MILWAUKEE AREA TECHNICAL COLLE

ADVANCED ALGEBRA

THOMAS J. McHALE
PAUL T. WITZKE
Milwaukee Area Technical College, Milwaukee, Wisconsin

ADDISON–WESLEY PUBLISHING COMPANY
Reading, Massachusetts
Menlo Park, California · London · Amsterdam · Don Mills, Ontario · Sydney

Milwaukee Area Technical College Mathematics Series

BASIC ALGEBRA

CALCULATION AND SLIDE RULE

BASIC TRIGONOMETRY

ADVANCED ALGEBRA

Reproduced by Addison-Wesley from camera-ready copy prepared by the authors.

Copyright © 1972 by Addison-Wesley Publishing Company, Inc. Philippines copyright 1972 by Addison-Wesley Publishing Company, Inc.

All rights reserved. No part of this publication may be reproduced, stored in a retrieval system, or transmitted, in any form or by any means, electronic, mechanical, photocopying, recording, or otherwise, without the prior written permission of the publisher. Printed in the United States of America. Published simultaneously in Canada.

ISBN 0-201-04633-4
IJKLMNOPQR-MU-89876543210

FOREWORD

The <u>Milwaukee</u> <u>Area</u> <u>Technical</u> <u>College</u> <u>Mathematics</u> <u>Series</u> is the product of a five-year project whose goal has been the development of a method of communicating the mathematics skills needed in basic science and technology to a wide range of students, including average and below-average students. This <u>Series</u> is not just a set of textbooks. It is a highly-organized, highly-assessed, and highly-successful system of instruction which has been designed to cope with the learning process of individual students by a combination of programmed instruction, continual diagnostic assessment, and tutoring. Though this system of instruction is different than conventional mathematics instruction, a deliberate effort has been made to keep it realistic in the following two ways:

1. It can be used in a regular classroom by regular teachers without any special training.
2. Its cost has been minimized by avoiding the use of educational hardware.

<u>History</u>. The five-year project was originally funded by the Carnegie Corporation of New York and subsequently funded by the Wisconsin State Board of Vocational, Technical, and Adult Education and the Milwaukee Area Technical College (MATC). The system of instruction offered in this <u>Series</u> was specifically developed for a two-semester Technical Mathematics course at MATC. It has been used in that capacity for the past five years with the following general results:

1. The dropout rate in the course has been reduced by 50%.
2. Average scores on equivalent final exams have increased from 55% to 85%.
3. The rate of absenteeism has decreased.
4. Student motivation and attitudes have been very favorable.

Besides its use in the Technical Mathematics course at MATC, parts or all of the system have been field tested in various other courses at MATC, in other technical institutes in Wisconsin and neighboring states, and in secondary schools in the Milwaukee area. These field tests with over 4,000 students have provided many constructive comments by teachers and students, plus a wealth of test data which has been item-analyzed and error-analyzed. On the basis of all this feedback, the textbooks and tests have been revised several times until a high level of learning is now achieved by a high percentage of students.

<u>Uses</u> <u>of</u> <u>the</u> <u>System</u>. Though specifically prepared for a two-semester Technical Mathematics course, the system of instruction can be used in a variety of contexts. At the college level, parts of it can be used in pre-technical, health occupations, apprentice, trade, and developmental programs, or in an intermediate algebra course. At the secondary school level, the whole system can be used in a two-year Technical Mathematics course in Grades 11 and 12; parts of it can be used in other courses or as supplementary materials in other courses. The system is highly flexible because it can be used either in a conventional classroom or in a learning center, with either a paced or self-paced schedule.

Written for students who have completed one year of algebra and one year of geometry in high school, the system has been used successfully with students without these prerequisites. Though not designed for students with serious deficiencies in arithmetic, it has been successfully used with students who were below average in arithmetic skills.

Three Main Elements in the System. The three main elements in the instructional system are programmed textbooks, diagnostic assessment, and a teacher. Though discussed in greater detail in the Teacher's Manual - MATC Mathematics Series, each main element is described briefly below:

1. Programmed Textbooks - The Series includes four programmed textbooks: BASIC ALGEBRA, CALCULATION AND SLIDE RULE, BASIC TRIGONOMETRY, and ADVANCED ALGEBRA. These textbooks include frequent short self-tests (with answers) which are an integral part of the instruction.

2. Diagnostic Assessment - Each textbook is accompanied by a test book which includes a diagnostic test for each assignment, chapter tests (three parallel forms), and a comprehensive final examination. Pre-tests in arithmetic and algebra are included in two of the test books, TESTS FOR BASIC ALGEBRA and TESTS FOR CALCULATION AND SLIDE RULE. A complete set of keys is provided in each test book for all tests in that book.

    Note: Test books are provided only to teachers, not to individual students. Copies of the tests for student use must be made by the Xerox process or some similar process.

3. A Teacher - A teacher is essential to the success of this system. Its success depends on the teacher's ability to diagnose learning difficulties and to remedy them by means of tutoring. Its success also depends on the teacher's skill in maintaining a suitable effort by each student. This new role for the teacher can be best described by calling him a "manager of the learning process". Teachers who use the system correctly find that this new role is a highly professional and satisfying one because it permits them to deal successfully with the wide range of individual differences which exist in any group of students.

Special Features. Some of the major features of the system of instruction are listed below:

Individual Attention - The system is designed so that the teacher can deal with the learning problems of individual students.

Learnability - When the system is used correctly, teachers can expect (and will obtain) a high level of learning in a high percentage of students.

Relevant Content - All of the content has been chosen on the basis of its relevance for basic science and technology. The content also provides a basis for the study of more advanced topics in mathematics.

Assessment - The many tests which are provided enable the teacher to maintain a constant assessment of each student's progress and the overall success of the system.

Student Motivation - Because students receive individual attention and have a high probability of success, motivation problems are minimized and student attitudes are generally quite positive.

Besides these main features, the system also provides a mechanism for dealing with absenteeism, and it lends itself quite easily to the use of paraprofessionals.

HOW TO USE THE SYSTEM

Though discussed in greater detail in the Teacher's Manual - MATC Mathematics Series, the procedure for using the system is briefly outlined below. All of the tests mentioned are available in the book TESTS FOR ADVANCED ALGEBRA. Both books are available from the Addison-Wesley Publishing Company, Reading, Massachusetts 01867.

1. Each chapter is covered in a number of assignments. Each assignment can be assessed by the diagnostic test which is provided. The assignments for ADVANCED ALGEBRA are listed at the bottom of this page. Though the best results have been obtained with a paced schedule, whether daily or otherwise, the assignments can also be covered on a self-paced schedule. Since the diagnostic tests are designed to take only 15 to 25 minutes, ample time is left for correction and tutoring within a normal class period. The diagnostic tests need not be graded since they are simply a teaching tool.

2. After all assignments for a chapter are completed, one of the three equivalent forms of the chapter test can be administered. Ordinarily, these tests should be graded. If the diagnostic tests are used in conjunction with the assignments, high scores should be obtained on the chapter tests, and grades can be assigned on a percentage basis.

3. When all of the chapters in a book are completed in the manner above, the comprehensive test for that book can be given. Or, if the teacher prefers, the items in that comprehensive test can be included in a final exam at the end of the semester.

ASSIGNMENTS FOR ADVANCED ALGEBRA

Ch. 1: #1 (pp. 1-19)
        #2 (pp. 19-38)
        #3 (pp. 38-60)
        #4 (pp. 60-75)
        #5 (pp. 76-92)

Ch. 2: #6 (pp. 93-120)
        #7 (pp. 121-142)
        #8 (pp. 143-161)
        #9 (pp. 161-182)
        #10 (pp. 182-196)

Ch. 3: #11 (pp. 197-214)
        #12 (pp. 215-230)
        #13 (pp. 231-247)
        #14 (pp. 248-269)
        #15 (pp. 270-282)

Ch. 4: #16 (pp. 283-304)
        #17 (pp. 304-321)
        #18 (pp. 322-338)
        #19 (pp. 338-353)
        #20 (pp. 353-372)
        #21 (pp. 372-389)

Ch. 5: #22 (pp. 390-409)
        #23 (pp. 410-431)
        #24 (pp. 431-451)
        #25 (pp. 452-475)
        #26 (pp. 475-497)

Ch. 6: #27 (pp. 498-522)
        #28 (pp. 523-537)
        #29 (pp. 537-551)
        #30 (pp. 551-570)
        #31 (pp. 570-587)
        #32 (pp. 587-599)

Ch. 7: #33 (pp. 600-617)
        #34 (pp. 618-632)
        #35 (pp. 633-649)
        #36 (pp. 650-660)

ADVANCED ALGEBRA

PREREQUISITES: A knowledge of the content of <u>BASIC ALGEBRA</u> is essential. Though a brief review of common logarithms is included, a more detailed introduction to common logarithms (and powers of ten) is probably necessary for most students. Chapter 4 (Powers of Ten) and Chapter 8 (Introduction to Logarithms) in <u>CALCULATION AND SLIDE RULE</u> are adequate for this purpose.

SEQUENCING: Chapters 1 and 2 should be studied in the order listed.

Chapters 1 and 2 are prerequisites for Chapters 3 and 4, but Chapters 3 and 4 can be studied in any order.

Chapters 5 through 7 have no prerequisites in this book, with the single exception that Chapter 6 (Common and Natural Logarithms) is a prerequisite for Chapter 7 (Logarithms and Exponentials: Laws and Formulas).

FEATURES: Squares, square root radicals, and related equations are extensively treated before the more general treatment of exponents and radicals.

The principles involved in the solution of radical and quadratic equations are generalized to the rearrangement of radical and quadratic formulas.

When the quadratic formula is used to solve quadratic equations, answers are reported both in radical form and as decimal numbers.

The relationship between fractional exponents and radicals is heavily emphasized.

After the concepts of slope, intercepts, and the slope-intercept form of linear equations are fully developed using "x" and "y" as the variables, these concepts are generalized to other variables.

Heavy emphasis is given to evaluating and rearranging logarithmic and base-"e" exponential formulas.

Semi-log, log-log, and exponential graphs are discussed.

ACKNOWLEDGMENTS

The Carnegie Corporation of New York made this Series possible by originally funding the project in 1965.

Dr. George A. Parkinson, now Director Emeritus of MATC, was instrumental in obtaining the original grant and totally supported the efforts of the project staff. Dr. William L. Ramsey, present Director of MATC, has continued this support.

Dr. Lawrence M. Stolurow, Professor and Director of Graduate Studies, Division of Educational Research and Development, State University of New York at Stony Brook, New York, served as the original chairman of the project.

Mr. Gail W. Davis of the project staff was largely responsible for the development of an efficient classroom procedure and the organization of a learning center. He also contributed many suggestions which helped to improve the instructional materials.

Mr. Keith J. Roberts, Mr. Allan A. Christenson, and Mr. Joseph A. Colla, members of the project staff, offered many constructive criticisms and helpful suggestions.

Many other teachers and administrators contributed to the implementation, field testing, and improvement of the instructional system.

Ms. Arleen A. D'Amore typed the camera-ready copy of the textbooks and tests.

ABOUT THE AUTHORS

THOMAS J. McHALE

He has been director of the MATC Mathematics Project since its beginning in June, 1965. He received his doctorate in experimental psychology at the University of Illinois, with major emphasis on the psychology of learning. He is currently a part-time member of the Psychology Department at Marquette University. He has taught mathematics at the secondary-school level.

PAUL T. WITZKE

He has been a member of the MATC Mathematics Project staff since its beginning. He has taught mathematics, physics, electronics, and other technical courses at the Milwaukee Area Technical College for 25 years. He has written numerous instructional manuals and has developed several televised courses in mathematics.

C O N T E N T S

CHAPTER 1   SQUARES AND SQUARE ROOT RADICALS   (Pages 1-92)
  1-1   The Squares Of Numbers   1
  1-2   The Square Roots Of Positive Numbers   4
  1-3   The Squares And Square Roots Of Letters   9
  1-4   The Square Roots Of The Indicated Squares Of Numbers And Letters   12
  1-5   Multiplying Square Root Radicals   15
  1-6   Multiplying Non-Radicals And Square Root Radicals   19
  1-7   The Square Roots Of Multiplications   23
  1-8   Division Of Square Root Radicals And The Square Roots Of Fractions   28
  1-9   The Squares Of Letter-Terms And Fractions   38
  1-10 The Squares Of Square Root Radicals   44
  1-11 The Squares Of Terms Containing Square Root Radicals   48
  1-12 Rationalizing Denominators Which Contain Square Root Radicals   53
  1-13 Squaring Binomial Expressions   60
  1-14 The Square Roots Of Perfect-Square Trinomials   70
  1-15 Addition And Subtraction Of Terms Containing Square Root Radicals   76
  1-16 The Square Roots Of Additions And Subtractions   81
  1-17 Operations Involving The Square Roots Of Binomials   85

CHAPTER 2   EQUATIONS AND FORMULAS CONTAINING SQUARES AND SQUARE ROOT RADICALS
  2-1   Solving Radical Equations By Direct Inspection   93   (Pages 93-196)
  2-2   The Squaring Principle For Radical Equations   96
  2-3   Isolating Radicals In Radical Equations   103
  2-4   Isolating Radicals And Solving Radical Equations   114
  2-5   The Squaring Principle And Radical Formulas   121
  2-6   Isolating Radicals And Rearranging Radical Formulas   130
  2-7   An Alternate Method For Solving Some Radical Equations And Formulas   143
  2-8   Squaring Binomials That Contain A Square Root Radical   150
  2-9   Limitations Of The Alternate Method For Solving Radical Equations And Formulas   156
  2-10 Solving Equations Containing A Squared Letter   161
  2-11 The Square Root Principle And Formulas Containing Squared Letters   165
  2-12 Radical Formulas Which Contain Squared Letters   177
  2-13 The Equivalence Method And Systems Of Formulas Involving Squared Letters And Square Root Radicals   182
  2-14 The Substitution Method And Systems Of Formulas Involving Squared Letters And Square Root Radicals   191

CHAPTER 3   QUADRATIC EQUATIONS AND FORMULAS   (Pages 197-282)
  3-1   The Meaning Of Quadratic Equations   197
  3-2   Multiplying Two Binomials By Means Of The Distributive Principle   199
  3-3   A Shorter Method For Multiplying Two Binomials   205
  3-4   The Three Types Of Quadratic Equations   211
  3-5   The Basic Principle Used In Solving Quadratic Equations By The Factoring Method   215
  3-6   Solving Incomplete Quadratic Equations By The Factoring Method   220
  3-7   Solving "Pure" Quadratics By The Factoring Method   224
  3-8   The Factoring Method And "Complete" Quadratics Of The Form: $1x^2 + bx + c = 0$ 231
  3-9   The Factoring Method And "Complete" Quadratics Of The Form: $ax^2 + bx + c = 0$ 241
  3-10 Perfect-Square Trinomials And The Factoring Method   246
  3-11 Solving Quadratic Equations By Means Of The Quadratic Formula   248
  3-12 Methods For Solving Quadratic Equations   256
  3-13 Quadratic Equations In Non-Standard Form   261
  3-14 Radical Equations And Extraneous Roots   270
  3-15 Some Applied Problems Involving Quadratic Equations   273
  3-16 Formula Rearrangement With Formulas In Quadratic Form   278

CHAPTER 4   EXPONENTS AND RADICALS   (Pages 283-389)

- 4-1   The Meaning Of Powers With Positive Whole-Number Exponents   283
- 4-2   Multiplication And Division Of Powers With Positive Whole-Number Exponents   288
- 4-3   The Meaning Of Powers Whose Exponents Are Either "1" Or "0"   292
- 4-4   The Meaning of Powers With Negative Whole-Number Exponents   297
- 4-5   More-Complicated Multiplications And Divisions Involving Powers   304
- 4-6   The Reciprocals Of Powers And Converting Divisions To Multiplications   309
- 4-7   Converting Multiplications Of Powers To Divisions   314
- 4-8   Equivalent Forms Of Fractions Containing Powers   317
- 4-9   Powers Of Letter-Terms And Other Indicated Multiplications   322
- 4-10 Powers Of Fractions   327
- 4-11 Powers Of Powers   332
- 4-12 Adding And Subtracting Letter-Terms Which Contain Powers   338
- 4-13 A Review Of Powers And The Laws Of Exponents   341
- 4-14 The Meaning Of Roots And Radicals   345
- 4-15 The Roots Of Multiplications And Divisions   349
- 4-16 The Roots Of Powers   353
- 4-17 Simplifying Radicals By Factoring Out Perfect Powers   359
- 4-18 Raising Radicals To Powers   366
- 4-19 The Meaning Of Powers With Fractional Exponents   372
- 4-20 Fractional Exponents And The Laws Of Exponents And Radicals   375
- 4-21 Powers With Decimal Exponents   386

CHAPTER 5   LINEAR GRAPHS AND SLOPE   (Pages 390-497)

- 5-1   Identifying Linear Equations In "x" And "y"   390
- 5-2   The Intercepts Of Linear Equations   395
- 5-3   Graphing Linear Equations By Plotting Intercepts   400
- 5-4   A Graphical Representation Of Changes In "y" Corresponding To Increases In "x"   403
- 5-5   The Slope Of A Line Expressed As A Signed Number   410
- 5-6   The Slope Of A Line As A Ratio Of Various Pairs Of Changes In "x" And "y"   417
- 5-7   The Slope-Intercept Form Of Linear Equations   425
- 5-8   Writing Linear Equations In Slope-Intercept Form   431
- 5-9   Determining The Slope Of A Line Through Two Known Points   438
- 5-10 Determining The Equation Of A Line Through Two Known Points   442
- 5-11 The Equations Of Lines Through The Origin   447
- 5-12 Slope And Parallel Lines   452
- 5-13 The Equations Of Horizontal And Vertical Lines   454
- 5-14 The Slope Of Horizontal And Vertical Lines   457
- 5-15 Identifying Linear Equations With Variables Other Than "x" And "y"   459
- 5-16 The Intercepts Of "Non-xy" Linear Equations   462
- 5-17 The Slope Of "Non-xy" Linear Equations   466
- 5-18 The Slope-Intercept Form Of "Non-xy" Equations   470
- 5-19 "Non-xy" Equations Of Lines Through The Origin   475
- 5-20 The Slope As A Ratio Of Various Pairs Of Changes In "Non-xy" Variables   480
- 5-21 Estimating The Equations Of Graphed Lines   484
- 5-22 The Slope Of Lines When The Scales On The Axes Are Different   491
- 5-23 The Slope Of Curves   493

CHAPTER 6   COMMON AND NATURAL LOGARITHMS   (Pages 498-599)

- 6-1   The Power-Of-Ten Form Of Numbers Greater Than 1   498
- 6-2   Common Logarithms Of Numbers Greater Than 1   504
- 6-3   The Power-Of-Ten Form Of Numbers Between 0 And 1   509
- 6-4   Common Logarithms Of Numbers Between 0 And 1   516
- 6-5   Evaluating "Log" Formulas For "Solved-For" Variables   523
- 6-6   Evaluating "Log" Formulas For "Non-Solved For" Variables   528
- 6-7   Evaluating "Log" Formulas When A Conversion To Power-Of-Ten Form Is Required   532
- 6-8   Logarithmic Scales   537
- 6-9   Semi-Log And Log-Log Graphs   542
- 6-10   Finding Common Logarithms On The Slide Rule   548
- 6-11   The Number "e" As A Base   551
- 6-12   Evaluations With Formulas Which Contain "e"   557
- 6-13   Natural (Base "e") Logarithms   563
- 6-14   Evaluating "Ln" Formulas   570
- 6-15   Evaluating "Ln" Formulas When A Conversion To Power-Of-"e" Form Is Required   575
- 6-16   Evaluating "e" Formulas When A Conversion To "Ln" Notation Is Required   579
- 6-17   A Review Of The Strategies Used In Evaluations With "Log", "e", And "Ln" Formulas   583
- 6-18   The Graphs Of Exponential Equations In "x" And "y"   587
- 6-19   The Graphs Of Exponential Formulas   595
- 6-20   A Note On "exp" Notation For Base-"e" Exponential Formulas   598

CHAPTER 7   LOGARITHMS AND EXPONENTIALS: LAWS AND FORMULAS   (Pages 600-660)

- 7-1   The Laws of Common Logarithms   600
- 7-2   Rearranging "Log" Formulas - Without Using The Laws Of Logarithms   605
- 7-3   Rearranging "Log" Formulas - Using The Laws Of Logarithms   609
- 7-4   The Relationship Between "Log" Formulas And Base-10 Exponential Formulas   614
- 7-5   The Laws Of Natural Logarithms   618
- 7-6   Rearranging "Ln" Formulas   620
- 7-7   The Relationship Between "Ln" Formulas And Base-"e" Exponential Formulas   624
- 7-8   Combined Use Of The Laws Of Logarithms   633
- 7-9   Reversing The Laws Of Logarithms   637
- 7-10   Rearranging Base-"e" Exponential Formulas   641
- 7-11   The Log Principle For Equations   650
- 7-12   Exponential Formulas With Other Bases   655

Chapter 1 SQUARES AND SQUARE ROOT RADICALS

Squares and square root radicals appear in many equations and formulas. In order to solve these equations and rearrange these formulas, a knowledge of the squaring operation and operations with square root radicals is necessary. Therefore, the general purpose of this chapter is to discuss the operations needed to handle equations and formulas of this type. These equations and formulas are discussed in the next chapter.

In this chapter, the squaring operation is applied to numbers, letters, letter-terms, fractions, square root radicals and terms containing these radicals. And though the addition and subtraction of square root radicals is considered, major emphasis is given to the multiplication and division of radicals. The treatment of squares and square root radicals in this chapter will serve as a foundation for a more general treatment of exponents and radicals in a later chapter.

1-1 THE SQUARES OF NUMBERS

In this section, we will discuss the meaning of the "squares" of numbers. In doing so, the meaning of the "squaring" operation and the notation used to indicate that operation will be introduced.

1. Squaring a number means "multiplying the number by itself". For example, both "5" and "3" are squared below.

    (5)(5) = 25   (3)(3) = 9

    Using the blanks at the right:

    (a) Square the number "4". (4)(4) = __16__

    (b) Square the number "6". (6)(6) = __36__

2. Both -3 and -8 are squared below. Notice that in each case the result is a positive number.

    (-3)(-3) = 9   (-8)(-8) = 64

    Using the blanks at the right:

    (a) Square the number "-2". (-2)(-2) = __4__

    (b) Square the number "-7". (-7)(-7) = __49__

---

a) (4)(4) = 16
b) (6)(6) = 36

a) (-2)(-2) = 4
b) (-7)(-7) = 49

2   Squares and Square Root Radicals

3. When a number is squared, the new number is called the "square" of the original number. For example:

   25 is called the "square of 5", since $(5)(5) = 25$
   49 is called the "square of -7", since $(-7)(-7) = 49$

   Using this fact, complete each of these:

   (a) The square of 9 is __81__.

   (b) The square of -8 is __64__.

---

4. Since the square of any number is <u>positive</u>, any signed number has the same square as its opposite. For example:

   25 is the square of both +5 and -5
   (<u>Note</u>: +5 and -5 are a pair of opposites since their sum is 0.)

   (a) 100 is the square of both __-10__ and __+10__.

   (b) 1 is the square of both __-1__ and __+1__.

   a) 81
   b) 64

---

5. Mathematicians use a special notation to indicate the squaring operation.

   Instead of writing: "Square 5" or "Square 7"
   they simply write: 5^2 or 7^2

   The small "2" indicates that "5" and "7" should be squared. Therefore:

   5^2 means $(5)(5)$
   7^2 means $(\_7\_)(\_7\_)$

   a) +10 and -10
   b) +1 and -1

---

6. When using the shorter notation to indicate that a <u>negative</u> number should be squared, <u>parentheses</u> <u>are</u> <u>written</u> <u>around</u> <u>the</u> <u>negative</u> <u>number</u>. That is:

   "Square -4" is written $(-4)^2$
   "Square -6" is written $(-6)^2$

   Therefore: (a) $(-5)^2$ means $(-5)(-5)$ (b) $(-1)^2$ means $(-1)(-1)$

   (7)(7)

---

7. The name "8 squared" is used for 8^2.

   Similarly: (a) 10^2 is called "__10 squared__"
               (b) $(-3)^2$ is called "__-3 squared__"

   a) (-5)(-5)
   b) (-1)(-1)

---

8. Complete each of these:

   (a) $4^2 = (4)(4) = \underline{16}$ (c) $2^2 = \underline{4}$
   (b) $(-7)^2 = (-7)(-7) = \underline{49}$ (d) $(-1)^2 = \underline{1}$

   a) "10 squared"
   b) "-3 squared"

---

a) 16
b) 49
c) 4
d) 1

Squares and Square Root Radicals   3

9. Though the actual square of 8 is 64, the "square of 8" is <u>indicated</u> by the notation 8^2. Therefore:

   64 is called the "<u>square</u>" of 8.
   8^2 is called the "<u>indicated square</u>" of 8.

   Using the terms "square" and "indicated square", complete these:
   (a) 16 is called the ___indicated___ of 4.
   (b) 4^2 is called the ___square___ of 4.
   (c) $(-6)^2$ is called the ___square___ of -6.
   (d) 36 is called the ___indicated___ of -6.

---

10. (a) The <u>square</u> of 3 is ___9___.
    (b) The <u>indicated square</u> of 10 is ___10^2___.
    (c) The <u>indicated square</u> of -4 is ___$(-4)^2$___.
    (d) The <u>square</u> of -2 is ___4___.

    a) square
    b) indicated square
    c) indicated square
    d) square

---

11. Using "indicated square" notation, complete each of these:

    25 = 5^2 and $(-5)^2$ (b) 49 = ___7___ and ___-7^2___
    (a) 81 = ___9___ and ___-9^2___   (c) 1 = ___1___ and ___-1^2___

    a) 9^2 and $(-9)^2$
    b) 7^2 and $(-7)^2$
    c) 1^2 and $(-1)^2$

---

12. To square the number "0", we simply multiply "0" by itself.

    Since $0^2 = (0)(0)$, the square of 0 is ___0___.

    0

---

13. To square a non-whole number, we also multiply the number by itself. That is:
    (a) $1.2^2 = (1.2)(1.2) =$ ___1.44___
    (b) $(-0.5)^2 = (-0.5)(-0.5) =$ ___.25___

    a) 1.44
    b) 0.25

---

14. Write the <u>indicated square</u> notation for each of these:
    (a) $(5.9)(5.9) =$ ___5.9^2___   (b) $(-2.66)(-2.66) =$ ___$(-2.66)^2$___

    a) 5.9^2
    b) $(-2.66)^2$

---

15. Complete each of these:
    If $x = 6^2$, $x = 36$ (b) If $t = 0^2$, $t =$ ___0___
    (a) If $y = (-2)^2$, $y =$ ___4___   (c) If $m = (-0.4)^2$, $m =$ ___.16___

    a) $y = 4$
    b) $t = 0$
    c) $m = 0.16$

4 Squares and Square Root Radicals

SELF-TEST 1 (Frames 1-15)

1. Square the number -9. __81__
2. The square of 15 is __225__.
3. 36 is the square of both __6__ and __-6__.
4. $11^2 = $ __121__   7. $(-4)^2 = $ __16__
5. $(-1)^2 = $ __1__   8. $2.5^2 = $ __6.25__
6. $100^2 = $ __10000__   9. $(-0.7)^2 = $ __.49__
10. The indicated square of 30 is __30^2__.
11. The square of 30 is __900__.
12. The indicated square of -1.4 is __-1.4^2__.
13. The square of -1.4 is __1.96__.
14. Complete: $9 = (3)^2$ and $(-3)^2$
15. Complete: $1 = (1)^2$ and $(-1)^2$

ANSWERS:
1. 81   4. 121   7. 16   10. 30^2 13. 1.96
2. 225   5. 1   8. 6.25   11. 900   14. $(3)^2$ and $(-3)^2$
3. +6 and -6   6. 10,000   9. 0.49   12. $(-1.4)^2$ 15. $(1)^2$ and $(-1)^2$

1-2   THE SQUARE ROOTS OF POSITIVE NUMBERS

In this section, we will discuss the meaning of the "square root" of a positive number. In doing so, the meaning of the "square root" operation and the notation used to indicate that operation will be introduced.

16. <u>A square root of positive number N is a number whose square is number N</u>. For example:

   4 is a square root of 16, since $4^2 = 16$
   -5 is a square root of 25, since $(-5)^2 = 25$

   (a) Since $3^2 = 9$, _____ is a square root of 9.
   (b) Since $(-6)^2 = 36$, _____ is a square root of 36.

17. Instead of the phrase "find the square root of", mathematicians use the symbol $\sqrt{\ }$ to indicate the square root operation. That is:

   $\sqrt{49}$ means "find the square root of 49".

   Write "find the square root of 36" in symbol form. _____

a) 3
b) -6

$\sqrt{36}$

18. There is some special terminology which is used in connection with expressions like $\sqrt{81}$.

   (1) The entire expression $\sqrt{81}$ is called a square root radical.
   (2) The $\sqrt{}$ symbol is called a radical sign.
   (3) The number "81" under the radical sign is called the radicand.

   (a) In $\sqrt{16}$, the radicand is _____.
   (b) In $\sqrt{100}$, the $\sqrt{}$ is called the _____.
   (c) Write the square root radical whose radicand is 64. _____

19. Any positive number has two square roots, one positive and one negative. For example:

   Since both 5^2 and $(-5)^2 = 25$:

   +5 is the positive square root of 25.
   −5 is the negative square root of 25.

   As you can see, the two square roots are a pair of opposites since their sum is 0.

   (a) The positive square root of 100 is _____.
   (b) The negative square root of 4 is _____.
   (c) The two square roots of 64 are _____ and _____.

20. The positive square root of a number is ordinarily called its "principal" square root.

   (a) The principal (or positive) square root of 49 is _____.
   (b) The negative square root of 49 is _____.
   (c) Does the principal square root of 49 equal the negative square root of 49? _____

21. Plus and minus signs are used to indicate whether the principal (positive) or negative square root of a number is desired.

   To indicate a principal square root, we write a "+" sign (or no sign) in front of the radical. That is:

   $+\sqrt{100} = +10$ and $\sqrt{100} = +10$

   To indicate a negative square root, we write a "−" sign in front of the radical. That is:

   $-\sqrt{100} = -10$

   Complete the following with the correct signed number:

   (a) $\sqrt{36} =$ _____   (b) $+\sqrt{36} =$ _____   (c) $-\sqrt{36} =$ _____

---

a) 16
b) radical sign
c) $\sqrt{64}$

---

a) +10
b) −2
c) +8 and −8

---

a) +7
b) −7
c) No. They are a pair of opposites.

---

a) +6
b) +6
c) −6

Squares and Square Root Radicals

22. (a) The "principal square root of 47" can be indicated in radical form in two different ways. They are _____ and _____.
    (b) The "negative square root of 89" can be indicated in radical form in only one way. It is _____.

23. (a) If $x = -\sqrt{49}$, $x = $ _____
    (b) If $t = \sqrt{81}$, $t = $ _____
    (c) If $a = +\sqrt{9}$, $a = $ _____

    a) $+\sqrt{47}$ or $\sqrt{47}$
    b) $-\sqrt{89}$

24. Write "true" or "false" in the blank after each of the following statements:
    (a) $-\sqrt{16} = +4$ _____
    (b) $-7 = \sqrt{49}$ _____
    (c) $\sqrt{64} = +8$ _____
    (d) $-12 = -\sqrt{144}$ _____

    a) -7
    b) $+9$
    c) $+3$

25. Though the principal square root of 36 is actually 6, the "principal square root of 36" is indicated by the radical $\sqrt{36}$. Therefore:

    6 is called the "principal square root" of 36.
    $\sqrt{36}$ is called the "indicated principal square root" of 36.

    Using the terms above, complete these:
    (a) 7 is called the _____ of 49.
    (b) $\sqrt{49}$ is called the _____ of 49.

    a) False, since:
       $-\sqrt{16} = -4$
    b) False, since:
       $\sqrt{49} = +7$
    c) True
    d) True

26. Though the negative square root of 16 is actually -4, the "negative square root of 16" is indicated by the radical $-\sqrt{16}$. Therefore:

    -4 is called the "negative square root" of 16.
    $-\sqrt{16}$ is called the "indicated negative square root" of 16.

    Using the terms above, complete these:
    (a) $-\sqrt{25}$ is called the _____ of 25.
    (b) -5 is called the _____ of 25.

    a) principal square root
    b) indicated principal square root

27. (a) The principal square root of 81 is _____.
    (b) The indicated principal square root of 100 is _____.
    (c) The indicated negative square root of 64 is _____.
    (d) The negative square root of 9 is _____.

    a) indicated negative square root
    b) negative square root

28. To indicate that both the principal and negative square roots of a number are desired, we write a "±" in front of the radical. The "±" symbol means both "+" and "−". That is:

    Since $\pm\sqrt{16}$ means both $+\sqrt{16}$ and $-\sqrt{16}$,
    $\pm\sqrt{16}$ means both $+4$ and _____.

    a) $+9$
    b) $+\sqrt{100}$ or $\sqrt{100}$
    c) $-\sqrt{64}$
    d) -3

    -4

Squares and Square Root Radicals 7

29. Instead of writing: $\pm\sqrt{25}$ = +5 and −5
we usually write: $\pm\sqrt{25} = \pm 5$

(Note: ±5 means "both +5 and −5")

Therefore: (a) $\pm\sqrt{81}$ = _____ (b) $\pm\sqrt{1}$ = _____

---

30. Find the roots of each of the following equations:

(a) If $y = -\sqrt{36}$, y = _____

(b) If $y = \sqrt{4}$, y = _____

(c) If $y = \pm\sqrt{144}$, y = _____

a) ±9 b) ±1

---

31. (a) Since $0^2 = 0$, $\sqrt{0}$ = _____

(b) Since both $(+1)^2$ and $(-1)^2 = 1$, $\pm\sqrt{1}$ = _____ and _____

a) −6
b) +2 or 2
c) ±12

---

32. (a) Since $(0.3)^2 = 0.09$, the <u>principal</u> square root of 0.09 is _____.

(b) Since $(-1.9)^2 = 3.61$, the <u>negative</u> square root of 3.61 is _____.

(c) Since $(2.5)^2 = 6.25$, $\pm\sqrt{6.25}$ = _____.

a) 0
b) +1 and −1

---

33. There are many numbers whose square roots can be obtained mentally. For example:

$\sqrt{25} = 5$ $\sqrt{144} = 12$

However, to find the square roots of most numbers, the long square root procedure or a table of square roots or the slide rule must be used. In most of these cases, the square roots are non-ending decimal numbers. For example:

$\sqrt{17} = 4.12310\ldots$

$\sqrt{8.69} = 2.94788\ldots$

Note: The three dots after each number mean that we could add more digits indefinitely.

When the square root is a non-ending decimal number, we usually round it to a certain number of digits. This rounded number is, of course, only a close approximation. For example, if we would round the square roots above to three digits, we would say that:

$\sqrt{17}$ is approximately 4.12

$\sqrt{8.69}$ is approximately _____.

a) +0.3
b) −1.9
c) ±2.5

2.95

8   Squares and Square Root Radicals

34. If the square root of a number is a whole number, we call the original number a "perfect square". For example:

   Since $\sqrt{25}$ is exactly 5, 25 is called a perfect square.

   Since $\sqrt{81}$ is exactly 9, 81 is called a perfect square.

   Name the three whole numbers between 25 and 81 which are perfect squares:

   _____, _____, and _____

35. Since the square of any number (positive or negative) is positive, there is no regular number which is the square root of a negative number. For example:

   $\sqrt{-25} \neq +5$, since $(+5)^2 = +25$.

   $\sqrt{-25} \neq -5$, since $(-5)^2 = +25$.   (Note: "$\neq$" means is not equal to.)

   Though mathematicians have developed a method of handling the square roots of negative numbers, we will not discuss that method in this chapter.

a) 36, since $\sqrt{36} = 6$
b) 49, since $\sqrt{49} = 7$
c) 64, since $\sqrt{64} = 8$

SELF-TEST 2 (Frames 16-35)

In the expression $\sqrt{21}$:

1. The radicand is _____.
2. The radical sign is _____.
3. The radical is _____.
4. The two square roots of 64 are _____ and _____.
5. The principal square root of 16 is _____.
6. The negative square root of 4 is _____.
7. The indicated negative square root of 25 is _____.
8. The indicated principal square root of 43 is _____ or _____.

9. The principal square root of 144 is _____.
10. Write $+\sqrt{87}$ and $-\sqrt{87}$ as a single radical expression. _____
11. Evaluate: $\pm\sqrt{49} =$ _____
12. Since $(0.2)^2 = 0.04$, the two square roots of 0.04 are _____ and _____.
13. Is -3 a square root of -9? _____
14. Which of these numbers are perfect squares?

   36   24   1   60   -36   81

ANSWERS:
1. 21
2. $\sqrt{}$
3. $\sqrt{21}$
4. +8 and -8
5. +4 (or 4)
6. -2
7. $-\sqrt{25}$
8. $+\sqrt{43}$ or $\sqrt{43}$
9. +12 (or 12)
10. $\pm\sqrt{87}$
11. ±7
12. +0.2 and -0.2
13. No.
14. 36, 1, and 81

Squares and Square Root Radicals   9

1-3   THE SQUARES AND SQUARE ROOTS OF LETTERS

In this section, we will discuss the squares and square roots of letters like "x" or "t". We will show that the same definitions, notation, and terminology which apply to the squares and square roots of numbers also apply to the squares and square roots of letters.

---

36. Just as numbers can be squared, letters can be squared. The same notation is used to indicate "squaring" for a letter. That is:

    x^2 means that "x" should be squared.

    To square a letter, we simply multiply it by itself. Therefore:

    x^2 means $(x)(x)$
    $(-t)^2$ means $(\ )(\ )$

---

37. Just as 7^2 is called "7 squared",
    y^2 is called "y squared"
    and $(-x)^2$ is called "_____".

    | $(-t)(-t)$

---

38. Write each of these in "indicated square" notation:

    (a) $(d)(d) = $ _____   (b) $(-y)(-y) = $ _____

    | "-x squared"

---

39. When the square of a <u>number</u> is indicated, we can actually obtain its "square" by performing the multiplication. For example:

    $4^2 = (4)(4) = 16$

    When the square of a <u>letter</u> is indicated, however, we cannot obtain its "square" by performing the multiplication. For example:

    $x^2 = (x)(x) = ?$

    Therefore, the square of a letter can only be written in indicated form. That is:

    The "square of x" can only be written as x^2.
    The "square of y" can only be written as _____.

    | a) d^2
    | b) $(-y)^2$

---

40. Though the square of a letter can only be written in indicated square notation, we can write both the indicated and the actual square of any number which is plugged in for the letter. For example:

    If we plug in 5 for x in x^2, $x^2 = 5^2 = 25$

    If we plug in -3 for t in t^2, $t^2 = $ _____ = _____

    | y^2

---

41. If $y = x^2$, find the value of "y" when:

    (a) $x = 9$ $y = $ _____
    (b) $x = -1$ $y = $ _____
    (c) $x = 0$ $y = $ _____

    | $(-3)^2 = 9$

| a) 81
| b) 1
| c) 0

10 Squares and Square Root Radicals

42. Just as 3x means (3)(x) or "3 times x",
    $3x^2$ means $(3)(x^2)$ or "3 times x^2".
    Therefore: If x = 4, $3x^2 = 3(4^2) = 3(16) = 48$
    If x = -5, $3x^2 = 3[(-5)^2] = 3(25) = 75$

    Following the steps above, complete these:
    (a) If x = 10, $2x^2 = $ _____.
    (b) If x = -2, $5x^2 = $ _____.

    a) 2(100) = 200
    b) 5(4) = 20

43. If $a = 4b^2$, find the value of "a" when:
    (a) b = 3   a = _____
    (b) b = -5   a = _____
    (c) b = 0.1   a = _____

    a) 4(9) = 36
    b) 4(25) = 100
    c) 4(0.01) = 0.04

44. The square root radical is also used to indicate the square root of a letter. That is:

    $+\sqrt{x}$ (or $\sqrt{x}$) indicates the <u>principal</u> (or positive) square root of x.

    $-\sqrt{x}$ indicates the <u>negative</u> square root of x.

    $\pm\sqrt{x}$ indicates <u>both</u> <u>the</u> <u>principal</u> <u>and</u> <u>negative</u> square roots of x.

    Write the radical (or radicals) which indicates:
    (a) The <u>negative</u> square root of "t". _____
    (b) The <u>principal</u> square root of "y". _____
    (c) <u>Both</u> <u>the</u> <u>principal</u> <u>and</u> <u>negative</u> square roots of "A". _____

45. In $\sqrt{m}$: (a) The symbol $\sqrt{}$ is called the _____.
    (b) The letter "m" is called the _____.

    a) $-\sqrt{t}$
    b) $+\sqrt{y}$ or $\sqrt{y}$
    c) $\pm\sqrt{A}$

46. When the square root of a <u>number</u> is indicated, we can actually obtain its square root. For example:
    $\sqrt{100} = +10$

    When the square root of a <u>letter</u> is indicated, however, we cannot obtain its square root. For example:
    $\sqrt{x} = ?$

    Therefore, the square root of a letter can only be written in its indicated form. That is:

    The "<u>principal</u> square root of t" can only be written as $+\sqrt{t}$ or $\sqrt{t}$.

    (a) The "<u>negative</u> square root of y" can only be written as _____.
    (b) "<u>Both</u> square roots of d" can only be written as _____.

    a) radical sign
    b) radicand

Squares and Square Root Radicals   11

47. Though the square root of a letter can only be written in indicated square root notation, we can write both the indicated and the actual square root of any number which is plugged in for the letter. For example:

$$\text{If } x = 9: \sqrt{x} = \sqrt{9} = +3$$
$$-\sqrt{x} = -\sqrt{9} = -3$$
$$\pm\sqrt{x} = \pm\sqrt{9} = \pm 3$$

(a) If $y = 16$: $\sqrt{y} = \sqrt{\underline{}} = \underline{}$
(b) If $t = 49$: $-\sqrt{t} = -\sqrt{\underline{}} = \underline{}$
(c) If $R = 81$: $\pm\sqrt{R} = \pm\sqrt{\underline{}} = \underline{}$

a) $-\sqrt{y}$
b) $\pm\sqrt{d}$

---

48. If $y = \sqrt{x}$, find the value of "y" when:

(a) $x = 25$ $y = \underline{}$
(b) $x = 100$ $y = \underline{}$

a) $\sqrt{16} = +4$
b) $-\sqrt{49} = -7$
c) $\pm\sqrt{81} = \pm 9$

---

49. If $m = -\sqrt{q}$, find the value of "m" when:

(a) $q = 16$ $m = \underline{}$
(b) $q = 36$ $m = \underline{}$

a) +5
b) +10

---

50. Complete each of these:

(a) If $a = \sqrt{b}$ and $b = 0$, $a = \underline{}$
(b) If $t = -\sqrt{R}$ and $R = 1$, $t = \underline{}$
(c) If $d = \pm\sqrt{S}$ and $S = 49$, $d = \underline{}$

a) -4
b) -6

---

51. Just as $3x^2$ means $(3)(x^2)$ or "3 times x^2",

$3\sqrt{x}$ means $(3)(\sqrt{x})$ or "3 times $\sqrt{x}$".

Therefore: If $x = 4$, $3\sqrt{x} = 3\sqrt{4} = 3(2) = 6$

If $x = 25$, $3\sqrt{x} = 3\sqrt{25} = 3() = \underline{}$

a) 0
b) -1
c) ± 7

---

52. If $t = 4\sqrt{m}$, find the value of "t" when:

(a) $m = 100$ $t = \underline{}$
(b) $m = 0$ $t = \underline{}$

$3(5) = 15$

---

a) $4(10) = 40$
b) $4(0) = 0$

SELF-TEST 3 (Frames 36-52)

1. In symbols, "-t squared" is _____.
2. Given: $h = t^2$. If $t = 0.3$, $h =$ _____.
3. Given: $w = 9a^2$. If $a = -1$, $w =$ _____.
4. Write the radical which indicates both the principal and negative square roots of R. _____
5. In $\sqrt{p}$, p is called the _____.
6. Given: $s = \sqrt{x}$. If $x = 144$, $s =$ _____.
7. Given: $A = -\sqrt{d}$. If $d = 36$, $A =$ _____.
8. Given: $v = \pm\sqrt{p}$. If $p = 100$, $v =$ _____.
9. Given: $P = 9\sqrt{r}$. If $r = 4$, $P =$ _____.
10. Given: $F = 20\sqrt{t}$. If $t = 0$, $F =$ _____.

ANSWERS:
1. $(-t)^2$
2. $h = 0.09$
3. $w = 9$
4. $\pm\sqrt{R}$
5. radicand
6. $s = 12$
7. $A = -6$
8. $v = \pm 10$
9. $P = 18$
10. $F = 0$

1-4 THE SQUARE ROOTS OF THE INDICATED SQUARES OF NUMBERS AND LETTERS

In this section, we will discuss the square roots of the indicated squares of numbers and letters, like 7^2 and x^2. We will show that $\sqrt{7^2}$ and $\sqrt{x^2}$ can be simplified to non-radical expressions since 7^2 and x^2 are perfect squares.

53. Up to this point, we have discussed the square roots of numbers and letters. We can also take the square roots of the <u>indicated squares of numbers</u>. The same square root notation is used. For example:

   $+\sqrt{6^2}$ (or $\sqrt{6^2}$) means "the <u>principal</u> square root of 6^2".

   $-\sqrt{6^2}$ means "the <u>negative</u> square root of 6^2".

   $\pm\sqrt{6^2}$ means "<u>both</u> square roots of 6^2".

   Write each of the following in <u>indicated</u> square root notation:

   (a) The <u>negative</u> square root of 7^2. _____
   (b) The <u>principal</u> square root of 3^2. _____
   (c) <u>Both</u> square roots of 5^2. _____

a) $-\sqrt{7^2}$
b) $+\sqrt{3^2}$ or $\sqrt{3^2}$
c) $\pm\sqrt{5^2}$

54. The same notation is used to indicate the square roots of squared letters. For example:

$+\sqrt{x^2}$ (or $\sqrt{x^2}$) means "the <u>principal</u> square root of x^2".

$-\sqrt{x^2}$ means "the <u>negative</u> square root of x^2".

$\pm\sqrt{x^2}$ means "<u>both</u> square roots of x^2".

Write each of the following in indicated square root notation:

(a) <u>Both</u> square roots of t^2. _____

(b) The <u>principal</u> square root of y^2. _____

(c) The <u>negative</u> square root of m^2. _____

---

55. In $\sqrt{10^2}$, the radicand is 10^2.

In $-\sqrt{d^2}$, the radicand is d^2.

(a) Write the principal square root radical whose radicand is 4^2. _____

(b) Write the negative square root radical whose radicand is b^2. _____

a) $\pm\sqrt{t^2}$

b) $+\sqrt{y^2}$ or $\sqrt{y^2}$

c) $-\sqrt{m^2}$

---

56. Since $6^2 = 36$, $+\sqrt{6^2} = +\sqrt{36} = +6$

$-\sqrt{6^2} = -\sqrt{36} = -6$

$\pm\sqrt{6^2} = \pm\sqrt{36} = \pm 6$

Since $9^2 = 81$, $+\sqrt{9^2} = +\sqrt{81} = +9$

(a) $-\sqrt{9^2} = -\sqrt{81} =$ _____

(b) $\pm\sqrt{9^2} = \pm\sqrt{81} =$ _____

a) $+\sqrt{4^2}$ or $\sqrt{4^2}$

b) $-\sqrt{b^2}$

---

57. (a) Since $3^2 = 9$, $-\sqrt{3^2} =$ _____

(b) Since $7^2 = 49$, $\sqrt{7^2} =$ _____

(c) Since $8^2 = 64$, $\pm\sqrt{8^2} =$ _____

(d) Since $1^2 = 1$, $+\sqrt{1^2} =$ _____

a) -9

b) ± 9

---

58. Complete each of these:

(a) $+\sqrt{4^2} =$ _____   (c) $\pm\sqrt{1^2} =$ _____

(b) $-\sqrt{8^2} =$ _____   (d) $\sqrt{0^2} =$ _____

a) -3

b) $+7$

c) ± 8

d) $+1$

---

59. Complete each of these:

(a) $-\sqrt{(0.6)^2} =$ _____   (c) $\pm\sqrt{(0.1)^2} =$ _____

(b) $+\sqrt{(1.9)^2} =$ _____   (d) $\sqrt{(9.87)^2} =$ _____

a) $+4$

b) -8

c) ± 1

d) 0

14 Squares and Square Root Radicals

60. Just as the two square roots of 5^2 are $+5$ and -5, the two square roots of x^2 are "$+x$" and "$-x$". That is:

$$+\sqrt{x^2} \text{ (or } \sqrt{x^2}\text{)} = +x$$
$$-\sqrt{x^2} = -x$$
$$\pm\sqrt{x^2} = \pm x$$

It is easy to see that "x^2" has both a positive and a negative square root if we plug in numbers for "x".

If $x = 7$, $\pm\sqrt{x^2} = \pm\sqrt{7^2} = \pm 7$

(a) If $x = 9$, $\pm\sqrt{x^2} = \pm\sqrt{9^2} = $ _____

(b) If $x = 1$, $\pm\sqrt{x^2} = \pm\sqrt{1^2} = $ _____

a) -0.6
b) $+1.9$
c) ± 0.1
d) $+9.87$

---

61. (a) The principal square root of t^2 is _____.

(b) The negative square root of a^2 is _____.

(c) The two square roots of y^2 are _____ and _____.

a) ± 9
b) ± 1

---

62. Complete each of these:

(a) $+\sqrt{d^2} = $ _____ (c) $\sqrt{R^2} = $ _____

(b) $-\sqrt{m^2} = $ _____ (d) $\pm\sqrt{p^2} = $ _____

a) t (or $+t$)
b) $-a$
c) y and $-y$

---

63. The indicated square of any number or letter is called a "<u>perfect square</u>" since its square root is a number or letter. That is:

7^2 is called a "perfect square" since $\sqrt{7^2} = 7$.

x^2 is called a "perfect square" since $\sqrt{x^2} = x$.

t^2 is called a "perfect square" since ____ = ____.

a) d (or $+d$)
b) $-m$
c) R (or $+R$)
d) $\pm p$

---

64. Though $\sqrt{x^2}$ can be simplified to "x", $\sqrt{x}$ cannot be simplified.

Which of the following expressions can be simplified? _____

(a) $\sqrt{t}$ (b) $-\sqrt{m^2}$ (c) $-\sqrt{R}$ (d) $\sqrt{A^2}$

$\sqrt{t^2} = t$

---

65. Similarly, $3\sqrt{x^2}$ can be simplified to $3x$, but $3\sqrt{x}$ cannot be simplified.

Simplify the expressions below if possible:

(a) $7\sqrt{y^2} = $ _____   (c) $5\sqrt{R} = $ _____

(b) $7\sqrt{y} = $ _____   (d) $4\sqrt{A^2} = $ _____

Only (b) and (d)

---

a) $7y$
b) Not possible.
c) Not possible.
d) $4A$

66. If $y = 4\sqrt{x^2}$, $y = 4x$
   (a) If $t = 3\sqrt{s^2}$, $t = $ _____
   (b) If $m = \sqrt{d^2}$, $m = $ _____

   a) $3s$
   b) d

SELF-TEST 4 (Frames 53-66)

Write each of the following in <u>indicated</u> <u>square</u> <u>root</u> <u>notation</u>:

1. Both square roots of v^2. _____
2. The principal square root of 19^2. _____
3. The negative square root whose radicand is E. _____

Simplify: 4. $\sqrt{h^2} = $ _____   5. $\pm\sqrt{3.9^2} = $ _____   6. $+\sqrt{r^2} = $ _____   7. $-\sqrt{(0.6)^2} = $ _____

8. Which of these are perfect squares?   (a) 12^2 (b) E^2 (c) R   (d) w

Simplify if possible: 9. $8\sqrt{H^2} = $ _____   10. $\pm 4\sqrt{t^2} = $ _____   11. $-\sqrt{a} = $ _____

ANSWERS: 1. $\pm\sqrt{v^2}$ 4. h 8. (a) and (b)
          2. $\sqrt{19^2}$ (or $+\sqrt{19^2}$)   5. ± 3.9 9. $8H$
          3. $-\sqrt{E}$ 6. $+r$ (or r)   10. $\pm 4t$
                                7. -0.6 11. Cannot be simplified.

1-5   MULTIPLYING SQUARE ROOT RADICALS

In this section, we will show the procedure for multiplying two or more square root radicals. Since the products of these multiplications are frequently radicals whose radicands are letter-terms, the meaning of such radicals is discussed.

> Though negative square roots are clearly a part of formal mathematics, negative square roots have little or no significance in basic science and technology. Therefore, in the discussion of square root radicals in the rest of this chapter, <u>only</u> <u>principal</u> <u>square</u> <u>roots</u> will be considered.

67. Various symbols can be used to indicate a multiplication of two radicals. For example:

$$\left.\begin{array}{c}\sqrt{5} \cdot \sqrt{7} \\ \sqrt{5}\,\sqrt{7} \\ \left(\sqrt{5}\right)\left(\sqrt{7}\right)\end{array}\right\} \text{ all mean "multiply } \sqrt{5} \text{ and } \sqrt{7}\text{".}$$

Of the three ways listed above, the top two are the most common.

Write "multiply $\sqrt{3}$ times $\sqrt{8}$" in the two most common ways:

_____ and _____

68. The following pattern is used to multiply two square root radicals.

$$\sqrt{\Box} \cdot \sqrt{\triangle} = \sqrt{\Box\,\triangle}$$

That is: <u>WHEN MULTIPLYING TWO RADICALS, THE PRODUCT IS A SINGLE RADICAL WHOSE RADICAND IS THE PRODUCT OF THE RADICANDS OF THE TWO FACTORS.</u> Here are some examples:

$$\sqrt{5} \cdot \sqrt{7} = \sqrt{(5)(7)}$$
$$\sqrt{3} \cdot \sqrt{x} = \sqrt{(3)(x)}$$
$$\sqrt{a} \cdot \sqrt{b} = \sqrt{(a)(b)}$$

Following the pattern above, complete these:

(a) $\sqrt{3} \cdot \sqrt{8} = \sqrt{()()}$

(b) $\sqrt{7} \cdot \sqrt{y} = \sqrt{()()}$

(c) $\sqrt{R} \cdot \sqrt{S} = \sqrt{()()}$

69. The sensibleness of the pattern used to multiply two radicals can be shown by some numerical examples.

In these two examples, the radicands are perfect squares:

$$\sqrt{9} \cdot \sqrt{4} = \sqrt{(9)(4)}$$
$$3 \cdot 2 = \sqrt{36}$$
$$6 = 6$$

$$\sqrt{16} \cdot \sqrt{25} = \sqrt{(16)(25)}$$
$$4 \cdot 5 = \sqrt{400}$$
$$20 = 20$$

In the two examples below, the radicands are not perfect squares:

$$\sqrt{8} \cdot \sqrt{3} = \sqrt{(8)(3)}$$
$$(2.83)(1.73) = \sqrt{24}$$
$$4.90 = 4.90$$

$$\sqrt{5} \cdot \sqrt{12} = \sqrt{(5)(12)}$$
$$(2.24)(3.46) = \sqrt{60}$$
$$7.75 = 7.75$$

Using the pattern for multiplying two radicals, complete these:

(a) $\sqrt{5} \cdot \sqrt{d} = \sqrt{()()}$ 　(b) $\sqrt{m} \cdot \sqrt{t} = \sqrt{()()}$

$\sqrt{3} \cdot \sqrt{8}$ and $\sqrt{3}\sqrt{8}$

a) $\sqrt{(3)(8)}$
b) $\sqrt{(7)(y)}$
c) $\sqrt{(R)(S)}$

Answer to Frame 69:　　a) $\sqrt{(5)(d)}$　　b) $\sqrt{(m)(t)}$

Squares and Square Root Radicals   17

70. When the radicand of each factor is a number, the radicand of the product can be written as a single number. For example:

$$\sqrt{5} \cdot \sqrt{7} = \sqrt{(5)(7)} = \sqrt{35}$$

(a) $\sqrt{3} \cdot \sqrt{11} = \sqrt{(3)(11)} = \sqrt{}$

(b) $\sqrt{10} \cdot \sqrt{17} = \sqrt{(10)(17)} = \sqrt{}$

---

71. When the radicand of one or both factors is a letter, the radicand of the product can only be written as an indicated multiplication. For example:

$$\sqrt{5} \cdot \sqrt{x} = \sqrt{(5)(x)} = \sqrt{5x}$$
$$\sqrt{d} \cdot \sqrt{R} = \sqrt{(d)(R)} = \sqrt{dR}$$

In each product above, the radicand is a letter-term. Remember that these letter-terms are really indicated multiplications. That is:

   5x means (5)(x) or "multiply 5 and x ".

   dR means (d)(R) or "multiply ___ and ___".

a) $\sqrt{33}$
b) $\sqrt{170}$

---

72. When multiplying radicals whose radicands are letters, the letters in the product are written in alphabetical order. That is:

$$\sqrt{H} \cdot \sqrt{t} = \sqrt{(H)(t)} = \sqrt{Ht}$$
$$\sqrt{y} \cdot \sqrt{x} = \sqrt{(y)(x)} = \sqrt{xy}$$
$$\sqrt{t} \cdot \sqrt{b} = \sqrt{(t)(b)} = \sqrt{}$$

d and R

---

73. Here are some examples of multiplications in which the radicand of one or both factors is itself a letter-term.

$$\sqrt{5} \cdot \sqrt{3x} = \sqrt{(5)(3x)} = \sqrt{15x}$$
$$\sqrt{2y} \cdot \sqrt{7t} = \sqrt{(2y)(7t)} = \sqrt{14ty}$$
$$\sqrt{x} \cdot \sqrt{ay} = \sqrt{(x)(ay)} = \sqrt{axy}$$

Write the product of each of the following:

(a) $\sqrt{6y} \cdot \sqrt{7} = \sqrt{}$ (c) $\sqrt{2x} \cdot \sqrt{3y} = \sqrt{}$

(b) $\sqrt{x} \cdot \sqrt{3m} = \sqrt{}$ (d) $\sqrt{ax} \cdot \sqrt{by} = \sqrt{}$

$\sqrt{bt}$

---

a) $\sqrt{42y}$
b) $\sqrt{3mx}$
c) $\sqrt{6xy}$
d) $\sqrt{abxy}$

18 Squares and Square Root Radicals

74. The same multiplication pattern is used when the radicands involve squared letters. For example:

$$\sqrt{5} \cdot \sqrt{t^2} = \sqrt{(5)(t^2)} = \sqrt{5t^2}$$
$$\sqrt{3a} \cdot \sqrt{7x^2} = \sqrt{(3a)(7x^2)} = \sqrt{21ax^2}$$
$$\sqrt{t^2} \cdot \sqrt{b^2} = \sqrt{(t^2)(b^2)} = \sqrt{b^2t^2}$$

In each product above, the radicand is a letter-term which indicates a multiplication. That is:

$5t^2$ means $(5)(t^2)$ or "multiply $\underline{5}$ and $\underline{t^2}$".

(a) $21ax^2$ means $(21)(a)(x^2)$ or "multiply ____ and ____ and ____".

(b) b^2t^2 means $(b^2)(t^2)$ or "multiply ____ and ____".

a) $\underline{21}$ and $\underline{a}$ and $\underline{x^2}$
b) $\underline{b^2}$ and $\underline{t^2}$

75. Write the product of each of the following:

(a) $\sqrt{d^2} \cdot \sqrt{7} = \sqrt{}$

(b) $\sqrt{2y^2} \cdot \sqrt{11} = \sqrt{}$

(c) $\sqrt{4x^2} \cdot \sqrt{3a} = \sqrt{}$

(d) $\sqrt{m^2} \cdot \sqrt{5c^2} = \sqrt{}$

a) $\sqrt{7d^2}$
b) $\sqrt{22y^2}$
c) $\sqrt{12ax^2}$
d) $\sqrt{5c^2m^2}$

76. In each factor below, the radicand contains the same letter.

$$\sqrt{3x} \cdot \sqrt{2x} = \sqrt{(3x)(2x)} = \sqrt{6xx}$$

Since "xx" means "multiply $\underline{x}$ and $\underline{x}$", it indicates a squaring of "x". This indicated squaring of "x" is ordinarily written as "x^2" since the latter form is shorter. Therefore:

$$\sqrt{3x} \cdot \sqrt{2x} = \sqrt{(3x)(2x)} = \sqrt{6x^2}$$

Perform each of these multiplications:

(a) $\sqrt{4y} \cdot \sqrt{5y} = \sqrt{}$

(b) $\sqrt{8t} \cdot \sqrt{7t} = \sqrt{}$

(c) $\sqrt{ax} \cdot \sqrt{7x} = \sqrt{}$

(d) $\sqrt{cd} \cdot \sqrt{cd} = \sqrt{}$

a) $\sqrt{20y^2}$
b) $\sqrt{56t^2}$
c) $\sqrt{7ax^2}$
d) $\sqrt{c^2d^2}$

77. The same general pattern is used to multiply three or more radicals. That is:

$$\sqrt{\square} \cdot \sqrt{\triangle} \cdot \sqrt{\bigcirc} = \sqrt{(\square)(\triangle)(\bigcirc)}$$

Here are some examples:

$$\sqrt{5} \cdot \sqrt{7} \cdot \sqrt{3} = \sqrt{(5)(7)(3)} = \sqrt{105}$$
$$\sqrt{11} \cdot \sqrt{a} \cdot \sqrt{b} = \sqrt{(11)(a)(b)} = \sqrt{11ab}$$
$$\sqrt{2} \cdot \sqrt{x} \cdot \sqrt{5} \cdot \sqrt{y^2} = \sqrt{(2)(x)(5)(y^2)} = \sqrt{10xy^2}$$

Write the product of each of the following:

(a) $\sqrt{2} \cdot \sqrt{t} \cdot \sqrt{3} = \sqrt{}$

(b) $\sqrt{a} \cdot \sqrt{x} \cdot \sqrt{5t} = \sqrt{}$

(c) $\sqrt{3} \cdot \sqrt{5a} \cdot \sqrt{ax} = \sqrt{}$

(d) $\sqrt{d^2} \cdot \sqrt{6} \cdot \sqrt{m} \cdot \sqrt{11m} = \sqrt{}$

Answer to Frame 77:   a) $\sqrt{6t}$ b) $\sqrt{5atx}$ c) $\sqrt{15a^2x}$ d) $\sqrt{66d^2m^2}$

SELF-TEST 5 (Frames 67-77)

Write each of the following as a single radical:

1. $\sqrt{3} \cdot \sqrt{5} =$ _____
2. $\sqrt{2p} \cdot \sqrt{7} =$ _____
3. $\sqrt{h} \cdot \sqrt{b} =$ _____
4. $\sqrt{6r} \cdot \sqrt{5d} =$ _____
5. $\sqrt{3t} \cdot \sqrt{2a^2} =$ _____
6. $\sqrt{7x} \cdot \sqrt{3x} =$ _____
7. $\sqrt{2rs^2} \cdot \sqrt{3ar} =$ _____
8. $\sqrt{8} \cdot \sqrt{3} \cdot \sqrt{5} =$ _____
9. $\sqrt{b} \cdot \sqrt{2d} \cdot \sqrt{bh} =$ _____

ANSWERS: 1. $\sqrt{15}$ 4. $\sqrt{30dr}$ 7. $\sqrt{6ar^2s^2}$
           2. $\sqrt{14p}$ 5. $\sqrt{6a^2t}$ 8. $\sqrt{120}$
           3. $\sqrt{bh}$ 6. $\sqrt{21x^2}$ 9. $\sqrt{2b^2dh}$

1-6   MULTIPLYING NON-RADICALS AND SQUARE ROOT RADICALS

In this section, we will examine multiplications involving both non-radicals and square root radicals. The procedure for converting a multiplication of non-radicals and radicals into a single radical is also discussed.

78.   The multiplication of a non-radical and a radical is indicated by writing the non-radical in front of the radical. For example:

   $2\sqrt{5}$ means "multiply 2 and $\sqrt{5}$".

   $3\sqrt{7x}$ means "multiply 3 and $\sqrt{7x}$".

   $a\sqrt{bc}$ means "multiply _____ and _____".

79.   When writing the multiplication of a non-radical and a radical, we always write the non-radical first. That is:

   "Multiply 3 and $\sqrt{5}$" is written $3\sqrt{5}$, not $\sqrt{5} \cdot 3$

We write the non-radical first so that it does not get confused with the radicand of the radical.

Write each of the following in the preferred form:

   (a) Multiply x and $\sqrt{5}$ _____
   (b) Multiply $\sqrt{7}$ and 2   _____
   (c) Multiply $\sqrt{2R}$ and 7t   _____
   (d) Multiply $\sqrt{cd}$ and R   _____

a and $\sqrt{bc}$

20 Squares and Square Root Radicals

80. In the multiplications below, two factors are non-radicals and one factor is a radical. Notice how the multiplication is simplified by multiplying the two non-radicals.

$$3\left(4\sqrt{7}\right) = 12\sqrt{7}$$
$$5 \cdot 2\sqrt{3} = 10\sqrt{3}$$

Simplify each of the following multiplications:

(a) $9\left(2\sqrt{5}\right) = $ _____

(b) $y\left(4\sqrt{2x}\right) = $ _____

(c) $3 \cdot 6\sqrt{5t^2} = $ _____

a) $x\sqrt{5}$
b) $2\sqrt{7}$
c) $7t\sqrt{2R}$
d) $R\sqrt{cd}$

81. In the multiplication below, one factor is a non-radical and two factors are radicals. Notice how the multiplication is simplified by multiplying the two radicals.

$$7\sqrt{3} \cdot \sqrt{2} = 7\sqrt{6}$$

Simplify each of the following three-factor multiplications:

(a) $5\sqrt{3} \cdot \sqrt{11} = $ _____

(b) $9\sqrt{7x^2} \cdot \sqrt{y} = $ _____

(c) $m\sqrt{2} \cdot \sqrt{3t} = $ _____

a) $18\sqrt{5}$
b) $4y\sqrt{2x}$
c) $18\sqrt{5t^2}$

82. In the multiplication below, two factors are non-radicals and two factors are radicals. Notice how the order of the factors is rearranged so that the multiplication can be simplified.

$$2\sqrt{3} \cdot 5\sqrt{7} = \underline{(2)(5)}\,\sqrt{3}\,\sqrt{7}$$
$$= 10 \quad \sqrt{21}$$

Simplify each of the following:

(a) $10\sqrt{11} \cdot 5\sqrt{2} = $ _____

(b) $3\sqrt{5d} \cdot a\sqrt{7} = $ _____

(c) $c\sqrt{b} \cdot d\sqrt{ab} = $ _____

a) $5\sqrt{33}$
b) $9\sqrt{7x^2y}$
c) $m\sqrt{6t}$

83. Simplify each of the following:

(a) $m\sqrt{v} \cdot \sqrt{t} = $ _____   (c) $5\sqrt{d^2} \cdot 4\sqrt{2F} = $ _____

(b) $8\left(5\sqrt{x}\right) = $ _____   (d) $a\left(3\sqrt{5b}\right) = $ _____

a) $50\sqrt{22}$
b) $3a\sqrt{35d}$
c) $cd\sqrt{ab^2}$

84. Simplify each of the following:

(a) $4a\sqrt{7m} \cdot 3\sqrt{10p} = $ _____

(b) $5\sqrt{2a} \cdot \sqrt{3a} = $ _____

a) $m\sqrt{tv}$ c) $20\sqrt{2d^2F}$
b) $40\sqrt{x}$ d) $3a\sqrt{5b}$

a) $12a\sqrt{70mp}$
b) $5\sqrt{6a^2}$

Squares and Square Root Radicals   21

85. <u>Any positive number is equal to the square root radical whose radicand is the square of the number.</u> For example:

$$3 = \sqrt{9} \qquad 7 = \sqrt{49} \qquad 9 = \sqrt{81}$$

Write the square root radical which is equal to each of these numbers:

(a) $2 = \sqrt{}$ (b) $5 = \sqrt{}$ (c) $10 = \sqrt{}$

a) $\sqrt{4}$
b) $\sqrt{25}$
c) $\sqrt{100}$

86. Since $\sqrt{4}$ can be substituted for "2", the multiplication below can be converted into a multiplication of two radicals.

$$2\sqrt{5} = \sqrt{4} \cdot \sqrt{5}$$

Convert each of the following multiplications into a multiplication of two radicals:

(a) $3\sqrt{8} = \sqrt{} \cdot \sqrt{8}$ (c) $9\sqrt{x} = \sqrt{} \cdot \sqrt{x}$
(b) $5\sqrt{2} = \sqrt{} \cdot \sqrt{2}$ (d) $4\sqrt{y} = \sqrt{} \cdot \sqrt{y}$

a) $\sqrt{9} \cdot \sqrt{8}$
b) $\sqrt{25} \cdot \sqrt{2}$
c) $\sqrt{81} \cdot \sqrt{x}$
d) $\sqrt{16} \cdot \sqrt{y}$

87. Any letter is equal to the square root radical whose radicand is the indicated square of the letter. For example:

$$x = \sqrt{x^2}$$

(a) $y = \sqrt{}$ (b) $R = \sqrt{}$ (c) $m = \sqrt{}$

a) $\sqrt{y^2}$
b) $\sqrt{R^2}$
c) $\sqrt{m^2}$

88. Since $\sqrt{x^2}$ can be substituted for "x", the multiplication below can be converted into a multiplication of two radicals.

$$x\sqrt{t} = \sqrt{x^2} \cdot \sqrt{t}$$

Convert each of the following multiplications into a multiplication of two radicals:

(a) $y\sqrt{5} = \sqrt{} \cdot \sqrt{5}$
(b) $R\sqrt{d} = \sqrt{} \cdot \sqrt{d}$
(c) $a\sqrt{bc} = \sqrt{} \cdot \sqrt{bc}$

a) $\sqrt{y^2} \cdot \sqrt{5}$
b) $\sqrt{R^2} \cdot \sqrt{d}$
c) $\sqrt{a^2} \cdot \sqrt{bc}$

89. After converting a multiplication of a non-radical and a radical into a multiplication of two radicals, the product can be written as a single radical. For example:

$$5\sqrt{2} = \sqrt{25} \cdot \sqrt{2} = \sqrt{50}$$
$$4\sqrt{x} = \sqrt{16} \cdot \sqrt{x} = \sqrt{16x}$$
$$a\sqrt{b} = \sqrt{a^2} \cdot \sqrt{b} = \sqrt{a^2 b}$$

Following the pattern above, write each of the following as a single radical:

(a) $3\sqrt{7} =$ _____   (c) $m\sqrt{d} =$ _____
(b) $8\sqrt{t} =$ _____   (d) $y\sqrt{7} =$ _____

22 Squares and Square Root Radicals

90. In each case below, write the right side of the equation as a single radical:

(a) If $y = 10\sqrt{7}$, $y = \sqrt{}$

(b) If $t = 6\sqrt{R}$, $t = \sqrt{}$

(c) If $x = b\sqrt{c}$, $x = \sqrt{}$

(d) If $V = m\sqrt{2y}$, $V = \sqrt{}$

a) $\sqrt{9} \cdot \sqrt{7} = \sqrt{63}$
b) $\sqrt{64} \cdot \sqrt{t} = \sqrt{64t}$
c) $\sqrt{m^2} \sqrt{d} = \sqrt{dm^2}$
d) $\sqrt{y^2} \sqrt{7} = \sqrt{7y^2}$

91. By two substitutions, each of the multiplications below is converted into a multiplication of three radicals and then written as a single radical.

$3x\sqrt{y} = \sqrt{9} \cdot \sqrt{x^2} \cdot \sqrt{y} = \sqrt{9x^2y}$

$cd\sqrt{a} = \sqrt{c^2} \cdot \sqrt{d^2} \cdot \sqrt{a} = \sqrt{ac^2d^2}$

Convert each of the following to a multiplication of three radicals and then write the product as a single radical:

(a) $5y\sqrt{2t} =$ _____ = _____

(b) $xy\sqrt{ab} =$ _____ = _____

a) $y = \sqrt{700}$
b) $t = \sqrt{36R}$
c) $x = \sqrt{b^2c}$
d) $V = \sqrt{2m^2y}$

92. By three substitutions, each multiplication below is written as a multiplication of four radicals and then written as a single radical.

$3ab\sqrt{d} = \sqrt{9} \cdot \sqrt{a^2} \cdot \sqrt{b^2} \cdot \sqrt{d} = \sqrt{9a^2b^2d}$

$axy\sqrt{3R} = \sqrt{a^2} \cdot \sqrt{x^2} \cdot \sqrt{y^2} \cdot \sqrt{3R} = \sqrt{3a^2Rx^2y^2}$

Convert each of these to a multiplication of four radicals and then write the product as a single radical:

(a) $7mp\sqrt{2F} =$ _____ = _____

(b) $cdt\sqrt{y} =$ _____ = _____

a) $\sqrt{25} \cdot \sqrt{y^2} \cdot \sqrt{2t} = \sqrt{50ty^2}$
b) $\sqrt{x^2} \cdot \sqrt{y^2} \cdot \sqrt{ab} = \sqrt{abx^2y^2}$

93. Convert each of the following multiplications to a single radical:

(a) $5\sqrt{x} =$ _____   (c) $2x\sqrt{3y} =$ _____

(b) $3\sqrt{7} =$ _____   (d) $4cd\sqrt{a} =$ _____

a) $\sqrt{49} \cdot \sqrt{m^2} \cdot \sqrt{p^2} \cdot \sqrt{2F}$
$= \sqrt{98Fm^2p^2}$
b) $\sqrt{c^2} \cdot \sqrt{d^2} \cdot \sqrt{t^2} \cdot \sqrt{y}$
$= \sqrt{c^2d^2t^2y}$

a) $\sqrt{25x}$ c) $\sqrt{12x^2y}$
b) $\sqrt{63}$ d) $\sqrt{16ac^2d^2}$

SELF-TEST 6 (Frames 78-93)

Multiply. Write each product in the form $\boxed{}\sqrt{\bigcirc}$:

1. $5\left(8\sqrt{2}\right) =$ _____
2. $2\sqrt{5} \cdot 3\sqrt{2} =$ _____
3. $t\sqrt{r} \cdot r\sqrt{t} =$ _____
4. $3b\sqrt{2w} \cdot ab\sqrt{3w} =$ _____

In Problems 5-7, find each radicand:

5. $4 = \sqrt{}$
6. $h = \sqrt{}$
7. $E = \sqrt{}$

Write each of the following as a single radical:

8. $3\sqrt{2} =$ _____
9. $2t\sqrt{3s} =$ _____
10. $b\sqrt{ad} =$ _____
11. $4pw\sqrt{5t} =$ _____

ANSWERS:
1. $40\sqrt{2}$
2. $6\sqrt{10}$
3. $rt\sqrt{rt}$
4. $3ab^2\sqrt{6w^2}$
5. $\sqrt{16}$
6. $\sqrt{h^2}$
7. $\sqrt{E^2}$
8. $\sqrt{18}$
9. $\sqrt{12st^2}$
10. $\sqrt{ab^2d}$
11. $\sqrt{80p^2tw^2}$

1-7 THE SQUARE ROOTS OF MULTIPLICATIONS

The square root of any multiplication can be factored into a multiplication of the square roots of each of its factors. This factoring process, which is simply the reverse of the process for multiplying radicals, will be discussed in this section. The meaning of "factoring out perfect squares" from a radical is also discussed.

94. Here is the pattern for multiplying two square root radicals.

$$\sqrt{\square} \cdot \sqrt{\triangle} = \sqrt{(\square)(\triangle)}$$

By examining the pattern, you can see this fact: When two square root radicals are multiplied, the product is a single square root radical whose radicand is a multiplication.

Therefore, when a square root radical contains a radicand which is a multiplication, we can reverse the procedure and factor the single radical into the two original square root radicals. Here is the pattern:

$$\sqrt{(\square)(\triangle)} = \sqrt{\square} \cdot \sqrt{\triangle}$$

(Continued on following page.)

24 Squares and Square Root Radicals

94. (Continued)

Here are some examples of this factoring procedure:
$$\sqrt{(7)(11)} = \sqrt{7} \cdot \sqrt{11}$$
$$\sqrt{3x} = \sqrt{3} \cdot \sqrt{x}$$
$$\sqrt{mp} = \sqrt{m} \cdot \sqrt{p}$$

Factor each of these single radicals into two radicals:

(a) $\sqrt{7y} = \sqrt{} \cdot \sqrt{}$

(b) $\sqrt{bc} = \sqrt{} \cdot \sqrt{}$

(c) $\sqrt{(5)(8)} = \sqrt{} \cdot \sqrt{}$

---

95. The same pattern for factoring applies if the radicand contains three or more factors. For example:
$$\sqrt{5mv} = \sqrt{5} \cdot \sqrt{m} \cdot \sqrt{v}$$
$$\sqrt{axy} = \sqrt{a} \cdot \sqrt{x} \cdot \sqrt{y}$$
$$\sqrt{3cde} = \sqrt{3} \cdot \sqrt{c} \cdot \sqrt{d} \cdot \sqrt{e}$$

Following the pattern above, factor each of these single radicals:

(a) $\sqrt{7pq} = $ _____

(b) $\sqrt{13RSV} = $ _____

a) $\sqrt{7} \cdot \sqrt{y}$
b) $\sqrt{b} \cdot \sqrt{c}$
c) $\sqrt{5} \cdot \sqrt{8}$

---

96. When the square root of a multiplication is factored, any radical factor containing a perfect square (or an indicated square) can be replaced by a non-radical. For example:
$$\sqrt{(16)(5)} = \sqrt{16} \cdot \sqrt{5} = 4\sqrt{5}$$
$$\sqrt{7R^2} = \sqrt{7} \cdot \sqrt{R^2} = R\sqrt{7}$$
$$\sqrt{9y^2} = \sqrt{9} \cdot \sqrt{y^2} = 3y$$

Factor each of the following and substitute non-radicals wherever possible:

(a) $\sqrt{(3)(49)} = $ _____ = _____

(b) $\sqrt{25x} = $ _____ = _____

(c) $\sqrt{36p^2} = $ _____ = _____

a) $\sqrt{7} \cdot \sqrt{p} \cdot \sqrt{q}$
b) $\sqrt{13} \cdot \sqrt{R} \cdot \sqrt{S} \cdot \sqrt{V}$

---

97. Factor each of the following and substitute non-radicals wherever possible:

(a) $\sqrt{a^2b^2} = $ _____ = _____

(b) $\sqrt{9xy^2} = $ _____ = _____

(c) $\sqrt{m^2pq^2} = $ _____ = _____

(d) $\sqrt{64ac^2F^2} = $ _____ = _____

a) $\sqrt{3} \cdot \sqrt{49} = 7\sqrt{3}$
b) $\sqrt{25} \cdot \sqrt{x} = 5\sqrt{x}$
c) $\sqrt{36} \cdot \sqrt{p^2} = 6p$

98. When the radicand of a square root of a multiplication contains a perfect square or an indicated square, we ordinarily factor it and substitute non-radicals wherever possible. We do so because the latter form is easier to evaluate. For example:

$a\sqrt{b}$ is easier to evaluate than $\sqrt{a^2 b}$,
since $a\sqrt{b}$ requires only a square root and a multiplication, whereas $\sqrt{a^2 b}$ requires a squaring, a multiplication, and a square root.

xy is easier to evaluate than $\sqrt{x^2 y^2}$,
since xy requires only a multiplication, whereas $\sqrt{x^2 y^2}$ requires two squarings, a multiplication, and a square root.

Write the right side of each equation below in the preferred form for evaluation:

(a) If $y = \sqrt{25x^2}$, y = _____
(b) If $t = \sqrt{4a^2 s}$, t = _____
(c) If $V = \sqrt{ad^2}$, V = _____
(d) If $R = \sqrt{9a^2 bc^2}$, R = _____

a) $\sqrt{a^2} \cdot \sqrt{b^2} = ab$
b) $\sqrt{9} \cdot \sqrt{x} \cdot \sqrt{y^2} = 3y\sqrt{x}$
c) $\sqrt{m^2} \cdot \sqrt{p} \cdot \sqrt{q^2} = mq\sqrt{p}$
d) $\sqrt{64} \cdot \sqrt{a} \cdot \sqrt{c^2} \cdot \sqrt{F^2}$
   $= 8cF\sqrt{a}$

---

99. The procedure we have been using with the square roots of multiplications is called "<u>factoring out perfect squares</u>". That is:

In writing $\sqrt{16x^2 y}$ as $4x\sqrt{y}$,
we have factored out "16" and "x^2".

(a) In writing $\sqrt{a^2 bc^2}$ as $ac\sqrt{b}$,
we have factored out _____ and _____.
(b) In writing $\sqrt{81xy^2 z^2}$ as $9yz\sqrt{x}$,
we have factored out _____, _____, and _____.

a) y = 5x
b) $t = 2a\sqrt{s}$
c) $V = d\sqrt{a}$
d) $R = 3ac\sqrt{b}$

---

100. "Factor out the perfect squares" in the right side of each equation below:

(a) If $a = \sqrt{x^2 y^2}$, a = _____
(b) If $t = \sqrt{9a^2 b^2}$, t = _____

a) a^2 and c^2
b) 81, y^2, and z^2

---

101. When multiplying radicals, we ordinarily "factor out the perfect squares" in the product. For example:

$\sqrt{7x} \cdot \sqrt{3x} = \sqrt{21x^2} = x\sqrt{21}$

$\sqrt{3a^2} \cdot \sqrt{12b} = \sqrt{36a^2 b} = 6a\sqrt{b}$

Perform the multiplications below and factor out the perfect squares in the products:

(a) $\sqrt{8} \cdot \sqrt{2F}$ = _____ = _____
(b) $\sqrt{5t^2} \cdot \sqrt{7y^2}$ = _____ = _____
(c) $\sqrt{32m} \cdot \sqrt{2x^2}$ = _____ = _____

a) xy
b) 3ab

Squares and Square Root Radicals   25

26   Squares and Square Root Radicals

102. When "factoring out perfect squares" from a radical, we leave all of the non-perfect-square factors <u>under</u> <u>a</u> <u>single</u> <u>radical</u>. That is:

   Instead of writing $\sqrt{a^2bc} = a\sqrt{b} \cdot \sqrt{c}$,
   we write $\sqrt{a^2bc} = a\sqrt{bc}$

   Instead of writing $\sqrt{16mpq^2} = 4q\sqrt{m} \cdot \sqrt{p}$,
   we write $\sqrt{16mpq^2} = $ _____

a) $\sqrt{16F} = 4\sqrt{F}$
b) $\sqrt{35t^2y^2} = ty\sqrt{35}$
c) $\sqrt{64mx^2} = 8x\sqrt{m}$

---

103. Factor out the perfect squares in the radicals below:

   (a) $\sqrt{36F_1F_2} = $ _____   (c) $\sqrt{81g^2hk} = $ _____
   (b) $\sqrt{RT^2S} = $ _____   (d) $\sqrt{abcd^2} = $ _____

$4q\sqrt{mp}$

---

104. When the radicals are multiplied in each case below, we can factor out a perfect square from their product.

   $a\sqrt{3x} \cdot \sqrt{2x} = a\sqrt{6x^2} = ax\sqrt{6}$
   $2\sqrt{3} \cdot \sqrt{12h} = 2\sqrt{36h} = 2 \cdot 6\sqrt{h} = 12\sqrt{h}$

   Do these, factoring out perfect squares whenever possible:

   (a) $3\sqrt{8a} \cdot \sqrt{2b} = $ _____   (c) $\sqrt{3x} \cdot 7\sqrt{5x} = $ _____
   (b) $R\sqrt{12y} \cdot \sqrt{3g} = $ _____   (d) $t\sqrt{2y} \cdot 6\sqrt{5y} = $ _____

a) $6\sqrt{F_1F_2}$ c) $9g\sqrt{hk}$
b) $T\sqrt{RS}$ d) $d\sqrt{abc}$

---

105. If the radicand is a <u>number</u>:

   The radical can be factored if the number is factorable. For example:
   $\sqrt{14} = \sqrt{(2)(7)} = \sqrt{2} \cdot \sqrt{7}$

   The radical cannot be factored if the number is not factorable. For example:
   $\sqrt{13}$, $\sqrt{17}$, $\sqrt{43}$ are not factorable.

   Complete the following factorings:
   (a) $\sqrt{6} = \sqrt{(3)(2)} = \sqrt{} \cdot \sqrt{}$
   (b) $\sqrt{15} = \sqrt{()()} = \sqrt{} \cdot \sqrt{}$
   (c) $\sqrt{39} = \sqrt{()()} = \sqrt{} \cdot \sqrt{}$
   (d) $\sqrt{11} = \sqrt{()()} = \sqrt{} \cdot \sqrt{}$

a) $12\sqrt{ab}$
b) $6R\sqrt{gy}$
c) $7x\sqrt{15}$
d) $6ty\sqrt{10}$

---

106. Whenever the radicand is a number (like 18) which can be factored in various ways, the radical can be factored in various ways. Always try to find a "perfect-square" factor so that the perfect square can be factored out. For example:

   Instead of factoring this way:
   $\sqrt{18} = \sqrt{(6)(3)} = \sqrt{6} \cdot \sqrt{3}$

   We factor this way so that we can factor out the perfect square:
   $\sqrt{18} = \sqrt{(9)(2)} = 3\sqrt{2}$

(Continued on following page.)

a) $\phantom{\sqrt{(0)(0)}} = \sqrt{3} \cdot \sqrt{2}$
b) $\sqrt{(5)(3)} = \sqrt{5} \cdot \sqrt{3}$
c) $\sqrt{(3)(13)} = \sqrt{3} \cdot \sqrt{13}$
d) Not factorable, since "11" cannot be factored.

Squares and Square Root Radicals   27

106. (Continued)

Which one of the following ways of factoring 12 allows us to factor out a perfect square?

(a) $\sqrt{12} = \sqrt{(6)(2)}$ (b) $\sqrt{12} = \sqrt{(4)(3)}$

(b) We get: $2\sqrt{3}$

107. Complete each of the following so that a perfect-square factor can be factored out:

(a) $\sqrt{27} = \sqrt{()()} = \underline{}$

(b) $\sqrt{75} = \sqrt{()()} = \underline{}$

(c) $\sqrt{500} = \sqrt{()()} = \underline{}$

108. Factor a perfect square out of each of the following:

(a) $\sqrt{8} = \underline{}$ (c) $\sqrt{50} = \underline{}$

(b) $\sqrt{20} = \underline{}$ (d) $\sqrt{28} = \underline{}$

a) $\sqrt{(9)(3)} = 3\sqrt{3}$
b) $\sqrt{(25)(3)} = 5\sqrt{3}$
c) $\sqrt{(100)(5)} = 10\sqrt{5}$

109. When factoring a numerical radicand, look for the <u>largest</u> perfect-square factor. Otherwise, a perfect square can still be factored out of the remaining radical. For example, since 16 is the largest perfect-square factor of 32, $\sqrt{32} = \sqrt{(16)(2)} = 4\sqrt{2}$.

If we factor $\sqrt{32}$ this way:

$\sqrt{32} = \sqrt{(4)(8)} = 2\sqrt{8}$

we can still factor a perfect square out of $\sqrt{8}$. We get:

$\sqrt{32} = \sqrt{(4)(8)} = 2\sqrt{8} = 2\sqrt{(4)(2)} = 4\sqrt{2}$

In factoring these, look for the largest perfect-square factor:

(a) $\sqrt{48} = \underline{}$ (c) $\sqrt{96} = \underline{}$

(b) $\sqrt{72} = \underline{}$ (d) $\sqrt{288} = \underline{}$

a) $2\sqrt{2}$ c) $5\sqrt{2}$
b) $2\sqrt{5}$ c) $2\sqrt{7}$

110. When the radicand contains a numerical coefficient which is not a perfect square, factor a perfect square out of the coefficient if possible. For example:

$\sqrt{48x} = \sqrt{(16)(3)(x)} = 4\sqrt{3x}$

$\sqrt{200h^2} = \sqrt{(2)(100)(h^2)} = 10h\sqrt{2}$

Factor out as many perfect squares as possible from these:

(a) $\sqrt{125q} = \underline{}$ (c) $\sqrt{56ab} = \underline{}$

(b) $\sqrt{15t} = \underline{}$ (d) $\sqrt{45x^2y^2} = \underline{}$

a) $4\sqrt{3}$, from $\sqrt{(16)(3)}$
b) $6\sqrt{2}$, from $\sqrt{(36)(2)}$
c) $4\sqrt{6}$, from $\sqrt{(16)(6)}$
d) $12\sqrt{2}$, from $\sqrt{(144)(2)}$

111. Factor out as many perfect squares as possible from these:

(a) $\sqrt{12xy} = \underline{}$ (c) $\sqrt{200p^2q^2} = \underline{}$

(b) $\sqrt{18a^2b} = \underline{}$ (d) $\sqrt{85dt^2} = \underline{}$

a) $5\sqrt{5q}$
b) No perfect square can be factored out.
c) $2\sqrt{14ab}$
d) $3xy\sqrt{5}$

112. Perform these multiplications and factor out the perfect squares from their products:

(a) $\sqrt{5ac^2} \cdot \sqrt{10b} =$ _____

(b) $4\sqrt{6h} \cdot \sqrt{2v} =$ _____

a) $5c\sqrt{2ab}$
b) $8\sqrt{3hv}$

SELF-TEST 7 (Frames 94–112)

1. Factor this radical into two radicals: $\sqrt{2H} =$ _____
2. Factor this radical into three radicals: $\sqrt{5rt} =$ _____

Factor out the perfect squares in the radicals below:

3. $\sqrt{25a} =$ _____ 4. $\sqrt{9r^2w^2} =$ _____ 5. $\sqrt{4mv^2} =$ _____

Do these multiplications, factoring out perfect squares wherever possible:

6. $\sqrt{3t} \cdot 2\sqrt{3h} =$ _____ 7. $\sqrt{2bp} \cdot \sqrt{18p} =$ _____

Factor out as many perfect squares as possible from each of these:

8. $\sqrt{300} =$ _____ 9. $\sqrt{24xy^2} =$ _____ 10. $\sqrt{63b^2c^2} =$ _____

Do these multiplications, factoring out perfect squares wherever possible:

11. $\sqrt{2st} \cdot \sqrt{10s} =$ _____ 12. $3\sqrt{6mv^2} \cdot \sqrt{12m} =$ _____

ANSWERS:
1. $\sqrt{2} \cdot \sqrt{H}$
2. $\sqrt{5} \cdot \sqrt{r} \cdot \sqrt{t}$
3. $5\sqrt{a}$
4. $3rw$
5. $2v\sqrt{m}$
6. $6\sqrt{ht}$
7. $6p\sqrt{b}$
8. $10\sqrt{3}$
9. $2y\sqrt{6x}$
10. $3bc\sqrt{7}$
11. $2s\sqrt{5t}$
12. $18mv\sqrt{2}$

1-8 DIVISION OF SQUARE ROOT RADICALS AND THE SQUARE ROOTS OF FRACTIONS

This section will begin with a brief discussion of the division of two square root radicals. Then we will discuss the square roots of fractions, showing how these square roots are related to the division of two radicals.

Squares and Square Root Radicals   29

113. Just as other divisions can be written as fractions, any division of two square root radicals can be written as a fraction. For example:

"Divide $\sqrt{36}$ by $\sqrt{9}$" is written $\dfrac{\sqrt{36}}{\sqrt{9}}$

"Divide $\sqrt{ab^2}$ by $\sqrt{c}$" is written $\dfrac{\sqrt{ab^2}}{\sqrt{c}}$

Therefore: (a) $\dfrac{\sqrt{12}}{\sqrt{3}}$ means "divide _____ by _____".

(b) $\dfrac{\sqrt{4x^2}}{\sqrt{9y}}$ means "divide _____ by _____".

114. A division of radicals can be performed in either of two ways. As an example, we will divide $\sqrt{36}$ by $\sqrt{9}$.

Method 1: $\dfrac{\sqrt{36}}{\sqrt{9}} = \dfrac{6}{3} = 2$

In Method 1, we found the square root in both the numerator and denominator and then divided.

Method 2: $\dfrac{\sqrt{36}}{\sqrt{9}} = \sqrt{\dfrac{36}{9}} = \sqrt{4} = 2$

In Method 2, we wrote the division of the two radicals as the square root of a fraction, simplified the fraction, and then found the square root.

Using the two methods above, divide $\sqrt{64}$ by $\sqrt{4}$.

(a) Method 1: $\dfrac{\sqrt{64}}{\sqrt{4}} =$ _____

(b) Method 2: $\dfrac{\sqrt{64}}{\sqrt{4}} =$ _____

a) $\sqrt{12}$ by $\sqrt{3}$
b) $\sqrt{4x^2}$ by $\sqrt{9y}$

115. The pattern used in Method 2 in the last frame is given below:

$$\dfrac{\sqrt{\square}}{\sqrt{\triangle}} = \sqrt{\dfrac{\square}{\triangle}}$$

This pattern says: A DIVISION OF TWO SQUARE ROOT RADICALS IS EQUAL TO THE SQUARE ROOT OF A FRACTION.

Using this pattern, complete the following:

(a) $\dfrac{\sqrt{12}}{\sqrt{3}} = \sqrt{\dfrac{12}{3}} = \sqrt{4} =$ _____

(b) $\dfrac{\sqrt{32}}{\sqrt{2}} = \sqrt{\dfrac{32}{2}} = \sqrt{} =$ _____

(c) $\dfrac{\sqrt{144}}{\sqrt{4}} = \sqrt{\dfrac{144}{4}} = \sqrt{} =$ _____

(d) $\dfrac{\sqrt{243}}{\sqrt{3}} = \sqrt{\dfrac{243}{3}} = \sqrt{} =$ _____

a) $\dfrac{\sqrt{64}}{\sqrt{4}} = \dfrac{8}{2} = 4$

b) $\dfrac{\sqrt{64}}{\sqrt{4}} = \sqrt{\dfrac{64}{4}} =$
    $= \sqrt{16} = 4$

30 Squares and Square Root Radicals

116. The same pattern can be used to convert <u>any</u> division of two square root radicals to the square root of a fraction. For example:

$$\frac{\sqrt{50}}{\sqrt{7}} = \sqrt{\frac{50}{7}} \qquad\qquad \frac{\sqrt{3b}}{\sqrt{xy}} = \sqrt{\frac{3b}{xy}}$$

Convert each of these to a square root of a fraction:

(a) $\dfrac{\sqrt{67}}{\sqrt{13}} = $ _____ (b) $\dfrac{\sqrt{VT}}{\sqrt{S}} = $ _____

a) ___ = 2
b) $\sqrt{16} = 4$
c) $\sqrt{36} = 6$
d) $\sqrt{81} = 9$

---

117. By interchanging the sides of the pattern, we get the new pattern below:

$$\sqrt{\frac{\square}{\triangle}} = \frac{\sqrt{\square}}{\sqrt{\triangle}}$$

This new form of the pattern is very useful in dealing with the square roots of fractions. It says:

<u>THE SQUARE ROOT OF ANY FRACTION IS EQUAL TO THE SQUARE ROOT OF ITS NUMERATOR DIVIDED BY THE SQUARE ROOT OF ITS DENOMINATOR.</u>

Here are some examples of this new form of the pattern:

$$\sqrt{\frac{93}{17}} = \frac{\sqrt{93}}{\sqrt{17}} \qquad\qquad \sqrt{\frac{3x}{5}} = \frac{\sqrt{3x}}{\sqrt{5}}$$

Complete these: (a) $\sqrt{\dfrac{7}{8}} = \dfrac{\sqrt{}}{\sqrt{}}$ (b) $\sqrt{\dfrac{c}{d}} = \dfrac{\sqrt{}}{\sqrt{}}$

a) $\sqrt{\dfrac{67}{13}}$

b) $\sqrt{\dfrac{VT}{S}}$

---

118. After converting the square root of a fraction into a division of two radicals, we can factor out any perfect-square factors in both the numerator and the denominator. For example:

$$\sqrt{\frac{3x^2}{16}} = \frac{\sqrt{3x^2}}{\sqrt{16}} = \frac{x\sqrt{3}}{4}$$

$$\sqrt{\frac{25y}{cd^2}} = \frac{\sqrt{25y}}{\sqrt{cd^2}} = \frac{5\sqrt{y}}{d\sqrt{c}}$$

$$\sqrt{\frac{a^2}{b^2}} = \frac{\sqrt{a^2}}{\sqrt{b^2}} = \frac{a}{b}$$

Factor out the perfect squares in these:

(a) $\sqrt{\dfrac{v^2 m}{p^2}} = \dfrac{\sqrt{v^2 m}}{\sqrt{p^2}} = $ _____

(b) $\sqrt{\dfrac{ab}{c^2}} = \dfrac{\sqrt{ab}}{\sqrt{c^2}} = $ _____

(c) $\sqrt{\dfrac{4t^2}{xy}} = \dfrac{\sqrt{4t^2}}{\sqrt{xy}} = $ _____

a) $\dfrac{\sqrt{7}}{\sqrt{8}}$

b) $\dfrac{\sqrt{c}}{\sqrt{d}}$

Squares and Square Root Radicals   31

119. Convert each of these square roots to a division of radicals and factor out perfect squares whenever possible:

(a) $\sqrt{\dfrac{m^2 q^2}{ab}} =$ _____

(b) $\sqrt{\dfrac{7x^2}{9y^2}} =$ _____

(c) $\sqrt{\dfrac{5t}{11d}} =$ _____

a) $\dfrac{v\sqrt{m}}{p}$

b) $\dfrac{\sqrt{ab}}{c}$

c) $\dfrac{2t}{\sqrt{xy}}$

---

120. Simplify each of these by converting to a division of two radicals and factoring out perfect squares:

(a) $\sqrt{\dfrac{s^2 t v^2}{cd^2}} =$ _____

(b) $\sqrt{\dfrac{4d^2}{g^2}} =$ _____

(c) $\sqrt{\dfrac{25b^2}{81a^2}} =$ _____

a) $\dfrac{mq}{\sqrt{ab}}$

b) $\dfrac{x\sqrt{7}}{3y}$

c) $\dfrac{\sqrt{5t}}{\sqrt{11d}}$

(No perfect squares can be factored out.)

---

121. To handle cases in which the numerator is "1", you must remember that "1" is a perfect square since $\sqrt{1} = 1$. That is:

$$\sqrt{\dfrac{1}{3x}} = \dfrac{\sqrt{1}}{\sqrt{3x}} = \dfrac{1}{\sqrt{3x}}$$

Convert each of these to a division and factor out perfect squares:

(a) $\sqrt{\dfrac{1}{5t}} =$ _____

(b) $\sqrt{\dfrac{1}{9x}} =$ _____

(c) $\sqrt{\dfrac{1}{16y^2}} =$ _____

(d) $\sqrt{\dfrac{1}{a^2 b^2}} =$ _____

a) $\dfrac{sv\sqrt{t}}{d\sqrt{c}}$

b) $\dfrac{2d}{g}$

c) $\dfrac{5b}{9a}$

---

122. When factoring out perfect squares, always check the numerical coefficients to see whether a perfect square can be factored out of them. For example:

$$\sqrt{\dfrac{32 g^2 h}{45 b^2}} = \dfrac{4g\sqrt{2h}}{3b\sqrt{5}}$$

Note: $\sqrt{32} = \sqrt{16} \cdot \sqrt{2}$ and $\sqrt{45} = \sqrt{9} \cdot \sqrt{5}$

a) $\dfrac{1}{\sqrt{5t}}$

b) $\dfrac{1}{3\sqrt{x}}$

c) $\dfrac{1}{4y}$

d) $\dfrac{1}{ab}$

(Continued on following page.)

32 Squares and Square Root Radicals

122. (Continued)

Simplify these by converting to division and factoring out perfect squares:

(a) $\sqrt{\dfrac{18a}{d^2}}$ = _____

(b) $\sqrt{\dfrac{7y}{50x^2}}$ = _____

(c) $\sqrt{\dfrac{98b^2}{27m}}$ = _____

(d) $\sqrt{\dfrac{1}{75R}}$ = _____

a) $\dfrac{3\sqrt{2a}}{d}$

b) $\dfrac{\sqrt{7y}}{5x\sqrt{2}}$

c) $\dfrac{7b\sqrt{2}}{3\sqrt{3m}}$

d) $\dfrac{1}{5\sqrt{3R}}$

---

123. The square root of a fraction is easier to evaluate than a division of two radicals. For example:

Evaluating $\sqrt{\dfrac{77}{39}}$ requires a division and <u>one</u> square root.

Evaluating $\dfrac{\sqrt{77}}{\sqrt{39}}$ requires <u>two</u> square roots and a division.

Therefore, if neither the numerator nor the denominator of the fraction contains a perfect square, we would not convert the square root of a fraction to a division of radicals.

Which of the following square roots would be left in their present form? _____

(a) $\sqrt{\dfrac{7}{11}}$ (b) $\sqrt{\dfrac{a^2 b}{y^2}}$ (c) $\sqrt{\dfrac{cd}{x}}$

Both (a) and (c)

---

124. However, when one or more perfect squares can be factored out of the numerator or denominator or both, converting to division and factoring out perfect squares usually simplifies the evaluation. For example:

Evaluating $\sqrt{\dfrac{a^2}{b}}$ requires a squaring, a division, and a square root.

Evaluating $\dfrac{a}{\sqrt{b}}$ requires only a square root and a division.

In which of the cases below should we convert to division and factor out perfect squares? _____

(a) $\sqrt{\dfrac{x^2 y}{m^2}}$ (b) $\sqrt{\dfrac{ab}{V}}$ (c) $\sqrt{\dfrac{7t}{19m}}$

Only in (a)

Squares and Square Root Radicals   33

125. After converting to a division of radicals and factoring out perfect squares, we sometimes obtain a fraction with a radical in both its numerator or denominator. In such cases, it is usually easier to evaluate if this division of two radicals is converted back to a square root of a fraction. For example:

Instead of $\dfrac{a\sqrt{w}}{\sqrt{t}}$, we write $a\sqrt{\dfrac{w}{t}}$, because:

Evaluating $\dfrac{a\sqrt{w}}{\sqrt{t}}$ requires <u>two</u> square roots, a multiplication, and a division.

Evaluating $a\sqrt{\dfrac{w}{t}}$ requires a division, only <u>one</u> square root, and a multiplication.

Write each of the following in the preferred form for evaluation:

(a) $\dfrac{3\sqrt{x}}{4\sqrt{y}} = $ _____

(b) $\dfrac{2x\sqrt{d}}{\sqrt{R}} = $ _____

126. There is a special problem with expressions like those below in which only the denominator contains a non-radical factor or factors.

$$\dfrac{\sqrt{a}}{b\sqrt{c}} \qquad \dfrac{\sqrt{xy}}{5\sqrt{R}} \qquad \dfrac{\sqrt{t}}{3y\sqrt{x}}$$

The problem is this: <u>What should be done with the non-radical factors when the two radicals are converted back to one radical?</u>

To handle this conversion, we insert "1" as a non-radical factor in the numerator. We get:

$$\dfrac{\sqrt{a}}{b\sqrt{c}} = \dfrac{1\sqrt{a}}{b\sqrt{c}} = \dfrac{1}{b}\sqrt{\dfrac{a}{c}}$$

(a) $\dfrac{\sqrt{xy}}{5\sqrt{R}} = \dfrac{1\sqrt{xy}}{5\sqrt{R}} = $ _____

(b) $\dfrac{\sqrt{t}}{3y\sqrt{x}} = \dfrac{1\sqrt{t}}{3y\sqrt{x}} = $ _____

127. Write each of these in the preferred form for evaluation:

(a) $\sqrt{\dfrac{a}{9b^2}} = $ _____

(b) $\sqrt{\dfrac{ab}{c}} = $ _____

(c) $\sqrt{\dfrac{m^2 t}{R}} = $ _____

a) $\dfrac{3}{4}\sqrt{\dfrac{x}{y}}$

b) $2x\sqrt{\dfrac{d}{R}}$

a) $\dfrac{1}{5}\sqrt{\dfrac{xy}{R}}$

b) $\dfrac{1}{3y}\sqrt{\dfrac{t}{x}}$

34 Squares and Square Root Radicals

128. Write each of these in the preferred form for evaluation:

(a) $\sqrt{\dfrac{11t^2}{vs^2}} = $ _____

(b) $\sqrt{\dfrac{t^2d^2}{V}} = $ _____

(c) $\sqrt{\dfrac{y}{9x}} = $ _____

a) $\dfrac{\sqrt{a}}{3b}$

b) $\sqrt{\dfrac{ab}{c}}$ (It is in the preferred form.)

c) $m\sqrt{\dfrac{t}{R}}$

---

129. Before doing anything with the square root of a fraction, always check to see that the fraction is in lowest terms. Here are some examples in which the numerical coefficients can be reduced to lowest terms.

$\sqrt{\dfrac{48t}{8b^2}} = \sqrt{\dfrac{6t}{b^2}} = \dfrac{\sqrt{6t}}{\sqrt{b^2}} = \dfrac{\sqrt{6t}}{b}$

$\sqrt{\dfrac{9x^2}{18y}} = \sqrt{\dfrac{x^2}{2y}} = \dfrac{\sqrt{x^2}}{\sqrt{2y}} = \dfrac{x}{\sqrt{2y}}$

Reduce the fractions below to lowest terms and then write each square root in the preferred form for evaluation:

(a) $\sqrt{\dfrac{6b}{18d^2}} = $ _____

(b) $\sqrt{\dfrac{72R^2}{8V}} = $ _____

a) $\dfrac{t}{s}\sqrt{\dfrac{11}{v}}$

b) $\dfrac{td}{\sqrt{V}}$

c) $\dfrac{1}{3}\sqrt{\dfrac{y}{x}}$

---

130. If a fraction contains the same letter in both its numerator and denominator, it can be reduced to lowest terms. For example:

$\dfrac{7ax}{11ay} = \left(\dfrac{7x}{11y}\right)\left(\dfrac{a}{a}\right) = \dfrac{7x}{11y}(1) = \dfrac{7x}{11y}$

Reduce each fraction to lowest terms and then write each square root in the preferred form for evaluation:

(a) $\sqrt{\dfrac{5ax}{7bx}} = $ _____

(b) $\sqrt{\dfrac{27ax}{3x}} = $ _____

(c) $\sqrt{\dfrac{ct}{cy^2}} = $ _____

(d) $\sqrt{\dfrac{b^2y}{cy}} = $ _____

a) $\dfrac{1}{d}\sqrt{\dfrac{b}{3}}$ $\left(\text{from } \dfrac{\sqrt{b}}{d\sqrt{3}}\right)$

b) $\dfrac{3R}{\sqrt{V}}$

---

a) $\sqrt{\dfrac{5a}{7b}}$

b) $3\sqrt{a}$

c) $\dfrac{\sqrt{t}}{y}$

d) $\dfrac{b}{\sqrt{c}}$

131. If a fraction contains the indicated square of the same letter in both its numerator and denominator, it can be reduced to lowest terms. For example:
$$\frac{ax^2}{bx^2} = \left(\frac{a}{b}\right)\left(\frac{x^2}{x^2}\right) = \left(\frac{a}{b}\right)(1) = \frac{a}{b}$$

Reduce each fraction to lowest terms and then write each square root in the preferred form for evaluation:

(a) $\sqrt{\dfrac{cy^2}{dy^2}} =$ _____

(b) $\sqrt{\dfrac{mp^2q^2}{ab^2q^2}} =$ _____

(c) $\sqrt{\dfrac{a^2b}{a^2cd^2}} =$ _____

132. If a fraction contains the indicated square of a letter in its numerator and the <u>same</u> letter in its denominator, it can be reduced to lowest terms. For example:
$$\frac{ax^2}{bx} = \frac{axx}{bx} = \frac{ax}{b}\left(\frac{x}{x}\right) = \frac{ax}{b}(1) = \frac{ax}{b}$$
$$\frac{y^2}{y} = \frac{yy}{y} = y\left(\frac{y}{y}\right) = y(1) = y$$

Reduce each fraction below to lowest terms and then write each square root in the preferred form for evaluation:

(a) $\sqrt{\dfrac{x^2}{x}} =$ _____

(b) $\sqrt{\dfrac{b^2y^2}{cy}} =$ _____

(c) $\sqrt{\dfrac{t^2}{mt}} =$ _____

a) $\sqrt{\dfrac{c}{d}}$

b) $\dfrac{p}{b}\sqrt{\dfrac{m}{a}}$

c) $\dfrac{1}{d}\sqrt{\dfrac{b}{c}}$

133. If a fraction contains a letter in its numerator and the indicated square of the <u>same</u> letter in its denominator, it can be reduced to lowest terms. For example:
$$\frac{cd}{ad^2} = \frac{cd}{add} = \left(\frac{c}{ad}\right)\left(\frac{d}{d}\right) = \left(\frac{c}{ad}\right)(1) = \frac{c}{ad}$$
$$\frac{x}{bx^2} = \frac{1x}{bxx} = \left(\frac{1}{bx}\right)\left(\frac{x}{x}\right) = \left(\frac{1}{bx}\right)(1) = \frac{1}{bx}$$

Reduce each fraction to lowest terms and then write each square root in the preferred form for evaluation:

(a) $\sqrt{\dfrac{ab}{b^2}} =$ _____

(b) $\sqrt{\dfrac{t}{pt^2}} =$ _____

a) $\sqrt{x}$

b) $b\sqrt{\dfrac{y}{c}}$

c) $\sqrt{\dfrac{t}{m}}$

36 Squares and Square Root Radicals

134. When reducing fractions to lowest terms, we can sometimes factor out more than one instance of $\frac{\Box}{\Box}$. For example:

$$\frac{abx^2}{acx} = \left(\frac{a}{a}\right)\left(\frac{x}{x}\right)\left(\frac{bx}{c}\right) = \frac{bx}{c}$$

$$\frac{6mt^2}{9pt^2} = \left(\frac{3}{3}\right)\left(\frac{t^2}{t^2}\right)\left(\frac{2m}{3p}\right) = \frac{2m}{3p}$$

$$\frac{14d^2y}{10f^2y^2} = \left(\frac{2}{2}\right)\left(\frac{y}{y}\right)\left(\frac{7d^2}{5f^2y}\right) = \frac{7d^2}{5f^2y}$$

Reduce each fraction below to lowest terms:

(a) $\dfrac{7ax^2}{8ax^2} =$ _____

(b) $\dfrac{24by}{8cy^2} =$ _____

(c) $\dfrac{atx^2}{amx} =$ _____

a) $\sqrt{\dfrac{a}{b}}$

b) $\dfrac{1}{\sqrt{pt}}$

135. Reduce each fraction to lowest terms and then write the square root in the preferred form for evaluation:

(a) $\sqrt{\dfrac{abt^2}{4bt}} =$ _____

(b) $\sqrt{\dfrac{8x^2y}{24xy}} =$ _____

(c) $\sqrt{\dfrac{5mq}{10b^2q}} =$ _____

a) $\dfrac{7}{8}$

b) $\dfrac{3b}{cy}$

c) $\dfrac{tx}{m}$

136. After a fraction is reduced to lowest terms, occasionally its square root can be written without a radical. For example:

$$\sqrt{\frac{ax^2}{a}} = \sqrt{x^2} = x$$

$$\sqrt{\frac{a^2y^2}{c^2y^2}} = \sqrt{\frac{a^2}{c^2}} = \frac{\sqrt{a^2}}{\sqrt{c^2}} = \frac{a}{c}$$

$$\sqrt{\frac{7}{7x^2}} = \sqrt{\frac{1}{x^2}} = \frac{\sqrt{1}}{\sqrt{x^2}} = \frac{1}{x}$$

Reduce each fraction to lowest terms and simplify:

(a) $\sqrt{\dfrac{9t^2}{25t^2}} =$ _____

(b) $\sqrt{\dfrac{mpx^2}{mp}} =$ _____

(c) $\sqrt{\dfrac{10x}{40xy^2}} =$ _____

a) $\dfrac{\sqrt{at}}{2}$

b) $\sqrt{\dfrac{x}{3}}$

c) $\dfrac{1}{b}\sqrt{\dfrac{m}{2}}$

137. Whenever perfect squares are factored out of a radical in a fraction, always check to see that the fraction is in lowest terms. For example:

$$\frac{\sqrt{t^2 w}}{2tw} = \frac{t\sqrt{w}}{2tw} = \left(\frac{t}{t}\right)\left(\frac{\sqrt{w}}{2w}\right) = \frac{\sqrt{w}}{2w}$$

Factor out perfect squares and reduce to lowest terms:

(a) $\dfrac{\sqrt{m^2 p^2 q}}{5mpq} = $ _____

(b) $\dfrac{w}{\sqrt{w^2 R}} = $ _____

(c) $\dfrac{ct}{\sqrt{c^2 t}} = $ _____

a) $\dfrac{\sqrt{q}}{5q}$

b) $\dfrac{1}{\sqrt{R}}$

c) $\dfrac{t}{\sqrt{t}}$

138. A radical divided by itself is also an instance of $\dfrac{\square}{\square}$. For example:

$$\frac{a\sqrt{x}}{b\sqrt{x}} = \left(\frac{a}{b}\right)\left(\frac{\sqrt{x}}{\sqrt{x}}\right) = \left(\frac{a}{b}\right)(1) = \frac{a}{b}$$

Reduce each fraction to lowest terms:

(a) $\dfrac{3b\sqrt{y}}{d\sqrt{y}} = $ _____

(b) $\dfrac{7\sqrt{mp}}{8\sqrt{mp}} = $ _____

a) $\dfrac{3b}{d}$

b) $\dfrac{7}{8}$

SELF-TEST 8 (Frames 113-138)

1. Write the following as the square root of a fraction:

   $\dfrac{\sqrt{5rt}}{\sqrt{2s}} = $ _____

2. Write the following as the division of two square roots:

   $\sqrt{\dfrac{3d}{h}} = $ _____

Convert each of the following square roots to a division of radicals and factor out perfect squares whenever possible:

3. $\sqrt{\dfrac{4at^2}{9s^2}} = $ _____

4. $\sqrt{\dfrac{1}{2h^2}} = $ _____

5. $\sqrt{\dfrac{8b}{d^2}} = $ _____

6. $\sqrt{\dfrac{3P}{500I^2R}} = $ _____

Write each of the following so that each contains *only* *one* radical:

7. $\dfrac{2\sqrt{h}}{r\sqrt{3t}} = $ _____

8. $\dfrac{\sqrt{w}}{p\sqrt{2d}} = $ _____

9. $\dfrac{5b\sqrt{ax}}{\sqrt{cr}} = $ _____

Simplify each of the following by factoring out perfect squares. Answers must be in lowest terms and should contain *only* *one* radical *or* *no* radical:

10. $\sqrt{\dfrac{mv^2}{4t}} = $ _____

11. $\sqrt{\dfrac{4b^2d}{8a^2b}} = $ _____

12. $\sqrt{\dfrac{20hp}{5hk^2p}} = $ _____

13. $\dfrac{s\sqrt{2r}}{rs\sqrt{8r}} = $ _____

ANSWERS:

1. $\sqrt{\dfrac{5rt}{2s}}$

2. $\dfrac{\sqrt{3d}}{\sqrt{h}}$

3. $\dfrac{2t\sqrt{a}}{3s}$

4. $\dfrac{1}{h\sqrt{2}}$

5. $\dfrac{2\sqrt{2b}}{d}$

6. $\dfrac{\sqrt{3P}}{10I\sqrt{5R}}$

7. $\dfrac{2}{r}\sqrt{\dfrac{h}{3t}}$

8. $\dfrac{1}{p}\sqrt{\dfrac{w}{2d}}$

9. $5b\sqrt{\dfrac{ax}{cr}}$

10. $\dfrac{v}{2}\sqrt{\dfrac{m}{t}}$

11. $\dfrac{1}{a}\sqrt{\dfrac{bd}{2}}$

12. $\dfrac{2}{k}$

13. $\dfrac{1}{2r}$

1-9 THE SQUARES OF LETTER-TERMS AND FRACTIONS

In this section, we will discuss the procedure for squaring both letter-terms and fractions. We will also identify the types of letter-terms and fractions which are called "perfect squares".

139. To square a letter-term, we simply multiply the letter-term by itself. For example:

$$(3x)^2 = (3x)(3x) = 9x^2$$

$$(ab)^2 = (ab)(ab) = a^2b^2$$

Complete each of the following squarings:

(a) $(7R)^2 = (7R)(7R) = $ _____

(b) $(mt)^2 = (mt)(mt) = $ _____

(c) $(4xy)^2 = (4xy)(4xy) = $ _____

---

140. It should be obvious that squaring a letter-term is the equivalent of squaring each factor in the letter-term. For example:

$$(5x)^2 = 5^2 x^2 = 25x^2$$

$$(3bd)^2 = 3^2 b^2 d^2 = 9b^2 d^2$$

Complete each of the following squarings:

(a) $(10y)^2 = $ _____ (c) $(2xy)^2 = $ _____

(b) $(ac)^2 = $ _____ (d) $(6mpq)^2 = $ _____

a) $49R^2$

b) $m^2 t^2$

c) $16x^2 y^2$

---

141. (a) The "square of 8t" is _____.

(b) The "square of xy" is _____.

(c) The "square of 9ad" is _____.

a) $100y^2$

b) $a^2 c^2$

c) $4x^2 y^2$

d) $36m^2 p^2 q^2$

---

142. Don't confuse $(3x)^2$ with $3x^2$.

$(3x)^2$ means $(3x)(3x)$ or "square 3x".
$3x^2$ means $(3)(x^2)$ or "multiply 3 and x^2".

(a) Does $4t^2$ or $(4t)^2$ mean "square 4t"? _____

(b) Does $4t^2 = 16t^2$? _____

(c) Does $(4t)^2 = 16t^2$? _____

a) $64t^2$

b) $x^2 y^2$

c) $81a^2 d^2$

---

143. If a letter-term is squared, its square is a second letter-term in which each factor is a perfect square. Therefore, if each factor in a letter-term is a perfect square, its square root is a letter-term. For example:

$$\sqrt{9x^2} = \sqrt{9} \cdot \sqrt{x^2} = 3x$$

$$\sqrt{a^2 b^2} = \sqrt{a^2} \cdot \sqrt{b^2} = ab$$

$$\sqrt{4x^2 y^2} = \sqrt{4} \cdot \sqrt{x^2} \cdot \sqrt{y^2} = 2xy$$

Find the following square roots:

(a) $\sqrt{16y^2} = $ _____ (c) $\sqrt{25a^2 b^2} = $ _____

(b) $\sqrt{m^2 t^2} = $ _____ (d) $\sqrt{36R^2} = $ _____

a) $(4t)^2$

b) No

c) Yes, since: $(4t)(4t) = 16t^2$

40 Squares and Square Root Radicals

144. A letter-term is called a "perfect square" only if its square root is a letter-term which does not contain a radical. That is:

$25b^2$ is a perfect square, since $\sqrt{25b^2} = 5b$.

$25b$ is not a perfect square, since $\sqrt{25b} = 5\sqrt{b}$.

$7b^2$ is not a perfect square, since $\sqrt{7b^2} = b\sqrt{7}$.

Which of the following letter-terms are perfect squares? _____

(a) $49t$ (b) $81x^2$ (c) $36a^2b$

a) 4y   c) 5ab
b) mt   d) 6R

145. Which of the following letter-terms are perfect squares? _____

(a) $4a^2x^2$ (b) $25d^2$ (c) $5y^2$

Only (b)

146. The square root of the indicated square of a letter-term is the letter-term itself. For example:

$\sqrt{(4x)^2} = 4x$, since $\sqrt{(4x)^2} = \sqrt{16x^2} = 4x$

$\sqrt{(ab)^2} = ab$, since $\sqrt{(ab)^2} = \sqrt{a^2b^2} = ab$

Find the following square roots:

(a) $\sqrt{(7t)^2} =$ _____   (b) $\sqrt{(mt)^2} =$ _____   (c) $\sqrt{(2xy)^2} =$ _____

Both (a) and (b)

147. Find these square roots:

(a) $\sqrt{(4x)^2} =$ _____   (c) $\sqrt{(cd)^2} =$ _____
(b) $\sqrt{16x^2} =$ _____   (d) $\sqrt{c^2d^2} =$ _____

a) 7t
b) mt
c) 2xy

148. Be careful with these because some of the radicands are not the indicated squares of letter-terms:

(a) $\sqrt{(9y)^2} =$ _____   (c) $\sqrt{(16b)^2} =$ _____
(b) $\sqrt{9y^2} =$ _____   (d) $\sqrt{16b^2} =$ _____

a) 4x   c) cd
b) 4x   d) cd

149. To square a fraction, we simply multiply the fraction by itself. That is:

$$\left(\frac{3}{4}\right)^2 = \left(\frac{3}{4}\right)\left(\frac{3}{4}\right) = \frac{9}{16}$$

(a) $\left(\frac{a}{b}\right)^2 = \left(\frac{a}{b}\right)\left(\frac{a}{b}\right) =$ _____

(b) $\left(\frac{2x}{3y}\right)^2 = \left(\frac{2x}{3y}\right)\left(\frac{2x}{3y}\right) =$ _____

a) 9y   c) 16b
b) 3y   d) 4b

a) $\frac{a^2}{b^2}$

b) $\frac{4x^2}{9y^2}$

150. Squaring a fraction is the equivalent of squaring <u>both its numerator and its denominator</u>. That is:
$$\left(\frac{4}{5}\right)^2 = \frac{4^2}{5^2} = \frac{16}{25}$$

(a) $\left(\frac{3x}{7}\right)^2 = \frac{(3x)^2}{7^2} = $ _____

(b) $\left(\frac{ab}{6m}\right)^2 = \frac{(ab)^2}{(6m)^2} = $ _____

151. Write the squares of each of these fractions:

(a) $\left(\frac{3}{7}\right)^2 = $ _____ (b) $\left(\frac{5}{8}\right)^2 = $ _____ (c) $\left(\frac{1}{3}\right)^2 = $ _____

a) $\frac{9x^2}{49}$

b) $\frac{a^2b^2}{36m^2}$

152. Find the following squares:

(a) $\left(\frac{4y}{3m}\right)^2 = $ _____ (b) $\left(\frac{1}{a}\right)^2 = $ _____ (c) $\left(\frac{ab}{cd}\right)^2 = $ _____

a) $\frac{9}{49}$

b) $\frac{25}{64}$

c) $\frac{1}{9}$

153. Since squaring a fraction is the equivalent of squaring both its numerator and its denominator, the <u>square</u> of any fraction is a fraction whose numerator and denominator are both "perfect squares". For example:

$\left(\frac{a}{b}\right)^2 = \frac{a^2}{b^2}$, and both "$a^2$" and "$b^2$" are perfect squares.

$\left(\frac{3t}{2x}\right)^2 = \frac{9t^2}{4x^2}$, and both _____ and _____ are perfect squares.

a) $\frac{16y^2}{9m^2}$

b) $\frac{1}{a^2}$

c) $\frac{a^2b^2}{c^2d^2}$

154. If both the numerator and denominator of a fraction are perfect squares, the square root of the fraction is a fraction which does not contain a radical. For example:

$$\sqrt{\frac{9}{25}} = \frac{\sqrt{9}}{\sqrt{25}} = \frac{3}{5}$$

(a) $\sqrt{\frac{x^2}{y^2}} = \frac{\sqrt{x^2}}{\sqrt{y^2}} = $ _____

(b) $\sqrt{\frac{4t^2}{49a^2}} = \frac{\sqrt{4t^2}}{\sqrt{49a^2}} = $ _____

$9t^2$ and $4x^2$

a) $\frac{x}{y}$

b) $\frac{2t}{7a}$

42   Squares and Square Root Radicals

155. A fraction is called a "perfect square" if <u>both</u> its numerator and denominator are perfect squares. For example:

$\frac{25}{49}$, $\frac{m^2}{p^2}$, and $\frac{a^2b^2}{c^2d^2}$ are called "perfect squares".

Which of the fractions below are perfect squares? _____

(a) $\frac{4}{9}$ (b) $\frac{d^2}{p}$ (c) $\frac{a^2c^2}{b^2}$

---

156. Which of the fractions below are perfect squares? _____   Both (a) and (c)

(a) $\frac{16x}{24t^2}$ (b) $\frac{9d^2}{49}$ (c) $\frac{25m^2p^2}{100b^2d^2}$

---

157. Find the square roots of each of the following "perfect-square" fractions: Both (b) and (c)

(a) $\sqrt{\frac{4}{25}} = $ _____   (b) $\sqrt{\frac{16}{81}} = $ _____   (c) $\sqrt{\frac{9}{100}} = $ _____

---

158. Find the square roots of each of these perfect-square fractions:

(a) $\sqrt{\frac{t^2}{p^2}} = $ _____   (b) $\sqrt{\frac{9x^2}{4y^2}} = $ _____   (c) $\sqrt{\frac{x^2y^2}{V^2S^2}} = $ _____

a) $\frac{2}{5}$

b) $\frac{4}{9}$

c) $\frac{3}{10}$

---

159. The square root of the indicated square of a fraction is the fraction itself. For example:

$\sqrt{\left(\frac{a}{b}\right)^2} = \frac{a}{b}$, since $\sqrt{\left(\frac{a}{b}\right)^2} = \sqrt{\frac{a^2}{b^2}} = \frac{a}{b}$

$\sqrt{\left(\frac{3x}{2y}\right)^2} = \frac{3x}{2y}$, since $\sqrt{\left(\frac{3x}{2y}\right)^2} = \sqrt{\frac{9x^2}{4y^2}} = \frac{3x}{2y}$

Find the following square roots:

(a) $\sqrt{\left(\frac{t}{m}\right)^2} = $ _____   (b) $\sqrt{\left(\frac{5d}{3}\right)^2} = $ _____

a) $\frac{t}{p}$

b) $\frac{3x}{2y}$

c) $\frac{xy}{VS}$

---

a) $\frac{t}{m}$

b) $\frac{5d}{3}$

160. Be careful with these because some of the radicands are <u>not</u> the indicated squares of fractions:

(a) $\sqrt{\left(\dfrac{c}{d}\right)^2} =$ ____

(b) $\sqrt{\dfrac{c^2}{d}} =$ ____

(c) $\sqrt{\left(\dfrac{5x}{9y}\right)^2} =$ ____

(d) $\sqrt{\dfrac{5x^2}{9y}} =$ ____

a) $\dfrac{c}{d}$ c) $\dfrac{5x}{9y}$

b) $\dfrac{c}{\sqrt{d}}$ d) $\dfrac{x}{3}\sqrt{\dfrac{5}{y}}$

SELF-TEST 9 (Frames 139-160)

Find the following squares:

1. $(9d)^2 =$ ____
2. $(4at)^2 =$ ____
3. $\left(\dfrac{7}{10}\right)^2 =$ ____
4. $\left(\dfrac{d}{2}\right)^2 =$ ____
5. $\left(\dfrac{1}{5h}\right)^2 =$ ____
6. $\left(\dfrac{3p}{4b}\right)^2 =$ ____

Find the following square roots:

7. $\sqrt{64t^2} =$ ____
8. $\sqrt{(64t)^2} =$ ____
9. $\sqrt{\dfrac{1}{4}} =$ ____
10. $\sqrt{\dfrac{4t^2}{9x^2}} =$ ____
11. $\sqrt{\left(\dfrac{s}{2}\right)^2} =$ ____
12. $\sqrt{\left(\dfrac{25a}{16b}\right)^2} =$ ____

Find these square roots:

13. $\sqrt{\left(\dfrac{9t}{4w}\right)^2} =$ ____
14. $\sqrt{\dfrac{9t^2}{4w^2}} =$ ____

ANSWERS:
1. $81d^2$
2. $16a^2t^2$
3. $\dfrac{49}{100}$
4. $\dfrac{d^2}{4}$
5. $\dfrac{1}{25h^2}$
6. $\dfrac{9p^2}{16b^2}$
7. $8t$
8. $64t$
9. $\dfrac{1}{2}$
10. $\dfrac{2t}{3x}$
11. $\dfrac{s}{2}$
12. $\dfrac{25a}{16b}$
13. $\dfrac{9t}{4w}$
14. $\dfrac{3t}{2w}$

44   Squares and Square Root Radicals

1-10   THE SQUARES OF SQUARE ROOT RADICALS

In this section, we will discuss the squares of radicals. To do so, we will examine the following pattern:

$$\left(\sqrt{\Box}\right)^2 = \sqrt{\Box} \cdot \sqrt{\Box} = \sqrt{(\Box)^2} = \Box$$

This pattern states that the square of any radical is its own radicand.

---

161. The following pattern says that the square root of the indicated square of a quantity is the quantity itself.

    $$\sqrt{(\Box)^2} = \Box$$

    We have seen that this pattern applies to numbers, letters, letter-terms, and fractions. That is:

    (a) $\sqrt{7^2} =$ _____   (c) $\sqrt{(5y)^2} =$ _____

    (b) $\sqrt{R^2} =$ _____   (d) $\sqrt{\left(\dfrac{a}{b}\right)^2} =$ _____

162. When two radicals with the same radicand are multiplied, the product is a radical containing the indicated square of that radicand. For example:

    $$\sqrt{7} \cdot \sqrt{7} = \sqrt{(7)(7)} = \sqrt{7^2}$$
    $$\sqrt{R} \cdot \sqrt{R} = \sqrt{(R)(R)} = \sqrt{R^2}$$
    $$\sqrt{3x} \cdot \sqrt{3x} = \sqrt{(3x)(3x)} = \sqrt{(3x)^2}$$
    $$\sqrt{\dfrac{y}{5}} \cdot \sqrt{\dfrac{y}{5}} = \sqrt{\left(\dfrac{y}{5}\right)\left(\dfrac{y}{5}\right)} = \sqrt{\left(\dfrac{y}{5}\right)^2}$$

    Therefore, when we multiply two radicals with the same radicand, the product is that radicand. That is:

    (a) $\sqrt{3} \cdot \sqrt{3} = \sqrt{3^2} =$ _____   (c) $\sqrt{cd} \cdot \sqrt{cd} = \sqrt{(cd)^2} =$ _____

    (b) $\sqrt{m} \cdot \sqrt{m} = \sqrt{m^2} =$ _____   (d) $\sqrt{\dfrac{x}{y}} \cdot \sqrt{\dfrac{x}{y}} = \sqrt{\left(\dfrac{x}{y}\right)^2} =$ _____

163. When two radicals with the same radicand are multiplied, the product is that radicand. The following numerical examples show the sensibleness of this principle:

    $$\sqrt{9} \cdot \sqrt{9} = 9 \text{ , since } \sqrt{9} \cdot \sqrt{9} = 3 \cdot 3 = 9$$
    $$\sqrt{25} \cdot \sqrt{25} = 25 \text{ , since } \sqrt{25} \cdot \sqrt{25} = 5 \cdot 5 = 25$$

    Using the principle above, complete these:

    $$\sqrt{16} \cdot \sqrt{16} = 16$$

    (a) $\sqrt{V} \cdot \sqrt{V} =$ _____

    (b) $\sqrt{4p} \cdot \sqrt{4p} =$ _____   (c) $\sqrt{\dfrac{3}{h}} \cdot \sqrt{\dfrac{3}{h}} =$ _____

Answers:

161. a) 7   b) R   c) 5y   d) $\dfrac{a}{b}$

162. a) 3   b) m   c) cd   d) $\dfrac{x}{y}$

Squares and Square Root Radicals   45

164. Here is the pattern that we have developed:

$$\sqrt{\Box} \cdot \sqrt{\Box} = \sqrt{(\Box)^2} = \Box$$

This pattern says two things:

(1) The <u>square root of the indicated square of a quantity is the quantity itself</u>. That is:

$$\sqrt{(\Box)^2} = \Box$$

(2) <u>If two radicals containing the same radicand are multiplied, the product is that radicand</u>. That is:

$$\sqrt{\Box} \cdot \sqrt{\Box} = \Box$$

Using the pattern above, complete these:

(a) $\sqrt{77} \cdot \sqrt{77} =$ _____   (c) $\sqrt{F} \cdot \sqrt{F} =$ _____

(b) $\sqrt{(3R)^2} =$ _____   (d) $\sqrt{\left(\dfrac{a}{bc}\right)^2} =$ _____

a) V
b) 4p
c) $\dfrac{3}{h}$

165. To square a radical, we multiply the radical by itself. Therefore, squaring a radical is merely a special case of multiplying two radicals. That is:

Just as $(5)^2$ means $5 \cdot 5$,

$\left(\sqrt{7}\right)^2$ means $\sqrt{7} \cdot \sqrt{7}$.

Complete these: (a) $\left(\sqrt{R}\right)^2$ means $\sqrt{} \cdot \sqrt{}$

(b) $\left(\sqrt{3t}\right)^2$ means $\sqrt{} \cdot \sqrt{}$

(c) $\left(\sqrt{\dfrac{x}{4}}\right)^2$ means $\sqrt{} \cdot \sqrt{}$

a) 77   c) F
b) 3R   d) $\dfrac{a}{bc}$

166. Since squaring a radical is the equivalent of multiplying two radicals with the same radicand, the "square" of a radical is the <u>original radicand</u>. That is:

$\left(\sqrt{11}\right)^2 = \sqrt{11} \cdot \sqrt{11} = 11$

(a) $\left(\sqrt{t}\right)^2 = \sqrt{t} \cdot \sqrt{t} =$ _____

(b) $\left(\sqrt{4x}\right)^2 = \sqrt{4x} \cdot \sqrt{4x} =$ _____

(c) $\left(\sqrt{\dfrac{y}{6}}\right)^2 = \sqrt{\dfrac{y}{6}} \cdot \sqrt{\dfrac{y}{6}} =$ _____

a) $\sqrt{R} \cdot \sqrt{R}$
b) $\sqrt{3t} \cdot \sqrt{3t}$
c) $\sqrt{\dfrac{x}{4}} \cdot \sqrt{\dfrac{x}{4}}$

a) t
b) 4x
c) $\dfrac{y}{6}$

46 Squares and Square Root Radicals

167. Complete: (a) $\left(\sqrt{47}\right)^2 = \sqrt{47} \cdot \sqrt{47} = $ _____
   (b) $\left(\sqrt{V}\right)^2 = \sqrt{} \cdot \sqrt{} = $ _____
   (c) $\left(\sqrt{ab}\right)^2 = \sqrt{} \cdot \sqrt{} = $ _____
   (d) $\left(\sqrt{\dfrac{s}{t}}\right)^2 = \sqrt{\phantom{\dfrac{s}{t}}} \cdot \sqrt{\phantom{\dfrac{s}{t}}} = $ _____

168. You should be able to write each "square" below immediately.

   (a) $\left(\sqrt{13}\right)^2 = $ ____   (b) $\left(\sqrt{7x}\right)^2 = $ ____   (c) $\left(\sqrt{\dfrac{3}{R}}\right)^2 = $ ____

a) _____ = 47
b) $\sqrt{V} \cdot \sqrt{V} = V$
c) $\sqrt{ab} \cdot \sqrt{ab} = ab$
d) $\sqrt{\dfrac{s}{t}} \cdot \sqrt{\dfrac{s}{t}} = \dfrac{s}{t}$

169. Here is the complete pattern which has been developed:

$$\left(\sqrt{\square}\right)^2 = \sqrt{\square} \cdot \sqrt{\square} = \sqrt{(\square)^2} = \square$$

This complete pattern says three things:

(1) The <u>square root</u> of the <u>indicated square</u> of <u>a quantity</u> is <u>the quantity itself</u>. That is:

$$\sqrt{(\square)^2} = \square$$

(2) If <u>two radicals</u> contain <u>the same radicand</u>, <u>their product is that radicand</u>. That is:

$$\sqrt{\square} \cdot \sqrt{\square} = \square$$

(3) The "square" <u>of any radical is the original radicand</u>. That is:

$$\left(\sqrt{\square}\right)^2 = \square$$

The pattern above says that $\left(\sqrt{5}\right)^2$, $\sqrt{5} \cdot \sqrt{5}$, and $\sqrt{5^2}$ are equivalent because each equals _____.

a) 13
b) 7x
c) $\dfrac{3}{R}$

170. The three radical expressions below are equivalent because each equals "y".

$\left(\sqrt{y}\right)^2$ $\sqrt{y} \cdot \sqrt{y}$ $\sqrt{y^2}$

(a) Write three equivalent radical expressions which equal "4m":

_____ _____ _____

(b) Write three equivalent radical expressions which equal $\dfrac{"a"}{b}$:

_____ _____ _____

5

171. (a) The sentence below is true because each side equals _____.

$$\left(\sqrt{101}\right)^2 = \sqrt{(101)^2}$$

(b) The sentence below is true because each side equals _____.

$$\left(\sqrt{ef}\right)^2 = \sqrt{ef} \cdot \sqrt{ef}$$

(c) The sentence below is true because each side equals _____.

$$\sqrt{\frac{m}{q}} \cdot \sqrt{\frac{m}{q}} = \sqrt{\left(\frac{m}{q}\right)^2}$$

a) $\left(\sqrt{4m}\right)^2$, $\sqrt{4m} \cdot \sqrt{4m}$, $\sqrt{(4m)^2}$

b) $\left(\sqrt{\frac{a}{b}}\right)^2$, $\sqrt{\frac{a}{b}} \cdot \sqrt{\frac{a}{b}}$, $\sqrt{\left(\frac{a}{b}\right)^2}$

172. Write a non-radical in each blank:

(a) $\sqrt{5d} \cdot \sqrt{5d} =$ ____ (b) $\sqrt{(707)^2} =$ ____ (c) $\left(\sqrt{\frac{t}{7}}\right)^2 =$ ____

a) 101
b) ef
c) $\dfrac{m}{q}$

173. Write a non-radical in each blank:

(a) $\left(\sqrt{cd}\right)^2 =$ ____ (b) $\sqrt{\left(\frac{2x}{3y}\right)^2} =$ ____ (c) $\sqrt{F_1} \cdot \sqrt{F_1} =$ ____

a) 5d
b) 707
c) $\dfrac{t}{7}$

Answer to Frame 173:   a) cd   b) $\dfrac{2x}{3y}$ c) F_1

SELF-TEST 10 (Frames 161-173)

1. $\sqrt{s^2} =$ _____

2. $\sqrt{(97)^2} =$ _____

3. $\sqrt{(5r)^2} =$ _____

4. $\sqrt{\left(\frac{d}{2}\right)^2} =$ _____

5. $\sqrt{100} \cdot \sqrt{100} =$ _____

6. $\sqrt{2} \cdot \sqrt{2} =$ _____

7. $\sqrt{A} \cdot \sqrt{A} =$ _____

8. $\sqrt{\frac{P}{R}} \cdot \sqrt{\frac{P}{R}} =$ _____

9. $\left(\sqrt{49}\right)^2 =$ _____

10. $\left(\sqrt{3}\right)^2 =$ _____

11. $\left(\sqrt{4x}\right)^2 =$ _____

12. $\left(\sqrt{\frac{2E}{m}}\right)^2 =$ _____

ANSWERS:
1. s
2. 97
3. 5r
4. $\dfrac{d}{2}$
5. 100
6. 2
7. A
8. $\dfrac{P}{R}$
9. 49
10. 3
11. 4x
12. $\dfrac{2E}{m}$

48 Squares and Square Root Radicals

1-11 THE SQUARES OF TERMS CONTAINING SQUARE ROOT RADICALS

Each of the following terms contains a square root radical:

$$3\sqrt{x}, \quad ab\sqrt{xy}, \quad \frac{2\sqrt{y}}{3}, \quad 2g\sqrt{\frac{a}{d}}$$

In this section, we will discuss the procedure for squaring terms of this type.

---

174. The usual symbols (parentheses and a small "2") are used to indicate the squaring of a term containing a radical. That is:

$$\left(3\sqrt{x}\right)^2 \quad \text{means} \quad \left(3\sqrt{x}\right)\left(3\sqrt{x}\right)$$

$$\left(\frac{\sqrt{y}}{5}\right)^2 \quad \text{means} \quad \left(\frac{\sqrt{y}}{5}\right)\left(\frac{\sqrt{y}}{5}\right)$$

$$\left(\frac{t}{\sqrt{ab}}\right)^2 \quad \text{means} \quad (\quad)(\quad)$$

---

175. Squaring a non-fractional term containing a radical leads to a multiplication of non-radicals and radicals. In this multiplication, we multiply each factor in the term by itself. For example:

$$\left(5\sqrt{x}\right)^2 = \left(5\sqrt{x}\right)\left(5\sqrt{x}\right) = \underline{5 \cdot 5} \cdot \underline{\sqrt{x} \cdot \sqrt{x}}$$
$$= 25 \cdot x \quad \text{or} \quad 25x$$

(a) $\left(y\sqrt{ab}\right)^2 = \left(y\sqrt{ab}\right)\left(y\sqrt{ab}\right) = \underline{y \cdot y} \cdot \underline{\sqrt{ab} \cdot \sqrt{ab}}$

$$= \underline{\quad} \cdot \underline{\quad} \quad \text{or} \quad \underline{\quad}$$

(b) $\left(4p\sqrt{m}\right)^2 = \left(4p\sqrt{m}\right)\left(4p\sqrt{m}\right) = \underline{4 \cdot 4} \cdot \underline{p \cdot p} \cdot \underline{\sqrt{m} \cdot \sqrt{m}}$

$$= \underline{\quad} \cdot \underline{\quad} \cdot \underline{\quad} \quad \text{or} \quad \underline{\quad}$$

Answer column: $\left(\dfrac{t}{\sqrt{ab}}\right)\left(\dfrac{t}{\sqrt{ab}}\right)$

---

176. Complete each of these squarings:

(a) $\left(8\sqrt{d}\right)^2 = 8 \cdot 8 \cdot \sqrt{d} \cdot \sqrt{d} = \underline{\quad}$

(b) $\left(ab\sqrt{c}\right)^2 = a \cdot a \cdot b \cdot b \cdot \sqrt{c} \cdot \sqrt{c} = \underline{\quad}$

(c) $\left(R\sqrt{3x}\right)^2 = R \cdot R \cdot \sqrt{3x} \cdot \sqrt{3x} = \underline{\quad}$

Answer column:
a) $y^2 \cdot ab = aby^2$
b) $16 \cdot p^2 \cdot m = 16mp^2$

---

177. As you can see, squaring a non-fractional term containing a radical is the equivalent of squaring each factor in the term. For example:

$$\left(7\sqrt{t}\right)^2 = 7 \cdot 7 \cdot \sqrt{t} \cdot \sqrt{t} = 7^2 \cdot \left(\sqrt{t}\right)^2$$

$$\left(3d\sqrt{q}\right)^2 = 3 \cdot 3 \cdot d \cdot d \cdot \sqrt{q} \cdot \sqrt{q} = 3^2 \cdot d^2 \cdot \left(\sqrt{q}\right)^2$$

$$\left(5\sqrt{3a}\right)^2 = 5 \cdot 5 \cdot \sqrt{3a} \cdot \sqrt{3a} = (\quad)^2 \cdot (\quad)^2$$

Answer column:
a) 64d
b) $a^2 b^2 c$
c) $3R^2 x$ (from $R^2 \cdot 3x$)

Answer column: $(5)^2 \cdot \left(\sqrt{3a}\right)^2$

Squares and Square Root Radicals   49

178. Complete each of the following:
$$\left(9\sqrt{y}\right)^2 = 9^2 \cdot \left(\sqrt{y}\right)^2 = 81y$$
  (a) $\left(6\sqrt{b}\right)^2 = 6^2 \cdot \left(\sqrt{b}\right)^2 = $ _____
  (b) $\left(5m\sqrt{t}\right)^2 = 5^2 \cdot m^2 \cdot \left(\sqrt{t}\right)^2 = $ _____
  (c) $\left(V\sqrt{2g}\right)^2 = V^2 \cdot \left(\sqrt{2g}\right)^2 = $ _____

179. Complete each of the following by squaring each factor in the term:
  (a) $\left(10\sqrt{y}\right)^2 = $ _____
  (b) $\left(3a\sqrt{x}\right)^2 = $ _____
  (c) $\left(t\sqrt{2b}\right)^2 = $ _____
  (d) $\left(7m\sqrt{cd}\right)^2 = $ _____

a) 36b
b) 25m²t
c) 2gV²

180. When squaring a non-fractional term containing a radical, a common error is <u>forgetting</u> <u>to</u> <u>square</u> <u>the</u> <u>non-radical</u> <u>factor</u> <u>or</u> <u>factors</u>. Here are some examples of that error:
$$\left(5\sqrt{x}\right)^2 = 5x \qquad \text{(error)}$$
$$\left(3a\sqrt{b}\right)^2 = 3ab \qquad \text{(error)}$$

If the terms above are squared properly:
  (a) $\left(5\sqrt{x}\right)^2 = $ _____
  (b) $\left(3a\sqrt{b}\right)^2 = $ _____

a) 100y
b) 9a²x
c) 2bt²
d) 49cdm²

181. Avoid the common error in squaring these:
  (a) $\left(9\sqrt{p}\right)^2 = $ _____
  (b) $\left(2d\sqrt{y}\right)^2 = $ _____
  (c) $\left(m\sqrt{2x}\right)^2 = $ _____
  (d) $\left(xy\sqrt{R}\right)^2 = $ _____

a) 25x
b) 9a²b

182. To square a <u>fraction</u> which contains a radical, we simply multiply the fraction by itself. For example:
  (a) $\left(\dfrac{\sqrt{t}}{3}\right)^2 = \left(\dfrac{\sqrt{t}}{3}\right)\left(\dfrac{\sqrt{t}}{3}\right) = \dfrac{\sqrt{t} \cdot \sqrt{t}}{3 \cdot 3} = $ _____
  (b) $\left(\dfrac{7}{\sqrt{y}}\right)^2 = \left(\dfrac{7}{\sqrt{y}}\right)\left(\dfrac{7}{\sqrt{y}}\right) = \dfrac{7 \cdot 7}{\sqrt{y} \cdot \sqrt{y}} = $ _____
  (c) $\left(\dfrac{\sqrt{3x}}{4}\right)^2 = \left(\dfrac{\sqrt{3x}}{4}\right)\left(\dfrac{\sqrt{3x}}{4}\right) = \dfrac{\sqrt{3x} \cdot \sqrt{3x}}{4 \cdot 4} = $ _____

a) 81p
b) 4d²y
c) 2m²x
d) Rx²y²

a) $\dfrac{t}{9}$
b) $\dfrac{49}{y}$
c) $\dfrac{3x}{16}$

50 Squares and Square Root Radicals

183. Complete each of these:

(a) $\left(\dfrac{2\sqrt{x}}{3}\right)^2 = \left(\dfrac{2\sqrt{x}}{3}\right)\left(\dfrac{2\sqrt{x}}{3}\right) = \dfrac{2 \cdot 2 \cdot \sqrt{x} \cdot \sqrt{x}}{3 \cdot 3} = \underline{\qquad}$

(b) $\left(\dfrac{a}{3\sqrt{b}}\right)^2 = \left(\dfrac{a}{3\sqrt{b}}\right)\left(\dfrac{a}{3\sqrt{b}}\right) = \dfrac{a \cdot a}{3 \cdot 3 \cdot \sqrt{b} \cdot \sqrt{b}} = \underline{\qquad}$

(c) $\left(\dfrac{M}{\sqrt{xy}}\right)^2 = \left(\dfrac{M}{\sqrt{xy}}\right)\left(\dfrac{M}{\sqrt{xy}}\right) = \dfrac{M \cdot M}{\sqrt{xy} \cdot \sqrt{xy}} = \underline{\qquad}$

a) $\dfrac{4x}{9}$

b) $\dfrac{a^2}{9b}$

c) $\dfrac{M^2}{xy}$

184. It should be obvious that squaring a fraction which contains a radical is the equivalent of squaring both its numerator and its denominator. That is:

(a) $\left(\dfrac{r}{\sqrt{a}}\right)^2 = \dfrac{(r)^2}{(\sqrt{a})^2} = \underline{\qquad}$

(b) $\dfrac{\sqrt{2d}}{5} = \dfrac{(\sqrt{2d})^2}{(5)^2} = \underline{\qquad}$

(c) $\left(\dfrac{7}{3\sqrt{m}}\right)^2 = \dfrac{(7)^2}{(3\sqrt{m})^2} = \underline{\qquad}$

a) $\dfrac{r^2}{a}$

b) $\dfrac{2d}{25}$

c) $\dfrac{49}{9m}$

185. Complete these:

(a) $\left(\dfrac{1}{\sqrt{x}}\right)^2 = \dfrac{(1)^2}{(\sqrt{x})^2} = \underline{\qquad}$

(b) $\left(\dfrac{1}{6\sqrt{y}}\right)^2 = \dfrac{(1)^2}{(6\sqrt{y})^2} = \underline{\qquad}$

(c) $\left(\dfrac{1}{2d\sqrt{rw}}\right)^2 = \dfrac{(1)^2}{(2d\sqrt{rw})^2} = \underline{\qquad}$

a) $\dfrac{1}{x}$

b) $\dfrac{1}{36y}$

c) $\dfrac{1}{4d^2 rw}$

186. Complete each of these:

(a) $(10p\sqrt{t})^2 = \underline{\qquad}$

(b) $\left(\dfrac{\sqrt{bc}}{d}\right)^2 = \underline{\qquad}$

(c) $(3\sqrt{4x})^2 = \underline{\qquad}$

(d) $\left(\dfrac{7}{\sqrt{2y}}\right)^2 = \underline{\qquad}$

a) $100p^2 t$ c) $36x$ (from $9 \cdot 4x$)

b) $\dfrac{bc}{d^2}$ d) $\dfrac{49}{2y}$

187. Complete each of these:

(a) $\left(\dfrac{1}{5\sqrt{R}}\right)^2 = \underline{\qquad}$

(b) $\left(\dfrac{1}{7\sqrt{2t}}\right)^2 = \underline{\qquad}$

a) $\dfrac{1}{25R}$

b) $\dfrac{1}{98t}$ (from $\dfrac{1}{49 \cdot 2t}$)

Squares and Square Root Radicals 51

188. Here is a term in which the radical contains a fraction. Watch the steps used to square it.

$$\left(2g\sqrt{\frac{a}{r}}\right)^2 = 2^2 \cdot g^2 \cdot \left(\sqrt{\frac{a}{r}}\right)^2 = 4g^2\left(\frac{a}{r}\right) = \frac{4ag^2}{r}$$

Following the steps above, complete each of these:

(a) $\left(5a\sqrt{\frac{m}{t}}\right)^2 = $ _____

(b) $\left(pq\sqrt{\frac{y}{5}}\right)^2 = $ _____

189. After squaring a fraction containing a radical, look for opportunities to reduce the new fraction to lowest terms. For example:

$$\left(\frac{p}{2\sqrt{2p}}\right)^2 = \frac{p^2}{4 \cdot 2p} = \frac{p^2}{8p} = \frac{p}{8}$$

a) $\dfrac{25a^2m}{t}$

b) $\dfrac{p^2q^2y}{5}$

Do these squarings and reduce to lowest terms if possible:

(a) $\left(\dfrac{b}{a\sqrt{b}}\right)^2 = $ _____

(b) $\left(\dfrac{3\sqrt{t}}{t}\right)^2 = $ _____

(c) $\left(\dfrac{c}{a\sqrt{bc}}\right)^2 = $ _____

(d) $\left(\dfrac{\sqrt{ab}}{ab}\right)^2 = $ _____

190. In the term $2\sqrt{2}$, the radicand is a number. The procedure for squaring terms of this type is the same. That is, we simply square each factor. For example:

$$\left(2\sqrt{2}\right)^2 = \left(2\sqrt{2}\right)\left(2\sqrt{2}\right) = 2^2 \cdot \left(\sqrt{2}\right)^2 = 4 \cdot 2 = 8$$

a) $\dfrac{b}{a^2}$

b) $\dfrac{9}{t}$

c) $\dfrac{c}{a^2b}$

d) $\dfrac{1}{ab}$

Complete these: (a) $\left(2\sqrt{3}\right)^2 = $ _____

(b) $\left(3\sqrt{3}\right)^2 = $ _____

(c) $\left(5\sqrt{5}\right)^2 = $ _____

(d) $\left(4\sqrt{2}\right)^2 = $ _____

a) 12 (from 4 · 3)

b) 27 (from 9 · 3)

c) 125 (from 25 · 5)

d) 32 (from 16 · 2)

52 Squares and Square Root Radicals

191. In the fractions below, either the numerator or denominator is a radical containing a number. To square fractions of this type, we use the same procedure. That is, we square both the numerator and the denominator. For example:

$$\left(\frac{\sqrt{2}}{3}\right)^2 = \frac{\left(\sqrt{2}\right)^2}{3^2} = \frac{2}{9}$$

$$\left(\frac{1}{\sqrt{3}}\right)^2 = \frac{1^2}{\left(\sqrt{3}\right)^2} = \frac{1}{3}$$

Find these squares: (a) $\left(\frac{\sqrt{5}}{2}\right)^2 =$ _____

(b) $\left(\frac{1}{\sqrt{7}}\right)^2 =$ _____

192. Be sure to reduce the "squares" to lowest terms if possible. For example:

$$\left(\frac{\sqrt{3}}{3}\right)^2 = \frac{3}{9} = \frac{1}{3}$$

$$\left(\frac{2}{\sqrt{2}}\right)^2 = \frac{4}{2} = 2$$

Complete each of the following. Reduce to lowest terms whenever possible:

(a) $\left(\frac{\sqrt{5}}{5}\right)^2 =$ _____

(b) $\left(\frac{\sqrt{3}}{2}\right)^2 =$ _____

(c) $\left(\frac{3}{\sqrt{3}}\right)^2 =$ _____

(d) $\left(\frac{\sqrt{2}}{2}\right)^2 =$ _____

a) $\frac{5}{4}$

b) $\frac{1}{7}$

193. Here is another case where we can reduce the "square" to lowest terms:

$$\left(\frac{2\sqrt{3}}{3}\right)^2 = \frac{\left(2\sqrt{3}\right)^2}{3^2} = \frac{4 \cdot 3}{9} = \frac{4}{3}$$

Complete these and reduce each answer to lowest terms:

(a) $\left(\frac{3\sqrt{5}}{5}\right)^2 =$ _____

(b) $\left(\frac{2}{3\sqrt{2}}\right)^2 =$ _____

a) $\frac{1}{5}$ (from $\frac{5}{25}$)

b) $\frac{3}{4}$

c) 3 (from $\frac{9}{3}$)

d) $\frac{1}{2}$ (from $\frac{2}{4}$)

a) $\frac{9}{5}$ (from $\frac{9 \cdot 5}{25}$)

b) $\frac{2}{9}$ (from $\frac{4}{9 \cdot 2}$)

Squares and Square Root Radicals 53

SELF-TEST 11 (Frames 174-193)

Complete the following squarings:

1. $\left(5\sqrt{x}\right)^2 =$ _____
2. $\left(4t\sqrt{r}\right)^2 =$ _____
3. $\left(d\sqrt{3h}\right)^2 =$ _____
4. $\left(ap\sqrt{2b}\right)^2 =$ _____
5. $\left(3\sqrt{2}\right)^2 =$ _____
6. $\left(4t\sqrt{2}\right)^2 =$ _____

Do these squarings. Where possible, reduce answers to lowest terms:

7. $\left(\dfrac{\sqrt{t}}{3}\right)^2 =$ _____
8. $\left(\dfrac{\sqrt{6p}}{2}\right)^2 =$ _____
9. $\left(\dfrac{1}{3\sqrt{2w}}\right)^2 =$ _____
10. $\left(\dfrac{2a}{\sqrt{2d}}\right)^2 =$ _____
11. $\left(\dfrac{p\sqrt{3r}}{r\sqrt{2p}}\right)^2 =$ _____
12. $\left(\dfrac{3\sqrt{2}}{2\sqrt{3}}\right)^2 =$ _____

After doing these squarings, reduce each answer to lowest terms:

13. $\left(\dfrac{t}{2}\sqrt{\dfrac{v}{2t}}\right)^2 =$ _____
14. $\left(3h\sqrt{\dfrac{2c}{h}}\right)^2 =$ _____

ANSWERS:
1. $25x$
2. $16rt^2$
3. $3d^2h$
4. $2a^2bp^2$
5. 18
6. $32t^2$
7. $\dfrac{t}{9}$
8. $\dfrac{3p}{2}$
9. $\dfrac{1}{18w}$
10. $\dfrac{2a^2}{d}$
11. $\dfrac{3p}{2r}$
12. $\dfrac{3}{2}$
13. $\dfrac{tv}{8}$
14. $18ch$

1-12 RATIONALIZING DENOMINATORS WHICH CONTAIN SQUARE ROOT RADICALS

When a fraction contains a radical in its denominator, it can be converted to an equivalent form <u>without a radical in the denominator</u>. For example:

$$\dfrac{2}{\sqrt{5}} \text{ can be converted to } \dfrac{2\sqrt{5}}{5}$$

$$\dfrac{m}{a\sqrt{b}} \text{ can be converted to } \dfrac{m\sqrt{b}}{ab}$$

This process of eliminating a radical from the denominator is called "<u>rationalizing a radical denominator</u>". The "rationalizing" process will be described in this section.

54   Squares and Square Root Radicals

194. "Rationalizing a radical denominator" is a method of obtaining an equivalent form of a fraction. As we have seen, the process for obtaining equivalent forms of fractions is based on these two principles:

$$\frac{\Box}{\Box} = 1 \quad \text{and} \quad \Box(1) = \Box$$

The principle $\dfrac{\Box}{\Box} = 1$ says this: <u>If any quantity is divided by itself, the quotient is "1"</u>. That is:

$$\frac{8}{8} = 1 \quad \frac{x}{x} = 1 \quad \frac{a+b}{a+b} = 1 \quad \frac{y^2}{y^2} = 1$$

When rationalizing denominators, we use instances of this principle in which a radical is divided by itself. That is:

$$\frac{\sqrt{7}}{\sqrt{7}} = 1 \quad \frac{\sqrt{x}}{\sqrt{x}} = 1 \quad \frac{\sqrt{3t}}{\sqrt{3t}} = 1 \quad \frac{\sqrt{ab}}{\sqrt{ab}} = 1$$

Based on the principle above, complete these:

(a) $\dfrac{\sqrt{b}}{\sqrt{b}} = $ _____   (b) $\dfrac{\sqrt{9}}{\Box} = 1$ (c) $\dfrac{\Box}{\sqrt{2R}} = 1$

a) 1

b) $\sqrt{9}$

c) $\sqrt{2R}$

195. The principle $\Box(1) = \Box$ says this: If a quantity is multiplied by "<u>1</u>", <u>the product is equivalent to the original quantity</u>.

To obtain equivalent forms of fractions, we substitute instances of $\dfrac{\Box}{\Box}$ for "1" in the principle above. For example:

$$\frac{3}{4} = \frac{3}{4}(1) = \frac{3}{4}\left(\frac{2}{2}\right) = \frac{6}{8}$$

$$\frac{a}{b} = \frac{a}{b}(1) = \frac{a}{b}\left(\frac{x}{x}\right) = \frac{ax}{bx}$$

When obtaining equivalent forms of fractions, we can also substitute instances of $\dfrac{\sqrt{}}{\sqrt{}}$ for "1" in the principle above since $\dfrac{\sqrt{}}{\sqrt{}} = \dfrac{\Box}{\Box}$.

For example:

$$\frac{3}{4} = \frac{3}{4}(1) = \frac{3}{4}\left(\frac{\sqrt{5}}{\sqrt{5}}\right) = \frac{3\sqrt{5}}{4\sqrt{5}}$$

$$\frac{a}{b} = \frac{a}{b}(1) = \frac{a}{b}\left(\frac{\sqrt{t}}{\sqrt{t}}\right) = \underline{}$$

$\dfrac{a\sqrt{t}}{b\sqrt{t}}$

196. The same procedure is used to obtain equivalent forms of fractions which contain a radical. For example:

$$\frac{3}{\sqrt{5}} = \frac{3}{\sqrt{5}}(1) = \frac{3}{\sqrt{5}}\left(\frac{2}{2}\right) = \frac{6}{2\sqrt{5}}$$

$$\frac{a}{\sqrt{b}} = \frac{a}{\sqrt{b}}(1) = \frac{a}{\sqrt{b}}\left(\frac{c}{c}\right) = \frac{ac}{c\sqrt{b}}$$

To obtain equivalent forms of fractions which contain a radical, we can also use instances of $\frac{\sqrt{}}{\sqrt{}}$. For example:

$$\frac{3}{\sqrt{5}} = \frac{3}{\sqrt{5}}(1) = \frac{3}{\sqrt{5}}\left(\frac{\sqrt{6}}{\sqrt{6}}\right) = \frac{3\sqrt{6}}{\sqrt{5}\cdot\sqrt{6}} = \frac{3\sqrt{6}}{\sqrt{30}}$$

$$\frac{a}{\sqrt{b}} = \frac{a}{\sqrt{b}}(1) = \frac{a}{\sqrt{b}}\left(\frac{\sqrt{t}}{\sqrt{t}}\right) = \frac{a\sqrt{t}}{\sqrt{b}\cdot\sqrt{t}} = \frac{a\sqrt{t}}{\sqrt{bt}}$$

Write the equivalent form of each fraction in the blank:

(a) $\dfrac{4}{\sqrt{2}}\left(\dfrac{\sqrt{7}}{\sqrt{7}}\right) = $ _____

(b) $\dfrac{1}{\sqrt{3}}\left(\dfrac{\sqrt{10}}{\sqrt{10}}\right) = $ _____

(c) $\dfrac{\sqrt{5}}{\sqrt{4}}\left(\dfrac{\sqrt{9}}{\sqrt{9}}\right) = $ _____

---

197. Write the equivalent form of each fraction in the blank:

(a) $\dfrac{m}{\sqrt{t}}\left(\dfrac{\sqrt{3}}{\sqrt{3}}\right) = $ _____ (c) $\dfrac{\sqrt{a}}{\sqrt{b}}\left(\dfrac{\sqrt{2x}}{\sqrt{2x}}\right) = $ _____

(b) $\dfrac{R}{\sqrt{V}}\left(\dfrac{\sqrt{t}}{\sqrt{t}}\right) = $ _____ (d) $\dfrac{1}{\sqrt{cd}}\left(\dfrac{\sqrt{y}}{\sqrt{y}}\right) = $ _____

a) $\dfrac{4\sqrt{7}}{\sqrt{14}}$

b) $\dfrac{\sqrt{10}}{\sqrt{30}}$

c) $\dfrac{\sqrt{45}}{\sqrt{36}}$

---

198. In order to "rationalize a denominator", we must obtain an equivalent form <u>in which there is no radical in the denominator</u>. To obtain an equivalent form, we must multiply by some instance of $\dfrac{\sqrt{}}{\sqrt{}}$.

Let's try to rationalize the denominator of $\dfrac{3}{\sqrt{7}}$.

(a) If we multiply by $\dfrac{\sqrt{5}}{\sqrt{5}}$, we get:

$$\frac{3}{\sqrt{7}}\left(\frac{\sqrt{5}}{\sqrt{5}}\right) = \frac{3\sqrt{5}}{\sqrt{35}}$$

Have we obtained an equivalent form without a radical in the denominator? _____

a) $\dfrac{m\sqrt{3}}{\sqrt{3t}}$ c) $\dfrac{\sqrt{2ax}}{\sqrt{2bx}}$

b) $\dfrac{R\sqrt{t}}{\sqrt{tV}}$ d) $\dfrac{\sqrt{y}}{\sqrt{cdy}}$

(Continued on following page.)

198. (Continued)

   (b) If we multiply by $\dfrac{\sqrt{7}}{\sqrt{7}}$, we get:

   $$\dfrac{3}{\sqrt{7}}\left(\dfrac{\sqrt{7}}{\sqrt{7}}\right) = \dfrac{3\sqrt{7}}{\sqrt{7}\cdot\sqrt{7}} = \dfrac{3\sqrt{7}}{7}$$

   Have we obtained an equivalent form without a radical in the denominator? _____

199. Though we can obtain an equivalent form of a fraction whose denominator contains a radical by multiplying by any instance of $\dfrac{\sqrt{}}{\sqrt{}}$, we can "rationalize the denominator" only by using the instance of $\dfrac{\sqrt{}}{\sqrt{}}$ which contains the radical in the denominator. That is:

   To rationalize $\dfrac{3}{\sqrt{5}}$, we must multiply by $\dfrac{\sqrt{5}}{\sqrt{5}}$.

   To rationalize $\dfrac{1}{\sqrt{13}}$, we must multiply by $\dfrac{\sqrt{13}}{\sqrt{13}}$.

   (a) To rationalize $\dfrac{a}{\sqrt{c}}$, we must multiply by _____.

   (b) To rationalize $\dfrac{1}{\sqrt{ab}}$, we must multiply by _____.

200. Here are some examples of "rationalizing denominators":

   $$\dfrac{5}{\sqrt{11}} = \dfrac{5}{\sqrt{11}}\left(\dfrac{\sqrt{11}}{\sqrt{11}}\right) = \dfrac{5\sqrt{11}}{11}$$

   $$\dfrac{m}{\sqrt{q}} = \dfrac{m}{\sqrt{q}}\left(\dfrac{\sqrt{q}}{\sqrt{q}}\right) = \dfrac{m\sqrt{q}}{q}$$

   $$\dfrac{1}{\sqrt{2x}} = \dfrac{1}{\sqrt{2x}}\left(\dfrac{\sqrt{2x}}{\sqrt{2x}}\right) = \dfrac{\sqrt{2x}}{2x}$$

   Rationalize the denominator of each fraction below:

   (a) $\dfrac{7}{\sqrt{15}} =$ _____

   (b) $\dfrac{c}{\sqrt{t}} =$ _____

   (c) $\dfrac{5}{\sqrt{3y}} =$ _____

   (d) $\dfrac{1}{\sqrt{ab}} =$ _____

---

a) No

b) Yes

---

a) $\dfrac{\sqrt{c}}{\sqrt{c}}$

b) $\dfrac{\sqrt{ab}}{\sqrt{ab}}$

Squares and Square Root Radicals   57

201. Rationalize the denominators of these fractions:

(a) $\dfrac{7}{\sqrt{2}} = $ _____

(b) $\dfrac{1}{\sqrt{m}} = $ _____

(c) $\dfrac{y}{\sqrt{5x}} = $ _____

(d) $\dfrac{R}{\sqrt{SV}} = $ _____

a) $\dfrac{7\sqrt{15}}{15}$

b) $\dfrac{c\sqrt{t}}{t}$

c) $\dfrac{5\sqrt{3y}}{3y}$

d) $\dfrac{\sqrt{ab}}{ab}$

202. In the fraction on the left below, the denominator contains both a non-radical and a radical factor. We rationalize the denominator in the usual way.

$$\dfrac{1}{5\sqrt{2}} = \dfrac{1}{5\sqrt{2}}\left(\dfrac{\sqrt{2}}{\sqrt{2}}\right) = \dfrac{\sqrt{2}}{5 \cdot 2} = \dfrac{\sqrt{2}}{10}$$

Following the steps above, rationalize each denominator below:

(a) $\dfrac{5}{7\sqrt{3}} = $ _____

(b) $\dfrac{1}{10\sqrt{7}} = $ _____

(c) $\dfrac{m}{a\sqrt{b}} = $ _____

(d) $\dfrac{R}{S\sqrt{tv}} = $ _____

a) $\dfrac{7\sqrt{2}}{2}$ c) $\dfrac{y\sqrt{5x}}{5x}$

b) $\dfrac{\sqrt{m}}{m}$ d) $\dfrac{R\sqrt{SV}}{SV}$

203. Rationalize each of the denominators below:

(a) $\dfrac{3}{5\sqrt{2x}} = $ _____

(b) $\dfrac{1}{7\sqrt{5x}} = $ _____

a) $\dfrac{5\sqrt{3}}{21}$ c) $\dfrac{m\sqrt{b}}{ab}$

b) $\dfrac{\sqrt{7}}{70}$ d) $\dfrac{R\sqrt{tv}}{Stv}$

a) $\dfrac{3\sqrt{2x}}{10x}$

b) $\dfrac{\sqrt{5x}}{35x}$

204. When denominators are rationalized, sometimes the new fraction can be reduced to lowest terms. Here are some examples:

$$\frac{5}{3\sqrt{5}} = \frac{5}{3\sqrt{5}}\left(\frac{\sqrt{5}}{\sqrt{5}}\right) = \frac{5\sqrt{5}}{3 \cdot 5} = \frac{\sqrt{5}}{3}$$

$$\frac{2}{\sqrt{2}} = \frac{2}{\sqrt{2}}\left(\frac{\sqrt{2}}{\sqrt{2}}\right) = \frac{2\sqrt{2}}{2} = \sqrt{2}$$

Rationalize each denominator and reduce the new fraction to lowest terms:

(a) $\dfrac{7}{4\sqrt{7}} =$ _____

(b) $\dfrac{11}{\sqrt{11}} =$ _____

205. In each case below, the new fraction can be reduced to lowest terms:

$$\frac{a}{b\sqrt{a}} = \frac{a}{b\sqrt{a}}\left(\frac{\sqrt{a}}{\sqrt{a}}\right) = \frac{a\sqrt{a}}{b \cdot a} = \frac{\sqrt{a}}{b}$$

$$\frac{x}{\sqrt{x}} = \left(\frac{x}{\sqrt{x}}\right)\left(\frac{\sqrt{x}}{\sqrt{x}}\right) = \frac{x\sqrt{x}}{x} = \sqrt{x}$$

Rationalize each denominator and reduce the new fraction to lowest terms:

(a) $\dfrac{3y}{t\sqrt{3y}} =$ _____

(b) $\dfrac{5t}{\sqrt{t}} =$ _____

(c) $\dfrac{ab}{\sqrt{b}} =$ _____

206. When the square root of a fraction has been simplified by factoring out perfect squares, we can frequently rationalize the denominator of the resulting fraction. For example:

$$\sqrt{\frac{a^2 b}{c^2 d}} = \frac{a\sqrt{b}}{c\sqrt{d}} = \frac{a\sqrt{b}}{c\sqrt{d}}\left(\frac{\sqrt{d}}{\sqrt{d}}\right) = \frac{a\sqrt{bd}}{cd}$$

Factor out perfect squares and then rationalize the denominator of the resulting fractions:

(a) $\sqrt{\dfrac{m^2 x^2}{ab^2}} =$ _____

(b) $\sqrt{\dfrac{1}{x^2 y^2 z}} =$ _____

---

a) $\dfrac{\sqrt{7}}{4}$

b) $\sqrt{11}$

---

a) $\dfrac{\sqrt{3y}}{t}$

b) $5\sqrt{t}$

c) $a\sqrt{b}$

---

a) $\dfrac{mx\sqrt{a}}{ab}$ b) $\dfrac{\sqrt{z}}{xyz}$

207. Even when we cannot factor out perfect squares from the square root of a fraction, we can still rationalize its denominator. For example:

$$\sqrt{\frac{2}{3}} = \frac{\sqrt{2}}{\sqrt{3}} = \frac{\sqrt{2}}{\sqrt{3}}\left(\frac{\sqrt{3}}{\sqrt{3}}\right) = \frac{\sqrt{6}}{3}$$

$$\sqrt{\frac{m}{p}} = \frac{\sqrt{m}}{\sqrt{p}} = \frac{\sqrt{m}}{\sqrt{p}}\left(\frac{\sqrt{p}}{\sqrt{p}}\right) = \frac{\sqrt{mp}}{p}$$

Rationalize the denominators of each of the following:

(a) $\sqrt{\dfrac{3}{5}} =$ _____

(b) $\sqrt{\dfrac{ab}{c}} =$ _____

208. First factor out perfect squares if possible and then rationalize the denominators:

(a) $\sqrt{\dfrac{r}{4s^2 t}} =$ _____

(b) $\sqrt{\dfrac{3h}{2p}} =$ _____

a) $\dfrac{\sqrt{15}}{5}$

b) $\dfrac{\sqrt{abc}}{c}$

209. Though denominators containing radicals are usually rationalized in formal mathematics, they are not always rationalized in science and technology. In science and technology, the decision to rationalize or not is usually based on the comparative ease of evaluating with the two forms.

Here are two examples from electronics in which the denominators are not rationalized:

(1) Relationship of current (I), power (P), and resistance (R).

We use this form: $I = \sqrt{\dfrac{P}{R}}$

instead of: $I = \dfrac{\sqrt{PR}}{R}$

(2) Relationship of resonant frequency (f), inductance (L), and capacitance (C).

We use this form: $f = \dfrac{1}{2\pi\sqrt{LC}}$

instead of: $f = \dfrac{\sqrt{LC}}{2\pi LC}$

Whether a denominator is rationalized or not is something that must be learned for each scientific or technical formula.

a) $\dfrac{\sqrt{rt}}{2st}$

b) $\dfrac{\sqrt{6hp}}{2p}$

60   Squares and Square Root Radicals

SELF-TEST 12 (Frames 194-209)

1. To rationalize the denominator of $\dfrac{r}{\sqrt{2a}}$, multiply the fraction by _____

Rationalize the denominator of each of the following fractions:

2. $\dfrac{1}{2\sqrt{3}} = $ _____

3. $\dfrac{v}{4\sqrt{a}} = $ _____

4. $\dfrac{t}{r\sqrt{2s}} = $ _____

Rationalize each denominator and write the resulting fraction in lowest terms:

5. $\dfrac{3}{\sqrt{3}} = $ _____

6. $\dfrac{bd}{2\sqrt{d}} = $ _____

7. $\dfrac{4t}{b\sqrt{2t}} = $ _____

Factor out perfect squares, if possible, and then rationalize each denominator:

8. $\sqrt{\dfrac{2}{5}} = $ _____

9. $\sqrt{\dfrac{h}{m}} = $ _____

10. $\sqrt{\dfrac{4mv^2}{9k}} = $ _____

11. $\sqrt{\dfrac{3r^2}{8ab^2}} = $ _____

ANSWERS:
1. $\dfrac{\sqrt{2a}}{\sqrt{2a}}$
2. $\dfrac{\sqrt{3}}{6}$
3. $\dfrac{v\sqrt{a}}{4a}$
4. $\dfrac{t\sqrt{2s}}{2rs}$
5. $\sqrt{3}$
6. $\dfrac{b\sqrt{d}}{2}$
7. $\dfrac{2\sqrt{2t}}{b}$
8. $\dfrac{\sqrt{10}}{5}$
9. $\dfrac{\sqrt{hm}}{m}$
10. $\dfrac{2v\sqrt{km}}{3k}$
11. $\dfrac{r\sqrt{6a}}{4ab}$

1-13   SQUARING BINOMIAL EXPRESSIONS

A binomial expression is an expression which contains <u>two terms</u>. Since a binomial expression is either an addition or a subtraction, squaring a binomial requires either a multiplication of two additions or a multiplication of two subtractions. For example:

$$(x + 5)^2 = (x + 5)(x + 5)$$
$$(y - 3)^2 = (y - 3)(y - 3)$$

Multiplications of this type are performed by means of the distributive principle. We will show the method in this section. And since squaring binomials by direct use of the distributive principle is a long process, we will show a shortcut procedure at the end of the section.

Squares and Square Root Radicals 61

210. Each of the additions and subtractions below contains <u>two terms</u>. Since the technical word for "two-term" is "<u>binomial</u>", these expressions are called "<u>binomial</u> expressions".

$x + 5$ $3y - 7$ $4a + b$ $c - d$

Which of the following are "binomial expressions"? _____

(a) $7x$ (b) $3x + 7$ (c) $4x + 3y + 6$ (d) $2a - c$

---

211. Each of the expressions below contains <u>three terms</u>. Since the technical word for "three-term" is "<u>trinomial</u>", they are called "<u>trinomial</u> expressions".

$3x^2 + 2x + 7$ $y^2 - 4y + 4$ $2a - 3b - 6$

Which of the following are "trinomial expressions"? _____

(a) $3y - 5$ (c) $x^2 - 6x + 9$
(b) $a^2 + 2ab + b^2$ (d) $a^2 - 6$

Only (b) and (d)

---

212. Write "binomial" or "trinomial" in the blank after each expression:

(a) $5a - 8$ _____   (c) $y^2 - 8y + 16$ _____
(b) $4t^2 + 20t + 25$ _____   (d) $7 - 2d$ _____

Only (b) and (c)

---

213. If an expression contains <u>only one term</u>, it is called a "<u>monomial</u>" expression. For example:

x^2, $5y$, and 6 are <u>monomial</u> expressions.

If an expression contains <u>more than three terms</u>, it is called a <u>polynomial</u> expression. And ordinarily, the number of terms in these expressions is identified. For example:

$3x^2 + 5x + 3y + 7$ is called a "<u>polynomial of four terms</u>".

$y^2 + 3a + 2cd + b^2 + 8$ is called a "<u>polynomial of five terms</u>".

(a) Which of the following are <u>monomial expressions</u>? _____

$10t$ $3x + 5$ x^2 $7b - 2$

(b) "$5d^2 - 4t - 3d + 1$" is a polynomial of _____ terms.

a) binomial
b) trinomial
c) trinomial
d) binomial

---

a) Only $10t$ and x^2
b) four

62   Squares and Square Root Radicals

214. In order to square a binomial which is an addition, we must multiply the addition by itself. For example:

$$(x + 5)^2 = (x + 5)(x + 5)$$

This type of multiplication in which both factors are additions is really an instance of the distributive principle.

Here is a general statement of the distributive principle:

$$(\triangle + \bigcirc)(\square) = \triangle(\square) + \bigcirc(\square)$$

By plugging in the factors $(x + 5)(x + 5)$, we can see that a multiplication of two additions is an instance of this principle:

$$(\triangle{x} + \bigcirc{5})(\boxed{x+5}) = \triangle{x}(\boxed{x+5}) + \bigcirc{5}(\boxed{x+5})$$

or

$$(x + 5)(x + 5) = x(x + 5) + 5(x + 5)$$

Notice these two points about the multiplication above:

(1) By using the distributive principle, the <u>first</u> addition is broken up.

(2) The right side contains two terms, $x(x + 5)$ and $5(x + 5)$, each of which is still an instance of the distributive principle. However, these two terms contain only one addition.

Using the same pattern, complete these:

(a) $(y + 4)(y + 4) = y(y + 4) + ()(y + 4)$

(b) $(2a + 7)(2a + 7) = 2a() + 7(2a + 7)$

(c) $(a + b)(a + b) = a() + b()$

---

215. Complete each of the following:

(a) $(3t + 2)(3t + 2) = 3t(3t + 2) + 2()$

(b) $(c + d)(c + d) = c() + d()$

(c) $(2m + p)(2m + p) = 2m() + p()$

a) 4

b) $2a + 7$

c) $a + b$ ... $a + b$

---

216. Complete these by applying the distributive principle:

(a) $(m + 6)(m + 6) = ()() + ()()$

(b) $(p + q)(p + q) = ()() + ()()$

(c) $(2t + a)(2t + a) = ()() + ()()$

a) $3t + 2$

b) $c + d$ ... $c + d$

c) $2m + p$ ... $2m + p$

---

a) $m(m + 6) + 6(m + 6)$

b) $p(p + q) + q(p + q)$

c) $2t(2t + a) + a(2t + a)$

Squares and Square Root Radicals   63

217. Three steps are needed to complete the squaring of a binomial which is an addition. Here is an example:

Step 1: $(x + 5)(x + 5) = x(x + 5) + 5(x + 5)$

Step 2: $= x^2 + 5x + 5x + 25$

Step 3: $= x^2 + 10x + 25$

In Step 1, we used the distributive principle to multiply the two additions. By doing so, the first addition was broken up and we obtained two terms, each of which contains only one addition.

In Step 2, we multiplied by the distributive principle in each term.

In Step 3, we combined like terms.

Following the steps above, complete this one:

$(y + 3)(y + 3) = y(y + 3) + 3(y + 3)$

= _____

= _____

---

218. Complete each of the following:

(a) $(t + 7)(t + 7) = t(t + 7) + 7(t + 7)$

= _____

= _____

(b) $(a + b)(a + b) = a(a + b) + b(a + b)$

= _____

= _____

Answers:
$= y^2 + 3y + 3y + 9$
$= y^2 + 6y + 9$

---

219. Complete each of the following:

(a) $(2x + 3)(2x + 3) = 2x(2x + 3) + 3(2x + 3)$

= _____

= _____

(b) $(4c + 5d)(4c + 5d) = 4c(4c + 5d) + 5d(4c + 5d)$

= _____

= _____

Answers:
a) $= t^2 + 7t + 7t + 49$
    $= t^2 + 14t + 49$
b) $= a^2 + ab + ab + b^2$
    $= a^2 + 2ab + b^2$

---

a) $= 4x^2 + 6x + 6x + 9$
    $= 4x^2 + 12x + 9$
b) $= 16c^2 + 20cd + 20cd + 25d^2$
    $= 16c^2 + 40cd + 25d^2$

Squares and Square Root Radicals

220. Complete each of the following:

(a) $(2y + 7)(2y + 7) = $ _____

$= $ _____

$= $ _____

(b) $(5t + 2s)(5t + 2s) = $ _____

$= $ _____

$= $ _____

a) $= 2y(2y+7) + 7(2y+7)$
$= 4y^2 + 14y + 14y + 49$
$= 4y^2 + 28y + 49$

b) $= 5t(5t+2s) + 2s(5t+2s)$
$= 25t^2 + 10st + 10st + 4s^2$
$= 25t^2 + 20st + 4s^2$

221. In order to square a binomial which is a subtraction, we must multiply the subtraction by itself. For example:

$$(x - 5)^2 = (x - 5)(x - 5)$$

By converting both subtractions to additions, this multiplication of two subtractions can also be performed by means of the distributive principle for multiplication over addition. That is:

$(x - 5)(x - 5) = \left[x + (-5)\right]\left[x + (-5)\right]$

$= x\left[x + (-5)\right] + (-5)\left[x + (-5)\right]$

$= x^2 + (-5x) + (-5x) + 25$

$= x^2 + (-10x) + 25$

or

$x^2 - 10x + 25$

Following the steps above, complete this one:

$(y - 4)(y - 4) = \left[y + (-4)\right]\left[y + (-4)\right]$

$= y\left[y + (-4)\right] + (-4)\left[y + (-4)\right]$

$= $ _____

$= $ _____

$= y^2 + (-4y) + (-4y) + 16$
$= y^2 + (-8y) + 16$
or
$y^2 - 8y + 16$

222. Complete each of the following:

(a) $(x - y)(x - y) = \left[x + (-y)\right]\left[x + (-y)\right]$

$= $ _____

$= $ _____

$= $ _____

(b) $(4t - 3)(4t - 3) = \left[4t + (-3)\right]\left[4t + (-3)\right]$

$= $ _____

$= $ _____

$= $ _____

Squares and Square Root Radicals   65

223. When squaring a binomial which is a subtraction, the second term of its square is ordinarily converted to subtraction form. That is:

   Instead of: $x^2 + (-6x) + 9$
   we write: $x^2 - 6x + 9$

Complete each of the following and write the "square" in the preferred form:

(a) $(m - t)(m - t) = $ _____
   = _____
   = _____
   = _____

(b) $(3b - 2a)(3b - 2a) = $ _____
   = _____
   = _____
   = _____

a) $= x[x+(-y)] + (-y)[x+(-y)]$
   $= x^2 + (-xy) + (-xy) + y^2$
   $= x^2 + (-2xy) + y^2$
   or
   $x^2 - 2xy + y^2$

b) $= 4t[4t+(-3)] + (-3)[4t+(-3)]$
   $= 16t^2 + (-12t) + (-12t) + 9$
   $= 16t^2 + (-24t) + 9$
   or
   $16t^2 - 24t + 9$

---

224. As we have seen in the preceding frames, the "square" of any binomial is a <u>trinomial</u>. Though the third term of the trinomial is added in all cases:

If the binomial is an <u>addition</u>, the second term of the trinomial is <u>added</u>. That is:

$$(x + 3)^2 = x^2 \overset{\downarrow}{+} 6x + 9$$

$$(2a + 5b)^2 = 4a^2 \overset{\downarrow}{+} 20ab + 25b^2$$

If the binomial is a <u>subtraction</u>, the second term of the trinomial is <u>subtracted</u>. That is:

$$(y - 5)^2 = y^2 \overset{\downarrow}{-} 10y + 25$$

$$(3t - 4m)^2 = 9t^2 \overset{\downarrow}{-} 24mt + 16m^2$$

Look at the operation before the second term to answer these:

(a) Which of the following is the "square" of $t + 6$? _____
   $t^2 + 12t + 36$ $t^2 - 12t + 36$

(b) Which of the following is the "square" of $y - 7$? _____
   $y^2 + 14y + 49$ $y^2 - 14y + 49$

a) $= [m + (-t)][m + (-t)]$
   $= m[m+(-t)] + (-t)[m+(-t)]$
   $= m^2 + (-mt) + (-mt) + t^2$
   $= m^2 - 2mt + t^2$

b) $= [3b+(-2a)][3b+(-2a)]$
   $= 3b[3b + (-2a)]$
   $+ (-2a)[3b + (-2a)]$
   $= 9b^2 + (-6ab) + (-6ab) + 4a^2$
   $= 9b^2 - 12ab + 4a^2$

---

a) $t^2 + 12t + 36$

b) $y^2 - 14y + 49$

66   Squares and Square Root Radicals

225. Multiplying by the distributive principle in order to square a binomial is a long process. Fortunately, there is a shortcut that can be used. Here is an example:

If we square $(x + 5)$ by means of the distributive principle, we get:

$$(x + 5)^2 = x^2 + 10x + 25$$

By examining the "square", we can see these facts:

(1) The <u>first term of the trinomial</u> (x^2) can be obtained by squaring "x", the first term of the binomial. That is:

$$(x)(x) = x^2$$

(2) The <u>second term of the trinomial</u> $(10x)$ can be obtained by doubling the multiplication of the two terms of the binomials. That is:
$$2(x)(5) = 10x$$

(3) The <u>third term of the trinomial</u> (25) can be obtained by squaring "5", the second term of the binomial. That is:

$$(5)(5) = 25$$

Let's use this new procedure to find the square of $(y + 7)$.

(a) Squaring the first term of the binomial, we get: $(y)(y) =$ _____

(b) Doubling the multiplication of the two terms of the binomial, we get: $(2)(y)(7) =$ _____

(c) Squaring the second term of the binomial, we get: $(7)(7) =$ _____

(d) Therefore, the square of $(y + 7)$ is the trinomial _____.

---

226. Let's square $(2x + 9)$ by the shorter process:

(a) Squaring the first term of the binomial, we get: $(2x)(2x) =$ _____

(b) Doubling the multiplication of the two terms of the binomial, we get: $2(2x)(9) =$ _____

(c) Squaring the second term of the binomial, we get: $(9)(9) =$ _____

(d) Therefore, the square of $(2x + 9)$ is the trinomial _____.

a) y^2
b) $14y$
c) 49
d) $y^2 + 14y + 49$

---

227. Let's square $(3a + 4b)$ by the shorter process:

(a) Squaring the first term of the binomial, we get: $(3a)(3a) =$ _____

(b) Doubling the multiplication of the two terms of the binomial, we get: $2(3a)(4b) =$ _____

(c) Squaring the second term of the binomial, we get: $(4b)(4b) =$ _____

(d) Therefore, the square of $(3a + 4b)$ is the trinomial _____.

a) $4x^2$
b) $36x$
c) 81
d) $4x^2 + 36x + 81$

---

a) $9a^2$
b) $24ab$
c) $16b^2$
d) $9a^2 + 24ab + 16b^2$

228. Using the shorter procedure, square each of these:

(a) $(t + 3)^2 =$ _____ + _____ + _____

(b) $(2y + 5)^2 =$ _____ + _____ + _____

(c) $(R + V)^2 =$ _____ + _____ + _____

(d) $(3m + 5d)^2 =$ _____ + _____ + _____

229. We can also use the same short procedure to find the square of a binomial which is a subtraction. However, when using it, <u>we must remember</u> <u>to subtract the second term</u>.

a) $t^2 + 6t + 9$
b) $4y^2 + 20y + 25$
c) $R^2 + 2RV + V^2$
d) $9m^2 + 30md + 25d^2$

Let's square $(y - 4)$ as an example:

(a) The first term of the trinomial is: $(y)(y) =$ _____

(b) The second term of the trinomial is: $2(y)(4) =$ _____

(c) The third term of the trinomial is: $(4)(4) =$ _____

(d) $(y - 4)^2 =$ _____ - _____ + _____

230. Let's square $(4t - 5b)$ by the short procedure:

(a) The first term of the square is: $(4t)(4t) =$ _____

(b) The second term of the square is: $(2)(4t)(5b) =$ _____

(c) The third term of the square is: $(5b)(5b) =$ _____

(d) $(4t - 5b)^2 =$ _____ - _____ + _____

a) y^2
b) $8y$
c) 16
d) $y^2 - 8y + 16$

231. Using the short procedure, square each of these:

(a) $(x - 2)^2 =$ _____

(b) $(3t - 4)^2 =$ _____

(c) $(p - q)^2 =$ _____

(d) $(5t - m)^2 =$ _____

a) $16t^2$
b) $40bt$
c) $25b^2$
d) $16t^2 - 40bt + 25b^2$

232. When using the short procedure, we must square both terms of the binomial. If these terms are letter-terms, be sure to square each factor of the letter-term. For example:

$(3y)^2$ is $9y^2$, not $3y$

$(4ab)^2$ is $16a^2b^2$, not $4ab^2$ or $4a^2b^2$

a) $x^2 - 4x + 4$
b) $9t^2 - 24t + 16$
c) $p^2 - 2pq + q^2$
d) $25t^2 - 10mt + m^2$

Square each of the following binomials:

(a) $(ab + cd)^2 =$ _____

(b) $(3mt - 2bd)^2 =$ _____

a) $a^2b^2 + 2abcd + c^2d^2$
b) $9m^2t^2 - 12bdmt + 4b^2d^2$

68 Squares and Square Root Radicals

233. When using the short procedure, remember that the second term of the trinomial is <u>double</u> the multiplication of the two terms. For example:

In the square of $(3x + 5y)$,
 the second term is $(2)(3x)(5y) = 30xy$.

Note: A common error is forgetting to <u>double</u> the multiplication of the two terms.

Write the <u>second</u> <u>term</u> of each of the following squares:

(a) $(a + b)^2$ _____ (c) $(5x + 7)^2$ _____
(b) $(2c - d)^2$ _____ (d) $(7y - 4x)^2$ _____

---

234. Be careful when one of the terms is "1". For example, in the square of $(x + 1)$:

The first   term is: $(x)(x) = x^2$
The second term is: $(2)(x)(1) = 2x$
The third   term is: $(1)(1) = 1$

Therefore: $(x + 1)^2 =$ _____

a) 2ab   c) 70x
b) 4cd   d) 56xy

---

235. Complete: (a) $(y - 1)^2 =$ _____
             (b) $(1 - d)^2 =$ _____
             (c) $(3b - 1)^2 =$ _____
             (d) $(1 + 5a)^2 =$ _____

$x^2 + 2x + 1$

---

a) $y^2 - 2y + 1$
b) $1 - 2d + d^2$
c) $9b^2 - 6b + 1$
d) $1 + 10a + 25a^2$

Squares and Square Root Radicals   69

SELF-TEST 13 (Frames 210-235)

State whether each of the following is a monomial, a binomial, or a trinomial:

1. $3d + 1$ _____   2. $9t^2$ _____   3. $4x^2 + 20x + 25$ _____

Using the distributive principle, fill in the following blanks:

4. $(r + 6)^2 = (r + 6)(r + 6) = (\ \ )(\ \ \ \ ) + (\ \ )(\ \ \ \ )$

Do these complete squarings by the long method using the distributive principle:

5. $(3p + 2t)^2 = (3p + 2t)(3p + 2t) = 3p(3p + 2t) + 2t(3p + 2t)$

   = _____

   = _____

6. $(5a - 4)^2 = [5a + (-4)][5a + (-4)] = 5a[5a + (-4)] + (-4)[5a + (-4)]$

   = _____

   = _____

Find the following squares using the shortcut method:

7. $(7x + 2)^2 =$ _____   10. $(1 - 6s)^2 =$ _____
8. $(1 - A)^2 =$ _____   11. $(2d + bt)^2 =$ _____
9. $(3h + 1)^2 =$ _____   12. $(4r - 3w)^2 =$ _____

ANSWERS:
1. binomial
2. monomial
3. trinomial
4. $(r)(r + 6) + (6)(r + 6)$
5. $= 9p^2 + 6pt + 6pt + 4t^2$
   $= 9p^2 + 12pt + 4t^2$
6. $= 25a^2 + (-20a) + (-20a) + 16$
   $= 25a^2 - 40a + 16$
7. $49x^2 + 28x + 4$
8. $1 - 2A + A^2$
9. $9h^2 + 6h + 1$
10. $1 - 12s + 36s^2$
11. $4d^2 + 4bdt + b^2t^2$
12. $16r^2 - 24rw + 9w^2$

1-14 THE SQUARE ROOTS OF PERFECT-SQUARE TRINOMIALS

In the last section we saw that the squares of binomials are trinomials. Trinomials of this type are called "perfect-square trinomials". Since any three-term expression is a trinomial, there are many trinomials which are not "perfect-square" trinomials. In this section, we will discuss the method of identifying perfect-square trinomials. After doing so, we will discuss the square roots of perfect-square trinomials.

236. Three tests are needed to identify a perfect-square trinomial. Since two of the terms of a perfect-square trinomial are obtained by squaring the two terms of a binomial, the first test is this:

A trinomial is a perfect-square trinomial only if two of its terms are perfect squares. (In the ordinary way of writing trinomials, these two terms are the first and last terms.)

We have discussed the following types of perfect-square terms:

(1) 1, 4, 9, 16, 25, etc. are perfect squares since their square roots are whole numbers.

(2) x^2, y^2, t^2, etc. are perfect squares since their square roots are letters.

(3) $4x^2$, $9y^2$, a^2b^2, and $49c^2d^2$ are perfect squares since their square roots are letter-terms which do not contain a radical.

In $t^2 + 5t + 9$: (a) Is t^2 a perfect square? _____

(b) Is 9 a perfect square? _____

In $25x^2 + 12xy + 14y^2$: (c) Is $25x^2$ a perfect square? _____

(d) Is $14y^2$ a perfect square? _____

---

237. Which of the following trinomials contain two perfect squares? _____

(a) $y^2 - 6y + 16$ (c) $9x^2 - 5x + 1$

(b) $c^2 + 2cd + d$ (d) $4a^2 + 10ab + 9b^2$

---

238. Which of the following trinomials contain two perfect squares? _____

(a) $x^2 - 7x + 11$ (c) $3y^2 - 8y + 7$

(b) $m^2 + 2mt + t^2$ (d) $5d^2 + 16dp + 49p^2$

---

a) Yes, since $\sqrt{t^2} = t$.

b) Yes, since $\sqrt{9} = 3$.

c) Yes, since $\sqrt{25x^2} = 5x$.

d) No, since $\sqrt{14y^2} = y\sqrt{14}$.

(a), (c), and (d)

Only (b)

239. The second test needed to identify a perfect-square trinomial involves its <u>middle</u> term. In squaring a binomial, the middle term of the trinomial is obtained by doubling the product of the two terms of the binomial. For example:

$$(x + 7)^2 = x^2 + 14x + 49$$
where $14x = (2)(x)(7)$

$$(3a - 4b)^2 = 9a^2 - 24ab + 16b^2$$
where $24ab = (2)(3a)(4b)$

Therefore, the second test for a perfect-square trinomial is this:

<u>A trinomial is a perfect-square trinomial only if its middle term is double the product of the square roots of the first and last terms.</u> That is:

$x^2 + 14x + 49$ is a perfect-square trinomial, since:
$$2\sqrt{x^2} \cdot \sqrt{49} = 2(x)(7) = 14x$$

$9a^2 - 24ab + 16b^2$ is a perfect-square trinomial, since:
$$2\sqrt{9a^2} \cdot \sqrt{16b^2} = 2(3a)(4b) = 24ab$$

(a) In $y^2 + 6y + 9$, does $2\sqrt{y^2} \cdot \sqrt{9} = 6y$? _____

(b) In $4x^2 + 10x + 25$, does $2\sqrt{4x^2} \cdot \sqrt{25} = 10x$? _____

(c) In $a^2 + 2ab + b^2$, does $2\sqrt{a^2} \cdot \sqrt{b^2} = 2ab$? _____

(d) In $9y^2 + 5ty + t^2$, does $2\sqrt{9y^2} \cdot \sqrt{t^2} = 5ty$? _____

240. Does the middle term equal twice the product of the square roots of the first and last terms of:

(a) $x^2 + 8x + 16$? _____   (c) $y^2 - y + 9$? _____

(b) $t^2 - 2t + 1$? _____   (d) $16b^2 + 40bd + 25d^2$? _____

a) Yes

b) No, since:
$2\sqrt{4x^2} \cdot \sqrt{25} = 2(2x)(5)$
$= 20x$

c) Yes

d) No, since:
$2\sqrt{9y^2} \cdot \sqrt{t^2} = 2(3y)(t)$
$= 6ty$

241. Does the middle term equal twice the product of the square roots of the first and last terms of:

(a) $x^2 + 2xy + y^2$? _____   (c) $4a^2 - 12ab + 9b^2$? _____

(b) $p^2 - pq + q^2$? _____   (d) $49c^2 + 42cd + 36d^2$? _____

a) Yes, since:
$2\sqrt{x^2} \cdot \sqrt{16} = 2(x)(4) = 8x$

b) Yes, since:
$2\sqrt{t^2} \cdot \sqrt{1} = 2(t)(1) = 2t$

c) No, since:
$2\sqrt{y^2} \cdot \sqrt{9} = 2(y)(3) = 6y$

d) Yes, since:
$2\sqrt{16b^2} \cdot \sqrt{25d^2}$
$= 2(4b)(5d) = 40bd$

<u>Answer to Frame 241</u>:   a) Yes, since: $2\sqrt{x^2} \cdot \sqrt{y^2} = 2xy$ c) Yes, since: $2\sqrt{4a^2} \cdot \sqrt{9b^2} = 2(2a)(3b) = 12ab$

b) No, since: $2\sqrt{p^2} \cdot \sqrt{q^2} = 2pq$ d) No, since: $2\sqrt{49c^2} \cdot \sqrt{36d^2} = 2(7c)(6d) = 84cd$

72   Squares and Square Root Radicals

242. The third test needed to identify a perfect-square trinomial is this:

> Though the second term of the trinomial can be either added or subtracted, the third term is always added.

For example, the following is not a perfect-square trinomial, since its third term is subtracted:

$$x^2 - 8x \overset{\downarrow}{-} 16$$

Which of the following cannot be perfect-square trinomials for the reason above? _____

(a) $x^2 - 4x - 4$
(b) $y^2 - 6y + 9$
(c) $m^2 + 12m - 36$
(d) $a^2 - 2ab - b^2$

| (a), (c), and (d) |

243. If either the first or last term of a trinomial is not a perfect square, you know immediately that the trinomial is not a perfect-square trinomial.

Which of the following are not perfect-square trinomials for the reason above? _____

(a) $x^2 + 5x + 7$
(b) $b^2 - 6b + 9$
(c) $a^2 + 2ab + b$
(d) $4m^2 + 16mt + 16t^2$

| Both (a) and (c) |

244. Even though both the first and last terms of a trinomial are perfect squares, the trinomial is still not a perfect-square trinomial if its middle term is not twice the product of the square roots of the first and last terms.

Which of the following trinomials are not perfect-square trinomials for the reason above? _____

(a) $x^2 + 4x + 4$
(b) $4y^2 - 8x + 9$
(c) $a^2 + ad + d^2$
(d) $9x^2 - 12xy + 4y^2$

| Both (b) and (c) |

245. Even though both the first and last terms are perfect squares and the middle term is twice the product of the square roots of the first and last terms, a trinomial is still not a perfect-square trinomial if its third term is subtracted.

Which of the following trinomials are not perfect-square trinomials for the reason above? _____

(a) $x^2 + 4x - 4$
(b) $9t^2 - 6t - 1$
(c) $a^2 - 2ab - b^2$
(d) $4t^2 + 12mt - 9m^2$

| All of them |

Squares and Square Root Radicals   73

246. Here are the three tests needed to identify a perfect-square trinomial:
   (1) The first and last terms must be perfect squares.
   (2) The middle term must be twice the product of the square roots of the first and last terms.
   (3) The last term must be added.

   Apply the three tests to decide which of the following are perfect-square trinomials. _____

   (a) $m^2 + 10m - 25$
   (b) $t^2 - 12t + 35$
   (c) $R^2 - 2RV + V^2$
   (d) $b^2 + 2bd + d$

247. Which of the following are perfect-square trinomials? _____

   (a) $4a^2 + 6ab + 9b^2$
   (b) $25x^2 - 20xy + 4y^2$
   (c) $9d^2 - 36dt + 36t$
   (d) $49r^2 + 126rs - 81s^2$

   Only (c)

248. Any perfect square trinomial is the square of a binomial. For example:

   $x^2 + 6x + 9$ is the square of $x + 3$

   $a^2 - 2ab + b^2$ is the square of $a - b$

   As you can see from the examples above:
   (1) The terms of the binomials are the square roots of the two perfect-square terms of the trinomials.
   (2) The binomial is an addition if the second term of the trinomial is added. The binomial is a subtraction if the second term of the trinomial is subtracted.

   Let's use the steps above to find the binomial whose square is $y^2 + 14y + 49$.
   (a) The first term of the binomial is _____.
   (b) The second term of the binomial is _____.
   (c) The binomial is an _____ (addition/subtraction).
   (d) The binomial is _____.

   Only (b)

249. Let's find the binomial whose square is $9a^2 - 30ab + 25b^2$.
   (a) The first term of the binomial is _____.
   (b) The second term of the binomial is _____.
   (c) The binomial is an _____ (addition/subtraction).
   (d) The binomial is _____.

   a) y, from $\sqrt{y^2}$
   b) 7, from $\sqrt{49}$
   c) addition
   d) y + 7

   a) 3a, from $\sqrt{9a^2}$
   b) 5b, from $\sqrt{25b^2}$
   c) subtraction
   d) 3a - 5b

74   Squares and Square Root Radicals

250. Use the same steps for these:
   (a) $a^2 + 4a + 4$ is the square of _____ .
   (b) $y^2 - 10y + 25$ is the square of _____ .
   (c) $b^2 + 2bt + t^2$ is the square of _____ .
   (d) $16m^2 - 24mx + 9x^2$ is the square of _____ .

251. Instead of saying: $x^2 + 12x + 36$ is the square of $x + 6$
   we can simply write: $x^2 + 12x + 36 = (x + 6)^2$

   Complete each of these:
   (a) $R^2 + 16R + 64 = ($ _____ $)^2$
   (b) $c^2 - 2cd + d^2 = ($ _____ $)^2$
   (c) $64x^2 + 144xy + 81y^2 = ($ _____ $)^2$
   (d) $49s^2 - 70st + 25t^2 = ($ _____ $)^2$

   a) $a + 2$
   b) $y - 5$
   c) $b + t$
   d) $4m - 3x$

252. Just as $\sqrt{9} = 3$,   since $3^2 = 9$
   and $\sqrt{16x^2} = 4x$,   since $(4x)^2 = 16x^2$,
   $\sqrt{x^2 + 6x + 9} = x + 3$, since $(x + 3)^2 = x^2 + 6x + 9$.

   Find the following square roots:
   (a) $\sqrt{y^2 + 4y + 4} = $ _____
   (b) $\sqrt{p^2 - 2pq + q^2} = $ _____
   (c) $\sqrt{9b^2 + 12bc + 4c^2} = $ _____
   (d) $\sqrt{16R^2 - 40RV + 25V^2} = $ _____

   a) $(R + 8)^2$ c) $(8x + 9y)^2$
   b) $(c - d)^2$ d) $(7s - 5t)^2$

253. Complete each of these:
   (a) $d^2 - 18d + 81 = ($ _____ $)^2$
   (b) $\sqrt{F^2 + 14F + 49} = $ _____
   (c) $m^2 + 4mt + 4t^2 = ($ _____ $)^2$
   (d) $\sqrt{y^2 - 2y + 1} = $ _____

   a) $y + 2$ c) $3b + 2c$
   b) $p - q$ d) $4R - 5V$

254. We have seen that the square root of the indicated square of a number, letter, or letter-term is the number, letter, or letter-term. For example:
   $\sqrt{7^2} = 7$ $\sqrt{x^2} = x$ $\sqrt{(3y)^2} = 3y$

   Similarly, the square root of the indicated square of a binomial is the binomial. That is:
   $\sqrt{(t + 5)^2} = \sqrt{t^2 + 10t + 25} = t + 5$
   (a) $\sqrt{(R - 1)^2} = \sqrt{R^2 - 2R + 1} = $ _____
   (b) $\sqrt{(2x + 3y)^2} = \sqrt{4x^2 + 12xy + 9y^2} = $ _____

   a) $(d - 9)^2$ c) $(m + 2t)^2$
   b) $F + 7$ d) $y - 1$

   a) $R - 1$
   b) $2x + 3y$

255. Complete each of these:

(a) $4d^2 + 8dm + 4m^2 = ($ ____ $)^2$ (d) $\sqrt{16x^2 - 8x + 1} =$ _____

(b) $\sqrt{t^2 - 2t + 1} =$ _____ (e) $\sqrt{(2F - 3h)^2} =$ _____

(c) $\sqrt{(a + b)^2} =$ _____ (f) $\sqrt{1 + 6y + 9y^2} =$ _____

a) $(2d + 2m)^2$
b) $t - 1$
c) $a + b$
d) $4x - 1$
e) $2F - 3h$
f) $1 + 3y$

256. If a trinomial is a perfect square, its square root can be replaced by a binomial. If a trinomial <u>is not</u> a perfect square, however, its square root <u>cannot be replaced</u> by a binomial.

Replace each square root with a binomial if possible:

(a) $\sqrt{R^2 + 10R + 25} =$ _____ (c) $\sqrt{y^2 - 3y + 9} =$ _____

(b) $\sqrt{a^2 - 2ab - b^2} =$ _____ (d) $\sqrt{p^2 + 2pq + q^2} =$ _____

<u>Answer</u> <u>to</u> <u>Frame</u> <u>256</u>:   a) $R + 5$ b) Not possible   c) Not possible   d) $p + q$

SELF-TEST 14 (Frames 236-256)

1. Which of the following are perfect-square trinomials? _____

(a) $x^2 - 2x + 1$ (c) $9y^2 + 18y + 36$ (e) $4a^2 - 12ab + 9b^2$
(b) $16d^2 - 8d - 1$ (d) $r^2 - 2rs - t^2$ (f) $100h^2 + 20hp + p^2$

Complete each of the following, if possible, by writing a binomial in the parentheses:

2. $A^2 + 4A + 4 = ($ ____ $)^2$ 3. $x^2 - 2xy + y^2 = ($ ____ $)^2$ 4. $a^2 + b^2 = ($ ____ $)^2$

Find the following square roots, if possible:

5. $\sqrt{(x + 5)^2} =$ _____   7. $\sqrt{a^2 + 2ab + b^2} =$ _____   9. $\sqrt{p^2 - 2p - 1} =$ _____

6. $\sqrt{(3r - t)^2} =$ _____   8. $\sqrt{25h^2 - 10h + 1} =$ _____   10. $\sqrt{1 + 4x + 4x^2} =$ _____

ANSWERS: 1. (a), (e), (f)   2. $(A + 2)^2$ 5. $x + 5$ 7. $a + b$ 9. Not possible.
3. $(x - y)^2$ 6. $3r - t$ 8. $5h - 1$ 10. $1 + 2x$
4. Not possible.

76 Squares and Square Root Radicals

1-15 ADDITION AND SUBTRACTION OF TERMS CONTAINING SQUARE ROOT RADICALS

In this section, we will discuss the addition and subtraction of terms containing square root radicals. We will see that these terms can be combined into one term only if they are "like" terms.

257. Two letter-terms are called "like" terms <u>if they contain the same letter</u>. For example:

$3x$ and $2x$ are "like" terms.
$4y$ and $5t$ are "unlike" terms.

Similarly, two terms which contain radicals are called "like" terms <u>if they contain the same radical</u>. For example:

$3\sqrt{y}$ and $2\sqrt{y}$ are "like" terms.
$4\sqrt{t}$ and $5\sqrt{p}$ are "unlike" terms.

Which pairs of terms below are "like" terms? _____

(a) $5\sqrt{x}$ and $4\sqrt{x}$ (c) $7\sqrt{5}$ and $3\sqrt{6}$
(b) $3\sqrt{m}$ and $2\sqrt{s}$ (d) $8\sqrt{7}$ and $2\sqrt{7}$

Only (a) and (d)

258. An addition of two letter-terms can be combined into one term only if they are "like" terms. Basically, they can be combined only because we can factor by the distributive principle and simplify. That is:

$3x + 2x$ can be combined into one term since:

$3x + 2x = (3 + 2)(x) = 5x$

$4y + 5t$ cannot be combined into one term <u>since they do not contain a common literal factor.</u>

Similarly, an addition of two radical terms can be combined into one term only if they are "like" terms. Again, they can be combined only because we can factor by the distributive principle and simplify. That is:

$3\sqrt{y} + 2\sqrt{y}$ can be combined into one term since:

$3\sqrt{y} + 2\sqrt{y} = (3 + 2)\sqrt{y} = 5\sqrt{y}$

$4\sqrt{p} + 5\sqrt{t}$ cannot be combined into one term <u>since they do not contain a common radical factor.</u>

Which of the following additions can be combined into one term? _____

(a) $5\sqrt{a} + 6\sqrt{b}$ (c) $8\sqrt{2} + 5\sqrt{2}$
(b) $7\sqrt{t} + 4\sqrt{t}$ (d) $2\sqrt{5} + 3\sqrt{6}$

Only (b) and (c)

Squares and Square Root Radicals   77

259. Complete each of the following:

(a) $5\sqrt{x} + 7\sqrt{x} = (5 + 7)\sqrt{x} = $ _____

(b) $3\sqrt{7} + 8\sqrt{7} = (3 + 8)\sqrt{7} = $ _____

a) $12\sqrt{x}$
b) $11\sqrt{7}$

---

260. Combining "like" terms by means of the distributive principle is the equivalent of simply adding the coefficients of the letters or radicals. For example:

Just as $7x + 3x = 10x$, $8\sqrt{y} + 4\sqrt{y} = 12\sqrt{y}$

Combine each of these by simply adding their coefficients:

(a) $6\sqrt{y} + 8\sqrt{y} = $ _____   (b) $2\sqrt{5} + 3\sqrt{5} = $ _____

a) $14\sqrt{y}$ b) $5\sqrt{5}$

---

261. Remember that an addition of radical terms can be combined into <u>one term only if the terms contain the same radical</u>. Combine the following into one term if possible:

(a) $7\sqrt{t} + 3\sqrt{t} = $ _____   (c) $3\sqrt{5} + 4\sqrt{5} = $ _____

(b) $10\sqrt{S} + 7\sqrt{R} = $ _____   (d) $5\sqrt{2} + 4\sqrt{3} = $ _____

a) $10\sqrt{t}$
b) Not possible.
c) $7\sqrt{5}$
d) Not possible.

---

262. If the coefficient of a letter or radical is not explicitly written, its coefficient is "1". That is:

Just as $x = 1x$, $\sqrt{x} = 1\sqrt{x}$

Therefore: (a) $5\sqrt{x} + \sqrt{x} = 5\sqrt{x} + 1\sqrt{x} = $ _____

(b) $\sqrt{7} + 3\sqrt{7} = 1\sqrt{7} + 3\sqrt{7} = $ _____

(c) $\sqrt{2} + \sqrt{2} = 1\sqrt{2} + 1\sqrt{2} = $ _____

a) $6\sqrt{x}$
b) $4\sqrt{7}$
c) $2\sqrt{2}$

---

263. Complete each of these:

(a) $10\sqrt{t} + \sqrt{t} = $ _____   (c) $\sqrt{13} + 7\sqrt{13} = $ _____

(b) $\sqrt{p} + \sqrt{p} = $ _____   (d) $\sqrt{27} + \sqrt{27} = $ _____

a) $11\sqrt{t}$ c) $8\sqrt{13}$
b) $2\sqrt{p}$ d) $2\sqrt{27}$

---

264. A subtraction of radical terms can be combined into one term only if the two terms are "like". In this case, the combining is possible because we can factor by the distributive principle <u>over subtraction</u> and simplify. For example:

$5\sqrt{3} - 2\sqrt{3}$ can be combined into one term since:

$5\sqrt{3} - 2\sqrt{3} = (5 - 2)\sqrt{3} = 3\sqrt{3}$

$7\sqrt{x} - 5\sqrt{y}$ cannot be combined into one term <u>since they do not contain a common radical factor</u>.

Which of the following subtractions can be combined into one term? _____

(a) $10\sqrt{m} - 7\sqrt{m}$ (b) $12\sqrt{5} - 5\sqrt{3}$

Only (a)

78 Squares and Square Root Radicals

265. As you can see from the examples below, factoring by the distributive principle over subtraction is the equivalent of <u>subtracting</u> coefficients.

$$10\sqrt{t} - 5\sqrt{t} = (10 - 5)\sqrt{t} = 5\sqrt{t}$$
$$9\sqrt{3} - 7\sqrt{3} = (9 - 7)\sqrt{3} = 2\sqrt{3}$$

Subtract coefficients to combine these:

(a) $20\sqrt{p} - 10\sqrt{p} =$ _____   (b) $15\sqrt{7} - 8\sqrt{7} =$ _____

266. To complete these, remember that the coefficient of a radical is "1" if it is not explicitly written.

(a) $3\sqrt{x} - \sqrt{x} =$ _____   (c) $\sqrt{7} - \sqrt{7} =$ _____

(b) $2\sqrt{5} - \sqrt{5} =$ _____

a) $10\sqrt{p}$ b) $7\sqrt{7}$

267. Perform these additions and subtractions:

(a) $\sqrt{3} + 5\sqrt{3} =$ _____   (c) $\sqrt{x} + \sqrt{x} =$ _____

(b) $7\sqrt{5} - \sqrt{5} =$ _____   (d) $\sqrt{x} - \sqrt{x} =$ _____

a) $2\sqrt{x}$
b) $1\sqrt{5}$ or $\sqrt{5}$
c) 0, since $0\sqrt{7} = 0$

268. Additions and subtractions of radical terms can also be combined into one term when their coefficients are letters. For example:

$$a\sqrt{x} + b\sqrt{x} = (a + b)\sqrt{x}$$
$$c\sqrt{y} - d\sqrt{y} = (c - d)\sqrt{y}$$

Combine each of these into one term:

(a) $m\sqrt{p} + n\sqrt{p} =$ _____

(b) $R\sqrt{V} - S\sqrt{V} =$ _____

(c) $a\sqrt{x} + \sqrt{x} =$ _____

(d) $b\sqrt{y} - \sqrt{y} =$ _____

a) $6\sqrt{3}$ c) $2\sqrt{x}$
b) $6\sqrt{5}$ d) $0\sqrt{x}$ or 0

269. Each of the additions below contains one radical term and one non-radical term:

$$3 + \sqrt{2} \qquad a + \sqrt{x}$$

Since these terms are "unlike", they cannot be combined into one term. That is:

$3 + \sqrt{2}$ does not equal $3\sqrt{2}$

$a + \sqrt{x}$ does not equal $a\sqrt{x}$

Perform the following additions or subtractions if possible:

(a) $2\sqrt{m} - \sqrt{m} =$ _____

(b) $5 + \sqrt{y} =$ _____

(c) $\sqrt{t} + b =$ _____

(d) $\sqrt{d} + 2\sqrt{d} =$ _____

a) $(m + n)\sqrt{p}$
b) $(R - S)\sqrt{V}$
c) $(a + 1)\sqrt{x}$
d) $(b - 1)\sqrt{y}$

270. There is a clear difference between the multiplication of radicals and the addition (or subtraction) of radicals.

In a multiplication, two radicals can be combined into one radical whether their radicands are the same or not. For example:
$$\sqrt{x} \cdot \sqrt{x} = \sqrt{x^2}$$
$$\sqrt{a} \cdot \sqrt{b} = \sqrt{ab}$$

In an addition (or subtraction), two radicals can be combined into one term only if their radicands are the same. For example:
$$\sqrt{x} + \sqrt{x} = 2\sqrt{x}$$
$$\sqrt{a} + \sqrt{b} \text{ cannot be combined into one term.}$$

Perform each of the following if possible:

(a) $\sqrt{t} + \sqrt{t} = $ _____
(b) $\sqrt{t} \cdot \sqrt{t} = $ _____
(c) $\sqrt{5} \cdot \sqrt{3} = $ _____
(d) $\sqrt{5} + \sqrt{3} = $ _____

a) $1\sqrt{m}$ or $\sqrt{m}$
b) Not possible
c) Not possible
d) $3\sqrt{d}$

271. In a multiplication of radicals, the product is obtained by multiplying their radicands.
$$\sqrt{a} \cdot \sqrt{b} = \sqrt{ab}$$

However, in an addition (or subtraction) of radicals, the sum is not obtained by adding their radicands.
$$\sqrt{a} + \sqrt{b} \text{ does not equal } \sqrt{a+b}$$

Which of the following statements are true? _____

(a) $\sqrt{t} + \sqrt{m} = \sqrt{t+m}$
(b) $\sqrt{R} \cdot \sqrt{S} = \sqrt{RS}$
(c) $\sqrt{7} \cdot \sqrt{5} = \sqrt{(7)(5)}$
(d) $\sqrt{11} - \sqrt{6} = \sqrt{11-6}$

a) $2\sqrt{t}$
b) $\sqrt{t^2}$ (or t)
c) $\sqrt{15}$
d) Not possible

Only (b) and (c)

SELF-TEST 15 (Frames 257-271)

Combine each of the following into one term, if possible:

1. $2\sqrt{2} + 3\sqrt{2} = $ _____
2. $5\sqrt{3} + \sqrt{3} = $ _____
3. $8\sqrt{A} + 6\sqrt{A} = $ _____
4. $3\sqrt{a} - 2\sqrt{a} = $ _____
5. $\sqrt{x} + \sqrt{x} = $ _____
6. $4\sqrt{t} - \sqrt{t} = $ _____
7. $2\sqrt{r} - 2\sqrt{r} = $ _____
8. $2 + \sqrt{w} = $ _____
9. $5\sqrt{b} - 3\sqrt{d} = $ _____

10. Does $\sqrt{4} \cdot \sqrt{9} = \sqrt{(4)(9)}$? _____
11. Does $\sqrt{4} + \sqrt{9} = \sqrt{4+9}$? _____
12. Does $\sqrt{r} + \sqrt{s} = \sqrt{r+s}$? _____
13. Does $\sqrt{r} \cdot \sqrt{s} = \sqrt{rs}$? _____

ANSWERS:
1. $5\sqrt{2}$
2. $6\sqrt{3}$
3. $14\sqrt{A}$
4. $\sqrt{a}$
5. $2\sqrt{x}$
6. $3\sqrt{t}$
7. 0
8. Not possible
9. Not possible
10. Yes
11. No
12. No
13. Yes

80   Squares and Square Root Radicals

1-16   THE SQUARE ROOTS OF ADDITIONS AND SUBTRACTIONS

In this section, we will discuss the square roots of additions and subtractions. Emphasis will be given to the square roots of binomials. The square roots of binomials will be contrasted with both the square roots of multiplications and the square roots of perfect-square trinomials.

272. In the radicals below, the radicands are either additions or subtractions.

$$\sqrt{x+3} \qquad \sqrt{x-3} \qquad \sqrt{a+b} \qquad \sqrt{a-b}$$

Radicals of this type can be evaluated by plugging in numbers for the letters. For example:

(a) If $x = 13$, $\sqrt{x+3} = \sqrt{13+3} = \sqrt{16} = $ _____

(b) If $a = 14$ and $b = 5$, $\sqrt{a-b} = \sqrt{14-5} = \sqrt{9} = $ _____

273. In the last section, we saw that an addition or subtraction of two radicals with different radicands <u>cannot</u> <u>be</u> <u>combined</u> <u>into</u> <u>one</u> <u>radical</u>. That is:

$$\sqrt{y} + \sqrt{5} \text{ does } \underline{\text{not}} \text{ equal } \sqrt{y+5}$$

$$\sqrt{m} - \sqrt{t} \text{ does } \underline{\text{not}} \text{ equal } \sqrt{m-t}$$

Similarly, the reverse procedure is not possible. That is, a radical which contains an addition or subtraction <u>cannot</u> <u>be</u> <u>broken</u> <u>up</u> <u>into</u> <u>an</u> <u>addition</u> <u>or</u> <u>subtraction</u> <u>of</u> <u>two</u> <u>radicals</u>. For example:

$$\sqrt{y+5} \text{ does } \underline{\text{not}} \text{ equal } \sqrt{y} + \sqrt{5}$$

$$\sqrt{m-t} \text{ does } \underline{\text{not}} \text{ equal } \sqrt{m} - \sqrt{t}$$

Are the following statements true or false?

(a) $\sqrt{x} - \sqrt{7} = \sqrt{x-7}$ _____

(b) $\sqrt{a+b} = \sqrt{a} + \sqrt{b}$ _____

274. The following statement says that the square root of an addition <u>is</u> <u>not</u> <u>equal</u> <u>to</u> the addition of the square roots of the terms.

$$\sqrt{x+y} \text{ does } \underline{\text{not}} \text{ equal } \sqrt{x} + \sqrt{y}$$

We can see that the statement above is true by plugging in some numbers for "x" and "y".

If $x = 16$ and $y = 9$:

(a) $\sqrt{x+y} = \sqrt{16+9} = \sqrt{25} = $ _____

(b) $\sqrt{x} + \sqrt{y} = \sqrt{16} + \sqrt{9} = 4 + 3 = $ _____

(c) Does $\sqrt{16+9} = \sqrt{16} + \sqrt{9}$? _____

---

a) +4

b) +3

Both are false.

---

a) 5

b) 7

c) No

275. The following statement says that the square root of a subtraction is not equal to the subtraction of the square roots of the terms.

$$\sqrt{x-y} \text{ does not equal } \sqrt{x} - \sqrt{y}$$

We can also see that the statement above is true by plugging in some numbers for "x" and "y".

If $x = 100$ and $y = 36$:

(a) $\sqrt{x-y} = \sqrt{100-36} = \sqrt{64} = $ _____

(b) $\sqrt{x} - \sqrt{y} = \sqrt{100} - \sqrt{36} = 10 - 6 = $ _____

(c) Does $\sqrt{100-36} = \sqrt{100} - \sqrt{36}$? _____

276. Square roots of additions and subtractions like those below are in their simplest form. That is, they cannot be written in any simpler radical form.

$$\sqrt{x+7} \qquad \sqrt{y-3} \qquad \sqrt{c+d} \qquad \sqrt{R-V}$$

a) 8

b) 4

c) No

Don't confuse the square root of an addition or subtraction with the square root of a multiplication.

If the radicand is a multiplication of two factors, the radical can be factored into a multiplication of two radicals. For example:

$$\sqrt{2x} = \sqrt{2} \cdot \sqrt{x}$$
$$\sqrt{ab} = \sqrt{a} \cdot \sqrt{b}$$

If the radicand is an addition or subtraction involving two terms, the radical cannot be broken down into an addition or subtraction of two radicals. That is:

$$\sqrt{x+2} \text{ does not equal } \sqrt{x} + \sqrt{2}$$
$$\sqrt{a-b} \text{ does not equal } \sqrt{a} - \sqrt{b}$$

Answer "true" or "false" for these:

(a) $\sqrt{7x} = \sqrt{7} \cdot \sqrt{x}$ _____

(b) $\sqrt{7+x} = \sqrt{7} + \sqrt{x}$ _____

(c) $\sqrt{b-d} = \sqrt{b} - \sqrt{d}$ _____

(d) $\sqrt{bd} = \sqrt{b} \cdot \sqrt{d}$ _____

277. Answer "true" or "false" for these:

(a) $\sqrt{mt} = \sqrt{m} \cdot \sqrt{t}$ _____

(b) $\sqrt{m+t} = \sqrt{m} + \sqrt{t}$ _____

(c) $\sqrt{m-t} = \sqrt{m} - \sqrt{t}$ _____

a) True

b) False

c) False

d) True

a) True

b) False

c) False

82 Squares and Square Root Radicals

278. The greatest temptation to break up the square root of an addition or subtraction into two radicals occurs when each of the terms is a perfect square. Here are some examples:

$$\sqrt{x^2 + 9} \qquad \sqrt{y^2 - 25} \qquad \sqrt{a^2 + b^2} \qquad \sqrt{c^2 - d^2}$$

Square root radicals of this type are also in their simplest form.
That is:

$$\sqrt{x^2 + 9} \text{ does \underline{not} equal } \sqrt{x^2} + \sqrt{9}$$
$$\sqrt{c^2 - d^2} \text{ does \underline{not} equal } \sqrt{c^2} - \sqrt{d^2}$$

We can see that the two statements above are true by plugging in some numbers for the letters:

If $x = 4$: (a) $\sqrt{x^2 + 9} = \sqrt{4^2 + 9} = \sqrt{16 + 9} = \sqrt{25} = $ _____

(b) $\sqrt{x^2} + \sqrt{9} = \sqrt{4^2} + \sqrt{9} = \sqrt{16} + \sqrt{9} = 4 + 3 = $ _____

If $c = 10$ and $d = 8$:

(c) $\sqrt{c^2 - d^2} = \sqrt{10^2 - 8^2} = \sqrt{100 - 64} = \sqrt{36} = $ _____

(d) $\sqrt{c^2} - \sqrt{d^2} = \sqrt{10^2} - \sqrt{8^2} = \sqrt{100} - \sqrt{64} = 10 - 8 = $ _____

a) 5
b) 7
c) 6
d) 2

279. Again, don't confuse the square root of a multiplication with the square root of an addition or subtraction.

Though $\sqrt{c^2 d^2}$ does equal $\sqrt{c^2} \cdot \sqrt{d^2}$,

$\sqrt{c^2 + d^2}$ does \underline{not} equal $\sqrt{c^2} + \sqrt{d^2}$, and

$\sqrt{c^2 - d^2}$ does \underline{not} equal $\sqrt{c^2} - \sqrt{d^2}$.

Answer "true" or "false" for each of these:

(a) $\sqrt{9x} = \sqrt{9} \cdot \sqrt{x}$ _____

(b) $\sqrt{x^2 + 9} = \sqrt{x^2} + \sqrt{9}$ _____

(c) $\sqrt{a^2 b^2} = \sqrt{a^2} \cdot \sqrt{b^2}$ _____

(d) $\sqrt{a^2 - b^2} = \sqrt{a^2} - \sqrt{b^2}$ _____

a) True
b) False
c) True
d) False

280. Both terms in the radicand of $\sqrt{x^2 + y^2}$ are perfect squares. The reason there is a temptation to break up such a radical incorrectly into $\sqrt{x^2} + \sqrt{y^2}$ is the fact that each simpler radical can then be replaced by a non-radical, since $\sqrt{x^2} + \sqrt{y^2}$ does equal $x + y$.

However, we know that the binomial "$x + y$" cannot be the square root of $x^2 + y^2$, since "$x + y$" is the square root of a perfect-square trinomial. That is:

Since $(x + y)^2 = x^2 + 2xy + y^2$, $x + y$ is the square root of $x^2 + 2xy + y^2$.

Complete each of the following:

(a) Is $x + 3$ the square root of $x^2 + 9$ or of $x^2 + 6x + 9$? _____

(b) Is $y - 5$ the square root of $y^2 - 25$ or of $y^2 - 10y + 25$? _____

(c) Is $a - b$ the square root of $a^2 - b^2$ or of $a^2 - 2ab + b^2$? _____

a) $x^2 + 6x + 9$
b) $y^2 - 10y + 25$
c) $a^2 - 2ab + b^2$

281. (a) Which one of the following statements is true? _____

$\sqrt{t^2 + 8t + 16} = t + 4$

$\sqrt{t^2 + 16} = t + 4$

(b) Which one of the following statements is true? _____

$\sqrt{d^2 - m^2} = d - m$

$\sqrt{d^2 - 2dm + m^2} = d - m$

a) $\sqrt{t^2 + 8t + 16} = t + 4$
b) $\sqrt{d^2 - 2dm + m^2} = d - m$

282. Answer "true" or "false" for each of these:

(a) $\sqrt{x^2 y^2} = xy$ _____

(b) $\sqrt{x^2 + y^2} = x + y$ _____

(c) $\sqrt{x^2 + 2xy + y^2} = x + y$ _____

(d) $x + y$ is the square root of $x^2 + y^2$. _____

(e) The radical $\sqrt{x^2 + y^2}$ is in its simplest form. _____

283. Answer "true" or "false" for each of these:

(a) $\sqrt{y^2 - 25} = y - 5$ _____

(b) $\sqrt{a^2 + 4a + 4} = a + 2$ _____

(c) $\sqrt{V^2 - 2VS + S^2} = V - S$ _____

(d) $\sqrt{R^2 F^2} = RF$ _____

(e) $\sqrt{t^2 + y^2} = t + y$ _____

a) True
b) False
c) True
d) False
e) True

84 Squares and Square Root Radicals

284. Don't confuse the square root of $a^2 + b^2$ with the square root of the indicated square of $a + b$.

$\sqrt{a^2 + b^2}$ does not equal $\sqrt{(a + b)^2}$,

since $\sqrt{(a + b)^2} = \sqrt{a^2 + 2ab + b^2} = a + b$

Answer "true" or "false" for each of these:

(a) $\sqrt{t^2 - s^2} = t - s$ _____

(b) $\sqrt{(t - s)^2} = \sqrt{t^2 - 2ts + s^2}$ _____

(c) $\sqrt{(t - s)^2} = t - s$ _____

(d) $\sqrt{t^2 - s^2} = \sqrt{(t - s)^2}$ _____

a) False
b) True
c) True
d) True
e) False

285. Answer "true" or "false" for each of these:

(a) $\sqrt{(x + 3)^2} = x + 3$ _____

(b) $\sqrt{9y^2} = 3y$ _____

(c) $\sqrt{p^2 - q^2} = p - q$ _____

(d) $\sqrt{4x^2 + 12xy + 9y^2} = 2x + 3y$ _____

(e) $\sqrt{16t^2 + 25s^2} = 4t + 5s$ _____

(f) $\sqrt{(9m - 4t)^2} = 9m - 4t$ _____

a) False
b) True
c) True
d) False

286. (a) Can we write the radical $\sqrt{x + 3}$ in a simpler form? _____

(b) Can we write the radical $\sqrt{a^2 - b^2}$ in a simpler form? _____

(c) Can we write the radical $\sqrt{(c + d)^2}$ in a simpler form? _____

a) True   d) True
b) True   e) False
c) False   f) True

287. Write each of the following in a simpler form if possible:

(a) $\sqrt{(y - 16)^2} =$ _____   (d) $\sqrt{R^2 + V^2} =$ _____

(b) $\sqrt{t^2 + 49} =$ _____   (e) $\sqrt{a^2 b^2} =$ _____

(c) $\sqrt{x^2 + 2xy + y^2} =$ _____   (f) $\sqrt{(m + 9)^2} =$ _____

a) No
b) No
c) Yes, since:
   $\sqrt{(c + d)^2} = c + d$

288. Just as the square root of a binomial cannot be broken up into an addition of the square roots of each term, the square roots of additions or subtractions with more than two terms cannot be broken up. That is:

$\sqrt{m^2 + p^2 + q^2}$ does not equal $\sqrt{m^2} + \sqrt{p^2} + \sqrt{q^2}$

$\sqrt{a^2 - b^2 - c^2}$ does not equal $\sqrt{a^2} - \sqrt{b^2} - \sqrt{c^2}$

a) $y - 16$
b) Not possible.
c) $x + y$
d) Not possible.
e) ab
f) $m + 9$

Squares and Square Root Radicals 85

SELF-TEST 16 (Frames 272-288)

1. Which of the following expressions are equal to "r + t"?

   (a) $\sqrt{(r+t)^2}$ (b) $\sqrt{r^2 + t^2}$ (c) $\sqrt{r^2 + 4rt + t^2}$ (d) $\sqrt{r^2 + 2rt + t^2}$

2. Does $\sqrt{16h^2} = \sqrt{16} \cdot \sqrt{h^2}$? _____
3. Does $\sqrt{16 + h^2} = \sqrt{16} + \sqrt{h^2}$? _____
4. Does $\sqrt{h^2 + 16} = h + 4$? _____
5. Does $\sqrt{h^2 + 16} = \sqrt{h^2 + 8h + 16}$? _____

Write each of the following in a simpler form if possible:

6. $\sqrt{a^2 - c^2} =$ _____
7. $\sqrt{a^2 + c^2} =$ _____
8. $\sqrt{(a - c)^2} =$ _____
9. $\sqrt{a^2 c^2} =$ _____
10. $\sqrt{a^2 + 2ac - c^2} =$ _____
11. $\sqrt{a^2 - 2ac + c^2} =$ _____

ANSWERS:
1. (a), (d)
2. Yes
3. No
4. No
5. No
6. Not possible.
7. Not possible.
8. a - c
9. ac
10. Not possible.
11. a - c

1-17 OPERATIONS INVOLVING THE SQUARE ROOTS OF BINOMIALS

In this section, we will show how the same principles apply in operations with square root radicals when the radicands are binomials. Therefore, this section contains a review and an extension of many of the principles taught in this chapter.

289. The pattern for multiplying two radicals is given below:

$$\sqrt{\Box} \cdot \sqrt{\bigcirc} = \sqrt{(\Box)(\bigcirc)}$$

This pattern also applies when one of the radicands is a binomial. For example:

$\sqrt{x} \cdot \sqrt{3x - 1} = \sqrt{(x)(3x - 1)} = \sqrt{3x^2 - x}$

$\sqrt{t^2} \cdot \sqrt{s + 1} = \sqrt{(t^2)(s + 1)} = \sqrt{st^2 + t^2}$

Complete each of these:

(a) $\sqrt{y^2} \cdot \sqrt{a + b} = \sqrt{(\quad)(\quad)} = \sqrt{\rule{2cm}{0.3pt}}$

(b) $\sqrt{5} \cdot \sqrt{d - 7} = \sqrt{(\quad)(\quad)} = \sqrt{\rule{2cm}{0.3pt}}$

| | a) $\sqrt{(y^2)(a + b)} = \sqrt{ay^2 + by^2}$ |
| | b) $\sqrt{(5)(d - 7)} = \sqrt{5d - 35}$ |

290. Just as $3\sqrt{x}$ means "multiply 3 and $\sqrt{x}$"
   and $R\sqrt{V}$ means "multiply R and $\sqrt{V}$",

   (a) $5\sqrt{x + 1}$ means "multiply _____ and _____".
   (b) $a\sqrt{b + c}$ means "multiply _____ and _____".

| | a) 5 and $\sqrt{x + 1}$ |
| | b) a and $\sqrt{b + c}$ |

86   Squares and Square Root Radicals

291. The same principle is used with multiplications involving non-radicals and radicals when one of the radicands is a binomial. That is, we simply multiply the non-radical factors and the radical factors. For example:

$$y\sqrt{x} \cdot \sqrt{x-3} = y\sqrt{(x)(x-3)} = y\sqrt{x^2 - 3x}$$
$$R\sqrt{b^2} \cdot V\sqrt{c+1} = RV\sqrt{(b^2)(c+1)} = RV\sqrt{b^2c + b^2}$$

Multiply: (a) $5\sqrt{d} \cdot \sqrt{d-4} =$ _____

(b) $m\sqrt{c+d} \cdot p\sqrt{t^2} =$ _____

a) $5\sqrt{(d)(d-4)} = 5\sqrt{d^2 - 4d}$
b) $mp\sqrt{(c+d)(t^2)}$
   $= mp\sqrt{ct^2 + dt^2}$

292. Any multiplication of a non-radical and a radical containing a binomial can be converted to a single radical. To do so, we replace the non-radical with the square root of its square, and then multiply. For example:

$$2\sqrt{x+3} = \sqrt{4} \cdot \sqrt{x+3} = \sqrt{4x+12}$$
$$a\sqrt{c+d} = \sqrt{a^2} \cdot \sqrt{c+d} = \sqrt{a^2c + a^2d}$$
$$3x\sqrt{y+1} = \sqrt{9x^2} \cdot \sqrt{y+1} = \sqrt{9x^2y + 9x^2}$$

Convert each of these to a single radical:

(a) $5\sqrt{t-4} =$ _____

(b) $m\sqrt{b-c} =$ _____

(c) $2y\sqrt{x+7} =$ _____

(d) $RV\sqrt{x-y} =$ _____

a) $\sqrt{25t - 100}$
b) $\sqrt{bm^2 - cm^2}$
c) $\sqrt{4xy^2 + 28y^2}$
d) $\sqrt{R^2V^2x - R^2V^2y}$

293. In the radicand below, "25" and "x - 7" are factors of a multiplication. Even though "x - 7" is a binomial, the radical can be factored and $\sqrt{25}$ can be replaced by 5. That is:

$$\sqrt{25(x-7)} = \sqrt{25} \cdot \sqrt{x-7} = 5\sqrt{x-7}$$

Factor each of these and replace the square roots of perfect squares with non-radicals:

(a) $\sqrt{16(y+1)} =$ _____

(b) $\sqrt{a^2(b+c)} =$ _____

(c) $\sqrt{9x^2(x+5)} =$ _____

(d) $\sqrt{c^2d^2(m+p)} =$ _____

a) $4\sqrt{y+1}$
b) $a\sqrt{b+c}$
c) $3x\sqrt{x+5}$
d) $cd\sqrt{m+p}$

Squares and Square Root Radicals   87

294. When the radicand is a binomial, look for opportunities to factor out a perfect square by using the distributive principle. For example:
$$\sqrt{4t + 4y} = \sqrt{4(t+y)} = 2\sqrt{t+y}$$
   (a) $\sqrt{9a - 18b} = \sqrt{9(a-2b)} =$ _____
   (b) $\sqrt{ax^2 + bx^2} = \sqrt{x^2(a+b)} =$ _____
   (c) $\sqrt{4p^2q + 8p^2t} = \sqrt{4p^2(q+2t)} =$ _____

295. Using the distributive principle, factor out the perfect squares from these:

   (a) $\sqrt{16d + 16f} =$ _____
   (b) $\sqrt{27R - 9V} =$ _____
   (c) $\sqrt{m^2p + m^2q} =$ _____
   (d) $\sqrt{a^2b^2c - a^2b^2d} =$ _____

a) $3\sqrt{a - 2b}$
b) $x\sqrt{a+b}$
c) $2p\sqrt{q + 2t}$

296. The square root of a binomial can be written in a simpler form <u>only if</u> <u>we</u> <u>can</u> <u>factor</u> <u>out</u> <u>a</u> <u>perfect</u> <u>square</u> <u>by</u> <u>means</u> <u>of</u> <u>the</u> <u>distributive</u> <u>principle</u>. That is:

   Though $\sqrt{ax^2 + bx^2}$ can be written as $x\sqrt{a+b}$
   and $\sqrt{a^2x^2 + b^2x^2}$ can be written as $x\sqrt{a^2 + b^2}$,
   $\sqrt{a+b}$ and $\sqrt{a^2 + b^2}$ cannot be written in a simpler form since we cannot factor out a perfect square from either.

   Write each of these in a simpler form if possible:
   (a) $\sqrt{16x^2 + 16y^2} =$ _____
   (b) $\sqrt{17t^2 + 11m^2} =$ _____
   (c) $\sqrt{V^2 - S^2} =$ _____
   (d) $\sqrt{25a^2 - 100b^2} =$ _____

a) $4\sqrt{d+f}$
b) $3\sqrt{3R - V}$
c) $m\sqrt{p+q}$
d) $ab\sqrt{c-d}$

297. Write each of these in a simpler form if possible:
   (a) $\sqrt{c^2y^2 + b^2y^2} =$ _____
   (b) $\sqrt{m^2p^2 + m^2x^2} =$ _____
   (c) $\sqrt{12c^2d^2 - 24b^2d^2} =$ _____
   (d) $\sqrt{f^2g^2h^2 - g^2h^2j^2} =$ _____

a) $4\sqrt{x^2 + y^2}$
b) Not possible.
c) Not possible.
d) $5\sqrt{a^2 - 4b^2}$

a) $y\sqrt{c^2 + b^2}$
b) $m\sqrt{p^2 + x^2}$
c) $2d\sqrt{3c^2 - 6b^2}$
d) $gh\sqrt{f^2 - j^2}$

298. In each square root below, either the numerator or denominator of the fraction is a binomial. Notice how we factor out perfect squares after converting the square root to a division of two radicals.

$$\sqrt{\frac{4p - 8w}{p^2}} = \frac{\sqrt{4(p - 2w)}}{\sqrt{p^2}} = \frac{2\sqrt{p - 2w}}{p} \text{ or } \frac{2}{p}\sqrt{p - 2w}$$

$$\sqrt{\frac{1}{9x - 36y}} = \frac{\sqrt{1}}{\sqrt{9(x - 4y)}} = \frac{1}{3\sqrt{x - 4y}}$$

Factor out the perfect squares from these:

(a) $\sqrt{\dfrac{16t + 32v}{m^2}} = $ _____

(b) $\sqrt{\dfrac{b^2 - 4ac}{4a^2}} = $ _____

(c) $\sqrt{\dfrac{1}{a^2x^2 + b^2x^2}} = $ _____

299. After factoring out the perfect squares below, notice how we convert the remaining division of radicals back to a single radical:

$$\sqrt{\frac{4b - 4c}{d}} = \frac{2\sqrt{b - c}}{\sqrt{d}} = 2\sqrt{\frac{b - c}{d}}$$

$$\sqrt{\frac{V^2 t}{m^2 p + m^2 q}} = \frac{V\sqrt{t}}{m\sqrt{p + q}} = \frac{V}{m}\sqrt{\frac{t}{p + q}}$$

Factor out the perfect squares and convert the remaining division of radicals back to a single radical:

(a) $\sqrt{\dfrac{49d - 98f}{g^2 h}} = $ _____

(b) $\sqrt{\dfrac{ab^2 - cb^2}{18x - 9y}} = $ _____

a) $\dfrac{4\sqrt{t + 2v}}{m}$ or $\dfrac{4}{m}\sqrt{t + 2v}$

b) $\dfrac{\sqrt{b^2 - 4ac}}{2a}$

c) $\dfrac{1}{x\sqrt{a^2 + b^2}}$

a) $\dfrac{7}{g}\sqrt{\dfrac{d - 2f}{h}}$

b) $\dfrac{b}{3}\sqrt{\dfrac{a - c}{2x - y}}$

300. We have already seen these two facts:

(1) If we multiply two radicals containing the same radicand, the product is the radicand. For example:

$$\sqrt{x} \cdot \sqrt{x} = x$$

$$\sqrt{\frac{a}{b}} \cdot \sqrt{\frac{a}{b}} = \frac{a}{b}$$

(2) If a radical is squared, its square is the radicand. For example:

$$\left(\sqrt{x}\right)^2 = x$$

$$\left(\sqrt{\frac{a}{b}}\right)^2 = \frac{a}{b}$$

These two facts also apply to radicals whose radicands are binomials. That is:

$$\sqrt{x+3} \cdot \sqrt{x+3} = x + 3$$

$$\left(\sqrt{a-b}\right)^2 = a - b$$

Complete each of the following:

(a) $\sqrt{y-5} \cdot \sqrt{y-5} = $ _____

(b) $\left(\sqrt{m+p}\right)^2 = $ _____

(c) $\left(\sqrt{2y+7}\right)^2 = $ _____

(d) $\sqrt{c^2 - d^2} \cdot \sqrt{c^2 - d^2} = $ _____

---

301. To square a term which has the square root of a binomial as one of its factors, we simply square each factor. That is:

$$\left(3\sqrt{x+5}\right)^2 = 3^2 \cdot \left(\sqrt{x+5}\right)^2 = 9(x+5) \text{ or } 9x + 45$$

$$\left(x\sqrt{c-d}\right)^2 = x^2 \cdot \left(\sqrt{c-d}\right)^2 = x^2(c-d) \text{ or } cx^2 - dx^2$$

Square each of the following terms:

(a) $\left(7\sqrt{y-1}\right)^2 = $ _____

(b) $\left(t\sqrt{x^2 + y^2}\right)^2 = $ _____

a) $y - 5$ c) $2y + 7$
b) $m + p$ d) $c^2 - d^2$

---

302. To square a fraction whose numerator or denominator contains the square root of a binomial, we simply square both the numerator and denominator. For example:

$$\left(\frac{\sqrt{t+4}}{5}\right)^2 = \frac{\left(\sqrt{t+4}\right)^2}{5^2} = \frac{t+4}{25}$$

$$\left(\frac{a}{b\sqrt{c-d}}\right)^2 = \frac{a^2}{b^2\left(\sqrt{c-d}\right)^2} = \frac{a^2}{b^2(c-d)} \text{ or } \frac{a^2}{b^2c - b^2d}$$

Square each of the following fractions:

(a) $\left(\dfrac{1}{\sqrt{c+d}}\right)^2 = $ _____

(b) $\left(\dfrac{3\sqrt{x+1}}{5}\right)^2 = $ _____

(c) $\left(\dfrac{R}{2\sqrt{x+y}}\right)^2 = $ _____

a) $49(y - 1)$ or $49y - 49$
b) $t^2(x^2 + y^2)$ or $t^2x^2 + t^2y^2$

90 Squares and Square Root Radicals

303. When any radical is squared, its square is the radicand. For example:

$$\left(\sqrt{5}\right)^2 = 5 \qquad \left(\sqrt{\frac{a}{b}}\right)^2 = \frac{a}{b} \qquad \left(\sqrt{c-d}\right)^2 = c - d$$

Similarly, if we square a radical whose radicand is a fraction with a binomial numerator or denominator, its square is the fraction. That is:

$$\left(\sqrt{\frac{y+5}{4}}\right)^2 = \frac{y+5}{4}$$

(a) $\left(\sqrt{\dfrac{1}{a+b}}\right)^2 = $ _____ (b) $\left(\sqrt{\dfrac{x-y}{c-d}}\right)^2 = $ _____

a) $\dfrac{1}{c+d}$

b) $\dfrac{9(x+1)}{25}$ or $\dfrac{9x+9}{25}$

c) $\dfrac{R^2}{4(x+y)}$ or $\dfrac{R^2}{4x+4y}$

---

304. As we saw earlier, we can rationalize the denominator of a fraction by multiplying the fraction by $\dfrac{\sqrt{}}{\sqrt{}}$, where $\sqrt{}$ is the denominator radical. For example:

$$\frac{5}{\sqrt{x}} = \frac{5}{\sqrt{x}}\left(\frac{\sqrt{x}}{\sqrt{x}}\right) = \frac{5\sqrt{x}}{x}$$

We can use this same method to rationalize a denominator when the radical in it contains a binomial. That is:

$$\frac{1}{\sqrt{p+q}} = \frac{1}{\sqrt{p+q}}\left(\frac{\sqrt{p+q}}{\sqrt{p+q}}\right) = \frac{\sqrt{p+q}}{p+q}$$

$$\frac{a}{b\sqrt{c-d}} = \frac{a}{b\sqrt{c-d}}\left(\frac{\sqrt{c-d}}{\sqrt{c-d}}\right) = \frac{a\sqrt{c-d}}{b(c-d)} \quad \text{or} \quad \frac{a\sqrt{c-d}}{bc-bd}$$

Rationalize each of the following denominators:

(a) $\dfrac{1}{\sqrt{a+b}} = $ _____

(b) $\dfrac{3t}{\sqrt{y-4}} = $ _____

a) $\dfrac{1}{a+b}$ b) $\dfrac{x-y}{c-d}$

---

305. Rationalize the denominators in each of these:

(a) $\dfrac{t}{2\pi\sqrt{f-k}} = $ _____

(b) $\dfrac{1}{p\sqrt{q+v}} = $ _____

a) $\dfrac{\sqrt{a+b}}{a+b}$

b) $\dfrac{3t\sqrt{y-4}}{y-4}$

---

a) $\dfrac{t\sqrt{f-k}}{2\pi(f-k)}$ or $\dfrac{t\sqrt{f-k}}{2\pi f - 2\pi k}$

b) $\dfrac{\sqrt{q+v}}{p(q+v)}$ or $\dfrac{\sqrt{q+v}}{pq+pv}$

Squares and Square Root Radicals   91

306. Here is a case where we can factor out a perfect square and reduce to lowest terms after rationalizing the denominator:

$$\frac{abd}{\sqrt{ad^2 - bd^2}} = \frac{abd}{d\sqrt{a-b}} = \frac{abd\sqrt{a-b}}{d(a-b)} = \frac{ab\sqrt{a-b}}{a-b}$$

Factor out perfect squares where possible and reduce to lowest terms after rationalizing the denominator:

(a) $\dfrac{3t}{\sqrt{18x - 9y}} =$ _____

(b) $\dfrac{b^2 y}{\sqrt{3b^2 + ab^2}} =$ _____

307. As we saw earlier, we can combine an addition or subtraction of two radical terms into one term if the radicals contain the same radicand. For example:

$$7\sqrt{x} + 5\sqrt{x} = 12\sqrt{x}$$
$$11\sqrt{3} - 6\sqrt{3} = 5\sqrt{3}$$

This same principle is also true when the common radicand is a binomial. For example:

$$8\sqrt{y+3} + 7\sqrt{y+3} = 15\sqrt{y+3}$$
$$6\sqrt{a-b} - 4\sqrt{a-b} = 2\sqrt{a-b}$$

Combine each of the following into one term:

(a) $3\sqrt{R^2 + V^2} + 4\sqrt{R^2 + V^2} =$ _____

(b) $10\sqrt{1 - d^2} - 6\sqrt{1 - d^2} =$ _____

308. If the radicals do not have a coefficient explicitly written, their coefficient is "1". That is:

(a) $\sqrt{x+3} + \sqrt{x+3} = 1\sqrt{x+3} + 1\sqrt{x+3} =$ _____

(b) $5\sqrt{c-d} - \sqrt{c-d} = 5\sqrt{c-d} - 1\sqrt{c-d} =$ _____

(c) $\sqrt{a+b} + \sqrt{a+b} = 1\sqrt{a+b} + 1\sqrt{a+b} =$ _____

---

a) $\dfrac{t\sqrt{2x-y}}{2x-y}$

b) $\dfrac{by\sqrt{3+a}}{3+a}$

---

a) $7\sqrt{R^2 + V^2}$

b) $4\sqrt{1-d^2}$

---

a) $2\sqrt{x+3}$

b) $4\sqrt{c-d}$

c) $2\sqrt{a+b}$

Squares and Square Root Radicals

SELF-TEST 17 (Frames 289-308)

Multiply:

1. $\sqrt{t-1} \cdot \sqrt{t} =$ _____

2. $r\sqrt{2b} \cdot 2d\sqrt{a+3b} =$ _____

Convert each of the following to a single radical:

3. $d\sqrt{p+2} =$ _____

4. $5h\sqrt{s-2b} =$ _____

Factor each radicand and replace square roots of perfect squares with non-radicals:

5. $\sqrt{8a^2 - 4b^2} =$ _____

6. $\sqrt{9r^2s^2 + 9s^2t^2} =$ _____

7. In the following, factor out perfect squares after converting the problem to a division of two radicals:

$\sqrt{\dfrac{16m + 8n}{d^2}} =$ _____

8. In the following, factor out perfect squares and convert the remaining division of radicals back to a single radical:

$\sqrt{\dfrac{mv^2}{4mt^2 - 4t^2v}} =$ _____

Do the following squarings:

9. $\left(2h\sqrt{r-s}\right)^2 =$ _____

10. $\left(\dfrac{h}{p\sqrt{x^2-1}}\right)^2 =$ _____

11. $\left(\sqrt{\dfrac{a+b}{a-b}}\right)^2 =$ _____

Rationalize each denominator:

12. $\dfrac{1}{2\sqrt{x^2+y^2}} =$ _____

13. $\dfrac{4rt}{\sqrt{4r^2t - 4r^2}} =$ _____

14. Combine into one term: $2\sqrt{Z^2 - R^2} - \sqrt{Z^2 - R^2} =$ _____

ANSWERS:

1. $\sqrt{t^2 - t}$
2. $2dr\sqrt{2ab + 6b^2}$
3. $\sqrt{d^2p + 2d^2}$
4. $\sqrt{25h^2s - 50bh^2}$
5. $2\sqrt{2a^2 - b^2}$
6. $3s\sqrt{r^2 + t^2}$
7. $\dfrac{2}{d}\sqrt{4m + 2n}$
8. $\dfrac{v}{2t}\sqrt{\dfrac{m}{m-v}}$
9. $4h^2r - 4h^2s$
10. $\dfrac{h^2}{p^2x^2 - p^2}$
11. $\dfrac{a+b}{a-b}$
12. $\dfrac{\sqrt{x^2+y^2}}{2x^2 + 2y^2}$
13. $\dfrac{2t\sqrt{t-1}}{t-1}$
14. $\sqrt{Z^2 - R^2}$

Chapter 2 EQUATIONS AND FORMULAS CONTAINING SQUARES AND SQUARE ROOT RADICALS

In the last chapter, we discussed operations involving squares and square root radicals. In this chapter, we will discuss equations and formulas which contain squared letters and square root radicals. We will show how the squaring principle is used both to solve radical equations and to rearrange radical formulas. And after a brief discussion of "pure" quadratic equations, we will show how the square root principle is used to rearrange formulas which contain a squared letter. At the end of the chapter, we will show how either the equivalence method or the substitution method can be used to eliminate a variable from a system of formulas involving a squared letter or a square root radical.

2-1 SOLVING RADICAL EQUATIONS BY DIRECT INSPECTION

In this section, we will show what is meant by a radical equation. Then we will solve some simple radical equations by the direct-inspection method. Though the direct-inspection method is only efficient when solving simple radical equations with whole-number roots, it is used here to show what is meant by a root of a radical equation.

1. There are two types of equations that contain square root radicals. In one type, the radicand is simply a whole number, numerical fraction, or decimal. For example:

   $2x = \sqrt{5}$ $t - \sqrt{\frac{7}{4}} = 1$ $y - 2 = \sqrt{5.94}$

   In these cases, we can solve the equation by simply replacing the radical with a regular number. That is:

   (a) Since $\sqrt{5} = 2.24$, $2x = \sqrt{5}$ is the same as $2x = 2.24$

   Therefore, x = _____

   (b) Since $\sqrt{\frac{7}{4}} = 1.32$, $t - \sqrt{\frac{7}{4}} = 1$ is the same as $t - 1.32 = 1$

   Therefore, t = _____

   (c) Since $\sqrt{5.94} = 2.44$, $y - 2 = \sqrt{5.94}$ is the same as $y - 2 = 2.44$

   Therefore, y = _____

a) $x = 1.12$
b) $t = 2.32$
c) $y = 4.44$

94 Equations and Formulas Containing Squares and Square Root Radicals

2. In the second type of equation that contains a square root radical, the radicand contains the variable (or letter). For example:

$\sqrt{x} = 3$ $\sqrt{m+4} = 5$ $\sqrt{\dfrac{y}{7}} = 2$

In these cases, the equations cannot be solved by simply replacing the radical with a regular number.

Which of the following equations cannot be solved by replacing the radical with a regular number? _____

(a) $2\sqrt{t} = 7$ (b) $3x = \sqrt{11}$ (c) $\sqrt{\dfrac{8}{R}} = 4$

---

3. If an equation contains a radical with a variable in its radicand, it cannot be solved by replacing the radical with a regular number. Equations of this type are called "radical equations".

Which of the following are called "radical equations"? _____

(a) $d = \sqrt{\dfrac{11}{7}}$ (b) $\sqrt{\dfrac{x}{5}} = 1$ (c) $m - \sqrt{4.62} = 5$ (d) $\sqrt{m+1} = 3$

(a) and (c)

---

4. Though "radical equations" cannot be solved by replacing the radical with a regular number, radical equations do have roots. Sometimes, we can determine their roots by simple inspection. For example:

The root of $\sqrt{x} = 3$ is "9", since $\sqrt{9} = 3$

The root of $\sqrt{y+1} = 4$ is "15", since $\sqrt{15+1} = 4$

The root of $\sqrt{\dfrac{t}{3}} = 1$ is "3", since $\sqrt{\dfrac{3}{3}} = 1$

By simple inspection, write the root of each of the following:

(a) $\sqrt{d} = 4$ d = _____

(b) $\sqrt{V+2} = 5$ V = _____

Only (b) and (d)

---

5. To check whether a specific number is or is not the root of a radical equation, we simply plug the number in for the letter and evaluate. For example:

"36" is the root of $2\sqrt{x} = 12$, since:

$2\sqrt{36} = 2(6) = 12$

"13" is not the root of $\sqrt{y+3} = 8$, since:

$\sqrt{13+3} = \sqrt{16} = 4$ (not 8)

If we plug in 28 for "t" in $\sqrt{\dfrac{t}{7}} = 2$:

(a) $\sqrt{\dfrac{t}{7}} = \sqrt{\dfrac{28}{7}} = \sqrt{4} =$ _____

(b) Is "28" the root of $\sqrt{\dfrac{t}{7}} = 2$? _____

a) d = 16, since:
$\sqrt{16} = 4$

b) V = 23, since:
$\sqrt{23+2} = 5$

Equations and Formulas Containing Squares and Square Root Radicals   95

6. If we plug in 16 for "x" in $2\sqrt{x} = 6$:

    (a) $2\sqrt{x} = 2\sqrt{16} = 2(4) =$ _____

    (b) Is 16 the root of $2\sqrt{x} = 6$? _____

a) 2

b) Yes, since $2 = 2$

---

7. If we plug in 44 for "y" in $\sqrt{y + 5} = 7$:

    (a) $\sqrt{y + 5} = \sqrt{44 + 5} = \sqrt{49} =$ _____

    (b) Is "44" the root of $\sqrt{y + 5} = 7$? _____

a) 8

b) No, since $8 \neq 6$

---

8. If we plug in 15 for "k" in $\sqrt{\dfrac{k + 1}{4}} = 3$:

    (a) $\sqrt{\dfrac{k + 1}{4}} = \sqrt{\dfrac{15 + 1}{4}} = \sqrt{\dfrac{16}{4}} = \sqrt{4} =$ _____

    (b) Is "15" the root of $\sqrt{\dfrac{k + 1}{4}} = 3$? _____

a) 7

b) Yes, since $7 = 7$

---

9. (a) Of the numbers 9, 16, and 25, which one is the root of $\sqrt{t} = 5$? _____

    (b) Of the numbers 1, 4, and 9, which one is the root of $2\sqrt{s} = 2$? _____

    (c) Of the numbers 2, 9, and 18, which one is the root of $\sqrt{p + 7} = 4$? _____

a) 25, since:
$\sqrt{25} = 5$

b) 1, since:
$2\sqrt{1} = 2(1) = 2$

c) 9, since:
$\sqrt{9 + 7} = \sqrt{16} = 4$

---

10. (a) Of the numbers 16, 64, and 100, which one is the root of $\sqrt{\dfrac{x}{4}} = 5$? _____

    (b) Of the numbers 2, 8, and 32, which one is the root of $\sqrt{\dfrac{128}{m}} = 4$? _____

    (c) Of the numbers 17, 31, and 49, which one is the root of $\sqrt{\dfrac{t + 1}{2}} = 3$? _____

a) 100, since:
$\sqrt{\dfrac{100}{4}} = \sqrt{25} = 5$

b) 8, since:
$\sqrt{\dfrac{128}{8}} = \sqrt{16} = 4$

c) 17, since:
$\sqrt{\dfrac{17 + 1}{2}} = \sqrt{9} = 3$

96   Equations and Formulas Containing Squares and Square Root Radicals

---

SELF-TEST 1 (Frames 1-10)

1. Which of the following equations are called "radical equations"? _____

   (a) $\sqrt{2t} = 5$ (b) $2\sqrt{3} = 7x$ (c) $18 = \sqrt{\dfrac{1}{h}}$ (d) $4p = \sqrt{\dfrac{2}{3}}$

By direct inspection, write the root of each of the following:

2. $\sqrt{y} = 11$ y = _____   3. $\sqrt{a - 2} = 7$ a = _____

4. Of the numbers 1, 3, 9, 18, and 27, which one is the root of $\sqrt{\dfrac{81}{p}} = 3$? _____

5. Of the numbers 1, 4, 12, 20, and 28, which one is the root of $\sqrt{\dfrac{w + 4}{2}} = 4$? _____

ANSWERS: 1. (a) and (c)   2. y = 121   4. 9
                              3. a = 51   5. 28

2-2   THE SQUARING PRINCIPLE FOR RADICAL EQUATIONS

In the last section, we solved some radical equations by the direct-inspection method. That method is efficient only with very simple radical equations with whole-number roots. In this section, we will discuss an algebraic method of solving a radical equation. The algebraic method involves the use of the "squaring principle for radical equations". In this section, the algebraic method will be used to solve only those radical equations in which the radical is already isolated on one side of the equation.

11. We have already seen that the "square" of a square root radical is a non-radical expression. That is:

$\left(\sqrt{2}\right)^2 = 2$ $\left(\sqrt{t + 3}\right)^2 = t + 3$

$\left(\sqrt{y}\right)^2 = y$ $\left(\sqrt{\dfrac{x + 1}{2}}\right)^2 = \dfrac{x + 1}{2}$

Notice in each case that the <u>radicand</u> is the "square" of the square root radical.

Write the "squares" of each of these:

(a) $\left(\sqrt{5x}\right)^2 =$ _____   (c) $\left(\sqrt{\dfrac{y}{6}}\right)^2 =$ _____

(b) $\left(\sqrt{3R + 5}\right)^2 =$ _____   (d) $\left(\sqrt{\dfrac{5}{p + 7}}\right)^2 =$ _____

a) 5x   c) $\dfrac{y}{6}$

b) 3R + 5   d) $\dfrac{5}{p + 7}$

Equations and Formulas Containing Squares and Square Root Radicals   97

12. The "squaring principle for equations" is based on this fact: **If two quantities are equal, their squares are equal.** That is:

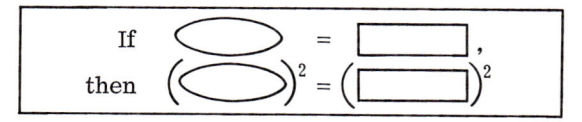

If ◯ = ▭,
then (◯)² = (▭)²

This principle is especially useful in solving radical equations because the square of any square root radical is a non-radical expression. Therefore, **by applying this squaring principle to a radical equation, we can eliminate the radical.** That is:

If we square both sides of $\sqrt{x} = 7$, we get:
$$\left(\sqrt{x}\right)^2 = 7^2$$
or $\quad x = 49$

If we square both sides of $\sqrt{t+3} = 5$, we get:
$$\left(\sqrt{t+3}\right)^2 = 5^2$$
or $\quad t + 3 = 25$

What non-radical equations do we get by squaring both sides of these equations?

(a) $\sqrt{3y+1} = 2$ $\qquad$ (b) $\sqrt{\dfrac{2x}{3}} = 3$

_____ = ____ $\qquad$ _____ = ____

---

13. If we square both sides of $\sqrt{2x} = 4$, we get:
$$\left(\sqrt{2x}\right)^2 = 4^2$$
or $\quad 2x = 16$

The root of the last equation is 8.
Is "8" also the root of the original radical equation? _____

a) $3y + 1 = 4$

b) $\dfrac{2x}{3} = 9$

---

14. If we square both sides of $\sqrt{t+3} = 5$, we get:
$$\left(\sqrt{t+3}\right)^2 = 5^2$$
or $\quad t + 3 = 25$

The root of the last equation is 22.
Is "22" the root of the original radical equation? _____

Yes, since:
$\sqrt{2(8)} = \sqrt{16} = 4$

Yes, since:
$\sqrt{22+3} = \sqrt{25} = 5$

98 Equations and Formulas Containing Squares and Square Root Radicals

15. Only two steps are needed to solve a radical equation <u>when the radical is isolated on one side</u>. They are:

   (1) Eliminate the radical by squaring both sides of the equation.
   (2) Then solve the resulting non-radical equation.

   An example of a solution is given on the right. Show that "80" is the root of the original radical equation, $\sqrt{y+1} = 9$

   $\sqrt{y+1} = 9$
   $\left(\sqrt{y+1}\right)^2 = 9^2$
   $y + 1 = 81$
   $y = 80$

   $\sqrt{80 + 1} = \sqrt{81} = 9$

16. When using the squaring principle for radical equations, <u>be sure to square both sides of the equation</u>.

   What non-radical equation is obtained by squaring both sides of each of these?

   (a) $\sqrt{y} = \dfrac{2}{5}$

   (c) $\sqrt{\dfrac{50}{m-1}} = 9$

   (b) $4 = \sqrt{2x+3}$

   (d) $\dfrac{1}{2} = \sqrt{\dfrac{3d}{5}}$

17. When using the squaring principle, <u>the most common error is forgetting to square the non-radical side</u>.

   In which cases below has this common error been made? _____

   (a) $\sqrt{4x+3} = 7$
       $4x + 3 = 7$

   (b) $\sqrt{5t} = \dfrac{2}{3}$
       $5t = \dfrac{4}{9}$

   (c) $8 = \sqrt{\dfrac{2b}{5}}$
       $8 = \dfrac{2b}{5}$

   a) $y = \dfrac{4}{25}$

   b) $16 = 2x + 3$

   c) $\dfrac{50}{m-1} = 81$

   d) $\dfrac{1}{4} = \dfrac{3d}{5}$

18. Let's solve: $4 = \sqrt{2R - 6}$

   (a) What non-radical equation do we get by applying the squaring principle?
       _____ = _____

   (b) The root of the non-radical equation is _____.

   (c) Show that this root satisfies the original radical equation.

   In both (a) and (c).

Equations and Formulas Containing Squares and Square Root Radicals 99

19. Let's solve: $\sqrt{\dfrac{x}{2}} = 4$

    (a) What non-radical equation do we get by applying the squaring principle?

        _____ = _____

    (b) The root of the non-radical equation is _____.

    (c) Show that this root satisfies the original radical equation.

a) $16 = 2R - 6$
b) $R = 11$
c) $4 = \sqrt{2(11) - 6}$
    $= \sqrt{22 - 6}$
    $= \sqrt{16}$
    $= 4$

---

20. Solve each of the following:

    (a) $\sqrt{\dfrac{3y}{4}} = 1$ (b) $\sqrt{\dfrac{2x+1}{3}} = 5$

        y = _____   x = _____

a) $\dfrac{x}{2} = 16$
b) $x = 32$
c) $\sqrt{\dfrac{32}{2}} = 4$
    $\sqrt{16} = 4$
    $4 = 4$

---

21. Solve each of the following:

    (a) $\sqrt{\dfrac{2x}{5}} = 4$ (b) $\dfrac{1}{2} = \sqrt{\dfrac{1}{t}}$

        x = _____   t = _____

a) $y = \dfrac{4}{3}$
b) $x = 37$

---

a) $x = 40$, from $\dfrac{2x}{5} = 16$

b) $t = 4$, from $\dfrac{1}{4} = \dfrac{1}{t}$

22. Solve each of the following:

    (a) $\dfrac{4}{3} = \sqrt{\dfrac{m+1}{3}}$ 　　　　(b) $\sqrt{\dfrac{100}{v+1}} = 5$

    m = _____ 　　　　　　　　　　v = _____

    a) $m = \dfrac{13}{3}$ or $4\dfrac{1}{3}$, from:
    $$\dfrac{16}{9} = \dfrac{m+1}{3}$$
    $$16 = 3(m+1)$$
    $$16 = 3m + 3$$

    b) $v = 3$, from:
    $$\dfrac{100}{v+1} = 25$$
    $$100 = 25(v+1)$$
    $$100 = 25v + 25$$

23. If we square both sides of $\sqrt{3x} = \dfrac{5}{2}$, we get: $3x = \dfrac{25}{4}$

    There are two ways of solving $3x = \dfrac{25}{4}$:

    (1) Clearing the fraction and then multiplying both sides by $\dfrac{1}{12}$. That is:
    $$3x = \dfrac{25}{4}$$
    $$12x = 25$$
    $$x = \dfrac{25}{12}$$

    (2) Simply multiplying both sides by $\dfrac{1}{3}$. That is:
    $$3x = \dfrac{25}{4}$$
    $$\dfrac{1}{3}(3x) = \dfrac{1}{3}\left(\dfrac{25}{4}\right)$$
    $$x = \dfrac{25}{12}$$

    Solve each equation below. Use either method above to solve the non-radical equation.

    (a) $\sqrt{2y} = \dfrac{4}{3}$ 　　(b) $\sqrt{5t} = \dfrac{2}{3}$ 　　(c) $\sqrt{3m} = \dfrac{1}{2}$

    y = _____ 　　　t = _____ 　　　m = _____

24. If we square both sides of $\sqrt{x-1} = \frac{2}{3}$, we get: $x - 1 = \frac{4}{9}$

    There are two ways of solving $x - 1 = \frac{4}{9}$:

    (1) Clearing the fraction and then solving the resulting non-fractional equation. That is:

    $$x - 1 = \frac{4}{9}$$
    $$9x - 9 = 4$$
    $$9x = 13$$
    $$x = \frac{13}{9} \text{ or } 1\frac{4}{9}$$

    (2) Simply using the addition axiom with the equation as it stands. That is:

    $$x - 1 = \frac{4}{9}$$
    $$x = \frac{4}{9} + 1$$
    $$x = 1\frac{4}{9} \text{ or } \frac{13}{9}$$

    Note: In this case, the second method seems easier.

    Solve each equation below. Use either method above to solve the non-radical equation.

    (a) $\sqrt{R+1} = \frac{3}{2}$ (b) $\sqrt{x - \frac{1}{2}} = 2$ (c) $\sqrt{2y - 1} = \frac{3}{4}$

    R = _____   x = _____   y = _____

a) $y = \frac{16}{18}$ or $\frac{8}{9}$

b) $t = \frac{4}{45}$

c) $m = \frac{1}{12}$

---

a) $R = \frac{5}{4}$ or $1\frac{1}{4}$

b) $x = 4\frac{1}{2}$ or $\frac{9}{2}$

c) $y = \frac{25}{32}$, from:

$$2y = \frac{25}{16}$$

102   Equations and Formulas Containing Squares and Square Root Radicals

SELF-TEST 2 (Frames 11-24)

Apply the squaring principle for equations to each of the following radical equations, and write the resulting equation. Do not solve.

1. $\sqrt{5x} = 12$

2. $\dfrac{4}{7} = \sqrt{\dfrac{3r}{5}}$

3. $\sqrt{\dfrac{2d-3}{8}} = \dfrac{3}{2}$

4. $6 = \sqrt{4w + \dfrac{3}{4}}$

Solve the following radical equations:

5. $\sqrt{\dfrac{2}{3x}} = 4$

   x = _____

6. $\sqrt{t - \dfrac{3}{8}} = \dfrac{3}{4}$

   t = _____

7. $5 = \sqrt{\dfrac{50}{2p - 3}}$

   p = _____

ANSWERS:
1. $5x = 144$
2. $\dfrac{16}{49} = \dfrac{3r}{5}$
3. $\dfrac{2d - 3}{8} = \dfrac{9}{4}$
4. $36 = 4w + \dfrac{3}{4}$
5. $x = \dfrac{1}{24}$
6. $t = \dfrac{15}{16}$
7. $p = \dfrac{5}{2}$ or $2\dfrac{1}{2}$

2-3 ISOLATING RADICALS IN RADICAL EQUATIONS

All radical equations are solved by applying the squaring principle. This principle is easiest to apply when the radical is isolated on one side of the equation. All of the radical equations solved in the last section were equations of that type. However, there are many radical equations in which the radical is not isolated on one side. Here are some examples:

$$\frac{2\sqrt{x}}{3} = 5 \qquad 4 + 2\sqrt{t} = 9 \qquad \frac{5 + \sqrt{y}}{7} = 1$$

The most general method for solving equations like those above is to isolate the radical and then square both sides of the new equation. In this section, we will discuss the steps needed to isolate radicals in radical equations.

25. Here is an equation in which the radical is not isolated on one side:

$$4 + \sqrt{x} = 9$$

In this case, we can isolate the radical by using the addition axiom. We add "-4" to both sides and get:

$$(-4) + 4 + \sqrt{x} = 9 + (-4)$$
$$0 + \sqrt{x} = 5$$
$$\sqrt{x} = 5$$

Isolate the radical in each equation below:

(a) $7 + \sqrt{m} = 10$ (b) $8 = \sqrt{3y} - 4$

26. When rearranging equations to isolate radicals, <u>the radicand is left intact</u> even if it is a more complicated expression than a letter or a letter term. That is, the radical is treated like a single quantity in all cases.

Here is an equation in which the radicand is a binomial. We have isolated the radical by adding "-2" to both sides.

$$2 + \sqrt{R+1} = 9$$
$$(-2) + 2 + \sqrt{R+1} = 9 + (-2)$$
$$\sqrt{R+1} = 7$$

Isolate the radical in each of these:

(a) $\sqrt{y+8} - 9 = 5$ (b) $17 = 6 + \sqrt{2t-1}$

---

a) $\sqrt{m} = 3$

b) $12 = \sqrt{3y}$

---

a) $\sqrt{y+8} = 14$

b) $11 = \sqrt{2t-1}$

104   Equations and Formulas Containing Squares and Square Root Radicals

27. Here is a case in which the radicand is a fraction. We have isolated the radical by adding "-3" to both sides.

$$\sqrt{\frac{4x}{5}} + 3 = 8$$

$$\sqrt{\frac{4x}{5}} + 3 + (-3) = 8 + (-3)$$

$$\sqrt{\frac{4x}{5}} = 5$$

Isolate the radical in each equation below:

(a) $7 + \sqrt{\frac{y}{9}} = 15$ (b) $20 = \sqrt{\frac{2t+5}{3}} - 10$

---

28. In the equation below, the radical has a numerical coefficient.

$$2\sqrt{x} = 5$$

To isolate the radical, we use the multiplication axiom, multiplying both sides by $\frac{1}{2}$. We get:

$$\underset{1}{\frac{1}{2}}\left(2\sqrt{x}\right) = \frac{1}{2}(5)$$

$$\sqrt{x} = \frac{5}{2}$$

or $\sqrt{x} = \frac{5}{2}$

Use the multiplication axiom to isolate the radical in each of these:

(a) $7\sqrt{3y} = 15$ (b) $13 = 4\sqrt{x+5}$ (c) $5\sqrt{\frac{d+7}{20}} = 9$

---

a) $\sqrt{\frac{y}{9}} = 8$

b) $30 = \sqrt{\frac{2t+5}{3}}$

---

a) $\sqrt{3y} = \frac{15}{7}$

b) $\frac{13}{4} = \sqrt{x+5}$

c) $\sqrt{\frac{d+7}{20}} = \frac{9}{5}$

29. In the equation below, we must use both the addition axiom and the multiplication axiom to isolate the radical.

$$3 + 5\sqrt{F} = 12$$

Adding "-3" to both sides, we get:

$$5\sqrt{F} = 9$$

Multiplying both sides by $\frac{"1"}{5}$, we get:

$$\sqrt{F} = \frac{9}{5}$$

Use both axioms to isolate the radicals in these:

(a) $10 + 7\sqrt{x} = 15$ (b) $14 = 3\sqrt{2t} - 6$

---

30. Use both axioms to isolate the radicals in these:

(a) $15 + 4\sqrt{R+8} = 24$ (b) $25 = 5\sqrt{\frac{x+1}{x}} - 10$

a) $\sqrt{x} = \frac{5}{7}$

b) $\sqrt{2t} = \frac{20}{3}$

---

31. If two radical terms contain the same radical, they are a pair of opposites if the coefficients of the radicals are a pair of opposites. For example:

$3\sqrt{x}$ and $-3\sqrt{x}$ are a pair of opposites, since:

$$3\sqrt{x} + \left(-3\sqrt{x}\right) = \left[3 + (-3)\right]\sqrt{x} = 0\sqrt{x} = 0$$

Write the opposites of each of these radical terms:

(a) $7\sqrt{t}$ _____   (b) $-4\sqrt{y}$ _____   (c) $-2\sqrt{5d}$ _____

a) $\sqrt{R+8} = \frac{9}{4}$

b) $7 = \sqrt{\frac{x+1}{x}}$, from:

$\frac{35}{5} = \sqrt{\frac{x+1}{x}}$

---

a) $-7\sqrt{t}$

b) $4\sqrt{y}$

c) $2\sqrt{5d}$

106   Equations and Formulas Containing Squares and Square Root Radicals

32. If the coefficient of a radical is not explicitly written, remember that its coefficient is either "+1" or "-1". That is:

$$\sqrt{m} = +1\sqrt{m}$$
$$-\sqrt{m} = -1\sqrt{m}$$

Therefore $\sqrt{m}$ and $-\sqrt{m}$ are a pair of opposites, since:

$$\sqrt{m} + (-\sqrt{m}) = 1\sqrt{m} + (-1\sqrt{m}) = [1 + (-1)]\sqrt{m} = 0\sqrt{m} = 0$$

Write the opposite of each of these:

(a) $\sqrt{R}$ _____ (b) $-\sqrt{d}$ _____ (c) $-\sqrt{2x}$ _____

---

33. Even with more-complicated radicands, two radical terms containing the same radical are a pair of opposites <u>if the coefficients of the radicals are a pair of opposites</u>. For example:

$5\sqrt{y+1}$ and $-5\sqrt{y+1}$ are a pair of opposites.

$\sqrt{\dfrac{50}{x+5}}$ and $-\sqrt{\dfrac{50}{x+5}}$ are a pair of opposites.

Write the opposite of each of these:

(a) $-3\sqrt{s-7}$ _____ (c) $4\sqrt{\dfrac{x}{3}}$ _____

(b) $-\sqrt{2R+1}$ _____ (d) $-\sqrt{\dfrac{m}{m+1}}$ _____

a) $-\sqrt{R}$
b) $\sqrt{d}$
c) $\sqrt{2x}$

---

34. A radical is isolated only if its coefficient is "+1". (Note: The "+1" is ordinarily not written.) Here is an equation in which the coefficient of the radical is "-1":

$$-\sqrt{x} = -5$$

To isolate the radical in this case, we simply apply the oppositing principle for equations. That is, we replace both sides of the equation by their opposites. That is:

$$\sqrt{x} = 5$$

Isolate the radical in each equation below by applying the oppositing principle:

(a) $-\sqrt{t} = -6$ (b) $-8 = -\sqrt{y}$ (c) $-\sqrt{3R} = -7$

_____ = _____   _____ = _____   _____ = _____

a) $3\sqrt{s-7}$
b) $\sqrt{2R+1}$
c) $-4\sqrt{\dfrac{x}{3}}$
d) $\sqrt{\dfrac{m}{m+1}}$

---

35. Isolate the radical in each equation below by applying the oppositing principle:

(a) $-\sqrt{2y+7} = -3$ (b) $-4 = -\sqrt{\dfrac{1}{x}}$ (c) $-\sqrt{\dfrac{40}{6-t}} = -1$

_____ = _____   _____ = _____   _____ = _____

a) $\sqrt{t} = 6$
b) $8 = \sqrt{y}$
c) $\sqrt{3R} = 7$

Equations and Formulas Containing Squares and Square Root Radicals 107

36. Two steps are needed to isolate the radical in the equation below.
$$4 - \sqrt{x} = 1$$
Step 1: Adding −4 to both sides, we get:
$$-\sqrt{x} = -3$$
Step 2: Applying the oppositing principle, we get:
$$\sqrt{x} = 3$$

Isolate the radical in each equation below:
(a) $10 - \sqrt{y} = 5$
(b) $7 = 8 - \sqrt{3x}$

a) $\sqrt{2y+7} = 3$

b) $4 = \sqrt{\dfrac{1}{x}}$

c) $\sqrt{\dfrac{40}{6-t}} = 1$

---

37. Isolate the radical in each equation below:
(a) $20 - \sqrt{V+4} = 10$
(b) $5 = 9 - \sqrt{\dfrac{y}{7}}$

a) $\sqrt{y} = 5$

b) $1 = \sqrt{3x}$

---

38. In the equation below, the radical has a negative coefficient.
$$-3\sqrt{y} = -5$$
We can isolate the radical in either of two ways:

(1) By multiplying both sides by $"-\dfrac{1}{3}"$. We get:
$$\left(-\dfrac{1}{3}\right)\left(-3\sqrt{y}\right) = \left(-\dfrac{1}{3}\right)(-5)$$
$$1\quad \sqrt{y} = +\dfrac{5}{3}$$
or $\sqrt{y} = \dfrac{5}{3}$

(2) By applying the oppositing principle and then multiplying both sides by $"\dfrac{1}{3}"$. We get:
$$3\sqrt{y} = 5$$
$$\dfrac{1}{3}\left(3\sqrt{y}\right) = \dfrac{1}{3}(5)$$
$$\sqrt{y} = \dfrac{5}{3}$$

(Continued on following page.)

a) $\sqrt{V+4} = 10$

b) $4 = \sqrt{\dfrac{y}{7}}$

108   Equations and Formulas Containing Squares and Square Root Radicals

38. (Continued)

Using either method, isolate the radicals in these:

(a) $-4\sqrt{x} = -7$ (b) $-1 = -5\sqrt{3F}$ (c) $-8\sqrt{y-5} = -9$

---

39. Two steps are needed to isolate the radical in the equation below.

$$7 - 3\sqrt{x} = 2$$

Adding -7 to both sides, we get:

$$-3\sqrt{x} = -5$$

Isolating the radical in this new equation, we get:

$$\sqrt{x} = \frac{5}{3}$$

Isolate the radical in each equation below:

(a) $40 - 7\sqrt{3t} = 10$ (b) $2 = 12 - 5\sqrt{V+1}$

a) $\sqrt{x} = \frac{7}{4}$

b) $\frac{1}{5} = \sqrt{3F}$

c) $\sqrt{y-5} = \frac{9}{8}$

---

40. Isolate the radical in each equation:

(a) $4\sqrt{m} - 7 = 14$ (b) $-6 = -\sqrt{t}$ (c) $8 - \sqrt{x} = 1$

a) $\sqrt{3t} = \frac{30}{7}$

b) $2 = \sqrt{V+1}$

---

a) $\sqrt{m} = \frac{21}{4}$

b) $6 = \sqrt{t}$

c) $\sqrt{x} = 7$

41. Isolate the radical in each equation:

(a) $\sqrt{\dfrac{3x}{x+1}} - 7 = 9$ (b) $10 = 30 - 2\sqrt{r-6}$

a) $\sqrt{\dfrac{3x}{x+1}} = 16$

b) $10 = \sqrt{r-6}$

---

42. In the equation below, the radical is part of the numerator of a fraction.

$$\dfrac{2\sqrt{x}}{3} = 5$$

To isolate the radical, we begin by clearing the fraction. We do so by multiplying both sides by "3". We get:

$$3\left(\dfrac{2\sqrt{x}}{3}\right) = 5(3)$$

$$2\sqrt{x} = 15$$

Now we can isolate the radical in the usual way. We get:

$$\sqrt{x} = \dfrac{15}{2}$$

Isolate the radical in each of these:

(a) $\dfrac{\sqrt{t}}{3} = 4$ (b) $\dfrac{3\sqrt{y}}{5} = 1$ (c) $8 = \dfrac{5\sqrt{2d}}{4}$

a) $\sqrt{t} = 12$

b) $\sqrt{y} = \dfrac{5}{3}$

c) $\sqrt{2d} = \dfrac{32}{5}$

---

43. Isolate the radical in each of these:

(a) $\dfrac{\sqrt{y-1}}{3} = 2$ (b) $\dfrac{2\sqrt{x+4}}{5} = 8$ (c) $1 = \dfrac{9\sqrt{m-7}}{8}$

a) $\sqrt{y-1} = 6$

b) $\sqrt{x+4} = 20$

c) $\sqrt{m-7} = \dfrac{8}{9}$

110   Equations and Formulas Containing Squares and Square Root Radicals

44. In the equation below, the radical is part of the denominator of a fraction.

$$\frac{2}{5\sqrt{m}} = 3$$

To isolate the radical, we again begin by clearing the fraction. We do so by multiplying both sides by "$5\sqrt{m}$". We get:

$$5\sqrt{m}\left(\frac{2}{5\sqrt{m}}\right) = 3\left(5\sqrt{m}\right)$$

$$2 = 15\sqrt{m}$$

Now we can isolate the radical in the usual way. We get:

$$\frac{2}{15} = \sqrt{m}$$

Isolate the radical in each of these:

(a) $\dfrac{5}{\sqrt{y}} = 4$ (b) $\dfrac{4}{3\sqrt{t}} = 1$ (c) $7 = \dfrac{1}{2\sqrt{3p}}$

45. Isolate the radical in each of these:

(a) $\dfrac{7}{\sqrt{x+1}} = 2$ (b) $\dfrac{1}{3\sqrt{2R-7}} = 5$ (c) $2 = \dfrac{5}{6\sqrt{t+3}}$

a) $\sqrt{y} = \dfrac{5}{4}$

b) $\sqrt{t} = \dfrac{4}{3}$

c) $\sqrt{3p} = \dfrac{1}{14}$

a) $\sqrt{x+1} = \dfrac{7}{2}$

b) $\sqrt{2R-7} = \dfrac{1}{15}$

c) $\sqrt{t+3} = \dfrac{5}{12}$

46. To clear the fractions in the equation below, we multiply both sides by both 2 and 3, and get:

$$\frac{5\sqrt{x}}{2} = \frac{1}{3}$$

$$(2)(3)\left(\frac{5\sqrt{x}}{2}\right) = (2)(3)\left(\frac{1}{3}\right)$$

$$3\left(5\sqrt{x}\right) = 2$$

$$15\sqrt{x} = 2$$

Clear the fractions in these equations and then isolate the radicals:

(a) $\dfrac{\sqrt{t}}{5} = \dfrac{2}{3}$ (b) $\dfrac{5}{4} = \dfrac{3\sqrt{2m}}{5}$ (c) $\dfrac{2\sqrt{p-1}}{3} = \dfrac{4}{9}$

---

47. To clear the fractions in the equation below, we multiply both sides by $3\sqrt{y}$ and 5, and get:

$$\frac{1}{3\sqrt{y}} = \frac{4}{5}$$

$$(5)\left(3\sqrt{y}\right)\left(\frac{1}{3\sqrt{y}}\right) = \frac{4}{5}(5)\left(3\sqrt{y}\right)$$

$$5(1) = 4\left(3\sqrt{y}\right)$$

$$5 = 12\sqrt{y}$$

Clear the fractions in each of these equations and then isolate the radical:

(a) $\dfrac{7}{\sqrt{x}} = \dfrac{5}{2}$ (b) $\dfrac{1}{2} = \dfrac{4}{3\sqrt{5t}}$ (c) $\dfrac{1}{5\sqrt{1-a}} = \dfrac{2}{3}$

---

a) $\sqrt{t} = \dfrac{10}{3}$, from:
$3\sqrt{t} = 10$

b) $\sqrt{2m} = \dfrac{25}{12}$, from:
$25 = 12\sqrt{2m}$

c) $\sqrt{p-1} = \dfrac{2}{3}$, from:
$6\sqrt{p-1} = 4$

112   Equations and Formulas Containing Squares and Square Root Radicals

48. To clear the fractions in the equation below, we multiply both sides by "2" and get:

$$\frac{3 + \sqrt{y}}{2} = 4$$

$$2\left(\frac{3 + \sqrt{y}}{2}\right) = 2(4)$$

$$3 + \sqrt{y} = 8$$

Clear the fraction in these and then isolate the radical:

(a) $\dfrac{7 + 2\sqrt{t}}{3} = 4$ (b) $10 = \dfrac{30 - \sqrt{x+1}}{2}$

a) $\sqrt{x} = \dfrac{14}{5}$, from:
$14 = 5\sqrt{x}$

b) $\sqrt{5t} = \dfrac{8}{3}$, from:
$3\sqrt{5t} = 8$

c) $\sqrt{1-a} = \dfrac{3}{10}$, from:
$3 = 10\sqrt{1-a}$

---

49. To clear the fraction in the equation below, we multiply both sides by the binomial $5 + \sqrt{d}$. Notice that $9\left(5 + \sqrt{d}\right)$ on the right side is an instance of the distributive principle.

$$\frac{4}{5 + \sqrt{d}} = 9$$

$$\left(5 + \sqrt{d}\right)\left(\frac{4}{5 + \sqrt{d}}\right) = 9\left(5 + \sqrt{d}\right)$$

$$4 = 45 + 9\sqrt{d}$$

Clear the fractions in these and then isolate the radical. In each case, you will have to multiply by the distributive principle on one side.

(a) $\dfrac{1}{3 - 2\sqrt{d}} = 4$ (b) $2 = \dfrac{5}{4 - \sqrt{2y+7}}$

a) $\sqrt{t} = \dfrac{5}{2}$, from:
$7 + 2\sqrt{t} = 12$

b) $\sqrt{x+1} = 10$, from:
$20 = 30 - \sqrt{x+1}$

---

a) $\sqrt{d} = \dfrac{11}{8}$, from:
$1 = 12 - 8\sqrt{d}$

b) $\sqrt{2y+7} = \dfrac{3}{2}$, from:
$8 - 2\sqrt{2y+7} = 5$

Equations and Formulas Containing Squares and Square Root Radicals   113

50. Isolate the radical in each of these:

(a) $\dfrac{2\sqrt{5t}}{3} = 1$ (b) $\dfrac{1}{\sqrt{x}} = \dfrac{5}{3}$ (c) $7 = \dfrac{1}{3\sqrt{R+1}}$

---

51. Isolate the radical in each of these:

(a) $\dfrac{3}{4\sqrt{2t}} = \dfrac{1}{5}$ (b) $1 = \dfrac{5 - 8\sqrt{y}}{2}$ (c) $9 = \dfrac{100}{10 + \sqrt{R}}$

a) $\sqrt{5t} = \dfrac{3}{2}$

b) $\sqrt{x} = \dfrac{3}{5}$

c) $\sqrt{R+1} = \dfrac{1}{21}$

---

a) $\sqrt{2t} = \dfrac{15}{4}$

b) $\sqrt{y} = \dfrac{3}{8}$

c) $\sqrt{R} = \dfrac{10}{9}$

SELF-TEST 3 (Frames 25-51)

Isolate the radical in each of the following equations:

1. $2 + 3\sqrt{5t} = 8$

2. $5 = 7 - 2\sqrt{r-1}$

3. $\dfrac{4\sqrt{x+7}}{5} = 12$

4. $\dfrac{3}{5\sqrt{2w}} = \dfrac{9}{10}$

5. $\dfrac{29}{7 + 6\sqrt{3h}} = 3$

6. $\dfrac{7}{3} = \dfrac{5}{3 - 2\sqrt{2w-1}}$

ANSWERS:
1. $\sqrt{5t} = 2$
2. $\sqrt{r-1} = 1$
3. $\sqrt{x+7} = 15$
4. $\sqrt{2w} = \dfrac{2}{3}$
5. $\sqrt{3h} = \dfrac{4}{9}$
6. $\sqrt{2w-1} = \dfrac{3}{7}$

2-4 ISOLATING RADICALS AND SOLVING RADICAL EQUATIONS

The general method for solving a radical equation is to isolate the radical, apply the squaring principle, and then solve the resulting non-radical equation. In the last section, we discussed the steps needed to isolate radicals. In this section, we will use the general method to solve radical equations.

Equations and Formulas Containing Squares and Square Root Radicals 115

52. The general method for solving radical equations involves three steps:
   (1) Isolating the radical.
   (2) Applying the squaring principle.
   (3) Solving the resulting non-radical equation.

   Let's use these three steps to solve: $3 + \sqrt{2x} = 9$

   (a) Isolate the radical.

   (b) Apply the squaring principle.   ___ = ___

   ___ = ___

   (c) Solve the resulting non-radical equation. $x = $ ___

   (d) The root of $3 + \sqrt{2x} = 9$ is ___.

53. Let's use the same three steps to solve: $7 + \sqrt{R+3} = 9$

   (a) Isolate the radical.

   (b) Apply the squaring principle.   ___ = ___

   ___ = ___

   (c) Solve the non-radical equation.

   $R = $ ___

   (d) The root of $7 + \sqrt{R+3} = 9$ is ___.

---

a) $\sqrt{2x} = 6$
b) $2x = 36$
c) $x = 18$
d) 18

---

a) $\sqrt{R+3} = 2$
b) $R + 3 = 4$
c) $R = 1$
d) 1

116   Equations and Formulas Containing Squares and Square Root Radicals

54. Sometimes the non-radical equation obtained by applying the squaring principle is already solved. Here is an example.

   Let's solve this equation: $3\sqrt{R} = 6$

   (a) Isolate the radical.

   _____ = _____

   (b) Apply the squaring principle.

   _____ = _____

   (c) The root of $3\sqrt{R} = 6$ is _____.

---

55.   Let's solve this equation: $5 = 15 - 2\sqrt{t}$

   (a) Isolate the radical.

   (b) Apply the squaring principle.

   _____ = _____

   (c) The root of $5 = 15 - 2\sqrt{t}$ is _____.

a) $\sqrt{R} = 2$
b) $R = 4$
c) 4

---

56.   Let's solve this equation: $\sqrt{\dfrac{6r}{r-1}} - 4 = 0$

   (a) Isolate the radical.

   (b) Apply the squaring principle.

   (c) Solve the non-radical equation.

   $r =$ _____

a) $\sqrt{t} = 5$
b) $t = 25$
c) 25

---

a) $\sqrt{\dfrac{6r}{r-1}} = 4$

b) $\dfrac{6r}{r-1} = 16$

c) $r = \dfrac{16}{10}$ or $\dfrac{8}{5}$,
   from: $6r = 16r - 16$

Equations and Formulas Containing Squares and Square Root Radicals 117

57. Solve each of these:

(a) $12 - 3\sqrt{x+1} = 6$

(b) $7 = 10 - \sqrt{\dfrac{w}{5}}$

x = _____   w = _____

58. When applying the squaring principle to a radical equation, <u>don't forget to square the non-radical side</u>.

Solve each equation below:

(a) $2 + \sqrt{2s-3} = 11$

(b) $\sqrt{7-3b} - 5 = 0$

a) $\sqrt{x+1} = 2$
$x+1 = 4$
$x = 3$

b) $\sqrt{\dfrac{w}{5}} = 3$

$\dfrac{w}{5} = 9$

$w = 45$

s = _____   b = _____

59. Let's solve this equation: $\dfrac{4\sqrt{2t}}{5} = 8$

(a) Isolate the radical.

(b) Apply the squaring principle.

(c) Solve the non-radical equation.   t = _____

a) s = 42
b) b = -6

a) $\sqrt{2t} = 10$
b) $2t = 100$
c) $t = 50$

118 Equations and Formulas Containing Squares and Square Root Radicals

60. Let's solve this equation: $\dfrac{1}{3\sqrt{m}} = 2$

(a) Isolate the radical.

_____

(b) Apply the squaring principle.

_____

(c) The root of the original radical equation is _____.

---

61. Solve each of the following equations:

(a) $\dfrac{14}{\sqrt{y+1}} = 7$ (b) $6 = \dfrac{3\sqrt{10 - 2d}}{5}$

y = _____   d = _____

a) $\sqrt{m} = \dfrac{1}{6}$

b) $m = \dfrac{1}{36}$

c) $\dfrac{1}{36}$

---

62. Let's solve this equation: $\dfrac{1}{5\sqrt{t}} = \dfrac{2}{3}$

a) y = 3

b) d = −45

(a) Isolate the radical.

_____

(b) Apply the squaring principle.

_____

(c) The root of the original equation is _____.

---

a) $\sqrt{t} = \dfrac{3}{10}$

b) $t = \dfrac{9}{100}$

c) $\dfrac{9}{100}$

Equations and Formulas Containing Squares and Square Root Radicals   119

63. Solve: (a) $\dfrac{\sqrt{5x}}{6} = \dfrac{1}{3}$ (b) $\dfrac{3}{8} = \dfrac{1}{2\sqrt{R}+1}$

x = _____   R = _____

| a) $x = \dfrac{4}{5}$ |
| b) $R = \dfrac{7}{9}$ |

64. Let's solve this equation: $\dfrac{16}{2+3\sqrt{y}} = 4$

(a) Isolate the radical.

(b) Apply the squaring principle.

(c) The root of the original equation is _____ .

| a) $\sqrt{y} = \dfrac{2}{3}$, from: $16 = 8 + 12\sqrt{y}$ |
| b) $y = \dfrac{4}{9}$ |
| c) $\dfrac{4}{9}$ |

65. Solve: (a) $\dfrac{1+2\sqrt{h}}{3} = 7$ (b) $\dfrac{3}{2+\sqrt{3v-2}} = 1$

h = _____   v = _____

a) $h = 100$
b) $v = 1$

120 Equations and Formulas Containing Squares and Square Root Radicals

66. Solve: (a) $\dfrac{2}{\sqrt{x-1}} = 3$ (b) $\dfrac{7}{\sqrt{2t}} = 4$

Answer to Frame 66: a) $x = 1\dfrac{4}{9}$ or $\dfrac{13}{9}$ b) $t = \dfrac{49}{32}$ or $1\dfrac{17}{32}$

SELF-TEST 4 (Frames 52-66)

Solve each of the following radical equations:

1. $5 - 4\sqrt{R} = 3$

    R = _____

2. $\dfrac{2\sqrt{3x}}{3} = 1$

    x = _____

3. $10 = 3\sqrt{2t+4} - 2$

    t = _____

4. $\dfrac{2 + \sqrt{5d}}{4} = 3$

    d = _____

5. $\dfrac{1}{6} = \dfrac{3}{2\sqrt{2h-1}}$

    h = _____

6. $\dfrac{1}{1 - \sqrt{3y-2}} = 2$

    y = _____

ANSWERS: 1. $R = \dfrac{1}{4}$ 3. $t = 6$ 5. $h = 41$

2. $x = \dfrac{3}{4}$ 4. $d = 20$ 6. $y = \dfrac{3}{4}$

2-5 THE SQUARING PRINCIPLE AND RADICAL FORMULAS

In each of the formulas below, a square root radical is isolated on one side.

$$t = \sqrt{\frac{2s}{g}} \qquad I = \sqrt{\frac{P}{R}} \qquad \sqrt{LC} = \frac{1}{2\pi f}$$

In order to solve for a variable which is part of the radicand in formulas of this type, we must apply the squaring principle. We will discuss the method in this section.

---

67. As we have seen, the square of any square root radical is its radicand. Therefore, its square is a non-radical expression. For example:

    $$\left(\sqrt{\frac{A}{\pi}}\right)^2 = \frac{A}{\pi} \qquad \left(\sqrt{\frac{2s}{g}}\right)^2 = \frac{2s}{g} \qquad \left(\sqrt{2gs}\right)^2 = 2gs$$

    Write the squares of each of the following:

    (a) $\left(\sqrt{\frac{P}{R}}\right)^2 = $ _____   (b) $\left(\sqrt{LC}\right)^2 = $ _____   (c) $\left(\sqrt{h+2r}\right)^2 = $ _____

68. If a radical is isolated on one side of a formula, we can eliminate the radical by applying the squaring principle. For example:

    Squaring both sides of $r = \sqrt{\frac{A}{\pi}}$, we get: $r^2 = \frac{A}{\pi}$

    Squaring both sides of $v = \sqrt{2gs}$, we get: $v^2 = 2gs$

    What non-radical equation do we get by squaring both sides of each formula below?

    (a) $d = \sqrt{\frac{A}{0.7854}}$ (b) $r = \sqrt{\frac{V}{\pi h}}$ (c) $s = \sqrt{A}$

    Answers:
    a) $\frac{P}{R}$
    b) LC
    c) $h + 2r$

69. When applying the squaring principle to a radical formula, don't forget to square the non-radical side.

    Apply the squaring principle to each equation below:

    (a) $p = \sqrt{b+c}$ (b) $d = \sqrt{\frac{I_A - I_B}{m}}$ (c) $v_0 = \sqrt{v_f^2 - 2gs}$

    Answers:
    a) $d^2 = \frac{A}{0.7854}$
    b) $r^2 = \frac{V}{\pi h}$
    c) $s^2 = A$

Answers:
a) $p^2 = b + c$
b) $d^2 = \frac{I_A - I_B}{m}$
c) $v_0^2 = v_f^2 - 2gs$

122 Equations and Formulas Containing Squares and Square Root Radicals

70. To solve for a letter in the radicand, we apply the squaring principle to obtain a non-radical equation. Then we complete the solution in the usual way. For example, we have applied the squaring principle and solved for "A" in the formula below.

$$r = \sqrt{\frac{A}{\pi}}$$

$$r^2 = \frac{A}{\pi}$$

$$A = \pi r^2$$

Let's solve for "R" in this formula: $I = \sqrt{\frac{P}{R}}$

(a) Applying the squaring principle, we get: _____

(b) Solving for "R" in this new equation, we get:

$$R = \underline{\qquad}$$

71. Let's solve for "s" in this formula: $v = \sqrt{2gs}$

(a) Applying the squaring principle, we get: _____

(b) Solving for "s", we get:

$$s = \underline{\qquad}$$

72. Apply the squaring principle and then solve for the indicated letter:

(a) $S = \sqrt{A}$ (b) $a = \sqrt{b+c}$ (c) $t = \sqrt{\frac{2s}{g}}$

A = _____   b = _____   g = _____

---

a) $I^2 = \frac{P}{R}$

b) $R = \frac{P}{I^2}$

---

a) $v^2 = 2gs$

b) $s = \frac{v^2}{2g}$

---

a) $A = S^2$

b) $b = a^2 - c$

c) $g = \frac{2s}{t^2}$

Equations and Formulas Containing Squares and Square Root Radicals   123

73. (a) Solve for "A":   (b) Solve for "P":   (c) Solve for "L":

$$d = \sqrt{\dfrac{A}{0.7854}} \qquad I = \sqrt{\dfrac{P}{R}} \qquad d = \sqrt{\dfrac{pL}{R}}$$

A = _____   P = _____   L = _____

74. When applying the squaring principle to a radical formula, the non-radical side can be more complicated than a single letter.

If the non-radical side <u>is a letter-term</u>, we square it by squaring each factor in the term. For example:

$$(ad)^2 = a^2 d^2 \qquad (2gs)^2 = 4g^2 s^2$$

If the non-radical side <u>is a fraction</u>, we square it by squaring both its numerator and its denominator. For example:

$$\left(\dfrac{ab}{c}\right)^2 = \dfrac{a^2 b^2}{c^2} \qquad \left(\dfrac{1}{2\pi f}\right)^2 = \dfrac{1}{4\pi^2 f^2}$$

Apply the squaring principle to each equation below:

(a) $\sqrt{L_1 L_2} = \dfrac{M}{K}$ (b) $\sqrt{cd} = ab$ (c) $\sqrt{R} = \dfrac{1}{3xy}$

a) $A = 0.7854 d^2$
b) $P = I^2 R$
c) $L = \dfrac{d^2 R}{p}$

75. Apply the squaring principle to each equation below:

(a) $\sqrt{h + 2r} = sd$ (b) $\sqrt{\dfrac{r}{a}} = \dfrac{t}{2\pi}$ (c) $\sqrt{d} = 2w$

a) $L_1 L_2 = \dfrac{M^2}{K^2}$
b) $cd = a^2 b^2$
c) $R = \dfrac{1}{9x^2 y^2}$

a) $h + 2r = s^2 d^2$
b) $\dfrac{r}{a} = \dfrac{t^2}{4\pi^2}$
c) $d = 4w^2$

76. Apply the squaring principle to each of these formulas:

(a) $\sqrt{Z^2 - R^2} = 2\pi fL$ (b) $\sqrt{\dfrac{Fr}{m}} = \dfrac{2\pi r}{t}$ (c) $\sqrt{\dfrac{T}{P}} = \dfrac{N}{P}$

_____ _____ _____ _____

a) $Z^2 - R^2 = 4\pi^2 f^2 L^2$

b) $\dfrac{Fr}{m} = \dfrac{4\pi^2 r^2}{t^2}$

c) $\dfrac{T}{P} = \dfrac{N^2}{P^2}$

77. Let's solve for "b" in this equation: $\sqrt{bc} = ad$

(a) Squaring both sides, we get: _____

(b) Solving for "b", we get:

$b = $ _____

a) $bc = a^2 d^2$

b) $b = \dfrac{a^2 d^2}{c}$

78. Apply the squaring principle and then solve for the indicated letter:

(a) $\sqrt{\dfrac{P}{K}} = mt$ (b) $\sqrt{h + 2r} = sd$

$K = $ _____ $r = $ _____

a) $K = \dfrac{P}{m^2 t^2}$

b) $r = \dfrac{s^2 d^2 - h}{2}$

79. (a) Solve for "a":   (b) Solve for "x":   (c) Solve for "m":

$\sqrt{a} = \dfrac{r}{t}$ $\sqrt{x} = \dfrac{bc}{a}$ $\sqrt{m} = xy$

$a = $ _____ $x = $ _____ $m = $ _____

a) $a = \dfrac{r^2}{t^2}$

b) $x = \dfrac{b^2 c^2}{a^2}$

c) $m = x^2 y^2$

Equations and Formulas Containing Squares and Square Root Radicals   125

80. If we square both sides of $\sqrt{L_1 L_2} = \dfrac{M}{K}$, we get: $L_1 L_2 = \dfrac{M^2}{K^2}$

We can use either of two methods to solve the non-radical equation for "L_2".

(1) Clearing the fraction and then multiplying both sides by $\dfrac{1}{K^2 L_1}$:

$$L_1 L_2 = \dfrac{M^2}{K^2}$$

$$K^2 L_1 L_2 = M^2$$

$$\dfrac{1}{K^2 L_1}(K^2 L_1 L_2) = M^2 \left(\dfrac{1}{K^2 L_1}\right)$$

$$L_2 = \dfrac{M^2}{K^2 L_1}$$

(2) Simply multiplying both sides by $\dfrac{1}{L_1}$:

$$L_1 L_2 = \dfrac{M^2}{K^2}$$

$$\dfrac{1}{L_1}(L_1 L_2) = \dfrac{1}{L_1}\left(\dfrac{M^2}{K^2}\right)$$

$$L_2 = \dfrac{M^2}{K^2 L_1}$$

Note: The second method is shorter and easier.

Let's solve for "L" in: $\sqrt{LC} = \dfrac{1}{2\pi f}$

(a) Squaring both sides, we get: _____

(b) Now using either method above, solve for "L":

L = _____

---

81. (a) Solve for "h": (b) Solve for "g_p": (c) Solve for "r":

$\sqrt{ht} = \dfrac{W}{F}$ $\sqrt{\dfrac{g_p}{g_o}} = \dfrac{r_o}{r_p}$ $\dfrac{t}{2\pi} = \sqrt{\dfrac{r}{a}}$

h = _____ g_p = _____   r = _____

a) $LC = \dfrac{1}{4\pi^2 f^2}$

b) $L = \dfrac{1}{4\pi^2 C f^2}$

126 Equations and Formulas Containing Squares and Square Root Radicals

82. We applied the squaring principle to the formula below and then solved for "a". Notice how we were able to reduce the fraction in the solution to lowest terms.

$$\sqrt{ar} = wr$$
$$ar = w^2r^2$$
$$a = \frac{w^2r^2}{r} = w^2r\left(\frac{r}{r}\right) = w^2r$$

Do these problems, and reduce the solutions to lowest terms if possible:

(a) Solve for "d":
$$\sqrt{cd} = 2cr$$

(b) Solve for "a":
$$\sqrt{ar} = \frac{2\pi r}{T}$$

d = _____   a = _____

a) $h = \dfrac{W^2}{F^2 t}$

b) $g_p = \dfrac{g_o r_o^2}{r_p^2}$

c) $r = \dfrac{at^2}{4\pi^2}$

---

83. In solving for "w" below, we were also able to reduce the fraction in the solution to lowest terms.

$$\sqrt{\frac{aR}{w}} = Rt$$
$$\frac{aR}{w} = R^2t^2$$
$$w = \frac{aR}{R^2t^2} = \frac{a}{Rt^2}\left(\frac{R}{R}\right) = \frac{a}{Rt^2}$$

In the following problems, reduce the solutions to lowest terms if possible:

(a) Solve for "b":
$$\sqrt{\frac{2a}{b}} = ac$$

(b) Solve for "m":
$$\sqrt{\frac{Fr}{m}} = \frac{2\pi r}{T}$$

b = _____   m = _____

a) $d = 4cr^2$, from:
$$d = \frac{4c^2r^2}{c}$$

b) $a = \dfrac{4\pi^2 r}{T^2}$, from:
$$a = \frac{4\pi^2 r^2}{rT^2}$$

84. Reduce each solution to lowest terms if possible:

    (a) Solve for "p":

$$\sqrt{\dfrac{dt}{p}} = 4t$$

    (b) Solve for "h":

$$\sqrt{\dfrac{ch}{3}} = \dfrac{c}{R}$$

p = _____ h = _____

a) $b = \dfrac{2}{ac^2}$, from:
$$b = \dfrac{2a}{a^2c^2}$$

b) $m = \dfrac{FT^2}{4\pi^2 r}$, from:
$$m = \dfrac{FrT^2}{4\pi^2 r^2}$$

---

85. (a) Solve for "r":

$$\sqrt{\dfrac{a}{r}} = \dfrac{w}{2g}$$

    (b) Solve for "F":

$$\sqrt{\dfrac{Fr}{m}} = \dfrac{2\pi r}{T}$$

    (c) Solve for "h":

$$\sqrt{h + 2r} = sd$$

r = _____ F = _____ h = _____

a) $p = \dfrac{d}{16t}$

b) $h = \dfrac{3c}{R^2}$

---

86. Here are some of the steps needed to solve for "s" in the formula below:

$$v_0 = \sqrt{v_f^2 - 2gs}$$
$$v_0^2 = v_f^2 - 2gs$$
$$v_0^2 - v_f^2 = -2gs$$

At this point in the solution, we can apply the opposing principle to the equation to eliminate the "-" in front of "2gs".

    Since the opposite of -2gs is 2gs
      and the opposite of $v_0^2 - v_f^2$ is $v_f^2 - v_0^2$,

        we get: $2gs = v_f^2 - v_0^2$

    and: s = _____

a) $r = \dfrac{4ag^2}{w^2}$

b) $F = \dfrac{4\pi^2 mr}{T^2}$

(Did you reduce to lowest terms?)

c) $h = s^2 d^2 - 2r$

$$s = \dfrac{v_f^2 - v_0^2}{2g}$$

128 Equations and Formulas Containing Squares and Square Root Radicals

87. When using the opposing principle in the last frame, we had to replace a binomial which contained a subtraction by its opposite. Remember:

   The opposite of "a - b" is "b - a".

   Solve for the indicated letter in each of these:

   (a) Solve for "I_c":
   $$d = \sqrt{\frac{I_a - I_c}{m}}$$

   (b) Solve for "e":
   $$v = \sqrt{w - e}$$

   I_c = _____   e = _____

---

88. Here are the steps needed to solve for "r" in the formula below:

   $$\frac{F}{2c} = \sqrt{b - r}$$

   $$\frac{F^2}{4c^2} = b - r$$

   $$\frac{F^2}{4c^2} - b = -r$$

   $$r = b - \frac{F^2}{4c^2}$$

   By converting "b" to a fraction whose denominator is "$4c^2$", we can write the solution in an alternate form. Do so.

   r = _____

---

a) $I_c = I_a - d^2 m$

   (from: $d^2 m - I_a = -I_c$)

b) $e = w - v^2$

   (from: $v^2 - w = -e$)

$$r = \frac{4bc^2 - F^2}{4c^2}, \text{ from:}$$

$$\frac{4bc^2}{4c^2} - \frac{F^2}{4c^2}$$

89. Let's solve for "h" in: $\sqrt{1 - \dfrac{H}{h}} = C$

   (a) Square both sides.

   (b) Clear the fraction.

   (c) Isolate the two terms which contain "h".

   (d) Factor by the distributive principle and complete the solution.

---

a) $1 - \dfrac{H}{h} = C^2$

b) $h\left(1 - \dfrac{H}{h}\right) = hC^2$

   $h - H = hC^2$

c) $h - hC^2 = H$
   (from: $-H = hC^2 - h$)

d) $h(1 - C^2) = H$

   $h = \dfrac{H}{1 - C^2}$

130 Equations and Formulas Containing Squares and Square Root Radicals

---

SELF-TEST 5 (Frames 67-89)

1. Solve for P:

   $E = \sqrt{PR}$

   P = _____

2. Solve for h:

   $p = \sqrt{H - h}$

   h = _____

3. Solve for t:

   $\sqrt{at} = 2ar$

   t = _____

4. Solve for d: $kp = \sqrt{\dfrac{2kw}{d}}$

   d = _____

5. Solve for m: $\sqrt{1 + \dfrac{M}{m}} = F$

   m = _____

ANSWERS: 1. $P = \dfrac{E^2}{R}$ 2. $h = H - p^2$ 3. $t = 4ar^2$ 4. $d = \dfrac{2w}{kp^2}$ 5. $m = \dfrac{M}{F^2 - 1}$

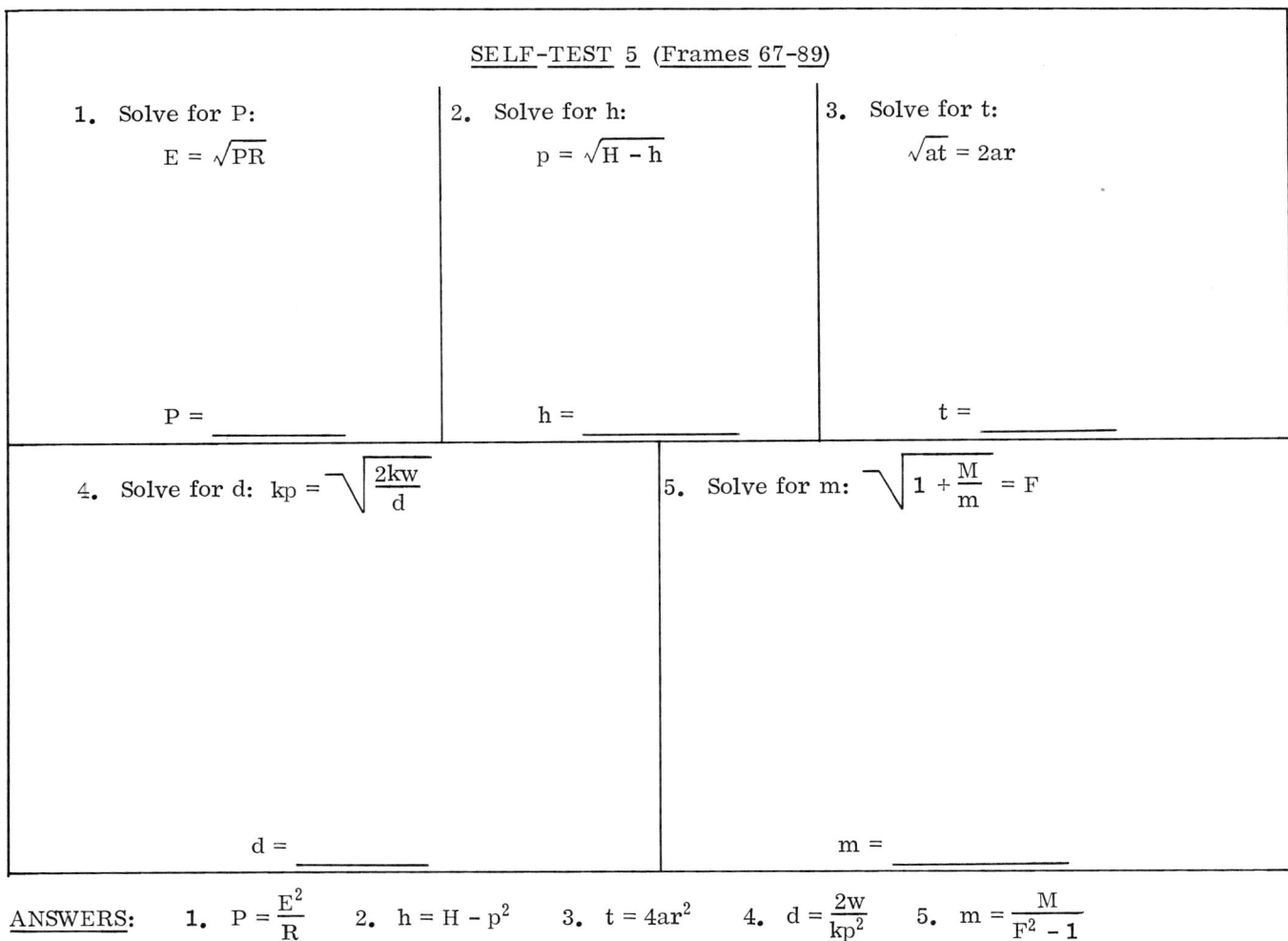

2-6 ISOLATING RADICALS AND REARRANGING RADICAL FORMULAS

In the last section, we solved for variables in radical formulas in which the radical was already isolated on one side. In this section, we will solve for variables in formulas in which the radical is not isolated on one side. To solve for a variable which is part of the radicand in these cases, the first step is isolating the radical. At the end of this section, we will also discuss the method of solving for variables which are not part of the radicand.

Equations and Formulas Containing Squares and Square Root Radicals   131

90. In the formula below, the radical is not isolated on one side. To solve for "a", the first step is isolating the radical. We get:

$$t = \frac{r}{\sqrt{a}}$$

$$t\sqrt{a} = r$$

$$\sqrt{a} = \frac{r}{t}$$

Now we can solve for "a" by applying the squaring principle. We get:

a = _____

---

91. To solve for a letter in a radical which is not isolated on one side, the first step is "isolating the radical".

Isolate the radical in each equation below:

(a) $R = P + \sqrt{Q}$ (b) $K = \dfrac{M}{\sqrt{L_1 L_2}}$ (c) $a = \dfrac{\sqrt{bc}}{d}$

$\sqrt{Q} = $ _____ $\sqrt{L_1 L_2} = $ _____ $\sqrt{bc} = $ _____

$a = \dfrac{r^2}{t^2}$

---

92. Isolate the radical in each equation below:

(a) $s = \dfrac{\sqrt{h + 2r}}{d}$ (b) $w = \dfrac{\sqrt{d}}{2}$ (c) $\dfrac{W}{\sqrt{ht}} = F$

$\sqrt{h + 2r} = $ _____ $\sqrt{d} = $ _____ $\sqrt{ht} = $ _____

a) $\sqrt{Q} = R - P$
b) $\sqrt{L_1 L_2} = \dfrac{M}{K}$
c) $\sqrt{bc} = ad$

---

93. Isolate the radical in each equation below:

(a) $\dfrac{a\sqrt{x}}{b} = c$ (b) $w = 2g\sqrt{\dfrac{a}{r}}$ (c) $f = \dfrac{1}{2\pi\sqrt{LC}}$

$\sqrt{x} = $ _____ $\sqrt{\dfrac{a}{r}} = $ _____ $\sqrt{LC} = $ _____

a) $\sqrt{h + 2r} = ds$
b) $\sqrt{d} = 2w$
c) $\sqrt{ht} = \dfrac{W}{F}$

132   Equations and Formulas Containing Squares and Square Root Radicals

94. Isolate the radical in each of these:

(a) $t = 2\pi\sqrt{\dfrac{r}{a}}$ (b) $p = m + r\sqrt{y}$ (c) $a = \dfrac{1}{\sqrt{p^2+w^2}}$

$\sqrt{\dfrac{r}{a}} =$ _____ $\sqrt{y} =$ _____ $\sqrt{p^2+w^2} =$ _____

a) $\sqrt{x} = \dfrac{bc}{a}$

b) $\sqrt{\dfrac{a}{r}} = \dfrac{w}{2g}$

c) $\sqrt{LC} = \dfrac{1}{2\pi f}$

---

95. After isolating the radical, we can solve for a letter in the radicand in the usual way.

Let's solve for "c" in: $\dfrac{\sqrt{cd}}{T} = R$

(a) Isolating the radical, we get:

(b) Squaring both sides, we get:

(c) Completing the solution, we get:

c = _____

a) $\sqrt{\dfrac{r}{a}} = \dfrac{t}{2\pi}$

b) $\sqrt{y} = \dfrac{p-m}{r}$

c) $\sqrt{p^2+w^2} = \dfrac{1}{a}$

---

96. Isolate the radical and then solve for the indicated letter:

(a) Solve for "d":   (b) Solve for "y":   (c) Solve for "L_1":

$w = \dfrac{\sqrt{d}}{2}$ $\dfrac{c\sqrt{y}}{t} = m$ $K = \dfrac{M}{\sqrt{L_1 L_2}}$

d = _____   y = _____ L_1 = _____

a) $\sqrt{cd} = RT$

b) $cd = R^2 T^2$

c) $c = \dfrac{R^2 T^2}{d}$

---

a) $d = 4w^2$

b) $y = \dfrac{m^2 t^2}{c^2}$

c) $L_1 = \dfrac{M^2}{K^2 L_2}$

Equations and Formulas Containing Squares and Square Root Radicals   133

97. (a) Solve for "h": (b) Solve for "a": (c) Solve for "g_o":

$$s = \frac{\sqrt{h + 2r}}{d}$$ $$w = 2g\sqrt{\frac{a}{r}}$$ $$r_o = r_p\sqrt{\frac{g_p}{g_o}}$$

h = _____   a = _____ g_o = _____

98. (a) Solve for "t": (b) Solve for "b": (c) Solve for "m":

$$\frac{W}{\sqrt{ht}} = F$$ $$F = \frac{\sqrt{b + r}}{2c}$$ $$\frac{1}{b\sqrt{m}} = t$$

a) $h = s^2 d^2 - 2r$

b) $a = \dfrac{rw^2}{4g^2}$

c) $g_o = \dfrac{g_p r_p^2}{r_o^2}$

t = _____   b = _____   m = _____

99. To solve for "Q" in the formula below, we must isolate the radical and then apply the squaring principle. We get:

$$R = P + \sqrt{Q}$$
$$\sqrt{Q} = R - P$$
$$Q = (R - P)^2$$

Since $(R - P)^2$ indicates the squaring of a binomial, we could also write the solution in this form:

$$Q = R^2 - 2RP + P^2$$

Ordinarily however, we leave the solution in the form $Q = (R - P)^2$ because it makes evaluation easier.

Let's solve for "b" in this one:

$$a - \sqrt{b} = c$$
$$-\sqrt{b} = c - a$$
$$\sqrt{b} = a - c$$

Note: In the last step, we applied the oppositing principle.

Complete the solution. Write it
in the preferred form.   b = _____

a) $t = \dfrac{W^2}{F^2 h}$

b) $b = 4c^2 F^2 - r$

c) $m = \dfrac{1}{b^2 t^2}$

134   Equations and Formulas Containing Squares and Square Root Radicals

100. We have shown the steps needed to solve for "y" in the equation below.

$$p = m + r\sqrt{y}$$
$$r\sqrt{y} = p - m$$
$$\sqrt{y} = \frac{p - m}{r}$$
$$y = \left(\frac{p - m}{r}\right)^2$$

Note: Though we could square the fraction on the right, we usually leave it in the form above because it makes evaluation easier.

Solve for "x" in this one. Write the solution in the preferred form:

$$t = b - c\sqrt{x}$$

$b = (a - c)^2$

---

101. To solve for a letter which is not part of the radicand, we do not have to eliminate the radical. Therefore, we don't need the squaring principle. Here is an example:

To solve for "r" below, we simply multiply both sides by $\sqrt{a}$. We get:

$$t = \frac{r}{\sqrt{a}}$$
$$\sqrt{a}(t) = \frac{r}{\sqrt{a}}\left(\sqrt{a}\right)$$
$$t\sqrt{a} = r$$
$$\text{or } r = t\sqrt{a}$$

Solve for the indicated letter in these:

(a) Solve for "M": (b) Solve for "d": (c) Solve for "b":

$$K = \frac{M}{\sqrt{L_1 L_2}} \qquad a = \frac{\sqrt{bc}}{d} \qquad \frac{a\sqrt{x}}{b} = c$$

M = _____   d = _____   b = _____

$x = \left(\dfrac{b - t}{c}\right)^2$, from:

$t - b = -c\sqrt{x}$

$b - t = c\sqrt{x}$

$\dfrac{b - t}{c} = \sqrt{x}$

Equations and Formulas Containing Squares and Square Root Radicals   135

102. To solve for "P" in the formula below, we simply add the opposite of $\sqrt{Q}$ to both sides. We get:

$$R = P + \sqrt{Q}$$
$$R + \left(-\sqrt{Q}\right) = P + \sqrt{Q} + \left(-\sqrt{Q}\right)$$
$$R - \sqrt{Q} = P$$
$$\text{or} \quad P = R - \sqrt{Q}$$

Solve for the indicated letter in each case below:

(a) Solve for "t":

$$m = t - \sqrt{R}$$

(b) Solve for "m":

$$p = m - r\sqrt{y}$$

t = _____   m = _____

a) $M = K\sqrt{L_1 L_2}$

b) $d = \dfrac{\sqrt{bc}}{a}$

c) $b = \dfrac{a\sqrt{x}}{c}$

103. The squaring principle is needed only to solve for variables <u>in the radicand</u> of a radical formula. That is:

In $K = \dfrac{M}{\sqrt{L_1 L_2}}$, the squaring principle is needed only to solve for L_1 or L_2.

In $s = \dfrac{\sqrt{h + 2r}}{d}$, the squaring principle is needed only to solve for _____.

a) $t = m + \sqrt{R}$

b) $m = p + r\sqrt{y}$

104. Do we need the squaring principle to solve for:

(a) "m" in $p = m + r\sqrt{y}$? _____

(b) "r" in $t = 2\pi\sqrt{\dfrac{r}{a}}$? _____

(c) "g" in $w = 2g\sqrt{\dfrac{a}{r}}$? _____

h or r

a) No
b) Yes
c) No

105. If we solve for "c" in the equation below, the solution is a fraction with a radical in its denominator.

$$c\sqrt{y} = d$$

$$c = \frac{d}{\sqrt{y}}$$

Frequently solutions are left in this non-rationalized form because it makes evaluation easier.

Do these without rationalizing the denominators of the solutions:

(a) Solve for "V":

$$\frac{V\sqrt{d}}{R} = S$$

(b) Solve for "r":

$$p = m + r\sqrt{t}$$

V = _____ r = _____

a) $V = \dfrac{RS}{\sqrt{d}}$

b) $r = \dfrac{p - m}{\sqrt{t}}$

106. Here are the solutions obtained in the last frame. The denominators were left in non-rationalized form.

$$c = \frac{d}{\sqrt{y}} \qquad V = \frac{RS}{\sqrt{d}} \qquad r = \frac{p-m}{\sqrt{t}}$$

If we want to rationalize the denominators, we multiply by the appropriate instance of $\dfrac{\sqrt{}}{\sqrt{}}$. For example:

$$c = \frac{d}{\sqrt{y}}\left(\frac{\sqrt{y}}{\sqrt{y}}\right) = \frac{d\sqrt{y}}{y}$$

Rationalize the denominator in each of these solutions:

(a) $V = \dfrac{RS}{\sqrt{d}}$

(b) $r = \dfrac{p - m}{\sqrt{t}}$

V = _____ r = _____

a) $V = \dfrac{RS\sqrt{d}}{d}$

b) $r = \dfrac{(p - m)\sqrt{t}}{t}$

107. If we solve for "m" in the equation below, the solution contains a square root of a fraction in its denominator.

$$t = m\sqrt{\frac{a}{b}}$$

$$m = \frac{t}{\sqrt{\frac{a}{b}}}$$

Solve for the indicated letter in each case:

(a) Solve for "r_p":

$$r_o = r_p \sqrt{\frac{g_p}{g_o}}$$

(b) Solve for "T":

$$\sqrt{\frac{Fr}{m}} = \frac{2\pi r}{T}$$

$r_p = $ _____

T = _____

---

a) $r_p = \dfrac{r_o}{\sqrt{\dfrac{g_p}{g_o}}}$

b) $T = \dfrac{2\pi r}{\sqrt{\dfrac{Fr}{m}}}$

---

108. Here is one of the solutions obtained in the last frame:

$$m = \frac{t}{\sqrt{\frac{a}{b}}}$$

The fraction on the right is really a division. It means: divide "t" by $\sqrt{\frac{a}{b}}$. By converting this division to a multiplication, we can simplify the solution. That is:

$$m = \frac{t}{\sqrt{\frac{a}{b}}} = t\left(\text{the reciprocal of } \sqrt{\frac{a}{b}}\right) = t\sqrt{\frac{b}{a}}$$

Note: The reciprocal of $\sqrt{\frac{a}{b}}$ is $\sqrt{\frac{b}{a}}$, since:

$$\sqrt{\frac{a}{b}} \cdot \sqrt{\frac{b}{a}} = \sqrt{\frac{ab}{ab}} = \sqrt{1} = 1$$

Notice that the conversion above is the equivalent of inverting the radicand and putting the radical in the numerator.

Write the reciprocal of each of these:

(a) $\sqrt{\dfrac{g_p}{g_o}}$ _____

(b) $\sqrt{\dfrac{a}{r}}$ _____

(c) $\sqrt{\dfrac{Fr}{m}}$ _____

138   Equations and Formulas Containing Squares and Square Root Radicals

109. Let's simplify the solution below by converting the division to a multiplication.

$$m = \dfrac{t}{\sqrt{\dfrac{a}{b}}} = t\left(\text{the reciprocal of } \sqrt{\dfrac{a}{b}}\right) = t\sqrt{\dfrac{b}{a}}$$

Notice that the conversion above is the equivalent of inverting the fractional radicand and moving the radical to the numerator.

Using the procedure above, write each solution below in the preferred form:

(a) $r_p = \dfrac{r_o}{\sqrt{\dfrac{g_p}{g_o}}}$ is written $r_p = $ _____

(b) $T = \dfrac{2\pi r}{\sqrt{\dfrac{Fr}{m}}}$ is written $T = $ _____

a) $\sqrt{\dfrac{g_o}{g_p}}$

b) $\sqrt{\dfrac{r}{a}}$

c) $\sqrt{\dfrac{m}{Fr}}$

---

110. Notice how we have simplified the solution below by inverting the radicand and moving the radical to the numerator.

$$g = \dfrac{w}{2\sqrt{\dfrac{a}{r}}} = \dfrac{w\sqrt{\dfrac{r}{a}}}{2} = \dfrac{w}{2}\sqrt{\dfrac{r}{a}}$$

Write each of the following solutions in the preferred form:

(a) $T = \dfrac{m}{b\sqrt{\dfrac{c}{d}}} = $ _____

(b) $F = \dfrac{2t}{R\sqrt{\dfrac{m}{s}}} = $ _____

a) $r_p = r_o\sqrt{\dfrac{g_o}{g_p}}$

b) $T = 2\pi r\sqrt{\dfrac{m}{Fr}}$

---

a) $T = \dfrac{m}{b}\sqrt{\dfrac{d}{c}}$

b) $F = \dfrac{2t}{R}\sqrt{\dfrac{s}{m}}$

111. If we solve for "P" in the formula below, we get:

$$PR = \sqrt{\frac{T}{N}}$$

$$P = \frac{\sqrt{\frac{T}{N}}}{R}$$

This solution can also be written in a simpler form by converting the division to a multiplication. That is:

$$P = \frac{\sqrt{\frac{T}{N}}}{R} = \sqrt{\frac{T}{N}\left(\frac{1}{R}\right)} = \frac{1}{R}\sqrt{\frac{T}{N}}$$

Notice that we write $\frac{1}{R}$ in front of the radical.

Write each of these solutions in the preferred form:

(a) $T = \dfrac{\sqrt{\frac{a}{b}}}{m}$ (b) $r = \dfrac{\sqrt{\frac{t}{m}}}{ab}$ (c) $S = \dfrac{\sqrt{\frac{2g}{a}}}{mt}$

T = _____   r = _____   S = _____

112. We have solved for "T" in the formula below.

$$\sqrt{ar} = \frac{2\pi r}{T}$$

$$T = \frac{2\pi r}{\sqrt{ar}}$$

Since there is an "r" in both the numerator and the radicand of the denominator, we can write the solution in a simpler form. Here are the steps:

(1) We factor the radical in the denominator.

$$T = \frac{2\pi r}{\sqrt{a}\sqrt{r}}$$

(2) Then rationalize the $\sqrt{r}$ in the denominator so that we can reduce to lowest terms.

$$T = \frac{2\pi r \cdot \sqrt{r}}{\sqrt{a} \cdot \sqrt{r} \cdot \sqrt{r}}$$

$$T = \frac{2\pi r\sqrt{r}}{\sqrt{a} \cdot r}$$

$$T = \frac{2\pi\sqrt{r}}{\sqrt{a}}$$

(Continued on following page.)

---

a) $T = \dfrac{1}{m}\sqrt{\dfrac{a}{b}}$

b) $r = \dfrac{1}{ab}\sqrt{\dfrac{t}{m}}$

c) $S = \dfrac{1}{mt}\sqrt{\dfrac{2g}{a}}$

140   Equations and Formulas Containing Squares and Square Root Radicals

112. (Continued)

(3) Then write the two radicals as a single radical.

$$T = 2\pi \sqrt{\dfrac{r}{a}}$$

Using the steps above, write each of these solutions in the preferred form:

(a) $R = \dfrac{cd}{\sqrt{ad}}$ (b) $m = \dfrac{2gt}{\sqrt{ty}}$

---

a) $R = c\sqrt{\dfrac{d}{a}}$

b) $m = 2g\sqrt{\dfrac{t}{y}}$

---

113. Here is a summary of what is generally done with fractional solutions which contain a radical only in the <u>numerator</u>.

(1) If the radicand is a <u>non-fraction</u>, we leave the solution in that form. For example:

$$r = \dfrac{\sqrt{ab}}{c}$$ is left in that form.

(2) If the radicand is a <u>fraction</u>, we convert the division to a multiplication. For example:

$$m = \dfrac{\sqrt{\dfrac{c}{d}}}{t}$$ is written: $m = \dfrac{1}{t}\sqrt{\dfrac{c}{d}}$

Write each of these solutions in the preferred form:

(a) $Q = \dfrac{\sqrt{PR}}{ST}$ (b) $F = \dfrac{\sqrt{\dfrac{2g}{a}}}{t}$ (c) $a = \dfrac{\sqrt{\dfrac{m}{t}}}{2g}$

Q = _____   F = _____   a = _____

---

a) $Q = \dfrac{\sqrt{PR}}{ST}$

b) $F = \dfrac{1}{t}\sqrt{\dfrac{2g}{a}}$

c) $a = \dfrac{1}{2g}\sqrt{\dfrac{m}{t}}$

114. Here is a summary of what is generally done with fractional solutions which contain a radical only in the denominator.

(1) If the radicand is a <u>fraction</u>, we invert the fraction and move the radical to the numerator. For example:

$$T = \frac{2\pi r}{\sqrt{\frac{Fr}{m}}} \text{ is written } T = 2\pi r \sqrt{\frac{m}{Fr}}$$

(2) If the radicand is a <u>non-fraction</u>:

(a) We partially rationalize the denominator if there is a common letter in both the numerator and the radicand of the denominator. For example, we rationalize the "b" below:

$$T = \frac{abd}{\sqrt{bt}} \text{ is written } ad\sqrt{\frac{b}{t}}$$

(b) We generally rationalize other denominators only if the new form makes evaluation easier. For example:

$$R = \frac{a}{\sqrt{c}} \text{ is left in that form.}$$

$$m = \frac{b}{\sqrt{b}} \text{ is written } m = \sqrt{b}$$

Write each of the following solutions in the preferred form:

(a) $P = \dfrac{cd}{\sqrt{dt}}$ (b) $m = \dfrac{Fr}{\sqrt{2p}}$ (c) $t = \dfrac{2g}{\sqrt{\dfrac{a}{H}}}$

P = _____   m = _____   t = _____

a) $P = c\sqrt{\dfrac{d}{t}}$

b) $m = \dfrac{Fr}{\sqrt{2p}}$

c) $t = 2g\sqrt{\dfrac{H}{a}}$

115. If the solution contains a radical in both its numerator and its denominator, we combine the two radicals into one. For example:

$$\frac{a\sqrt{b}}{\sqrt{c}} \text{ is written: } a\sqrt{\frac{b}{c}}$$

(a) $\dfrac{2g\sqrt{h}}{\sqrt{wr}}$ is written: _____

(b) $\dfrac{V\sqrt{aR}}{\sqrt{2P}}$ is written: _____

a) $2g\sqrt{\dfrac{h}{wr}}$

b) $V\sqrt{\dfrac{aR}{2P}}$

SELF-TEST 6 (Frames 90-115)

1. Solve for w:

    $t = \dfrac{h\sqrt{w}}{2}$

    w = _____

2. Solve for b:

    $P = \dfrac{EN}{\sqrt{ab}}$

    b = _____

3. Solve for G:

    $H = \dfrac{\sqrt{A - G}}{B}$

    G = _____

In Problems 4 to 6, put each expression in the preferred form:

4. $\dfrac{\sqrt{\dfrac{E}{e}}}{2r}$

5. $\dfrac{p}{d\sqrt{\dfrac{s}{t}}}$

6. $\dfrac{tw}{\sqrt{pt}}$

7. Solve for K:

    $H = P - K\sqrt{d}$

8. Solve for v_2:

    $v_1 = 2v_2 \sqrt{\dfrac{M}{m}}$

9. Solve for F:

    $\dfrac{Ah}{F} = a\sqrt{ht}$

ANSWERS:

1. $w = \left(\dfrac{2t}{h}\right)^2$ or $\dfrac{4t^2}{h^2}$

2. $b = \dfrac{1}{a}\left(\dfrac{EN}{P}\right)^2$ or $\dfrac{E^2 N^2}{aP^2}$

3. $G = A - (BH)^2$ or $A - B^2 H^2$

4. $\dfrac{1}{2r}\sqrt{\dfrac{E}{e}}$

5. $\dfrac{p}{d}\sqrt{\dfrac{t}{s}}$

6. $w\sqrt{\dfrac{t}{p}}$

7. $K = \dfrac{P - H}{\sqrt{d}}$

8. $v_2 = \dfrac{v_1}{2}\sqrt{\dfrac{m}{M}}$

9. $F = \dfrac{A}{a}\sqrt{\dfrac{h}{t}}$

2-7 AN ALTERNATE METHOD FOR SOLVING SOME RADICAL EQUATIONS AND FORMULAS

The method of solving radical equations and formulas which we have used up to this point is a general method which works in all cases. In this general method, the radical is isolated before the squaring principle is applied. There is an alternate method which can be used to solve some radical equations and formulas. In this alternate method, the radical is not isolated before the squaring principle is applied. We will discuss the alternate method in this section.

116. To square a radical term, we simply square each factor in the term. For example:

$$(3\sqrt{x})^2 = 3^2 \cdot (\sqrt{x})^2 = 9x$$
$$(2a\sqrt{b})^2 = 2^2 \cdot a^2 \cdot (\sqrt{b})^2 = 4a^2b$$
$$(at\sqrt{2m})^2 = a^2 \cdot t^2 \cdot (\sqrt{2m})^2 = a^2t^2(2m) = 2a^2mt^2$$

Notice in each case that the radical was eliminated in the squaring process.

What non-radical term do we obtain by squaring each of the following?

(a) $(5\sqrt{y})^2 = $ _____
(b) $(a\sqrt{x})^2 = $ _____
(c) $(4c\sqrt{t})^2 = $ _____
(d) $(p\sqrt{3q})^2 = $ _____

117. Since the radical in a radical term is eliminated by the squaring process, we can eliminate the radical in the equation below by applying the squaring principle to the equation as it stands.

$$2\sqrt{x} = 5$$
$$(2\sqrt{x})^2 = 5^2$$
$$4x = 25$$

Apply the squaring principle to each equation below as it stands:

(a) $3\sqrt{m} = 4$ (b) $a\sqrt{x} = b$ (c) $4\sqrt{ar} = t$

_____ = _____   _____ = _____   _____ = _____

a) 25y   c) $16c^2t$
b) a^2x d) $3p^2q$

118. To square a fraction which contains a radical, we simply square both its numerator and its denominator. For example:

$$\left(\frac{\sqrt{x}}{3}\right)^2 = \frac{(\sqrt{x})^2}{3^2} = \frac{x}{9}$$

$$\left(\frac{P}{\sqrt{R}}\right)^2 = \frac{P^2}{(\sqrt{R})^2} = \frac{P^2}{R}$$

$$\left(\frac{a\sqrt{y}}{b}\right)^2 = \frac{a^2 \cdot (\sqrt{y})^2}{b^2} = \frac{a^2y}{b^2}$$

Notice again that the radical was eliminated in the squaring process in each case.

(Continued on following page.)

a) $9m = 16$
b) $a^2x = b^2$
c) $16ar = t^2$

144   Equations and Formulas Containing Squares and Square Root Radicals

118. (Continued)

What non-radical expression do we obtain by squaring each of the following?

(a) $\left(\dfrac{\sqrt{r}}{t}\right)^2 =$ _____

(b) $\left(\dfrac{4}{\sqrt{m}}\right)^2 =$ _____

(c) $\left(\dfrac{3\sqrt{p}}{5}\right)^2 =$ _____

(d) $\left(\dfrac{R}{P\sqrt{Q}}\right)^2 =$ _____

a) $\dfrac{r}{t^2}$ c) $\dfrac{9p}{25}$

b) $\dfrac{16}{m}$ d) $\dfrac{R^2}{P^2 Q}$

119. In the equation below, the radical can be eliminated by applying the squaring principle to the equation as it stands.

$$\dfrac{\sqrt{t}}{3} = 2$$

$$\left(\dfrac{\sqrt{t}}{3}\right)^2 = 2^2$$

$$\dfrac{t}{9} = 4$$

Apply the squaring principle to each equation below as it stands:

(a) $\dfrac{d}{\sqrt{r}} = m$

(b) $\dfrac{2\sqrt{y}}{5} = 1$

(c) $\dfrac{a}{b\sqrt{c}} = x$

_____ = _____   _____ = _____   _____ = _____

120. When applying the squaring principle to an equation as it stands, <u>be sure to square every number and letter which is not part of the radicand.</u>

Apply the squaring principle to each equation below as it stands:

(a) $t = \dfrac{r}{\sqrt{a}}$

(b) $a = \dfrac{\sqrt{bc}}{d}$

(c) $f = \dfrac{1}{2\pi\sqrt{LC}}$

a) $\dfrac{d^2}{r} = m^2$

b) $\dfrac{4y}{25} = 1$

c) $\dfrac{a^2}{b^2 c} = x^2$

_____   _____   _____

121. Apply the squaring principle to each equation below as it stands:

(a) $s = \dfrac{\sqrt{h + 2r}}{d}$

(b) $w = \dfrac{\sqrt{d}}{2}$

(c) $a = \dfrac{1}{\sqrt{p^2 + w^2}}$

a) $t^2 = \dfrac{r^2}{a}$

b) $a^2 = \dfrac{bc}{d^2}$

c) $f^2 = \dfrac{1}{4\pi^2 LC}$

_____   _____   _____

Equations and Formulas Containing Squares and Square Root Radicals 145

122. Each of the terms below contains a square root of a fraction as one of its factors. To square the terms, we simply square each factor. That is:

$$\left(a\sqrt{\frac{b}{c}}\right)^2 = a^2 \cdot \left(\sqrt{\frac{b}{c}}\right)^2 = a^2\left(\frac{b}{c}\right) = \frac{a^2 b}{c}$$

$$\left(2\pi\sqrt{\frac{r}{a}}\right)^2 = 2^2 \cdot \pi^2 \cdot \left(\sqrt{\frac{r}{a}}\right)^2 = 4\pi^2\left(\frac{r}{a}\right) = \frac{4\pi^2 r}{a}$$

Notice again that the radical is eliminated by the squaring process in each case.

What non-radical expression do we get by squaring each of the following terms?

(a) $\left(4\sqrt{\frac{x}{3}}\right)^2 = $ _____

(b) $\left(t\sqrt{\frac{p}{q}}\right)^2 = $ _____

(c) $\left(2g\sqrt{\frac{a}{r}}\right)^2 = $ _____

(d) $\left(3\sqrt{\frac{5}{y}}\right)^2 = $ _____

a) $s^2 = \dfrac{h + 2r}{d^2}$

b) $w^2 = \dfrac{d}{4}$

c) $a^2 = \dfrac{1}{p^2 + w^2}$

123. The radical in the equation below can be eliminated by applying the squaring principle to the equation as it stands. That is:

$$t = 2\pi\sqrt{\frac{r}{a}}$$

$$t^2 = \left(2\pi\sqrt{\frac{r}{a}}\right)^2$$

$$t^2 = \frac{4\pi^2 r}{a}$$

Apply the squaring principle to each equation below as it stands:

(a) $r_o = r_p\sqrt{\dfrac{g_p}{g_o}}$

(b) $w = 2g\sqrt{\dfrac{a}{r}}$

a) $\dfrac{16x}{3}$ c) $\dfrac{4ag^2}{r}$

b) $\dfrac{pt^2}{q}$ d) $\dfrac{45}{y}$

a) $r_o^2 = \dfrac{r_p^2 g_p}{g_o}$

b) $w^2 = \dfrac{4ag^2}{r}$

146   Equations and Formulas Containing Squares and Square Root Radicals

124. The radical in the equation below can be eliminated by applying the squaring principle to the equation as it stands.

$$\frac{2\sqrt{x}}{3} = 5$$

Therefore, we can solve the equation by either of two methods:

(1) Squaring both sides of the equation as it stands.

$$\left(\frac{2\sqrt{x}}{3}\right)^2 = 5^2$$

$$\frac{4x}{9} = 25$$

$$4x = 225$$

$$x = \frac{225}{4}$$

(2) Isolating the radical first and then squaring both sides.

$$\frac{2\sqrt{x}}{3} = 5$$

$$2\sqrt{x} = 15$$

$$\sqrt{x} = \frac{15}{2}$$

$$x = \left(\frac{15}{2}\right)^2 \text{ or } \frac{225}{4}$$

Using either method, the root of the original radical equation is _____.

---

125. Solve each of the following by applying the squaring principle to the equation as it stands:

(a) $\dfrac{\sqrt{y}}{2} = 5$ (b) $\dfrac{1}{\sqrt{x}} = 6$ (c) $\dfrac{\sqrt{t-1}}{3} = 2$

y = _____   x = _____   t = _____

$\dfrac{225}{4}$

---

126. Solve each of these by applying the squaring principle to the equation as it stands:

(a) $\dfrac{2}{\sqrt{x+3}} = 1$ (b) $\dfrac{4\sqrt{d}}{3} = 1$ (c) $1 = \dfrac{5}{3\sqrt{b}}$

x = _____   d = _____   b = _____

a) y = 100, from:

$\dfrac{y}{4} = 25$

b) $x = \dfrac{1}{36}$, from:

$\dfrac{1}{x} = 36$

c) t = 37, from:

$\dfrac{t-1}{9} = 4$

127. The radical in the formula below can be eliminated by applying the squaring principle to the formula as it stands.

$$t = \frac{r}{\sqrt{a}}$$

Therefore, we can solve for "a" in either of two ways:

(1) Squaring both sides as they stand.

$$t^2 = \left(\frac{r}{\sqrt{a}}\right)^2$$

$$t^2 = \frac{r^2}{a}$$

$$at^2 = r^2$$

$$a = \frac{r^2}{t^2}$$

(2) Isolating the radical first and then squaring both sides.

$$t = \frac{r}{\sqrt{a}}$$

$$t\sqrt{a} = r$$

$$\sqrt{a} = \frac{r}{t}$$

$$a = \left(\frac{r}{t}\right)^2 \text{ or } \frac{r^2}{t^2}$$

Did we obtain the same solution in both methods? _____

a) $x = 1$, from:

$$\frac{4}{x+3} = 1$$

b) $d = \frac{9}{16}$, from:

$$\frac{16d}{9} = 1$$

c) $b = \frac{25}{9}$, from:

$$1 = \frac{25}{9b}$$

---

128. We have solved for "L" in the formula below by isolating the radical first. In the space on the right, solve for "L" by squaring both sides of the formula as it stands.

$$f = \frac{1}{2\pi\sqrt{LC}}$$

$$2\pi f\sqrt{LC} = 1$$

$$\sqrt{LC} = \frac{1}{2\pi f}$$

$$LC = \frac{1}{4\pi^2 f^2}$$

$$L = \frac{1}{4\pi^2 C f^2}$$

$$f = \frac{1}{2\pi\sqrt{LC}}$$

L = _____

Yes

---

129. Solve for the indicated letter in each case by squaring both sides of the formula as it stands:

(a) Solve for "L_2":

$$K = \frac{M}{\sqrt{L_1 L_2}}$$

(b) Solve for "r":

$$s = \frac{\sqrt{h + 2r}}{d}$$

$L_2 =$ _____

r = _____

$$f^2 = \frac{1}{4\pi^2 LC}$$

$$4\pi^2 f^2 LC = 1$$

$$L = \frac{1}{4\pi^2 C f^2}$$

148   Equations and Formulas Containing Squares and Square Root Radicals

130. Solve for the indicated letter in each case by squaring both sides of the formula as it stands:

(a) Solve for "a":

$$w = 2g\sqrt{\frac{a}{r}}$$

(b) Solve for "g_o":

$$r_o = r_p\sqrt{\frac{g_p}{g_o}}$$

a) $L_2 = \dfrac{M^2}{K^2 L_1}$, from:

$$K^2 = \frac{M^2}{L_1 L_2}$$

b) $r = \dfrac{s^2 d^2 - h}{2}$, from:

$$s^2 = \frac{h + 2r}{d^2}$$

a = _____ g_o = _____

Answer to Frame 130:   a) $a = \dfrac{rw^2}{4g^2}$, from: $w^2 = \dfrac{4g^2 a}{r}$ b) $g_o = \dfrac{r_p^2 g_p}{r_o^2}$, from: $r_o^2 = \dfrac{r_p^2 g_p}{g_o}$

The general method for solving radical equations and formulas requires "isolating the radical" as the first step. The alternate method discussed in this section avoids that step by squaring both sides of the equation or formula as it stands. However, since it is much easier to make mistakes in squaring when using the alternate method, it is usually preferable to use the general method.

SELF-TEST 7 (Frames 116–130)

1. $\left(\dfrac{h\sqrt{d}}{p}\right)^2 =$ _____

2. $\left(2t\sqrt{\dfrac{s}{r}}\right)^2 =$ _____

3. $\left(\dfrac{w}{d\sqrt{a^2+b^2}}\right)^2 =$ _____

In Problems 4 to 9, apply the squaring principle to each equation as it stands, and then solve for the indicated letter.

4. $\dfrac{3\sqrt{r}}{2} = 1$

   r = _____

5. $\dfrac{2}{\sqrt{2h}} = 5$

   h = _____

6. $\dfrac{3}{2\sqrt{2x-1}} = \dfrac{1}{3}$

   x = _____

7. $r = \dfrac{m}{\sqrt{pt}}$

   t = _____

8. $s = \dfrac{b\sqrt{a-d}}{2}$

   d = _____

9. $v = a\sqrt{\dfrac{T}{t}}$

   t = _____

ANSWERS:

1. $\dfrac{dh^2}{p^2}$

2. $\dfrac{4st^2}{r}$

3. $\dfrac{w^2}{d^2(a^2+b^2)}$

4. $r = \dfrac{4}{9}$

5. $h = \dfrac{2}{25}$

6. $x = \dfrac{85}{8}$

7. $t = \dfrac{m^2}{pr^2}$

8. $d = \dfrac{ab^2 - 4s^2}{b^2}$

9. $t = \dfrac{a^2 T}{v^2}$

150   Equations and Formulas Containing Squares and Square Root Radicals

2-8   SQUARING BINOMIALS THAT CONTAIN A SQUARE ROOT RADICAL

In this section, we will extend the procedure for squaring binomials to binomials that contain a radical in one term. The ability to square binomials of this type is needed in order to see the limitations of the alternate method for solving radical equations. The limitations of that method will be discussed in the next section.

131. In an earlier section, we saw that the distributive principle is used to square a binomial. For example:

$$(c + 5)^2 = (c + 5)(c + 5) = c(c + 5) + 5(c + 5)$$
$$= c^2 + 5c + 5c + 25$$
$$= c^2 + 10c + 25$$

We also use the distributive principle to square a binomial in which one term contains a radical. For example:

$$\left(3 + \sqrt{x}\right)^2 = \left(3 + \sqrt{x}\right)\left(3 + \sqrt{x}\right) = 3\left(3 + \sqrt{x}\right) + \sqrt{x}\left(3 + \sqrt{x}\right)$$
$$= 9 + 3\sqrt{x} + 3\sqrt{x} + \left(\sqrt{x}\right)^2$$
$$= 9 + 6\sqrt{x} + x$$

Following the steps above, complete this one:

$$\left(a + \sqrt{y}\right)^2 = \left(a + \sqrt{y}\right)\left(a + \sqrt{y}\right) = \underline{\hspace{3cm}}$$
$$= \underline{\hspace{3cm}}$$
$$= \underline{\hspace{3cm}}$$

132. To square a binomial which contains a subtraction, we convert the subtractions to additions first. For example:

$$(x - 3)^2 = (x - 3)(x - 3) = [x + (-3)][x + (-3)]$$
$$= x[x + (-3)] + (-3)[x + (-3)]$$
$$= x^2 + (-3x) + (-3x) + 9$$
$$= x^2 + (-6x) + 9$$
$$= x^2 - 6x + 9$$

This same procedure is used when one of the terms contains a radical. Here is an example:

$$\left(5 - \sqrt{t}\right)^2 = \left(5 - \sqrt{t}\right)\left(5 - \sqrt{t}\right) = \left[5 + (-\sqrt{t})\right]\left[5 + (-\sqrt{t})\right]$$
$$= 5\left[5 + (-\sqrt{t})\right] + (-\sqrt{t})\left[5 + (-\sqrt{t})\right]$$
$$= 25 + (-5\sqrt{t}) + (-5\sqrt{t}) + (-\sqrt{t})^2$$
$$= 25 + (-10\sqrt{t}) + t$$
$$= 25 - 10\sqrt{t} + t$$

When squaring $5 - \sqrt{t}$, notice that we encounter $\left(-\sqrt{t}\right)^2$ in the third step. The square of $-\sqrt{t}$ is $+1t$ or t. It is easy to see this fact if we write the "-1" coefficient explicitly. That is:

$$\left(-\sqrt{t}\right)^2 = \left(-1\sqrt{t}\right)^2 = (-1)^2 \cdot \left(\sqrt{t}\right)^2 = +1t \text{ or } t$$

Similarly, the square of any radical term with a negative coefficient is a positive quantity. For example:

$$\left(-2\sqrt{x}\right)^2 = (-2)^2\left(\sqrt{x}\right)^2 = +4x \text{ or } 4x$$
$$\left(-a\sqrt{b}\right)^2 = (-a)^2\left(\sqrt{b}\right)^2 = a^2 b$$

(Continued on following page.)

$= a\left(a + \sqrt{y}\right) + \sqrt{y}\left(a + \sqrt{y}\right)$
$= a^2 + a\sqrt{y} + a\sqrt{y} + \left(\sqrt{y}\right)^2$
$= a^2 + 2a\sqrt{y} + y$

132. (Continued)

Write the squares of each of the following radical terms:

(a) $\left(-5\sqrt{y}\right)^2 =$ _____ (b) $\left(-c\sqrt{d}\right)^2 =$ _____ (c) $\left(-2R\sqrt{P}\right)^2 =$ _____

---

133. Following the steps in the last frame, complete this one:

$\left(a - \sqrt{x}\right)^2 = \left(a - \sqrt{x}\right)\left(a - \sqrt{x}\right) = \left[a + \left(-\sqrt{x}\right)\right]\left[a + \left(-\sqrt{x}\right)\right]$

= _____

= _____

= _____

= _____

a) $25y$

b) $c^2 d$

c) $4R^2 P$

---

134. Here are the results we obtained in the last few frames. The square in each case is a trinomial:

$\left(3 + \sqrt{x}\right)^2 = 9 + 6\sqrt{x} + x$

$\left(a + \sqrt{y}\right)^2 = a^2 + 2a\sqrt{y} + y$

$\left(5 - \sqrt{t}\right)^2 = 25 - 10\sqrt{t} + t$

$\left(a - \sqrt{x}\right)^2 = a^2 - 2a\sqrt{x} + x$

By examining these examples, we can see that the same shortcut can be used to square a binomial when one of the terms is a radical. That is:

(1) To obtain the <u>first</u> **term** of the trinomial, we square the first term of the binomial.

(2) To obtain the <u>last</u> **term** of the trinomial, we square the second term of the binomial.

(3) To obtain the <u>middle</u> **term**, we double the product of the first and last terms of the binomial.

Let's use the shortcut to square $7 + \sqrt{m}$.

(a) The square of 7 is _____.

(b) Double the product of 7 and $\sqrt{m}$ is _____.

(c) The square of $\sqrt{m}$ is _____.

(d) Therefore: $\left(7 + \sqrt{m}\right)^2 =$ _____

$= a\left[a + \left(-\sqrt{x}\right)\right]$
$\quad + \left(-\sqrt{x}\right)\left[a + \left(-\sqrt{x}\right)\right]$

$= a^2 + \left(-a\sqrt{x}\right) + \left(-a\sqrt{x}\right)$
$\quad + \left(-\sqrt{x}\right)^2$

$= a^2 + \left(-2a\sqrt{x}\right) + x$

$= a^2 - 2a\sqrt{x} + x$

---

a) 49

b) $14\sqrt{m}$

c) m

d) $49 + 14\sqrt{m} + m$

152 Equations and Formulas Containing Squares and Square Root Radicals

135. Let's use the shortcut to square $8 - \sqrt{b}$. Remember that the second term of the trinomial product is subtracted when the binomial is a subtraction.

    (a) The square of 8 is _____.

    (b) Double the product of 8 and $\sqrt{b}$ is _____.

    (c) The square of $\sqrt{b}$ is _____.

    (d) $\left(8 - \sqrt{b}\right)^2 = $ _____

---

136. Below we have squared $5 + 2\sqrt{t}$ by means of the distributive principle. Notice that its square is also a trinomial.

$$\left(5 + 2\sqrt{t}\right)^2 = \left(5 + 2\sqrt{t}\right)\left(5 + 2\sqrt{t}\right) = 5\left(5 + 2\sqrt{t}\right) + 2\sqrt{t}\left(5 + 2\sqrt{t}\right)$$
$$= 25 + 10\sqrt{t} + 10\sqrt{t} + \left(2\sqrt{t}\right)^2$$
$$= 25 + 20\sqrt{t} + 4t$$

Let's use the shortcut to square $5 + 2\sqrt{t}$ to see if we get the same result.

    (a) The square of 5 is _____.

    (b) Double the product of 5 and $2\sqrt{t}$ is _____.

    (c) The square of $2\sqrt{t}$ is _____.

    (d) $\left(5 + 2\sqrt{t}\right)^2 = $ _____

    (e) Is this the same square we obtained above? _____

a) 64

b) $16\sqrt{b}$

c) b

d) $64 - 16\sqrt{b} + b$

Note: Did you forget to subtract the second term?

---

137. We have squared $7 - 3\sqrt{y}$ by means of the distributive principle below.

$$\left(7 - 3\sqrt{y}\right)^2 = \left(7 - 3\sqrt{y}\right)\left(7 - 3\sqrt{y}\right) = \left[7 + \left(-3\sqrt{y}\right)\right]\left[7 + \left(-3\sqrt{y}\right)\right]$$
$$= 7\left[7 + \left(-3\sqrt{y}\right)\right] + \left(-3\sqrt{y}\right)\left[7 + \left(-3\sqrt{y}\right)\right]$$
$$= 49 + \left(-21\sqrt{y}\right) + \left(-21\sqrt{y}\right) + \left(-3\sqrt{y}\right)^2$$
$$= 49 + \left(-42\sqrt{y}\right) + 9y$$
$$= 49 - 42\sqrt{y} + 9y$$

Let's use the shortcut to square $7 - 3\sqrt{y}$ to see if we get the same result.

    (a) The square of 7 is _____.

    (b) Double the product of 7 and $3\sqrt{y}$ is _____.

    (c) The square of $3\sqrt{y}$ is _____.

    (d) $\left(7 - 3\sqrt{y}\right)^2 = $ _____

    (e) Is this the same square we obtained above? _____

a) 25

b) $20\sqrt{t}$

c) 4t

d) $25 + 20\sqrt{t} + 4t$

e) Yes

---

a) 49

b) $42\sqrt{y}$

c) 9y

d) $49 - 42\sqrt{y} + 9y$

e) Yes

138. Let's use the shortcut to square $m + r\sqrt{y}$.

   (a) The square of "m" is _____.

   (b) Double the product of "m" and "$r\sqrt{y}$" is _____.

   (c) The square of "$r\sqrt{y}$" is _____.

   (d) $(m + r\sqrt{y})^2 = $ _____

---

139. Let's use the shortcut to square $d - p\sqrt{t}$.

   (a) The first term of the trinomial is _____.

   (b) The middle term of the trinomial is _____.

   (c) The last term of the trinomial is _____.

   (d) $(d - p\sqrt{t})^2 = $ _____

a) m^2
b) $2mr\sqrt{y}$
c) $r^2 y$
d) $m^2 + 2mr\sqrt{y} + r^2 y$

---

140. Here is an example of squaring a binomial in which the radicand is also a binomial.

$$(2 + \sqrt{r+1})^2 = (2 + \sqrt{r+1})(2 + \sqrt{r+1})$$
$$= 2(2 + \sqrt{r+1}) + \sqrt{r+1}(2 + \sqrt{r+1})$$
$$= 4 + 2\sqrt{r+1} + 2\sqrt{r+1} + (\sqrt{r+1})^2$$
$$= 4 + 4\sqrt{r+1} + r + 1$$
$$= 5 + 4\sqrt{r+1} + r$$

Notice this point: When $(\sqrt{r+1})^2$ is replaced by $r+1$, we obtain a four-term expression. The "like" terms must then be combined to obtain the final trinomial.

Following the steps above, complete this one. Combine like terms to write the square as a trinomial.

$$(3 + \sqrt{t-2})^2 = (3 + \sqrt{t-2})(3 + \sqrt{t-2})$$
= _____
= _____
= _____
= _____

a) d^2
b) $2dp\sqrt{t}$
c) $p^2 t$
d) $d^2 - 2dp\sqrt{t} + p^2 t$

$= 3(3+\sqrt{t-2}) + \sqrt{t-2}(3+\sqrt{t-2})$
$= 9 + 3\sqrt{t-2} + 3\sqrt{t-2} + (\sqrt{t-2})^2$
$= 9 + 6\sqrt{t-2} + t - 2$
$= 7 + 6\sqrt{t-2} + t$

154   Equations and Formulas Containing Squares and Square Root Radicals

141. In the last frame, we squared the binomial below and obtained the trinomial shown.
$$\left(3 + \sqrt{t-2}\right)^2 = 7 + 6\sqrt{t-2} + t$$

We can square the same binomial by the shortcut method. However, when we do so, we do not obtain the trinomial above directly. We obtain a four-term expression which must be simplified to obtain the trinomial. Here is the process:

Applying the shortcut method to $\left(3 + \sqrt{t-2}\right)^2$, we get:
$$3^2 + 2(3)\sqrt{t-2} + \left(\sqrt{t-2}\right)^2$$
or $\quad 9 + 6\sqrt{t-2} + t - 2$

The last expression which contains four terms simplifies to:
$$7 + 6\sqrt{t-2} + t$$

Let's square $5 + \sqrt{y+4}$ by the shortcut method.

(a) The square of 5 is _____.

(b) Double the product of 5 and $\sqrt{y+4}$ is _____.

(c) The square of $\sqrt{y+4}$ is _____.

(d) The four-term expression is _____.

(e) The trinomial is _____.

---

142. Using the shortcut method, square each of the following:

(a) $\left(\sqrt{t} + 4\right)^2 =$ _____   (c) $\left(2\sqrt{y} + 1\right)^2 =$ _____

(b) $\left(\sqrt{x} - 3\right)^2 =$ _____   (d) $\left(3\sqrt{m} - 2\right)^2 =$ _____

a) 25
b) $10\sqrt{y+4}$
c) $y + 4$
d) $25 + 10\sqrt{y+4} + y + 4$
e) $29 + 10\sqrt{y+4} + y$

---

143. Using the shortcut method, square each of these:

(a) $\left(R + \sqrt{P}\right)^2 =$ _____   (c) $\left(F + G\sqrt{H}\right)^2 =$ _____

(b) $\left(y - \sqrt{x}\right)^2 =$ _____   (d) $\left(a - b\sqrt{c}\right)^2 =$ _____

a) $t + 8\sqrt{t} + 16$
b) $x - 6\sqrt{x} + 9$
c) $4y + 4\sqrt{y} + 1$
d) $9m - 12\sqrt{m} + 4$

---

144. Using the shortcut method, square each of these:

(a) $\left(7 + \sqrt{R+10}\right)^2 =$ _____
    = _____

(b) $\left(9 + \sqrt{y-50}\right)^2 =$ _____
    = _____

a) $R^2 + 2R\sqrt{P} + P$
b) $y^2 - 2y\sqrt{x} + x$
c) $F^2 + 2FG\sqrt{H} + G^2 H$
d) $a^2 - 2ab\sqrt{c} + b^2 c$

Equations and Formulas Containing Squares and Square Root Radicals 155

145. The numerator of the fraction below is a binomial which contains a radical. To square the fraction, we simply square both its numerator and its denominator. The shortcut process can be used to square the binomial in the numerator. That is:

$$\left(\frac{3+\sqrt{y}}{2}\right)^2 = \frac{(3+\sqrt{y})^2}{2^2} = \frac{9+6\sqrt{y}+y}{4}$$

Square each of the following fractions:

(a) $\left(\dfrac{5}{4+\sqrt{t}}\right)^2 =$ _____

(b) $\left(\dfrac{1}{\sqrt{m}+2}\right)^2 =$ _____

a) $= 49 + 14\sqrt{R+10} + R + 10$
$= 59 + 14\sqrt{R+10} + R$

b) $= 81 + 18\sqrt{y-50} + y - 50$
$= 31 + 18\sqrt{y-50} + y$

146. Square each fraction:

(a) $\left(\dfrac{\sqrt{p}-1}{3}\right)^2 =$ _____

(b) $\left(\dfrac{1+2\sqrt{h}}{6}\right)^2 =$ _____

a) $\dfrac{25}{16 + 8\sqrt{t} + t}$

b) $\dfrac{1}{m + 4\sqrt{m} + 4}$

147. Square each fraction:

(a) $\left(\dfrac{m+\sqrt{y}}{t}\right)^2 =$ _____

(b) $\left(\dfrac{a}{b+c\sqrt{d}}\right)^2 =$ _____

a) $\dfrac{p - 2\sqrt{p} + 1}{9}$

b) $\dfrac{1 + 4\sqrt{h} + 4h}{36}$

148. Square the fraction below and write the denominator as a trinomial:

$$\left(\dfrac{1}{2+\sqrt{3V-2}}\right)^2 =$$ _____

$=$ _____

a) $\dfrac{m^2 + 2m\sqrt{y} + y}{t^2}$

b) $\dfrac{a^2}{b^2 + 2bc\sqrt{d} + c^2 d}$

$= \dfrac{1}{4 + 4\sqrt{3V-2} + 3V - 2}$

$= \dfrac{1}{2 + 4\sqrt{3V-2} + 3V}$

156 Equations and Formulas Containing Squares and Square Root Radicals

---

SELF-TEST 8 (Frames 131-148)

Do these squarings, preferably by the shortcut method.

1. $(3 + \sqrt{x})^2 = $ _____
2. $(5 - 2\sqrt{R})^2 = $ _____
3. $(2\sqrt{d} + 7)^2 = $ _____
4. $(\sqrt{t} - s)^2 = $ _____
5. $(h + p\sqrt{a})^2 = $ _____
6. $(c\sqrt{r} - p)^2 = $ _____

7. $(4 + \sqrt{x + 3})^2 = $ _____
8. $(7 - \sqrt{3w - 2})^2 = $ _____

9. $\left(\dfrac{a - b\sqrt{c}}{h}\right)^2 = $ _____

10. $\left(\dfrac{2}{3 + \sqrt{5x - 3}}\right)^2 = $ _____

ANSWERS:

1. $9 + 6\sqrt{x} + x$
2. $25 - 20\sqrt{R} + 4R$
3. $4d + 28\sqrt{d} + 49$
4. $t - 2s\sqrt{t} + s^2$
5. $h^2 + 2hp\sqrt{a} + ap^2$
6. $c^2r - 2cp\sqrt{r} + p^2$
7. $19 + 8\sqrt{x + 3} + x$
8. $47 - 14\sqrt{3w - 2} + 3w$
9. $\dfrac{a^2 - 2ab\sqrt{c} + b^2c}{h^2}$
10. $\dfrac{4}{6 + 6\sqrt{5x - 3} + 5x}$

2-9 LIMITATIONS OF THE ALTERNATE METHOD FOR SOLVING RADICAL EQUATIONS AND FORMULAS

In an earlier section, we discussed an alternate method of solving some radical equations and formulas. In that method, the squaring principle is applied to the equation as it stands without isolating the radical first. There are limitations to the use of this alternate method. That is, the method cannot be used to solve some radical equations and formulas. We will discuss its limitations in this section.

Equations and Formulas Containing Squares and Square Root Radicals 157

149. In order to solve for a variable in the radicand of a radical equation, we must eliminate the radical. To do so, we use the squaring principle.

   (1) If the radical is already isolated on one side, we can <u>always</u> eliminate it by applying the squaring principle directly. For example:

$$\sqrt{2x} = 5 \qquad\qquad \sqrt{R+1} = 3$$
$$(\sqrt{2x})^2 = 5^2 \qquad\qquad (\sqrt{R+1})^2 = 3^2$$
$$2x = 25 \qquad\qquad R+1 = 9$$

   (2) If the radical is not isolated on one side, we can <u>sometimes</u> eliminate it by applying the squaring principle directly. For example:

$$3\sqrt{t} = 7 \qquad\qquad \frac{2\sqrt{y}}{5} = 4$$
$$(3\sqrt{t})^2 = 7^2 \qquad\qquad \left(\frac{2\sqrt{y}}{5}\right)^2 = 4^2$$
$$9t = 49 \qquad\qquad \frac{4y}{25} = 16$$

Eliminate the radicals in these equations by applying the squaring principle directly:

(a) $\sqrt{cd} = t$ (b) $R\sqrt{S} = P$ (c) $\dfrac{a}{b\sqrt{c}} = d$

---

150. Here is an equation in which the radical is one term of the binomial on the left side.

$$3 + \sqrt{x} = 5$$

If we square both sides of the equation as it stands, we get:

$$(3 + \sqrt{x})^2 = 5^2$$
$$9 + 6\sqrt{x} + x = 25$$

Did we eliminate the radical by applying the squaring principle directly?

a) $cd = t^2$

b) $R^2 S = P^2$

c) $\dfrac{a^2}{b^2 c} = d^2$

---

151. Apply the squaring principle to each equation below directly and write the resulting equation.

(a) $\sqrt{t} - 7 = 1$ (b) $1 - 3\sqrt{y} = 2$

(c) Did we eliminate the radical in either (a) or (b)? _____

No. The middle term on the left side contains $\sqrt{x}$.

158   Equations and Formulas Containing Squares and Square Root Radicals

152. Apply the squaring principle to each formula below directly and write the resulting equation.

(a) $F = G + \sqrt{H}$

(b) $E + K\sqrt{F} = R$

(c) Did we eliminate the radical in either (a) or (b)? _____

a) $t - 14\sqrt{t} + 49 = 1$
b) $1 - 6\sqrt{y} + 9y = 4$
c) No. The middle term of each trinomial contains a radical.

153. We have applied the squaring principle directly to the equation below.

$$3 + \sqrt{x+5} = 1$$
$$\left(3 + \sqrt{x+5}\right)^2 = 1^2$$
$$9 + 6\sqrt{x+5} + x + 5 = 1$$
$$14 + 6\sqrt{x+5} + x = 1$$

Did we eliminate the radical by applying the squaring principle directly? _____

a) $F^2 = G^2 + 2G\sqrt{H} + H$
b) $E^2 + 2EK\sqrt{F} + K^2 F = R^2$
c) No

154. The numerator of the fraction on the left side of the equation below is a binomial containing a radical in one term. We have applied the squaring principle to the equation as it stands.

$$\frac{6 + \sqrt{m}}{5} = 2$$
$$\left(\frac{6 + \sqrt{m}}{5}\right)^2 = 2^2$$
$$\frac{36 + 12\sqrt{m} + m}{25} = 4$$

Did we eliminate the radical by applying the squaring principle directly? _____

No

155. Apply the squaring principle directly to each equation below.

(a) $\dfrac{1}{5 - 3\sqrt{x}} = 4$

(b) $t = \dfrac{m + p\sqrt{q}}{R}$

(c) Did we eliminate the radical by applying the squaring principle in either (a) or (b)? _____

No. There is still a radical in the middle term of the numerator.

a) $\dfrac{1}{25 - 30\sqrt{x} + 9x} = 16$

b) $t^2 = \dfrac{m^2 + 2mp\sqrt{q} + p^2 q}{R^2}$

c) No

156. If a radical is not already isolated in an equation:

   (1) We can eliminate the radical by applying the squaring principle directly if the radical is part of a radical term or part of a simple fraction. Here are some examples:

   $$2\sqrt{x} = 5 \qquad \frac{3\sqrt{x}}{7} = 1 \qquad \frac{a}{b\sqrt{c}} = d$$

   (2) We cannot eliminate the radical by applying the squaring principle directly if the radical is in one term of a binomial. Here are some examples:

   $$3 + \sqrt{x} = 7 \qquad \frac{a + b\sqrt{c}}{d} = S$$

   In which equations below can we eliminate the radical by squaring both sides as they stand? _____

   (a) $5 = \dfrac{\sqrt{x}}{3}$ \qquad (c) $7 - \sqrt{R+1} = 9$

   (b) $1 = 8\sqrt{t}$ \qquad (d) $\dfrac{4}{2 + 3\sqrt{5t}} = 1$

157. In which formulas below can we eliminate the radical by applying the squaring principle directly? _____

   (a) $h + a\sqrt{m} = d$ \qquad (c) $f = \dfrac{1}{2\pi\sqrt{LC}}$

   (b) $\dfrac{1}{R + \sqrt{Q}} = P$ \qquad (d) $w = 2g\sqrt{\dfrac{a}{r}}$

Only in (a) and (b).

158. In the alternate method for solving radical equations, the squaring principle is applied to the equation as it stands. This alternate method, however, cannot be used in instances where direct use of the squaring principle does not eliminate the radical.

   In which equations below can we solve for the variable in the radicand by using the alternate method? _____

   (a) $R = w\sqrt{t}$ \qquad (c) $S = a + b\sqrt{c}$

   (b) $\dfrac{3 + \sqrt{t}}{7} = 1$ \qquad (d) $5 = 3\sqrt{\dfrac{y}{4}}$

Only in (c) and (d).

Only in (a) and (d).

160   Equations and Formulas Containing Squares and Square Root Radicals

159. When the alternate method cannot be used, we must use the general method. That is, we must isolate the radical <u>before</u> applying the squaring principle. For example:

$$3 + \sqrt{x} = 7$$
$$\sqrt{x} = 4$$
$$\left(\sqrt{x}\right)^2 = 4^2$$
$$x = 16$$

Use the general method to solve each of these:

(a) $5 + 2\sqrt{y} = 8$ (b) $\dfrac{3 + 2\sqrt{2w}}{3} = 2$

y = _____   w = _____

---

160. Use the general method to solve each of these:

(a) $3 + \sqrt{t-1} = 7$ (b) $\dfrac{10}{1 + 3\sqrt{h}} = 5$

t = _____   h = _____

a) $y = \dfrac{9}{4}$, from:

$\sqrt{y} = \dfrac{3}{2}$

b) $w = \dfrac{9}{8}$, from:

$\sqrt{2w} = \dfrac{3}{2}$

---

161. Use the general method to solve for "x" in each of these:

(a) $y + \sqrt{x} = a$ (b) $d = R - m\sqrt{x}$

x = _____   x = _____

a) $t = 17$, from:

$\sqrt{t-1} = 4$

b) $h = \dfrac{1}{9}$, from:

$\sqrt{h} = \dfrac{5}{15}$ or $\dfrac{1}{3}$

Though the alternate method for solving radical equations and formulas works in some cases, we feel that you will be safer if you use the general method for these two reasons:

(1) Since the general method works in all cases, you don't have to decide whether it will work or not.

(2) Applying the squaring principle is usually easier if the radical is isolated.

a) $x = (a - y)^2$

b) $x = \left(\dfrac{R - d}{m}\right)^2$ or $\dfrac{(R - d)^2}{m^2}$

SELF-TEST 9 (Frames 149-161)

1. In which of the following equations can the radical be eliminated by squaring the equation as it stands? _____

   (a) $15 = \dfrac{3}{8\sqrt{5x}}$ (b) $\dfrac{2}{5 - 7\sqrt{3x}} = 1$ (c) $\dfrac{2h\sqrt{at}}{r} = T$ (d) $t = \dfrac{v - \sqrt{2as}}{d}$

2. Solve for P:

   $\dfrac{1}{3 - 2\sqrt{2P}} = 2$

   P = _____

3. Solve for F:

   $h = 2r - d\sqrt{F}$

   F = _____

4. Solve for r:

   $\dfrac{1 - \sqrt{ar}}{p} = s$

   r = _____

ANSWERS: 1. (a) and (c)   2. $P = \dfrac{25}{32}$ 3. $F = \left(\dfrac{2r - h}{d}\right)^2$ 4. $r = \dfrac{(1 - ps)^2}{a}$

2-10   SOLVING EQUATIONS CONTAINING A SQUARED LETTER

All of the equations below contain a squared letter.

$x^2 = 36$ $4y^2 = 9$ $t^2 - 16 = 0$

Equations of this type have two roots. In this section, we will briefly discuss the method of finding the roots of such equations.

162. The equation below contains a squared letter.

$$x^2 = 36$$

Intuitively, you can see that both "+6" and "−6" are roots of the equation, since:

$$(+6)(+6) = 36 \quad \text{and} \quad (-6)(-6) = 36$$

(a) The two roots of $y^2 = 100$ are _____ and _____.

(b) The two roots of $t^2 = 64$ are _____ and _____.

---

163. When an equation with a squared letter is written in this form:

$$y^2 = 16$$

its two roots are the principal and negative square roots of the number on the right. That is:

$$y = \pm\sqrt{16} = +4 \text{ and } -4$$

Similarly: (a) If $d^2 = 9$, $d = \pm\sqrt{9}$ = _____ and _____.

(b) If $m^2 = 49$, $m = \pm\sqrt{49}$ = _____ and _____.

(c) If $R^2 = 144$, $R = \pm\sqrt{144}$ = _____ and _____.

a) 10 and −10

b) 8 and −8

---

164. In the last frame, we saw this fact:

$$\boxed{\text{If } \square^2 = \triangle, \text{ then } \square = \pm\sqrt{\triangle}}$$

That is: If $x^2 = 17$, then $x = \pm\sqrt{17}$.

(a) If $b^2 = 89$, then $b = \pm\sqrt{}$.

(b) If $P^2 = 413$, then $P = \pm\sqrt{}$.

a) +3 and −3

b) +7 and −7

c) +12 and −12

---

165. The equation below is not in the form $\square^2 = \triangle$. However, we can easily write it in that form by solving for the squared letter. We get:

$$y^2 - 25 = 0$$
$$y^2 = 25$$

Therefore, the two roots of the equation are:

$$y = \pm\sqrt{25} = +5 \text{ and } -5$$

Show that these two roots satisfy the original equation:

_____   _____

a) $\pm\sqrt{89}$

b) $\pm\sqrt{413}$

$$\begin{array}{ll} y^2 - 25 = 0 & y^2 - 25 = 0 \\ (+5)^2 - 25 = 0 & (-5)^2 - 25 = 0 \\ 25 - 25 = 0 & 25 - 25 = 0 \end{array}$$

Equations and Formulas Containing Squares and Square Root Radicals 163

166. Solve for the squared letter and then write the two roots of each equation:

    (a) $x^2 - 81 = 0$ (b) $q^2 - 1 = 0$

    $x^2 = $ _____ $q^2 = $ _____

    $x = $ _____ and _____ $q = $ _____ and _____

---

167. Solve for the squared letter and then write the two roots of each equation:

    (a) $x^2 - 13 = 36$ (b) $F^2 - 1 = 8$

    $x^2 = $ _____ $F^2 = $ _____

    $x = $ _____ and _____ $F = $ _____ and _____

a) $x^2 = 81$
   $x = +9$ and -9

b) $q^2 = 1$
   $q = +1$ and -1

---

168. The fraction on the right below is a perfect square.

$$x^2 = \frac{9}{4}$$

Therefore, it is easy to find the two roots of the equation:

$$x = \pm\sqrt{\frac{9}{4}} = \frac{3}{2} \text{ and } -\frac{3}{2}$$

Write the two roots of each equation below:

    (a) $y^2 = \frac{25}{36}$ (b) $F^2 = \frac{81}{49}$

    $y = $ _____ and _____ $F = $ _____ and _____

a) $x^2 = 49$
   $x = 7$ and -7

b) $F^2 = 9$
   $F = 3$ and -3

---

169. If we solve for the squared letter in the equation below, we can easily write its two roots. That is:

$$9x^2 = 4$$

$$x^2 = \frac{4}{9}$$

$$x = \frac{2}{3} \text{ and } -\frac{2}{3}$$

Show that $\frac{2}{3}$ and $-\frac{2}{3}$ satisfy the original equation:

a) $\frac{5}{6}$ and $-\frac{5}{6}$

b) $\frac{9}{7}$ and $-\frac{9}{7}$

_____ _____

$9x^2 = 4$ $9x^2 = 4$

$9\left(\frac{2}{3}\right)^2 = 4$ $9\left(-\frac{2}{3}\right)^2 = 4$

$9\left(\frac{4}{9}\right) = 4$ $9\left(\frac{4}{9}\right) = 4$

$4 = 4$ $4 = 4$

164   Equations and Formulas Containing Squares and Square Root Radicals

170. Solve for the squared letter and then write the two roots of each equation:

   (a) $64t^2 = 9$ (b) $16x^2 - 81 = 0$ (c) $25d^2 - 1 = 0$

   $t^2 =$ _____ $x^2 =$ _____ $d^2 =$ _____

   $t^2 =$ _____ and _____ $x^2 =$ _____ and _____ $d^2 =$ _____ and _____

---

a) $t^2 = \dfrac{9}{64}$

$t = \dfrac{3}{8}$ and $-\dfrac{3}{8}$

b) $x^2 = \dfrac{81}{16}$

$x = \dfrac{9}{4}$ and $-\dfrac{9}{4}$

c) $d^2 = \dfrac{1}{25}$

$d = \dfrac{1}{5}$ and $-\dfrac{1}{5}$

---

171. If we apply the squaring principle to the radical equation below, we obtain an equation that contains a squared letter. That is:

$$\sqrt{p^2 - 4} = 0$$
$$\left(\sqrt{p^2 - 4}\right)^2 = 0^2$$
$$p^2 - 4 = 0$$

The two roots of the equation are _____ and _____.

+2 and −2

---

172. Write the two roots of each equation below:

   (a) $\sqrt{t^2 - 16} = 0$ (b) $\sqrt{y^2 - 36} = 8$

   $t =$ _____ and _____ $y =$ _____ and _____

---

a) $t = +4$ and -4
   from $t^2 = 16$

b) $y = +10$ and -10
   from $y^2 = 100$

---

173. Write the two roots of each equation below:

   (a) $\sqrt{9m^2 - 36} = 0$ (b) $\sqrt{4d^2 - 9} = 4$

   $m =$ _____ and _____ $d =$ _____ and _____

---

a) $m = 2$ and -2

b) $d = \dfrac{5}{2}$ and $-\dfrac{5}{2}$

SELF-TEST 10 (Frames 162-173)

Find the two roots of each of the following equations:

1. $9x^2 - 16 = 0$

2. $t^2 + 4 = 5$

3. $\sqrt{2h^2 - 7} = 5$

x = _____ and _____   t = _____ and _____   h = _____ and _____

ANSWERS: 1. $x = \frac{4}{3}$ and $-\frac{4}{3}$ 2. $t = 1$ and -1 3. $h = 4$ and -4

2-11 THE SQUARE ROOT PRINCIPLE AND FORMULAS CONTAINING SQUARED LETTERS

Each of the formulas below contains a squared letter.

$$P = I^2R \qquad s = \frac{1}{2}at^2 \qquad H = (1 - C^2)h$$

In this section, we will discuss the method of solving for letters which are squared in formulas. We will begin by showing how the square root principle for equations can be used to eliminate squared letters.

174. In the last section, we saw that equations like $x^2 = 49$ have two roots, one positive and one negative. If the letter stands for a scientific variable, however, the negative root does not ordinarily make sense. For example:

If $s^2 = 100$, where "s" stands for "distance", "-10" does not ordinarily make sense as a value for distance.

If $v^2 = 81$, where "v" stands for "velocity", ordinarily only one of the two roots (+9 and -9) makes sense as a value for "velocity". Which one? _____

+9

166 Equations and Formulas Containing Squares and Square Root Radicals

175. Squared letters which appear in scientific formulas stand for quantities such as length, mass, time, force, velocity, resistance, pressure, and so forth. Because such quantities are usually positive, we can use the "<u>square root principle for equations</u>" to solve for the letter which is squared. This principle applies to <u>positive</u> quantities and avoids negative square roots and negative values.

The "<u>square root principle for equations</u>" says this:

> IF TWO POSITIVE QUANTITIES ARE EQUAL, THEIR <u>PRINCIPAL</u> (<u>POSITIVE</u>) SQUARE ROOTS ARE EQUAL.
>
> In symbol form, the principle says:
>
> If □ = ○ ,
>
> then √□ = √○ .

We have used the square root principle to eliminate v^2 in the formula below. Notice that by doing so, we have solved for "v".

$$\text{If } v^2 = 2as,$$
$$\text{then } \sqrt{v^2} = \sqrt{2as}$$
$$\text{or } v = \sqrt{2as}$$

Since the square root principle avoids negative square roots, it avoids the following solution for v:

$$v = -\sqrt{2as}$$

Since "v" stands for "velocity", why do we want to avoid this negative square root solution? _____

176. Use the square root principle to complete each of these:

(a) If $E^2 = PR$,   (b) If $s^2 = A$,   (c) If $V^2 = \dfrac{Fr}{M}$,

E = _____   s = _____   V = _____

Because a negative value for "velocity" does not ordinarily make sense.

a) $E = \sqrt{PR}$

b) $s = \sqrt{A}$

c) $V = \sqrt{\dfrac{Fr}{M}}$

177. When using the square root principle to solve for a letter which is squared, <u>remember to take the square root of the entire right side of the formula</u>. For example:

If $a^2 = \dfrac{b+c}{d}$,

$a = \sqrt{\dfrac{b+c}{d}}$

If $a^2 = b^2 + c^2$,

$a = \sqrt{b^2 + c^2}$

Use the square root principle to complete each of these:

(a) If $d^2 = \dfrac{I_A - I_C}{m}$,   (b) If $c^2 = b^2 - a^2$,   (c) If $v_0^2 = v_f^2 - 2gs$,

d = _____   c = _____ v_0 = _____

---

178. To solve for "w" in: $a = w^2 r$, two steps are needed.

(1) We isolate w^2 and get: $w^2 = \dfrac{a}{r}$

(2) We apply the square root principle and get: $w = \sqrt{\dfrac{a}{r}}$

Using these two steps, solve for the indicated letter in each formula below:

(a) Solve for "I":   (b) Solve for "d":   (c) Solve for "v":

$P = I^2 R$ $A = 0.7854 d^2$ $F = \dfrac{mv^2}{r}$

I^2 = _____ d^2 = _____ v^2 = _____

I = _____   d = _____   v = _____

Answers:

a) $d = \sqrt{\dfrac{I_A - I_C}{m}}$

b) $c = \sqrt{b^2 - a^2}$

c) $v_0 = \sqrt{v_f^2 - 2gs}$

a) $I^2 = \dfrac{P}{R}$

$I = \sqrt{\dfrac{P}{R}}$

b) $d^2 = \dfrac{A}{0.7854}$

$d = \sqrt{\dfrac{A}{0.7854}}$

c) $v^2 = \dfrac{Fr}{m}$

$v = \sqrt{\dfrac{Fr}{m}}$

168 Equations and Formulas Containing Squares and Square Root Radicals

179. Solve for the indicated letter in each case:

(a) Solve for "D": $H = \dfrac{D^2 N}{2.5}$

(b) Solve for "d": $R = \dfrac{pL}{d^2}$

(c) Solve for "t": $P = \dfrac{rw}{2t^2}$

D = _____   d = _____   t = _____

a) $D = \sqrt{\dfrac{2.5H}{N}}$

b) $d = \sqrt{\dfrac{pL}{R}}$

c) $t = \sqrt{\dfrac{rw}{2P}}$

180. The formula on the left below is equivalent to the one on the right.

$s = \dfrac{1}{2}at^2 \qquad s = \dfrac{at^2}{2}$

Therefore, to clear the fraction, we simply multiply both sides by "2". Do so and then solve for "t".

t = _____

$t = \sqrt{\dfrac{2s}{a}}$

181. (a) Solve for "v": $E = \dfrac{1}{2}mv^2$

(b) Solve for "r": $\dfrac{1}{2}dr^2 = h$

(c) Solve for "d": $F = \dfrac{m_1 m_2}{rd^2}$

v = _____   r = _____   d = _____

a) $v = \sqrt{\dfrac{2E}{m}}$

b) $r = \sqrt{\dfrac{2h}{d}}$

c) $d = \sqrt{\dfrac{m_1 m_2}{Fr}}$

182. (a) Solve for "r": (b) Solve for "d": (c) Solve for "a":

$$\frac{E}{d} = \frac{2r^2}{a}$$ $$I_A = I_C + md^2$$ $$a^2 + 1 = \frac{b}{c}$$

r = _____   d = _____   a = _____

183. Though the formula below does not contain a squared letter, we obtain a squared letter by clearing the fractions.

$$\frac{V}{G} = \frac{G}{R}$$

$$RV = G^2$$

and $G = \sqrt{RV}$

Solve each of these for the indicated letter:

(a) Solve for "k":   (b) Solve for "d":

$$\frac{k}{m} = \frac{t}{k}$$ $$\frac{ab}{d} = \frac{cd}{t}$$

k = _____   d = _____

---

a) $r = \sqrt{\dfrac{aE}{2d}}$

b) $d = \sqrt{\dfrac{I_A - I_C}{m}}$

c) $a = \sqrt{\dfrac{b}{c} - 1}$

or

$a = \sqrt{\dfrac{b - c}{c}}$

---

a) $k = \sqrt{mt}$

b) $d = \sqrt{\dfrac{abt}{c}}$

184. When solving for a letter which is squared, the solution is always a square root radical. If the radical contains a perfect square, we always factor it out. For example, if we solve for "V" in the formula below, we can factor out a perfect square.

$$h = \frac{V^2}{64}$$

$$V = \sqrt{64h}$$

$$V = 8\sqrt{h}$$

Solve for "c" in: $81a = \dfrac{c^2}{b}$

c = _____

---

185. We have solved for D_1 below. Notice that the radical in the solution contains a squared letter.

$$\frac{F_1}{F_2} = \frac{(D_1)^2}{(D_2)^2}$$

$$(D_1)^2 = \frac{F_1(D_2)^2}{F_2}$$

$$D_1 = \sqrt{\frac{F_1(D_2)^2}{F_2}}$$

Since $(D_2)^2$ is a perfect square, we can factor it out of the radical. We get:

$$D_1 = \sqrt{(D_2)^2 \left(\frac{F_1}{F_2}\right)} = \sqrt{D_2^{\,2}} \cdot \sqrt{\frac{F_1}{F_2}} = D_2\sqrt{\frac{F_1}{F_2}}$$

Factor out perfect squares if possible in these:

(a) Solve for "a":

$$\frac{a^2}{b^2} = \frac{p}{s}$$

(b) Solve for "D_2":

$$\frac{F_1}{F_2} = \frac{(D_1)^2}{(D_2)^2}$$

a = _____ D_2 = _____

$c = 9\sqrt{ab}$, from:
$c = \sqrt{81ab}$

Equations and Formulas Containing Squares and Square Root Radicals 171

186. In the formula on the left below, the fraction $\frac{m}{t}$ is squared. To solve for either "m" or "t", we must square the fraction as we have done on the right.

$$k = \left(\frac{m}{t}\right)^2 \qquad\qquad k = \frac{m^2}{t^2}$$

Using the formula on the right, solve for "m".

m = _____

a) $a = b\sqrt{\dfrac{p}{s}}$

b) $D_2 = D_1\sqrt{\dfrac{F_2}{F_1}}$

---

187. The formula on the right contains a squared fraction.

$$\frac{g_o}{g_p} = \left(\frac{r_p}{r_o}\right)^2$$

(a) Solve for "r_p": (b) Solve for "r_o":

$r_p =$ _____ $r_o =$ _____

$m = t\sqrt{k}$

(Did you factor out the perfect square?)

---

188. The formula below contains a squared fraction. We have written it in two equivalent forms.

$$k = 2a\left(\frac{b}{c}\right)^2$$

$$k = 2a\left(\frac{b^2}{c^2}\right)$$

$$k = \frac{2ab^2}{c^2}$$

Using the last form, it is easy to solve for either "b" or "c".

(a) Solve for "b": (b) Solve for "c":

b = _____   c = _____

a) $r_p = r_o\sqrt{\dfrac{g_o}{g_p}}$

b) $r_o = r_p\sqrt{\dfrac{g_p}{g_o}}$

---

a) $b = c\sqrt{\dfrac{k}{2a}}$

b) $c = b\sqrt{\dfrac{2a}{k}}$

189. Notice how we obtain a "1" in the numerator when we factor out "a^2" from the radical below:

$$\sqrt{\frac{a^2}{b}} = \sqrt{\frac{a^2(1)}{b}} = a\sqrt{\frac{1}{b}}$$

Factor out the perfect squares in these solutions:

(a) Solve for "t":

$$k = \left(\frac{m}{t}\right)^2$$

(b) Solve for "P":

$$CH = \left(\frac{RS}{P}\right)^2$$

t = _____   P = _____

---

190. Always check your solutions to make sure that a fractional radicand is in lowest terms. For example:

If $w^2r^2 = ar$, $w = \sqrt{\frac{ar}{r^2}} = \sqrt{\frac{a}{r}}$

Solve for the indicated letter in each of these:

(a) Solve for "h":
$dh^2 = bd^2$

(b) Solve for "m":
$am^2 = aR^2T$

(c) Solve for "t":
$c^2t^2 = 2cy^2$

h = _____   m = _____   t = _____

---

a) $t = m\sqrt{\dfrac{1}{k}}$

b) $P = RS\sqrt{\dfrac{1}{CH}}$

---

a) $h = \sqrt{bd}$

b) $m = R\sqrt{T}$

c) $t = y\sqrt{\dfrac{2}{c}}$

191. If we clear the fraction in the formula on the left below, we obtain the formula on the right.

$$2dp^2 = \frac{1}{2dR} \qquad 4d^2p^2R = 1$$

Watch how we can factor out perfect squares when we solve for "d".

$$d = \sqrt{\frac{1}{4p^2R}} = \frac{\sqrt{1}}{\sqrt{4p^2R}} = \frac{1}{2p\sqrt{R}}$$

Factor out the perfect squares in each of these solutions:

(a) Solve for "s":

$$a^2bs = \frac{1}{bcs}$$

(b) Solve for "f":

$$2\pi fL = \frac{1}{2\pi fC}$$

s = _____   f = _____

192. Notice how we insert a "1" in the numerator to enable us to factor out the perfect square from the denominator of the radicand below.

$$\sqrt{\frac{a}{b^2c}} = \sqrt{\frac{(1)(a)}{b^2c}} = \frac{1}{b}\sqrt{\frac{a}{c}}$$

Factor out the perfect squares from each of these radicals:

(a) $\sqrt{\dfrac{3b}{16a^2d}}$ = _____

(b) $\sqrt{\dfrac{RT}{P^2S^2Y}}$ = _____

a) $s = \dfrac{1}{ab\sqrt{c}}$

b) $f = \dfrac{1}{2\pi\sqrt{LC}}$

193. If you solve for "f" in the formula at the right, you can both reduce to lowest terms and factor out perfect squares. Do so.

$$\frac{gr}{h} = 4\pi^2 f^2 r$$

f = _____

a) $\dfrac{1}{4a}\sqrt{\dfrac{3b}{d}}$

b) $\dfrac{1}{PS}\sqrt{\dfrac{RT}{Y}}$

$f = \dfrac{1}{2\pi}\sqrt{\dfrac{g}{h}}$

194. We can solve for "R" in $A = \pi(R^2 - r^2)$ in two different ways:

   (1) By multiplying by the distributive principle first:

   $A = \pi(R^2 - r^2)$

   $A = \pi R^2 - \pi r^2$

   $\pi R^2 = A + \pi r^2$

   $R^2 = \dfrac{A + \pi r^2}{\pi}$

   $R = \sqrt{\dfrac{A + \pi r^2}{\pi}}$

   (2) By multiplying by $\dfrac{1}{\pi}$ first:

   $A = \pi(R^2 - r^2)$

   $\dfrac{A}{\pi} = R^2 - r^2$

   $R^2 = \dfrac{A}{\pi} + r^2$

   $R = \sqrt{\dfrac{A}{\pi} + r^2}$

   Show that the two radicands are equivalent.

   Either:
   $\dfrac{A + \pi r^2}{\pi} = \dfrac{A}{\pi} + \dfrac{\pi r^2}{\pi} = \dfrac{A}{\pi} + r^2$

   or:
   $\dfrac{A}{\pi} + r^2 = \dfrac{A}{\pi} + \dfrac{\pi r^2}{\pi} = \dfrac{A + \pi r^2}{\pi}$

195. We can solve for "C" in $H = (1 - C^2)h$ in two different ways, as shown below. Notice how we have to apply the opposing principle in each method.

   (1) By multiplying by the distributive principle first:

   $H = (1 - C^2)h$

   $H = h - C^2 h$

   $H - h = -C^2 h$

   $h - H = C^2 h$

   $C^2 = \dfrac{h - H}{h}$

   $C = \sqrt{\dfrac{h - H}{h}}$

   (2) By multiplying by $\dfrac{1}{h}$ first:

   $H = (1 - C^2)h$

   $\dfrac{H}{h} = 1 - C^2$

   $\dfrac{H}{h} - 1 = -C^2$

   $1 - \dfrac{H}{h} = C^2$

   $C = \sqrt{1 - \dfrac{H}{h}}$

   Show that the two radicands are equivalent.

   Either:
   $\dfrac{h - H}{h} = \dfrac{h}{h} - \dfrac{H}{h} = 1 - \dfrac{H}{h}$

   or:
   $1 - \dfrac{H}{h} = \dfrac{h}{h} - \dfrac{H}{h} = \dfrac{h - H}{h}$

196. Use either of the two methods to solve these:

   (a) Solve for "d":

   $m = (d^2 - 1)t$

   (b) Solve for "r":

   $A = \pi(R^2 - r^2)$

   d = _____   r = _____

197. (a) Solve for "V_1":

$$H = \frac{M(V_1^2 - V_2^2)}{1{,}100gt}$$

(b) Solve for "t":

$$p = \frac{k(t^2 - b^2)}{ar}$$

$V_1 = $ _____

$t = $ _____

a) $d = \sqrt{\dfrac{m}{t} + 1}$ or $d = \sqrt{\dfrac{m + t}{t}}$

b) $r = \sqrt{R^2 - \dfrac{A}{\pi}}$ or $r = \sqrt{\dfrac{\pi R^2 - A}{\pi}}$

---

198. (a) Solve for "V_2":

$$H = \frac{M(V_1^2 - V_2^2)}{1{,}100gt}$$

(b) Solve for "b":

$$p = \frac{k(t^2 - b^2)}{ar}$$

$V_2 = $ _____

$b = $ _____

a) $V_1 = \sqrt{\dfrac{1{,}100gHt}{M} + V_2^2}$

or $V_1 = \sqrt{\dfrac{1{,}100gHt + MV_2^2}{M}}$

b) $t = \sqrt{\dfrac{apr}{k} + b^2}$

or $t = \sqrt{\dfrac{apr + b^2k}{k}}$

---

199. One method of solving for "v" in the formula below is shown on the left. In this method, notice that we did not clear the fraction first.

$$\boxed{P = \frac{mv^2}{r} - mg}$$

$P + mg = \dfrac{mv^2}{r}$

$r(P + mg) = mv^2$

$v^2 = \dfrac{r(P + mg)}{m}$

$v = \sqrt{\dfrac{r(P + mg)}{m}}$

$$\boxed{P = \frac{mv^2}{r} - mg}$$

In the answer, notice that we left the numerator of the radicand in factored form. Using the space provided on the right above, show the steps involved in solving for "v" if we clear the fraction first.

a) $V_2 = \sqrt{V_1^2 - \dfrac{1{,}100gHt}{M}}$

or $V_2 = \sqrt{\dfrac{MV_1^2 - 1{,}100gHt}{M}}$

b) $b = \sqrt{t^2 - \dfrac{apr}{k}}$

or $b = \sqrt{\dfrac{kt^2 - apr}{k}}$

Answer to Frame 199: $Pr = mv^2 - mgr$

$Pr + mgr = mv^2$

$v^2 = \dfrac{Pr + mgr}{m}$

$v = \sqrt{\dfrac{Pr + mgr}{m}}$

$v = \sqrt{\dfrac{r(P + mg)}{m}}$

SELF-TEST 11 (Frames 174-199)

1. Solve for h:

   $h^2 = pt$

   h = _____

2. Solve for r:

   $A = \pi r^2$

   r = _____

3. Solve for X:

   $Z^2 = X^2 + R^2$

   X = _____

4. Solve for d: $E = \dfrac{kQ_1 Q_2}{d^2}$

   d = _____

5. Solve for r: $2ars^2 = \dfrac{at}{2r}$

   r = _____

6. Solve for v_1:

   $\dfrac{t_1}{t_2} = \left(\dfrac{v_2}{v_1}\right)^2$

   v_1 = _____

7. Solve for a:

   $\dfrac{a}{bc} = \dfrac{b}{ad}$

   a = _____

8. Solve for d:

   $A = 0.785(D^2 - d^2)$

   d = _____

ANSWERS:

1. $h = \sqrt{pt}$

2. $r = \sqrt{\dfrac{A}{\pi}}$

3. $X = \sqrt{Z^2 - R^2}$

4. $d = \sqrt{\dfrac{kQ_1 Q_2}{E}}$

5. $r = \dfrac{\sqrt{t}}{2s}$

   or $r = \dfrac{1}{2s}\sqrt{t}$

6. $v_1 = v_2 \sqrt{\dfrac{t_2}{t_1}}$

7. $a = b\sqrt{\dfrac{c}{d}}$

8. $d = \sqrt{D^2 - \dfrac{A}{0.785}}$

   or $d = \sqrt{\dfrac{0.785 D^2 - A}{0.785}}$

2-12 RADICAL FORMULAS WHICH CONTAIN SQUARED LETTERS

Each of the radical formulas below contains at least one squared letter in its radicand.

$$c = \sqrt{a^2 + b^2} \qquad v_o = \sqrt{v_f^2 - 2gs} \qquad h = \frac{t}{\sqrt{s^2 + 1}}$$

In this section, we will discuss the method of solving for the squared letters in this type of formula. We will show that both the squaring principle and the square root principle are needed in the solutions.

200. To solve for "R" in $Z = \sqrt{R^2 + X^2}$, we need both the squaring principle (to eliminate the radical) and the square root principle (to solve for the squared letter). Here are the steps:

$$Z = \sqrt{R^2 + X^2}$$

(1) Applying the squaring principle, we get:
$$Z^2 = R^2 + X^2$$

(2) Isolating R^2, we get:
$$R^2 = Z^2 - X^2$$

(3) Applying the square root principle, we get:
$$R = \sqrt{Z^2 - X^2}$$

Using the steps above, solve for "b" in:

$$c = \sqrt{a^2 + b^2}$$

b = _____

201. Notice how we use the oppositing principle in solving for "m" below:

$$t = \sqrt{r^2 - m^2}$$
$$t^2 = r^2 - m^2$$
$$t^2 - r^2 = -m^2$$
$$r^2 - t^2 = m^2$$
$$m = \sqrt{r^2 - t^2}$$

Following the steps above, solve for the indicated letter in each of these:

(a) Solve for "b":
$$c = \sqrt{a^2 - b^2}$$

(b) Solve for "y":
$$P = \sqrt{x - y^2}$$

b = _____   y = _____

$b = \sqrt{c^2 - a^2}$

178   Equations and Formulas Containing Squares and Square Root Radicals

202. (a) Solve for "v_f":

$v_o = \sqrt{v_f^2 - 2gs}$

(b) Solve for "R":

$\sqrt{Z^2 - R^2} = 2\pi fL$

a) $b = \sqrt{a^2 - c^2}$

b) $y = \sqrt{x - P^2}$

$v_f =$ _____ $R =$ _____

---

203. To solve for "p" in the formula at the right, we must isolate the radical first.

$a = \dfrac{1}{\sqrt{p^2 + w^2}}$

(a) Isolate the radical:

$\sqrt{p^2 + w^2} =$ _____

(b) Square both sides: _____

(c) Isolate "p^2":

$p^2 =$ _____

(d) Apply the square root principle: $p =$ _____

a) $v_f = \sqrt{v_o^2 + 2gs}$

b) $R = \sqrt{Z^2 - 4\pi^2 f^2 L^2}$

or $R = \sqrt{Z^2 - (2\pi fL)^2}$

---

204. To solve for "s" in the formula on the right, we must also isolate the radical first. Do so and complete the solution.

$h = \dfrac{t}{\sqrt{s^2 + 1}}$

$s =$ _____

a) $\sqrt{p^2 + w^2} = \dfrac{1}{a}$

b) $p^2 + w^2 = \dfrac{1}{a^2}$

c) $p^2 = \dfrac{1}{a^2} - w^2$

d) $p = \sqrt{\dfrac{1}{a^2} - w^2}$

---

205. We saw earlier that the square root of an addition is not equal to the addition of the square roots of each term. That is:

$\sqrt{c^2 + d^2}$ does not equal $\sqrt{c^2} + \sqrt{d^2}$

Can $\sqrt{c^2 + d^2}$ be written in a simpler form? _____

$\sqrt{s^2 + 1} = \dfrac{t}{h}$

$s^2 + 1 = \left(\dfrac{t}{h}\right)^2$

$s^2 = \left(\dfrac{t}{h}\right)^2 - 1$

$s = \sqrt{\left(\dfrac{t}{h}\right)^2 - 1}$

No

206. Answer "true" or "false" for each of the following:

_____ (a) $\sqrt{R^2 + X^2} = \sqrt{R^2} + \sqrt{X^2}$ _____ (c) $\sqrt{1 - \frac{H}{h}} = \sqrt{1} - \sqrt{\frac{H}{h}}$

_____ (b) $\sqrt{c^2 - a^2} = \sqrt{c^2} - \sqrt{a^2}$ _____ (d) $\sqrt{\frac{A}{\pi} + r^2} = \sqrt{\frac{A}{\pi}} + \sqrt{r^2}$

---

207. Answer "true" or "false" for each of these:

_____ (a) If $a = \sqrt{c^2 - b^2}$, $a = c - b$

_____ (b) If $v_f = \sqrt{v_o^2 - 2gs}$, $v_f = v_o - \sqrt{2gs}$

_____ (c) If $p = \sqrt{\frac{1}{a^2} - w^2}$, $p = \frac{1}{a} - w$

_____ (d) If $r = \sqrt{R^2 - \frac{A}{\pi}}$, $r = R - \sqrt{\frac{A}{\pi}}$

All are "false".

---

208. Which of the following solutions can be written in a simpler form?

(a) $R = \sqrt{Z^2 - X^2}$ (c) $d = \sqrt{\frac{m}{t} + 1}$

(b) $s = \sqrt{\left(\frac{t}{h}\right)^2 - 1}$ (d) $t = \sqrt{\frac{apr}{k} + b^2}$

All are "false".

---

209. To solve for "P" in the formula on the right, we must apply the squaring principle because P is in the radicand. The resulting equation is shown.

$\sqrt{\frac{T}{P}} = \frac{N}{P}$

$\frac{T}{P} = \frac{N^2}{P^2}$

Though it looks as if we will need the square root principle to eliminate P^2, that principle is not needed because the equation obtained by clearing the fractions does not contain P^2.

To clear the fractions in the equation above, we multiply both sides by P^2. The steps are shown at the right.

$P^2\left(\frac{T}{P}\right) = \frac{N^2}{P^2}(P^2)$

$\left(\frac{P^2}{P}\right)(T) = N^2\left(\frac{P^2}{P^2}\right)$

$(P)T = N^2(1)$

$PT = N^2$

Therefore, P = _____

None of them.

---

210. What quantity would we multiply both sides by to clear the fractions in each of these?

(a) $\frac{2s}{a} = \frac{V^2}{a^2}$ _____ (b) $\frac{GM}{r_o} = \frac{v^2}{r_o^2}$ _____ (c) $\frac{K}{am} = \frac{V^2}{2a}$ _____

$P = \frac{N^2}{T}$

a) a^2

b) r_o^2

c) $2am$

Equations and Formulas Containing Squares and Square Root Radicals

211. Clear the fractions in each formula below:

(a) $\dfrac{2s}{a} = \dfrac{V^2}{a^2}$ (b) $\dfrac{GM}{r_o} = \dfrac{v^2}{r_o^2}$ (c) $\dfrac{K}{am} = \dfrac{V^2}{2a}$

a) $2as = V^2$
b) $GMr_o = v^2$
c) $2K = mV^2$

212. (a) Solve for "T": (b) Solve for "a": (c) Solve for "M":

$\sqrt{\dfrac{T}{P}} = \dfrac{N}{P}$ $\sqrt{\dfrac{2s}{a}} = \dfrac{V}{a}$ $\sqrt{\dfrac{GM}{r_o}} = \dfrac{v}{r_o}$

T = _____   a = _____   M = _____

a) $T = \dfrac{N^2}{P}$
b) $a = \dfrac{V^2}{2s}$
c) $M = \dfrac{v^2}{Gr_o}$

213. (a) Solve for "K": (b) Solve for "b": (c) Solve for "y":

$\dfrac{K}{am} = \dfrac{V^2}{2a}$ $\sqrt{\dfrac{ab}{c}} = \dfrac{t}{c}$ $\sqrt{\dfrac{x}{y}} = \dfrac{a}{y}$

K = _____   b = _____   y = _____

a) $K = \dfrac{mV^2}{2}$
b) $b = \dfrac{t^2}{ac}$
c) $y = \dfrac{a^2}{x}$

214. To solve for "R" in the formula at the right, we do not need the squaring principle. We can simply multiply both sides by "t". The resulting equation is shown.

$$\sqrt{\frac{V}{t}} = \frac{R}{t}$$

$$t\sqrt{\frac{V}{t}} = R$$

$$\text{or } R = t\sqrt{\frac{V}{t}}$$

Since there is a "t" outside the radical and in the denominator of the radical, we can simplify the solution by moving the "t" under the radical. To do so, we substitute $\sqrt{t^2}$ for "t" and simplify, as shown.

$$R = t\sqrt{\frac{V}{t}}$$

$$R = \sqrt{t^2}\sqrt{\frac{V}{t}}$$

$$R = \sqrt{\frac{t^2 V}{t}}$$

$$R = \sqrt{tV}$$

Using the steps above, solve for the indicated letter in each of the following formulas and write each solution in the preferred form.

(a) Solve for "N":

$$\sqrt{\frac{a}{b}} = \frac{N}{b}$$

(b) Solve for "c":

$$\sqrt{\frac{m}{r}} = \frac{c}{r}$$

N = _____

c = _____

a) $N = \sqrt{ab}$

b) $c = \sqrt{mr}$

SELF-TEST 12 (Frames 200-214)

1. Solve for d:

   $h = \sqrt{a^2 + d^2}$

   d = _____

2. Solve for X:

   $R = \sqrt{Z^2 - X^2}$

   X = _____

3. Solve for p:

   $\sqrt{\dfrac{t}{2p}} = \dfrac{r}{p}$

   p = _____

4. Solve for t: $r = \dfrac{a}{\sqrt{t^2 + 1}}$

   t = _____

5. Solve for H: $\sqrt{\dfrac{2w}{s}} = \dfrac{H}{s}$

   H = _____

ANSWERS: 1. $d = \sqrt{h^2 - a^2}$ 2. $X = \sqrt{Z^2 - R^2}$ 3. $p = \dfrac{2r^2}{t}$ 4. $t = \sqrt{\dfrac{a^2}{r^2} - 1}$ 5. $H = \sqrt{2sw}$

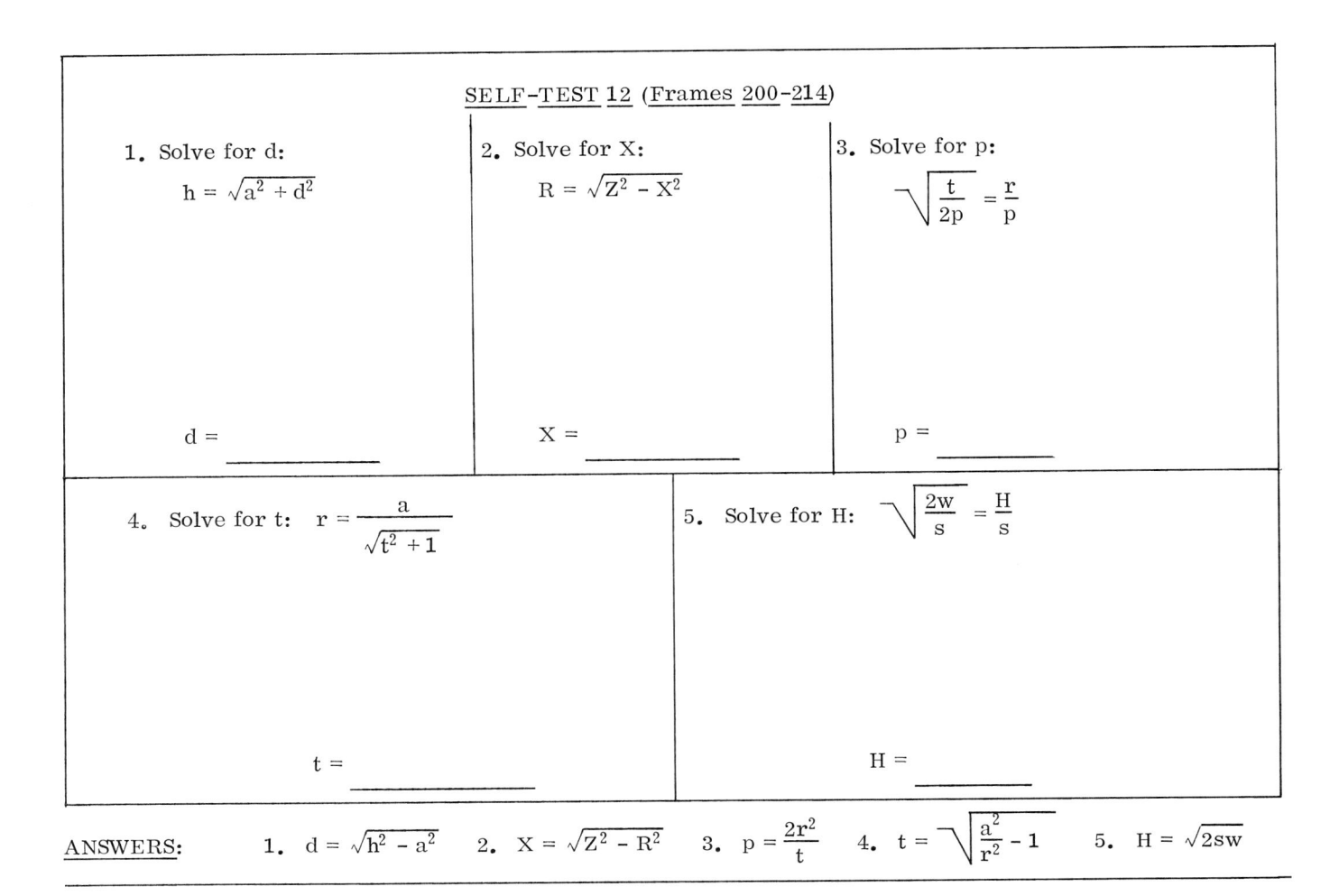

2-13 THE EQUIVALENCE METHOD AND SYSTEMS OF FORMULAS INVOLVING SQUARED LETTERS AND SQUARE ROOT RADICALS

In earlier work in solving systems of formulas, we discussed the use of the <u>equivalence method</u> to derive a new formula by eliminating a variable from a system of formulas. At that time, we avoided derivations which required the use of the squaring and square root principles. In this section, we will use the equivalence method to derive new formulas from systems which involve squared letters or square root radicals. Derivations of this type require the use of the squaring and square root principles.

Equations and Formulas Containing Squares and Square Root Radicals   183

215. As we saw earlier, the equivalence method can be used with a system of formulas to derive a new relationship of variables by eliminating one or more variables. The equivalence method involves two basic steps. We will review these two steps in the process of eliminating "E" and deriving a new relationship from the system shown at the right.

$$E = IR$$
$$P = EI$$

　　Step 1: Solving for "E" in each formula, we get:　　　$E = IR$

$$E = \frac{P}{I}$$

　　Step 2: Applying the equivalence principle, we get:　　$IR = \frac{P}{I}$

　　Solve for "P" in the new relationship:　　　$P = \underline{\qquad}$

---

216. After applying the equivalence principle in the last frame, we obtained this formula:　　$IR = \frac{P}{I}$

$P = I^2R$

In order to solve for I in this formula, we must clear the fraction. We get:　　$I^2R = P$

Now to solve for "I", we must use the square root principle. Do so.　　$I^2 = \underline{\qquad}$

$I = \underline{\qquad}$

---

217. Here is the same system. Let's eliminate "I" and then solve for "E" in the new formula. Notice that we need the square root principle to solve for "E".

$$E = IR$$
$$P = EI$$

$I^2 = \frac{P}{R}$

$I = \sqrt{\dfrac{P}{R}}$

(a) Solve for "I" in each formula: $I = \underline{\qquad}$ and $I = \underline{\qquad}$

(b) Apply the equivalence principle: $\underline{\qquad} = \underline{\qquad}$

(c) Solve for "E" in this new formula:

$E = \underline{\qquad}$

---

218. Let's eliminate "a" and then solve for "V" in this system.

$$F = ma$$
$$a = \frac{V^2}{r}$$

a) $I = \dfrac{E}{R}$ and $I = \dfrac{P}{E}$

b) $\dfrac{E}{R} = \dfrac{P}{E}$

c) $E = \sqrt{PR}$, from:
$E^2 = PR$

(a) Solve for "a" in both formulas: $a = \underline{\qquad}$ and $a = \underline{\qquad}$

(b) Apply the equivalence principle: $\underline{\qquad} = \underline{\qquad}$

(c) Solve for "V":

$V = \underline{\qquad}$

184 Equations and Formulas Containing Squares and Square Root Radicals

219. Eliminate "F" from this system and then solve for "v".

$$F = mg$$
$$Fr = mv^2$$

v = _____

a) $a = \dfrac{F}{m}$

$a = \dfrac{V^2}{r}$

b) $\dfrac{F}{m} = \dfrac{V^2}{r}$

c) $V = \sqrt{\dfrac{Fr}{m}}$

---

220. Eliminate "t" from this system and then solve for "r".

$$E = dt$$
$$t = \dfrac{2r^2}{a}$$

r = _____

$v = \sqrt{gr}$

(Did you reduce to lowest terms?)

---

221. Eliminate "s" from this system and then solve for "V".

$$K = mas$$
$$s = \dfrac{V^2}{2a}$$

V = _____

$r = \sqrt{\dfrac{aE}{2d}}$, from:

$\dfrac{E}{d} = \dfrac{2r^2}{a}$

---

222. Eliminate "m" from this system and then solve for "v".

$$F = \dfrac{mv^2}{r}$$
$$m = \dfrac{W}{g}$$

v = _____

$V = \sqrt{\dfrac{2K}{m}}$, from:

$\dfrac{K}{ma} = \dfrac{V^2}{2a}$

$v = \sqrt{\dfrac{Fgr}{W}}$, from:

$\dfrac{Fr}{v^2} = \dfrac{W}{g}$

Equations and Formulas Containing Squares and Square Root Radicals 185

223. We want to eliminate "G" from the system on the right. Notice that we need the square root principle to solve for "G" in the top formula.

$$T = G^2N$$
$$G = PR$$

Solving for "G" in both formulas, we get: $G = \sqrt{\dfrac{T}{N}}$ and $G = PR$

Applying the equivalence principle, we get: $\sqrt{\dfrac{T}{N}} = PR$

(a) We <u>need</u> the squaring principle to solve for either _____ or _____ in the new formula.

(b) We <u>do not need</u> the squaring principle to solve for either _____ or _____ in the new formula.

(c) Solve for "T".

T = _____

---

224. By applying the equivalence principle in the last frame, we obtained this equation:

$$\sqrt{\dfrac{T}{N}} = PR$$

If we solve this equation for "P", we get:

$$P = \dfrac{\sqrt{\dfrac{T}{N}}}{R}$$

This solution can be written in a simpler form by converting the division to a multiplication. Do so.

P = _____

Answers:
a) T or N
b) P or R
c) $T = N(PR)^2$ or $T = NP^2R^2$

---

225. We want to eliminate "G" from this system.

$$T = G^2P$$
$$P = \dfrac{N}{G}$$

After solving for "G" in each formula and then applying the equivalence principle, we get:

$$\sqrt{\dfrac{T}{P}} = \dfrac{N}{P}$$

To solve for "P", we must apply the squaring principle. We get:

$$\dfrac{T}{P} = \dfrac{N^2}{P^2}$$

(a) To clear the fractions, we should multiply both sides by _____.

(b) Do so and solve for P.

P = _____

Answer: $P = \dfrac{1}{R}\sqrt{\dfrac{T}{N}}$

Answers:
a) P^2
b) $P = \dfrac{N^2}{T}$

186   Equations and Formulas Containing Squares and Square Root Radicals

226. After applying the equivalence principle in the last frame, we obtained this formula:

$$\sqrt{\dfrac{T}{P}} = \dfrac{N}{P}$$

Solve for "N" and write the solution in the preferred form.

N = _____

---

227. Eliminate "t" from this system and then solve for "V". Write the solution in the preferred form.

$$V = at$$
$$s = \dfrac{1}{2}at^2$$

V = _____

$N = \sqrt{PT}$, from:

$$N = P\sqrt{\dfrac{T}{P}}$$

$$N = \sqrt{P^2}\sqrt{\dfrac{T}{P}}$$

$$N = \sqrt{\dfrac{P^2 T}{P}}$$

$$N = \sqrt{PT}$$

---

228. Here is the same system. Eliminate "t" and then solve for "a".

$$V = at$$
$$s = \dfrac{1}{2}at^2$$

a = _____

$V = \sqrt{2as}$, from:

$$\dfrac{V}{a} = \sqrt{\dfrac{2s}{a}}$$

$$V = a\sqrt{\dfrac{2s}{a}}$$

$$V = \sqrt{\dfrac{2a^2 s}{a}}$$

$$V = \sqrt{2as}$$

---

229. If we eliminate "v" from the system at the right, we get this new formula:

$$\sqrt{ar} = \dfrac{2\pi r}{T}$$

$$a = \dfrac{v^2}{r}$$
$$v = \dfrac{2\pi r}{T}$$

In order to solve for "a", we must apply the squaring principle. Do so and solve for "a".

a = _____

$$a = \dfrac{V^2}{2s}$$

$$a = \dfrac{4\pi^2 r}{T^2}$$

(Did you reduce to lowest terms?)

Equations and Formulas Containing Squares and Square Root Radicals 187

230. Eliminate "v" and then solve for "F".

$$F = \frac{mv^2}{r}$$
$$v = \frac{2\pi r}{T}$$

F = _____

---

231. Eliminate "g_o" and then solve for "v".

$$v^2 = g_o r_o$$
$$g_o = \frac{GM}{r_o^2}$$

v = _____

$F = \dfrac{4\pi^2 mr}{T^2}$

(Did you reduce to lowest terms?)

---

232. Let's eliminate "v" and then solve for "T".

$$a = \frac{v^2}{r}$$
$$v = \frac{2\pi r}{T}$$

Solving for "v" in each formula, we get:

$v = \sqrt{ar}$ and $v = \dfrac{2\pi r}{T}$

Applying the equivalence principle, we get:

$\sqrt{ar} = \dfrac{2\pi r}{T}$

Solving for "T", we get:

$T = \dfrac{2\pi r}{\sqrt{ar}}$

Since there is an "r" in both the numerator and the radicand of the denominator, we can simplify this solution by partially rationalizing the denominator. Do so.

T = _____

$v = \sqrt{\dfrac{GM}{r_o}}$, from:

$\dfrac{v^2}{r_o} = \dfrac{GM}{r_o^2}$

$T = 2\pi \sqrt{\dfrac{r}{a}}$, from:

$T = \dfrac{2\pi r \sqrt{r}}{\sqrt{a} \cdot \sqrt{r} \cdot \sqrt{r}}$

$\quad = \dfrac{2\pi r \sqrt{r}}{\sqrt{a} \cdot r}$

$\quad = 2\pi \sqrt{\dfrac{r}{a}}$

188 Equations and Formulas Containing Squares and Square Root Radicals

233. If we eliminate "v" and solve for "T" in the system on the right, we get:

$$\sqrt{\frac{Fr}{m}} = \frac{2\pi r}{T}$$

$$T = \frac{2\pi r}{\sqrt{\dfrac{Fr}{m}}}$$

$$F = \frac{mv^2}{r}$$

$$v = \frac{2\pi r}{T}$$

(a) This solution can be written in a simpler form by converting the division to a multiplication. Do so.

T = _____

(b) The solution in (a) can be put in a simpler form as follows: Put the "r" (in $2\pi r$) under the radical (as r^2), and then simplify the radicand. Do so.

T = _____

a) $T = 2\pi r\sqrt{\dfrac{m}{Fr}}$

b) $T = 2\pi\sqrt{\dfrac{mr}{F}}$, from:

$T = 2\pi r\sqrt{\dfrac{m}{Fr}}$

$T = 2\pi\sqrt{\dfrac{mr^2}{Fr}}$

$T = 2\pi\sqrt{\dfrac{mr}{F}}$

234. Eliminate "v" and solve for "T". Write your solution in the preferred form.

$$v^2 = \frac{GM}{r_0}$$

$$v = \frac{2\pi r_0}{T}$$

T = _____

$T = 2\pi r_0\sqrt{\dfrac{r_0}{GM}}$, from:

$T = \dfrac{2\pi r_0}{\sqrt{\dfrac{GM}{r_0}}}$

235. Since "X" is solved for in each formula in this system, it can easily be eliminated. We get:

$$2\pi fL = \frac{1}{2\pi fC}$$

Clearing the fraction, we get:

$X = 2\pi fL$

$X = \dfrac{1}{2\pi fC}$

$4\pi^2 f^2 LC = 1$

Solve for "f" and factor out the perfect squares in the denominator of the solution.

f = _____

$f = \dfrac{1}{2\pi\sqrt{LC}}$, from:

$f = \dfrac{1}{\sqrt{4\pi^2 LC}}$

236. In the last frame, we obtained the solution on the left below. Frequently, this solution is written in the equivalent form on the right.

$$f = \frac{1}{2\pi\sqrt{LC}} \qquad f = \frac{0.159}{\sqrt{LC}}$$

The equivalent form is obtained in these steps:

$$f = \frac{1}{2\pi\sqrt{LC}} = \left(\frac{1}{2\pi}\right)\left(\frac{1}{\sqrt{LC}}\right) = (0.159)\left(\frac{1}{\sqrt{LC}}\right) = \frac{0.159}{\sqrt{LC}}$$

Note: Since $\pi = 3.14$, $\frac{1}{2\pi} = \frac{1}{6.28} = 0.159$

Using the steps above, write each of these expressions with a decimal number in the numerator:

(a) $\dfrac{3}{4\sqrt{ab}} =$ _____ (b) $\dfrac{5}{2\sqrt{cd}} =$ _____ (c) $\dfrac{9}{10\sqrt{RT}} =$ _____

---

237. We want to eliminate "X" from this system and then solve for "L". To solve for "X" in the top formula, we must use both the squaring principle and the square root principle.

$Z = \sqrt{R^2 + X^2}$
$X = 2\pi fL$

(a) What equation do you get when you apply the equivalence principle? _____

(b) Solve for "L".

L = _____

a) $\dfrac{0.75}{\sqrt{ab}}$

b) $\dfrac{2.5}{\sqrt{cd}}$

c) $\dfrac{0.9}{\sqrt{RT}}$

---

238. Here is the same system. Let's eliminate "X" and solve for "Z".

$Z = \sqrt{R^2 + X^2}$
$X = 2\pi fL$

Z = _____

a) $\sqrt{Z^2 - R^2} = 2\pi fL$

b) $L = \dfrac{\sqrt{Z^2 - R^2}}{2\pi f}$

$Z = \sqrt{4\pi^2 f^2 L^2 + R^2}$

190 Equations and Formulas Containing Squares and Square Root Radicals

239. The solution from the last frame is given below.
$$Z = \sqrt{4\pi^2 f^2 L^2 + R^2}$$
Can this solution be written in a simpler form? _____

---

240. Let's eliminate "e" and solve for "w" in this system.

$$v = \sqrt{w - e}$$
$$w = \frac{e}{v}$$

No.

Solving for "e" in each formula, we get:

$e = w - v^2$ and $e = wv$

Applying the equivalence principle, we get:

$w - v^2 = wv$

We want to solve for "w". Since "w" appears in two terms, we must isolate those terms and factor by the distributive principle. Do so and complete the solution.

w = _____

---

241. Eliminate "V" and solve for "v" in this system.

$$mv = (m + M)V$$
$$V^2 = 2gh$$

$w = \dfrac{v^2}{1 - v}$, from:

$w - wv = v^2$

$w(1 - v) = v^2$

$w = \dfrac{v^2}{1 - v}$

v = _____

$v = \dfrac{(m + M)\sqrt{2gh}}{m}$

or

$\left(\dfrac{m + M}{m}\right)\sqrt{2gh}$

or

$\left(1 + \dfrac{M}{m}\right)\sqrt{2gh}$

SELF-TEST 13 (Frames 215-241)

In Problems 1 to 3, write the right side of each formula in a simpler form:

1. $w = \dfrac{\sqrt{\dfrac{s}{a}}}{h}$

   $w = $ _____

2. $D = r\sqrt{\dfrac{2F}{r}}$

   $D = $ _____

3. $T = \dfrac{kt}{\sqrt{\dfrac{t}{p}}}$

   $T = $ _____

Refer to the system of formulas shown at the right.

$F = GP$
$A = FG$

4. Eliminate F and solve for G.
5. Eliminate G and solve for F.

   $G = $ _____ $F = $ _____

6. Eliminate b and solve for w.

   $b = \sqrt{c^2 - a^2}$
   $w = br$

   $w = $ _____

7. Eliminate d and solve for t.

   $h^2 = t - d$
   $t = \dfrac{d}{h}$

   $t = $ _____

8. Referring to the formula at the right, write its right side with a decimal number in the numerator.

   $Q = \dfrac{1}{2.5R} = $ _____

ANSWERS:

1. $w = \dfrac{1}{h}\sqrt{\dfrac{s}{a}}$
2. $D = \sqrt{2Fr}$
3. $T = k\sqrt{pt}$
4. $G = \sqrt{\dfrac{A}{P}}$
5. $F = \sqrt{AP}$
6. $w = r\sqrt{c^2 - a^2}$
7. $t = \dfrac{h^2}{1 - h}$
8. $Q = \dfrac{0.40}{R}$

2-14 THE SUBSTITUTION METHOD AND SYSTEMS OF FORMULAS INVOLVING SQUARED LETTERS AND SQUARE ROOT RADICALS

In earlier work in solving systems of formulas, we discussed the use of the substitution method for deriving a new formula by eliminating a variable from a system of formulas. In this section, we will use the substitution method with systems which involve squared letters and square root radicals. Of course, the new formulas obtained by the substitution method are identical to those obtained by the equivalence method.

192 Equations and Formulas Containing Squares and Square Root Radicals

242. Since "E" is solved for in the top formula of the system on the right, we can eliminate "E" by substituting "IR" for "E" in the bottom formula.

$E = IR$
$P = EI$

We get: $P = EI$
$P = (IR)I$
$P = I^2R$

Solve for "I" in the new formula: $I = $ _____

$I = \sqrt{\dfrac{P}{R}}$

---

243. Since "a" is solved for in the bottom formula on the right, it is easy to eliminate "a" from this system by the substitution method.

$F = ma$
$a = \dfrac{V^2}{r}$

(a) Eliminate "a" by substituting $\dfrac{"V^2"}{r}$ for "a" in the top formula: _____

(b) Now solve for "V" in this new formula: $V = $ _____

a) $F = \dfrac{mV^2}{r}$, from:
$F = m\left(\dfrac{V^2}{r}\right)$

b) $V = \sqrt{\dfrac{Fr}{m}}$

---

244. In this system, eliminate "t" by substituting $\dfrac{"2r^2"}{a}$ for "t" in the top formula. Then solve the new formula for "r".

$E = dt$
$t = \dfrac{2r^2}{a}$

$r = $ _____

$r = \sqrt{\dfrac{aE}{2d}}$, from:
$E = \dfrac{2dr^2}{a}$

---

245. In this system, eliminate "F" by substituting "mg" for "F" in the bottom formula. Then solve for "v". Be sure to reduce the solution to lowest terms.

$F = mg$
$Fr = mv^2$

$v = $ _____

$v = \sqrt{gr}$, from:
$mgr = mv^2$

246. If we substitute $\frac{\text{"GM"}}{r_o^2}$ for "g_o" in the top formula of this system, we get:

$$v^2 = \left(\frac{GM}{r_o^2}\right)r_o = \frac{GMr_o}{r_o^2} = \frac{GM}{r_o}$$

Notice how we reduced the fraction to lowest terms. Now solve for "v" in the new formula.

$v^2 = g_o r_o$
$g_o = \frac{GM}{r_o^2}$

v = _____

$v = \sqrt{\frac{GM}{r_o}}$

---

247. (a) What new formula do we get if we substitute "$\frac{V^2}{2a}$" for "s" in the top formula of this system? Be sure the new formula is in lowest terms.

$K = mas$
$s = \frac{V^2}{2a}$

_____

(b) Solve for "V" in this new formula.

V = _____

a) $K = \frac{mV^2}{2}$

b) $V = \sqrt{\frac{2K}{m}}$

---

248. If the variable which we want to eliminate is not already solved for in one formula, we must do so first. For example, to eliminate "I" from the system at the right:

$E = IR$
$P = EI$

(1) We can solve for "I" in the top formula: $I = \frac{E}{R}$

(2) Then substitute this value for "I" in the bottom formula: $P = E\left(\frac{E}{R}\right)$

$P = \frac{E^2}{R}$

Solve for "E" in this new formula: E = _____

$E = \sqrt{PR}$

---

249. Let's eliminate "m" from this system and then solve for "v".

$F = mg$
$Fr = mv^2$

(a) Solve for "m" in the top formula: m = _____

(b) Substitute this value for "m" in the bottom formula: _____

(c) Now solve for "v": v = _____

194 Equations and Formulas Containing Squares and Square Root Radicals

250. In the system at the right, "v" is solved for in the top formula. Therefore, to eliminate "v", we can substitute "wr" for "v" in the bottom formula. We get:

$$v^2 = ar$$
$$(wr)^2 = ar$$
$$w^2 r^2 = ar$$

Notice that we squared "wr" since "v" was squared in the bottom formula.

Solve for "a" in the new formula:

$\boxed{\begin{array}{c} v = wr \\ v^2 = ar \end{array}}$

a = _____

a) $m = \dfrac{F}{g}$

b) $Fr = \dfrac{Fv^2}{g}$

c) $v = \sqrt{gr}$

(Did you reduce to lowest terms?)

---

251. To eliminate "G" from this system:

(1) We can solve for "G" in the bottom formula, and get:

$$G = \dfrac{N}{P}$$

(2) We then substitute this value for "G" in the top formula, $T = G^2 P$. We get:

$$T = G^2 P = \left(\dfrac{N}{P}\right)^2 P = \left(\dfrac{N^2}{P^2}\right) P = \dfrac{N^2 P}{P^2} = \dfrac{N^2}{P}$$

Notice that we squared $\dfrac{N}{P}$ and then reduced to lowest terms.

Solve for N in this new formula: $T = \dfrac{N^2}{P}$

N = _____

$\boxed{\begin{array}{c} T = G^2 P \\ P = \dfrac{N}{G} \end{array}}$

$a = w^2 r$

(Did you reduce to lowest terms?)

---

252. Let's eliminate "t" from this system and then solve for "V".

(a) Solve for "t" in the top formula: t = _____

(b) Substitute this value for "t" in the bottom formula and then simplify:

(c) Now solve for "V":   V = _____

$\boxed{\begin{array}{c} V = at \\ s = \dfrac{1}{2} at^2 \end{array}}$

$N = \sqrt{PT}$

Equations and Formulas Containing Squares and Square Root Radicals   195

253. We can eliminate "m" from this system by substituting $\frac{W}{g}$ for "m" in the top formula. We get:

$$F = \frac{mv^2}{r} = \frac{\left(\frac{W}{g}\right)v^2}{r} = \frac{\frac{Wv^2}{g}}{r} = \frac{Wv^2}{g}\left(\frac{1}{r}\right) = \frac{Wv^2}{gr}$$

$\boxed{F = \frac{mv^2}{r} \\ m = \frac{W}{g}}$

Notice how we converted the division of $\frac{Wv^2}{g}$ by "r" to a multiplication by multiplying $\frac{Wv^2}{g}$ by $\frac{1}{r}$, the reciprocal of "r".

Solve for "v" in the new formula: $F = \frac{Wv^2}{gr}$

v = _____

a) $t = \frac{V}{a}$

b) $s = \frac{V^2}{2a}$, from:

$$s = \frac{1}{2}a\left(\frac{V}{a}\right)^2$$
$$= \frac{1}{2}a\left(\frac{V^2}{a^2}\right)$$
$$= \frac{aV^2}{2a^2} = \frac{V^2}{2a}$$

b) $V = \sqrt{2as}$

254. We can eliminate "v" from this system by substituting $\frac{2\pi r}{T}$ for "v" in the top formula. We get:

$$a = \frac{v^2}{r} = \frac{\left(\frac{2\pi r}{T}\right)^2}{r} = \frac{\frac{4\pi^2 r^2}{T^2}}{r} = \left(\frac{4\pi^2 r^2}{T^2}\right)\left(\frac{1}{r}\right) = \frac{4\pi^2 r^2}{T^2 r} = \frac{4\pi^2 r}{T^2}$$

$\boxed{a = \frac{v^2}{r} \\ v = \frac{2\pi r}{T}}$

Notice how we squared $\frac{2\pi r}{T}$, converted the division of $\frac{4\pi^2 r^2}{T^2}$ by "r" to a multiplication, and then reduced the fraction to lowest terms.

Solve for "T" in the new formula: $a = \frac{4\pi^2 r}{T^2}$

T = _____

$v = \sqrt{\frac{Fgr}{W}}$

255. Let's eliminate "v" from this system and then solve for "T".

$\boxed{F = \frac{mv^2}{r} \\ v = \frac{2\pi r}{T}}$

(a) Substitute $\frac{2\pi r}{T}$ for "v" in the top formula and simplify.

(b) Now solve for "T".

T = _____

$T = 2\pi\sqrt{\frac{r}{a}}$, from:

$$T = \sqrt{\frac{4\pi^2 r}{a}}$$
$$= \frac{2\pi\sqrt{r}}{\sqrt{a}} = 2\pi\sqrt{\frac{r}{a}}$$

196   Equations and Formulas Containing Squares and Square Root Radicals

256. Let's eliminate "e" from this system and then solve for "w".

$v = \sqrt{w - e}$
$w = \dfrac{e}{v}$

a) $F = \dfrac{4\pi^2 mr}{T^2}$

b) $T = 2\pi\sqrt{\dfrac{mr}{F}}$

(a) Solving for "e" in the bottom formula, and then substituting that value for "e" in the top formula, we get:

_____

(b) Now to solve for "w", you must use the squaring principle and then factor by the distributive principle. Do so.

w = _____

257. To eliminate "X" in this system, we can substitute "$2\pi fL$" for "X" in the top formula. Do so and then solve for "L".

$Z = \sqrt{R^2 + X^2}$
$X = 2\pi fL$

a) $v = \sqrt{w - vw}$

b) $w = \dfrac{v^2}{1 - v}$, from:

$v^2 = w - vw$
$v^2 = w(1 - v)$

L = _____

Either the equivalence method or the substitution method can be used to eliminate a variable from a system of formulas. The two methods produce identical results. In some cases, the substitution method is easier. In other cases in which the substitution method requires operations with complicated fractions, the equivalence method is easier to use.

$L = \dfrac{\sqrt{Z^2 - R^2}}{2\pi f}$

SELF-TEST 14 (Frames 242-257)

1. Using the substitution method, eliminate b and solve for r.

$b = rh$
$p = br$

r = _____

2. Using the substitution method, eliminate E and solve for d.

$E = kt$
$aE = kd^2$

d = _____

ANSWERS: 1. $r = \sqrt{\dfrac{p}{h}}$ 2. $d = \sqrt{at}$

Chapter 3   QUADRATIC EQUATIONS AND FORMULAS

Any one-letter equation which has two (and only two) roots is called a "quadratic" equation. There are three types of quadratic equations - "pure", "incomplete", and "complete". Various methods are available for solving each type. In this chapter, we will discuss both the factoring method and the quadratic-formula method. In conjunction with the discussion of the factoring method, the procedure for multiplying two binomials and factoring trinomials will be examined. The chapter is concluded with some applied problems and a brief discussion of formula rearrangement with formulas in quadratic form.

3-1   THE MEANING OF QUADRATIC EQUATIONS

There are various types of one-letter equations which have two roots. All of these "two-root" equations are called "quadratic" equations. We will briefly discuss the meaning of quadratic equations in this section.

1.   In earlier work, we saw that equations like those below have <u>two roots</u>.

   $x^2 - 81 = 0$ $4y^2 - 9 = 0$

   To find the two roots of each, we solved for the squared letter and then took both the positive and negative square roots of the quantity on the right. That is:

   $x^2 = 81$ $y^2 = \dfrac{9}{4}$

   $x = \pm\sqrt{81}$ $y = \pm\sqrt{\dfrac{9}{4}}$

   $x = +9$ and -9 $y = +\dfrac{3}{2}$ and $-\dfrac{3}{2}$

   Write the two roots of each of these equations:

   (a) $t^2 - 16 = 0$ $t =$ _____ and _____
   (b) $9x^2 - 25 = 0$ $x =$ _____ and _____

---

a) $t = +4$ and -4

b) $x = +\dfrac{5}{3}$ and $-\dfrac{5}{3}$

198   Quadratic Equations and Formulas

2. Any one-letter equation which has two and only two roots is called a "quadratic" equation.

   Which of the following are "quadratic" equations because they have two roots? _____

   (a) $m^2 = 36$
   (b) $3m = 12$
   (c) $16d^2 - 25 = 0$
   (d) $16d - 25 = 0$

---

3. The equation below is a "quadratic" equation. Its two roots are "0" and "3".

   $$x^2 - 3x = 0$$

   Show that both "0" and "3" satisfy the equation:

   Only (a) and (c)

---

4. The equation below is also a "quadratic" equation. Its two roots are "2" and "4".

   $$y^2 - 6y + 8 = 0$$

   Show that both "2" and "4" satisfy the equation:

   $0^2 - 3(0) = 0$
   $0 -\ \ 0\ = 0$
   $\ \ \ \ \ \ \ \ \ 0\ = 0$

   $3^2 - 3(3) = 0$
   $9 -\ \ 9\ = 0$
   $\ \ \ \ \ \ \ \ \ 0\ = 0$

---

5. We have seen that the three equations below are called "quadratic" equations since each has two roots.

   $y^2 - 6y + 8 = 0 \qquad x^2 - 3x = 0 \qquad t^2 - 16 = 0$

   Notice that each of the equations contains a squared letter in one term. The presence of a squared letter can be used as a clue to identify quadratic equations.

   Which of the following are quadratic equations? _____

   (a) $4b^2 - 81 = 0$
   (b) $3y + 8 = 0$
   (c) $m^2 - 5m + 6 = 0$
   (d) $R^2 + 5R = 0$

   $2^2 - 6(2) + 8 = 0$
   $4 - 12\ \ + 8 = 0$
   $\ \ \ \ \ \ \ \ \ \ \ \ \ \ \ \ 0 = 0$

   $4^2 - 6(4) + 8 = 0$
   $16 - 24\ \ + 8 = 0$
   $\ \ \ \ \ \ \ \ \ \ \ \ \ \ \ \ 0 = 0$

| The rest of this chapter will be devoted to discussing the principles and methods used to find the two roots of quadratic equations. |

(a), (c), and (d)

Quadratic Equations and Formulas   199

SELF-TEST 1 (Frames 1-5)

1. State which of the following are quadratic equations. _____

   (a) $x^2 + 3x - 4 = 0$ (b) $9R - 16 = 0$ (c) $4P^2 - 1 = 0$ (d) $36y + 25 = 0$

2. State which of the following are roots of: $\boxed{t^2 + 2t - 3 = 0}$ _____

   (a) $t = 0$ (b) $t = 1$ (c) $t = -1$ (d) $t = 3$ (e) $t = -3$

ANSWERS: 1. (a) and (c)   2. (b) and (e)

3-2 MULTIPLYING TWO BINOMIALS BY MEANS OF THE DISTRIBUTIVE PRINCIPLE

The procedure for multiplying two binomials is one of the basic principles needed to generate and solve quadratic equations. In this section, we will show how the distributive principle is used to multiply two different binomials. The use of the distributive principle for these multiplications is identical to its use in squaring binomials.

6. In symbol form, here is the distributive principle for multiplication over addition:

$$(\triangle + \bigcirc)\square = \triangle(\square) + \bigcirc(\square)$$

   As we have seen, this principle can be used to multiply a binomial (two-term) expression times a monomial (one-term) expression. For example:

   $(x + 2)(3) = x(3) + 2(3) = 3x + 6$

   If both the binomial and the monomial contain the same letter, the product contains that letter squared. For example:

   $(y + 4)(y) = y(y) + 4(y) = y^2 + 4y$

   Use the distributive principle to perform these multiplications:

   (a) $(t + 1)t = $ _____

   (b) $(2x + 3)(4x) = $ _____

---

7. The distributive principle for multiplication over addition can be used directly if the binomial is an addition. However, to use it when the binomial is a subtraction, we must convert the subtraction to addition first. For example:

   $(m - 2)(m) = \bigl[m + (-2)\bigr](m) = m(m) + (-2)(m) = m^2 + (-2m)$

   Complete each of the following multiplications:

   (a) $(t - 4)(t) = \bigl[t + (-4)\bigr](t) = $ _____

   (b) $(2y - 5)(3y) = \bigl[2y + (-5)\bigr](3y) = $ _____

a) $t^2 + t$

b) $8x^2 + 12x$

200   Quadratic Equations and Formulas

8. In the last frame, we obtained these products:

$$(t - 4)(t) = t^2 + (-4t)$$
$$(2y - 5)(3y) = 6y^2 + (-15y)$$

Ordinarily, the products on the right are converted back to subtraction form. That is:

$$t^2 + (-4t) \text{ is written } t^2 - 4t$$
$$6y^2 + (-15y) \text{ is written } 6y^2 - 15y$$

Perform each multiplication below and write each product in the preferred form:

(a) $(x - 7)(2x) =$ _____

(b) $(5m - 1)(2m) =$ _____

a) $t^2 + (-4t)$
b) $6y^2 + (-15y)$

---

9. Here is the distributive principle in symbol form again:

$$(\triangle + \bigcirc)(\square) = \triangle(\square) + \bigcirc(\square)$$

As we have seen earlier, this principle is used to square a binomial. For example:

$$(x + 4)(x + 4) = x(x + 4) + 4(x + 4)$$

Similarly, it is used to multiply two different binomials. For example:

$$(y + 2)(y + 3) = y(y + 3) + 2(y + 3)$$

Notice these points about both multiplications above:

(1) By applying the distributive principle to the two binomials, the <u>first</u> binomial is broken up into two monomials.

(2) The two terms on the right in each case are still instances of the distributive principle, but they contain <u>only one</u> binomial.

Following the pattern above, complete these:

(a) $(b + 5)(b + 7) = \triangle(\square) + \bigcirc(\square)$

(b) $(2m + 1)(5m + 3) = \triangle(\square) + \bigcirc(\square)$

a) $2x^2 - 14x$
b) $10m^2 - 2m$

---

a) $b(b + 7) + 5(b + 7)$
b) $2m(5m + 3) + 1(5m + 3)$

10. Two steps are required to perform a multiplication of two binomials. Here is an example:

$$\text{Step 1:} \quad (x + 2)(y + 3) = x(y + 3) + 2(y + 3)$$
$$\text{Step 2:} \quad xy + 3x + 2y + 6$$

In Step 1: We broke up the first binomial to obtain two terms which contain only one binomial.

In Step 2: We multiplied by the distributive principle in each term.

Using the steps above, complete this multiplication:

(a + 5)(b + 4) = _____

= _____

---

11. Here is a multiplication in which we can combine "like" terms in the product:

$$(y + 3)(y + 4) = y(y + 4) + 3(y + 4)$$
$$= y^2 + 4y + 3y + 12$$
$$= y^2 + 7y + 12$$

Following the steps above, complete this multiplication and combine like terms in the product:

(2t + 1)(t + 5) = _____

= _____

= _____

Answer box:
= a(b + 4) + 5(b + 4)
= ab + 4a + 5b + 20

---

12. When one (or both) of the binomials is a subtraction, we must convert the subtraction to an addition before using the distributive principle for multiplication <u>over</u> <u>addition</u>. This conversion of the subtraction to addition is a big help for minimizing "sign" errors.

In the multiplication below, the second binomial is a subtraction. Notice how we converted this subtraction to an addition first.

$$(a + 6)(t - 4) = (a + 6)\left[t + (-4)\right]$$
$$= a\left[t + (-4)\right] + 6\left[t + (-4)\right]$$
$$= at + (-4a) + 6t + (-24)$$
$$= at - 4a + 6t - 24$$

Notice also how we converted the addition of negative terms back to subtractions in the final form of the product.

Answer box:
= 2t(t + 5) + 1(t + 5)
= 2t² + 10t + t + 5
= 2t² + 11t + 5

(Continued on following page.)

202 Quadratic Equations and Formulas

12. (Continued)

Following the steps above, complete this multiplication. Be sure to convert the additions of negative terms back to subtractions in the final form of the product.

$(2R + 5)(3V - 7) = (2R + 5)\left[3V + (-7)\right]$

= _____

= _____

= _____

$= 2R\left[3V+(-7)\right] + 5\left[3V+(-7)\right]$
$= 6RV + (-14R) + 15V + (-35)$
$= 6RV - 14R + 15V - 35$

---

13. When multiplying two binomials, there is a preferred form for the product. In this preferred form:

(1) Like terms are combined.
(2) All additions of negative terms are converted to subtraction form.

The product in the multiplication below is written in the preferred form:

$(m + 1)(m - 4) = (m + 1)\left[m + (-4)\right]$

$\qquad = m\left[m + (-4)\right] + 1\left[m + (-4)\right]$

$\qquad = m^2 + (-4m) + 1m + (-4)$

$\qquad = m^2 - 3m - 4$

Following the steps above, complete this multiplication and write the product in the preferred form:

$(3x + 2)(x - 5) = (3x + 2)\left[x + (-5)\right]$

= _____

= _____

= _____

$= 3x\left[x+(-5)\right] + 2\left[x+(-5)\right]$
$= 3x^2 + (-15x) + 2x + (-10)$
$= 3x^2 - 13x - 10$

---

14. In the multiplication below, the first binomial is a subtraction. Notice how we converted this subtraction to an addition first to minimize "sign" errors.

$(R - 8)(S + 2) = \left[R + (-8)\right](S + 2)$

$\qquad = R(S + 2) + (-8)(S + 2)$

$\qquad = RS + 2R + (-8S) + (-16)$

$\qquad = RS + 2R - 8S - 16$

Following the steps above, complete this multiplication and write the product in the preferred form:

$(x - 2)(x + 5) = \left[x + (-2)\right](x + 5)$

= _____

= _____

= _____

Quadratic Equations and Formulas 203

15. In the multiplication below, both binomials are subtractions. Notice how we convert both subtractions to additions first to minimize "sign" errors.

$(p - 3)(t - 4) = \left[p + (-3)\right]\left[t + (-4)\right]$
$= p\left[t + (-4)\right] + (-3)\left[t + (-4)\right]$
$= pt + (-4p) + (-3t) + 12$
$= pt - 4p - 3t + 12$

$= x(x + 5) + (-2)(x + 5)$
$= x^2 + 5x + (-2x) + (-10)$
$= x^2 + 3x - 10$

Following the steps above, complete this multiplication and write the product in the preferred form:

$(d - 6)(d - 7) = \left[d + (-6)\right]\left[d + (-7)\right]$
= _____
= _____
= _____

16. Show all of the steps in these. Write each product in the preferred form:

(a) $(2y - 1)(5y + 4) = \left[2y + (-1)\right](5y + 4)$
= _____
= _____
= _____

$= d\left[d+(-7)\right] + (-6)\left[d+(-7)\right]$
$= d^2 + (-7d) + (-6d) + 42$
$= d^2 - 13d + 42$

(b) $(2x - 4)(3x - 1) = \left[2x + (-4)\right]\left[3x + (-1)\right]$
= _____
= _____
= _____
= _____

17. When we combine like terms in the product below, the product simplifies to two terms since (3x) and (-3x) are a pair of opposites.

$(x + 3)(x - 3) = (x + 3)\left[x + (-3)\right]$
$= x\left[x + (-3)\right] + 3\left[x + (-3)\right]$
$= x^2 + \underline{(-3x) + 3x} + (-9)$
$= x^2 + \quad 0 \quad + (-9)$
$= x^2 + (-9)$ or $x^2 - 9$

a) $=2y(5y+4)+(-1)(5y+4)$
$=10y^2 + 8y + (-5y) + (-4)$
$= 10y^2 + 3y - 4$

b) $=2x\left[3x+(-1)\right]+(-4)\left[3x+(-1)\right]$
$= 6x^2 + (-2x) + (-12x) + 4$
$= 6x^2 - 14x + 4$

Following the steps above, complete this multiplication. The product also simplifies to two terms.

$(2t - 7)(2t + 7) = \left[2t + (-7)\right](2t + 7)$
= _____
= _____
= _____

204 Quadratic Equations and Formulas

Answer to Frame 17:
$= 2t(2t + 7) + (-7)(2t + 7)$
$= 4t^2 + 14t + (-14t) + (-49)$
$= 4t^2 - 49$

SELF-TEST 2 (Frames 6-17)

Do these multiplications by means of the distributive principle. Write each product in the preferred form.

1. $(a + 3)(2b + 5)$

2. $(2x + 1)(3x - 5)$

3. $(5s - 1)(5s + 1)$

4. $(3t - 2)(2t - 5)$

ANSWERS:
1. $2ab + 5a + 6b + 15$
2. $6x^2 - 7x - 5$
3. $25s^2 - 1$
4. $6t^2 - 19t + 10$

Quadratic Equations and Formulas   205

3-3 A SHORTER METHOD FOR MULTIPLYING TWO BINOMIALS

In the last section, we showed how two binomials are multiplied by means of the distributive principle. The formal distributive principle method requires a number of steps. There is a shorter method based on the distributive principle method which avoids one of these steps. We will discuss the shorter method in this section.

18. Here are the formal steps needed to multiply $(a + 3)$ and $(b + 2)$ by means of the distributive principle:

$$(a + 3)(b + 2) = a(b + 2) + 3(b + 2)$$
$$= ab + 2a + 3b + 6$$

There are four terms in the product in the second step above. We can skip the first step and write these four terms directly as we have done below:

$$(a + 3)(b + 2) = ab + 2a + 3b + 6$$

Notice these points:

(1) The first two terms of the product are obtained by multiplying the first term of the first binomial times both terms of the second binomial.

(2) The last two terms of the product are obtained by multiplying the second term of the first binomial times both terms of the second binomial.

Following the arrows in each case, write the four terms of the product directly:

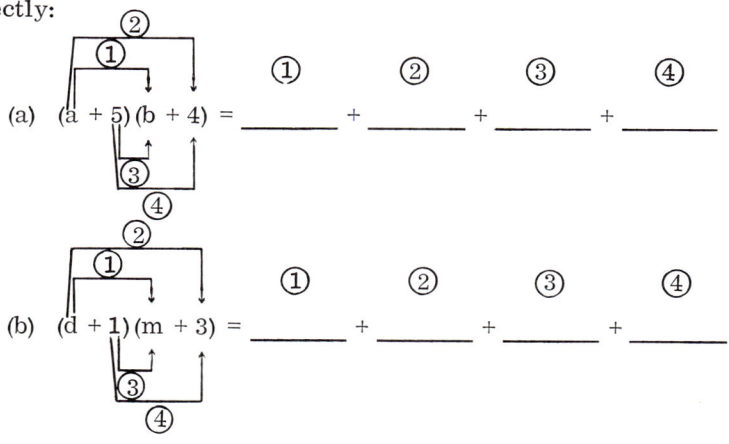

(a) $(a + 5)(b + 4) = $ _____ + _____ + _____ + _____

(b) $(d + 1)(m + 3) = $ _____ + _____ + _____ + _____

a) $ab + 4a + 5b + 20$

b) $dm + 3d + m + 3$

206   Quadratic Equations and Formulas

19. Here are the formal steps needed to multiply (a + 5) and (a - 2). Notice that we converted the subtraction to addition first.

$$(a + 5)(b - 2) = (a + 5)\left[b + (-2)\right] = a\left[b + (-2)\right] + 5\left[b + (-2)\right]$$
$$= ab + (-2a) + 5b + (-10)$$

After converting the subtraction to addition, we can write the four-term product directly. That is:

$$(a + 5)(b - 2) = (a + 5)\left[b + (-2)\right] = ab + (-2a) + 5b + (-10)$$

Again, we multiplied the first term of the first binomial times both terms of the second, and then multiplied the second term of the first binomial times both terms of the second.

Following the steps above, write the four-term product for each of these:

(a) $(m + 5)(t - 4) = (m + 5)\left[t + (-4)\right] = \underline{} + \underline{} + \underline{} + \underline{}$

(b) $(x + 7)(y - 6) = (x + 7)\left[y + (-6)\right] = \underline{} + \underline{} + \underline{} + \underline{}$

---

a) mt + (-4m) + 5t + (-20)

b) xy + (-6x) + 7y + (-42)

20. Here are the steps needed to multiply $(p - 4)$ and $(q + 9)$. Notice that we converted the subtraction to addition first.

$$(p - 4)(q + 9) = [p + (-4)](q + 9) = p(q + 9) + (-4)(q + 9)$$
$$= pq + 9p + (-4q) + (-36)$$

After converting the subtraction to addition, we can write the four-term product directly. That is:

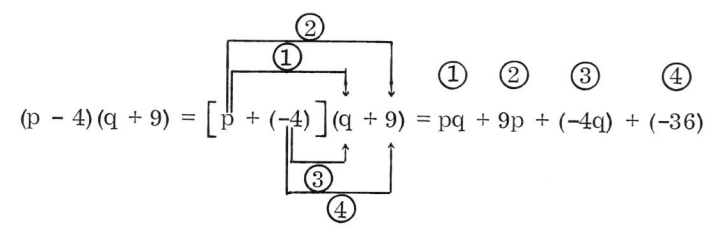

$$(p - 4)(q + 9) = [p + (-4)](q + 9) = pq + 9p + (-4q) + (-36)$$

Following the steps above, write the four-term product for each of these:

(a) $(a - 7)(c + 2) = [a + (-7)](c + 2) = \underline{} + \underline{} + \underline{} + \underline{}$

(b) $(h - 1)(p + 8) = [h + (-1)](p + 8) = \underline{} + \underline{} + \underline{} + \underline{}$

---

a) $ac + 2a + (-7c) + (-14)$

b) $hp + 8h + (-p) + (-8)$

208   Quadratic Equations and Formulas

21. Here are the formal steps needed to multiply (a − 5) and (t − 6). Notice that we converted both subtractions to additions first.

$$(a - 5)(t - 6) = [a + (-5)][t + (-6)] = a[t + (-6)] + (-5)[t + (-6)]$$
$$= at + (-6a) + (-5t) + 30$$

After converting both subtractions to additions, we can write the four-term product directly. That is:

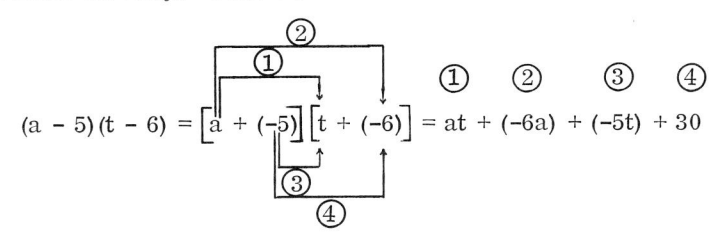

$$(a - 5)(t - 6) = [a + (-5)][t + (-6)] = at + (-6a) + (-5t) + 30$$
$$\ \ ①\ \ \ \ ②\ \ \ \ ③\ \ \ \ ④$$

Following the steps above, write the four-term product for each of these:

(a) $(k - 3)(m - 6) = [k + (-3)][m + (-6)] = \underline{} + \underline{} + \underline{} + \underline{}$

(b) $(a - 1)(b - 2) = [a + (-1)][b + (-2)] = \underline{} + \underline{} + \underline{} + \underline{}$

---

22. In the shorter method, we multiply both terms in the second binomial:

(1) by the <u>first</u> <u>term</u> of the <u>first</u> binomial,
and then (2) by the <u>second</u> <u>term</u> of the <u>first</u> binomial.

Using the shorter method, write each four-term product:

(a) $(c + 7)(d + 9) = \underline{} + \underline{} + \underline{} + \underline{}$

(b) $(R + 1)(S + 5) = \underline{} + \underline{} + \underline{} + \underline{}$

---

a) km + (−6k) + (−3m) + 18
b) ab + (−2a) + (−b) + 2

---

a) cd + 9c + 7d + 63
b) RS + 5R + S + 5

23. Before using the shorter method, any binomial which is a subtraction should be converted to an addition. These conversions help to minimize "sign" errors.

Write the four-term product for each of these:

(a) $(d + 2)(f - 8) = (d + 2)\left[f + (-8)\right]$ = _____ + _____ + _____ + _____

(b) $(c - 3)(t + 8) = \left[c + (-3)\right](t + 8)$ = _____ + _____ + _____ + _____

(c) $(R - 7)(V - 3) = \left[R + (-7)\right]\left[V + (-3)\right]$ = _____ + _____ + _____ + _____

---

24. In the preferred form of the product, any addition of a negative term is converted to a subtraction. Write the four-term product for each of these, and then write the product in the preferred form.

a) $df + (-8d) + 2f + (-16)$
b) $ct + 8c + (-3t) + (-24)$
c) $RV + (-3R) + (-7V) + 21$

(a) $(2p + 5)(3q - 1) = (2p + 5)\left[3q + (-1)\right]$ = _____ + _____ + _____ + _____

= _____

(b) $(a - 7)(5b + 4) = \left[a + (-7)\right](5b + 4)$ = _____ + _____ + _____ + _____

= _____

(c) $(2c - 1)(3d - 2) = \left[2c + (-1)\right]\left[3d + (-2)\right]$ = _____ + _____ + _____ + _____

= _____

---

25. After writing the four-term products for each of these, we can combine like terms. Do so.

a) $6pq + (-2p) + 15q + (-5)$
    $6pq - 2p + 15q - 5$
b) $5ab + 4a + (-35b) + (-28)$
    $5ab + 4a - 35b - 28$
c) $6cd + (-4c) + (-3d) + 2$
    $6cd - 4c - 3d + 2$

(a) $(x + 8)(x + 9)$ = _____ + _____ + _____ + _____

= _____

(b) $(2c + 1)(4c + 5)$ = _____ + _____ + _____ + _____

= _____

---

26. After writing the four-term product for each of these, write each product in the preferred form:

a) $x^2 + 9x + 8x + 72$
    $x^2 + 17x + 72$
b) $8c^2 + 10c + 4c + 5$
    $8c^2 + 14c + 5$

(a) $(y + 5)(y - 9) = (y + 5)\left[y + (-9)\right]$ = _____ + _____ + _____ + _____

= _____

(b) $(t - 4)(t + 3) = \left[t + (-4)\right](t + 3)$ = _____ + _____ + _____ + _____

= _____

(c) $(a - 7)(a - 9) = \left[a + (-7)\right]\left[a + (-9)\right]$ = _____ + _____ + _____ + _____

= _____

210 Quadratic Equations and Formulas

27. After writing each four-term product, write the product in the preferred form:

  (a) $(R + 7)(R - 7) = (R + 7)\bigl[R + (-7)\bigr]$ = ____ + ____ + ____ + ____

  = _____

  (b) $(m - 1)(m + 1) = \bigl[m + (-1)\bigr](m + 1)$ = ____ + ____ + ____ + ____

  = _____

a) $y^2 + (-9y) + 5y + (-45)$
   $y^2 - 4y - 45$

b) $t^2 + 3t + (-4t) + (-12)$
   $t^2 - t - 12$

c) $a^2 + (-9a) + (-7a) + 63$
   $a^2 - 16a + 63$

28. Write each final product in the preferred form:

  (a) $(2m + 3)(4m + 1) =$ _____

  (b) $(2t + 1)(3t - 4) =$ _____

  (c) $(5d - 2)(d + 3) =$ _____

  (d) $(3x - 1)(4x - 1) =$ _____

a) $R^2 + (-7R) + 7R + (-49)$
   $R^2 - 49$

b) $m^2 + m + (-m) + (-1)$
   $m^2 - 1$

29. Write each final product in the preferred form:

  (a) $(a + b)(c + d) =$ _____

  (b) $(p + q)(p - q) =$ _____

  (c) $(c - d)(p + q) =$ _____

  (d) $(a - b)(m - t) =$ _____

a) $8m^2 + 14m + 3$

b) $6t^2 - 5t - 4$

c) $5d^2 + 13d - 6$

d) $12x^2 - 7x + 1$

a) $ac + ad + bc + bd$

b) $p^2 - q^2$

c) $cp + cq - dp - dq$

d) $am - at - bm + bt$

35. In the blank in front of each quadratic equation below, identify whether it is "complete", "incomplete", or "pure".

    _____ (a) $y^2 - 4 = 0$

    _____ (b) $t^2 - 4t + 3 = 0$

    _____ (c) $R^2 - 3R = 0$

    _____ (d) $3m^2 + 14m + 8 = 0$

36. In the blank in front of each quadratic equation below, identify its type:

    _____ (a) $3x^2 - 7x = 0$

    _____ (b) $9d^2 - 64 = 0$

    _____ (c) $p^2 - 12p + 35 = 0$

    _____ (d) $V^2 + 6V = 0$

    a) Pure
    b) Complete
    c) Incomplete
    d) Complete

37. We have generated a quadratic equation below by multiplying the monomial and the binomial on the left side.

    $$x(x - 3) = 0$$
    $$x^2 - 3x = 0$$

    What type of quadratic have we generated? _____

    a) Incomplete
    b) Pure
    c) Complete
    d) Incomplete

38. We have generated a quadratic equation below by multiplying the two binomials on the left side.

    $$(x + 3)(x + 2) = 0$$
    $$x^2 + 5x + 6 = 0$$

    What type of quadratic have we generated? _____

    "Incomplete" quadratic

39. We have generated a quadratic equation below by multiplying the two binomials on the left side.

    $$(x + 4)(x - 4) = 0$$
    $$x^2 - 16 = 0$$

    What type of quadratic have we generated? _____

    "Complete" quadratic

40. In each case below, perform the multiplication on the left side and then identify the type of quadratic which is generated:

    (a) $(y - 5)(y + 5) = 0$

    _____ $= 0$ _____

    (b) $t(t - 6) = 0$

    _____ $= 0$ _____

    (c) $(p + 2)(p - 3) = 0$

    _____ $= 0$ _____

    "Pure" quadratic

214 Quadratic Equations and Formulas

41. In each case below, perform the multiplication on the left side and then identify the type of quadratic which is generated:

(a) $(2f + 3)(f - 1) = 0$

_____ $= 0$ _____

(b) $x(2x - 1) = 0$

_____ $= 0$ _____

(c) $(2t + 1)(2t - 1) = 0$

_____ $= 0$ _____

a) $y^2 - 25 = 0$
   "Pure" quadratic

b) $t^2 - 6t = 0$
   "Incomplete" quadratic

c) $p^2 - p - 6 = 0$
   "Complete" quadratic

Answer to Frame 41:   a) $2f^2 + f - 3 = 0$ b) $2x^2 - x = 0$ c) $4t^2 - 1 = 0$
                         "Complete" quadratic   "Incomplete" quadratic   "Pure" quadratic

---

SELF-TEST 4 (Frames 31-41)

State whether each quadratic equation is "complete", "incomplete", or "pure".

1. $2t^2 - 5t = 0$	2. $4d^2 - 9 = 0$	3. $y^2 - 7y + 10 = 0$
_____	_____	_____

In each problem below, do the multiplication. Then state whether the resulting quadratic equation is "complete", "incomplete", or "pure".

4. $(3R + 2)(3R - 2) = 0$	5. $(3s - 1)(2s + 3) = 0$	6. $x(4x + 5) = 0$
_____ $= 0$	_____ $= 0$	_____ $= 0$
_____	_____	_____

ANSWERS: 1. Incomplete   4. $9R^2 - 4 = 0$ 5. $6s^2 + 7s - 3 = 0$ 6. $4x^2 + 5x = 0$
           2. Pure   Pure   Complete   Incomplete
           3. Complete

3-5 THE BASIC PRINCIPLE USED IN SOLVING QUADRATIC EQUATIONS BY THE FACTORING METHOD

All three types of quadratic equations can be solved by the factoring method. There is one basic principle involved in the factoring method of solving all three types. We will discuss this basic principle in this section.

42. The term "ab" means: multiply "a" and "b".

    (a) If $a = 5$ and $b = 0$, $ab = (5)(0) =$ _____

    (b) If $a = 0$ and $b = 7$, $ab = (0)(7) =$ _____

43. The term "pq" means: multiply "p" and "q".

    (a) If $p = 0$ and $q = 10$, $pq =$ _____   (c) If $p = 0$ and $q = -5$, $pq =$ _____

    (b) If $p = 0$ and $q = 5$, $pq =$ _____   (d) If $p = 0$ and $q = -10$, $pq =$ _____

a) 0
b) 0

44. The term "dk" means: multiply "d" and "k".

    (a) If $d = 8$ and $k = 0$, $dk =$ _____   (c) If $d = -4$ and $k = 0$, $dk =$ _____

    (b) If $d = 4$ and $k = 0$, $dk =$ _____   (d) If $d = -8$ and $k = 0$, $dk =$ _____

a) 0   c) 0
b) 0   d) 0

45. In the multiplication "RS":

    (a) If we plug in "0" for R, no matter what number we plug in for S, the product will be _____.

    (b) If we plug in "0" for S, no matter what number we plug in for R, the product will be _____.

a) 0   c) 0
b) 0   d) 0

46. The equation below says this: if we multiply (a) and (b), their product is "0".

$$ab = 0$$

    (a) If we plug in "7" for "a", what number must we plug in for "b" to make their product equal "0"? _____

    (b) If we plug in "13" for "b", what number must we plug in for "a" to make their product equal "0"? _____

a) 0
b) 0

a) 0
b) 0

216   Quadratic Equations and Formulas

47. The point we are trying to make is this:

   <u>If</u> <u>the</u> <u>product</u> <u>of</u> <u>two</u> <u>factors</u> <u>is</u> "<u>0</u>", <u>then</u> <u>at</u> <u>least</u> <u>one</u> <u>of</u> <u>the</u> <u>two</u> <u>factors</u> <u>must</u> <u>equal</u> "<u>0</u>".

   This basic principle is written in general form below:

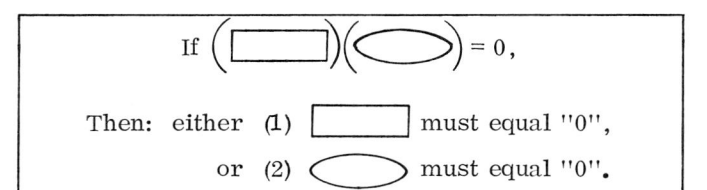

   If (☐)(◯) = 0,

   Then: either (1) ☐ must equal "0",

   or (2) ◯ must equal "0".

   That is, if PQ = 0, then either _____ or _____ must equal "0".

48. Here is another example of two factors whose product is "0". The factors in this case are binomials.

   $\boxed{x - 2}\ \widehat{x - 3} = 0$

   If we want their product to be "0",

   either (1) $\boxed{x - 2}$ must equal "0".

   or (2) $\widehat{x - 3}$ must equal "0".

   (a) What value of "x" will make the factor (x − 2) equal "0"? _____
   (b) What value of "x" will make the factor (x − 3) equal "0"? _____

   | | P or Q |

49. Both factors on the left below are binomials.

   (y + 5)(y − 4) = 0

   Since their product is "0": either (y + 5) must equal "0",

   or (y − 4) must equal "0".

   (a) If y + 5 = 0, y = _____.
   (b) If y − 4 = 0, y = _____.

   | | a) +2 |
   | | b) +3 |

50. Both factors on the left below are binomials:

   (3m − 7)(4m + 5) = 0

   Since their product is "0": either (3m − 7) must equal "0",

   or (4m + 5) must equal "0".

   (a) If 3m − 7 = 0, m = _____.
   (b) If 4m + 5 = 0, m = _____.

   | | a) −5 |
   | | b) 4 |

   | | a) $\frac{7}{3}$ |
   | | b) $-\frac{5}{4}$ |

Quadratic Equations and Formulas 217

51. The equation below is true if either (x − 5) or (x − 7) is "0".

$$(x - 5)(x - 7) = 0$$

(x − 5) is "0" when "x" equals 5.
(x − 7) is "0" when "x" equals 7.

Therefore, the two roots of the equation are _____ and _____.

---

52. The equation below is true if either (3y − 2) or (2y + 5) is "0".

$$(3y - 2)(2y + 5) = 0$$

(3y − 2) is "0" when $y = \frac{2}{3}$.

(2y + 5) is "0" when $y = -\frac{5}{2}$.

Therefore, the two roots of the equation are _____ and _____.

5 and 7

---

53. The equation below is true if either (t + 7) or (t − 5) is "0".

$$(t + 7)(t - 5) = 0$$

(a) t + 7 = 0 when t = _____   (b) t − 5 = 0 when t = _____

(c) Therefore, the two roots of the equation are _____ and _____.

$\frac{2}{3}$ and $-\frac{5}{2}$

---

54. The equation below is true if either (3p + 5) or (4p − 7) is "0".

$$(3p + 5)(4p - 7) = 0$$

(a) 3p + 5 = 0 when p = _____   (b) 4p − 7 = 0 when p = _____

(c) Therefore, the two roots of the equation are _____ and _____.

a) −7   b) 5
c) −7 and 5

---

55. Using the same principle, write the two roots for each equation:

(a) (t + 4)(t − 4) = 0   t = _____ and _____

(b) (3R + 8)(9R − 1) = 0   R = _____ and _____

(c) (4w − 9)(5w + 3) = 0   w = _____ and _____

a) $-\frac{5}{3}$ b) $\frac{7}{4}$

c) $-\frac{5}{3}$ and $\frac{7}{4}$

218 Quadratic Equations and Formulas

56. In the equation below, the first factor (y) is a monomial; the second factor (y - 2) is a binomial.

$$y(y - 2) = 0$$

Their product can be "0" only if: either (1) (y) is "0",
or (2) (y - 2) is "0".

Therefore, the two roots of the equation are _____ and _____.

| a) -4 and 4
| b) $-\frac{8}{3}$ and $\frac{1}{9}$
| c) $\frac{9}{4}$ and $-\frac{3}{5}$

---

57. In the last frame, we saw that "0" is one of the roots of: y(y - 2) = 0. This "0" root is frequently overlooked.

Write the two roots of each of these equations:

(a) S(S + 9) = 0   S = _____ and _____

(b) v(v - 10) = 0   v = _____ and _____

| 0 and 2

---

58. Write the two roots of each of these equations:

(a) t(5t + 11) = 0   t = _____ and _____

(b) x(3x - 4) = 0   x = _____ and _____

| a) 0 and -9
| b) 0 and 10

---

59. In the equation below, the first factor (3x) is a letter-term.

$$3x(x + 2) = 0$$

The product of the factors can be "0" only if:

either (1) 3x = 0,
or (2) x + 2 = 0.

Therefore, the two roots of the equation are _____ and _____.

| a) 0 and $-\frac{11}{5}$
| b) 0 and $\frac{4}{3}$

---

60. Here is the principle we have been using, in symbol form.

If (▢)(◯) = 0,
Then: either (1) ▢ must equal "0",
or (2) ◯ must equal "0".

This principle applies only when the product of the two factors is "0". IF THEIR PRODUCT IS NOT "0", THE PRINCIPLE DOES NOT APPLY. For example:

If (x - 3)(x - 9) = 5, we cannot say that either (x - 3) or (x - 9) must equal "5".

In which cases below does the principle apply? _____

(a) (t + 5)(t - 9) = 0   (c) y(y - 3) = 8

(b) (3x - 1)(2x + 3) = 9   (d) m(m + 1) = 0

| 0 and -2

61. (a) If (t + 3)(t - 4) = 0, can we say that either (t + 3) or (t - 4) must equal "0"? _____

(b) If (R - 1)(R + 1) = 9, can we say that either (R - 1) or (R + 1) must equal "9"? _____

(c) If x(x - 7) = 0, can we say that either (x) or (x - 7) must equal "0"? _____

(d) If V(V + 6) = 8, can we say that either (V) or (V + 6) must equal "8"? _____

Only in (a) and (d)

Answer to Frame 61:   a) Yes   b) No   c) Yes   d) No

SELF-TEST 5 (Frames 42-61)

Given: (x - 7)(2x + 5) = 0

1. What value of "x" will make the factor (x - 7) equal "0"? _____
2. What value of "x" will make the factor (2x + 5) equal "0"? _____
3. The two roots of the equation are _____ and _____.

Find the two roots of each of the following equations:

4. (y + 1)(y - 4) = 0

    y = _____ and _____

5. E(E + 6) = 0

    E = _____ and _____

6. (2t - 1)(3t + 6) = 0

    t = _____ and _____

State whether each of the following is "True" or "False":

_____ 7. To find the roots of (w + 2)(w - 5) = 7, we can say that either (w + 2) or (w - 5) must equal "7".

_____ 8. To find the roots of (3R + 4)(R - 1) = 0, we can say that either (3R + 4) or (R - 1) must equal "0".

ANSWERS:
1. 7
2. $-\frac{5}{2}$
3. 7 and $-\frac{5}{2}$
4. -1 and 4
5. 0 and -6
6. $\frac{1}{2}$ and -2
7. False
8. True

220 Quadratic Equations and Formulas

3-6 SOLVING INCOMPLETE QUADRATIC EQUATIONS BY THE FACTORING METHOD

In the last section, we discussed the basic principle needed to solve quadratic equations by the factoring method. In this section, we will show how the factoring method can be used to solve "incomplete" quadratic equations.

62. The equation below is an "incomplete" quadratic equation:
$$x^2 - 7x = 0$$
Using the distributive principle, we can factor the left side. We get:
$$x(x - 7) = 0$$
The two roots of this last equation are 0 and 7. Show that these two roots satisfy the original equation, $x^2 - 7x = 0$.

63. Here is another incomplete quadratic equation: $T^2 + 5T = 0$

Factoring the left side, we get: $T(T + 5) = 0$

(a) The two roots of this last equation are ____ and ____.

(b) Therefore, the two roots of the original equation must be ____ and ____.

$0^2 - 7(0) = 0$
$0 - 0 = 0$
$0 = 0$
$7^2 - 7(7) = 0$
$49 - 49 = 0$
$0 = 0$

64. Let's solve: $y^2 + y = 0$

(a) Factoring the left side, we get: _____ = 0

(b) The two roots of this new equation are ____ and ____.

(c) The two roots of the original equation must be ____ and ____.

a) 0 and -5
b) 0 and -5

65. Let's solve: $x^2 - x = 0$

(a) Factoring the left side, we get: _____ = 0

(b) The two roots of this new equation are ____ and ____.

(c) The two roots of the original equation must be ____ and ____.

a) $y(y + 1) = 0$
b) 0 and -1
c) 0 and -1

66. Let's solve: $t^2 - 9t = 0$

(a) Factoring the left side, we get: _____ = 0

(b) The two roots of the original equation are ____ and ____.

a) $x(x - 1) = 0$
b) 0 and +1
c) 0 and +1

a) $t(t - 9) = 0$
b) 0 and +9

Quadratic Equations and Formulas 221

67. Use the factoring method to find the two roots of each of these:

(a) $m^2 - 8m = 0$ m = _____ and _____

(b) $d^2 + 10d = 0$ d = _____ and _____

---

68. The equation at the right is also an incomplete quadratic equation: $3V^2 - 10V = 0$

If we factor the left side, we get: $V(3V - 10) = 0$

The two roots of this new equation are 0 and $\frac{10}{3}$. Show that these two roots satisfy the original equation, $3V^2 - 10V = 0$.

a) 0 and +8, from:
   $m(m - 8) = 0$

b) 0 and −10, from:
   $d(d + 10) = 0$

---

69. Let's solve: $5t^2 - 2t = 0$

(a) Factoring the left side, we get: _____ = 0

(b) The two roots are _____ and _____.

$3(0^2) - 10(0) = 0$
$0 - 0 = 0$
$0 = 0$

$3\left(\frac{10}{3}\right)^2 - 10\left(\frac{10}{3}\right) = 0$

$3\left(\frac{100}{9}\right) - \frac{100}{3} = 0$

$\frac{100}{3} - \frac{100}{3} = 0$

$0 = 0$

---

70. Let's solve: $9y^2 + 7y = 0$

(a) Factoring the left side, we get: _____ = 0

(b) The two roots are _____ and _____.

a) $t(5t - 2) = 0$

b) 0 and $\frac{2}{5}$

---

71. Use the factoring method to find the two roots of each of these:

(a) $6s^2 - 5s = 0$ (b) $3p^2 + 14p = 0$

s = _____ and _____   p = _____ and _____

a) $y(9y + 7) = 0$

b) 0 and $-\frac{7}{9}$

222 Quadratic Equations and Formulas

72. We can factor both a number and a letter out of the left side of the incomplete quadratic equation below. We get:

$$16t^2 - 32t = 0$$
$$16t(t - 2) = 0$$

The new equation is true if either: (1) $16t = 0$
or (2) $t - 2 = 0$

Therefore, the two roots of the equation are _____ and _____.

a) 0 and $\frac{5}{6}$, from:

$s(6s - 5) = 0$

b) 0 and $-\frac{14}{3}$, from:

$p(3p + 14) = 0$

---

73. Let's solve: $2x^2 + 6x = 0$

(a) Factoring both a number and a letter out of the left side, we get: _____ = 0

(b) The two roots are _____ and _____.

0 and 2

---

74. Use the factoring method to solve each quadratic equation below:

(a) $4y^2 - 20y = 0$ (b) $3d^2 + 12d = 0$

y = _____ and _____   d = _____ and _____

a) $2x(x + 3) = 0$

b) 0 and −3

---

75. When factoring an "incomplete" quadratic equation, we always get a letter (or letter-term) as one of the factors. For example:

$x^2 + 7x = 0$ $5t^2 - 15t = 0$

$x(x + 7) = 0$ $5t(t - 3) = 0$

Since we always get a letter (or letter-term) as one of the factors, what number is always one of the roots of an incomplete quadratic equation? _____

a) y = 0 and 5

b) d = 0 and −4

---

76. Since the equation below contains a squared-letter term and a letter-term, it is an "incomplete" quadratic equation. However, the two terms are on opposite sides of the equal sign.

$$x^2 = 3x$$

To solve this equation by the factoring method, we must:

(1) Use the addition axiom to get both terms on one side, leaving a "0" on the other side.

$$x^2 - 3x = 0$$

(2) Factor the left side as usual.

$$x(x - 3) = 0$$

The two roots of the equation are _____ and _____.

0

---

0 and +3

77. Following the steps in the last frame, find the two roots of each of these equations:

(a) $m^2 = 8m$ (b) $2R^2 = 8R$ (c) $5t^2 = -3t$

m = ___ and ___   R = ___ and ___   t = ___ and ___

Answer to Frame 77:   a) 0 and +8   b) 0 and +4   c) 0 and $-\frac{3}{5}$

SELF-TEST 6 (Frames 62-77)

Solve each of these incomplete quadratic equations:

1. $h^2 - 5h = 0$

   h = ___ and ___

2. $2E^2 + E = 0$

   E = ___ and ___

3. $12x^2 - 8x = 0$

   x = ___ and ___

4. $4y^2 + 7y = 0$

   y = ___ and ___

5. $2t^2 = 5t$

   t = ___ and ___

6. $3r^2 = -6r$

   r = ___ and ___

ANSWERS:
1. h = 0 and 5
2. E = 0 and $-\frac{1}{2}$
3. x = 0 and $\frac{2}{3}$
4. y = 0 and $-\frac{7}{4}$
5. t = 0 and $\frac{5}{2}$
6. r = 0 and -2

224 Quadratic Equations and Formulas

3-7 SOLVING "PURE" QUADRATICS BY THE FACTORING METHOD

In an earlier chapter, we solved "pure" quadratic equations by isolating the squared letter first. In this section, we will show the factoring method of solving the same type of equations. To do so, we must discuss the procedure for factoring a subtraction of two perfect squares. The section is concluded with a brief discussion of "pure" quadratic equations whose roots are not ordinary numbers.

78. The equation below is a "pure" quadratic equation.

$$x^2 - 9 = 0$$

In earlier work, we solved an equation of this type by:

(1) Isolating the squared letter:

$$x^2 = 9$$

(2) And then taking both the positive and negative square root of the number:

$$x = \pm\sqrt{9} \quad \text{or} \quad x = +3 \text{ and } -3$$

Following the steps above, solve each of these "pure" quadratic equations:

(a) $y^2 - 25 = 0$ (b) $4t^2 - 9 = 0$ (c) $9d^2 - 1 = 0$

y = ____ and ____   t = ____ and ____   d = ____ and ____

79. "Pure" quadratic equations can also be solved by the factoring method. Before doing so, however, we must discuss the method of factoring binomials like $x^2 - 9$ and $25t^2 - 16$.

We have performed a multiplication of two binomials below. Notice that the four-term product reduces to a binomial since two of the terms are a pair of opposites.

$$(x + 4)(x - 4) = x^2 - 16$$

Let's examine both sides of the above equation:

<u>On the left:</u>   (1) Both factors contain an "x" and a "4".
(2) However, one factor is an addition and the other is a subtraction.

<u>On the right:</u> (1) x^2 is the square of "x".
(2) 16 is the square of "4".
(3) These two perfect squares are <u>subtracted</u>.

Following the pattern above, complete these:

(a) $(y + 3)(y - 3) =$ _____ - _____

(b) $(m + 5)(m - 5) =$ _____ - _____

a) +5 and −5

b) $+\dfrac{3}{2}$ and $-\dfrac{3}{2}$

c) $+\dfrac{1}{3}$ and $-\dfrac{1}{3}$

a) $\underline{y^2} - \underline{9}$
b) $\underline{m^2} - \underline{25}$

Quadratic Equations and Formulas 225

80. Here is another multiplication of an addition and a subtraction containing the same terms:

$$(2s + 5)(2s - 5) = 4s^2 - 25$$

Notice these points about the binomial product:

(1) $4s^2$ is the square of "2s".
(2) 25 is the square of "5".
(3) These two perfect squares are <u>subtracted</u>.

Following the pattern above, complete these:

(a) $(3y + 2)(3y - 2) = $ _____ - _____
(b) $(4t + 7)(4t - 7) = $ _____ - _____

---

81. In the last two frames, we showed the following pattern:

$$(\triangle + \bigcirc)(\triangle - \bigcirc) = (\triangle)^2 - (\bigcirc)^2$$

Following the pattern above, write each of the following products:

(a) $(t + 6)(t - 6) = $ _____   (c) $(5d + 3)(5d - 3) = $ _____
(b) $(m + 1)(m - 1) = $ _____   (d) $(9x + 1)(9x - 1) = $ _____

a) $9y^2 - 4$
b) $16t^2 - 49$

---

82. Using the same pattern, write each product below:

(a) $(c + d)(c - d) = $ _____
(b) $(pq + m)(pq - m) = $ _____

a) $t^2 - 36$ c) $25d^2 - 9$
b) $m^2 - 1$ d) $81x^2 - 1$

---

83. $(x + 3)(x - 3) = x^2 - 9$

We have interchanged the product and the factors below:

$$x^2 - 9 = (x + 3)(x - 3)$$

Since factoring is a process of getting from a product back to the original factors, the last equation above represents a factoring of $x^2 - 9$. Notice these points about the factoring:

(1) "x" is the principal square root of "x^2".
(2) "3" is the principal square root of "9".
(3) One factor is an addition; the other is a subtraction.

Following the pattern above, complete these factorings:

(a) $y^2 - 25 = (y + 5)($ $)$
(b) $t^2 - 64 = ($ $)(t - 8)$
(c) $p^2 - 1 = ($ $)($ $)$

a) $c^2 - d^2$
b) $p^2q^2 - m^2$

---

a) $(y - 5)$
b) $(t + 8)$
c) $(p + 1)(p - 1)$

226 Quadratic Equations and Formulas

84. The equation below shows a factoring of $9m^2 - 36$.

$$9m^2 - 36 = (3m + 6)(3m - 6)$$

Notice these points about the factoring:

(1) "3m" is the principal square root of "$9m^2$".
(2) "6" is the principal square root of "36".
(3) One factor is an addition; the other is a subtraction.

Following the pattern above, complete these factorings:

(a) $4x^2 - 9 = (2x + 3)($ $)$

(b) $16t^2 - 49 = ($ $)(4t - 7)$

(c) $(25R^2 - 1) = ($ $)($ $)$

---

85. In the last few frames, we showed the following pattern of factoring a subtraction of two perfect squares.

$$(\triangle)^2 - (\bigcirc)^2 = (\triangle + \bigcirc)(\triangle - \bigcirc)$$

Using this pattern, factor each of the following:

(a) $F^2 - 81 = ($ $)($ $)$ (c) $49y^2 - 100 = ($ $)($ $)$

(b) $c^2 - 1 = ($ $)($ $)$ (d) $64x^2 - 1 = ($ $)($ $)$

a) $(2x - 3)$
b) $(4t + 7)$
c) $(5R + 1)(5R - 1)$

---

86. Using the same pattern, factor these. <u>Don't forget to write parentheses around each binomial factor.</u>

(a) $V^2 - t^2 =$ _____

(b) $9m^2 - 16b^2 =$ _____

(c) $a^2b^2 - 49d^2 =$ _____

a) $(F + 9)(F - 9)$
b) $(c + 1)(c - 1)$
c) $(7y + 10)(7y - 10)$
d) $(8x + 1)(8x - 1)$

---

87. Factor each of these binomials:

(a) $25 - t^2 =$ _____

(b) $1 - 4x^2 =$ _____

a) $(V + t)(V - t)$
b) $(3m + 4b)(3m - 4b)$
c) $(ab + 7d)(ab - 7d)$

---

88. The "pure" quadratic equation below can be solved by the factoring method.

$$x^2 - 16 = 0$$

To do so, we use the pattern to factor the left side. We get:

$$(x + 4)(x - 4) = 0$$

The product of $(x + 4)$ and $(x - 4)$ can be "0" only if one of the binomials equals "0". Therefore, the two roots of the bottom equation are -4 and $+4$.

Show that these two roots satisfy the original equation, $x^2 - 16 = 0$.

a) $(5 + t)(5 - t)$
b) $(1 + 2x)(1 - 2x)$

89. Let's solve $S^2 - 64 = 0$ by the factoring method.

    (a) Factor the left side: (   )(   ) = 0

    (b) The two roots are _____ and _____.

$(-4)^2 - 16 = 0$
$16 - 16 = 0$
$(+4)^2 - 16 = 0$
$16 - 16 = 0$

---

90. Let's solve $4x^2 - 9 = 0$ by the factoring method.

    (a) Factoring the left side, we get: (   )(   ) = 0

    (b) The two roots are _____ and _____.

a) $(S + 8)(S - 8) = 0$
b) -8 and $+8$

---

91. Let's solve $25 - t^2 = 0$ by the factoring method.

    (a) Factoring the left side, we get: (   )(   ) = 0

    (b) The two roots are _____ and _____.

a) $(2x + 3)(2x - 3) = 0$
b) $-\frac{3}{2}$ and $+\frac{3}{2}$

---

92. Let's solve $25 - 100d^2 = 0$ by the factoring method.

    (a) Factoring the left side, we get: (   )(   ) = 0

    (b) The two roots are _____ and _____.

a) $(5 + t)(5 - t) = 0$
b) -5 and $+5$

---

93. Solve each of these by the factoring method.

    (a) $16T^2 - 81 = 0$ (b) $25 - 49x^2 = 0$

    T = _____ and _____   x = _____ and _____

a) $(5 + 10d)(5 - 10d) = 0$
b) $-\frac{1}{2}$ and $+\frac{1}{2}$

(Did you reduce to lowest terms?)

---

94. The equation below is also a "pure" quadratic equation since it contains a squared-letter term and a number-term. However, the terms are on opposite sides.

$$y^2 = 49$$

Though we can easily solve it by taking both square roots of 49, we can also solve it by the factoring method. To do so, we:

    (1) Use the addition axiom to get both terms on one side: $y^2 - 49 = 0$

    (2) Factor the left side as usual: $(y + 7)(y - 7) = 0$

The two roots of the equation are _____ and _____.

a) $T = -\frac{9}{4}$ and $+\frac{9}{4}$
b) $x = -\frac{5}{7}$ and $+\frac{5}{7}$

-7 and $+7$

228 Quadratic Equations and Formulas

95. Show the steps needed to solve the "pure" quadratic at the right by the factoring method.

$4m^2 = 25$

---

96. The factoring pattern we have been using is generally used only when both terms in the subtraction are perfect squares. That is:

It <u>is</u> <u>used</u> with binomials like $y^2 - 49$ or $9t^2 - 64$.

It <u>is</u> <u>not</u> <u>used</u> with binomials like $y^2 - 17$ or $7t^2 - 64$.

The pattern would ordinarily be used to factor which binomials below? _____

(a) $8x^2 - 25$ (b) $4t^2 - 81$ (c) $16 - y^2$ (d) $81 - 11m^2$

$4m^2 - 25 = 0$

$(2m + 5)(2m - 5) = 0$

$m = -\frac{5}{2}$ or $+\frac{5}{2}$

---

97. Since the factoring pattern is not generally used with binomials like those on the left side of each equation below, "pure" quadratics of this type are not solved by the factoring method.

$t^2 - 17 = 0$ $\qquad$ $3x^2 - 4 = 0$

To solve "pure" quadratics of this type, we isolate the squared letter and then take both square roots of the quantity on the right side. We get:

$t^2 = 17$ $\qquad$ $x^2 = \frac{4}{3}$

$t = \pm\sqrt{17}$ $\qquad$ $x = \pm\sqrt{\frac{4}{3}}$

(a) Since $\sqrt{17}$ is 4.12, the two roots of $t^2 - 17 = 0$ are _____ and _____.

(b) Since $\sqrt{\frac{4}{3}} = 1.15$, the two roots of $3x^2 - 4 = 0$ are _____ and _____.

Only (b) and (c)

---

98. Here is the same factoring pattern:

$$\boxed{(\triangle)^2 - (\bigcirc)^2 = (\triangle + \bigcirc)(\triangle - \bigcirc)}$$

This pattern applies only to a subtraction of two perfect squares. It <u>does</u> <u>not</u> <u>apply</u> <u>to</u> <u>an</u> <u>addition</u> <u>of</u> <u>two</u> <u>perfect</u> <u>squares</u>.

The pattern can be used to factor which binomials below? _____

(a) $x^2 - 9$ (b) $x^2 + 9$ (c) $4y^2 - 25$ (d) $4y^2 + 25$

a) +4.12 and −4.12
b) +1.15 and −1.15

Only (a) and (c)

99. Which of the following statements are true? _____

    (a) $x^2 + 4 = (x + 2)(x - 2)$

    (b) $x^2 - 4 = (x + 2)(x - 2)$

100. Which of the following statements are true? _____

    (a) $9v^2 - 64 = (3v + 8)(3v - 8)$

    (b) $9v^2 + 64 = (3v + 8)(3v - 8)$

Only (b)

101. The equation below is a "pure" quadratic. However, since the binomial on the left side is an addition, the equation cannot be solved by the factoring method which we have been using.

$$t^2 + 25 = 0$$

If we try to solve it by isolating the squared letter and then taking both square roots of the quantity on the right, we get:

$$t^2 = -25$$

$$t = \pm\sqrt{-25}$$

However, we have already seen that no ordinary positive or negative number is the square root of a negative number like -25. That is:

Since $(+5)(+5) = +25$, +5 is not the square root of -25.

Since $(-5)(-5) = +25$, -5 is not the square root of -25.

Since no ordinary positive or negative number is the square root of -25, does $t^2 + 25 = 0$ have two ordinary numbers as its roots? _____

Only (a)

102. We also cannot solve the "pure" quadratic below by the factoring method we have been using since $9m^2 + 4$ is an addition.

$$9m^2 + 4 = 0$$

If we try to solve the equation by isolating the squared letter, we get:

$$m^2 = -\frac{4}{9}$$

$$m = \pm\sqrt{-\frac{4}{9}}$$

(a) Is $+\frac{2}{3}$ a square root of $-\frac{4}{9}$? _____

(b) Is $-\frac{2}{3}$ a square root of $-\frac{4}{9}$? _____

(c) Does $9m^2 + 4 = 0$ have two ordinary numbers as its roots? _____

No

103. Which of the following quadratic equations do not have two ordinary numbers as their roots? _____

    (a) $x^2 - 16 = 0$ (c) $t^2 = 17$ (e) $81m^2 + 1 = 0$

    (b) $y^2 + 64 = 0$ (d) $16d^2 - 25 = 0$ (f) $49R^2 = 36$

a) No, since: $\left(+\frac{2}{3}\right)\left(+\frac{2}{3}\right) = +\frac{4}{9}$

b) No, since: $\left(-\frac{2}{3}\right)\left(-\frac{2}{3}\right) = +\frac{4}{9}$

c) No

Answer to Frame 103:   (b) and (e)

SELF-TEST 7 (Frames 78-103)

Multiply:

1. $(w + 7)(w - 7) = $ _____

2. $(3t + 5)(3t - 5) = $ _____

Factor:

3. $9 - x^2 = $ _____

4. $4a^2 - 49 = $ _____

Solve each of these quadratic equations by means of the factoring method:

5. $9d^2 - 4 = 0$

     d = _____ and _____

6. $49 - 4r^2 = 0$

     r = _____ and _____

7. $16x^2 = 25$

     x = _____ and _____

8. Find the two roots of: $4y^2 = 17$, if $\sqrt{\frac{17}{4}} = 2.06$.   y = _____ and _____

9. Which of the following quadratic equations do not have two ordinary numbers as their roots? _____

    (a) $p^2 - 1 = 0$ (b) $s^2 + 1 = 0$ (c) $2t^2 = 35$ (d) $4x^2 + 25 = 0$ (e) $9 = 16r^2$

ANSWERS:

1. $w^2 - 49$
2. $9t^2 - 25$
3. $(3 + x)(3 - x)$
4. $(2a + 7)(2a - 7)$
5. $d = -\frac{2}{3}$ and $+\frac{2}{3}$
6. $r = -\frac{7}{2}$ and $+\frac{7}{2}$
7. $x = -\frac{5}{4}$ and $+\frac{5}{4}$
8. $y = +2.06$ and -2.06
9. (b) and (d)

Quadratic Equations and Formulas   231

3-8   THE FACTORING METHOD AND "COMPLETE" QUADRATICS OF THE FORM: $1x^2 + bx + c = 0$

Each of the equations below is a "complete" quadratic. The coefficient of the squared letter in each case is "1".

$$x^2 + 5x + 6 = 0 \qquad y^2 - 7y + 12 = 0 \qquad t^2 + t - 6 = 0$$

In this section, we will solve quadratic equations of this type by the factoring method. In order to do so, we must discuss the procedure for factoring trinomials like those in the equations above.

---

104. Each binomial below contains the letter "x" as one term and a number as the other term. We have written the four-terms of their product.

$$(x + 2)(x + 3) = x^2 + 3x + 2x + 6$$

The two "like" terms in the product can be combined since we can factor by the distributive principle and simplify. The formal steps are given below.

$$(x + 2)(x + 3) = x^2 + (3 + 2)x + 6$$
$$= x^2 + 5\ x + 6$$

Notice these points about the terms in the trinomial product:

(1) Since the coefficient of each "x" in the original factors is "1", the coefficient of the "x^2" term is "1".

(2) The coefficient of the "x" term is "5". This coefficient is obtained <u>by adding the numbers in each original binomial</u>.

(3) The number-term is "6". This number term is obtained <u>by multiplying the numbers in each original binomial</u>.

By using the facts above, we can immediately write the trinomial product for this multiplication.

$$(y + 3)(y + 4)$$

(a) Since the coefficient of each "y" is "1", the coefficient of the "y^2" term of the product is _____.

(b) Since $3 + 4 = 7$, the coefficient of the "y" term of the product must be _____.

(c) Since $(3)(4) = 12$, the number-term of the product must be _____.

(d) Therefore, $(y + 3)(y + 4) =$ _____ + _____ + _____

---

105. Using the procedure in the last frame, write the trinomial product for each of these:

(a) $(d + 1)(d + 4) =$ _____ + _____ + _____

(b) $(R + 6)(R + 2) =$ _____ + _____ + _____

a) 1
b) 7
c) 12
d) $y^2 + 7y + 12$
   or
   $1y^2 + 7y + 12$

a) $d^2 + 5d + 4$
b) $R^2 + 8R + 12$

232   Quadratic Equations and Formulas

106. In the multiplication below, the second binomial factor is a subtraction. We have converted this subtraction to an addition and then performed the multiplication.

$$(t + 2)(t - 4) = (t + 2)\left[t + (-4)\right] = t^2 + (-4t) + 2t + (-8)$$
$$= t^2 + \left[(-4) + 2\right]t + (-8)$$
$$= t^2 + \quad (-2t) \quad + (-8)$$

Notice the same points about the terms in the trinomial product.

(1) Since the coefficient of each "t" in the factors is "1", the coefficient of the "t^2" term is "1".

(2) Since $(-4) + 2 = -2$, the coefficient of the "t" term is "-2".

(3) Since $(2)(-4) = -8$, the number-term is "-8".

By using the facts above, we can immediately write the trinomial product for the multiplication below. We have converted the subtraction-binomial to an addition because the facts above <u>apply only to two additions</u>.

$$(m + 4)(m - 5) = (m + 4)\left[m + (-5)\right]$$

(a) Since the coefficient of each "m" is "1", the coefficient of the "m^2" term of the product is _____.

(b) Since $4 + (-5) = -1$, the coefficient of the "m" term of the product is _____.

(c) Since $(4)(-5) = -20$, the number-term of the product is _____.

(d) Therefore: $(m + 4)(m - 5) = $ _____ + _____ + _____

---

107. Using the procedure of the last frame, write the trinomial product for each of these:

(a) $(x + 7)(x - 1) = (x + 7)\left[x + (-1)\right] = $ _____ + _____ + _____

(b) $(S + 2)(S - 5) = (S + 2)\left[S + (-5)\right] = $ _____ + _____ + _____

a) 1
b) -1
c) -20
d) $m^2 + (-m) + (-20)$
   or
   $1m^2 + (-1m) + (-20)$

---

108. In the multiplication below, the first binomial factor is a subtraction. We have converted this subtraction to an addition and then performed the multiplication.

$$(p - 3)(p + 5) = \left[p + (-3)\right](p + 5) = p^2 + 5p + (-3p) + (-15)$$
$$= p^2 + \left[5 + (-3)\right]p + (-15)$$
$$= p^2 + \quad 2p \quad + (-15)$$

a) $x^2 + 6x + (-7)$
b) $S^2 + (-3S) + (-10)$

(Continued on following page.)

108. Continued

Notice these same points about the terms in the trinomial product.

(1) The coefficient of the "p^2" term is 1.

(2) The coefficient of the "p" term is "2", since $5 + (-3) = 2$.

(3) The number-term is -15, since $(5)(-3) = -15$.

Using these facts, let's find the trinomial product for the multiplication below. We have converted the first factor to an addition since the facts above apply only to two additions.

$$(x - 7)(x + 4) = [x + (-7)](x + 4)$$

(a) The coefficient of the "x^2" term is _____.

(b) The coefficient of the "x" term is _____.

(c) The number-term is _____.

(d) Therefore: $(x - 7)(x + 4) =$ _____ + _____ + _____

---

109. Using the procedure of the last frame, write the trinomial product for each of these:

(a) $(y - 7)(y + 2) = [y + (-7)](y + 2) =$ _____ + _____ + _____

(b) $(V - 1)(V + 5) = [V + (-1)](V + 5) =$ _____ + _____ + _____

a) 1

b) -3, since $4 + (-7) = -3$

c) -28, since $(-7)(4) = -28$

d) $x^2 + (-3x) + (-28)$

---

110. Both binomial factors are subtractions in the multiplication below. We have converted each to an addition and then performed the multiplication.

$$(x - 6)(x - 3) = [x + (-6)][x + (-3)] = x^2 + (-3x) + (-6x) + 18$$
$$= x^2 + [(-3) + (-6)]x + 18$$
$$= x^2 + (-9x) + 18$$

a) $y^2 + (-5y) + (-14)$

b) $V^2 + 4V + (-5)$

Notice the same points about the terms in the trinomial product.

(1) The coefficient of the "x^2" term is "1".

(2) The coefficient of the "x" term is "-9", since $(-3) + (-6) = -9$.

(3) The number-term is 18, since $(-6)(-3) = 18$.

Using these facts, let's find the trinomial product for the multiplication below. Both factors have been converted to additions since the facts above apply only to two additions.

$$(R - 3)(R - 8) = [R + (-3)][R + (-8)]$$

(a) The coefficient of the "R^2" term is _____.

(b) The coefficient of the "R" term is _____.

(c) The number-term is _____.

(d) Therefore: $(R - 3)(R - 8) =$ _____ + _____ + _____

234 Quadratic Equations and Formulas

111. Using the procedure of the last frame, write the trinomial product for each of these:

(a) $(y - 1)(y - 3) = [y + (-1)][y + (-3)] = \underline{} + \underline{} + \underline{}$

(b) $(t - 7)(t - 3) = [t + (-7)][t + (-3)] = \underline{} + \underline{} + \underline{}$

a) 1
b) -11, since $(-8)+(-3)=-11$
c) 24, since $(-3)(-8) = 24$
d) $R^2 + (-11R) + 24$

---

112. Since the procedure we have been using applies only to multiplying two binomials which are <u>additions</u>, we must convert any subtraction-binomial to an addition <u>first</u>. When this is done, the trinomial product always contains at least one addition of a negative term. This product must then be written in the preferred form by converting any addition of a negative term to a subtraction. For example:

$(b - 5)(b + 1) = [b + (-5)](b + 1) = b^2 + (-4b) + (-5)$
$ = b^2 - 4b - 5$

Write the trinomial product for each of these and then convert the product to the preferred form:

(a) $(x + 7)(x - 5) = (x + 7)[x + (-5)] = \underline{} + \underline{} + \underline{}$
$= \underline{}$

(b) $(y - 9)(y + 2) = [y + (-9)](y + 2) = \underline{} + \underline{} + \underline{}$
$= \underline{}$

(c) $(a - 6)(a - 7) = [a + (-6)][a + (-7)] = \underline{} + \underline{} + \underline{}$
$= \underline{}$

a) $y^2 + (-4y) + 3$
b) $t^2 + (-10t) + 21$

---

113. We have used the shortcut procedure to write the trinomial product below:

$(x + 4)(x + 2) = x^2 + 6x + 8$

If we interchange the two sides of this equation, we get:

$x^2 + 6x + 8 = (x + 4)(x + 2)$

This new equation says that the trinomial $x^2 + 6x + 8$ can be factored into the two binomials $(x + 4)$ and $(x + 2)$. Other than the fact that the coefficient of "x" in each binomial is "1", notice these two essential facts about the numbers (4 and 2) in the two binomials:

(1) Their <u>product</u> is "8", the number-term of the trinomial.

(2) Their <u>sum</u> is "6", the coefficient of the "x" term of the trinomial.

a) $x^2 + 2x + (-35)$
$x^2 + 2x - 35$
b) $y^2 + (-7y) + (-18)$
$y^2 - 7y - 18$
c) $a^2 + (-13a) + 42$
$a^2 - 13a + 42$

Continued on following page.

113. Continued

    Let's check these two essential facts about the numbers in the binomials below to see whether the factoring is correct.

    $$y^2 + 7y + 12 = (y + 3)(y + 4)$$

    (a) Does the <u>product</u> of 3 and 4 equal the number-term of the trinomial? _____

    (b) Does the <u>sum</u> of 3 and 4 equal the coefficient of the "y" term of the trinomial? _____

    (c) Can $y^2 + 7y + 12$ be factored into $(y + 3)$ and $(y + 4)$? _____

---

114. Here are the steps needed to factor the trinomial below into two binomials:

    $$m^2 + 5m + 6$$

    (1) Since the <u>product</u> of the numbers in the two binomials must be "6", we begin by listing all possible pairs of factors of "6". The possible pairs are either (6 and 1) or (3 and 2).

    (2) Since the <u>sum</u> of the numbers in the two binomials must be "5", the correct pair of factors is (3 and 2).

    Therefore: $m^2 + 5m + 6 = ($ ___ + ___ $)($ ___ + ___ $)$.

a) Yes
b) Yes
c) Yes

---

115. Let's factor $t^2 + 5t + 4$. The possible pairs of factors of "4" are (4 and 1) and (2 and 2).

    (a) The sum of which of these pairs of factors is "5"? _____

    (b) Therefore: $t^2 + 5t + 4 = ($ ___ + ___ $)($ ___ + ___ $)$

$(m + 3)(m + 2)$

---

116. Let's factor $x^2 + 8x + 12$. The possible pairs of factors for "12" are (12 and 1), (6 and 2), and (4 and 3).

    (a) The sum of which pair of factors is "8"? _____

    (b) Therefore: $x^2 + 8x + 12 = ($ ___ + ___ $)($ ___ + ___ $)$

a) 4 and 1
b) $(t + 4)(t + 1)$

---

117. The trinomial below contains a subtraction. Before attempting to factor it, we have converted the subtraction to an addition.

    $$y^2 - 7y + 10 = y^2 + (-7y) + 10$$

    Since the product of the numbers in the two binomials must be +10 and their sum must be -7, we can see that <u>both numbers must be negative</u>. The possible pairs of negative factors of +10 are (-10 and -1) and (-5 and -2).

    (a) The sum of which pair of factors is -7? _____

    (b) Therefore: $y^2 - 7y + 10 = [$ ___ $+ ($ ___ $)][$ ___ $+ ($ ___ $)]$ or $($ ___ $-$ ___ $)($ ___ $-$ ___ $)$

a) 6 and 2
b) $(x + 6)(x + 2)$

236　Quadratic Equations and Formulas

118. Here is another case in which the numbers in both binomials must be negative:
$$t^2 - 9t + 14 = t^2 + (-9t) + 14$$

The possible pairs of negative factors of +14 are (-14 and -1) and (-7 and -2).

(a) The sum of which pair of factors is -9? _____

(b) Therefore: $t^2 - 9t + 14 = [\quad + (\quad)][\quad + (\quad)]$ or
$(\quad - \quad)(\quad - \quad)$

a) -5 and -2
b) $[y + (-5)][y + (-2)]$
or
$(y - 5)(y - 2)$

---

119. The trinomial below also contains a subtraction. Before attempting to factor it, we have converted the subtraction to an addition.
$$x^2 + 5x - 6 = x^2 + 5x + (-6)$$

Since the product of the numbers in the two binomials must be -6, we can see that <u>one number must be positive and the other negative</u>. The possible pairs of factors of -6 are (+6 and -1), (-6 and +1), (+3 and -2), and (-3 and +2).

(a) The sum of the two numbers in the binomials must be +5. Which pair of factors above has +5 as its sum? _____

(b) Therefore: $x^2 + 5x - 6 = [\quad + (\quad)][\quad + (\quad)]$ or
$(\quad)(\quad)$

a) -7 and -2
b) $[t + (-7)][t + (-2)]$
or $(t - 7)(t - 2)$

---

120. We have converted both subtractions to additions in the trinomial below.
$$m^2 - 2m - 8 = m^2 + (-2m) + (-8)$$

Since the product of the numbers in the two binomials must be -8, one number must be positive and the other negative. The possible pairs of factors of -8 are (+8 and -1), (-8 and +1), (+4 and -2), and (-4 and +2).

(a) The sum of the numbers in the binomials must be -2. Which pair of factors above has -2 as its sum? _____

(b) Therefore: $m^2 - 2m - 8 = [\quad + (\quad)][\quad + (\quad)]$ or
$(\quad)(\quad)$

a) +6 and -1
b) $[x + (+6)][x + (-1)]$
or $(x + 6)(x - 1)$

---

121. If the number-term of a trinomial is <u>positive</u>, the numbers in the two binomials must be either <u>both positive</u> or <u>both negative</u>. To decide whether they are <u>both positive</u> or <u>both negative</u>, we must examine the sign of the letter-term (the middle term).

If the letter-term is <u>positive</u>, the numbers in both binomials must be <u>positive</u>.

If the letter-term is <u>negative</u>, the numbers in both binomials must be <u>negative</u>.

Using the facts above, factor each of these trinomials.

(a) $d^2 + 8d + 7 = (\quad)(\quad)$

(b) $R^2 - 4R + 3 = [\quad + (\quad)][\quad + (\quad)]$ or $(\quad)(\quad)$

a) -4 and +2
b) $[m + (-4)][m + (+2)]$
or $(m - 4)(m + 2)$

Quadratic Equations and Formulas   237

122. If the number-term of a trinomial is <u>negative</u>, the numbers in the two binomials must have <u>different signs</u>. That is, one must be positive and the other negative.

Using the fact above, factor each of these trinomials:
(a) $x^2 + 4x - 5 = [\ \ +(\ \ )][\ \ +(\ \ )]$ or $(\ \ )(\ \ )$
(b) $y^2 - 2y - 3 = [\ \ +(\ \ )][\ \ +(\ \ )]$ or $(\ \ )(\ \ )$

a) $(d+7)(d+1)$
b) $[R+(-3)][R+(-1)]$
or $(R-3)(R-1)$

123. After factoring a trinomial, always check to see whether the factoring is correct. To do so, multiply the two binomials to see whether you obtain the original trinomial. For example:

$x^2 + 8x + 15 = (x+3)(x+5)$ is a correct factoring, since:
$(x+3)(x+5) = x^2 + 8x + 15$

$y^2 + 7y - 18 = (y-9)(y+2)$ is <u>not</u> a correct factoring, since:
$(y-9)(y+2) = y^2 - 7y - 18$

Check each factoring below and state whether it is "correct" or "incorrect".

(a) $m^2 + 8m + 12 = (m+3)(m+4)$ _____

(b) $t^2 - 7t + 10 = (t-2)(t-5)$ _____

(c) $p^2 + 2p - 15 = (p-3)(p+5)$ _____

(d) $V^2 - 5V - 14 = (V-2)(V+7)$ _____

a) $[x+(+5)][x+(-1)]$
or $(x+5)(x-1)$
b) $[y+(-3)][y+(+1)]$
or $(y-3)(y+1)$

124. Let's solve the following quadratic equation by the factoring method.
$x^2 + 11x + 10 = 0$

Factoring the trinomial on the left, we get:
$(x+10)(x+1) = 0$

Setting each factor equal to "0", we get:
$x + 10 = 0$ or $x = -10$
$x + 1 = 0$ or $x = -1$

Therefore, the two roots are −10 and −1.
Show that these two roots satisfy the original equation, $x^2 + 11x + 10 = 0$.

a) Incorrect, since:
$(m+3)(m+4) = m^2+7m+12$
b) Correct.
c) Correct.
d) Incorrect, since:
$(V-2)(V+7) = V^2+5V-14$

238   Quadratic Equations and Formulas

125. Let's solve $m^2 - 5m + 6 = 0$ by the factoring method.

    (a) Factoring the trinomial on the left side, we get:

        (   )(   ) = 0

    (b) Therefore, the two roots are _____ and _____.

$(-10)^2 + 11(-10) + 10 = 0$
$100 \ + (-110) + 10 = 0$
$\qquad\qquad\qquad 0 = 0$

$(-1)^2 + 11(-1) + 10 = 0$
$1 \ + (-11) + 10 = 0$
$\qquad\qquad 0 = 0$

---

126. Let's solve $t^2 + 4x - 12 = 0$ by the factoring method.

    (a) Factoring the trinomial on the left, we get:

        (   )(   ) = 0

    (b) Therefore, t = _____ and _____.

a) $(m - 2)(m - 3) = 0$

b) +2 and +3

---

127. Solve each of the following by the factoring method.

    (a) $x^2 + 13x + 12 = 0$ (b) $y^2 - 8y - 9 = 0$

    x = _____ and _____   y = _____ and _____

a) $(t + 6)(t - 2) = 0$

b) −6 and +2

---

128. There are some trinomials that cannot be factored into two binomials in which the numbers are whole numbers. The trinomial below is an example.
$$x^2 + 9x + 10$$
The pairs of factors of +10 are (10 and 1) and (5 and 2). Neither pair of factors has 9 as its sum.

Which of the following trinomials <u>cannot</u> be factored into binomials which contain whole numbers? _____

(a) $x^2 + 9x + 14$ (b) $t^2 - 11x + 16$ (c) $R^2 + R - 11$ (d) $q^2 - 2q - 15$

a) x = −12 and −1, from:
   $(x + 12)(x + 1) = 0$

b) y = +9 and −1, from:
   $(y - 9)(y + 1) = 0$

---

129. When a "complete" quadratic contains a trinomial which cannot be factored into two binomials with whole numbers, we ordinarily do not solve it by the factoring method.

Which of the following quadratics <u>would</u> <u>not</u> <u>ordinarily</u> be solved by the factoring method? _____

    (a) $t^2 + 7t + 3 = 0$ (c) $d^2 + d - 12 = 0$

    (b) $p^2 - 11p + 18 = 0$ (d) $F^2 - F - 7 = 0$

Both (b) and (c)

(a) can be factored:
   $(x + 7)(x + 2)$

(d) can be factored:
   $(q - 5)(q + 3)$

130. The coefficients and numbers in each quadratic equation below are decimals.

$$y^2 + 0.87y + 1.63 = 0 \qquad b^2 - 47.8b - 85.6 = 0$$

(a) Can the trinomials above be factored into binomials containing whole numbers? _____

(b) Would we ordinarily solve the quadratics above by the factoring method? _____

---

131. The numbers in the equations below are quite large.

$$x^2 + 37x + 312 = 0 \qquad y^2 - 30y - 216 = 0$$

These quadratics can be solved by **the** factoring method. That is:

$$(x + 13)(x + 24) = 0 \qquad (y + 6)(y - 36) = 0$$

$$x = -13 \text{ and } -24 \qquad y = -6 \text{ and } +36$$

However, because of the difficulty in factoring trinomials of this type, we do not ordinarily solve such equations by the factoring method.

To see the difficulty involved in trying to solve a quadratic containing large whole numbers by the factoring method, try to solve the quadratic equation below by that method.

$$x^2 - 44x + 288 = 0$$

Both (a) and (d)

(b) can be factored:
   $(p - 2)(p - 9) = 0$

(c) can be factored:
   $(d + 4)(d - 3) = 0$

---

a) No

b) No

$x = 8$ and 36, from:

$(x - 8)(x - 36) = 0$

Quadratic Equations and Formulas 239

SELF-TEST 8 (Frames 104-131)

Multiply the following binomials:

1. $(x - 8)(x + 3) = $ _____

2. $(t - 7)(t - 5) = $ _____

Factor the following trinomials:

3. $r^2 + 9r + 20 = $ _____

4. $a^2 - 9a + 8 = $ _____

5. Which of the following are factored <u>incorrectly</u>? _____

    (a) $x^2 + 3x - 10 = (x + 5)(x - 2)$ (c) $b^2 + 5b - 6 = (b + 1)(b - 6)$

    (b) $h^2 - 2h + 2 = (h - 1)(h - 2)$ (d) $E^2 - 5E - 24 = (E - 8)(E + 3)$

Solve each of the following quadratic equations by the factoring method:

6. $x^2 + x - 2 = 0$

7. $w^2 - 11w + 24 = 0$

8. $P^2 - P - 12 = 0$

    x = ____ and ____   w = ____ and ____   P = ____ and ____

ANSWERS:
1. $x^2 - 5x - 24$
2. $t^2 - 12t + 35$
3. $(r + 5)(r + 4)$
4. $(a - 1)(a - 8)$
5. (b), (c)
6. x = +1 and -2
7. w = +3 and +8
8. P = +4 and -3

Quadratic Equations and Formulas   241

3-9   THE FACTORING METHOD AND "COMPLETE" QUADRATIC EQUATIONS OF THE FORM: $ax^2 + bx + c = 0$

Each of the equations below is a "complete" quadratic. The coefficient of the squared letter in each case is a number other than "1".

$2x^2 + 9x + 7 = 0$ $6y^2 - 14y + 4 = 0$ $8t^2 - 2t - 3 = 0$

In this section, we will briefly discuss the method of solving quadratics of this type by the factoring method.

---

132. We have shown the steps needed to multiply $(4x + 3)$ and $(x + 2)$ below.

$(4x + 3)(x + 2) = 4x^2 + 8x + 3x + 6 = 4x^2 + 11x + 6$

Let's examine the trinomial product on the right:

"$4x^2$" is the product of "$4x$" and "x", the letter-terms in the binomials.

"6" is the product of "3" and "2", the number-terms in the binomials.

Based on the facts above, if we multiplied $(2y + 4)$ and $(3y + 1)$:

(a) The first term of the trinomial product would be _____.

(b) The last term of the trinomial product would be _____.

133. Here are the steps needed to factor the trinomial below:

$5x^2 + 14x + 8$

(1) Since the product of the letter-terms of the binomials must be "$5x^2$", the only possible pair of letter-terms is ($5x$ and x).

(2) Since the product of the number-terms of the binomials must be "8", the possible pairs of number-terms are (8 and 1) and (4 and 2).

(3) Generating all possible pairs of binomials from these letter-terms and number-terms, we get:

A: $(5x + 8)(x + 1)$
B: $(5x + 1)(x + 8)$
C: $(5x + 4)(x + 2)$
D: $(5x + 2)(x + 4)$

(4) Only one of the four pairs of binomial factors above is correct. The correct pair is the one whose trinomial product has "$14x$" as its middle term.

(a) For the multiplications shown above, write the middle term of the trinomial product for each pair of binomial factors.

A: _____   B: _____   C: _____   D: _____

(b) Therefore: $5x^2 + 14x + 8 = ($ __ + __ $)($ __ + __ $)$

a) $6y^2$, from $(2y)(3y)$

b) 4, from $(4)(1)$

242   Quadratic Equations and Formulas

134. Let's factor the trinomial below by the same method. We have converted the subtraction to an addition on the right.

$$2y^2 - 11y + 15 = 2y^2 + (-11y) + 15$$

(1) The only possible pair of letter-terms is (2y and y).

(2) The possible pairs of number-terms are (-15 and -1) and (-5 and -3).

(3) The possible pairs of binomial factors are:

A: $[2y + (-15)][y + (-1)]$
B: $[2y + (-1)][y + (-15)]$
C: $[2y + (-5)][y + (-3)]$
D: $[2y + (-3)][y + (-5)]$

The correct pair of binomial factors is the one whose trinomial product has "-11y" as its middle term.

(a) Write the middle term of the trinomial product for each pair of binomial factors shown above:

A: _____   B: _____   C: _____   D: _____

(b) Therefore: $2y^2 - 11y + 15 = [\quad + (\quad)][\quad + (\quad)]$ or
$(\quad - \quad)(\quad - \quad)$

a) A: 13x
   B: 41x
   C: 14x
   D: 22x
b) (5x + 4)(x + 2)

---

135. Let's factor the trinomial below. We have converted the subtraction to an addition on the right.

$$3t^2 + 2t - 5 = 3t^2 + 2t + (-5)$$

(1) The only possible pair of letter-terms is (3t and t).

(2) Since -5 is negative, the possible pairs of number-terms are (+5 and -1) and (-5 and +1).

(3) The possible pairs of binomial factors are:

A: $[3t + (+5)][t + (-1)]$
B: $[3t + (-1)][t + (+5)]$
C: $[3t + (-5)][t + (+1)]$
D: $[3t + (+1)][t + (-5)]$

(a) Which one of these pairs of binomial factors has "2t" as the middle term of its product? Pair _____

(b) Therefore: $3t^2 + 2t - 5 = [\quad + (\quad)][\quad + (\quad)]$ or
$(\quad)(\quad)$

a) A: -17y
   B: -31y
   C: -11y
   D: -13y
b) $= [2y + (-5)][y + (-3)]$
   or (2y - 5)(y - 3)

a) Pair A
b) $[3t + (+5)][t + (-1)]$
   or (3t + 5)(t - 1)

136. Let's factor the trinomial below.

$$5p^2 - 7p - 6 = 5p^2 + (-7p) + (-6)$$

(1) The only possible pair of letter-terms is (5p and p).

(2) The possible pairs of number-terms are (+6 and -1), (-6 and +1), (+3 and -2), and (-3 and +2).

(3) The possible pairs of binomial factors are:

A: $[5p + (+6)][p + (-1)]$ E: $[5p + (+3)][p + (-2)]$
B: $[5p + (-1)][p + (+6)]$ F: $[5p + (-2)][p + (+3)]$
C: $[5p + (-6)][p + (+1)]$ G: $[5p + (-3)][p + (+2)]$
D: $[5p + (+1)][p + (-6)]$ H: $[5p + (+2)][p + (-3)]$

(a) Which pair of binomial factors has "-7p" as the middle term of its product? Pair _____

(b) Therefore: $5p^2 - 7p - 6 = [\_\_ + (\_)][\_\_ + (\_)]$ or

$(\_\_\_\_)(\_\_\_\_)$

a) Pair E

b) $[5p + (+3)][p + (-2)]$
or $(5p + 3)(p - 2)$

---

137. Here are the steps needed to solve the quadratic equation below by the factoring method.

$$6x^2 - 13x + 5 = 0$$

(1) Factoring the trinomial on the left side, we get:

$(3x - 5)(2x - 1) = 0$

(2) Setting each factor equal to "0", we get:

$3x - 5 = 0 \qquad 2x - 1 = 0$

(3) Therefore, the two roots are _____ and _____.

$\dfrac{5}{3}$ and $\dfrac{1}{2}$

---

138. Solve each quadratic below by the factoring method.

(a) $3y^2 + 10y + 7 = 0$ (b) $2p^2 - 5p - 12 = 0$

y = _____ and _____   p = _____ and _____

a) $y = -\dfrac{7}{3}$ and -1, from:
$(3y + 7)(y + 1) = 0$

b) $p = -\dfrac{3}{2}$ and $+4$, from:
$(2p + 3)(p - 4) = 0$

244   Quadratic Equations and Formulas

139. Let's try to solve the quadratic below by the factoring method.

$$2m^2 + 6m + 3 = 0$$

The possible pairs of factors for the trinomial are:

$$(2m + 3)(m + 1)$$
$$(2m + 1)(m + 3)$$

(a) Does either pair of factors above have "6m" as the middle term of its product? _____

(b) Can we factor the trinomial into two binomials which contain whole numbers? _____

(c) Would we ordinarily attempt to solve the equation above by the factoring method? _____

---

140. The coefficients and numbers in the quadratic equations below are decimals.

$$0.56x^2 + 1.93x + 1.37 = 0 \qquad 31.7y^2 + 46.0y + 29.6 = 0$$

(a) Can we factor either trinomial into two binomials which contain whole numbers? _____

(b) Would we ordinarily attempt to solve the quadratics above by the factoring method? _____

a) No
b) No
c) No

---

141. The numbers in the "complete" quadratic below are quite small. However, because of the large number of possible pairs of binomial factors, a fair amount of trial-and-error is needed to solve the equation by the factoring method.

$$6x^2 - 19x - 20 = 0$$

Try to solve the equation by the factoring method.

x = _____ and _____

a) No
b) No

---

142. The numbers in the quadratic equations below are quite large.

$$100t^2 + 123t + 36 = 0 \qquad 50y^2 - 495y + 144 = 0$$

These quadratics can be solved by the factoring method. That is:

$$(25t + 12)(4t + 3) = 0 \qquad (5y - 48)(10y - 3) = 0$$
$$t = -\frac{12}{25} \text{ and } -\frac{3}{4} \qquad y = \frac{48}{5} \text{ and } \frac{3}{10}$$

However, because of the difficulty of factoring trinomials like those above, the factoring method of solving such equations is very inefficient. Therefore, we would not ordinarily attempt to solve them by the factoring method.

$x = -\frac{5}{6}$ and +4, from:

$(6x + 5)(x - 4) = 0$

SELF-TEST 9 (Frames 132-142)

Factor each of the following trinomials:

1. $2x^2 + x - 3 =$ _____

2. $3h^2 - 2h - 5 =$ _____

Solve each of the following quadratic equations by the factoring method:

3. $2y^2 + 5y + 3 = 0$

    y = _____ and _____

4. $3t^2 + 5t - 2 = 0$

    t = _____ and _____

5. $5w^2 - 8w + 3 = 0$

    w = _____ and _____

ANSWERS:
1. $(2x + 3)(x - 1)$
2. $(3h - 5)(h + 1)$
3. $y = -\frac{3}{2}$ and -1
4. $t = +\frac{1}{3}$ and -2
5. $w = +\frac{3}{5}$ and $+1$

246 Quadratic Equations and Formulas

3-10 PERFECT-SQUARE TRINOMIALS AND THE FACTORING METHOD

In this section, we will use the factoring method to solve quadratic equations in which the trinomial is a perfect-square trinomial. We will see that the two roots are identical for equations of this type.

---

143. We have already discussed the criteria for identifying a perfect-square trinomial. They are:

    (1) The first and last terms must be perfect-squares.

    (2) The middle term must be double the product of the square roots of the first and last terms.

    For example: $4x^2 + 12x + 9$ is a perfect-square trinomial since:

    (1) $4x^2$ and 9 are perfect squares.

    (2) $12x$ equals $2\sqrt{4x^2} \cdot \sqrt{9}$ or $2(2x)(3)$.

    Which of the following are perfect-square trinomials? _____

    (a) $t^2 + 6t + 9$ (b) $R^2 - 5R + 25$ (c) $16d^2 + 42d + 49$ (d) $9y^2 - 30y + 25$

---

144. The trinomial in the quadratic equation below is a perfect-square trinomial.
$$x^2 + 8x + 16 = 0$$

    If we factor the trinomial, we get: $(x + 4)(x + 4) = 0$

    If we set each factor equal to "0" to find the two roots of the equation, we get −4 and −4.

    Are the two roots in this case identical or different? _____

Only (a) and (d)

---

145. The trinomial in the quadratic equation below is a perfect-square trinomial.
$$4t^2 - 12t + 9 = 0$$

    If we factor the trinomial, we get: $(2t - 3)(2t - 3) = 0$

    (a) Setting each factor equal to "0", we get: t = _____ and _____

    (b) Are the two roots identical or different? _____

Identical

---

146. Whenever the trinomial in a quadratic equation is a perfect-square trinomial, the two roots of the equation are identical.

    Find the two roots of each quadratic equation below:

    (a) $m^2 + 10m + 25 = 0$ (b) $d^2 - 4d + 4 = 0$

    m = _____ and _____   d = _____ and _____

a) $\dfrac{3}{2}$ and $\dfrac{3}{2}$

b) Identical

147. Find the two roots of each equation below:

(a) $9x^2 + 12x + 4 = 0$ (b) $16p^2 - 40p + 25 = 0$

a) -5 and -5, from: $(m + 5)(m + 5) = 0$

b) $+2$ and $+2$, from: $(d - 2)(d - 2) = 0$

x = _____ and _____   p = _____ and _____

Answer to Frame 147: a) $-\frac{2}{3}$ and $-\frac{2}{3}$, from: $(3x + 2)(3x + 2) = 0$ b) $\frac{5}{4}$ and $\frac{5}{4}$, from: $(4p - 5)(4p - 5) = 0$

SELF-TEST 10 (Frames 143-147)

Solve each of the following quadratic equations by the factoring method:

1. $x^2 - 2x + 1 = 0$ 2. $9F^2 + 6F + 1 = 0$ 3. $4r^2 - 20r + 25 = 0$

x = _____ and _____   F = _____ and _____   r = _____ and _____

ANSWERS: 1. $x = +1$ and $+1$ 2. $F = -\frac{1}{3}$ and $-\frac{1}{3}$ 3. $r = +\frac{5}{2}$ and $+\frac{5}{2}$

248 Quadratic Equations and Formulas

3-11 SOLVING QUADRATIC EQUATIONS BY MEANS OF THE QUADRATIC FORMULA

As we have seen, the usefulness and efficiency of solving "complete" quadratic equations by the factoring method is limited to those equations in which the trinomial can easily be factored into binomials which contain whole numbers. Another method is needed to solve quadratics which exceed the limits of the factoring method. This "other" method involves the use of the quadratic formula.

In this section, we will discuss the quadratic formula and show how it is used to solve quadratic equations. We will see that the quadratic-formula method is a general method which can be used to solve any quadratic equation of any type. Before discussing the quadratic formula, however, we must briefly discuss the meaning of the "standard form" of quadratic equations.

148. In order to use the quadratic formula to solve a quadratic equation, the equation must be written in "standard form". The four quadratic equations below are written in "standard form".

(1) $3x^2 + 4x + 9 = 0$

(2) $2y^2 + 3y + (-7) = 0$

(3) $7t^2 + (-2)t + 8 = 0$

(4) $(-3)d^2 + 7d + 5 = 0$

The general pattern is:
$a\square^2 + b\square + c = 0$

where (1) $\square$ stands for the variable,
and (2) "a", "b", and "c" are the constants.

In equation (1): a = 3   b = 4   c = 9

(a) What is "a" in equation (2)? a = _____

(b) What is "b" in equation (3)? b = _____

(c) What is "c" in equation (4)? c = _____

149. The general pattern for a quadratic equation in standard form is:

$a\square^2 + b\square + c = 0$

Notice that all of the operations are additions. There are no subtractions. Therefore, the quadratic equation below is not in standard form because it contains subtractions.

$5x^2 - 7x - 9 = 0$

To write it in standard form, we must convert all subtractions to additions.

$5x^2 + (-7)x + (-9) = 0$

Having done this, we can see that:

a = _____   b = _____   c = _____

a) a = +2
b) b = -2
c) c = +5

a = 5
b = -7
c = -9

Quadratic Equations and Formulas

150. (a) Write this quadratic equation in standard form: $12y^2 - 9y + 20 = 0$

_____

(b) a = _____   b = _____   c = _____

---

151. Here is the standard-form pattern for a quadratic equation again:

$$\boxed{a\square^2 + b\square + c = 0}$$

In this standard-form pattern, the left side is a trinomial which contains a squared-letter term $(a\square^2)$, a letter term $(b\square)$, and a number term (c). The number term (c) is missing in the "incomplete" quadratic below.

$$2x^2 + 3x = 0$$

To put this equation in standard form, we must insert a "0" for the missing "c" term. We get:

$$2x^2 + 3x + 0 = 0$$

Therefore: a = _____   b = _____   c = _____

a) $12y^2 + (-9)y + 20 = 0$

b) a = 12
   b = -9
   c = 20

---

152. A letter term $(b\square)$ is missing in the "pure" quadratic below.

$$4t^2 + 9 = 0$$

To put the equation in standard form, we must insert a "0t" for the missing "$b\square$" term. We get:

$$4t^2 + 0t + 9 = 0$$

Therefore: a = _____   b = _____   c = _____

a = 2
b = 3
c = 0

---

153. (a) Write this "incomplete" quadratic in standard form: $3y^2 - 2y = 0$

_____

(b) Therefore: a = _____   b = _____   c = _____

a = 4
b = 0
c = 9

---

154. (a) Write this "pure" quadratic in standard form: $7F^2 - 9 = 0$

_____

(b) Therefore: a = _____   b = _____   c = _____

a) $3y^2 + (-2)y + 0 = 0$
b) a = 3
   b = -2
   c = 0

---

155. Here is an equation in which the squared-letter term $(a\square^2)$ is missing:

$$3x + 7 = 0$$

(a) If the "$a\square^2$" term is missing, is the equation a quadratic equation? _____

(b) How many roots does the equation above have? _____

a) $7F^2 + (0)F + (-9) = 0$
b) a = 7
   b = 0
   c = -9

---

a) No

b) Only <u>one</u>: $x = -\dfrac{7}{3}$

250   Quadratic Equations and Formulas

156. In the equation below, the coefficient of the squared-letter term ("a") and the coefficient of the letter-term ("b") are not explicitly written.

$$x^2 + x + 7 = 0$$

Since each of these coefficients is "1", the standard form of the equation is:

$$1x^2 + 1x + 7 = 0$$

Therefore: a = _____   b = _____   c = _____

---

157. (a) Write this "incomplete" quadratic in standard form: $x^2 - 2x = 0$ _____

(b) Therefore: a = _____   b = _____   c = _____

a = 1
b = 1
c = 7

---

158. (a) Write this "pure" quadratic in standard form: $y^2 - 14 = 0$ _____

(b) Therefore: a = _____   b = _____   c = _____

a) $1x^2 + (-2)x + 0 = 0$
b) a = 1
   b = -2
   c = 0

Answer to Frame 158:   a) $1y^2 + 0y + (-14) = 0$ b) a = 1   b = 0   c = -14

159. The "quadratic formula" for finding the two roots of a quadratic equation is given below. In this formula, the letters "a", "b", and "c" stand for the comparable numbers in the standard form of quadratic equations. Since any quadratic equation has two roots, there are two parallel forms of the quadratic formula.

$$\text{First root} = \frac{(-b) \;\downarrow\!\!+\; \sqrt{b^2 - 4ac}}{2a}$$

$$\text{Second root} = \frac{(-b) \;\downarrow\!\!-\; \sqrt{b^2 - 4ac}}{2a}$$

The only difference between these two forms is the "+" and "−" between the two terms in the numerator. We have drawn arrows over them.

Rather than writing two formulas, we can combine them into a single formula by using the "±" symbol as we do below. The "±" symbol means "add the radical for the first root, and subtract the radical for the second root".

$$\text{Both roots} = \frac{(-b) \pm \sqrt{b^2 - 4ac}}{2a}$$

Let's use the quadratic formula to solve this quadratic equation: $x^2 + 5x + (-14) = 0$ or $1x^2 + 5x + (-14) = 0$

Since the equation is in standard form: a = 1   b = 5   c = -14

Let's plug these values into the quadratic formula. We get:

$$\text{Both roots} = \frac{(-b) \pm \sqrt{b^2 - 4ac}}{2a}$$

$$= \frac{(-5) \pm \sqrt{(5)^2 - 4(1)(-14)}}{2(1)} = \frac{(-5) \pm \sqrt{25 - (-56)}}{2} = \frac{(-5) \pm \sqrt{81}}{2} = \frac{(-5) \pm 9}{2}$$

(Continued on following page.)

Quadratic Equations and Formulas   251

159. Continued

Therefore, the two roots are: First root $= \dfrac{(-5) + 9}{2} = \dfrac{4}{2} = 2$

Second root $= \dfrac{(-5) - 9}{2} = \dfrac{-14}{2} = -7$

Show that these two roots satisfy the original equation, $x^2 + 5x + (-14) = 0$.

(a) The <u>first</u> <u>root</u> <u>is</u> "<u>2</u>".   (b) The <u>second</u> <u>root</u> <u>is</u> "<u>-7</u>".

---

160. Let's solve another one: $y^2 + 4y - 45 = 0$

The standard form is: $1y^2 + 4y + (-45) = 0$

Therefore: $a = 1$ $b = 4$ $c = -45$

Plugging these values into the quadratic formula, we get:

Both roots $= \dfrac{(-b) \pm \sqrt{b^2 - 4ac}}{2a}$

$= \dfrac{(-4) \pm \sqrt{(+4)^2 - 4(1)(-45)}}{2(1)} = \dfrac{(-4) \pm \sqrt{16 + 180}}{2}$

$= \dfrac{(-4) \pm \sqrt{196}}{2} = \dfrac{(-4) \pm 14}{2}$

(a) The first root = _____ .   (b) The second root = _____ .

Answers:

a) $x^2 + 5x + (-14) = 0$
   $(2)^2 + 5(2) + (-14) = 0$
   $4 + 10 + (-14) = 0$
   $0 = 0$

b) $x^2 + 5x + (-14) = 0$
   $(-7)^2 + 5(-7) + (-14) = 0$
   $49 + (-35) + (-14) = 0$
   $0 = 0$

a) 5, from:

$\dfrac{(-4) + 14}{2} = \dfrac{10}{2} = 5$

b) -9, from:

$\dfrac{(-4) - 14}{2} = \dfrac{-18}{2} = -9$

252 Quadratic Equations and Formulas

161. There are a few points you should notice about plugging values into the quadratic formula.

$$\text{Both roots} = \frac{(-b) \pm \sqrt{b^2 - 4ac}}{2a}$$

(1) (-b) means "the opposite of b". If b = 4, -b = -4
 If b = -5, -b = +5

(2) b^2 (under the radical) is always a positive number. If b = 4, $b^2 = (4)^2 = +16$
 If b = -5, $b^2 = (-5)^2 = +25$

(3) Watch your signs with $b^2 - 4ac$ under the radical.

"4ac" is a set of three factors. Perform the complete multiplication of these three factors before handling the subtraction. That is:

If a = 1 $b^2 - 4ac = (-10)^2 - [4(1)(-6)]$
 b = -10 $= 100 - [-24]$
 c = -6 $= 100 + [+24] = 124$

If a = -3 $b^2 - 4ac = (12)^2 - [4(-3)(-9)]$
 b = 12 $= 144 - [+108]$
 c = -9 $= 144 + [-108]$
           $= 36$

---

162. Given this equation: $2x^2 + (-7)x + (-3) = 0$

(a) -b = _____ (b) b^2 = _____ (c) $b^2 - 4ac$ = _____

| Go to next frame. |

---

163. Given this equation: $7d^2 + (-9)d + (-1) = 0$

(a) -b = _____ (b) b^2 = _____ (c) $b^2 - 4ac$ = _____

| a) +7 b) 49 c) 73 |

---

164. The quadratic formula is a general method for solving all types of quadratic equations. That is, it can also be used to solve "incomplete" and "pure" quadratic equations. Here is an example:

$$2x^2 + 3x = 0$$

If we solve the "incomplete" quadratic equation above by the factoring method, we get:

x(2x + 3) = 0   The roots are "0" and "$-\frac{3}{2}$".

Though it is not an efficient method to use in this case, let's solve the same equation $(2x^2 + 3x = 0)$ by means of the quadratic formula. In the equation:

 a = 2   b = 3   c = 0

Plugging these values into the quadratic formula, we get:

$$\text{Both roots} = \frac{(-b) \pm \sqrt{b^2 - 4ac}}{2a}$$

$$= \frac{(-3) \pm \sqrt{(3)^2 - 4(2)(0)}}{2(2)} = \frac{(-3) \pm \sqrt{9 - 0}}{4} = \frac{(-3) \pm \sqrt{9}}{4}$$

(a) The first root = _____.   (b) The second root = _____.

(c) Do we get the same two roots with the quadratic formula that we did with the factoring method? _____

| a) +9 b) 81 c) 109 |

Quadratic Equations and Formulas   253

165. Here is another example: $h^2 - 144 = 0$

If we solve the "pure" quadratic equation above by the factoring method, we get:
$$(h + 12)(h - 12) = 0$$

The two roots are "-12" and "+12".

Though it is not an efficient method to use in this case, let's solve the quadratic equation above by means of the quadratic formula. In the equation $h^2 - 144 = 0$:

$a = 1 \qquad b = 0 \qquad c = -144$

Plugging these values into the quadratic formula, we get:

$$\text{Both roots} = \frac{(-0) \pm \sqrt{(0)^2 - 4(1)(-144)}}{2(1)}$$

$$= \frac{0 \pm \sqrt{0 + 576}}{2} = \frac{0 \pm 24}{2}$$

(a) The first root = _____ .   (b) The second root = _____ .

(b) Do we get the same two roots with the formula method that we did with the factoring method? _____

a) 0, from:
$$\frac{(-3) + 3}{4} = \frac{0}{4} = 0$$

b) $-\frac{3}{2}$, from:
$$\frac{(-3) - 3}{4} = \frac{-6}{4} = \frac{-3}{2}$$

c) Yes

---

166. We would not ordinarily attempt to solve the quadratic equation below by the factoring method.
$$t^2 + 7t + 3 = 0$$

We can solve it, however, by using the quadratic formula. In the equation:

$a = 1 \qquad b = 7 \qquad c = 3$

Plugging these values into the quadratic formula, we get:

$$\text{Both roots} = \frac{(-7) \pm \sqrt{7^2 - 4(1)(3)}}{2(1)}$$

$$= \frac{(-7) \pm \sqrt{49 - 12}}{2} = \frac{(-7) \pm \sqrt{37}}{2}$$

(a) The first root = _____ .   (b) The second root = _____ .

a) 12, from:
$$\frac{0 + 24}{2} = \frac{24}{2} = 12$$

b) -12, from:
$$\frac{0 - 24}{2} = \frac{-24}{2} = -12$$

c) Yes

Answer to Frame 166:   a) -0.46, from: $\frac{(-7) + 6.08}{2} = \frac{-0.92}{2}$ b) -6.54, from: $\frac{(-7) - 6.08}{2} = \frac{-13.08}{2}$

In the last frame, we reported "-0.46" and "-6.54" as the two roots of the quadratic equation $t^2 + 7t + 3 = 0$. To do so, we had to evaluate $\sqrt{37}$. Since evaluations of this type are tedious for students who cannot use the slide rule or logarithms or who do not have a table of square roots available, some teachers will accept the non-evaluated expression below as the solution of the equation.
$$\frac{(-7) \pm \sqrt{37}}{2}$$

In this chapter, we will always report the roots of quadratic equations in both forms. Your instructor will tell you the form which you will be required to use.

If your instructor will accept the non-evaluated form, you can report $\frac{(-7) \pm \sqrt{37}}{2}$.

If your instructor will not accept the non-evaluated form, you must report "-0.46" and "-6.54".

254   Quadratic Equations and Formulas

167. We would not ordinarily attempt to solve the quadratic below by the factoring method.

$$2m^2 - 9m + 3 = 0$$

However, we can solve it by means of the quadratic formula. In the equation:

a = 2   b = -9   c = 3

Plugging these values into the formula, we get the expression below. Complete the solution.

$$\text{Both roots} = \frac{(+9) \pm \sqrt{(-9)^2 - 4(2)(3)}}{2(2)}$$

m = _____ and _____

---

168. Here is another equation which we would not attempt to solve by the factoring method.

$$1.2t^2 - 3.1t - 4.5 = 0$$

Let's use the quadratic formula to solve it. In the equation:

a = 1.2   b = -3.1   c = -4.5

Plugging these values into the formula, we get the expression below. Complete the solution.

$$\text{Both roots} = \frac{(+3.1) \pm \sqrt{(-3.1)^2 - 4(1.2)(-4.5)}}{2(1.2)}$$

t = _____ and _____

Either:
m = 4.14 and 0.36

From: $\frac{9 \pm 7.55}{4}$

Or: $m = \frac{9 \pm \sqrt{57}}{4}$

---

169. The equation below would not ordinarily be solved by the factoring method.

$$5x^2 + 3x + 2 = 0$$

Let's use the quadratic formula. Plugging in the proper values, we get:

$$\text{Both roots} = \frac{(-3) \pm \sqrt{3^2 - 4(5)(2)}}{2(5)}$$

$$= \frac{(-3) \pm \sqrt{9 - 40}}{10} = \frac{(-3) \pm \sqrt{-31}}{10}$$

Notice that the number in the radical is negative. Since there are no ordinary numbers which equal $\sqrt{-31}$, there are no ordinary numbers which are the roots of the equation above.

Note: In the remainder of this chapter, all quadratic equations studied will have roots which are ordinary numbers.

Either:
t = 3.62 and -1.04

From: $\frac{3.1 \pm 5.59}{2.4}$

Or: $t = \frac{3.1 \pm \sqrt{31.21}}{2.4}$

Quadratic Equations and Formulas 255

SELF-TEST 11 (Frames 148-169)

The general quadratic equation is: $a\square^2 + b\square + c = 0$ Its two roots are: $\dfrac{(-b) \pm \sqrt{b^2 - 4ac}}{2a}$

In each of the following quadratic equations, list the numerical values of "a", "b", and "c". Then use the quadratic formula to find the two roots of each.

$2x^2 + 7x = 0$

1. a = _____   b = _____   c = _____

2. x = _____ and _____

$4w^2 - 25 = 0$

3. a = _____   b = _____   c = _____

4. w = _____ and _____

$3t^2 - 5t - 2 = 0$

5. a = _____   b = _____   c = _____

6. t = _____ and _____

$r^2 + 3r - 2 = 0$

7. a = _____   b = _____   c = _____

8. r = _____ and _____

ANSWERS:

1. a = 2
   b = 7
   c = 0

2. x = 0 and $-\dfrac{7}{2}$

3. a = 4
   b = 0
   c = -25

4. w = $\dfrac{5}{2}$ and $-\dfrac{5}{2}$

5. a = 3
   b = -5
   c = -2

6. t = 2 and $-\dfrac{1}{3}$

7. a = 1
   b = 3
   c = -2

8. r = 0.56 and -3.56

   Or: r = $\dfrac{-3 \pm \sqrt{17}}{2}$

256 Quadratic Equations and Formulas

3-12 METHODS FOR SOLVING QUADRATIC EQUATIONS

Up to this point, we have discussed the various methods for solving the three types of quadratic equations. In this section, we will briefly review these methods and make some additional comments about them.

170. Let's begin with "pure" quadratics. Since both terms in the binomial subtractions below are perfect squares, we can solve each equation in three ways.

$$x^2 - 9 = 0 \qquad\qquad 4y^2 - 25 = 0$$

Method 1: Isolating the squared-letter term and then taking both square roots of the number. That is:

$$x^2 = 9 \qquad\qquad y^2 = \frac{25}{4}$$

$$x = \pm\sqrt{9} \qquad\qquad y = \pm\sqrt{\frac{25}{4}}$$

$$x = \pm 3 \qquad\qquad y = \pm\frac{5}{2} \text{ (or } \pm 2.5)$$

Method 2: Using the factoring method. That is:

$$(x+3)(x-3) = 0 \qquad\qquad (2y+5)(2y-5) = 0$$

$$x = -3, \text{ from } x + 3 = 0 \qquad\qquad y = \frac{-5}{2}, \text{ from } 2y + 5 = 0$$

$$x = +3, \text{ from } x - 3 = 0 \qquad\qquad y = +\frac{5}{2}, \text{ from } 2y - 5 = 0$$

Method 3: Using the quadratic formula. That is:

$$x = \frac{(-0) \pm \sqrt{0^2 - 4(1)(-9)}}{2(1)} \qquad\qquad y = \frac{(-0) \pm \sqrt{0^2 - 4(4)(-25)}}{2(4)}$$

$$= \frac{0 \pm \sqrt{36}}{2} \qquad\qquad = \frac{0 \pm \sqrt{400}}{8}$$

$$= \frac{0 \pm 6}{2} = \pm\frac{6}{2} = \pm 3 \qquad\qquad = \frac{0 \pm 20}{8} = \pm\frac{20}{8} = \pm\frac{5}{2}$$

Since shorter methods are available, we ordinarily would not use Method 3 (the quadratic formula) to solve a "pure" quadratic equation.

Of Methods 1 and 2, which one would we ordinarily use to solve:

(a) $x^2 = 36$ _____ (b) $9w^2 - 16 = 0$ _____

---

a) Method 1

b) Method 2 (or Method 1 if it is rearranged first to $9w^2 = 16$).

Quadratic Equations and Formulas 257

171. Most "pure" quadratic equations are like those below in which at least one term is not a perfect square.

$$x^2 - 29 = 0 \qquad 3d^2 - 11 = 0$$

With this type of "pure" quadratic, we ordinarily do not use the factoring method. We solve this type by isolating the squared letter. That is:

$$x^2 = 29 \qquad d^2 = \frac{11}{3}$$

$$x = \pm\sqrt{29} \qquad d = \pm\sqrt{\frac{11}{3}}$$

Though the roots are frequently left in the radical form above in many mathematics courses, this radical form would not be satisfactory in science or technology. In science or technology, the radicals would have to be evaluated. Therefore, in science or technology:

(a) Since $\sqrt{29} = 5.39$, the roots of $x^2 - 29 = 0$ would be reported as _____ and _____.

(b) Since $\sqrt{\frac{11}{3}} = \sqrt{3.67} = 1.92$, the roots of $3d^2 - 11 = 0$ would be reported as _____ and _____.

---

172. There are some "pure" quadratics which do not have ordinary numbers as their roots. For example, if we isolate the squared letter in each case below, we get:

$$y^2 + 37 = 0 \qquad 4p^2 + 21 = 0$$

$$y^2 = -37 \qquad p^2 = -\frac{21}{4} \text{ or } -5.25$$

$$y = \pm\sqrt{-37} \qquad p = \pm\sqrt{-\frac{21}{4}} \text{ or } \pm\sqrt{-5.25}$$

Since there are no ordinary numbers which are the square roots of negative numbers, neither equation above has ordinary numbers as its roots.

Which of the following equations do not have ordinary numbers as their roots? _____

(a) $t^2 - 101 = 0$ (c) $7.6d^2 - 34.6 = 0$

(b) $t^2 + 101 = 0$ (d) $7.6d^2 + 34.6 = 0$

---

a) +5.39 and -5.39
b) +1.92 and -1.92

Both (b) and (d)

258   Quadratic Equations and Formulas

173. We can solve the "incomplete" quadratic below in either of two ways.

$$x^2 - 6x = 0$$

Method 1: By the factoring method. That is:

$$x(x - 6) = 0$$

$$x = 0$$

$$x = +6 \quad \text{(From: } x - 6 = 0\text{)}$$

Method 2: By the quadratic formula. That is:

$$x = \frac{+6 \pm \sqrt{(-6)^2 - 4(1)(0)}}{2(1)}$$

$$= \frac{+6 \pm \sqrt{36}}{2} = \frac{+6 \pm 6}{2} = \frac{12}{2} \text{ or } 6 \text{ and } \frac{0}{2} \text{ or } 0$$

Since the factoring method is much simpler, however, we do not solve "incomplete" quadratics by means of the quadratic formula.

Solve each "incomplete" quadratic below by the factoring method.

(a) $y^2 + 4y = 0$ (b) $3t^2 - 7t = 0$

y = _____ and _____   t = _____ and _____

---

174. Though the quadratic formula is not used to solve "pure" and "incomplete" quadratics, it is the general method for solving "complete" quadratics. However, the factoring method can be used to solve "complete" quadratics when they are easily factorable.

Solve each "complete" quadratic below by the factoring method.

(a) $x^2 - 10x + 16 = 0$ (b) $2y^2 - y - 15 = 0$

x = _____ and _____   y = _____ and _____

a) y = 0 and −4

b) t = 0 and $\frac{7}{3}$

---

a) x = 2 and 8, from:
   $(x - 2)(x - 8) = 0$

b) y = $-\frac{5}{2}$ and 3, from:
   $(2y + 5)(y - 3) = 0$

175. Though the numbers in the "complete" quadratics below are reasonably small, neither trinomial can be factored into binomials which contain whole numbers.

$$t^2 - 14t + 27 = 0 \qquad 3p^2 + 8p - 5 = 0$$

(a) Would we solve either equation above by the factoring method? _____

(b) What method would we use to solve them?

_____

176. The complete quadratics below are typical of those which are encountered in science and technology.

$$16.1t^2 + 80.0t - 300 = 0$$

$$0.8w^2 - 3.5w + 1.27 = 0$$

(a) Would we attempt to solve either equation above by the factoring method? _____

(b) What method would we use to solve the equations above?

_____

a) No

b) The quadratic formula.

177. Here is one of the equations from the last frame.

$$16.1t^2 + 80.0t - 300 = 0$$

Solving this equation by the quadratic formula, we get:

$$t = \frac{(-80.0) \pm \sqrt{(80.0)^2 - 4(16.1)(-300)}}{2(16.1)}$$

$$= \frac{(-80.0) \pm \sqrt{6{,}400 + 19{,}300}}{32.2} = \frac{(-80.0) \pm \sqrt{25{,}700}}{32.2}$$

Though the solution is frequently left in the radical form above in many mathematics courses, this form would not be acceptable in science or technology. Since $\sqrt{25{,}700} \doteq 160$, how would the roots be reported in science or technology?

$$t = \underline{\hspace{1cm}} \text{ and } \underline{\hspace{1cm}}$$

a) No

b) The quadratic formula.

t = 2.48 and -7.45

Quadratic Equations and Formulas

178. There are some "complete" quadratics which do not have ordinary numbers as their roots. Here is an example:

$$3t^2 - 2t + 5 = 0$$

Plugging into the quadratic formula, we get:

$$t = \frac{(+2) \pm \sqrt{(-2)^2 - 4(3)(5)}}{2(3)} = \frac{2 \pm \sqrt{-56}}{6}$$

Since there are no ordinary numbers which are the square roots of negative quantities like "-56", the equation above does not have ordinary numbers as its roots.

You should be able to see that a quadratic does not have ordinary numbers as its roots when "$b^2 - 4ac$" in the quadratic formula is a _____ (negative/positive) quantity.

---

179. Since the quadratic formula is used to solve most "realistic" complete quadratic equations, we feel that you should memorize the formula. It is given below.

If: $a\square^2 + b\square + c = 0$,

Then: $\square = \dfrac{-b \pm \sqrt{b^2 - 4ac}}{2a}$

Answer to 178: negative

SELF-TEST 12 (Frames 170-179)

1. Which of the following equations would be solved by the quadratic formula method? _____
   (a) $3x^2 = 7$ (b) $5t^2 + 2t = 0$ (c) $h^2 + 7h - 1 = 0$ (d) $r^2 - 6r = 0$ (e) $8 - y^2 = 0$

2. Which of the following quadratic equations do not have ordinary numbers as their roots? _____
   (a) $w^2 + 4 = 0$ (b) $p^2 + 9p = 0$ (c) $9s^2 - 1 = 0$ (d) $x^2 + 2x + 3 = 0$ (e) $x^2 - 2x - 3 = 0$

Solve each of these quadratic equations:

3. $5r^2 - 9 = 0$	4. $4E^2 + 25E = 0$	5. $4w^2 - 2w - 1 = 0$
r = _____ and _____	E = _____ and _____	w = _____ and _____

ANSWERS: 1. Only (c)   2. (a) and (d)   3. $r = \pm 1.34$ or $\pm\sqrt{\dfrac{9}{5}}$ 4. $E = 0$ and $\dfrac{25}{4}$ (or 6.25)   5. $w = 0.81$ or $w = -0.31$ $\dfrac{2 \pm \sqrt{20}}{8}$

3-13 QUADRATIC EQUATIONS IN NON-STANDARD FORM

The quadratic-formula method of solving "complete" quadratic equations presupposes that the equation is written in standard form. However, when applied problems lead to quadratic equations, the original equations are usually not in standard form. Therefore, they must be rearranged to standard form before the quadratic formula can be used. In this section, we will discuss rearrangements of this type and then solve a few quadratic equations which require rearrangements. Some methods of simplifying the calculations required when using the quadratic formula are also discussed.

180. Here is the standard form of a quadratic equation:

$$a\square^2 + b\square + c = 0$$

In this standard form, all of the non-zero terms appear on one side with the squared-letter term first and the number-term last. Use of the quadratic formula requires this form.
Here is an equation which is not in standard form: $3x^2 - 2x = 5x + 9 - 7x^2$

Before using the quadratic formula to solve it, we must rearrange it to put it in standard form. We can use the addition axiom to do so.

(a) Write the equation in standard form: _____

(b) Therefore: a = _____   b = _____   c = _____

181. Here is another quadratic equation which is not in standard form:

$$9 - 3s^2 = 2s - 5s^2$$

(a) Write it in standard form: _____

(b) Therefore: a = _____   b = _____   c = _____

a) $10x^2 + (-7x) + (-9) = 0$
b) a = 10
   b = -7
   c = -9

Answer to Frame 181:   a) $2s^2 + (-2s) + 9 = 0$ b) a = 2   b = -2   c = 9

182. When rearranging a quadratic equation to put it in standard form, we can obtain a negative coefficient for the squared-letter term. For example:

$$(-3x^2) + 7x + 9 = 0$$

Since the denominator of the quadratic formula is "2a", the denominator would be "-6" for the equation above. Mathematicians like to avoid negative signs in this denominator since they are easy to overlook. Therefore, they always try to write the standard form so that "a" is positive.

In order to obtain an equation, equivalent to the one above, in which "a" is positive, we can replace each side of the equation by its opposite.

On the left side: The opposite of the trinomial is obtained by replacing each term with its opposite.

On the right side: The opposite of "0" is "0".

Therefore, we get the new equation:

$$3x^2 + (-7x) + (-9) = 0$$

Continued on following page.

262 Quadratic Equations and Formulas

182. Continued

For each of these, write the equivalent equation in which "a" is positive:

(a) $(-9x^2) + 7x + 11 = 0$ (b) $(-4q^2) + (-5q) + 10 = 0$ (c) $(-6p^2) + (-7p) + (-8) = 0$

---

183. One of the terms on the left side of this equation is an instance of the distributive principle.

$7 + x(x + 1) = 2$

(a) To put the equation in standard form, we must multiply by the distributive principle first. Do so.

(b) Now put the equation in standard form.

a) $9x^2 + (-7x) + (-11) = 0$
b) $4q^2 + 5q + (-10) = 0$
c) $6p^2 + 7p + 8 = 0$

---

184. The equation at the right includes a squaring of the binomial $(m + 2)$ on the left.

$(m + 2)^2 = 3$

(a) Square the binomial on the left side.

(b) Now put the equation in standard form.

a) $7 + x^2 + x = 2$
b) $x^2 + x + 5 = 0$

---

185. The equation at the right includes a multiplication of two binomials on the left side.

$(p + 3)(p - 3) = 4p$

(a) Perform the multiplication on the left side.

(b) Now write the equation in standard form.

a) $m^2 + 4m + 4 = 3$
b) $m^2 + 4m + 1 = 0$

---

186. (a) Perform the multiplications on both sides of the equation at the right.

$3y(y - 7) = (y + 4)(y - 3)$

(b) Now write the equation in standard form.

a) $p^2 - 9 = 4p$
b) $p^2 + (-4p) + (-9) = 0$

---

a) $3y^2 - 21y = y^2 + y - 12$
b) $2y^2 + (-22y) + 12 = 0$

Quadratic Equations and Formulas   263

187. Some fractional equations are really quadratic equations. The fact that they are quadratic equations becomes obvious <u>when the fractions are cleared</u>. Here is an example: $x + \frac{2}{x} = 3$

(a) To clear the fraction, we multiply both sides by "x". Do so.

(b) Now write the equation in standard form.

---

188. Let's put the equation at the right in standard form. $\frac{5}{t} = \frac{t}{7} + 2$

(a) To clear the fractions, we must multiply both sides by "7t". Do so.

(b) Now write the equation in standard form.

a) $x^2 + 2 = 3x$

b) $x^2 + (-3x) + 2 = 0$

---

189. To clear the fraction in the top equation at the right, we multiply both sides by $(w + 1)$ and get the bottom equation.

$w + 3 = \dfrac{7}{w + 1}$

$(w + 1)(w + 3) = 7$

(a) Perform the multiplication on the left side.

(b) Now write the equation in standard form.

a) $35 = t^2 + 14t$

b) $t^2 + 14t + (-35) = 0$

---

a) $w^2 + 4w + 3 = 7$

b) $w^2 + 4w + (-4) = 0$

264 Quadratic Equations and Formulas

190. Let's put the equation at the right in standard form. $\frac{1}{b} + \frac{1}{b+3} = 5$

(a) To clear the fractions, we must multiply both sides by "b" and "b + 3". Do so.

(b) Now write the equation in standard form.

---

191. Write each equation below in standard form:

(a) $t(2t - 3) = 6t - 2$ (b) $3I = 8 - \frac{7}{I}$

a) $b + 3 + b = 5b^2 + 15b$
   or
   $2b + 3 = 5b^2 + 15b$

b) $5b^2 + 13b + (-3) = 0$

---

192. Write each equation below in standard form:

(a) $\frac{F}{3} = \frac{7}{F} + 1$ (b) $\frac{5}{y+1} = 3 + \frac{1}{y}$

a) $2t^2 + (-9t) + 2 = 0$
b) $3I^2 + (-8I) + 7 = 0$

---

a) $F^2 + (-3F) + (-21) = 0$
b) $3y^2 + (-y) + 1 = 0$

193. Let's solve the equation at the right. $\dfrac{4}{E} = \dfrac{2E}{3} - 7$

(a) Clear the fractions and write the resulting quadratic equation in standard form. (Be sure that "a" is positive.)

(b) Now use the quadratic formula to find the two roots.

E = _____ and _____

---

194. Using the quadratic formula involves a fair amount of calculation. After an equation is written in standard form, it is sometimes possible to simplify the calculation. We can use the equation below as an example:

$$70x^2 + 60x + 10 = 0$$

In order to reduce the size of the numbers, we can replace the equation above with a simpler equivalent equation. Here are the steps:

(1) <u>Factoring the left side</u>, we get: $10(7x^2 + 6x + 1) = 0$

(2) <u>Multiplying both sides by the reciprocal of "10"</u>, we get:

$$\left(\dfrac{1}{10}\right)(10)(7x^2 + 6x + 1) = \left(\dfrac{1}{10}\right)0$$

(1) $(7x^2 + 6x + 1) = 0$

$7x^2 + 6x + 1 = 0$

Using this same procedure, write the following quadratic equations in the simplest standard form:

(a) $100p^2 + 700p + 1{,}000 = 0$ (b) $150t^2 - 240t - 560 = 0$

Answers:

a) $2E^2 + (-21E) + (-12) = 0$

b) $E = 11.0$ and -0.55
or
$E = \dfrac{21 \pm \sqrt{537}}{4}$

---

195. $3x^2 + 12x + 9 = 0$

(a) What is the common factor on the left side above? _____

(b) Write the equation in the simplest standard form.

Answers:

a) $p^2 + 7p + 10 = 0$

b) $15t^2 + (-24t) + (-56) = 0$

266   Quadratic Equations and Formulas

196. Write the following equations in the simplest standard form: (a) $25y^2 - 15y - 120 = 0$ (b) $4F^2 - 24F - 32 = 0$	a) 3 b) $x^2 + 4x + 3 = 0$
197. Even though "a", "b", and "c" are large numbers, the standard form <u>cannot</u> <u>always</u> be simplified. Here is an example: $$27x^2 + (-13x) + 17 = 0$$ Since there is no common factor on the left side, the equation cannot be simplified. It is in the simplest standard form. Can the following equation be simplified? _____ $$27x^2 + 56x + 117 = 0$$	a) $5y^2 + (-3y) + (-24) = 0$ b) $F^2 + (-6F) + (-8) = 0$
198. Simplify the following equations if possible: (a) $51x^2 + (-18x) + 36 = 0$ (b) $27V^2 + 36V + (-44) = 0$	No
199. In the following equation, "a", "b", and "c" are decimals. $$1.73x^2 + 0.95x + 0.21 = 0$$ We can eliminate the decimals by multiplying both sides of the equation by 100. We get: $$100(1.73x^2 + 0.95x + 0.21) = 100(0)$$ $$100(1.73x^2) + 100(0.95x) + 100(0.21) = 0$$ $$173x^2 + 95x + 21 = 0$$ (Since the procedure above is a straight-forward use of the multiplication axiom, the roots of the new equation are identical to the roots of the original equation.) Using this procedure, eliminate the decimals in these equations: (a) $1.7t^2 + 3.2t + 0.9 = 0$ (b) $8.56R^2 + (-9.98R) + (-0.51) = 0$	a) $17x^2 + (-6x) + 12 = 0$ ("3" was the common factor.) b) It cannot be simplified.
	a) $17t^2 + 32t + 9 = 0$ b) $856R^2 + (-998R) + (-51) = 0$

200. After the decimals are eliminated, the equation can sometimes be simplified further. Here is an example:

$1.4x^2 + 3.5x + 0.7 = 0$

Multiplying both sides by 10, we get:

$14x^2 + 35x + 7 = 0$

This last equation can also be simplified. Do so.

―――――――――――――――

201. Eliminate the decimals in the following, and then simplify further if possible.

(a) $1.7x^2 + 4.2x + (-6.7) = 0$ (b) $2.25x^2 + 6.75x + 1.25 = 0$

$2x^2 + 5x + 1 = 0$

―――――――――――――――

202. ┌───┐
    │ Eliminating the decimal coefficients does not always simplify │
    │ the calculations in using the quadratic formula. Sometimes   │
    │ it is easier to work with the decimals themselves. Use the   │
    │ numbers which are easiest for you.   │
    └───┘

Let's solve this quadratic equation: $200x(x + 4) - 700 = 0$

(a) Write the equation in standard form.

(b) Now use the quadratic formula to find its roots.

x = _____ and _____

a) $17x^2 + 42x + (-67) = 0$
Not possible to simplify further.

b) $9x^2 + 27x + 5 = 0$
("25" was the common factor.)

―――――――――――――――

a) $200x^2 + 800x + (-700) = 0$
or
$2x^2 + 8x + (-7) = 0$

b) $x = 0.74$ and -4.74
or
$x = \dfrac{-8 \pm \sqrt{120}}{4}$

268   Quadratic Equations and Formulas

203. Let's solve this quadratic equation: $\frac{2.5}{F} + 4.5F = 8.5$

(a) Write the equation in standard form.

_____

(b) Now use the quadratic formula to find its roots.

F = _____ and _____

---

204. When rearranging an equation to put it in standard form, we occasionally obtain a "pure" quadratic equation. In these cases, we do not use the quadratic formula since there is a simpler method.

Solve these equations. Each equation simplifies to a "pure" quadratic equation.

(a) $(F + 4)(F - 4) = 7$ (b) $\frac{4x}{5} = \frac{7}{x}$

F = _____ and _____   x = _____ and _____

a) $4.5F^2 + (-8.5F) + 2.5 = 0$
   or
   $45F^2 + (-85F) + 25 = 0$
   or
   $9F^2 + (-17F) + 5 = 0$

b) F = 1.52 and 0.364
   or
   $F = \frac{17 \pm \sqrt{109}}{18}$

---

205. The equation at the right looks like a quadratic equation. $y^2 - 4y = (y + 7)(y - 7)$

(a) What equation do we get if we put it in standard form?

_____

(b) Is this new equation a quadratic equation? _____

a) $F = \pm 4.80$
   or
   $F = \pm\sqrt{23}$

b) $x = \pm 2.96$
   or
   $x = \pm\sqrt{8.75}$

---

a) $0y^2 + (-4y) + 49 = 0$
   or
   $(-4y) + 49 = 0$
   or
   $4y - 49 = 0$

b) No. It has only one root.

SELF-TEST 13 (Frames 180-205)

Write each of the following quadratic equations in standard form. **Do not solve**.

1. $2x - x(x - 3) = 3$

2. $\dfrac{p}{2} + \dfrac{2}{p+1} = 3$

3. $\dfrac{20}{w} - 60 = \dfrac{30w}{w-2}$

Solve each of the following quadratic equations.

4. $5R + \dfrac{2}{R} = 8$

5. $54 - 27h = 9h^2$

6. $1.25t(2 - t) = 0.75$

R = _____ and _____

h = _____ and _____

t = _____ and _____

ANSWERS:

1. $x^2 + (-5x) + 3 = 0$
2. $p^2 + (-5p) + (-2) = 0$
3. $9w^2 + (-14w) + 4 = 0$

4. R = 1.29 and 0.31
   or
   $R = \dfrac{8 \pm \sqrt{24}}{2}$

5. h = 1.37 and -4.37
   or
   $h = \dfrac{-3 \pm \sqrt{33}}{2}$

6. t = 1.63 and 0.37
   or
   $t = \dfrac{10 \pm \sqrt{40}}{10}$

270   Quadratic Equations and Formulas

3-14   RADICAL EQUATIONS AND EXTRANEOUS ROOTS

When the squaring principle is used to eliminate the radical in a radical equation, the resulting non-radical equation is occasionally a quadratic equation. Though the quadratic equation has two roots, frequently one of these roots does not satisfy the original radical equation. A root of this type is called an "extraneous" root. We will briefly discuss "extraneous" roots in this section.

206. In order to solve the radical equation below, we must use the squaring principle to eliminate the radical. When we do so, the resulting non-radical equation is a quadratic equation. That is:

$$\sqrt{y + 5} = y - 1$$
$$\left(\sqrt{y + 5}\right)^2 = (y - 1)^2$$
$$y + 5 = y^2 - 2y + 1$$

or

$$y^2 - 3y - 4 = 0$$

The two roots of the quadratic equation are "4" and "-1". Let's check these two roots in the original radical equation.

Checking "4", we get:

$$\sqrt{4 + 5} = 4 - 1$$
$$\sqrt{9} = 3$$
$$3 = 3$$

Checking "-1", we get:

$$\sqrt{(-1) + 5} = (-1) - 1$$
$$\sqrt{4} = -2$$
$$+2 \neq -2$$

Only one of the two roots satisfies the original equation. Which one? _____

207. Let's summarize what we found in the last frame.

(1) When we squared both sides of the original equation, we obtained a quadratic equation.

(2) Of the two roots (+4 and -1) of this quadratic equation, only +4 satisfied the original radical equation.

When a quadratic equation is obtained by squaring both sides of a radical equation, frequently one of the roots of the quadratic equation does not satisfy the original equation. This root is called an "extraneous" root.

In the last frame, "-1" was an _____ root.

Only "4"

extraneous

208. An extraneous root is occasionally introduced by applying the squaring principle to a radical equation. The extraneous root is usually one of the two roots of a quadratic equation. The <u>only</u> <u>way</u> <u>to</u> <u>detect</u> <u>an</u> <u>extraneous</u> <u>root</u> <u>is</u> <u>to</u> <u>check</u> <u>the</u> <u>roots</u> <u>in</u> <u>the</u> <u>original</u> <u>radical</u> <u>equation.</u>

Let's solve this radical equation: $\sqrt{2y+5} = 2y - 1$

(a) Squaring both sides, we get:

_____ = _____

(b) Write the resulting quadratic equation in standard form.

_____

(c) The two roots of the quadratic equation are _____ and _____ .

(d) Check both roots in the original equation, $\sqrt{2y+5} = 2y - 1$

(e) Which one of the two roots is an extraneous root? _____

(f) The only root which satisfies the original equation is _____ .

---

209. Solve this radical equation: $t - 5 = \sqrt{4t+1}$

(a) The extraneous root is _____ .

(b) The only root of the original equation is _____ .

---

a) $2y + 5 = 4y^2 - 4y + 1$

b) $4y^2 + (-6y) + (-4) = 0$
or
$2y^2 + (-3y) + (-2) = 0$

c) 2 and $-\frac{1}{2}$

d) $\sqrt{2(2)+5} = 2(2) - 1$
$\sqrt{4+5} = 4 - 1$
$\sqrt{9} = 3$
$3 = 3$

$\sqrt{2\left(-\frac{1}{2}\right)+5} = 2\left(-\frac{1}{2}\right)-1$
$\sqrt{(-1)+5} = (-1) - 1$
$\sqrt{4} = -2$
$2 \neq -2$

e) $-\frac{1}{2}$ f) 2

---

a) 2   b) 12

272 Quadratic Equations and Formulas

210. When we apply the squaring principle to a radical equation and obtain a quadratic equation, one of the roots is not always extraneous. That is, sometimes both roots satisfy the original radical equation. Here is an example:

We have squared both sides of the equation below.

$$p - 1 = \sqrt{3p - 5}$$
$$(p - 1)^2 = \left(\sqrt{3p - 5}\right)^2$$
$$p^2 - 2p + 1 = 3p - 5$$
$$p^2 - 5p + 6 = 0$$

The two roots of the quadratic equation are +2 and +3. Show that both of these roots satisfy the original radical equation, $p - 1 = \sqrt{3p - 5}$.

When solving a radical equation, always check for an extraneous root if you obtain a quadratic equation by applying the squaring principle. <u>An extraneous root, of course, is not a root of the original equation.</u> In an applied problem, you can usually detect an extraneous root quite easily because the value generally makes no sense in the context of the problem.

(2) $-1 = \sqrt{3(2) - 5}$
$ 1 = \sqrt{1}$
$ 1 = 1$

(3) $-1 = \sqrt{3(3) - 5}$
$ 2 = \sqrt{4}$
$ 2 = 2$

SELF-TEST 14 (Frames 206-210)

Find the root or roots of each of the following radical equations:

1. $\sqrt{t + 3} = t + 1$

2. $R + 2 = \sqrt{7R + 2}$

ANSWERS: 1. $t = +1$
(Note: -2 is not a root.)

2. $R = +1$ and $+2$
(Both roots satisfy the original equation.)

3-15 SOME APPLIED PROBLEMS INVOLVING QUADRATIC EQUATIONS

Though there are many applied problems which involve "pure" quadratic equations, applied problems which involve "incomplete" or "complete" quadratics are not encountered very often. In this section, we will solve a few applied problems involving quadratic equations.

211. Since there are many formulas which contain squared letters, "pure" quadratic equations are frequently encountered in applied problems. The "pure" quadratics arise when specific values are plugged in for the variables. Here is an example:

> The formula below states a relationship between the distance traveled (s), acceleration (a), and time traveled (t) for an object starting from rest.
>
> $$s = \frac{1}{2}at^2$$
>
> If the distance traveled (s) is 400 feet and the object's acceleration (a) is 32 feet per second, we can find the time traveled (t) by plugging into the formula. Notice that we get a "pure" quadratic equation.
>
> $$400 = \frac{1}{2}(32)(t^2)$$
>
> $$400 = 16t^2$$
>
> $$t^2 = \frac{400}{16}$$
>
> $$t^2 = 25$$

If we solve the "pure" quadratic, we get:

$\qquad$ t = 5 seconds $\quad$ or $\quad$ t = -5 seconds

Which of these two values does not make sense in the context of the problem? _____

212. When solving a "pure" quadratic equation in an applied context, the negative root of the equation ordinarily does not make sense. Here is another example.

> The formula below states the relationship between the power (P), current (I), and resistance (R) in an electrical circuit.
>
> $$P = I^2R$$
>
> If the power (P) is 150 watts and the resistance (R) is 96.1 ohms, we can find the corresponding current (I) by plugging into the formula. We get:
>
> $$150 = I^2(96.1)$$
>
> $$I^2 = \frac{150}{96.1}$$
>
> $$I^2 = 1.56$$

Continued on following page.

---

211. -5 seconds

274   Quadratic Equations and Formulas

212. Continued
　　(a) Since $\sqrt{1.56} = 1.25$, the two roots of the equation are "_____ amperes" and "_____ amperes".

　　(b) Which of these two roots does not make sense in the context of the problem? _____

---

a) +1.25 amperes and −1.25 amperes

b) −1.25 amperes

---

213. Though "pure" quadratics occur frequently in applied problems, the two other types of quadratics occur infrequently. In the next few frames, we will give a few examples of applied problems which involve "complete" quadratic equations.

The formula below states the relationship between distance traveled (s), initial velocity (v_o) and time traveled (t) for an object which is thrown vertically downward.

$$s = v_o t + 16.1 t^2$$

If the initial velocity (v_o) is 80.0 feet per second, we can find the amount of time (t) needed for it to travel a distance (s) of 300 feet by plugging into the formula. Notice that we get a "complete" quadratic in which "t" is the variable.

$$300 = 80.0t + 16.1t^2$$
or
$$16.1t^2 + 80.0t + (-300) = 0$$

(a) Use the quadratic formula to find the roots of the last equation.

　　　　　　　　　t = _____ seconds and _____ seconds

(b) Only one of these roots makes sense in the context of the problem. Which one? _____

---

a) t = 2.49 seconds and −7.47 seconds

From:

$$t = \frac{-80 \pm \sqrt{25,720}}{32.2}$$

b) t = 2.49 seconds

214. Here is another problem which leads to a "complete" quadratic equation.

The figure at the right is a cross-section of an L-shaped steel bar. The outside dimensions are 2.00 inches and 1.50 inches. Because of the load which the bar must support, the area of the cross-section must be 1.20 square inches. We must find the width "w".

To solve this problem, we break the figure into two rectangles, #1 and #2, as shown on the right. Then:

$$\boxed{\text{Area \#1} + \text{Area \#2} = 1.20}$$

Area #1 = (2.00)(w) = 2w
Area #2 = (1.50 - w)(w) = 1.5w - w²

Since the sum of these two areas is 1.20, we can set up this equation:

$$\boxed{2w + (1.5w - w^2) = 1.20}$$

(a) Write this equation in standard form. _____

(b) Solve for "w".

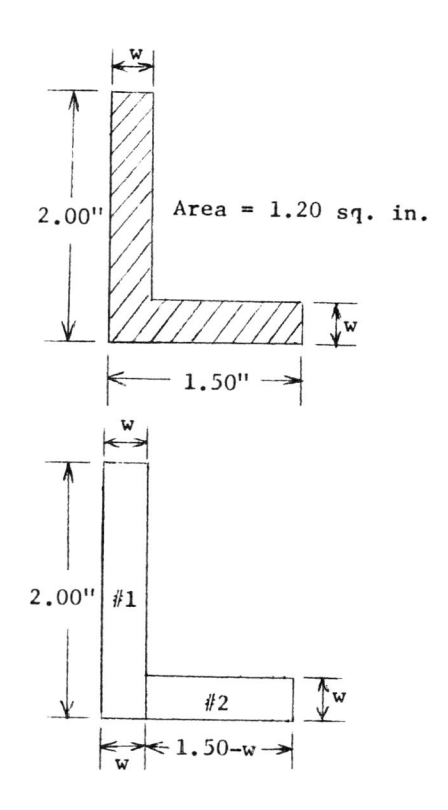

w = _____ and _____

(c) Which value of "w" is the correct answer to the problem? _____

---

215. In an electrical circuit, we must connect two resistors in parallel so that their total resistance is 20 ohms. One of the two resistors must be 10 ohms greater than the other. Let's find the size of the two resistors.

The formula relating total resistance to two resistances in parallel is:

$$\frac{1}{R_t} = \frac{1}{R_1} + \frac{1}{R_2}$$

If we let R_1 be the smaller resistor, then $R_2 = R_1 + 10$. Also, $R_t = 20$. Therefore, we get this new formula in which R_1 is the only unknown:

$$\frac{1}{20} = \frac{1}{R_1} + \frac{1}{R_1 + 10}$$

Continued on following page.

---

a) $w^2 + (-3.5w) + 1.20 = 0$

b) w = 0.385" and 3.12"

From:
$$w = \frac{3.5 \pm \sqrt{7.45}}{2}$$

c) w = 0.385"
Though the other root (3.12") satisfies the quadratic equation, it makes no sense in the context of this cross-section.

276   Quadratic Equations and Formulas

215. Continued

Clearing the fractions and writing the equation in standard form, we get:

$$R_1^2 + (-30R_1) + (-200) = 0$$

Using the quadratic formula to solve this last equation, we get:

$$R_1 = 35.6 \text{ ohms and } -5.6 \text{ ohms}$$

(a) Which of these two values makes no sense in the context of the problem? _____

(b) If $R_1 = 35.6$ ohms and $R_2 = R_1 + 10$, $R_2 =$ _____ ohms.

---

216. Quadratic equations sometimes arise in problems involving geometric shapes. Here is an example.

The diagram at the right shows a metal part which is circular in shape and which has a portion cut out. It is necessary to calculate "r", the radius of the circle.

To find "r", note that triangle EOF is a <u>right</u> triangle whose sides are:
EO = r - 1.00, OF = r, and EF = 4.00.

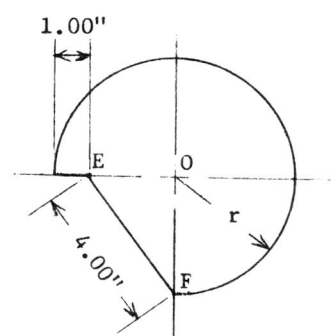

From the Pythagorean Theorem: $(EO)^2 + (OF)^2 = (EF)^2$ or $(r-1)^2 + r^2 = 4^2$

Simplifying the last equation, we get: $2r^2 - 2r - 15 = 0$

Using the quadratic formula to solve this equation, we get:

$$r = \frac{2 \pm \sqrt{124}}{4} \quad \text{or} \quad r = 3.28 \text{ and } -2.28$$

In the context of the problem, only one of these two answers makes sense. It is: r = _____

---

a) -5.6 ohms

b) 45.6 ohms

---

r = 3.28"

217. The diagram at the right shows an L-shaped inspection gage. It is desired to check the correctness of the gage by placing a cylindrical plug which will be tangent to two surfaces, as shown, and which will also touch the gage at point P. Calculate the radius of the cylindrical plug which should be used.

To find "r", <u>right</u> triangle 0NP has been drawn. Its sides are: $0N = r - 2.00$, $NP = 6.00 - r$, and $0P = r$.

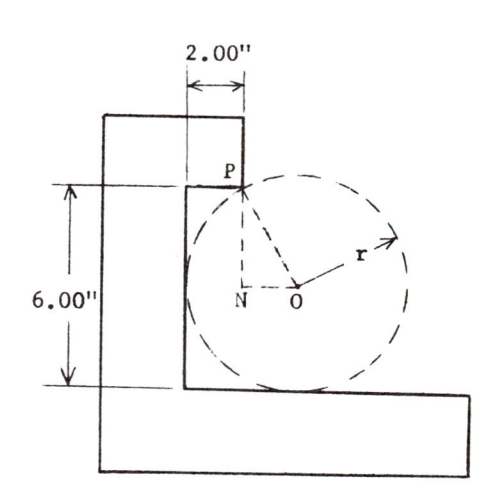

From the Pythagorean Theorem: $(0N)^2 + (NP)^2 = (0P)^2$ or $(r-2)^2 + (6-r)^2 = r^2$

Simplifying the last equation, we get: $r^2 - 16r + 40 = 0$

Solving by means of the quadratic formula, we get:

$$r = \frac{16 \pm \sqrt{96}}{2} \quad \text{or} \quad r = 12.90 \text{ and } 3.10$$

Only one of these two answers for "r" makes sense. It is: r = _____

---

218. A quadratic equation occurs in the following chemistry problem. It is desired to find the number of moles of thallium chloride that will dissolve in one liter of an aqueous solution of 0.100 molar potassium chloride.

Using principles of chemistry, the following quadratic equation can be set up, where "x" is the number of moles of thallium chloride:

$$0.000200 = x(0.100 + x)$$

The standard form is: $x^2 + 0.100x - 0.000200 = 0$

Solving by means of the quadratic formula, we get:

$$x = \frac{-0.100 \pm \sqrt{0.0108}}{2} \quad \text{or} \quad x = 0.00196 \text{ and } -0.102$$

In the context of this problem, "x" must be positive. Therefore: x = _____ mole.

---

r = 3.10"

x = 0.00196 mole

278 Quadratic Equations and Formulas

SELF-TEST 15 (Frames 211-218)

1. Calculate radius "r" in the diagram at the right. The circle touches point H and is tangent to the horizontal and vertical lines. Note: In right triangle OBH: OH = r
    OB = r - 2.00
    BH = r - 1.00

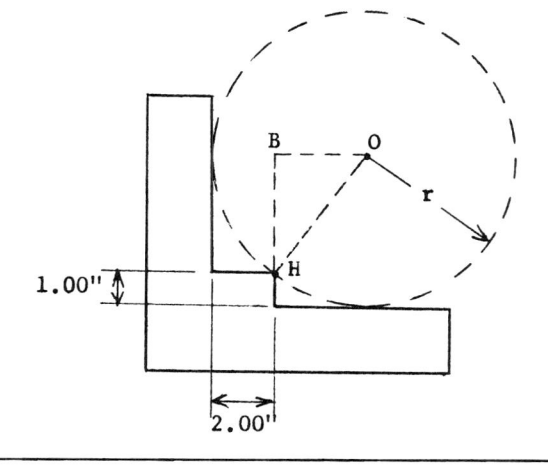

2. Calculate dimension "x" in the diagram at the right. The radius is 6.00", distance AB is 7.00", and AOB is a <u>right</u> triangle. Note: OA = 6.00 - x

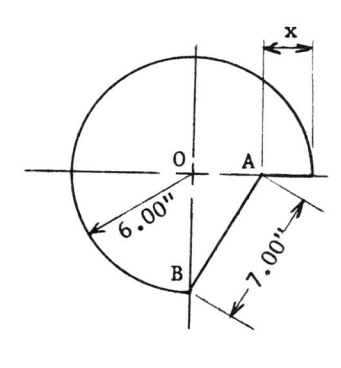

ANSWERS: 1. r = 5.00"   2. x = 2.39" or 2.40"

3-16 FORMULA REARRANGEMENT WITH FORMULAS IN QUADRATIC FORM

There are many formulas that contain squared letters. Since formulas of this type are similar to either pure, incomplete, or complete quadratic equations, we say that they are "in quadratic form". In this section, we will rearrange formulas which are in quadratic form. Some of the rearrangements require the use of the quadratic formula.

219. Since each of the formulas below contains a squared letter, we say that each formula is "in quadratic form". If we think of the squared letters as the variables and the other variables as constants, the formulas are similar to "pure" quadratic equations.

$$v^2 = 2as \qquad P = I^2R \qquad F = \frac{mv^2}{r}$$

The procedure for solving for the letter which is squared in formulas of this type was discussed earlier. The two basic steps are:

(1) Isolate the squared letter (if necessary).
(2) Then apply the square-root principle.

Using this procedure, solve for the letter which is squared in each formula below:

(a) $v^2 = 2as$ \qquad (b) $P = I^2R$ \qquad (c) $F = \frac{mv^2}{r}$

v = _____ \qquad I = _____ \qquad v = _____

220. Since the formula below contains a "w^2", we say that it is "in quadratic form". And since it contains a "w^2" in one term, a "w" in the other, and no other terms, it is similar to an "incomplete" quadratic with "w" as the variable.

$$aw^2 = bw$$

If we solve it like an incomplete quadratic by the factoring method, we get:

$$aw^2 - bw = 0$$
$$w(aw - b) = 0$$
$$w = 0 \text{ and } \frac{b}{a}$$

However, we would ordinarily solve it by multiplying both sides by $\frac{"1"}{w}$ to eliminate the square of the letter. We get:

$$\frac{1}{w}(aw^2) = bw\left(\frac{1}{w}\right)$$
$$aw = b$$
$$w = \frac{b}{a}$$

When we multiplied both sides by $\frac{"1"}{w}$ to get "$aw = b$", we eliminated one of the two roots of the original formula.

(a) Which root was eliminated? _____

(b) Why is it acceptable to eliminate this root? _____

---

a) $v = \sqrt{2as}$

b) $I = \sqrt{\dfrac{P}{R}}$

c) $v = \sqrt{\dfrac{Fr}{m}}$

280   Quadratic Equations and Formulas

221. To solve for "t" in the formula at the right, we begin by multiplying both sides by $\frac{"1"}{t}$.

Doing so eliminates the square of the letter and also eliminates the "0" root which is ordinarily meaningless in the context of formulas. Solve for "t".

$ct^2 = Rt$

t = _____

a) 0

b) Because "w = 0" makes little or no sense as the result of a formula rearrangement.

---

222. We have squared each formula below to eliminate the radical.

$\sqrt{ar} = wr$ $\qquad$ $\sqrt{ar} = \frac{2\pi r}{T}$

$ar = w^2 r^2$ $\qquad$ $ar = \frac{4\pi^2 r^2}{T^2}$

Notice that each new formula is similar to an "incomplete" quadratic with "r" as the variable. Solve for "r" in each case.

(a) $ar = w^2 r^2$ $\qquad$ (b) $ar = \frac{4\pi^2 r^2}{T^2}$

r = _____ $\qquad$ r = _____

$t = \frac{R}{c}$, from:

$ct = R$

---

223. To eliminate "v" from the system of formulas on the left by the equivalence method, we first solve for "v" in each formula as we have done on the right.

$\boxed{F = \frac{mv^2}{r}}$ $\qquad$ $\boxed{v = \sqrt{\frac{Fr}{m}}}$

$\boxed{v = \frac{2\pi r}{T}}$ $\qquad$ $\boxed{v = \frac{2\pi r}{T}}$

Equating the two expressions for "v", we get the equation at the right. Solve for "r" in this new equation.

$\sqrt{\frac{Fr}{m}} = \frac{2\pi r}{T}$

r = _____

a) $r = \frac{a}{w^2}$, from:

$a = w^2 r$

b) $r = \frac{aT^2}{4\pi^2}$, from:

$a = \frac{4\pi^2 r}{T^2}$

$r = \frac{FT^2}{4\pi^2 m}$, from:

$\frac{F}{m} = \frac{4\pi^2 r}{T^2}$

224. Since the formula below contains a "t^2", we say that it is "in quadratic form". And since it contains a "t^2" in one term, a "t" in a second term, and one other term, it is similar to a "complete" quadratic with "t" as the variable.

$$\tfrac{1}{2}gt^2 + vt - s = 0$$

To solve for "t" in this formula, we can use the quadratic formula. Two steps are needed:

(1) We write the formula in standard form:

$$\tfrac{1}{2}gt^2 + vt + (-s) = 0 \quad \text{where: } a = \tfrac{1}{2}g \text{ or } \tfrac{g}{2}$$

$$b = v$$

$$c = -s$$

(2) We plug these values in for "a", "b", and "c" in the quadratic formula and get:

$$t = \frac{(-b) \pm \sqrt{b^2 - 4ac}}{2a}$$

$$t = \frac{(-v) \pm \sqrt{(-v)^2 - 4\left(\tfrac{g}{2}\right)(-s)}}{2\left(\tfrac{g}{2}\right)}$$

Simplifying the solution, we get:

$$t = \underline{}$$

$$t = \frac{-v \pm \sqrt{v^2 + 2gs}}{g}$$

225. The formula at the right is similar to a "complete" quadratic in which "R" is the variable.

$$pR^2 - 2kR + p = 0$$

Writing the formula in standard form, we get:

$$pR^2 + (-2kR) + p = 0, \text{ where:}$$

$$a = p, \quad b = -2k \quad c = p$$

Solve for "R" by plugging these values in the quadratic formula.

$$R = \underline{}$$

$$R = \frac{k \pm \sqrt{k^2 - p^2}}{p}$$

from:

$$R = \frac{2k \pm \sqrt{4k^2 - 4p^2}}{2p}$$

282 Quadratic Equations and Formulas

226. To eliminate "e" from the system on the left below by the equivalence method, we solve for "e" in each formula as we have done on the right.

$$\boxed{\begin{array}{l} v = \sqrt{w - e} \\ w = \dfrac{e}{v} \end{array}} \qquad \boxed{\begin{array}{l} e = w - v^2 \\ e = wv \end{array}}$$

Now if we equate the two solutions, we obtain the formula below which is similar to a "complete" quadratic in which "v" is the variable.

$$w - v^2 = wv$$

(a) Write this formula in standard form with a positive coefficient for "v^2". _____

(b) Therefore: a = ____, b = ____, c = ____

(c) Plugging these values in the quadratic formula, solve for "v".

v = _____

Answer to Frame 226:   a) $v^2 + wv + (-w) = 0$ b) $a = 1, b = w, c = -w$ c) $v = \dfrac{-w \pm \sqrt{w^2 + 4w}}{2}$

SELF-TEST 16 (Frames 219-226)

1. Solve for "D": $\boxed{I = \dfrac{WD^2}{2}}$

   D = _____

2. Solve for "v_0": $\boxed{v^2 = v_0^2 + 2as}$

   v_0 = _____

3. Solve for "t": $\boxed{(h + r)t^2 = rt}$

   t = _____

4. Solve for "w": $\boxed{kw^2 - w + d = 0}$

   w = _____

5. Solve for "p": $\boxed{h(2p + m) = p^2}$

   p = _____

ANSWERS:

1. $D = \sqrt{\dfrac{2I}{W}}$ 2. $v_0 = \sqrt{v^2 - 2as}$ 3. $t = \dfrac{r}{h + r}$ 4. $w = \dfrac{1 \pm \sqrt{1 - 4dk}}{2k}$ 5. $p = h \pm \sqrt{h^2 + hm}$

Chapter 4 EXPONENTS AND RADICALS

In an earlier chapter, we discussed squares and square-root radicals. In this chapter, we will extend that discussion to a general treatment of powers and radicals. The meaning of powers and radicals and the laws of exponents and radicals will be emphasized. The laws of exponents will be generalized to powers with whole-number, fractional, and decimal exponents. After defining powers with fractional exponents, the laws of radicals will be related to the corresponding laws of exponents.

4-1 THE MEANING OF POWERS WITH POSITIVE WHOLE-NUMBER EXPONENTS

In this section, we will discuss the meaning of powers with positive whole-number exponents of "2" or more. In doing so, the meanings of the words "base" and "exponent" are explained. The language used in conjunction with powers is also introduced.

---

1. Expressions like 2^3 and 7^4 are in <u>base-exponent</u> form.

    The regular number on the bottom left is called the <u>base</u>.
    The small number on the upper right is called the <u>exponent</u>.

    For example: In 2^3, 2 is called the <u>base</u>.
                              3 is called the <u>exponent</u>.

    In 7^4, (a) 7 is called the _____.

             (b) 4 is called the _____.

---

2. In 8^2: (a) The <u>base</u> is ____.

           (b) The <u>exponent</u> is ____.

    a) base
    b) exponent

---

3. Write the base-exponent expression:

    (a) whose base is 5 and whose exponent is 6. _____

    (b) whose exponent is 3 and whose base is 10. _____

    a) 8
    b) 2

---

a) 5^6
b) 10^3

283

284   Exponents and Radicals

4. When the exponent is a positive whole number greater than "1", a base-exponent expression is shorthand for a <u>multiplication</u> <u>of</u> <u>identical</u> <u>factors</u>. For example:

5^3 is shorthand for (5)(5)(5).

The <u>base</u> "5" tells you what number to use for the factors.

The <u>exponent</u> "3" tells you how many times the base appears as a factor.

(a) 6^2 is shorthand for what multiplication? _____

(b) 7^4 is shorthand for what multiplication? _____

a) (6)(6)	
b) (7)(7)(7)(7)	

5. Write each of the following multiplications in <u>base-exponent</u> form.

(a) (5)(5) = _____   (b) (9)(9)(9)(9) = _____   (c) (8)(8)(8) = _____

a) 5^2 b) 9^4 c) 8^3

6. The base in a base-exponent expression can be a letter. For example:

In x^2, the base is "x".   In y^3, the base is _____ .

"y"

7. Write the base-exponent expression:

(a) whose base is "m" and whose exponent is "5". _____

(b) whose exponent is "7" and whose base is "a". _____

a) m^5
b) a^7

8. When the base is a letter and the exponent is a positive whole-number greater than "1", the base-exponent expression is shorthand for <u>a multiplication in which each factor is that letter</u>. For example:

x^4 is shorthand for (x)(x)(x)(x).

Note: The exponent again tells you how often the letter appears as a factor.

(a) y^3 is shorthand for what multiplication? _____

(b) t^5 is shorthand for what multiplication? _____

a) (y)(y)(y)
b) (t)(t)(t)(t)(t)

9. Write each of the following multiplications in base-exponent form:

(a) (d)(d) = _____   (b) (p)(p)(p)(p)(p) = _____   (c) (V)(V)(V) = _____

a) d^2 b) p^5 c) V^3

Exponents and Radicals   285

10. A base-exponent expression is called a "power".

   If the base is "8", it is called a "power of eight".
   8^2, 8^3, and 8^5 are all "powers of eight".

   If the base is "x", it is called a "power of x".
   x^3, x^4, and x^7 are all "powers of x".

   (a) 5^4 is a power of _____.   (b) y^3 is a power of _____.

---

11. (a) 7^8 is the power of _____ whose exponent is _____. | a) five (or 5)   b) y
    (b) t^4 is the power of _____ whose exponent is _____.

---

12. Write the base-exponent expression for each of the following: | a) seven 8
    (a) The "power of three" whose exponent is "2". _____ | b) t 4
    (b) The "power of y" whose exponent is "4". _____
    (c) The "power of F" whose exponent is "10". _____

---

13. If the base of a power is a number, the power can be converted to a regular number by performing the multiplication. For example: | a) 3^2
                                                                                                                                                | b) y^4
    $$4^3 = (4)(4)(4) = 64$$ | c) F^{10}

   (a) $3^4 = (3)(3)(3)(3) = $ _____   (b) $2^5 = (2)(2)(2)(2)(2) = $ _____

---

14. If the base of a power is a letter, we can write the power as a multiplication but we cannot convert it to a regular number. For example: | a) 81   b) 32

   Though $x^3 = (x)(x)(x)$, we cannot convert x^3 to a regular number.

   Which of the following can be converted to a regular number? _____
   (a) 3^8 (b) y^2 (c) t^7 (d) 10^4

---

15. Write each of the following as a regular number if possible: | Only (a) and (d)
    (a) $m^4 = $ _____   (b) $2^4 = $ _____   (c) $5^3 = $ _____   (d) $p^5 = $ _____

---

16. There are also powers of negative numbers. For example: | a) Not possible
                                                              | b) 16
   $(-2)^2 = (-2)(-2) = +4$ | c) 125
   $(-2)^3 = (-2)(-2)(-2) = -8$ | d) Not possible
   $(-2)^4 = (-2)(-2)(-2)(-2) = +16$

   Notice that we write parentheses around the base when it is negative.

   Convert each of the following to a signed number:
   (a) $(-3)^2 = (-3)(-3) = $ _____
   (b) $(-3)^3 = (-3)(-3)(-3) = $ _____
   (c) $(-3)^4 = (-3)(-3)(-3)(-3) = $ _____

286 Exponents and Radicals

17.	Though any power of a <u>positive</u> number is <u>positive</u>, the sign of a power of a <u>negative</u> number depends on the exponent of the power. That is: If the exponent is <u>even</u>, the power is <u>positive</u>. $\quad(-3)^4 = 81 \qquad (-2)^6 = 64$ If the exponent is <u>odd</u>, the power is <u>negative</u>. $\quad(-5)^3 = -125 \qquad (-2)^5 = -32$ Convert each of the following to a signed number: (a) $(-4)^3 = $ _____ (b) $5^4 = $ _____ (c) $(-10)^5 = $ _____	a) +9 b) −27 c) +81
18.	There is a standard language which is used to distinguish the various powers of the same number or letter. For example: $\quad 5^2$ is called "5 to the <u>second</u> power". $\quad 5^3$ is called "5 to the <u>third</u> power". $\quad 5^6$ is called "5 to the _____ power".	a) −64 b) 625 c) −100,000
19.	(a) 7^4 is called "7 to the _____ power". (b) x^5 is called "x to the _____ power". (c) t^7 is called "t to the _____ power".	sixth
20.	Write each of the following in base-exponent form: (a) 7 to the <u>fourth</u> power = _____ (c) y to the <u>fifth</u> power = _____ (b) 7 to the <u>sixth</u> power = _____ (d) y to the <u>eighth</u> power = _____	a) fourth b) fifth c) seventh
21.	Write each of the following as a multiplication of identical factors: (a) 5 to the third power = _____ (b) x to the fifth power = _____	a) 7^4 c) y^5 b) 7^6 d) y^8
22.	Write each of the following in base-exponent form and then as a multiplication of identical factors: $\quad -3$ to the second power $= (-3)^2 = (-3)(-3)$ (a) −7 to the third power = _____ = _____ (b) −9 to the fourth power = _____ = _____	a) (5)(5)(5) b) (x)(x)(x)(x)(x)
23.	Convert each of the following to a signed number: (a) 4 to the third power = _____ (b) −5 to the second power = _____ (c) −6 to the third power = _____	a) $(-7)^3 = (-7)(-7)(-7)$ b) $(-9)^4 = (-9)(-9)(-9)(-9)$

Exponents and Radicals 287

24. There are special names for the <u>second</u> and <u>third</u> power of any base. That is:

4^2 is called both "4 to the <u>second</u> power" and "4 <u>squared</u>".
6^3 is called both "6 to the <u>third</u> power" and "6 <u>cubed</u>".

(a) Another name for "x to the <u>second</u> power" is "_____".
(b) Another name for "t to the <u>third</u> power" is "_____".

a) +64, from 4^3
b) +25, from $(-5)^2$
c) -216, from $(-6)^3$

---

25. Write each of the following in base-exponent form:
   (a) F squared = _____   (c) F cubed = _____
   (b) 9 squared = _____   (d) 9 cubed = _____

a) "x squared"
b) "t cubed"

---

26. Convert each of the following to a signed number:
   (a) 3 squared = _____   (c) -2 squared = _____
   (b) 4 cubed = _____   (d) -5 cubed = _____

a) F^2 c) F^3
b) 9^2 d) 9^3

---

27. Instead of the instruction "raise 5 to the seventh power", we can simply write the symbol 5^7.

Write the symbol for each of the following instructions:
   (a) "Raise 3 to the fourth power". _____
   (b) "Raise -5 to the third power". _____
   (c) "Raise V to the eighth power". _____

a) +9   c) +4
b) +64   d) -125

---

28. Instead of the instruction "square 10", we can simply write the symbol 10^2.

Write the symbol for each of the following instructions:
   (a) "Square 7" _____   (c) "Cube -8" _____
   (b) "Square x" _____   (d) "Cube t" _____

a) 3^4
b) $(-5)^3$
c) V^8

---

29. Notice how we carried out the instructions below and obtained a signed number:

"Raise 2 to the fourth power" = 2^4 = +16
"Cube -4" = $(-4)^3$ = -64

Write the signed number you get by carrying out each instruction below:
   (a) "Raise 2 to the fifth power" _____
   (b) "Cube -4" _____   (c) "Square -10" _____

a) 7^2 c) $(-8)^3$
b) x^2 d) t^3

---

a) +32
b) -64
c) +100

288   Exponents and Radicals

SELF-TEST 1 (Frames 1-29)

Write the base-exponent expression: 1. Whose base is 4 and whose exponent is 7. _____

2. Whose exponent is 5 and whose base is t. _____

Write each of the following in base-exponent form:

3. $(12)(12)(12)(12)(12) =$ ____ | 4. $(h)(h)(h) =$ ____ | 5. p squared = ____ | 6. x cubed = ____

Find the numerical value of each of the following:

7. $2^5 =$ ____ | 8. $(-3)^4 =$ ____ | 9. $(-1)^7 =$ ____ | 10. 3 to the fourth power = ____

11. Write the "power of w" whose exponent is 6. _____

ANSWERS: 1. 4^7 3. 12^5 5. p^2 7. 32   9. -1   11. w^6
          2. t^5 4. h^3 6. x^3 8. 81   10. 81

4-2 MULTIPLICATION AND DIVISION OF POWERS WITH POSITIVE WHOLE-NUMBER EXPONENTS

In this section, we will discuss the laws of exponents which are used to multiply and divide powers with positive whole-number exponents of "2" or more. We will show that these laws apply only to powers with the same base. The division of powers is limited to those in which the exponent in the numerator is larger than the exponent in the denominator.

---

30. Each of the expressions below means "multiply 5^2 and 5^3".

$5^2 \times 5^3$ $5^2 \cdot 5^3$ $5^2(5^3)$ $5^2 5^3$

Therefore: (a) $7^3 \cdot 7^4$ means "multiply _____ and _____".

(b) $b^2 b^4$ means "multiply _____ and _____".

a) 7^3 and 7^4
b) b^2 and b^4

31. In the multiplication below, each factor is a power with "7" as its base.

$$7^2 \cdot 7^3 = (7)(7) \cdot (7)(7)(7) = 7^5$$

The multiplication was performed in two steps:

(1) Each factor was converted to a multiplication of 7's.
(2) The number of these 7's was then counted to obtain the exponent of the product.

Write each of the following products as powers:

(a) $5^4 \cdot 5^2 = (5)(5)(5)(5) \cdot (5)(5) = $ _____

(b) $x^3 \cdot x^5 = (x)(x)(x) \cdot (x)(x)(x)(x)(x) = $ _____

---

32. Here are the three multiplications which were performed in the last frame:

$$7^2 \cdot 7^3 = 7^5 \qquad 5^4 \cdot 5^2 = 5^6 \qquad x^3 \cdot x^5 = x^8$$

Notice in each case that <u>the exponent of the product is the sum of the exponents of the factors</u>. This fact can be stated in the general form below. It is called the "<u>law of exponents for multiplication</u>".

$$\boxed{a^m \cdot a^n = a^{m+n}}$$

Using this law, complete each of these:

(a) $6^3 \cdot 6^4 = $ _____   (b) $8^5 \cdot 8^6 = $ _____   (c) $y^5 \cdot y^4 = $ _____

a) 5^6
b) x^8

---

33. The law of exponents for multiplying powers can be used <u>only if the powers have the same base</u>. For example, the law does not apply in the multiplications below because the bases are different.

$$4^3 \cdot 7^2 = (4)(4)(4) \cdot (7)(7)$$

Note: The multiplication on the right cannot be written as a single power since all of the factors are not identical.

Use the law of exponents to write the product in those cases in which it applies:

(a) $3^5 \cdot 3^3 = $ _____   (c) $x^2 \cdot y^7 = $ _____

(b) $5^3 \cdot 6^4 = $ _____   (d) $m^5 \cdot m^5 = $ _____

a) 6^7, from 6^{3+4}
b) 8^{11}, from 8^{5+6}
c) y^9, from y^{5+4}

---

34. If possible, use the law of exponents for multiplication to write the right side of each equation below in a simpler form:

(a) If $y = x^3 \cdot x^4$,   (b) If $t = s^2 d^5$,   (c) If $F = v^7 \cdot v^3$,

   $y = $ _____ $t = $ _____ $F = $ _____

a) 3^8
b) Does not apply.
c) Does not apply.
d) m^{10}

---

a) $y = x^7$
b) Not possible.
c) $F = v^{10}$

290   Exponents and Radicals

35. A division of powers is ordinarily written as a fraction. For example:

"Divide 2^6 by 2^4" is written $\dfrac{2^6}{2^4}$.

"Divide x^{10} by x^7" is written _____ .

---

36. In the division of powers below, each power has "7" as its base.

$$\dfrac{7^5}{7^3} = \dfrac{7 \cdot 7 \cdot 7 \cdot 7 \cdot 7}{7 \cdot 7 \cdot 7} = \left(\dfrac{7 \cdot 7 \cdot 7}{7 \cdot 7 \cdot 7}\right)(7 \cdot 7) = (1)(7 \cdot 7) = 7 \cdot 7 = 7^2$$

Notice the steps used to perform the division:

(1) First we wrote both the numerator and denominator as a multiplication of 7's.

(2) Then we factored out $\left(\dfrac{7 \cdot 7 \cdot 7}{7 \cdot 7 \cdot 7}\right)$ which is an instance of _____ and substituted "1" for it.

(3) Then we obtained the exponent of the quotient by simply counting the remaining 7's.

Following the steps above, write each quotient below as a power:

(a) $\dfrac{3^5}{3^2} = \dfrac{3 \cdot 3 \cdot 3 \cdot 3 \cdot 3}{3 \cdot 3} = \left(\dfrac{3 \cdot 3}{3 \cdot 3}\right)(3 \cdot 3 \cdot 3) = (1)(3 \cdot 3 \cdot 3) =$ _____

(b) $\dfrac{x^7}{x^3} = \dfrac{x \cdot x \cdot x \cdot x \cdot x \cdot x \cdot x}{x \cdot x \cdot x} = \left(\dfrac{x \cdot x \cdot x}{x \cdot x \cdot x}\right)(x \cdot x \cdot x \cdot x) = (1)(x \cdot x \cdot x \cdot x)$
$=$ _____

---

37. Here are the three divisions which were performed in the last frame:

$\dfrac{7^5}{7^3} = 7^2 \qquad \dfrac{3^5}{3^2} = 3^3 \qquad \dfrac{x^7}{x^3} = x^4$

Notice in each case that <u>the exponent of the quotient can be obtained by subtracting the exponent of the denominator from the exponent of the numerator</u>. This fact can be stated in the general form below. It is called the "<u>law of exponents for division</u>".

$$\boxed{\dfrac{a^m}{a^n} = a^{m-n}}$$

Using this law, complete each of these:

(a) $\dfrac{7^6}{7^2} = 7^{6-2} =$ _____   (b) $\dfrac{10^{10}}{10^4} = 10^{10-4} =$ _____   (c) $\dfrac{y^7}{y^2} = y^{7-2} =$ _____

$\dfrac{x^{10}}{x^7}$

a) 3^3

b) x^4

a) 7^4 b) 10^6 c) y^5

38. The law of exponents for dividing powers can be used <u>only if the powers have the same base</u>. For example, the law does not apply to the division below because the bases are different.

$$\frac{t^4}{m^2} = \frac{t \cdot t \cdot t \cdot t}{m \cdot m}$$

Notice that we cannot factor out an instance of $\frac{\square}{\square}$ from the fraction on the right.

Use the law of exponents to write the quotient in those cases in which it applies:

(a) $\dfrac{4^7}{6^3} =$ _____ (b) $\dfrac{d^9}{d^5} =$ _____ (c) $\dfrac{8^8}{8^5} =$ _____ (d) $\dfrac{R^5}{P^2} =$ _____

---

39. If possible, use the law of exponents for division to write the right side of each equation below in a simpler form:

(a) If $m = \dfrac{p^{12}}{p^7}$, (b) If $R = \dfrac{h^6}{d^5}$, (c) If $q = \dfrac{t^{15}}{t^9}$,

m = _____   R = _____   q = _____

a) Does not apply.
b) d^4
c) 8^3
d) Does not apply.

---

40. Complete each of the following:

(a) $4^8 \cdot 4^3 =$ _____   (c) $x^7 \cdot x^5 =$ _____

(b) $\dfrac{4^8}{4^3} =$ _____   (d) $\dfrac{x^7}{x^5} =$ _____

a) $m = p^5$
b) Not possible
c) $q = t^6$

Answer to Frame 40:   a) 4^{11} b) 4^5 c) x^{12} d) x^2

SELF-TEST 2 (Frames 30-40)

Using the law of exponents for multiplication, do the following problems:

1. $x^6 \cdot x^3 =$ _____
2. $9^7 \cdot 9^5 =$ _____
3. $a^m \cdot a^n =$ _____
4. $b^2 \cdot c^4 =$ _____

Using the law of exponents for division, do the following problems:

5. $\dfrac{10^9}{10^2} =$ _____
6. $\dfrac{y^7}{y^4} =$ _____
7. $\dfrac{r^8}{t^6} =$ _____
8. $\dfrac{a^m}{a^n} =$ _____

ANSWERS: 1. x^9 3. a^{m+n} 5. 10^7 7. Law does not apply.
                2. 9^{12} 4. Law does not apply.   6. y^3 8. a^{m-n}

292 Exponents and Radicals

4-3 THE MEANING OF POWERS WHOSE EXPONENTS ARE EITHER "1" OR "0"

In this section, we will define the meaning of powers whose exponents are either "1" or "0". We will show that these definitions are consistent with the laws of exponents for multiplication and division.

41. Any power whose exponent is "1" is equal to its base. For example:

$7^1 = 7$ $\qquad$ $(-5)^1 = -5$ $\qquad$ $x^1 = x$

This fact fits the pattern for powers that we have already seen. That is:

$7^3 = (7)(7)(7)$ $\qquad$ $(-5)^3 = (-5)(-5)(-5)$ $\qquad$ $x^3 = (x)(x)(x)$
$7^2 = (7)(7)$ $\qquad$ $(-5)^2 = (-5)(-5)$ $\qquad$ $x^2 = (x)(x)$
$7^1 = (7)$ or 7 $\qquad$ $(-5)^1 = (-5)$ or -5 $\qquad$ $x^1 = (x)$ or x

Based on the definition above, complete these:

(a) $5^1 = $ _____ (b) $(-10)^1 = $ _____ (c) $y^1 = $ _____

42. The definition of a power whose exponent is "1" is consistent with the law of exponents for multiplying powers. To show this fact, we have performed the same multiplications in two different ways below.

(1) By using the law of exponents for multiplication.

$5^2 \cdot 5^1 = 5^{2+1} = 5^3$ $\qquad$ $x^1 \cdot x^3 = x^{1+3} = x^4$

(2) By substituting the meaning of each power for the power.

$5^2 \cdot 5^1 = (5)(5) \cdot (5) = 5^3$ $\qquad$ $x^1 \cdot x^3 = (x) \cdot (x)(x)(x) = x^4$

Use the law of exponents to complete each of these multiplications:

(a) $3^1 \cdot 3^5 = $ _____ (b) $(-4)^4 \cdot (-4)^1 = $ _____ (c) $t^1 \cdot t^9 = $ _____

a) 5

b) −10

c) y

43. The definition of a power whose exponent is "1" is also consistent with the law of exponents for dividing powers. To show this fact, we have performed the same divisions in two different ways below.

(1) By using the law of exponents for division.

$\dfrac{4^3}{4^1} = 4^{3-1} = 4^2$ $\qquad$ $\dfrac{t^4}{t^1} = t^{4-1} = t^3$

(2) By substituting the meaning of each power for the power.

$\dfrac{4^3}{4^1} = \dfrac{4 \cdot 4 \cdot 4}{4} = \left(\dfrac{4}{4}\right)(4 \cdot 4) = (1)(4 \cdot 4) = 4 \cdot 4 = 4^2$

$\dfrac{t^4}{t^1} = \dfrac{t \cdot t \cdot t \cdot t}{t} = \left(\dfrac{t}{t}\right)(t \cdot t \cdot t) = (1)(t \cdot t \cdot t) = t \cdot t \cdot t = t^3$

Use the law of exponents to complete each of these divisions:

(a) $\dfrac{8^5}{8^1} = $ _____ (b) $\dfrac{y^6}{y^1} = $ _____ (c) $\dfrac{V^{10}}{V^1} = $ _____

a) 3^6

b) $(-4)^5$

c) t^{10}

a) 8^4 b) y^5 c) V^9

44. By reversing the definition of a power whose exponent is "1", we can write any number or letter as a power. For example:

$$4 = 4^1 \qquad -3 = (-3)^1 \qquad m = m^1$$

This fact can be used to convert each of the multiplications below into a multiplication of two powers. That is:

$$4 \cdot 4^2 = 4^1 \cdot 4^2 = 4^3$$

(a) $3^8 \cdot 3 = 3^8 \cdot 3^1 =$ _____ (b) $x \cdot x^5 = x^1 \cdot x^5 =$ _____

---

45. The same fact can also be used to convert each division below into a division of two powers. That is:

$$\frac{9^3}{9} = \frac{9^3}{9^1} = 9^2$$

(a) $\dfrac{4^7}{4} = \dfrac{4^7}{4^1} =$ _____

(b) $\dfrac{t^5}{t} = \dfrac{t^5}{t^1} =$ _____

a) 3^9 \qquad b) x^6

---

46. Use the laws of exponents to complete each of these:

(a) $2^7 \cdot 2 =$ _____ (b) $\dfrac{11^4}{11} =$ _____ (c) $a \cdot a^5 =$ _____ (d) $\dfrac{b^9}{b} =$ _____

a) 4^6
b) t^4

---

47. When the exponent of a power is "1", the "1" is not ordinarily written explicitly. Therefore, the quotient below is ordinarily written as 5 instead of 5^1.

$$\frac{5^3}{5^2} = 5 \quad \text{(from } 5^1\text{)}$$

Write each quotient below in the ordinary form:

(a) $\dfrac{8^7}{8^6} =$ _____ (b) $\dfrac{t^4}{t^3} =$ _____ (c) $\dfrac{m^2}{m} =$ _____

a) 2^8
b) 11^3
c) a^6
d) b^8

---

48. When the exponent of a power is "0", the power equals +1. That is:

$$8^0 = 1 \qquad 5^0 = 1 \qquad (-3)^0 = 1 \qquad t^0 = 1$$

Based on the definition above, complete these:

(a) $7^0 =$ _____ (b) $49^0 =$ _____ (c) $(-2)^0 =$ _____ (d) $F^0 =$ _____

a) 8, not 8^1
b) t, not t^1
c) m, not m^1

---

49. The definition of any power whose exponent is "0" is consistent with the law of exponents for multiplication. To show this fact, we have performed the same multiplications in two different ways below:

(1) By using the law of exponents for multiplication.

$$4^0 \cdot 4^3 = 4^{0+3} = 4^3 \qquad x^2 \cdot x^0 = x^{2+0} = x^2$$

(2) By substituting the meaning of each power for the power.

$$4^0 \cdot 4^3 = (1) \cdot (4)(4)(4) = 4^3 \qquad x^2 \cdot x^0 = (x)(x) \cdot (1) = x^2$$

Use the law of exponents to complete each of these multiplications:

(a) $7^0 \cdot 7^6 =$ _____ (b) $(-2)^4 \cdot (-2)^0 =$ _____ (c) $y^0 \cdot y^5 =$ _____

The answer is +1 (or 1) for all.

294   Exponents and Radicals

50. The definition of any power whose exponent is "0" is also consistent with the law of exponents for dividing powers. To show this fact, we have performed the same divisions in two different ways below:

(1) By using the law of exponents for division.

$$\frac{6^4}{6^0} = 6^{4-0} = 6^4 \qquad \frac{x^3}{x^0} = x^{3-0} = x^3$$

(2) By substituting "1" for the "0" power and using the principle $\frac{\Box}{1} = \Box$.

$$\frac{6^4}{6^0} = \frac{6^4}{1} = 6^4 \qquad \frac{x^3}{x^0} = \frac{x^3}{1} = x^3$$

Use the law of exponents to complete each of these divisions:

(a) $\frac{5^2}{5^0} = $ _____ (b) $\frac{y^7}{y^0} = $ _____ (c) $\frac{P^{10}}{P^0} = $ _____

a) 7^6
b) $(-2)^4$
c) y^5

---

51. Complete each of the following:

(a) $t \cdot t^4 = $ _____ (b) $\frac{y^3}{y^0} = $ _____ (c) $\frac{8^4}{8} = $ _____ (d) $6^3 \cdot 6^0 = $ _____

a) 5^2
b) y^7
c) P^{10}

---

52. There is a standard language used to name powers whose exponents are either "1" or "0".

x^1 is called "x to the first power".
x^0 is called "x to the zero power".

Write the symbol which stands for each of the following names:

(a) "5 to the zero power" = _____   (b) "7 to the first power" = _____

a) t^5
b) y^3
c) 8^3
d) 6^3

---

53. Following the example below, complete each of the following:

"6 to the zero power" = 6^0 = 1

(a) "y to the zero power" = _____ = _____
(b) "4 to the first power" = _____ = _____

a) 5^0 b) 7^1

---

54. By reversing the law of exponents for multiplication, any power can be factored into two powers in various ways. For example:

$x^4 = x^4 \cdot x^0 \qquad x^4 = x^3 \cdot x^1 \qquad x^4 = x^2 \cdot x^2$

Notice in each case that the sum of the exponents of the factors equals "4".

Complete each of the following factorings:

(a) $4^5 = 4^2 \cdot 4^{\Box}$ (c) $y^9 = y^1 \cdot y^{\Box}$
(b) $8^7 = 8^5 \cdot 8^{\Box}$ (d) $m^5 = m^5 \cdot m^{\Box}$

a) $y^0 = 1$
b) $4^1 = 4$

55. By using the law of exponents for division and the definition of any zero power, we can show that the principle $\dfrac{\boxed{}}{\boxed{}} = 1$ also applies to powers. For example:

$$\dfrac{5^4}{5^4} = 5^{4-4} = 5^0 = 1 \qquad \dfrac{x^3}{x^3} = x^{3-3} = x^0 = 1$$

Since $\dfrac{7^8}{7^8}$ and $\dfrac{t^5}{t^5}$ are instances of $\dfrac{\boxed{}}{\boxed{}}$, they both equal _____.

a) $4^2 \cdot 4^{\boxed{3}}$

b) $8^5 \cdot 8^{\boxed{2}}$

c) $y^1 \cdot y^{\boxed{8}}$

d) $m^5 \cdot m^{\boxed{0}}$

---

56. If we use the law of exponents to divide x^5 by x^3, we get x^2 as the quotient. This fact can be confirmed by reducing the fraction to lowest terms. To do so, we factor out an instance of $\dfrac{\boxed{}}{\boxed{}}$.

$$\dfrac{x^5}{x^3} = \dfrac{x^3 \cdot x^2}{x^3} = \left(\dfrac{x^3}{x^3}\right)x^2 = (1)(x^2) = x^2$$

If we use the law of exponents to divide y^6 by y^2, we get y^4 as the quotient. Confirm this fact by reducing the fraction to lowest terms. Show all steps.

$$\dfrac{y^6}{y^2} = \underline{} = \left(\underline{}\right)(\;\;) = (\;\;)(\;\;) = \underline{}$$

1

---

57. We know that two quantities are a pair of reciprocals if their product is +1. Since $x^3\left(\dfrac{1}{x^3}\right) = \dfrac{x^3}{x^3} = 1$:

(a) The reciprocal of x^3 is _____. (b) The reciprocal of $\dfrac{1}{x^3}$ is _____.

$\dfrac{y^2 \cdot y^4}{y^2} = \left(\dfrac{y^2}{y^2}\right)(y^4) = (1)(y^4) = y^4$

---

58. Since $\left(\dfrac{1}{t^5}\right)(t^5) = \dfrac{t^5}{t^5} = 1$:

(a) The reciprocal of $\dfrac{1}{t^5}$ is _____. (b) The reciprocal of t^5 is _____.

a) $\dfrac{1}{x^3}$ b) x^3

---

59. Write the reciprocals of each of the following:

(a) 6^2 _____   (b) $\dfrac{1}{m^7}$ _____   (c) p^{10} _____   (d) $\dfrac{1}{7^8}$ _____

a) t^5 b) $\dfrac{1}{t^5}$

---

60. Any division can be converted to a multiplication by multiplying the numerator by the reciprocal of the denominator. For example:

$$\dfrac{\frac{1}{y^2}}{\frac{1}{y^5}} = \left(\dfrac{1}{y^2}\right)\left(\text{the reciprocal of } \dfrac{1}{y^5}\right) = \left(\dfrac{1}{y^2}\right)(y^5) = \dfrac{y^5}{y^2} = y^3$$

a) $\dfrac{1}{6^2}$

b) m^7

c) $\dfrac{1}{p^{10}}$

d) 7^8

Continued on following page.

Exponents and Radicals   297

4-4 THE MEANING OF POWERS WITH NEGATIVE WHOLE-NUMBER EXPONENTS

In this section, we will define what is meant by powers with negative whole-number exponents. We will then show that this definition is consistent with the laws of exponents for both multiplication and division.

62. We have applied the law of exponents for division to the division of two powers below. Notice that the exponent of the quotient is negative.

$$\frac{x^3}{x^5} = x^{3-5} = x^{3+(-5)} = x^{-2}$$

Using the law of exponents for division, find each quotient below:

(a) $\frac{3^4}{3^8} = $ _____   (b) $\frac{y^2}{y^5} = $ _____   (c) $\frac{t^5}{t^6} = $ _____

---

63. In the last frame, each quotient was a power with a negative whole-number exponent. The definition of such powers is given in the box below.

$$\boxed{x^{-n} = \frac{1}{x^n}}$$

That is: $x^{-1} = \frac{1}{x^1}$, $x^{-2} = \frac{1}{x^2}$, $x^{-3} = \frac{1}{x^3}$, and so on.

Using the definition above, write the equivalent fractional form of each power below:

(a) $5^{-4} = $ ___   (b) $6^{-5} = $ ___   (c) $y^{-8} = $ ___   (d) $t^{-10} = $ ___

a) 3^{-4}
b) y^{-3}
c) t^{-1}

---

64. Using the definition in the last frame, write each fraction below as a power with a negative exponent:

(a) $\frac{1}{8^2} = $ ___   (b) $\frac{1}{3^5} = $ ___   (c) $\frac{1}{y^7} = $ ___   (d) $\frac{1}{a^3} = $ ___

a) $\frac{1}{5^4}$ c) $\frac{1}{y^8}$
b) $\frac{1}{6^5}$ d) $\frac{1}{t^{10}}$

---

65. By applying the definition of a power with a negative exponent, we have converted the following powers to ordinary fractions.

$$2^{-3} = \frac{1}{2^3} = \frac{1}{8} \qquad 4^{-2} = \frac{1}{4^2} = \frac{1}{16}$$

Following the examples above, convert each of these powers to an ordinary fraction:

(a) $9^{-2} = $ ___ = ___   (b) $5^{-3} = $ ___ = ___

a) 8^{-2} c) y^{-7}
b) 3^{-5} d) a^{-3}

---

a) $= \frac{1}{9^2} = \frac{1}{81}$

b) $= \frac{1}{5^3} = \frac{1}{125}$

298   Exponents and Radicals

66. A power whose exponent is "-1" can easily be converted to an ordinary fraction. For example:

$$7^{-1} = \frac{1}{7^1} = \frac{1}{7} \qquad x^{-1} = \frac{1}{x^1} = \frac{1}{x}$$

Similarly, any ordinary fraction whose numerator is "1" can easily be converted to a power. For example:

$$\frac{1}{6} = \frac{1}{6^1} = 6^{-1} \qquad \frac{1}{b} = \frac{1}{b^1} = b^{-1}$$

Convert each power to a fraction and each fraction to a power:

(a) $3^{-1} =$ _____   (b) $\frac{1}{9} =$ _____   (c) $\frac{1}{y} =$ _____   (d) $P^{-1} =$ _____

---

67. Powers with negative whole-number exponents also have names. For example:

x^{-1} is called "x to the <u>minus-one</u> power".

x^{-2} is called "x to the <u>minus-two</u> power".

(a) x^{-3} is called "x to the _____ power".

(b) x^{-5} is called "x to the _____ power".

a) $\frac{1}{3}$ c) y^{-1}

b) 9^{-1} d) $\frac{1}{P}$

---

68. Following the example below, write each name in symbol form:

3 to the minus-four power = 3^{-4}

(a) 7 to the minus-eight power = _____

(b) y to the minus-one power = _____

(c) t to the minus-ten power = _____

a) minus-three

b) minus-five

---

69. Following the example below, convert each name to symbol form and then convert the power to a regular fraction:

2 to the minus-four power = $2^{-4} = \frac{1}{2^4} = \frac{1}{16}$

(a) 8 to the minus-two power = _____ = _____ = _____

(b) 10 to the minus-three power = _____ = _____ = _____

(c) 7 to the minus-one power = _____ = _____ = _____

a) 7^{-8}

b) y^{-1}

c) t^{-10}

---

a) $8^{-2} = \frac{1}{8^2} = \frac{1}{64}$

b) $10^{-3} = \frac{1}{10^3} = \frac{1}{1,000}$

c) $7^{-1} = \frac{1}{7^1} = \frac{1}{7}$

Exponents and Radicals 299

70. The definition of a power with a negative whole-number exponent is consistent with the law of exponents for division. To show this consistency, we have performed the division $x^3 \div x^5$ in two ways below.

   Applying the law of exponents, we get:
   $$\frac{x^3}{x^5} = x^{3-5} = x^{-2}$$

   Reducing the fraction to lowest terms, we get:
   $$\frac{x^3}{x^5} = \frac{x^3 \cdot 1}{x^3 \cdot x^2} = \left(\frac{x^3}{x^3}\right)\left(\frac{1}{x^2}\right) = (1)\left(\frac{1}{x^2}\right) = \frac{1}{x^2}$$

   Since $\frac{x^3}{x^5} = x^{-2}$ and $\frac{x^3}{x^5} = \frac{1}{x^2}$, the definition $x^{-2} = \frac{1}{x^2}$ makes sense.

   Use the law of exponents to write each quotient below as a power. Then convert each quotient to its equivalent fractional form.

   (a) $\frac{5^2}{5^6} = $ _____ = _____

   (b) $\frac{10^4}{10^5} = $ _____ = _____

   (c) $\frac{x^5}{x^7} = $ _____ = _____

   (d) $\frac{m^1}{m^4} = $ _____ = _____

71. The definition of a power with a negative exponent is also consistent with the law of exponents in the division below. To show the consistency, we have performed the division in two different ways.

   Applying the law of exponents, we get: $\frac{x^0}{x^3} = x^{0-3} = x^{-3}$

   Substituting "1" for x^0, we get: $\frac{x^0}{x^3} = \frac{1}{x^3}$

   Since $\frac{x^0}{x^3} = x^{-3}$ and $\frac{x^0}{x^3} = \frac{1}{x^3}$, the definition $\left(x^{-3} = \frac{1}{x^3}\right)$ makes sense.

   Use the law of exponents to write each quotient below as a power. Then write each quotient in its equivalent fractional form.

   (a) $\frac{4^0}{4^2} = $ _____ = _____

   (b) $\frac{y^0}{y^1} = $ _____ = _____

72. Since the number "1" is equal to the zero-power of any base, we can convert the fractions below into a division of powers. We get:

   $$\frac{1}{x^3} = \frac{x^0}{x^3} = x^{-3}$$

   (a) $\frac{1}{5^7} = $ _____ = _____

   (b) $\frac{1}{y^{10}} = $ _____ = _____

---

a) $5^{-4} = \frac{1}{5^4}$ c) $x^{-2} = \frac{1}{x^2}$

b) $10^{-1} = \frac{1}{10^1}$ d) $m^{-3} = \frac{1}{m^3}$

---

a) $4^{-2} = \frac{1}{4^2}$

b) $y^{-1} = \frac{1}{y^1}\left(\text{or }\frac{1}{y}\right)$

---

a) $= \frac{5^0}{5^7} = 5^{-7}$

b) $= \frac{y^0}{y^{10}} = y^{-10}$

300   Exponents and Radicals

73. The definition of a power with a negative exponent is also consistent with the law of exponents for multiplication. To show this fact, three examples are worked in two different ways below.

Applying the law of exponents for multiplication, we get:

$$x^{-3} \cdot x^5 = x^{(-3)+5} = x^2$$

$$x^2 \cdot x^{-6} = x^{2+(-6)} = x^{-4}$$

$$x^{-1} \cdot x^{-4} = x^{(-1)+(-4)} = x^{-5}$$

Substituting the equivalent fraction for each power with a negative exponent, we get:

$$x^{-3} \cdot x^5 = \left(\frac{1}{x^3}\right)(x^5) = \frac{x^5}{x^3} = x^2$$

$$x^2 \cdot x^{-6} = (x^2)\left(\frac{1}{x^6}\right) = \frac{x^2}{x^6} = x^{-4}$$

$$x^{-1} \cdot x^{-4} = \left(\frac{1}{x^1}\right)\left(\frac{1}{x^4}\right) = \frac{1}{x^5} = \frac{x^0}{x^5} = x^{-5}$$

Use the law of exponents to find each product below. If the product contains a negative exponent, also write it in an equivalent fractional form.

(a) $4^6 \cdot 4^{-3} = $ _____   (b) $m^{-7} \cdot m^5 = $ _____   (c) $p^{-2} \cdot p^{-3} = $ _____

a) 4^3

b) m^{-2} or $\frac{1}{m^2}$

c) p^{-5} or $\frac{1}{p^5}$

74. The definition of a power with a negative exponent is consistent with the law of exponents in the divisions below. To show the consistency, we have performed the same divisions in two different ways below:

Applying the law of exponents, we get:

$$\frac{x^3}{x^{-5}} = x^{3-(-5)} = x^{3+5} = x^8$$

$$\frac{x^{-4}}{x^{-7}} = x^{(-4)-(-7)} = x^{(-4)+7} = x^3$$

Substituting the equivalent fraction for each power with a negative exponent, we get:

$$\frac{x^3}{x^{-5}} = \frac{x^3}{\frac{1}{x^5}} = x^3\left(\text{the reciprocal of } \frac{1}{x^5}\right) = x^3(x^5) = x^8$$

$$\frac{x^{-4}}{x^{-7}} = \frac{\frac{1}{x^4}}{\frac{1}{x^7}} = \frac{1}{x^4}\left(\text{the reciprocal of } \frac{1}{x^7}\right) = \frac{1}{x^4}(x^7) = \frac{x^7}{x^4} = x^3$$

Use the law of exponents to write each quotient below as a power. Be careful with the signs.

(a) $\frac{3^2}{3^{-5}} = $ _____   (b) $\frac{7^3}{7^{-1}} = $ _____   (c) $\frac{t^{-1}}{t^{-2}} = $ _____

Exponents and Radicals   301

75. The definition of a power with a negative exponent is also consistent with the law of exponents in the divisions below. To show the consistency, both divisions are performed in two different ways.

Applying the law of exponents, we get:

$$\frac{x^{-3}}{x^5} = x^{(-3)-(5)} = x^{(-3)+(-5)} = x^{-8} \quad \left(\text{or } \frac{1}{x^8}\right)$$

$$\frac{x^{-7}}{x^{-4}} = x^{(-7)-(-4)} = x^{(-7)+4} = x^{-3} \quad \left(\text{or } \frac{1}{x^3}\right)$$

Substituting the equivalent fraction for each power with a negative exponent, we get:

$$\frac{x^{-3}}{x^5} = \frac{\frac{1}{x^3}}{x^5} = \left(\frac{1}{x^3}\right)(\text{the reciprocal of } x^5) = \left(\frac{1}{x^3}\right)\left(\frac{1}{x^5}\right) = \frac{1}{x^8} \text{ (or } x^{-8}\text{)}$$

$$\frac{x^{-7}}{x^{-4}} = \frac{\frac{1}{x^7}}{\frac{1}{x^4}} = \left(\frac{1}{x^7}\right)\left(\text{the reciprocal of } \frac{1}{x^4}\right) = \left(\frac{1}{x^7}\right)(x^4) = \frac{x^4}{x^7} = x^{-3} \left(\text{or } \frac{1}{x^3}\right)$$

Use the law of exponents to write each quotient below as a power. Be careful with the signs.

(a) $\dfrac{3^{-1}}{3^4} =$ _____   (b) $\dfrac{8^{-3}}{8^2} =$ _____   (c) $\dfrac{y^{-5}}{y^{-1}} =$ _____

a) 3^{-5}
b) 8^{-5}
c) y^{-4}

76. As we have seen, the definition of powers with negative whole-number exponents is consistent with the law of exponents for multiplying powers. Therefore, we can use that law to perform multiplications involving powers with negative exponents. For example:

$$8^{-7} \cdot 8^3 = 8^{(-7)+3} = 8^{-4}$$

Use the law of exponents for multiplication to write each product below as a power:

(a) $5^{-2} \cdot 5^7 =$ _____   (c) $x^0 \cdot x^{-7} =$ _____
(b) $8^{-5} \cdot 8^3 =$ _____   (d) $y^{-1} \cdot y^{-2} =$ _____

a) 5^5 c) x^{-7}
b) 8^{-2} d) y^{-3}

77. Before performing the multiplication below, we substitute 5^1 for 5 and get:

$$5 \cdot 5^{-3} = 5^1 \cdot 5^{-3} = 5^{1+(-3)} = 5^{-2}$$

Write each product below as a power:

(a) $7^{-4} \cdot 7 =$ _____   (b) $x \cdot x^{-5} =$ _____   (c) $y^{-2} \cdot y =$ _____

a) 7^{-3}
b) x^{-4}
c) y^{-1}

a) 3^7
b) 7^4
c) t^1 or t

302   Exponents and Radicals

78. If the product is a base <u>to the first power</u>, remember that such a power equals the base. For example:

$$x^4 \cdot x^{-3} = x^{4+(-3)} = x^1 \text{ or } x$$

If the product is a base <u>to the zero power</u>, remember that such a power equals "1". For example:

$$5^4 \cdot 5^{-4} = 5^{4+(-4)} = 5^0 \text{ or } 1$$

Use the law of exponents to perform each multiplication below:

(a) $7^2 \cdot 7^{-1} = $ _____   (c) $m^3 \cdot m^{-3} = $ _____

(b) $y^{-4} \cdot y^5 = $ _____   (d) $4 \cdot 4^{-1} = $ _____

---

79. As we have seen, the definition of powers with negative whole-number exponents is consistent with the law of exponents for dividing powers. Therefore, we can use that law to perform divisions involving powers with negative exponents.

a) 7^1 or 7 c) m^0 or 1
b) y^1 or y d) 4^0 or 1

When using the law of exponents for division, <u>sign errors can be avoided by converting the subtractions of exponents to equivalent additions</u>. For example:

$$\frac{x^2}{x^5} = x^{2-5} = x^{2+(-5)} = x^{-3}$$

$$\frac{y^{-4}}{y^6} = y^{(-4)-6} = y^{(-4)+(-6)} = y^{-10}$$

$$\frac{t^{-5}}{t^{-3}} = t^{(-5)-(-3)} = t^{(-5)+3} = t^{-2}$$

Use the law of exponents to write each quotient as a power. <u>Be careful with the signs</u>.

(a) $\frac{5^1}{5^4} = $ _____   (c) $\frac{t^4}{t^{-3}} = $ _____

(b) $\frac{7^{-5}}{7^2} = $ _____   (d) $\frac{m^{-7}}{m^{-2}} = $ _____

---

80. Before performing the division below, we substituted 6^1 for 6.

a) 5^{-3} c) t^7
b) 7^{-7} d) m^{-5}

$$\frac{6}{6^3} = \frac{6^1}{6^3} = 6^{1-3} = 6^{1+(-3)} = 6^{-2}$$

Write each quotient below as a power:

(a) $\frac{5^{-2}}{5} = $ _____   (c) $\frac{y}{y^5} = $ _____

(b) $\frac{7}{7^{-4}} = $ _____   (d) $\frac{p^{-5}}{p} = $ _____

---

a) 5^{-3} c) y^{-4}
b) 7^5 d) p^{-6}

81. Use the law of exponents to write each quotient below as a power:

   (a) $\dfrac{3^{-2}}{3^{-3}} =$ _____

   (b) $\dfrac{10^{-5}}{10^{-5}} =$ _____

   (c) $\dfrac{R^{-1}}{R^{-1}} =$ _____

   (d) $\dfrac{d^{-8}}{d^{-9}} =$ _____

---

82. Write each product or quotient below as a power:

   (a) $8^{-5} \cdot 8^{4} =$ _____

   (b) $\dfrac{8^{-5}}{8^{4}} =$ _____

   (c) $b^{-2} \cdot b^{-4} =$ _____

   (d) $\dfrac{b^{-2}}{b^{-4}} =$ _____

   a) 3^1 or 3
   b) 10^0 or 1
   c) R^0 or 1
   d) d^1 or d

---

83. Use the laws of exponents to find each product or quotient below:

   (a) $6^5 \cdot 6^{-5} =$ _____

   (b) $\dfrac{9^{-1}}{9^{-2}} =$ _____

   (c) $v^{-4} \cdot v^{5} =$ _____

   (d) $\dfrac{q^{-7}}{q^{-7}} =$ _____

   a) 8^{-1} c) b^{-6}
   b) 8^{-9} d) b^{2}

---

84. It should be obvious that the principle $\dfrac{\boxed{}}{\boxed{}} = 1$ applies to powers with either positive or negative exponents. We can see this fact by applying the law of exponents for division. That is:

   $$\dfrac{x^5}{x^5} = x^{5-5} = x^0 = +1$$

   $$\dfrac{y^{-2}}{y^{-2}} = y^{(-2)-(-2)} = y^{(-2)+2} = y^0 = +1$$

   Therefore, all of the fractions below equal what number? _____

   $\dfrac{5^3}{5^3}$ $\dfrac{9^{-4}}{9^{-4}}$ $\dfrac{m^2}{m^2}$ $\dfrac{b^{-1}}{b^{-1}}$

   a) 6^0 or 1 c) v^1 or v
   b) 9^1 or 9 d) q^0 or 1

+1

304 Exponents and Radicals

SELF-TEST 4 (Frames 62-84)

Write each of the following as a fraction containing a positive exponent:

1. $a^{-3} =$ _____   2. $3^{-5} =$ _____   3. $t^{-1} =$ _____   4. $10^{-2} =$ _____   5. $x^{-n} =$ _____

Using the laws of exponents for multiplication and division, do the following problems:

6. $\dfrac{r^2}{r^7} =$ _____   8. $4^{-2} \cdot 4 =$ _____   10. $\dfrac{p^0}{p^{-3}} =$ _____   12. $10^{-3} \cdot 10^3 =$ _____

7. $d^2 \cdot d^{-5} =$ _____   9. $\dfrac{3}{3^2} =$ _____   11. $5^0 \cdot 5^{-4} =$ _____   13. $\dfrac{2^{-5}}{2^{-1}} =$ _____

ANSWERS: 1. $\dfrac{1}{a^3}$ 3. $\dfrac{1}{t^1}$ 5. $\dfrac{1}{x^n}$ 6. r^{-5} 10. p^3
           2. $\dfrac{1}{3^5}$ 4. $\dfrac{1}{10^2}$ 7. d^{-3} 11. 5^{-4}
                                                                   8. 4^{-1} 12. 10^0 or 1
                                                                   9. 3^{-1} 13. 2^{-4}

4-5 MORE-COMPLICATED MULTIPLICATIONS AND DIVISIONS INVOLVING POWERS

In this section, we will examine more-complicated multiplications and divisions of powers. We will show how the laws of exponents can be used to simplify these more-complicated expressions.

---

85. The law of exponents for multiplication can be extended to more than two powers, <u>provided</u> <u>that</u> <u>each</u> <u>power</u> <u>has</u> <u>the</u> <u>same</u> <u>base</u>. For example:
$$x^2 \cdot x^3 \cdot x^4 = x^{2+3+4} = x^9$$

Use the law of exponents to write each product below:

(a) $3^4 \cdot 3^6 \cdot 3^5 =$ _____   (b) $F^2 \cdot F^4 \cdot F^6 =$ _____   (c) $y^3 \cdot y \cdot y^5 =$ _____

---

86. Though we cannot multiply all three powers below, we can simplify the multiplication by multiplying x^3 and x^5. We get:
$$x^3 \cdot a^4 \cdot x^5 = a^4 \cdot x^8$$

Simplify each multiplication below as much as possible:

(a) $y^2 \cdot y^3 \cdot y^5 =$ _____   (b) $b^4 \cdot m^2 \cdot m^7 =$ _____   (c) $a^2 \cdot b^3 \cdot c^4 =$ _____

a) 3^{15}
b) F^{12}
c) y^9

---

a) y^{10}
b) $b^4 \cdot m^9$
c) Cannot be simplified.

87. The multiplication below can be performed by multiplying the numerical factors and multiplying the power factors. That is:

$$(3x^2)(5x^4) = (3 \cdot 5)(x^2 \cdot x^4) = 15x^6$$

Following the procedure above, write each product below:

(a) $4(5x^3) =$ _____ (b) $(2d^5)(5d^{-2}) =$ _____ (c) $(8R)(9R^3) =$ _____

---

88. The multiplication below can be performed by multiplying the power factors with the same base. That is:

$$(a^5b^{-3})(a^{-1}b^2) = (a^5 \cdot a^{-1})(b^{-3} \cdot b^2) = a^4b^{-1}$$

Using the procedure above, do the following multiplications:

(a) $(x^4y)(x^3y^2) =$ _____ (b) $(m^{-4}p^3)(m^5p^{-7}) =$ _____ (c) $(c^2d^{-1})(c^{-2}d^4) =$ _____

a) $20x^3$
b) $10d^3$
c) $72R^4$

---

89. The multiplication below can be performed by multiplying the numerical factors and by multiplying the power factors with the same base. That is:

$$(3x^2y^4)(2xy^{-2}) = (3 \cdot 2)(x^2 \cdot x)(y^4 \cdot y^{-2}) = 6x^3y^2$$

Write the product of each of these:

(a) $(10d^2t)(5d^{-3}t^2) =$ _____ (b) $(3a^3)(6a^4y^{-1}) =$ _____

a) x^7y^3
b) m^1p^{-4} or mp^{-4}
c) d^3, from: $c^0d^3 = 1d^3$

---

90. In the fraction below, the base of each power is "x". Since the powers have the same base, the fraction can be simplified by using the laws of exponents. We get:

$$\frac{x^2 \cdot x^5}{x^3} = \frac{x^7}{x^3} = x^4$$

Note: (1) First we performed the multiplication in the numerator.
(2) Then we performed the division.

Following the steps above, simplify each fraction below:

(a) $\frac{4^{-3} \cdot 4^5}{4^7} =$ _____ = _____ (b) $\frac{y^{-7} \cdot y^2}{y^{-3}} =$ _____ = _____

a) $50d^{-1}t^3$
b) $18a^7y^{-1}$

---

91. To simplify the fraction below, we perform the multiplication in the denominator before dividing. We get:

$$\frac{y^3}{y^2 \cdot y^4} = \frac{y^3}{y^6} = y^{-3}$$

Following the steps above, simplify each fraction below:

(a) $\frac{5^7}{5^{-1} \cdot 5^6} =$ _____ = _____ (b) $\frac{a^{-2}}{a^{-9} \cdot a^4} =$ _____ = _____

a) $\frac{4^2}{4^7} = 4^{-5}$
b) $\frac{y^{-5}}{y^{-3}} = y^{-2}$

---

a) $\frac{5^7}{5^5} = 5^2$
b) $\frac{a^{-2}}{a^{-5}} = a^3$

306   Exponents and Radicals

92. To simplify the fraction below, we perform the multiplications in both numerator and denominator before performing the division. We get:

$$\frac{b^2 \cdot b^3}{b^4 \cdot b^{-1}} = \frac{b^5}{b^3} = b^2$$

Using the steps above, simplify each of these fractions:

(a) $\dfrac{2^3 \cdot 2^4}{2^6 \cdot 2^{-3}} =$ _____ = _____   (b) $\dfrac{t^{-5} \cdot t^2}{t^4 \cdot t^{-9}} =$ _____ = _____

a) $\dfrac{2^7}{2^3} = 2^4$

b) $\dfrac{t^{-3}}{t^{-5}} = t^2$

93. In the fraction below, only two of the powers have the same base. Therefore, we can simplify the fraction only by performing the multiplication in the numerator. We get:

$$\frac{x^2 \cdot x^3}{y^5} = \frac{x^5}{y^5}$$

Simplify each of the following fractions:

(a) $\dfrac{a^{-1}a^{-3}}{b^2} =$ _____   (b) $\dfrac{p^5}{q^{-1}q^7} =$ _____   (c) $\dfrac{m^5 m^{-2}}{t^{-4} t^6} =$ _____

a) $\dfrac{a^{-4}}{b^2}$

b) $\dfrac{p^5}{q^6}$

c) $\dfrac{m^3}{t^2}$

94. The fraction below can be simplified by performing two divisions.

$$\frac{a^7 y^5}{a^2 y^3} = \left(\frac{a^7}{a^2}\right)\left(\frac{y^5}{y^3}\right) = a^5 y^2$$

Simplify each of these fractions by performing two divisions:

(a) $\dfrac{b^{-4} t^{-2}}{b^{-6} t} =$ _____   (b) $\dfrac{x^3 y^4}{x^2 y^{-1}} =$ _____

a) $b^2 t^{-3}$ b) xy^5

95. The following fraction can be simplified by performing two divisions.

$$\frac{8x^5}{4x^2} = \left(\frac{8}{4}\right)\left(\frac{x^5}{x^2}\right) = 2x^3$$

Simplify each of these fractions:

(a) $\dfrac{15y^6}{5y^8} =$ _____   (b) $\dfrac{24a^{-4} b^3}{4a^{-1} b^2} =$ _____   (c) $\dfrac{42t^{-5}}{6t^{-6}} =$ _____

a) $3y^{-2}$

b) $6a^{-3}b$

c) $7t$, from $7t^1$

96. Notice the steps used to simplify the fraction below:

$$\frac{6x^9}{8x^4} = \left(\frac{6}{8}\right)\left(\frac{x^9}{x^4}\right) = \left(\frac{3}{4}\right)(x^5) = \frac{3x^5}{4}$$

Following the procedure above, simplify each of these fractions:

(a) $\dfrac{4m^3}{6m} =$ _____   (b) $\dfrac{8R^2}{10R^5} =$ _____   (c) $\dfrac{14x^2 y^{-3}}{35x^{-3} y^2} =$ _____

a) $\dfrac{2m^2}{3}$

b) $\dfrac{4R^{-3}}{5}$

c) $\dfrac{2x^5 y^{-5}}{5}$

Exponents and Radicals   309

SELF-TEST 5 (Frames **85-103**)

Do the following multiplications:

1. $(b^2c)(cd^3)(bd) = $ _____
2. $(rt^{-1})(3ar^2t) = $ _____
3. $(4x^2y^{-2})(6x^{-1}y) = $ _____

Do the following combined multiplications and divisions:

4. $\dfrac{h^{-2}p^3}{h^{-3}p^2} = $ _____

5. $\dfrac{18x^4y^2}{12xy^{-2}} = $ _____

6. $\dfrac{a^4b^{-3}}{bc^2} = $ _____

7. $\dfrac{F^{-2}F}{F^{-1}} = $ _____

8. $\dfrac{r^{-2}}{d^3r^{-2}} = $ _____

9. $\dfrac{15c}{9c^{-2}w^3} = $ _____

ANSWERS:
1. $b^3c^2d^4$
2. $3ar^3$
3. $24xy^{-1}$
4. hp
5. $\dfrac{3x^3y^4}{2}$
6. $\dfrac{a^4b^{-4}}{c^2}$
7. 1 (From: F^0)
8. $\dfrac{1}{d^3}$
9. $\dfrac{5c^3}{3w^3}$

4-6 THE RECIPROCALS OF POWERS AND CONVERTING DIVISIONS TO MULTIPLICATIONS

In this section, we will show that the reciprocal of any power can be written as a power. We will also show that the reciprocal of any multiplication of powers can be written as a multiplication of powers. After discussing these reciprocals, we will show how any division of powers can be converted to a multiplication of powers.

104. Since two quantities are a pair of reciprocals if their product is +1, the equation below is really a definition of reciprocals.

$$x^4 \cdot x^{-4} = x^0 = +1$$

Therefore: (a) The reciprocal of x^4 is _____.

(b) The reciprocal of x^{-4} is _____.

a) x^{-4}

b) x^4

310 Exponents and Radicals

105. From the example in the last frame, it should be obvious that two powers with the same base are a pair of reciprocals <u>if their exponents are a pair of opposites.</u> That is:

5^3 and 5^{-3} are a pair of reciprocals, since $5^3 \cdot 5^{-3} = 5^0 = +1$.

y^{-6} and y^6 are a pair of reciprocals, since $y^{-6} \cdot y^6 = y^0 = +1$.

Using this fact, write the reciprocal of each of the following:

(a) 4^7 _____   (b) 8^{-2} _____   (c) t^4 _____   (d) x^{-1} _____

---

106. Since any number or letter is a base to the first power, its reciprocal in power form is the same number or letter <u>to the minus-one power.</u> That is:

The reciprocal of 5 (or 5^1) is 5^{-1}.
The reciprocal of t (or t^1) is t^{-1}.

Write the reciprocal of each of these numbers and letters in power form:

(a) 7 _____   (b) x _____   (c) F _____   (d) 15 _____

a) 4^{-7}
b) 8^2
c) t^{-4}
d) x^1 or x

---

107. Since the zero-power of any base equals +1, its reciprocal is itself. For example:

The reciprocal of 3^0 is 3^0, since $3^0 \cdot 3^0 = 1 \cdot 1 = 1$.
The reciprocal of t^0 is t^0, since $t^0 \cdot t^0 = 1 \cdot 1 = 1$.

Write the reciprocal of each of the following as a power:

(a) 8^0 _____   (b) 4^3 _____   (c) m^{-7} _____   (d) b _____

a) 7^{-1}
b) x^{-1}
c) F^{-1}
d) 15^{-1}

---

108. Up to this point, we have used the following law of exponents to divide powers. When using this law, "b" is <u>subtracted</u> from "a".

$$\boxed{\frac{x^a}{x^b} = x^{a-b}}$$

However, we can also perform these divisions by converting them to multiplications first. To do so, we multiply the numerator by the reciprocal of the denominator. For example:

(a) $\dfrac{5^6}{5^2} = (5^6)$(the reciprocal of 5^2) $= (5^6)(5^{-2}) = $ _____

(b) $\dfrac{t^4}{t^7} = (t^4)$(the reciprocal of t^7) $= (t^4)(t^{-7}) = $ _____

(c) $\dfrac{m^3}{m^{-2}} = (m^3)$(the reciprocal of m^{-2}) $= (m^3)(m^2) = $ _____

a) 8^0
b) 4^{-3}
c) m^7
d) b^{-1}

---

a) 5^4
b) t^{-3}
c) m^5

Exponents and Radicals   311

109. Complete the following conversions of divisions to multiplications:

(a) $\dfrac{7^5}{7^2} = (7^5)()$

(b) $\dfrac{3^{-4}}{3^{-2}} = (3^{-4})()$

(c) $\dfrac{y^4}{y^{-5}} = (y^4)()$

(d) $\dfrac{p^{-1}}{p^3} = (p^{-1})()$

---

110. Complete these divisions after converting them to multiplications first:

(a) $\dfrac{w^1}{w^6} = ()() = \underline{}$

(b) $\dfrac{6^2}{6^{-7}} = ()() = \underline{}$

(c) $\dfrac{t^{-2}}{t^{-1}} = ()() = \underline{}$

(d) $\dfrac{4^{-5}}{4^{-1}} = ()() = \underline{}$

a) $(7^5)(7^{-2})$
b) $(3^{-4})(3^2)$
c) $(y^4)(y^5)$
d) $(p^{-1})(p^{-3})$

---

111. You now know two ways to perform a division like $\dfrac{x^4}{x^6}$. The two ways are:

$$\boxed{\dfrac{x^4}{x^6} = x^{4-6} = x^{-2}}$$ $$\boxed{\dfrac{x^4}{x^6} = (x^4)(x^{-6}) = x^{-2}}$$

Both ways are perfectly correct. Use the method of your choice to perform the divisions below.

(a) $\dfrac{b^{-1}}{b^3} = \underline{}$

(b) $\dfrac{m^{-2}}{m^{-5}} = \underline{}$

(c) $\dfrac{a^{-7}}{a^{-4}} = \underline{}$

(d) $\dfrac{t^4}{t^{-2}} = \underline{}$

a) $(w^1)(w^{-6}) = w^{-5}$
b) $(6^2)(6^7) = 6^9$
c) $(t^{-2})(t^1) = t^{-1}$
d) $(4^{-5})(4^1) = 4^{-4}$

---

112. In an earlier section, we simplified fractions like the one below by performing the multiplication in the numerator first. For example:

$$\dfrac{x^5 \cdot x^4}{x^3} = \dfrac{x^9}{x^3} = x^6$$

We can also simplify the same fraction by converting the division to a multiplication first and then applying the law of exponents for multiplication. We get:

$$\dfrac{x^5 \cdot x^4}{x^3} = (x^5 \cdot x^4)(\text{the reciprocal of } x^3) = x^5 \cdot x^4 \cdot x^{-3} = x^6$$

Simplify each of the following fractions by converting the division to a multiplication first:

(a) $\dfrac{7^{-1} \cdot 7^5}{7^{-2}} = (7^{-1} \cdot 7^5)() = \underline{}$

(b) $\dfrac{y^2 \cdot y^{-4}}{y^6} = (y^2 \cdot y^{-4})() = \underline{}$

a) b^{-4} c) a^{-3}
b) m^3 d) t^6

---

a) $(7^{-1} \cdot 7^5)(7^2) = 7^6$
b) $(y^2 \cdot y^{-4})(y^{-6}) = y^{-8}$

312   Exponents and Radicals

113. Though the law of exponents cannot be applied to the division below since the bases are different, we can convert the division to a multiplication. To do so, we again multiply the numerator by the reciprocal of the denominator. We get:

$$\frac{x^2}{y^3} = (x^2)(\text{the reciprocal of } y^3) = x^2 y^{-3}$$

Convert each division below to a multiplication:

(a) $\dfrac{a^4}{b^5} = $ _____ (b) $\dfrac{s^{-1}}{t^{-2}} = $ _____ (c) $\dfrac{R^6}{Q^{-3}} = $ _____

---

114. Though we cannot simplify the fraction below by using the laws of exponents, we can convert the division to a multiplication. That is:

$$\frac{a^3 b^4}{c^5} = (a^3 b^4)(\text{the reciprocal of } c^5) = a^3 b^4 c^{-5}$$

Convert each division to a multiplication:

(a) $\dfrac{m^{-1} p^2}{t^4} = $ _____ (b) $\dfrac{F^2 M^{-3}}{R^{-5}} = $ _____ (c) $\dfrac{a^2 b^3}{y} = $ _____

Answers:
a) $a^4 b^{-5}$
b) $s^{-1} t^2$
c) $R^6 Q^3$

---

115. Write the right side of each equation below as a multiplication of powers:

(a) If $y = \dfrac{a^4}{x^{-2}}$,   (b) If $F = \dfrac{C^2}{T}$,   (c) If $p = \dfrac{a^4 b^{-1}}{d^4}$

y = _____   F = _____   p = _____

Answers:
a) $m^{-1} p^2 t^{-4}$
b) $F^2 M^{-3} R^5$
c) $a^2 b^3 y^{-1}$

---

116. We can show that $x^2 y^3$ and $x^{-2} y^{-3}$ are a pair of reciprocals by showing that their product is +1.

$$(x^2 y^3)(x^{-2} y^{-3}) = (x^2 \cdot x^{-2})(y^3 \cdot y^{-3}) = x^0 \cdot y^0 = 1 \cdot 1 = 1$$

Show that $a^{-4} b^5$ and $a^4 b^{-5}$ are a pair of reciprocals by showing that their product is +1:

$(a^{-4} b^5)(a^4 b^{-5}) = $ _____

Answers:
a) $y = a^4 x^2$
b) $F = C^2 T^{-1}$
c) $p = a^4 b^{-1} d^{-4}$

---

117. As we can see from the last frame, two multiplications of powers are a pair of reciprocals if each factor in one is the reciprocal of a factor in the other. That is:

$x^2 y^3$ and $x^{-2} y^{-3}$ are a pair of reciprocals,

since: x^2 and x^{-2} are a pair of reciprocals
and: y^3 and y^{-3} are a pair of reciprocals.

Using the fact above, write the reciprocal of each multiplication below:

(a) $a^4 b^{-5}$ _____ (b) $3^{-7} \cdot 4^8$ _____ (c) $m^{-1} p^4$ _____

Answers:
$= (a^{-4} \cdot a^4)(b^5 \cdot b^{-5})$
$= a^0 b^0 = 1 \cdot 1 = 1$

---

a) $a^{-4} b^5$
b) $3^7 4^{-8}$
c) $m^1 p^{-4}$ or mp^{-4}

118. Write the reciprocal of each of these multiplications in power form:

   (a) xy^2 _____   (b) $c^{-1}d$ _____   (c) $3x^5$ _____

---

a) $x^{-1}y^{-2}$
b) $c^1 d^{-1}$ or cd^{-1}
c) $3^{-1}x^{-5}$

---

119. In an earlier section, we simplified fractions like the one below by performing the multiplication in the denominator first. For example:

$$\frac{x^7}{x^2 \cdot x^3} = \frac{x^7}{x^5} = x^2$$

We can also simplify the same fraction by converting the division to a multiplication first. We get:

$$\frac{x^7}{x^2 \cdot x^3} = x^7 \text{(the reciprocal of } x^2 \cdot x^3\text{)} = x^7 x^{-2} x^{-3} = x^2$$

Simplify each of the following fractions by converting the division to a multiplication first:

(a) $\dfrac{5^4}{5^{-1}5^6} = 5^4 ($ _____ $) =$ _____   (b) $\dfrac{m^{-3}}{m^{-1}m^{-7}} = m^{-3} ($ _____ $) =$ _____

---

120. In an earlier section, we simplified fractions like those below by performing the multiplications in both the numerator and denominator first. For example:

$$\frac{x^9 \cdot x^5}{x^2 \cdot x^7} = \frac{x^{14}}{x^9} = x^5$$

We can also simplify the same fraction by converting the division to a multiplication first. We get:

$$\frac{x^9 \cdot x^5}{x^2 \cdot x^7} = x^9 \cdot x^5 \text{(the reciprocal of } x^2 \cdot x^7\text{)} = x^9 \cdot x^5 (x^{-2} \cdot x^{-7}) = x^5$$

Simplify each of the following fractions by converting the division to a multiplication first:

(a) $\dfrac{y^{-1}y^4}{y^{-3}y^2} = y^{-1}y^4 ($ _____ $) =$ _____   (b) $\dfrac{a^5 a^{-3}}{a^4 a^{-7}} = a^5 a^{-3} ($ _____ $) =$ _____

---

a) $5^4(5^1 5^{-6}) = 5^{-1}$
b) $m^{-3}(m^1 m^7) = m^5$

---

121. We cannot use the laws of exponents to simplify the fraction below because each of the bases is different. However, we can convert the division to a multiplication. That is:

$$\frac{a^2}{b^4 c^{-3}} = a^2 \text{(the reciprocal of } b^4 c^{-3}\text{)} = a^2(b^{-4}c^3) = a^2 b^{-4} c^3$$

By converting the division to a multiplication, write the right side of each equation below as a multiplication of powers:

(a) If $m = \dfrac{C^2}{F^{-1}R^4}$,   (b) If $t = \dfrac{a^2 b^3}{pq^5}$,

   m = _____   t = _____

---

a) $y^{-1}y^4(y^3 y^{-2}) = y^4$
b) $a^5 a^{-3}(a^{-4}a^7) = a^5$

---

a) $m = C^2 F^1 R^{-4}$ or $C^2 F R^{-4}$
b) $t = a^2 b^3 p^{-1} q^{-5}$

314   Exponents and Radicals

SELF-TEST 6 (Frames 104-121)

Write the <u>reciprocal</u> of each of the following, in power form:

1. a^5 _____   2. r^{-2} _____   3. t^0 _____   4. P _____   5. $d^{-1}h^2$ _____

Convert each of the divisions below to a multiplication:

6. $\dfrac{a^2}{b^3} =$ _____   7. $\dfrac{rs^2}{t^{-1}} =$ _____   8. $\dfrac{k^{-3}}{p^{-2}w} =$ _____

Simplify each of the following fractions:

9. $\dfrac{v^{-2}v}{v^{-1}} =$ _____   10. $\dfrac{A^2}{A^{-3}A} =$ _____   11. $\dfrac{2 \cdot 2^0}{2^5 \cdot 2^{-1}} =$ _____

ANSWERS:
1. a^{-5} 4. P^{-1} 6. $a^2 b^{-3}$ 9. $v^0 = 1$
2. r^2 5. dh^{-2} 7. $rs^2 t$ 10. A^4
3. t^0 8. $k^{-3}p^2 w^{-1}$ 11. 2^{-3}

4-7 CONVERTING MULTIPLICATIONS OF POWERS TO DIVISIONS

In this section, we will show how multiplications of powers can be converted to equivalent fractions. Conversions of this type are ordinarily made in order to obtain an expression in which each power has a positive exponent.

---

122. Based on the definition of a power with a negative exponent, any such power can be written as a fraction. For example:

$$x^{-2} = \frac{1}{x^2} \qquad b^{-3} = \frac{1}{b^3} \qquad t^{-4} = \frac{1}{t^4}$$

Notice in each case that the numerator is "1" and the denominator is a power <u>with the same base but whose exponent is the opposite of the original exponent.</u>

The definition of a power with a negative exponent is consistent with the definition of converting a division to a multiplication. To show this fact, we can convert the fractions above back to the original powers. That is:

$$\frac{1}{x^2} = 1 \,(\text{the reciprocal of } x^2) = 1\,(x^{-2}) = x^{-2}$$

(a) $\dfrac{1}{b^3} = 1\,(\text{the reciprocal of } b^3) = 1(\quad) =$ _____

(b) $\dfrac{1}{t^4} = 1\,(\text{the reciprocal of } t^4) = 1(\quad) =$ _____

a) $1\,(b^{-3}) = b^{-3}$
b) $1\,(t^{-4}) = t^{-4}$

Exponents and Radicals   315

123. We can also write a fraction which is equivalent to any power with a positive exponent. For example:

$$m^2 = \frac{1}{m^{-2}} \qquad p^3 = \frac{1}{p^{-3}} \qquad y^4 = \frac{1}{y^{-4}}$$

Notice in each case that the numerator is "1" and the denominator is a power <u>with the same base but whose exponent is the opposite of the original exponent</u>.

The fact that the fractions and powers above are equivalent can be shown by converting each fraction back to the original power. We get:

$$\frac{1}{m^{-2}} = 1\,(\text{the reciprocal of } m^{-2}) = 1\,(m^2) = m^2$$

(a) $\frac{1}{p^{-3}} = 1\,(\text{the reciprocal of } p^{-3}) = 1(\quad) = $ _____

(b) $\frac{1}{y^{-4}} = 1\,(\text{the reciprocal of } y^{-4}) = 1(\quad) = $ _____

---

124. The following two equations are general statements of the facts we saw in the last two frames:

$$\boxed{x^{-n} = \frac{1}{x^n}} \qquad \boxed{x^n = \frac{1}{x^{-n}}}$$

Using these facts, write the fractions which are equivalent to each of the following powers:

(a) $y^2 = $ _____   (b) $m^{-3} = $ _____   (c) $V^{-4} = $ _____   (d) $p^5 = $ _____

a) $1(p^3) = p^3$

b) $1(y^4) = y^4$

---

125. The law of exponents does not apply to the multiplication $x^{-2}y^3$ because the bases are different. However, we can convert the multiplication to a fraction (or division) in various ways. To do so, we simply substitute the equivalent fraction for one of the powers. For example:

(1) Substituting $\frac{1}{x^2}$ for x^{-2}, we get:

$$x^{-2}y^3 = \left(\frac{1}{x^2}\right)y^3 = \frac{y^3}{x^2}$$

(2) Substituting $\frac{1}{y^{-3}}$ for y^3, we get:

$$x^{-2}y^3 = x^{-2}\left(\frac{1}{y^{-3}}\right) = \frac{x^{-2}}{y^{-3}}$$

Convert $a^{-5}b^4$ to a fraction as follows:

(a) Substitute $\frac{1}{a^5}$ for a^{-5}: $a^{-5}b^4 = \left(\text{―――}\right)(\quad) = $ _____

(b) Substitute $\frac{1}{b^{-4}}$ for b^4: $a^{-5}b^4 = (\quad)\left(\text{―――}\right) = $ _____

a) $\frac{1}{y^{-2}}$

b) $\frac{1}{m^3}$

c) $\frac{1}{V^4}$

d) $\frac{1}{p^{-5}}$

316 Exponents and Radicals

126. Convert $c^4 d^{-8}$ to a fraction by:

   (a) Substituting $\dfrac{1}{c^{-4}}$ for c^4. $c^4 d^{-8} =$ _____

   (b) Substituting $\dfrac{1}{d^8}$ for d^{-8}. $c^4 d^{-8} =$ _____

a) $\dfrac{1}{a^5}(b^4) = \dfrac{b^4}{a^5}$

b) $(a^{-5})\left(\dfrac{1}{b^{-4}}\right) = \dfrac{a^{-5}}{b^{-4}}$

---

127. We can convert the multiplication $p^{-3} q^4$ to a fraction whose numerator is "1" by:

   (1) substituting $\dfrac{1}{p^3}$ for p^{-3},

   and (2) substituting $\dfrac{1}{q^{-4}}$ for q^4.

   We get: $p^{-3} q^4 = \left(\dfrac{1}{p^3}\right)\left(\dfrac{1}{q^{-4}}\right) = \dfrac{1}{p^3 q^{-4}}$

   Make a double substitution to convert each of the following multiplications to a fraction whose numerator is "1":

   (a) $m^5 y^{-2} = \left(\dfrac{}{}\right)\left(\dfrac{}{}\right) =$ _____   (b) $F^{-3} R^{-4} = \left(\dfrac{}{}\right)\left(\dfrac{}{}\right) =$ _____

a) $\dfrac{d^{-8}}{c^{-4}}$

b) $\dfrac{c^4}{d^8}$

---

128. Multiplications like $d^{-3} f^5$ are not always converted to fractions. However, when such conversions are made, their purpose is to obtain an expression in which each power has a positive exponent. Therefore, we substitute to eliminate the power with the negative exponent.

   (a) To obtain a fraction in which each power has a positive exponent, should we substitute for d^{-3} or f^5 in $d^{-3} f^5$? _____

   (b) Do so. $d^{-3} f^5 = \left(\right)\left(\right) =$ _____

a) $\left(\dfrac{1}{m^{-5}}\right)\left(\dfrac{1}{y^2}\right) = \dfrac{1}{m^{-5} y^2}$

b) $\left(\dfrac{1}{F^3}\right)\left(\dfrac{1}{R^4}\right) = \dfrac{1}{F^3 R^4}$

---

129. To obtain a fraction in which each power has a positive exponent:

   (a) Should we substitute for x^4 or y^{-6} in $x^4 y^{-6}$? _____

   (b) Do so. $x^4 y^{-6} = \left(\right)\left(\right) =$ _____

a) d^{-3}

b) $\left(\dfrac{1}{d^3}\right)(f^5) = \dfrac{f^5}{d^3}$

---

130. To obtain a fraction in which each power has a positive exponent:

   (a) Should we substitute for a^{-1} or b^{-7} in $a^{-1} b^{-7}$? _____

   (b) Do so. $a^{-1} b^{-7} = \left(\right)\left(\right) =$ _____

a) y^{-6}

b) $x^4 \left(\dfrac{1}{y^6}\right) = \dfrac{x^4}{y^6}$

---

131. Convert each multiplication below to a fraction in which each power has a positive exponent:

   (a) $x^{-7} b^4 =$ _____   (b) $a t^{-3} =$ _____   (c) $F^{-2} R^{-3} =$ _____

a) For both.

b) $\left(\dfrac{1}{a^1}\right)\left(\dfrac{1}{b^7}\right) = \dfrac{1}{a^1 b^7}$

$\left(\text{or } \dfrac{1}{ab^7}\right)$

132. Write the right side of each equation below as a fraction in which each power has a positive exponent:

(a) If $y = b^2 x^{-4}$, $y =$ _____ (b) If $P = Q^{-2} R^5$, $P =$ _____

a) $\dfrac{b^4}{x^7}$ b) $\dfrac{a}{t^3}$

c) $\dfrac{1}{F^2 R^3}$

133. Write the right side of each equation below as a fraction in which each power has a positive exponent:

(a) If $m = a^{-1} c^{-4}$, $m =$ _____ (b) If $y = kx^{-1}$, $y =$ _____

a) $y = \dfrac{b^2}{x^4}$

b) $P = \dfrac{R^5}{Q^2}$

Answer to Frame 133:   a) $m = \dfrac{1}{ac^4}$ b) $y = \dfrac{k}{x}$

SELF-TEST 7 (Frames 122-133)

Write each of the following as a fraction whose numerator is "1":

1. $x^{-2} =$ _____
2. $y^3 =$ _____
3. $a^2 b^{-1} =$ _____
4. $ht^4 =$ _____

Write each multiplication below as a fraction in which each power has a positive exponent:

5. $p^2 w^{-3} =$ _____
6. $d^{-4} h^{-2} =$ _____
7. $c^{-2} k =$ _____
8. $rt^{-1} =$ _____

ANSWERS: 1. $\dfrac{1}{x^2}$ 3. $\dfrac{1}{a^{-2}b}$ 5. $\dfrac{p^2}{w^3}$ 7. $\dfrac{k}{c^2}$

2. $\dfrac{1}{y^{-3}}$ 4. $\dfrac{1}{h^{-1}t^{-4}}$ 6. $\dfrac{1}{d^4 h^2}$ 8. $\dfrac{r}{t}$

4-8 EQUIVALENT FORMS OF FRACTIONS CONTAINING POWERS

In this section, we will show how fractions can be written in equivalent forms by moving powers from the numerator to the denominator and vice versa. This procedure is ordinarily used to obtain expressions in which each power has a **positive** exponent.

318   Exponents and Radicals

134. We can interchange the powers in the numerator and denominator of the fraction below in two steps by:

(1) Converting the division to a multiplication:
$$\frac{a^3}{b^{-5}} = a^3 b^5$$

(2) Underline{Substituting $\frac{1}{a^{-3}}$ for a^3 in the multiplication}:
$$\frac{a^3}{b^{-5}} = a^3 b^5 = \left(\frac{1}{a^{-3}}\right) b^5 = \frac{b^5}{a^{-3}}$$

Interchange the powers in the numerator and denominator of the fraction below by converting the division to a multiplication and then substituting $\frac{1}{x^4}$ for x^{-4}.

$$\frac{x^{-4}}{y^2} = \underline{\qquad} = (\quad)(\quad) = \underline{\qquad}$$

---

135. Interchange the powers in the numerator and denominator of the fraction below by converting the division to a multiplication and then substituting $\frac{1}{t^3}$ for t^{-3}.

$$\frac{t^{-3}}{m^{-6}} = \underline{\qquad} = (\quad)(\quad) = \underline{\qquad}$$

$x^{-4} y^{-2} = \left(\frac{1}{x^4}\right) y^{-2} = \frac{y^{-2}}{x^4}$

---

136. In the last two frames, we interchanged the powers in the numerators and denominators of various fractions. We obtained these results:

$$\frac{a^3}{b^{-5}} = \frac{b^5}{a^{-3}} \qquad \frac{x^{-4}}{y^2} = \frac{y^{-2}}{x^4} \qquad \frac{t^{-3}}{m^{-6}} = \frac{m^6}{t^3}$$

As you can see from the examples above, when the powers in the numerator and denominator of a fraction are interchanged, the exponents are replaced by their opposites. Using this fact, interchange the powers in the numerator and denominator of each fraction below.

(a) $\frac{c^{-4}}{d^7} = \underline{\qquad}$ (b) $\frac{F^2}{R^{-1}} = \underline{\qquad}$ (c) $\frac{h^{-1}}{k^{-2}} = \underline{\qquad}$

$t^{-3} m^6 = \left(\frac{1}{t^3}\right) m^6 = \frac{m^6}{t^3}$

---

137. We have interchanged the numerator and denominator below by a two-step process.

$$\frac{a^{-2} b^3}{c^4} = a^{-2} b^3 c^{-4} = \left(\frac{1}{a^2}\right)\left(\frac{1}{b^{-3}}\right) c^{-4} = \frac{c^{-4}}{a^2 b^{-3}}$$

Notice that all of the exponents were changed to their opposites in this interchange. Using this fact, interchange the numerator and denominator in each fraction below:

(a) $\frac{d^5 f^{-3}}{t^{-4}} = \underline{\qquad}$ (b) $\frac{F^{-3} S^{-1}}{R^{-5}} = \underline{\qquad}$

a) $\frac{d^{-7}}{c^4}$

b) $\frac{R^1}{F^{-2}}$ or $\frac{R}{F^{-2}}$

c) $\frac{k^2}{h^1}$ or $\frac{k^2}{h}$

138. We have interchanged the numerator and denominator below by a two-step process.

$$\frac{x^{-4}}{a^5 y^{-3}} = x^{-4}a^{-5}y^3 = \left(\frac{1}{x^4}\right)a^{-5}y^3 = \frac{a^{-5}y^3}{x^4}$$

Notice again that each exponent was changed to its opposite in this interchange. Using this fact, interchange the numerator and denominator in each fraction below:

(a) $\dfrac{t^3}{s^{-7}y} = $ _____

(b) $\dfrac{c^{-6}}{d^{-7}p^{-8}} = $ _____

a) $\dfrac{t^4}{d^{-5}f^3}$

b) $\dfrac{R^5}{F^3 S^1}$ or $\dfrac{R^5}{F^3 S}$

139. We have interchanged the numerator and denominator below in a two-step process.

$$\frac{a^{-2}x^5}{b^3 y^{-4}} = a^{-2}x^5 b^{-3} y^4 = \left(\frac{1}{a^2}\right)\left(\frac{1}{x^{-5}}\right)b^{-3}y^4 = \frac{b^{-3}y^4}{a^2 x^{-5}}$$

Note that each exponent was changed to its opposite in the interchange. Use this fact to interchange the numerator and denominator in each fraction below:

(a) $\dfrac{c^4 t^{-6}}{d^{-5} s^7} = $ _____

(b) $\dfrac{D^{-2} F^{-3}}{G^{-4} T^{-5}} = $ _____

a) $\dfrac{s^7 y^{-1}}{t^{-3}}$ b) $\dfrac{d^7 p^8}{c^6}$

140. The point we are making is this:

> A power can be moved from the numerator of a fraction to its denominator (or vice versa) by simply changing the exponent to its opposite.

Here are some examples of the principle above:

(1) In the fraction below, we moved "a^3" from the numerator to the denominator and wrote it as "a^{-3}".

$$\frac{a^3 b^4}{c^5} = \frac{b^4}{a^{-3} c^5}$$

(2) In the fraction below, we moved "y^{-3}" from the denominator to the numerator and wrote it as "y^3".

$$\frac{d^5}{x^2 y^{-3}} = \frac{d^5 y^3}{x^2}$$

(a) Move "h^{-5}" from the numerator to the denominator of the fraction on the right. $\quad \dfrac{p^4 h^{-5}}{a^3} = $ _____

(b) Move "t^4" from the denominator to the numerator of the fraction on the right. $\quad \dfrac{q^2}{b^{-3} t^4} = $ _____

a) $\dfrac{d^5 s^{-7}}{c^{-4} t^6}$ b) $\dfrac{G^4 T^5}{D^2 F^3}$

a) $\dfrac{p^4}{a^3 h^5}$

b) $\dfrac{q^2 t^{-4}}{b^{-3}}$

320   Exponents and Radicals

141. Ordinarily, we move a power from the numerator to the denominator (or vice versa) in order to obtain an expression in which each exponent is positive. For example:

$$\frac{a^{-3}b^4}{c^2} \text{ is written as } \frac{b^4}{a^3 c^2}$$

$$\frac{x^2}{d^4 y^{-3}} \text{ is written as } \frac{x^2 y^3}{d^4}$$

Eliminate the negative exponent by moving a power in each of the following:

(a) $\dfrac{t^{-4} y^3}{c^2} =$ _____ (b) $\dfrac{R^5}{F^{-1} G^8} =$ _____ (c) $\dfrac{a^{-3} x^2}{b^4 y^5} =$ _____

---

142. Sometimes we have to move more than one power to obtain an expression with only positive exponents. For example:

$$\frac{c^{-4}}{d^{-3}} \text{ is written as } \frac{d^3}{c^4}$$

$$\frac{m^{-3} p^5}{d^{-4} q^{-2}} \text{ is written as } \frac{d^4 p^5 q^2}{m^3}$$

Write each of the following in an equivalent form which contains only positive exponents:

(a) $\dfrac{x^{-5}}{y^{-1}} =$ _____   (c) $\dfrac{h^{-8}}{g^{-4} p^2} =$ _____

(b) $\dfrac{a^4 c^{-3}}{b^{-2}} =$ _____   (d) $\dfrac{F^{-2} T^{-6}}{B^3 P^{-5}} =$ _____

a) $\dfrac{y^3}{c^2 t^4}$

b) $\dfrac{F^1 R^5}{G^8}$ or $\dfrac{F R^5}{G^8}$

c) $\dfrac{x^2}{a^3 b^4 y^5}$

---

143. When moving powers with negative exponents from the denominator to the numerator, we sometimes obtain a non-fractional expression. For example:

$$\frac{x^4}{y^{-5}} \text{ is written as } x^4 y^5$$

$$\frac{a^3}{b^{-2} c^{-4}} \text{ is written as } a^3 b^2 c^4$$

Notice that we have equivalently converted the divisions above to multiplication form.

Write each of the following in an equivalent form which contains only positive exponents:

(a) $\dfrac{c^5}{d^{-4}} =$ _____   (b) $\dfrac{h^2}{p^{-1} q^{-3}} =$ _____   (c) $\dfrac{x^2 y^3}{b^{-2} t^{-4}} =$ _____

a) $\dfrac{y^1}{x^5}$ or $\dfrac{y}{x^5}$

b) $\dfrac{a^4 b^2}{c^3}$

c) $\dfrac{g^4}{h^8 p^2}$

d) $\dfrac{P^5}{B^3 F^2 T^6}$

---

a) $c^5 d^4$

b) $h^2 p q^3$

c) $b^2 t^4 x^2 y^3$

144. When moving powers with negative exponents from the numerator to the denominator, we occasionally obtain a fraction whose numerator is "1". For example:

$$\frac{a^{-3}}{b^2} \left(\text{or } \frac{1a^{-3}}{b^2} \right) \text{ is written as } \frac{1}{a^3 b^2}$$

$$\frac{a^{-2}x^{-1}}{y^3} \left(\text{or } \frac{1a^{-2}x^{-1}}{y^3} \right) \text{ is written as } \frac{1}{a^2 x y^3}$$

Write each of the following in an equivalent form which contains only positive exponents:

(a) $\dfrac{t^{-5}}{c^2} =$ _____ (b) $\dfrac{a^{-1}b^{-2}}{d^4} =$ _____ (c) $\dfrac{x^{-2}y^{-3}}{a^4 b^5} =$ _____

---

145. Earlier we showed that any negative exponent in a multiplication can be eliminated by converting the multiplication to a fraction. If we use the principle $\square = \dfrac{\square}{1}$ to make a substitution, as we have done below, we can see that this conversion is equivalent to moving a power from the numerator to the denominator of a fraction.

$$p^{-2}q^4 = \frac{p^{-2}q^4}{1} = \frac{q^4}{p^2}$$

$$x^{-3}y^{-4} = \frac{1 x^{-3}y^{-4}}{1} = \frac{1}{x^3 y^4}$$

a) $\dfrac{1}{c^2 t^5}$

b) $\dfrac{1}{ab^2 d^4}$

c) $\dfrac{1}{a^4 b^5 x^2 y^3}$

Write each multiplication below as a fraction which contains only positive exponents:

(a) $c^{-3}d^5 =$ _____ (b) $m^{-5}t =$ _____ (c) $a^{-5}b^{-2} =$ _____

Answer to Frame 145:   a) $\dfrac{d^5}{c^3}$ b) $\dfrac{t}{m^5}$ c) $\dfrac{1}{a^5 b^2}$

SELF-TEST 8 (Frames 134-145)

1. Which of the following expressions are equivalent to: $\dfrac{a^{-1}r^2}{t}$ _____

(a) $a^{-1}r^2 t$ (b) $\dfrac{1}{ar^{-2}t}$ (c) $\dfrac{r^2 t^{-1}}{a}$ (d) $\dfrac{a^{-1}t}{r^2}$ (e) $\dfrac{a}{r^2 t}$ (f) $\dfrac{r^2}{at}$

Write each of the following in an equivalent form which contains only positive exponents:

2. $\dfrac{d^{-2}s}{h^2 p^{-3}} =$ _____

3. $\dfrac{r^{-1}}{w^{-2}} =$ _____

4. $\dfrac{b^3}{a^{-2}c^{-1}} =$ _____

5. $\dfrac{k^{-2}t^{-1}}{d^{-1}r} =$ _____

6. $p^2 s^{-3} w =$ _____

7. $x^{-2}y^{-1} =$ _____

ANSWERS: 1. (b), (c), (f)   2. $\dfrac{p^3 s}{d^2 h^2}$ 3. $\dfrac{w^2}{r}$ 4. $a^2 b^3 c$ 5. $\dfrac{d}{k^2 rt}$ 6. $\dfrac{p^2 w}{s^3}$ 7. $\dfrac{1}{x^2 y}$

322 Exponents and Radicals

4-9 POWERS OF LETTER-TERMS AND OTHER INDICATED MULTIPLICATIONS

In this section, we will discuss the powers of indicated multiplications, with special emphasis on letter-terms. We will show that there is a special law of exponents which applies to letter terms and other indicated multiplications.

---

146. Any power with a positive whole-number exponent greater than "1" stands for a multiplication of identical factors. For example:

$$8^3 = (8)(8)(8) \qquad x^5 = (x)(x)(x)(x)(x)$$

This meaning of a positive whole-number exponent greater than "1" also applies to powers whose base is a letter-term. For example:

$$(2y)^3 = (2y)(2y)(2y) \qquad (ab)^5 = (ab)(ab)(ab)(ab)(ab)$$

Write each of the following powers as a multiplication of identical factors:

(a) $(4t)^2 =$ _____ (b) $(xy)^4 =$ _____

---

147. Write each of the following multiplications as a power:

(a) $(7b)(7b)(7b) =$ _____ (b) $(cd)(cd)(cd)(cd)(cd) =$ _____

a) $(4t)(4t)$

b) $(xy)(xy)(xy)(xy)$

---

148. Don't confuse expressions like "$2x^3$" and $(2x)^3$.

In $2x^3$, only "x" is raised to the third power.

Therefore: $2x^3$ means $2 \cdot x \cdot x \cdot x$

In $(2x)^3$, the parentheses means that "2x" is raised to the third power.

Therefore: $(2x)^3$ means $(2x)(2x)(2x)$

Write each of the following expressions in multiplication form:

(a) $xy^4 =$ _____ (b) $(xy)^4 =$ _____

a) $(7b)^3$ b) $(cd)^5$

---

149. Write each of the following multiplications in power form:

(a) $5 \cdot t \cdot t \cdot t =$ _____ (b) $(5t)(5t)(5t) =$ _____

a) $x \cdot y \cdot y \cdot y \cdot y$

b) $(xy)(xy)(xy)(xy)$

---

150. As you know, any letter-term is a multiplication. That is:

"2x" means $(2)(x)$ \qquad "ab" means $(a)(b)$

Raising a letter-term to a power is the equivalent of raising each factor in the letter-term to that power. For example:

$$(2x)^3 = (2x)(2x)(2x) = (2 \cdot 2 \cdot 2)(x \cdot x \cdot x) = 2^3 x^3$$

$$(ab)^4 = (ab)(ab)(ab)(ab) = (a \cdot a \cdot a \cdot a)(b \cdot b \cdot b \cdot b) = a^4 b^4$$

Following the pattern above, complete these:

(a) $(xy)^3 = (xy)(xy)(xy) = (x \cdot x \cdot x)(y \cdot y \cdot y) = x^{\square} y^{\square}$

(b) $(7m)^5 = (7m)(7m)(7m)(7m)(7m) = (7 \cdot 7 \cdot 7 \cdot 7 \cdot 7)(m \cdot m \cdot m \cdot m \cdot m) = 7^{\square} \cdot m^{\square}$

a) $5t^3$ b) $(5t)^3$

Exponents and Radicals   323

151. In the last frame, we saw the following law of exponents for letter-terms.

$$(ab)^n = a^n b^n$$

That is: <u>any letter-term can be raised to a power by raising each factor in the letter-term to that power</u>.

Using the law above, complete each of these:

(a) $(4t)^3 = 4^{\square} \cdot t^{\square}$ (b) $(7x)^6 = 7^{\square} \cdot x^{\square}$ (c) $(pq)^9 = p^{\square} q^{\square}$

a) $x^{\boxed{3}} y^{\boxed{3}}$

b) $7^{\boxed{5}} \cdot m^{\boxed{5}}$

152. Using the law of exponents for a letter-term, we can write each power below in an equivalent form. That is:

$$(3m)^3 = 3^3 \cdot m^3 = 27m^3$$

(a) $(2h)^4 = 2^4 \cdot h^4 = $ _____   (b) $(2a)^5 = 2^5 \cdot a^5 = $ _____

a) $4^{\boxed{3}} \cdot t^{\boxed{3}}$

b) $7^{\boxed{6}} \cdot x^{\boxed{6}}$

c) $p^{\boxed{9}} q^{\boxed{9}}$

153. Any letter-term to the first power is equal to the letter-term. For example:

$$(5R)^1 = 5^1 \cdot R^1 = 5R \qquad (cd)^1 = c^1 \cdot d^1 = cd$$

Using the fact above, complete each of these:

(a) $(2y)^1 = $ _____   (b) $(10d)^1 = $ _____   (c) $(FR)^1 = $ _____

a) $16h^4$ b) $32a^5$

154. Like any other base to the zero power, a letter-term to the zero power equals +1. For example:

$$(9t)^0 = 9^0 \cdot t^0 = 1 \cdot 1 = 1 \qquad (hp)^0 = h^0 \cdot p^0 = 1 \cdot 1 = 1$$

Complete each of the following:

(a) $(8b)^0 = $ _____   (b) $(8b)^1 = $ _____   (c) $(qr)^0 = $ _____   (d) $(qr)^1 = $ _____

a) 2y
b) 10d
c) FR

155. The definition of a power with a negative whole-number exponent is given below.

$$\square^{-n} = \frac{1}{\square^n}$$

This definition also applies when the base is a letter-term. For example:

$$(5x)^{-3} = \frac{1}{(5x)^3} \qquad (ab)^{-5} = \frac{1}{(ab)^5}$$

Using the definition above, write an equivalent fraction for each of the following:

(a) $(3d)^{-4} = $ _____   (b) $(xy)^{-6} = $ _____   (c) $(RS)^{-10} = $ _____

a) 1
b) 8b
c) 1
d) qr

a) $\frac{1}{(3d)^4}$ c) $\frac{1}{(RS)^{10}}$

b) $\frac{1}{(xy)^6}$

324 Exponents and Radicals

156. Write each of the following as a power with a negative whole-number exponent:

(a) $\dfrac{1}{(9p)^2} =$ _____ (b) $\dfrac{1}{(pq)^5} =$ _____ (c) $\dfrac{1}{(ab)^9} =$ _____

---

157. Following the example at the right, complete each of the following: $(cd)^{-2} = \dfrac{1}{(cd)^2} = \dfrac{1}{c^2 d^2}$

(a) $(ax)^{-3} = \dfrac{1}{(ax)^3} =$ _____ (b) $(by)^{-7} = \dfrac{1}{(by)^7} =$ _____

a) $(9p)^{-2}$
b) $(pq)^{-5}$
c) $(ab)^{-9}$

---

158. Following the example at the right, complete each of the following: $(3t)^{-2} = \dfrac{1}{(3t)^2} = \dfrac{1}{3^2 t^2} = \dfrac{1}{9t^2}$

(a) $(5m)^{-3} = \dfrac{1}{(5m)^3} = \dfrac{1}{5^3 m^3} =$ _____ (b) $(2h)^{-5} = \dfrac{1}{(2h)^5} = \dfrac{1}{2^5 \cdot h^5} =$ _____

a) $\dfrac{1}{a^3 x^3}$ b) $\dfrac{1}{b^7 y^7}$

---

159. We can also convert a letter-term with a negative exponent to a fraction by applying the law of exponents for a letter-term first. We get:

$$(xy)^{-3} = x^{-3} y^{-3} = \left(\dfrac{1}{x^3}\right)\left(\dfrac{1}{y^3}\right) = \dfrac{1}{x^3 y^3}$$

(a) $(2t)^{-4} = 2^{-4} \cdot t^{-4} = \left(\dfrac{}{}\right)\left(\dfrac{}{}\right) =$ _____

(b) $(cp)^{-7} = c^{-7} p^{-7} = \left(\dfrac{}{}\right)\left(\dfrac{}{}\right) =$ _____

a) $\dfrac{1}{125 m^3}$ b) $\dfrac{1}{32 h^5}$

---

160. The laws of exponents for multiplication and division also apply to powers whose bases are letter-terms. For example:

$$(xy)^2 \cdot (xy)^5 = (xy)^{2+5} = (xy)^7$$

$$\dfrac{(cd)^8}{(cd)^5} = (cd)^{8-5} = (cd)^3$$

Using the laws of exponents, complete each of these:

(a) $(8t)^3 \cdot (8t)^{-1} =$ _____   (c) $(ab)^{-7} \cdot (ab)^4 =$ _____

(b) $\dfrac{(7m)^3}{(7m)^5} =$ _____   (d) $\dfrac{(ft)^{-3}}{(ft)^{-5}} =$ _____

a) $\left(\dfrac{1}{2^4}\right)\left(\dfrac{1}{t^4}\right) = \dfrac{1}{2^4 t^4}$

b) $\left(\dfrac{1}{c^7}\right)\left(\dfrac{1}{p^7}\right) = \dfrac{1}{c^7 p^7}$

---

161. The law of exponents for a letter-term is given below.

$$\boxed{(ab)^n = a^n b^n}$$

Since a letter-term is really a multiplication, the law of exponents for a letter-term applies to any multiplication. For example:

$(2 \times 3)^4 = 2^4 \times 3^4$ $(4 \times 5)^{-2} = 4^{-2} \times 5^{-2}$

Using the law above, complete each of these:

(a) $(7 \times 8)^3 =$ _____ $\times$ _____   (b) $(2 \times 10)^{-5} =$ _____ $\times$ _____

a) $(8t)^2$ c) $(ab)^{-3}$
b) $(7m)^{-2}$ d) $(ft)^2$

---

a) $7^3 \times 8^3$ b) $2^{-5} \times 10^{-5}$

162. Because the law of exponents below applies to any multiplication, it is usually called "the law of exponents for an indicated multiplication".

$$(ab)^n = a^n b^n$$

Let's use the law above to evaluate each expression below:

$(3 \times 5)^2 = 3^2 \times 5^2 = 9 \times 25 = 225$

(a) $(4 \times 10)^3 = 4^3 \times 10^3 = $ _____ x _____ = _____

(b) $(2 \times 10)^4 = 2^4 \times 10^4 = $ _____ x _____ = _____

---

163. Let's use the same law to evaluate each expression below:

$(7 \times 8)^{-1} = 7^{-1} \times 8^{-1} = \dfrac{1}{7^1} \times \dfrac{1}{8^1} = \dfrac{1}{7} \times \dfrac{1}{8} = \dfrac{1}{56}$

(a) $(5 \times 4)^{-2} = 5^{-2} \times 4^{-2} = \dfrac{1}{5^2} \times \dfrac{1}{4^2} = $ _____ x _____ = _____

(b) $(2 \times 3)^{-3} = 2^{-3} \times 3^{-3} = \dfrac{1}{2^3} \times \dfrac{1}{3^3} = $ _____ x _____ = _____

a) $64 \times 1{,}000 = 64{,}000$

b) $16 \times 10{,}000 = 160{,}000$

---

164. If we apply the law of exponents to each of the following, we get:

$(6 \times 5)^1 = 6^1 \times 5^1 = 6 \times 5 = 30$ $(9 \times 4)^0 = 9^0 \times 4^0 = 1 \times 1 = 1$

Based on the examples above, you should be able to answer these immediately:

(a) $(9 \times 10)^1 = $ _____   (c) $(4 \times 11)^0 = $ _____

(b) $(9 \times 10)^0 = $ _____   (d) $(4 \times 11)^1 = $ _____

a) $\dfrac{1}{25} \times \dfrac{1}{16} = \dfrac{1}{400}$

b) $\dfrac{1}{8} \times \dfrac{1}{27} = \dfrac{1}{216}$

---

165. The law of exponents for an indicated multiplication also applies to bases which contain more than two factors. For example:

$(7xy)^3 = 7^3 \cdot x^3 \cdot y^3$ $(abc)^4 = a^4 b^4 c^4$

Apply the law to each of the following:

(a) $(mpq)^5 = $ _____   (b) $(2FR)^7 = $ _____

a) 90   c) 1
b) 1   d) 44

---

166. If we reverse the law of exponents for an indicated multiplication, we get:

$$a^n b^n = (ab)^n$$

In this reversed form, the law says: <u>Any multiplication in which each factor is raised to the same power can be written as a letter-term (or indicated multiplication) to that power.</u> That is:

$2^3 \cdot x^3 = (2x)^3$ $p^5 q^5 = (pq)^5$

Using this reversed form of the law, complete each of these:

(a) $7^4 \cdot y^4 = (\quad)^4$ (c) $a^9 b^9 = (\quad)^9$

(b) $8^{-6} \cdot t^{-6} = (\quad)^{-6}$ (d) $h^{-1} m^{-1} = (\quad)^{-1}$

a) $m^5 p^5 q^5$ b) $2^7 \cdot F^7 \cdot R^7$

167. Of course, this reversed form of the law also applies to multiplications in which each base is a number. For example:

$$3^5 \times 7^5 = (3 \times 7)^5$$

(a) $8^4 \times 9^4 = (\quad \times \quad)^4$ (b) $2^{-3} \times 4^{-3} = (\quad \times \quad)^{-3}$

a) $(7y)^4$ c) $(ab)^9$
b) $(8t)^{-6}$ d) $(hm)^{-1}$

---

168. This reversed form of the law also applies to multiplications which contain more than two factors. For example:

$$c^2 d^2 f^2 = (cdf)^2$$

(a) $a^7 x^7 y^7 = (\quad)^7$ (b) $6^{-4} \cdot b^{-4} \cdot t^{-4} = (\quad)^{-4}$

a) $(8 \times 9)^4$ b) $(2 \times 4)^{-3}$

---

169. The reversed form of the law can be applied to the denominator in each fraction below:

(a) $\dfrac{1}{x^2 y^2} = \dfrac{1}{(\quad)^2}$ (b) $\dfrac{1}{p^5 q^5} = \dfrac{1}{(\quad)^5}$

a) $(axy)^7$ b) $(6bt)^{-4}$

---

170. After applying the reversed form of the law to the denominator of each fraction below, we can convert the fraction to a power with a negative exponent. That is:

(a) $\dfrac{1}{3^4 \cdot t^4} = \dfrac{1}{(3t)^4} = $ _____ (b) $\dfrac{1}{k^8 m^8} = \dfrac{1}{(km)^8} = $ _____

a) $\dfrac{1}{(xy)^2}$ b) $\dfrac{1}{(pq)^5}$

a) $(3t)^{-4}$ b) $(km)^{-8}$

Exponents and Radicals   327

SELF-TEST 9 (Frames 146-170)

1. State which of the following are equal to: $27t^3$ _____

   (a) $(3t)^{-3}$ (b) $(3t)^3$ (c) $3^3 \cdot t^3$ (d) $3t^3$ (e) $(27t^3)^0$ (f) $(27t^3)^1$ (g) $(27t)^3$

Using the laws of exponents, write each of the following in an equivalent non-fractional form:

2. $c^n d^n =$ _____ | 3. $(st)^{-n} =$ _____ | 4. $(hk)^n =$ _____ | 5. $(pr)^m \cdot (pr)^n =$ _____

Using the laws of exponents, find the numerical value of each of the following:

6. $(3 \times 2)^{-3} =$ _____ | 7. $2^5 \cdot 5^5 =$ _____ | 8. $2^{-5} \cdot 5^{-5} =$ _____ | 9. $\dfrac{(3 \times 8)^4}{(3 \times 8)^4} =$ _____

Using the laws of exponents, write each of the following in another way:

10. $(2w)^5 =$ _____ | 11. $(4d)^{-3} =$ _____ | 12. $8c^3 r^3 =$ _____

State whether each of the following is "True" or "False":

13. _____ $F^{-1} P^{-1} = (FP)^{-1}$ | 14. _____ $(ab)^5 = a^5 b^5$ | 15. _____ $(a+b)^2 = a^2 + b^2$

ANSWERS:

1. (b), (c), (f)
2. $(cd)^n$
3. $s^{-n} t^{-n}$
4. $h^n k^n$
5. $(pr)^{m+n}$
6. $\dfrac{1}{216}$
7. $100,000$
8. $\dfrac{1}{100,000}$
9. 1
10. $32w^5$
11. $4^{-3} d^{-3}$ or $\dfrac{1}{64d^3}$
12. $(2cr)^3$
13. True
14. True
15. False

4-10   POWERS OF FRACTIONS

In this section, we will discuss the powers of fractions. We will show that there is a special law of exponents which applies to fractions. We will also show that there is a pattern by which any fraction to a negative power can be converted to a fraction to a positive power.

171. When a power has a positive whole-number exponent greater than "1", it stands for a multiplication of identical factors <u>even</u> <u>if</u> <u>the</u> <u>base</u> <u>is</u> <u>a</u> <u>fraction</u>. For example:

$$\left(\frac{4}{5}\right)^3 = \left(\frac{4}{5}\right)\left(\frac{4}{5}\right)\left(\frac{4}{5}\right) \qquad \left(\frac{a}{b}\right)^4 = \left(\frac{a}{b}\right)\left(\frac{a}{b}\right)\left(\frac{a}{b}\right)\left(\frac{a}{b}\right)$$

Write each of the following powers as a multiplication of identical factors:

(a) $\left(\dfrac{8}{3}\right)^2 =$ _____   (b) $\left(\dfrac{x}{y}\right)^5 =$ _____

328   Exponents and Radicals

172. Write each of the following multiplications as a power:

(a) $\left(\dfrac{5}{3}\right)\left(\dfrac{5}{3}\right)\left(\dfrac{5}{3}\right) =$ _____   (b) $\left(\dfrac{s}{t}\right)\left(\dfrac{s}{t}\right)\left(\dfrac{s}{t}\right)\left(\dfrac{s}{t}\right) =$ _____

a) $\left(\dfrac{8}{3}\right)\left(\dfrac{8}{3}\right)$

b) $\left(\dfrac{x}{y}\right)\left(\dfrac{x}{y}\right)\left(\dfrac{x}{y}\right)\left(\dfrac{x}{y}\right)\left(\dfrac{x}{y}\right)$

---

173. Don't confuse expressions like $\dfrac{a^3}{b}$ and $\left(\dfrac{a}{b}\right)^3$.

In $\dfrac{a^3}{b}$, only "a" is raised to the third power.

Therefore: $\dfrac{a^3}{b}$ means $\dfrac{a \cdot a \cdot a}{b}$

In $\left(\dfrac{a}{b}\right)^3$, the parentheses mean that "$\dfrac{a}{b}$" is raised to the third power.

Therefore: $\left(\dfrac{a}{b}\right)^3$ means $\dfrac{a \cdot a \cdot a}{b \cdot b \cdot b}$

(a) Does $\dfrac{x^4}{5}$ or $\left(\dfrac{x}{5}\right)^4$ mean $\dfrac{x \cdot x \cdot x \cdot x}{5}$? _____

(b) Does $\dfrac{p^3}{q}$ or $\left(\dfrac{p}{q}\right)^3$ mean $\dfrac{p \cdot p \cdot p}{q \cdot q \cdot q}$? _____

a) $\left(\dfrac{5}{3}\right)^3$ b) $\left(\dfrac{s}{t}\right)^4$

---

174. Raising a fraction to a power is the equivalent of raising both the numerator and denominator of the fraction to that power. For example:

$$\left(\dfrac{7}{2}\right)^3 = \left(\dfrac{7}{2}\right)\left(\dfrac{7}{2}\right)\left(\dfrac{7}{2}\right) = \dfrac{7 \cdot 7 \cdot 7}{2 \cdot 2 \cdot 2} = \dfrac{7^3}{2^3}$$

(a) $\left(\dfrac{p}{q}\right)^4 = \left(\dfrac{p}{q}\right)\left(\dfrac{p}{q}\right)\left(\dfrac{p}{q}\right)\left(\dfrac{p}{q}\right) = \dfrac{p \cdot p \cdot p \cdot p}{q \cdot q \cdot q \cdot q} = \dfrac{p^{\Box}}{q^{\Box}}$

(b) $\left(\dfrac{x}{4}\right)^5 = \left(\dfrac{x}{4}\right)\left(\dfrac{x}{4}\right)\left(\dfrac{x}{4}\right)\left(\dfrac{x}{4}\right)\left(\dfrac{x}{4}\right) = \dfrac{x \cdot x \cdot x \cdot x \cdot x}{4 \cdot 4 \cdot 4 \cdot 4 \cdot 4} = \dfrac{x^{\Box}}{4^{\Box}}$

a) $\dfrac{x^4}{5}$ b) $\left(\dfrac{p}{q}\right)^3$

---

175. In the last frame, we saw the following law of exponents for fractions:

$$\boxed{\left(\dfrac{a}{b}\right)^n = \dfrac{a^n}{b^n}}$$

That is: <u>Any fraction can be raised to a power by raising both the numerator and denominator of the fraction to that power.</u>

Using the law above, complete each of these:

(a) $\left(\dfrac{6}{7}\right)^3 = \dfrac{6^{\Box}}{7^{\Box}}$ (b) $\left(\dfrac{2}{y}\right)^4 = \dfrac{2^{\Box}}{y^{\Box}}$ (c) $\left(\dfrac{m}{t}\right)^8 = \dfrac{m^{\Box}}{t^{\Box}}$

a) $\dfrac{p^{4}}{q^{4}}$

b) $\dfrac{x^{5}}{4^{5}}$

---

a) $\dfrac{6^{3}}{7^{3}}$

b) $\dfrac{2^{4}}{y^{4}}$

c) $\dfrac{m^{8}}{t^{8}}$

176. Using the law of exponents for fractions, we can write each fraction below in equivalent forms, as shown in the example.

$$\left(\frac{2}{x}\right)^5 = \frac{2^5}{x^5} = \frac{32}{x^5}$$

(a) $\left(\frac{a}{10}\right)^4 = \frac{a^4}{10^4} = $ _____   (b) $\left(\frac{4}{3}\right)^3 = \frac{4^3}{3^3} = $ _____

---

177. The law of exponents for fractions also applies to fractions in which the numerator or denominator (or both) is a letter-term. That is:

$$\left(\frac{3x}{2}\right)^4 = \frac{(3x)^4}{2^4}$$

(a) $\left(\frac{x}{5y}\right)^3 = $ _____   (b) $\left(\frac{ab}{pq}\right)^5 = $ _____

a) $\dfrac{a^4}{10{,}000}$ b) $\dfrac{64}{27}$

---

178. Following the example below, use the laws of exponents for fractions and letter-terms to write each power below in an equivalent form:

$$\left(\frac{5y}{m}\right)^2 = \frac{(5y)^2}{m^2} = \frac{5^2 \cdot y^2}{m^2} = \frac{25y^2}{m^2}$$

(a) $\left(\dfrac{c}{3d}\right)^3 = \dfrac{c^3}{(3d)^3} = \dfrac{c^3}{3^3 \cdot d^3} = $ _____

(b) $\left(\dfrac{2x}{5y}\right)^4 = \dfrac{(2x)^4}{(5y)^4} = \dfrac{2^4 \cdot x^4}{5^4 \cdot y^4} = $ _____

a) $\dfrac{x^3}{(5y)^3}$ b) $\dfrac{(ab)^5}{(pq)^5}$

---

179. Any fraction to the first power is equal to the fraction. For example:

$$\left(\frac{7}{8}\right)^1 = \frac{7^1}{8^1} = \frac{7}{8} \qquad \left(\frac{R}{F}\right)^1 = \frac{R^1}{F^1} = \frac{R}{F}$$

Any fraction to the zero power equals +1. For example:

$$\left(\frac{3}{4}\right)^0 = \frac{3^0}{4^0} = \frac{1}{1} = 1 \qquad \left(\frac{x}{y}\right)^0 = \frac{x^0}{y^0} = \frac{1}{1} = 1$$

Using the facts above, complete each of these:

(a) $\left(\dfrac{11}{7}\right)^1 = $ _____   (b) $\left(\dfrac{11}{7}\right)^0 = $ _____   (c) $\left(\dfrac{t}{d}\right)^0 = $ _____   (d) $\left(\dfrac{t}{d}\right)^1 = $ _____

a) $\dfrac{c^3}{27d^3}$ b) $\dfrac{16x^4}{625y^4}$

---

180. The facts in the last frame also apply to fractions in which the numerator or denominator (or both) is a letter-term. For example:

$$\left(\frac{5t}{s}\right)^1 = \frac{(5t)^1}{s^1} = \frac{5t}{s} \qquad \left(\frac{7}{3y}\right)^0 = \frac{7^0}{(3y)^0} = \frac{1}{1} = 1$$

Using the facts above, complete these:

(a) $\left(\dfrac{3x}{4}\right)^1 = $ _____   (b) $\left(\dfrac{3x}{4}\right)^0 = $ _____   (c) $\left(\dfrac{xy}{pq}\right)^1 = $ _____   (d) $\left(\dfrac{xy}{pq}\right)^0 = $ _____

a) $\dfrac{11}{7}$

b) 1

c) 1

d) $\dfrac{t}{d}$

---

a) $\dfrac{3x}{4}$ c) $\dfrac{xy}{pq}$

b) 1   d) 1

330   Exponents and Radicals

181. The law of exponents for fractions has been reversed below:

$$\frac{a^n}{b^n} = \left(\frac{a}{b}\right)^n$$

The reversed form of the law says this: <u>Any fraction whose numerator and denominator are raised to the same power can be written as a fraction to that power.</u> For example:

$$\frac{t^2}{5^2} = \left(\frac{t}{5}\right)^2 \qquad \frac{4^3}{m^3} = \left(\frac{4}{m}\right)^3 \qquad \frac{b^7}{a^7} = \left(\frac{b}{a}\right)^7$$

Using this reversed form of the law, complete each of these:

(a) $\dfrac{R^4}{7^4} =$ ____   (b) $\dfrac{8^9}{d^9} =$ ____   (c) $\dfrac{V^2}{F^2} =$ ____   (d) $\dfrac{(ab)^5}{c^5} =$ ____

182. By using the reversed form of the laws of exponents for both letter-terms and fractions, we can write each expression below as a fraction to a power. That is:

$$\frac{a^3b^3}{c^3} = \frac{(ab)^3}{c^3} = \left(\frac{ab}{c}\right)^3$$

(a) $\dfrac{m^4}{p^4q^4} = \dfrac{m^4}{(pq)^4} =$ ____   (b) $\dfrac{c^6d^6}{x^6y^6} = \dfrac{(cd)^6}{(xy)^6} =$ ____

a) $\left(\dfrac{R}{7}\right)^4$ c) $\left(\dfrac{V}{F}\right)^2$

b) $\left(\dfrac{8}{d}\right)^9$ d) $\left(\dfrac{ab}{c}\right)^5$

183. The definition of a power with a negative whole-number exponent is given below.

$$\Box^{-n} = \frac{1}{\Box^n}$$

This definition also applies <u>when the base is a fraction.</u> That is:

$$\left(\frac{3}{4}\right)^{-2} = \frac{1}{\left(\frac{3}{4}\right)^2} \qquad \left(\frac{x}{y}\right)^{-4} = \frac{1}{\left(\frac{x}{y}\right)^4}$$

Using the definition above, complete these:

(a) $\left(\dfrac{t}{5}\right)^{-3} =$ (b) $\left(\dfrac{c}{d}\right)^{-5} =$ (c) $\left(\dfrac{2}{p}\right)^{-1} =$

a) $\left(\dfrac{m}{pq}\right)^4$ b) $\left(\dfrac{cd}{xy}\right)^6$

a) $\dfrac{1}{\left(\dfrac{t}{5}\right)^3}$

b) $\dfrac{1}{\left(\dfrac{c}{d}\right)^5}$

c) $\dfrac{1}{\left(\dfrac{2}{p}\right)^1}$ or $\dfrac{1}{\dfrac{2}{p}}$

184. In the following example, notice the steps used to convert a fraction with a negative exponent to a fraction with a positive exponent.

$$\left(\frac{2}{3}\right)^{-4} = \frac{1}{\left(\frac{2}{3}\right)^4} = \frac{1}{\frac{2^4}{3^4}} = 1\left(\text{the reciprocal of } \frac{2^4}{3^4}\right) = 1\left(\frac{3^4}{2^4}\right) = \frac{3^4}{2^4} = \left(\frac{3}{2}\right)^4$$

Following the steps above, complete each of these conversions to a fraction with a positive exponent:

(a) $\left(\dfrac{x}{5}\right)^{-3} = \dfrac{1}{\left(\dfrac{x}{5}\right)^3} = \dfrac{1}{\dfrac{x^3}{5^3}} = 1\left(\text{the reciprocal of } \dfrac{x^3}{5^3}\right) = 1(\underline{}) = \underline{} = \underline{}$

(b) $\left(\dfrac{m}{t}\right)^{-5} = \dfrac{1}{\left(\dfrac{m}{t}\right)^5} = \dfrac{1}{\dfrac{m^5}{t^5}} = 1\left(\text{the reciprocal of } \dfrac{m^5}{t^5}\right) = 1(\underline{}) = \underline{} = \underline{}$

---

185. Here are the three conversions we made in the last frame:

$$\left(\frac{2}{3}\right)^{-4} = \left(\frac{3}{2}\right)^4 \qquad \left(\frac{x}{5}\right)^{-3} = \left(\frac{5}{x}\right)^3 \qquad \left(\frac{m}{t}\right)^{-5} = \left(\frac{t}{m}\right)^5$$

All of the conversions above fit the following pattern:

$$\boxed{\left(\frac{a}{b}\right)^{-n} = \left(\frac{b}{a}\right)^n}$$

That is: any fraction to a negative power can be converted to a fraction with a positive power by:

    (1) interchanging numerator and denominator,
and (2) replacing the exponent with its opposite.

Using the pattern above, convert each power below to a power with a positive exponent:

(a) $\left(\dfrac{8}{5}\right)^{-3} = $ _____

(b) $\left(\dfrac{t}{2}\right)^{-4} = $ _____

(c) $\left(\dfrac{x}{y}\right)^{-9} = $ _____

(d) $\left(\dfrac{a}{b}\right)^{-1} = $ _____

Answers:

a) $1\left(\dfrac{5^3}{x^3}\right) = \dfrac{5^3}{x^3} = \left(\dfrac{5}{x}\right)^3$

b) $1\left(\dfrac{t^5}{m^5}\right) = \dfrac{t^5}{m^5} = \left(\dfrac{t}{m}\right)^5$

a) $\left(\dfrac{5}{8}\right)^3$ c) $\left(\dfrac{y}{x}\right)^9$

b) $\left(\dfrac{2}{t}\right)^4$ d) $\left(\dfrac{b}{a}\right)^1$ or $\dfrac{b}{a}$

186. Using the same pattern, convert each power below to a power with a positive exponent:

(a) $\left(\dfrac{3x}{4}\right)^{-2} =$ _____   (b) $\left(\dfrac{c}{4d}\right)^{-5} =$ _____   (c) $\left(\dfrac{pq}{ab}\right)^{-7} =$ _____

Answer to Frame 186:   a) $\left(\dfrac{4}{3x}\right)^{2}$ b) $\left(\dfrac{4d}{c}\right)^{5}$ c) $\left(\dfrac{ab}{pq}\right)^{7}$

SELF-TEST 10 (Frames 171-186)

Using the law of exponents for fractions, raise each fraction to the power shown:

1. $\left(\dfrac{2}{3}\right)^{4} =$ _____   2. $\left(\dfrac{3r}{5d}\right)^{3} =$ _____   3. $\left(\dfrac{h}{2t}\right)^{5} =$ _____   4. $\left(\dfrac{a}{b}\right)^{0} =$ _____   5. $\left(\dfrac{a}{b}\right)^{w} =$ _____

Write each of the following as a fraction raised to a single power:

6. $\dfrac{p^3}{s^3} =$ _____   7. $\dfrac{h^5}{4^5 \cdot b^5} =$ _____

Write each of the following as a fraction raised to a positive power:

8. $\left(\dfrac{k}{2}\right)^{-4} =$ _____   9. $\left(\dfrac{2d}{3v}\right)^{-1} =$ _____

ANSWERS: 1. $\dfrac{64}{81}$ 2. $\dfrac{27r^3}{125d^3}$ 3. $\dfrac{h^5}{32t^5}$ 4. 1   5. $\dfrac{a^w}{b^w}$ 6. $\left(\dfrac{p}{s}\right)^{3}$ 7. $\left(\dfrac{h}{4b}\right)^{5}$ 8. $\left(\dfrac{2}{k}\right)^{4}$ 9. $\left(\dfrac{3v}{2d}\right)^{1}$ or $\dfrac{3v}{2d}$

4-11 POWERS OF POWERS

In this section, we will discuss the procedure for raising a power to a power. We will show that there is a special law of exponents which can be used to do so.

187. The expression below is a power in which the base is also a power. That is:

In $(x^3)^2$: the base is x^3.
           : the exponent is "2".

An expression like the one above is called a "power of a power". Which of the following are "powers of powers"? _____

(a) y^4 (b) $(y^2)^4$ (c) $(t^5)^{-3}$ (d) t^{-3}

Both (b) and (c).

188. When a power is raised to a whole-number power greater than "1", it can also be written as a multiplication of identical factors. That is:

$$(x^3)^2 = (x^3)(x^3)$$
$$(y^{-2})^3 = (y^{-2})(y^{-2})(y^{-2})$$
$$(t^5)^4 = (t^5)(t^5)(t^5)(t^5)$$

Write each of the following as a multiplication of identical factors:

(a) $(h^6)^3 =$ _____   (b) $(m^{-2})^5 =$ _____

---

189. Write each of the following as a "power of a power":

(a) $(3^4)(3^4) =$ _____   (b) $(7^{-1})(7^{-1})(7^{-1})(7^{-1}) =$ _____

a) $(h^6)(h^6)(h^6)$
b) $(m^{-2})(m^{-2})(m^{-2})(m^{-2})(m^{-2})$

---

190. Any power can be raised to a whole-number power greater than "1" by converting it to a multiplication and applying the law of exponents for multiplication. That is:

$$(7^3)^2 = (7^3)(7^3) = 7^{3+3} = 7^6$$

(a) $(4^{-5})^3 = (4^{-5})(4^{-5})(4^{-5}) = 4^{(-5)+(-5)+(-5)} =$ _____

(b) $(m^2)^4 = (m^2)(m^2)(m^2)(m^2) = m^{2+2+2+2} =$ _____

(c) $(p^{-1})^5 = (p^{-1})(p^{-1})(p^{-1})(p^{-1})(p^{-1}) = p^{(-1)+(-1)+(-1)+(-1)+(-1)} =$ _____

a) $(3^4)^2$ b) $(7^{-1})^4$

---

191. Here are the completed problems from the last frame:

$$(7^3)^2 = 7^6 \quad (4^{-5})^3 = 4^{-15} \quad (m^2)^4 = m^8 \quad (p^{-1})^5 = p^{-5}$$

As you can see from the examples above, <u>any power can be raised to a power by multiplying the exponent of the base by the desired power.</u> That is:

(a) $(8^5)^2 = 8^{(5)(2)} =$ _____   (c) $(2^{-3})^4 = 2^{(-3)(4)} =$ _____

(b) $(y^4)^3 = y^{(4)(3)} =$ _____   (d) $(q^{-4})^5 = q^{(-4)(5)} =$ _____

a) 4^{-15}
b) m^8
c) p^{-5}

---

192. The law of exponents for raising a power to a power is given in general form below.

$$\boxed{(a^m)^n = a^{mn}}$$

That is: <u>any power can be raised to a power by multiplying the exponent of the original power by the desired power.</u>

Using the law above, complete each of these:

(a) $(x^7)^2 =$ _____   (c) $(y^{-2})^5 =$ _____

(b) $(5^4)^7 =$ _____   (d) $(8^{-1})^{10} =$ _____

a) 8^{10} c) 2^{-12}
b) y^{12} d) q^{-20}

---

a) x^{14}, from $x^{(7)(2)}$
b) 5^{28}, from $5^{(4)(7)}$
c) y^{-10}, from $y^{(-2)(5)}$
d) 8^{-10}, from $8^{(-1)(10)}$

Exponents and Radicals 333

334   Exponents and Radicals

193. Any power raised to the first power equals the original power. For example:

$$(x^3)^1 = x^{(3)(1)} = x^3 \qquad (y^{-2})^1 = y^{(-2)(1)} = y^{-2}$$

Any power raised to the zero power equals +1. For example:

$$(t^4)^0 = t^{(4)(0)} = t^0 = 1 \qquad (m^{-3})^0 = m^{(-3)(0)} = m^0 = 1$$

Using the facts above, complete each of these:

(a) $(p^5)^1 =$ _____ (b) $(p^5)^0 =$ _____ (c) $(h^{-7})^0 =$ _____ (d) $(h^{-7})^1 =$ _____

---

194. The definition of a power with a negative whole-number exponent is given below.

$$\boxed{\Box^{-n} = \frac{1}{\Box^n}}$$

This definition also applies to powers which are raised to negative whole-number exponents. That is:

$$(x^4)^{-2} = \frac{1}{(x^4)^2} \qquad (y^5)^{-3} = \frac{1}{(y^5)^3} \qquad (8^{-3})^{-4} = \frac{1}{(8^{-3})^4}$$

Apply the definition above to each of the following:

(a) $(m^5)^{-4} =$ (b) $(7^4)^{-5} =$ (c) $(5^3)^{-1} =$

a) p^5
b) 1
c) 1
d) h^{-7}

---

195. Write each of the following as a power of a power:

(a) $\dfrac{1}{(x^2)^3} =$ (b) $\dfrac{1}{(4^3)^5} =$ (c) $\dfrac{1}{(m^{-4})^2} =$

a) $\dfrac{1}{(m^5)^4}$ b) $\dfrac{1}{(7^4)^5}$

c) $\dfrac{1}{(5^3)^1}$ or $\dfrac{1}{5^3}$

---

196. A power with a positive exponent can be raised to a negative power in three steps. For example:

$$(4^3)^{-2} = \frac{1}{(4^3)^2} = \frac{1}{4^6} = 4^{-6}$$

Note: (1) We obtained a fraction by applying the definition of a negative power.
(2) We simplified the denominator by applying the law of exponents for raising a power to a power.
(3) We converted back to a negative power by again applying the definition of a negative power.

a) $(x^2)^{-3}$
b) $(4^3)^{-5}$
c) $(m^{-4})^{-2}$

Continued on following page.

Exponents and Radicals   335

196. Continued

Following the preceding steps, raise each of the following to a negative power:

(a) $(x^4)^{-3} = \dfrac{1}{(x^4)^3} = $ _____ $= $ _____

(b) $(y^2)^{-4} = \dfrac{1}{(y^2)^4} = $ _____ $= $ _____

(c) $(m^5)^{-1} = \dfrac{1}{(m^5)^1} = $ _____ $= $ _____

197. A power with a negative exponent can also be raised to a negative power in three steps. For example:

$$(x^{-2})^{-3} = \dfrac{1}{(x^{-2})^3} = \dfrac{1}{x^{-6}} = x^6$$

Notice in the last step that we converted back to power form by using the principle $\dfrac{1}{x^{-n}} = x^n$.

Following the steps above, raise each of the following to a negative power:

(a) $(c^{-5})^{-2} = \dfrac{1}{(c^{-5})^2} = $ _____ $= $ _____

(b) $(k^{-4})^{-3} = \dfrac{1}{(k^{-4})^3} = $ _____ $= $ _____

(c) $(R^{-7})^{-1} = \dfrac{1}{(R^{-7})^1} = $ _____ $= $ _____

a) $\dfrac{1}{x^{12}} = x^{-12}$

b) $\dfrac{1}{y^8} = y^{-8}$

c) $\dfrac{1}{m^5} = m^{-5}$

198. Here are the examples from the last two frames:

$(x^4)^{-3} = x^{-12}$ $(y^2)^{-4} = y^{-8}$ $(m^5)^{-1} = m^{-5}$
$(c^{-5})^{-2} = c^{10}$ $(k^{-4})^{-3} = k^{12}$ $(R^{-7})^{-1} = R^7$

As you can see, the law of exponents for raising a power to a power also applies when we are raising powers to negative powers. That is, <u>we can simply multiply the exponent of the original power by the desired power</u>.

Using this fact, complete each of these:

(a) $(x^6)^{-2} = x^{(6)(-2)} = $ _____   (c) $(2^4)^{-4} = 2^{(4)(-4)} = $ _____

(b) $(y^{-3})^{-5} = y^{(-3)(-5)} = $ _____   (d) $(7^{-1})^{-9} = 7^{(-1)(-9)} = $ _____

a) $\dfrac{1}{c^{-10}} = c^{10}$

b) $\dfrac{1}{k^{-12}} = k^{12}$

c) $\dfrac{1}{R^{-7}} = R^7$

a) x^{-12} c) 2^{-16}

b) y^{15} d) 7^9

336   Exponents and Radicals

199. The law of exponents for raising a power to a power is given below. It applies when we are raising a power to either a positive or negative power.

$$(a^m)^n = a^{mn}$$

Using the law above, complete each of these:

(a) $(F^4)^6 = $ _____   (c) $(m^2)^{-8} = $ _____

(b) $(t^3)^7 = $ _____   (d) $(y^3)^{-3} = $ _____

200. Use the law of exponents for raising a power to a power to complete each of these:

(a) $(x^{-3})^9 = $ _____   (c) $(T^{-4})^8 = $ _____

(b) $(y^{-1})^{-10} = $ _____   (d) $(q^{-5})^{-5} = $ _____

a) F^{24} c) m^{-16}
b) t^{21} d) y^{-9}

201. We have already seen this law: raising a letter-term to a power is the equivalent of raising each factor in the letter-term to that power. For example:

$$(3x)^4 = 3^4 \cdot x^4 \qquad (ab)^3 = a^3 \cdot b^3$$

This same law applies when one (or both) of the factors in the letter-term is itself a power. For example:

$$(5y^2)^3 = (5)^3 \cdot (y^2)^3 \qquad (p^2q^3)^4 = (p^2)^4 \cdot (q^3)^4$$

We can simplify the right side of each equation above by using the law of exponents for raising a power to a power. That is:

$$(5y^2)^3 = 5^3 \cdot (y^2)^3 = 5^3 \cdot y^6 \text{ (or } 125y^6)$$
$$(p^2q^3)^4 = (p^2)^4 \cdot (q^3)^4 = p^8q^{12}$$

Following the steps above, raise each of the following letter-terms to the indicated power:

(a) $(3x^2)^5 = (3)^5 \cdot (x^2)^5 = $ _____

(b) $(c^4d^2)^3 = (c^4)^3 \cdot (d^2)^3 = $ _____

(c) $(b^{-2}y^5)^{-4} = (b^{-2})^{-4} \cdot (y^5)^{-4} = $ _____

a) x^{-27} c) T^{-32}
b) y^{10} d) q^{25}

202. Here are two of the completed problems from the last frame:

$$(c^4d^2)^3 = c^{12}d^6 \qquad (b^{-2}y^5)^{-4} = b^8y^{-20}$$

As you can see from the examples above, we can raise a letter-term to a desired power by multiplying the exponent of each of its factors by that power. That is:

(a) $(a^2b^3)^5 = a^{(2)(5)} \cdot b^{(3)(5)} = $ _____

(b) $(x^4y^{-1})^{-2} = x^{(4)(-2)} \cdot y^{(-1)(-2)} = $ _____

a) $3^5 \cdot x^{10}$ (or $243x^{10}$)
b) $c^{12}d^6$
c) b^8y^{-20}

a) $a^{10}b^{15}$
b) $x^{-8}y^2$

Exponents and Radicals   337

203. If the exponent of a factor is not explicitly written, remember that its exponent is "1". That is:

(a) $(ab^3)^2 = (a^1 b^3)^2 = a^{(1)(2)} \cdot b^{(3)(2)} = $ _____

(b) $(5t^3)^{-4} = (5^1 \cdot t^3)^{-4} = 5^{(1)(-4)} \cdot t^{(3)(-4)} = $ _____

204. Complete each of the following:

(a) $(2x^2)^6 = $ _____ (c) $(5y^2)^{-1} = $ _____

(b) $(m^{-1}t^3)^4 = $ _____ (d) $(a^4 b^{-1})^{-3} = $ _____

a) $a^2 b^6$
b) $5^{-4} \cdot t^{-12}$

205. We have already seen this law: <u>raising a fraction to a power is the equivalent of raising both its numerator and its denominator to that power</u>. For example:

$$\left(\frac{3}{x}\right)^4 = \frac{3^4}{x^4} \qquad \left(\frac{a}{b}\right)^3 = \frac{a^3}{b^3}$$

This same law applies when either the numerator or denominator (or both) is itself a power. For example:

$$\left(\frac{x^2}{y}\right)^3 = \frac{(x^2)^3}{y^3} \qquad \left(\frac{a}{b^3}\right)^4 = \frac{a^4}{(b^3)^4} \qquad \left(\frac{c^2}{d^2}\right)^{-3} = \frac{(c^2)^{-3}}{(d^2)^{-3}}$$

By applying the law of exponents for raising a power to a power, we can simplify the right side of each equation above. That is:

$$\left(\frac{x^2}{y}\right)^3 = \frac{(x^2)^3}{y^3} = \frac{x^6}{y^3}$$

(a) $\left(\dfrac{a}{b^3}\right)^4 = \dfrac{a^4}{(b^3)^4} = $ _____

(b) $\left(\dfrac{c^2}{d^2}\right)^{-3} = \dfrac{(c^2)^{-3}}{(d^2)^{-3}} = $ _____

a) $2^6 \cdot x^{12}$ c) $5^{-1} \cdot y^{-2}$
b) $m^{-4} t^{12}$ d) $a^{-12} b^3$

206. Using the procedure in the last frame, raise each fraction below to the indicated power:

(a) $\left(\dfrac{y^2}{5}\right)^3 = $ _____ (c) $\left(\dfrac{R^4}{7}\right)^{-1} = $ _____

(b) $\left(\dfrac{m}{p^3}\right)^5 = $ _____ (d) $\left(\dfrac{a^{-2}}{b^3}\right)^{-4} = $ _____

a) $\dfrac{a^4}{b^{12}}$ b) $\dfrac{c^{-6}}{d^{-6}}$

207. Using the same procedure, raise each fraction below to the indicated power:

(a) $\left(\dfrac{x^2 y}{b^3}\right)^4 = $ _____ (b) $\left(\dfrac{a^{-1} b^2}{c^{-2} d^3}\right)^{-2} = $ _____

a) $\dfrac{y^6}{5^3}$ c) $\dfrac{R^{-4}}{7^{-1}}$

b) $\dfrac{m^5}{p^{15}}$ d) $\dfrac{a^8}{b^{-12}}$

a) $\dfrac{x^8 y^4}{b^{12}}$ b) $\dfrac{a^2 b^{-4}}{c^4 d^{-6}}$

SELF-TEST 11 (Frames 187-207)

Raise each expression to the power indicated:

| 1. $(x^{-3})^2 = $ _____ | 2. $(h^{-1})^{-4} = $ _____ | 3. $(2r^{-3})^0 = $ _____ | 4. $(ab^{-2})^3 = $ _____ |

Raise each fraction to the power indicated:

| 5. $\left(\dfrac{t^3}{2s}\right)^4 = $ _____ | 6. $\left(\dfrac{x^3 y^{-1}}{d^2}\right)^{-2} = $ _____ | 7. $\left(\dfrac{b^2 c^{-3}}{h^{-2} t}\right)^3 = $ _____ |

ANSWERS: 1. x^{-6} 3. 1 5. $\dfrac{t^{12}}{16s^4}$ 6. $\dfrac{x^{-6} y^2}{d^{-4}}$ 7. $\dfrac{b^6 c^{-9}}{h^{-6} t^3}$
              2. h^4 4. $a^3 b^{-6}$

4-12 ADDING AND SUBTRACTING LETTER-TERMS WHICH CONTAIN POWERS

In this section, we will briefly discuss the addition and subtraction of letter-terms which contain powers. We will also discuss the proper order for terms in additions which contain various powers of the same letter.

208. Two letter-terms which contain the same letter to the same power can be added (or combined into one term) because we can factor by means of the distributive principle <u>over</u> <u>addition</u>. That is:

$$3x^2 + 5x^2 = (3 + 5)x^2 = 8x^2$$
$$7y^3 + 2y^3 = (7 + 2)y^3 = 9y^3$$

As you can see from the example above, factoring by the distributive principle is the equivalent of <u>adding the numerical coefficients</u>.

Perform each addition below by adding the numerical coefficients:

(a) $6m^4 + 5m^4 = $ _____   (b) $3R^3 + 9R^3 = $ _____

a) $11m^4$ b) $12R^3$

Exponents and Radicals 339

209. We cannot combine either pair of terms below because they do not contain the same powers, and therefore they cannot be factored by means of the distributive principle.

$$3x^2 + 4y^2 \qquad\qquad 6x^3 + 5x^7$$

Perform each addition below, if possible:

(a) $8a^3 + 5b^3 = $ _____ (b) $11m^4 + 4m^4 = $ _____ (c) $2y^2 + 3y^3 = $ _____

---

210. If a power does not have an explicit coefficient, remember that its coefficient is "1". For example:

$$x^2 + 4x^2 = 1x^2 + 4x^2 = 5x^2$$
$$5b^4 + b^4 = 5b^4 + 1b^4 = 6b^4$$

Perform each of these additions:

(a) $t^3 + 3t^3 = $ _____ (b) $2y^5 + y^5 = $ _____ (c) $x^4 + x^4 = $ _____

a) Not possible.
b) $15m^4$
c) Not possible.

---

211. Two letter-terms which contain the same power can also be subtracted (or combined into one term) because we can factor by means of the distributive principle <u>over</u> <u>subtraction</u>. That is:

$$5m^4 - 3m^4 = (5 - 3)m^4 = 2m^4$$
$$7d^2 - 2d^2 = (7 - 2)d^2 = 5d^2$$

As you can see from the examples above, factoring by the distributive principle is the equivalent of <u>subtracting the numerical coefficients</u>.

Perform each subtraction below by subtracting the numerical coefficients:

(a) $9x^3 - 5x^3 = $ _____ (b) $7y^5 - 4y^5 = $ _____

a) $4t^3$
b) $3y^5$
c) $2x^4$

---

212. Before performing the subtraction below, we have written the coefficient of x^3 explicitly.

$$5x^3 - x^3 = 5x^3 - 1x^3 = 4x^3$$

Perform each subtraction below:

(a) $3y^4 - y^4 = $ _____ (b) $x^6 - x^6 = $ _____

a) $4x^3$ b) $3y^5$

---

213. The addition below contains a number of terms, each of which has a power of "x" as one of its factors.

$$3x^4 + 5x^3 + 2x^2 + 7x$$

Notice that we have written the terms with the powers of "x" in order, with the largest power on the left.

Rewrite each addition below, putting the terms in the proper order:

(a) $9x + 3x^3 + 5x^2 = $ _____

(b) $x^8 + 5x^2 + 7x^4 = $ _____

a) $2y^4$
b) 0, from $0x^6$

---

a) $3x^3 + 5x^2 + 9x$
b) $x^8 + 7x^4 + 5x^2$

214. In the addition below, we have written the "7" at the far right since "7" is the equivalent of "$7x^0$" which equals $7 \cdot 1$ or 7.

$$4x^3 + 5x^2 + 2x + 7$$

Rewrite the addition below, putting the terms in the proper order:

$2y^5 + 9 + 4y + 8y^2 = $ _____

---

215. Each of the parentheses below contains a trinomial:

$$(x^2 + 5x + 7) + (x^5 + 4x + 2)$$

To perform the addition of the trinomials, we can simply drop the parentheses, add "like" terms, and then write the terms in their proper order. Do so.

$x^2 + 5x + 7 + x^5 + 4x + 2 = $ _____

$2y^5 + 8y^2 + 4y + 9$

---

216. Each of the parentheses below also contains a trinomial:

$$(4y^3 + 5y^2 + 7) - (2y^3 + 2y^2 + 3)$$

We can convert the subtraction of the second trinomial to an addition by adding the opposite of each term. We get:

$$4y^3 + 5y^2 + 7 + (-2y^3) + (-2y^2) + (-3)$$

Now add like terms and write the resulting terms in their proper order:

_____

$x^5 + x^2 + 9x + 9$

---

217. Perform each operation below. Write the resulting terms in their proper order:

(a) $(x^3 + 4x^2 + 2x) + (x^5 + 2x^2 + 1) = $ _____

(b) $(3m^2 + 7m + 11) - (m^2 + m + 1) = $ _____

$2y^3 + 3y^2 + 4$

Answer to Frame 217:   a) $x^5 + x^3 + 6x^2 + 2x + 1$ b) $2m^2 + 6m + 10$

SELF-TEST 12 (Frames 208-217)

1. Do this addition and write the resulting terms in their proper order:

$(x^4 + 3x^2 + 8) + (4x^3 + x^2 + 5x + 2) = $ _____

2. Do this subtraction and write the resulting terms in their proper order:

$(4y^3 + 3y^2 + 7y + 5) - (3y^3 + 2y^2 + 7y) = $ _____

ANSWERS: 1. $x^4 + 4x^3 + 4x^2 + 5x + 10$ 2. $y^3 + y^2 + 5$

4-13 A REVIEW OF POWERS AND THE LAWS OF EXPONENTS

Up to this point, we have discussed the meaning of powers with both positive and negative whole-number exponents. We have also introduced the following laws of exponents for powers.

$$a^m \cdot a^n = a^{m+n}$$

$$\frac{a^m}{a^n} = a^{m-n} = a^{m+(-n)}$$

$$(ab)^n = a^n b^n$$

$$\left(\frac{a}{b}\right)^n = \frac{a^n}{b^n}$$

$$(a^m)^n = a^{mn}$$

In this section, we will briefly review the meaning of powers, the laws of exponents, and a few other key points about powers.

218. Any power with a positive whole-number exponent greater than 1 is shorthand for a multiplication of identical factors. For example:

$$4^3 = (4)(4)(4) \qquad x^4 = (x)(x)(x)(x)$$

This meaning of powers with whole-number exponents greater than 1 also applies when the base is a letter-term, a fraction, or a power. That is:

(a) $(5x)^2 = (\quad)(\quad)$

(b) $(ab)^3 = (\quad)(\quad)(\quad)$

(c) $\left(\dfrac{x}{y}\right)^4 = (\quad)(\quad)(\quad)(\quad)$

(d) $(x^2)^5 = (\quad)(\quad)(\quad)(\quad)(\quad)$

219. Any power whose exponent is 1 equals its base. For example:

$$5^1 = 5 \qquad x^1 = x \qquad R^1 = R$$

This definition also applies when the base is a letter-term, a fraction, or a power. Therefore:

(a) $(3t)^1 = $ _____

(b) $\left(\dfrac{y}{7}\right)^1 = $ _____

(c) $(5^4)^1 = $ _____

220. Any power whose exponent is 0 equals +1. For example:

$$7^0 = +1 \qquad m^0 = +1 \qquad F^0 = +1$$

This definition also applies when the base is a letter-term, a fraction, or a power. Therefore:

(a) $(cd)^0 = $ _____

(b) $\left(\dfrac{p}{q}\right)^0 = $ _____

(c) $(x^7)^0 = $ _____

Answers:

a) $(5x)(5x)$

b) $(ab)(ab)(ab)$

c) $\left(\dfrac{x}{y}\right)\left(\dfrac{x}{y}\right)\left(\dfrac{x}{y}\right)\left(\dfrac{x}{y}\right)$

d) $(x^2)(x^2)(x^2)(x^2)(x^2)$

---

a) $3t$

b) $\dfrac{y}{7}$

c) 5^4

---

a) +1

b) +1

c) +1

342   Exponents and Radicals

221. The definition of a power with a negative whole-number exponent is given below.

$$\Box^{-n} = \frac{1}{\Box^n}$$

That is: $3^{-5} = \frac{1}{3^5}$ and $x^{-3} = \frac{1}{x^3}$

The definition above also applies when the base is a letter-term, a fraction, or a power. Therefore:

(a) $(2y)^{-4} =$ _____   (b) $\left(\frac{m}{2}\right)^{-3} =$ _____   (c) $(y^5)^{-2} =$ _____

222. A fraction raised to a negative power can be converted to a fraction raised to a positive power by means of the following principle:

$$\left(\frac{a}{b}\right)^{-n} = \left(\frac{b}{a}\right)^n$$

Therefore: $\left(\frac{y}{3}\right)^{-2} = \left(\frac{3}{y}\right)^2$ and $\left(\frac{m}{t}\right)^{-4} =$ _____

a) $\dfrac{1}{(2y)^4}$

b) $\dfrac{1}{\left(\frac{m}{2}\right)^3}$

c) $\dfrac{1}{(y^5)^2}$

223. The law of exponents below can be used to multiply two (or more) powers with the same base.

$$a^m \cdot a^n = a^{m+n}$$

Using the law above, write each product below:

(a) $3^7 \cdot 3^5 =$ _____   (c) $y^{-3} \cdot y^4 =$ _____
(b) $x^4 \cdot x^{-5} \cdot x^{-3} =$ _____   (d) $d^2 \cdot d^{-5} \cdot d^3 =$ _____

$\left(\dfrac{t}{m}\right)^4$

224. The law of exponents below can be used to divide two powers with the same base.

$$\frac{a^m}{a^n} = a^{m-n} = a^{m+(-n)}$$

Using the law above, write each quotient below:

(a) $\dfrac{x^2}{x^5} =$ _____   (b) $\dfrac{y^{-1}}{y^4} =$ _____   (c) $\dfrac{t^{-3}}{t^{-7}} =$ _____

a) 3^{12}

b) x^{-4} or $\dfrac{1}{x^4}$

c) y^1 or y

d) d^0 or 1

225. Using the laws of exponents for multiplication and division, write a single power which is equivalent to each fraction below:

(a) $\dfrac{x^4 \cdot x^2}{x^7} =$ _____   (b) $\dfrac{a^{-3}}{a^5 \cdot a^{-1}} =$ _____   (c) $\dfrac{y^{-2} y^3}{y^2 \cdot y^{-7}} =$ _____

a) x^{-3}

b) y^{-5}

c) t^4

a) x^{-1} b) a^{-7} c) y^6

226. We have seen that any power can be moved from the numerator of a fraction to its denominator (or vice versa) by simply changing the exponent to its opposite. This moving of powers is ordinarily done in order to obtain an expression in which each power has a positive exponent.

Write the right side of each equation below as a fraction which contains only positive powers:

(a) If $y = \dfrac{x^{-2}}{b^{-2}}$, (b) If $m = \dfrac{t^{-4}}{a^3}$, (c) If $t = \dfrac{a^2 b^{-3}}{c^{-4}}$,

y = m = t =

---

227. When moving a power from the denominator of a fraction to the numerator in order to eliminate a negative exponent, we sometimes obtain a multiplication instead of a fraction. For example:

$$\dfrac{a^2}{b^{-3}} = a^2 b^3 \qquad \dfrac{c^4}{d^{-2} f^{-3}} = c^4 d^2 f^3$$

a) $y = \dfrac{b^2}{x^2}$

b) $m = \dfrac{1}{a^3 t^4}$

c) $t = \dfrac{a^2 c^4}{b^3}$

Write the right side of each equation below as an expression which contains only positive powers:

(a) If $p = \dfrac{d^4}{q^{-2}}$, (b) If $t = \dfrac{x^4}{b^{-1} y^{-5}}$, (c) If $y = \dfrac{a^{-3}}{b^{-2} x^{-4}}$,

p = t = y =

---

228. If we think of multiplications as the numerators of fractions whose denominators are "1", we can use the same procedure to eliminate negative powers in multiplications. For example:

$$x^{-3} y^4 = \dfrac{x^{-3} y^4}{1} = \dfrac{y^4}{x^3} \qquad p^{-2} y^{-5} = \dfrac{p^{-2} y^{-5}}{1} = \dfrac{1}{p^2 y^5}$$

a) $p = d^4 q^2$

b) $t = bx^4 y^5$

c) $y = \dfrac{b^2 x^4}{a^3}$

Write the right side of each equation below as an expression which contains only positive powers:

(a) If $y = a^3 x^{-2}$, (b) If $t = c^{-1} d^{-3}$, (c) If $R = \dfrac{F^{-1}}{S^{-2}}$,

y = t = R =

---

229. The law of exponents at the right can be used to raise any letter-term (or indicated multiplication) to a power. Therefore:

$$(ab)^n = a^n b^n$$

(a) $(3x)^2 = 3^\square \cdot x^\square$ (b) $(xy)^{-4} = x^\square \cdot y^\square$ (c) $(3 \times 4)^5 = 3^\square \times 4^\square$

a) $y = \dfrac{a^3}{x^2}$

b) $t = \dfrac{1}{cd^3}$

c) $R = \dfrac{S^2}{F}$

---

a) $3^2 \cdot x^2$ or $9x^2$

b) $x^{-4} \cdot y^{-4}$

c) $3^5 \times 4^5$

Exponents and Radicals 343

344 Exponents and Radicals

230. The law of exponents at the right can be used to raise any fraction to a power. Therefore:

$$\left(\frac{a}{b}\right)^n = \frac{a^n}{b^n}$$

(a) $\left(\frac{1}{2}\right)^3 =$ _____   (b) $\left(\frac{t}{s}\right)^5 =$ _____   (c) $\left(\frac{p}{q}\right)^{-2} =$ _____

---

231. The law of exponents below can be used to raise any power to a power:

$$(a^m)^n = a^{mn}$$

Therefore: (a) $(3^4)^2 = 3^{\Box}$ (b) $(x^{-3})^5 = x^{\Box}$ (c) $(y^2)^{-6} = y^{\Box}$

a) $\frac{1^3}{2^3}$ or $\frac{1}{8}$

b) $\frac{t^5}{s^5}$

c) $\frac{p^{-2}}{q^{-2}}$ or $\frac{q^2}{p^2}$

---

232. Use the laws of exponents for letter-terms and for raising a power to a power to complete these:

(a) $(2x^3)^4 = 2^{\Box} x^{\Box}$ (b) $(a^4 b^2)^{-3} = a^{\Box} b^{\Box}$

a) 3^{8}

b) x^{-15}

c) y^{-12}

---

233. Use the laws of exponents to reduce each fraction below to a lowest-term form with only positive exponents:

(a) $\frac{y^6}{a^2 y^3} =$ _____   (b) $\frac{x^3}{b^3 x^5} =$ _____   (c) $\frac{a^3 x^2}{x^4} =$ _____

a) $2^4 x^{12}$

b) $a^{-12} b^{-6}$

---

234. Use the laws of exponents to reduce each fraction below to a lowest-term form with only positive exponents:

(a) $\frac{x^5}{(xy)^2} =$ _____   (b) $\frac{(a^2 b)^3}{a^4 c^5} =$ _____

a) $\frac{y^3}{a^2}$

b) $\frac{1}{b^3 x^2}$

c) $\frac{a^3}{x^2}$

---

a) $\frac{x^3}{y^2}$

b) $\frac{a^2 b^3}{c^5}$

> **SELF-TEST 13** (Frames 218-234)
>
> Using the laws of exponents, do the following problems:
>
> 1. $\dfrac{x}{x^2 x^{-5}} =$ _____
> 2. $(4at^2)^3 =$ _____
> 3. $\left(\dfrac{p^2}{p^{-3}}\right)^0 =$ _____
> 4. $\dfrac{2rs^3}{r^{-2}s} =$ _____
>
> Write each of the following as an expression which contains only positive exponents:
>
> 5. $\dfrac{k^3 p^{-1}}{d^{-4} h} =$ _____
> 6. $\dfrac{t^2}{r^{-3}} =$ _____
> 7. $\dfrac{b^{-5}}{w} =$ _____
> 8. $\left(\dfrac{2t}{5p}\right)^{-3} =$ _____
>
> Use the laws of exponents to reduce each fraction to a lowest-term form with only positive exponents:
>
> 9. $\dfrac{x^2}{rx^4} =$ _____
> 10. $\dfrac{(vw)^{-1}}{kv} =$ _____
> 11. $\dfrac{(bc^2)^3}{(ac^3)^2} =$ _____

ANSWERS:
1. x^4
2. $64a^3 t^6$
3. 1
4. $2r^3 s^2$
5. $\dfrac{d^4 k^3}{hp}$
6. $r^3 t^2$
7. $\dfrac{1}{b^5 w}$
8. $\left(\dfrac{5p}{2t}\right)^3$
9. $\dfrac{1}{rx^2}$
10. $\dfrac{1}{kv^2 w}$
11. $\dfrac{b^3}{a^2}$

4-14 THE MEANING OF ROOTS AND RADICALS

In this section, we will define what is meant by roots of various orders. We will also show that radical notation is used to express the "root" concept.

235. The "square root" concept was discussed earlier. We saw this fact: to find the <u>square root</u> (or <u>second</u> root) of a number, we must find one of <u>two</u> equal factors whose product is that number. For example: Since $(3)(3) = 9$, the square root of 9 is 3. Since $(7)(7) = 49$, the square root of 49 is _____.	
236. The concept of "roots" can be extended beyond square roots. For example, to find the <u>cube root</u> (or <u>third</u> root) of a number, we must find one of <u>three</u> equal factors whose product is that number. That is: Since $(2)(2)(2) = 8$, the <u>cube</u> root of 8 is 2. (a) Since $(3)(3)(3) = 27$, the <u>cube</u> root of 27 is _____. (b) Since $(5)(5)(5) = 125$, the <u>cube</u> root of 125 is _____.	7
	a) 3 b) 5

346 Exponents and Radicals

237. Though the <u>second</u> root of a number has the special name "<u>square</u>" root and the <u>third</u> root of a number has the special name "<u>cube</u>" root, roots beyond the third root do not have special names. They are simply called <u>fourth</u> root, <u>fifth</u> root, <u>sixth</u> root, and so on.

To find the <u>fourth</u> root of a number, we must find one of <u>four</u> equal factors whose product is that number. That is:

Since (2)(2)(2)(2) = 16, the <u>fourth</u> root of 16 is 2.

(a) Since (3)(3)(3)(3) = 81, the <u>fourth</u> root of 81 is _____.

(b) Since (5)(5)(5)(5) = 625, the <u>fourth</u> root of 625 is _____.

---

238. (a) To find the <u>fifth</u> root of a number, we must find one of <u>five</u> equal factors whose product is that number.

Since (3)(3)(3)(3)(3) = 243, the <u>fifth</u> root of 243 is _____.

(b) To find the <u>seventh</u> root of a number, we must find one of <u>seven</u> equal factors whose product is that number.

Since (2)(2)(2)(2)(2)(2)(2) = 128, the <u>seventh</u> root of 128 is _____.

a) 3
b) 5

---

239. (a) Since (4)(4)(4) = 64, the _____ root of 64 is 4.

(b) Since (2)(2)(2)(2)(2) = 32, the _____ root of 32 is 2.

(c) Since (3)(3)(3)(3)(3)(3)(3)(3) = 6,561, the _____ root of 6,561 is 3.

a) 3
b) 2

---

240. We have seen that the square root operation can be expressed in radical form. That is:

"Find the square root of 64" is written $\sqrt{64}$.

Similarly, the cube root operation can be expressed in radical form. That is:

"Find the cube root of 27" is written $\sqrt[3]{27}$.

There is some terminology about this radical form which you should know.

In $\sqrt[3]{27}$: (1) The entire expression $\sqrt[3]{27}$ is called a <u>cube root radical</u>.

(2) The small "3" is called the <u>index</u>.

(3) The symbol $\sqrt{}$ is called the <u>radical sign</u>.

(4) The number "27" is called the <u>radicand</u>.

(a) In $\sqrt[3]{81}$, the "3" is called the _____.

(b) Write the cube root radical whose radicand is 64. _____

a) cube
b) fifth
c) eighth

---

a) index
b) $\sqrt[3]{64}$

241. The same type of radical form is used to express roots higher than cube roots. The exact root desired is indicated by the <u>index</u> of the radical. That is:

$\sqrt[4]{81}$ means: find the <u>fourth</u> root of 81.

$\sqrt[5]{32}$ means: find the <u>fifth</u> root of 32.

$\sqrt[7]{1,089}$ means: find the _____ root of 1,089.

---

242. Though the index of a square root is actually "2", the "2" is not usually written explicitly. That is:

Instead of $\sqrt[2]{64}$, we simply write $\sqrt{64}$.

Write each of the following in radical form:

(a) The cube root of 17. _____

(b) The ninth root of 45. _____

(c) The square root of 37. _____

| | seventh |

---

243. $\sqrt{64}$ means: find one of <u>two</u> equal factors whose product is 64.

(a) $\sqrt[3]{81}$ means: find one of _____ equal factors whose product is 81.

(b) $\sqrt[7]{119}$ means: find one of _____ equal factors whose product is 119.

| | a) $\sqrt[3]{17}$
 b) $\sqrt[9]{45}$
 c) $\sqrt{37}$ (not $\sqrt[2]{37}$) |

---

244. (a) Since $(4)(4)(4) = 64$, $\sqrt[3]{64} = $ _____

(b) Since $(3)(3)(3)(3) = 81$, $\sqrt[4]{81} = $ _____

(c) Since $(2)(2)(2)(2)(2)(2) = 64$, $\sqrt[6]{64} = $ _____

| | a) three <br> b) seven |

---

245. Write the proper radical in the blank:

(a) Since $(7)(7)(7) = 343$, _____ = 7.

(b) Since $(2)(2)(2)(2)(2) = 32$, _____ = 2.

(c) Since $(10)(10)(10)(10) = 10,000$, _____ = 10.

| | a) 4 <br> b) 3 <br> c) 2 |

---

246. No matter how many times we multiply "1" by itself, the product is always "1". That is:

$(1)(1) = 1$ $(1)(1)(1) = 1$ $(1)(1)(1)(1) = 1$

Therefore: (a) $\sqrt{1} = $ _____ (b) $\sqrt[4]{1} = $ _____ (c) $\sqrt[6]{1} = $ _____

| | a) $\sqrt[3]{343}$
 b) $\sqrt[5]{32}$
 c) $\sqrt[4]{10,000}$ |

---

247. No matter how many times we multiply "0" by itself, the product is "0". That is:

$(0)(0) = 0$ $(0)(0)(0) = 0$ $(0)(0)(0)(0) = 0$

Therefore: (a) $\sqrt[3]{0} = $ _____ (b) $\sqrt[5]{0} = $ _____ (c) $\sqrt[7]{0} = $ _____

| | The answer is "1" for all. |

| | The answer is "0" for all. |

348 Exponents and Radicals

248. If a particular root of a number is a <u>small</u> whole number, we can sometimes find the root mentally. For example:

$\sqrt[3]{64} = 4 \qquad \sqrt[4]{81} = 3 \qquad \sqrt[5]{32} = 2$

However, most roots of numbers cannot be found mentally. Other techniques are needed. The major techniques are:

(1) Long arithmetic methods or the slide rule to find square roots and cube roots.
(2) Logarithms to find all roots.

Each of these roots can be completed mentally. Do so.

(a) $\sqrt[3]{125} = $ _____   (b) $\sqrt[4]{16} = $ _____   (c) $\sqrt[5]{243} = $ _____

249. We can also express the roots of <u>letters</u> in radical form. For example:

$\sqrt[3]{x}$ means: find the cube root of "x".

Of course, a radical like $\sqrt[3]{x}$ cannot be evaluated until we plug in some number for "x". That is:

If $x = 8$, $\sqrt[3]{x} = \sqrt[3]{8} = 2$.

(a) If $x = 27$, $\sqrt[3]{x} = \sqrt[3]{27} = $ _____.
(b) If $x = 64$, $\sqrt[3]{x} = \sqrt[3]{64} = $ _____.

a) 5
b) 2
c) 3

250. Write each of the following in radical form:

(a) The cube root of "y" = _____.
(b) The fourth root of "t" = _____.
(c) The seventh root of "m" = _____.

a) 3
b) 4

Answer to Frame 250:   a) $\sqrt[3]{y}$ b) $\sqrt[4]{t}$ c) $\sqrt[7]{m}$

SELF-TEST 14 (Frames 235-250)

1. $\sqrt[5]{218}$ means: find one of _____ equal factors whose product is _____.

Write each of the following in radical form:

2. The sixth root of 73. _____
3. The cube root of 497. _____
4. The second root of "N". _____
5. The ninth root of "x". _____

Find the numerical value of each of the following:

6. $\sqrt[3]{1,000} = $ _____   7. $\sqrt[7]{1} = $ _____   8. $\sqrt[9]{0} = $ _____   9. $\sqrt[6]{64} = $ _____

In $\sqrt[4]{159}$: 10. The "index" is _____.   11. The "radicand" is _____.

ANSWERS: 1. five ... 218   2. $\sqrt[6]{73}$ 4. $\sqrt{N}$ 6. 10   8. 0   10. 4
                                3. $\sqrt[3]{497}$ 5. $\sqrt[9]{x}$ 7. 1   9. 2   11. 159

4-15 THE ROOTS OF MULTIPLICATIONS AND DIVISIONS

In this section, we will briefly discuss the roots of multiplications and the roots of divisions (or fractions). The laws of radicals for multiplication and division will be emphasized.

251. In earlier work, we saw that two square root radicals can be multiplied by simply multiplying their radicands. For example:

$$\sqrt{9} \cdot \sqrt{4} = \sqrt{(9)(4)} = \sqrt{36}$$

Similarly, two cube root radicals can be multiplied by simply multiplying their radicands. That is:

$$\sqrt[3]{8} \cdot \sqrt[3]{27} = \sqrt[3]{(8)(27)} = \sqrt[3]{216}$$

We can see that the multiplication above makes sense by converting the radicals to regular numbers:

$$\sqrt[3]{8} \cdot \sqrt[3]{27} = \sqrt[3]{216}$$
$$\downarrow \quad \downarrow \quad \downarrow$$
$$2 \cdot 3 = 6$$

The multiplication below has been performed by multiplying the radicands. Show that it makes sense by converting the radicals to regular numbers.

$$\sqrt[3]{8} \cdot \sqrt[3]{64} = \sqrt[3]{512}$$
$$\downarrow \quad \downarrow \quad \downarrow$$
$$\underline{} \cdot \underline{} = \underline{}$$

252. We can also multiply two fourth root radicals by multiplying their radicands. For example:

$$\sqrt[4]{16} \cdot \sqrt[4]{81} = \sqrt[4]{(16)(81)} = \sqrt[4]{1,296}$$

Show that the multiplication above makes sense by converting the radicals to regular numbers:

$$\sqrt[4]{16} \cdot \sqrt[4]{81} = \sqrt[4]{1,296}$$
$$\downarrow \quad \downarrow \quad \downarrow$$
$$\underline{} \cdot \underline{} = \underline{}$$

$2 \cdot 4 = 8$

253. In general, we can multiply any two radicals <u>of the same order</u> by simply multiplying their radicands. This "<u>law of radicals for multiplication</u>" is expressed in general symbols below.

$$\boxed{\sqrt[n]{a} \cdot \sqrt[n]{b} = \sqrt[n]{ab}}$$

Using the law of radicals above, complete each of these:

(a) $\sqrt[3]{2} \cdot \sqrt[3]{8} = \sqrt[3]{}$ (c) $\sqrt[5]{7} \cdot \sqrt[5]{x} = \sqrt[5]{}$

(b) $\sqrt[4]{7} \cdot \sqrt[4]{5} = \sqrt[4]{}$ (d) $\sqrt[7]{m} \cdot \sqrt[7]{t} = \sqrt[7]{}$

$2 \cdot 3 = 6$

a) $\sqrt[3]{16}$ c) $\sqrt[5]{7x}$
b) $\sqrt[4]{35}$ d) $\sqrt[7]{mt}$

Exponents and Radicals   349

350　Exponents and Radicals

254. The law of radicals for multiplication applies only to radicals <u>of the</u> <u>same</u> <u>order</u>. That is, it cannot be used to perform either multiplication below.

$$\sqrt{5} \cdot \sqrt[3]{7} \qquad \sqrt[4]{x} \cdot \sqrt[6]{y}$$

If it applies, use the law to perform the multiplications below:

(a) $\sqrt[5]{m} \cdot \sqrt[5]{d} =$ _____   (c) $\sqrt[6]{R} \cdot \sqrt{V} =$ _____
(b) $\sqrt[3]{8} \cdot \sqrt[4]{y} =$ _____   (d) $\sqrt[8]{a} \cdot \sqrt[8]{b} =$ _____

---

255. When the law of radicals for multiplication is used, two radicals <u>of the</u> <u>same</u> <u>order</u> are combined into one radical <u>of the</u> <u>same</u> <u>order</u>. For example:

$$\sqrt[3]{x} \cdot \sqrt[3]{y} = \sqrt[3]{xy} \qquad \sqrt[5]{c} \cdot \sqrt[5]{d} = \sqrt[5]{cd}$$

Which of the following statements are true? _____

(a) $\sqrt[4]{F} \cdot \sqrt[4]{P} = \sqrt{FP}$ (c) $\sqrt[3]{p} \cdot \sqrt[4]{q} = \sqrt[3]{pq}$
(b) $\sqrt[6]{S} \cdot \sqrt[6]{Y} = \sqrt[6]{SY}$ (d) $\sqrt[8]{m} \cdot \sqrt[8]{x} = \sqrt[16]{mx}$

a) $\sqrt[5]{dm}$
b) Does not apply.
c) Does not apply.
d) $\sqrt[8]{ab}$

---

256. The law of radicals for multiplication can also be used to combine more than two radicals <u>of the</u> <u>same</u> <u>order</u>. For example:

$$\sqrt[3]{4} \cdot \sqrt[3]{x} \cdot \sqrt[3]{y} = \sqrt[3]{4xy}$$

Use the law to combine each of the following into a single radical:

(a) $\sqrt[4]{a} \cdot \sqrt[4]{b} \cdot \sqrt[4]{c} =$ _____   (b) $\sqrt[7]{10} \cdot \sqrt[7]{3} \cdot \sqrt[7]{2} =$ _____

Only (b)

---

257. If we reverse the law of radicals for multiplication, we get:

$$\boxed{\sqrt[n]{ab} = \sqrt[n]{a} \cdot \sqrt[n]{b}}$$

This reversed form of the law can be used to factor a "root of a multiplication". That is:

$$\sqrt[3]{4x} = \sqrt[3]{4} \cdot \sqrt[3]{x}$$

Notice that the cube root of the multiplication "4x" has been factored into a multiplication of the cube roots of each of its factors, "4" and "x".

Using this reversed form of the law, factor each of the roots below:

(a) $\sqrt[4]{7t} =$ _____   (b) $\sqrt[6]{2cd} =$ _____

a) $\sqrt[4]{abc}$ b) $\sqrt[7]{60}$

---

a) $\sqrt[4]{7} \cdot \sqrt[4]{t}$
b) $\sqrt[6]{2} \cdot \sqrt[6]{c} \cdot \sqrt[6]{d}$

258. We have seen that the square root of a fraction (or division) is equal to the square root of its numerator divided by the square root of its denominator. For example:

$$\sqrt{\frac{36}{9}} = \frac{\sqrt{36}}{\sqrt{9}}$$

Similarly, the cube root of a fraction (or division) is equal to the cube root of its numerator divided by the cube root of its denominator. That is:

$$\sqrt[3]{\frac{216}{8}} = \frac{\sqrt[3]{216}}{\sqrt[3]{8}}$$

We can see that this principle makes sense by converting the radicals above to regular numbers.
Since $\sqrt[3]{216} = 6$, $\sqrt[3]{8} = 2$, and $\sqrt[3]{27} = 3$:

(a) $\sqrt[3]{\frac{216}{8}} = \sqrt[3]{27} =$ _____   (b) $\frac{\sqrt[3]{216}}{\sqrt[3]{8}} = \frac{6}{2} =$ _____

259. The fourth root of a fraction (or division) is also equal to the fourth root of its numerator divided by the fourth root of its denominator. That is:

$$\sqrt[4]{\frac{1,296}{16}} = \frac{\sqrt[4]{1,296}}{\sqrt[4]{16}}$$

The sensibleness of the principle above can also be seen by converting each radical to a regular number.
Since $\sqrt[4]{1,296} = 6$, $\sqrt[4]{16} = 2$, and $\sqrt[4]{81} = 3$:

(a) $\sqrt[4]{\frac{1,296}{16}} = \sqrt[4]{81} =$ _____   (b) $\frac{\sqrt[4]{1,296}}{\sqrt[4]{16}} = \frac{6}{2} =$ _____

a) 3   b) 3

260. In general, any root of a fraction is equal to that root of its numerator divided by that root of its denominator. This "law of radicals for a fraction (or division)" is expressed in general symbols below:

$$\boxed{\sqrt[n]{\frac{a}{b}} = \frac{\sqrt[n]{a}}{\sqrt[n]{b}}}$$

Using the law of radicals above, complete each of these:

(a) $\sqrt[3]{\frac{5}{x}} =$ _____   (b) $\sqrt[4]{\frac{2m}{y}} =$ _____   (c) $\sqrt[7]{\frac{ab}{cd}} =$ _____

a) 3   b) 3

a) $\dfrac{\sqrt[3]{5}}{\sqrt[3]{x}}$

b) $\dfrac{\sqrt[4]{2m}}{\sqrt[4]{y}}$

c) $\dfrac{\sqrt[7]{ab}}{\sqrt[7]{cd}}$

352   Exponents and Radicals

261. When a root of a fraction is converted into a division of two radicals, the two new radicals must have the same index (or order) as the original radical. For example, the statements below are false.

$$\sqrt[3]{\frac{p}{q}} = \frac{\sqrt[3]{p}}{\sqrt{q}} \qquad \sqrt[5]{\frac{3m}{t}} = \frac{\sqrt[5]{3m}}{\sqrt[3]{t}}$$

Which statements below are false? _____

(a) $\sqrt[4]{\frac{x}{y}} = \frac{\sqrt[4]{x}}{\sqrt[4]{y}}$ (b) $\sqrt[6]{\frac{cd}{t}} = \frac{\sqrt{cd}}{\sqrt[6]{t}}$ (c) $\sqrt[9]{\frac{F_1}{F_2}} = \frac{\sqrt[3]{F_1}}{\sqrt[3]{F_2}}$

Both (b) and (c)

---

262. If we reverse the law of radicals for a fraction (or division), we get:

$$\boxed{\frac{\sqrt[n]{a}}{\sqrt[n]{b}} = \sqrt[n]{\frac{a}{b}}}$$

This reversed form of the law says that a division of two radicals of the same order can be converted into a single radical of the same order. For example:

$$\frac{\sqrt[3]{7}}{\sqrt[3]{2}} \text{ can be converted into } \sqrt[3]{\frac{7}{2}}$$

Using this reversed form of the law, convert each of the following divisions to a single radical.

(a) $\frac{\sqrt[3]{t}}{\sqrt[3]{11}} =$ (b) $\frac{\sqrt[4]{R}}{\sqrt[4]{P}} =$ (c) $\frac{\sqrt[7]{3d}}{\sqrt[7]{2m}} =$

---

263. Of course, a division of two radicals can be converted into a single radical by means of this law <u>only if the two radicals have the same order</u>.

If it applies, use the law to convert each division below into a single radical:

(a) $\frac{\sqrt[3]{x}}{\sqrt{y}} =$ (b) $\frac{\sqrt[5]{37}}{\sqrt[5]{a}} =$ (c) $\frac{\sqrt[4]{5p}}{\sqrt[6]{4d}} =$

a) $\sqrt[3]{\frac{t}{11}}$

b) $\sqrt[4]{\frac{R}{P}}$

c) $\sqrt[7]{\frac{3d}{2m}}$

---

264. Using the laws of radicals, convert each of the following into a single radical if possible:

(a) $\sqrt{x} \cdot \sqrt[4]{y} =$ (c) $\sqrt[3]{a} \cdot \sqrt[3]{bt} =$

(b) $\frac{\sqrt[5]{pq}}{\sqrt[5]{r}} =$ (d) $\frac{\sqrt[7]{F_1}}{\sqrt{F_2}} =$

a) Does not apply.

b) $\sqrt[5]{\frac{37}{a}}$

c) Does not apply.

---

a) Not possible.

b) $\sqrt[5]{\frac{pq}{r}}$

c) $\sqrt[3]{abt}$

d) Not possible.

Exponents and Radicals 353

SELF-TEST 15 (Frames 251-264)

Write each of the following as a single radical:

1. $\sqrt[4]{P} \cdot \sqrt[4]{5R} = $ _____
2. $\dfrac{\sqrt[8]{6t}}{\sqrt[8]{12}} = $ _____
3. $\sqrt[3]{5} \cdot \sqrt[3]{d} \cdot \sqrt[3]{2b} = $ _____
4. $\dfrac{\sqrt[5]{w} \cdot \sqrt[5]{8}}{\sqrt[5]{3r}} = $ _____

Write each of the following as two or more radicals:

5. $\sqrt{hk} = $ _____
6. $\sqrt[3]{\dfrac{a}{7}} = $ _____
7. $\sqrt[6]{\dfrac{y}{3x}} = $ _____
8. $\sqrt[7]{2rt} = $ _____

9. Which of the following are *false*? _____

(a) $\sqrt[3]{n} \cdot \sqrt{m} = \sqrt[6]{mn}$ (b) $\sqrt[6]{\dfrac{a}{9}} = \dfrac{\sqrt{a}}{3}$ (c) $\dfrac{\sqrt[6]{b}}{\sqrt[3]{c}} = \sqrt[6]{\dfrac{b}{c}}$ (d) $\sqrt[7]{2t} = \sqrt[7]{2} \cdot \sqrt[7]{t}$ (e) $\dfrac{\sqrt[5]{2x}}{\sqrt[5]{x}} = \sqrt[5]{2}$

ANSWERS:
1. $\sqrt[4]{5PR}$
2. $\sqrt[8]{\dfrac{t}{2}}$
3. $\sqrt[3]{10bd}$
4. $\sqrt[5]{\dfrac{8w}{3r}}$
5. $\sqrt{h} \cdot \sqrt{k}$
6. $\dfrac{\sqrt[3]{a}}{\sqrt[3]{7}}$
7. $\dfrac{\sqrt[6]{y}}{\sqrt[6]{3} \cdot \sqrt[6]{x}}$
8. $\sqrt[7]{2} \cdot \sqrt[7]{r} \cdot \sqrt[7]{t}$
9. (a) and (c)

4-16 THE ROOTS OF POWERS

As we shall see in this section, the definition of roots and radical notation can be extended to include the roots of powers. The law of radicals for powers is introduced. The meaning of "perfect powers" of a particular root is discussed.

265. The definition of roots can be extended to include the roots of powers. For example:

The *square* root of 2^6 is one of *two* equal factors whose product is 2^6.

Since $(2^3)(2^3) = 2^6$, the *square* root of 2^6 is 2^3.

The *cube* root of y^{12} is one of *three* equal factors whose product is y^{12}.

Since $(y^4)(y^4)(y^4) = y^{12}$, the *cube* root of y^{12} is y^4.

The *fourth* root of t^{20} is one of *four* equal factors whose product is t^{20}.

Since $(t^5)(t^5)(t^5)(t^5) = t^{20}$, the *fourth* root of t^{20} is _____ .

266. (a) Since $(5^2)(5^2) = 5^4$, the *square* root of 5^4 is _____ . (b) Since $(7^2)(7^2)(7^2) = 7^6$, the *cube* root of 7^6 is _____ . (c) Since $(x^3)(x^3)(x^3)(x^3)(x^3) = x^{15}$, the *fifth* root of x^{15} is _____ .	t^5
	a) 5^2 b) 7^2 c) x^3

354   Exponents and Radicals

267. Write the correct word in each blank:
   (a) Since $(y^8)(y^8) = y^{16}$, y^8 is the _____ root of y^{16}.
   (b) Since $(2^6)(2^6)(2^6)(2^6) = 2^{24}$, 2^6 is the _____ root of 2^{24}.
   (c) Since $(t^2)(t^2)(t^2)(t^2)(t^2)(t^2) = t^{12}$, t^2 is the _____ root of t^{12}.

268. The roots of powers can also be written in radical notation. That is:
   $\sqrt{2^4}$ means the square root of 2^4.
   (a) $\sqrt[3]{x^6}$ means the _____ root of x^6.
   (b) $\sqrt[5]{y^{15}}$ means the _____ root of y^{15}.

   a) square
   b) fourth
   c) sixth

269. Write the correct radical in each blank:
   Since $(x^4)(x^4) = x^8$, $\sqrt{x^8} = x^4$.
   (a) Since $(m^3)(m^3)(m^3) = m^9$, _____ $= m^3$.
   (b) Since $(3^5)(3^5)(3^5)(3^5)(3^5) = 3^{25}$, _____ $= 3^5$.

   a) cube
   b) fifth

270. The concept of roots also applies to powers with negative exponents. For example:
   Since $(x^{-3})(x^{-3}) = x^{-6}$, $\sqrt{x^{-6}} = x^{-3}$.
   (a) Since $(y^{-4})(y^{-4})(y^{-4}) = y^{-12}$, $\sqrt[3]{y^{-12}} =$ _____.
   (b) Since $(R^{-2})(R^{-2})(R^{-2})(R^{-2}) = R^{-8}$, $\sqrt[4]{R^{-8}} =$ _____.

   a) $\sqrt[3]{m^9}$
   b) $\sqrt[5]{3^{25}}$

271. To find the square root of a power, we can simply divide its exponent by 2. That is:
   $\sqrt{y^8} = y^{\frac{8}{2}}$ or y^4, since $(y^4)(y^4) = y^8$
   $\sqrt{t^{-10}} = t^{\frac{-10}{2}}$ or t^{-5}, since $(t^{-5})(t^{-5}) = t^{-10}$

   Using the fact above, write each square root below in power form:
   (a) $\sqrt{3^4} =$ _____   (b) $\sqrt{7^{-6}} =$ _____   (c) $\sqrt{x^{20}} =$ _____

   a) y^{-4}
   b) R^{-2}

272. To find the cube root of a power, we can simply divide its exponent by 3. That is:
   $\sqrt[3]{8^6} = 8^{\frac{6}{3}}$ or 8^2, since $(8^2)(8^2)(8^2) = 8^6$
   $\sqrt[3]{m^{-12}} = m^{\frac{-12}{3}}$ or m^{-4}, since $(m^{-4})(m^{-4})(m^{-4}) = m^{-12}$

   Using the fact above, write each cube root below in power form:
   (a) $\sqrt[3]{5^9} =$ _____   (b) $\sqrt[3]{x^{-15}} =$ _____   (c) $\sqrt[3]{y^{-24}} =$ _____

   a) 3^2
   b) 7^{-3}
   c) x^{10}

   a) 5^3
   b) x^{-5}
   c) y^{-8}

273. To find the <u>fourth</u> root of a power, we can simply divide its exponent by 4. That is:

$\sqrt[4]{2^{20}} = 2^{\frac{20}{4}}$ or 2^5, since $(2^5)(2^5)(2^5)(2^5) = 2^{20}$

$\sqrt[4]{x^{-8}} = x^{\frac{-8}{4}}$ or x^{-2}, since $(x^{-2})(x^{-2})(x^{-2})(x^{-2}) = x^{-8}$

Using the fact above, write each fourth root below in power form:
(a) $\sqrt[4]{7^{12}} =$ _____ (b) $\sqrt[4]{R^{-16}} =$ _____ (c) $\sqrt[4]{y^{28}} =$ _____

---

274. The general law of radicals for finding roots of powers is given in symbols below:

$$\boxed{\sqrt[n]{a^m} = a^{\frac{m}{n}}}$$

a) 7^3
b) R^{-4}
c) y^7

That is, to find any root of a power, <u>we simply divide the exponent of the power by the desired root</u>. (That is, we divide by the index.)

Using the law above, complete each of these:
(a) $\sqrt[5]{3^{10}} = 3^{\frac{10}{5}}$ or ____ (b) $\sqrt[6]{y^{-18}} = y^{\frac{-18}{6}}$ or ____ (c) $\sqrt[7]{m^{-14}} = m^{\frac{-14}{7}}$ or ____

---

275. Use the general law to find each root below:
(a) $\sqrt{t^{-14}} =$ ____ (b) $\sqrt[3]{R^{15}} =$ ____ (c) $\sqrt[4]{m^{24}} =$ ____ (d) $\sqrt[7]{h^{-35}} =$ ____

a) 3^2
b) y^{-3}
c) m^{-2}

---

276. Use the general law to find each root below:
(a) $\sqrt[6]{p^{36}} =$ ____ (b) $\sqrt[9]{F^{-27}} =$ ____ (c) $\sqrt[3]{y^{-33}} =$ ____ (d) $\sqrt{x^4} =$ ____

a) t^{-7} c) m^6
b) R^5 d) h^{-5}

---

277. The square root of any power is found by dividing its exponent by 2.

If the exponent of the power is a multiple of 2, the exponent of its square root is a <u>whole number</u>. For example:

$\sqrt{x^{10}} = x^5$ $\sqrt{y^{-8}} = y^{-4}$

If the exponent of the power is not a multiple of 2, the exponent of its square root is a <u>fraction</u>. For example:

$\sqrt{t^7} = t^{\frac{7}{2}}$ $\sqrt{p^{-5}} = p^{\frac{-5}{2}}$

a) p^6 c) y^{-11}
b) F^{-3} d) x^2

In which cases below is the square root a power with a whole-number exponent? _____
(a) $\sqrt{R^{12}}$ (b) $\sqrt{q^{-7}}$ (c) $\sqrt{h^{15}}$ (d) $\sqrt{F^{-6}}$

Only in (a) and (d).

356 Exponents and Radicals

278. If the square root of a power has a whole-number exponent, the power is called a perfect square. For example:

y^8 is a perfect square, since $\sqrt{y^8} = y^4$

y^{-5} is not a perfect square, since $\sqrt{y^{-5}} = y^{\frac{-5}{2}}$

Which of the following powers are perfect squares? _____

(a) x^{10} (b) t^7 (c) d^{-3} (d) F^{-14}

Only (a) and (d).

279. The cube root of any power is found by dividing its exponent by 3.

If the exponent of the power is a multiple of 3, the exponent of its cube root is a whole number. For example:

$\sqrt[3]{7^9} = 7^3$ $\sqrt[3]{b^{-18}} = b^{-6}$

If the exponent of the power is not a multiple of 3, the exponent of its cube root is a fraction. For example:

$\sqrt[3]{m^8} = m^{\frac{8}{3}}$ $\sqrt[3]{2^{-11}} = 2^{\frac{-11}{3}}$

If the cube root of a power has a whole-number exponent, the power is called a perfect cube.

Which of the powers below are perfect cubes? _____

(a) 10^{12} (b) 6^{-10} (c) t^{-6} (d) r^4

Only (a) and (c).

280. If the exponent of a power is a multiple of 4, its fourth root has a whole-number exponent. For example:

$\sqrt[4]{x^{12}} = x^3$ $\sqrt[4]{y^{-20}} = y^{-5}$

Therefore, powers like x^{12} and y^{-20} are called perfect fourths.

Which of the following powers are perfect fourths? _____

(a) m^{13} (b) t^{-8} (c) R^{16} (d) h^{-5}

Only (b) and (c).

281. Any power is called a "perfect" power of some particular root if that root of the power has a whole-number exponent. For example:

x^{15} is called a perfect fifth, since $\sqrt[5]{x^{15}} = x^3$

y^{-24} is called a perfect sixth, since $\sqrt[6]{y^{-24}} = y^{-4}$

t^{40} is called a perfect eighth, since $\sqrt[8]{t^{40}} = t^5$

(a) In order to be a perfect fifth, the exponent of a power must be a multiple of _____.

(b) In order to be a perfect eighth, the exponent of a power must be a multiple of _____.

a) 5
b) 8

Exponents and Radicals 357

282. (a) Is t^{-12} a perfect sixth? _____
(b) Is m^{11} a perfect fifth? _____
(c) Is R^{18} a perfect ninth? _____

---

283. The root of a power can be a power whose exponent is "1". For example:

Since $(x)(x) = x^2$, $\sqrt{x^2} = x^1$ or x

Since $(y)(y)(y) = y^3$, $\sqrt[3]{y^3} = y^1$ or y

Since $(a)(a)(a)(a)(a) = a^5$, $\sqrt[5]{a^5} = a^1$ or a

The examples above are consistent with the law of radicals for powers. That is:

$\sqrt{x^2} = x^{\frac{2}{2}} = x^1$ or x

$\sqrt[3]{y^3} = y^{\frac{3}{3}} = y^1$ or y

$\sqrt[5]{a^5} = a^{\frac{5}{5}} = a^1$ or a

It should be obvious that the exponent of a root is "1" whenever the index and the exponent of the original power are the same.

Therefore: (a) $\sqrt[4]{7^4} =$ _____ (b) $\sqrt[6]{R^6} =$ _____ (c) $\sqrt[10]{t^{10}} =$ _____

a) Yes, since $\sqrt[6]{t^{-12}} = t^{-2}$

b) No, since $\sqrt[5]{m^{11}} = m^{\frac{11}{5}}$

c) Yes, since $\sqrt[9]{R^{18}} = R^2$

---

284. The root of a power can be a power whose exponent is "-1". For example:

Since $(b^{-1})(b^{-1}) = b^{-2}$, $\sqrt{b^{-2}} = b^{-1}$

Since $(t^{-1})(t^{-1})(t^{-1})(t^{-1}) = t^{-4}$, $\sqrt[4]{t^{-4}} = t^{-1}$

The examples above are consistent with the law of radicals for powers. That is:

$\sqrt{b^{-2}} = b^{\frac{-2}{2}} = b^{-1}$ $\sqrt[4]{t^{-4}} = t^{\frac{-4}{4}} = t^{-1}$

It should be obvious that the exponent of a root is "-1" whenever the index and the exponent of the original power are a pair of opposites.

Therefore: (a) $\sqrt[3]{x^{-3}} =$ _____ (b) $\sqrt[5]{y^{-5}} =$ _____ (c) $\sqrt[7]{3^{-7}} =$ _____

a) 7^1 or 7
b) R^1 or R
c) t^1 or t

---

285. A power is also a "perfect power" of a particular root if the exponent of that root is either "1" or "-1". That is:

x^2 is a perfect square, since $\sqrt{x^2} = x^1$ or x

y^{-2} is a perfect square, since $\sqrt{y^{-2}} = y^{-1}$

(a) b^3 is a perfect _____, since $\sqrt[3]{b^3} = b^1$ or b

(b) d^{-4} is a perfect _____, since $\sqrt[4]{d^{-4}} = d^{-1}$

a) x^{-1}
b) y^{-1}
c) 3^{-1}

---

a) cube
b) fourth

358 Exponents and Radicals

286. Any root of a zero power is "1", since any zero power equals "1". That is:
$$\sqrt{t^0} = \sqrt{1} = 1$$
(a) $\sqrt[3]{p^0} = \sqrt[3]{1} = \underline{}$ (b) $\sqrt[7]{R^0} = \sqrt[7]{1} = \underline{}$

287. The law of radicals for the roots of powers can be generalized to powers whose bases are multiplications. For example:
$$\sqrt{(5x)^4} = (5x)^{\frac{4}{2}} = (5x)^2 = 5^2 \cdot x^2 \text{ (or } 25x^2)$$
$$\sqrt[3]{(ct)^9} = (ct)^{\frac{9}{3}} = (ct)^3 = c^3 t^3$$

Following the steps above, apply the law to each of these:
(a) $\sqrt{(2y)^6} = \underline{} = \underline{} = \underline{}$
(b) $\sqrt[3]{(ax)^6} = \underline{} = \underline{} = \underline{}$
(c) $\sqrt[4]{(bd)^{16}} = \underline{} = \underline{} = \underline{}$

a) 1 b) 1

288. The law of radicals for the roots of powers can also be generalized to powers whose bases are fractions. For example:
$$\sqrt{\left(\frac{x}{3}\right)^4} = \left(\frac{x}{3}\right)^{\frac{4}{2}} = \left(\frac{x}{3}\right)^2 = \frac{x^2}{3^2} \text{ or } \frac{x^2}{9}$$
$$\sqrt[3]{\left(\frac{p}{q}\right)^{12}} = \left(\frac{p}{q}\right)^{\frac{12}{3}} = \left(\frac{p}{q}\right)^4 = \frac{p^4}{q^4}$$

Following the steps above, apply the law to each of these:
(a) $\sqrt{\left(\frac{m}{t}\right)^8} = \underline{} = \underline{} = \underline{}$
(b) $\sqrt[3]{\left(\frac{2}{y}\right)^6} = \underline{} = \underline{} = \underline{}$
(c) $\sqrt[5]{\left(\frac{d}{h}\right)^{10}} = \underline{} = \underline{} = \underline{}$

a) $(2y)^{\frac{6}{2}} = (2y)^3 = 2^3 \cdot y^3$
   (or $8y^3$)
b) $(ax)^{\frac{6}{3}} = (ax)^2 = a^2 x^2$
c) $(bd)^{\frac{16}{4}} = (bd)^4 = b^4 d^4$

289. Apply the law to each of the following radicals:
(a) $\sqrt[4]{\left(\frac{a}{b}\right)^{12}} = \underline{}$ (c) $\sqrt[5]{\left(\frac{p}{h}\right)^{20}} = \underline{}$
(b) $\sqrt{(cd)^{10}} = \underline{}$ (d) $\sqrt[3]{(st)^{21}} = \underline{}$

a) $\left(\frac{m}{t}\right)^{\frac{8}{2}} = \left(\frac{m}{t}\right)^4 = \frac{m^4}{t^4}$
b) $\left(\frac{2}{y}\right)^{\frac{6}{3}} = \left(\frac{2}{y}\right)^2 = \frac{2^2}{y^2}$ or $\frac{4}{y^2}$
c) $\left(\frac{d}{h}\right)^{\frac{10}{5}} = \left(\frac{d}{h}\right)^2 = \frac{d^2}{h^2}$

a) $\left(\frac{a}{b}\right)^3$ or $\frac{a^3}{b^3}$
b) $(cd)^5$ or $c^5 d^5$
c) $\left(\frac{p}{h}\right)^4$ or $\frac{p^4}{h^4}$
d) $(st)^7$ or $s^7 t^7$

290. Apply the law to each of the following radicals:

(a) $\sqrt{\left(\frac{7}{m}\right)^2} =$ _____   (c) $\sqrt[3]{\left(\frac{p}{t}\right)^3} =$ _____

(b) $\sqrt{(3x)^2} =$ _____   (d) $\sqrt[4]{(RS)^4} =$ _____

Answer to Frame 290:   a) $\left(\frac{7}{m}\right)^1$ or $\frac{7}{m}$ b) $(3x)^1$ or $3x$ c) $\left(\frac{p}{t}\right)^1$ or $\frac{p}{t}$ d) $(RS)^1$ or RS

SELF-TEST 16 (Frames 265-290)

Find each of the following roots:

1. $\sqrt{r^{-8}} =$ _____   3. $\sqrt[3]{(3a)^0} =$ _____   5. $\sqrt[4]{\left(\frac{b}{w}\right)^{12}} =$ _____   7. $\sqrt[6]{\left(\frac{c}{d}\right)^{-6}} =$ _____

2. $\sqrt[4]{x^4} =$ _____   4. $\sqrt[5]{h^{-20}} =$ _____   6. $\sqrt[3]{(2r)^{15}} =$ _____   8. $\sqrt[n]{a^m} =$ _____

Given these powers: (a) t^{-6} (b) a^5 (c) p^{12} (d) s^{-8} (e) h^{-3} (f) r^{15}

9. Which of the above are "perfect cubes"? _____
10. Which of the above are "perfect fifths"? _____

ANSWERS: 1. r^{-4} 3. 1 5. $\left(\frac{b}{w}\right)^3$ or $\frac{b^3}{w^3}$ 7. $\left(\frac{c}{d}\right)^{-1}$ or $\frac{d}{c}$ 9. (a), (c), (e), (f)

2. x 4. h^{-4} 6. $(2r)^5$ or $32r^5$ 8. $a^{\frac{m}{n}}$ 10. (b), (f)

4-17 SIMPLIFYING RADICALS BY FACTORING OUT PERFECT POWERS

In this section, we will show how radicals can be simplified by factoring out perfect powers. This "factoring out" process will be used to simplify radicals whose radicands are single powers, multiplications of powers, or fractions.

360   Exponents and Radicals

291. Each of the square root radicals below has been simplified by factoring and replacing the perfect-square factors with their roots.

$$\sqrt{x^2 y^4} = \sqrt{x^2} \cdot \sqrt{y^4} = x \cdot y^2 \text{ or } xy^2$$
$$\sqrt{a^6 b^{10}} = \sqrt{a^6} \cdot \sqrt{b^{10}} = a^3 \cdot b^5 \text{ or } a^3 b^5$$

Using the procedure above, complete each of these:

(a) $\sqrt{4d^8} = \sqrt{4} \cdot \sqrt{d^8} =$ _____ (b) $\sqrt{25m^{12}p^2} = \sqrt{25} \cdot \sqrt{m^{12}} \cdot \sqrt{p^2} =$ _____

---

292. Each of the cube root radicals below has been simplified by factoring and then replacing the perfect-cube factors with their roots.

$$\sqrt[3]{8x^6} = \sqrt[3]{8} \cdot \sqrt[3]{x^6} = 2x^2$$
$$\sqrt[3]{27a^3 b^9} = \sqrt[3]{27} \cdot \sqrt[3]{a^3} \cdot \sqrt[3]{b^9} = 3ab^3$$

Using the procedure above, complete each of these:

(a) $\sqrt[3]{x^9 y^{12}} = \sqrt[3]{x^9} \cdot \sqrt[3]{y^{12}} =$ _____

(b) $\sqrt[3]{64p^6 q^3} = \sqrt[3]{64} \cdot \sqrt[3]{p^6} \cdot \sqrt[3]{q^3} =$ _____

a) $2d^4$ b) $5m^6 p$

---

293. The fourth root radical below has been simplified by factoring and replacing the perfect-fourth factors with their roots.

$$\sqrt[4]{16x^4 y^{20}} = \sqrt[4]{16} \cdot \sqrt[4]{x^4} \cdot \sqrt[4]{y^{20}} = 2xy^5$$

Using the procedure above, complete each of these:

(a) $\sqrt[4]{a^8 b^{12}} = \sqrt[4]{a^8} \cdot \sqrt[4]{b^{12}} =$ _____

(b) $\sqrt[4]{81c^4 d^{16}} = \sqrt[4]{81} \cdot \sqrt[4]{c^4} \cdot \sqrt[4]{d^{16}} =$ _____

a) $x^3 y^4$
b) $4p^2 q$

---

294. The procedure we have been using to simplify radicals is called "factoring out perfect powers".

(a) Factor out the perfect-fifths from this radical:
$$\sqrt[5]{x^{10} y^{25}} = \sqrt[5]{x^{10}} \cdot \sqrt[5]{y^{25}} =$$ _____

(b) Factor out the perfect-sevenths from this radical:
$$\sqrt[7]{a^7 b^{21}} = \sqrt[7]{a^7} \cdot \sqrt[7]{b^{21}} =$$ _____

a) $a^2 b^3$
b) $3cd^4$

---

295. We can also factor out perfect powers from radicals which contain negative powers. That is:

$$\sqrt{x^{-6} y^{-10}} = \sqrt{x^{-6}} \cdot \sqrt{y^{-10}} = x^{-3} y^{-5}$$

(a) $\sqrt[3]{a^3 b^{-9}} = \sqrt[3]{a^3} \cdot \sqrt[3]{b^{-9}} =$ _____

(b) $\sqrt[4]{16p^{-8} q^{16}} = \sqrt[4]{16} \cdot \sqrt[4]{p^{-8}} \cdot \sqrt[4]{q^{16}} =$ _____

(c) $\sqrt[5]{F^{-5} R^{-10}} = \sqrt[5]{F^{-5}} \cdot \sqrt[5]{R^{-10}} =$ _____

a) $x^2 y^5$
b) ab^3

---

a) ab^{-3}
b) $2p^{-2} q^4$
c) $F^{-1} R^{-2}$

296. Applying the law of radicals for the roots of powers to the radical below, we get:

$$\sqrt{(bd)^4} = (bd)^{\frac{4}{2}} = (bd)^2 = b^2d^2$$

By raising the letter-term to the fourth power first and then factoring out perfect squares, we can confirm the same fact. That is:

$$\sqrt{(bd)^4} = \sqrt{b^4d^4} = b^2d^2$$

(a) Apply the law of radicals for the roots of powers to this radical:

$$\sqrt[3]{(xy)^9} = \underline{\qquad} = \underline{\qquad} = \underline{\qquad}$$

(b) Now confirm this result by raising the letter-term to the ninth power first and then factoring out perfect cubes:

$$\sqrt[3]{(xy)^9} = \underline{\qquad} = \underline{\qquad}$$

---

297. Since the radicand in $\sqrt{y^7}$ is not a perfect square, we cannot factor out a perfect square from the radical as it stands. However, by factoring the radicand properly, we can factor out a perfect square. That is:

$$\sqrt{y^7} = \sqrt{y^6 \cdot y} = \sqrt{y^6} \cdot \sqrt{y} = y^3\sqrt{y}$$

Following the procedure above, complete each of these:

(a) $\sqrt{x^5} = \sqrt{x^4 \cdot x} = \sqrt{x^4} \cdot \sqrt{x} = \underline{\qquad}$

(b) $\sqrt{t^9} = \sqrt{t^8 \cdot t} = \sqrt{t^8} \cdot \sqrt{t} = \underline{\qquad}$

a) $(xy)^{\frac{9}{3}} = (xy)^3 = x^3y^3$

b) $\sqrt[3]{x^9y^9} = x^3y^3$

---

298. By factoring the radicand below properly, we can factor out a <u>perfect cube</u> from the radical. That is:

$$\sqrt[3]{a^5} = \sqrt[3]{a^3 \cdot a^2} = \sqrt[3]{a^3} \cdot \sqrt[3]{a^2} = a\sqrt[3]{a^2}$$

Following the procedure above, complete each of these:

(a) $\sqrt[3]{y^7} = \sqrt[3]{y^6 \cdot y} = \sqrt[3]{y^6} \cdot \sqrt[3]{y} = \underline{\qquad}$

(b) $\sqrt[3]{m^{11}} = \sqrt[3]{m^9 \cdot m^2} = \sqrt[3]{m^9} \cdot \sqrt[3]{m^2} = \underline{\qquad}$

a) $x^2\sqrt{x}$

b) $t^4\sqrt{t}$

---

299. By factoring the radicand below, we can factor out a <u>perfect fourth</u> from the radical. That is:

$$\sqrt[4]{x^7} = \sqrt[4]{x^4 \cdot x^3} = \sqrt[4]{x^4} \cdot \sqrt[4]{x^3} = x\sqrt[4]{x^3}$$

Following the procedure above, complete each of these:

(a) $\sqrt[4]{a^9} = \sqrt[4]{a^8 \cdot a} = \sqrt[4]{a^8} \cdot \sqrt[4]{a} = \underline{\qquad}$

(b) $\sqrt[4]{b^{14}} = \sqrt[4]{b^{12} \cdot b^2} = \sqrt[4]{b^{12}} \cdot \sqrt[4]{b^2} = \underline{\qquad}$

a) $y^2\sqrt[3]{y}$

b) $m^3\sqrt[3]{m^2}$

---

a) $a^2\sqrt[4]{a}$

b) $b^3\sqrt[4]{b^2}$

362 Exponents and Radicals

300. A perfect square can be factored out of $\sqrt{a^7}$ in three ways:

(1) $\sqrt{a^7} = \sqrt{a^6} \cdot \sqrt{a} = a^3\sqrt{a}$

(2) $\sqrt{a^7} = \sqrt{a^4} \cdot \sqrt{a^3} = a^2\sqrt{a^3}$

(3) $\sqrt{a^7} = \sqrt{a^2} \cdot \sqrt{a^5} = a\sqrt{a^5}$

However, when factoring out a perfect power from a radical, <u>we always factor out the largest perfect power</u>. Otherwise, we can still simplify the remaining radical by factoring out a perfect power.

Which of the three ways above is the proper way of factoring out a perfect square from $\sqrt{a^7}$? _____

301. Which of the three ways below is the proper way of factoring out a perfect cube from $\sqrt[3]{x^{10}}$? _____

(1) $\sqrt[3]{x^{10}} = \sqrt[3]{x^3} \cdot \sqrt[3]{x^7} = x\sqrt[3]{x^7}$

(2) $\sqrt[3]{x^{10}} = \sqrt[3]{x^6} \cdot \sqrt[3]{x^4} = x^2\sqrt[3]{x^4}$

(3) $\sqrt[3]{x^{10}} = \sqrt[3]{x^9} \cdot \sqrt[3]{x} = x^3\sqrt[3]{x}$

302. Simplify each radical below by factoring out the largest possible perfect power:

(a) $\sqrt[4]{t^{11}} = $ _____ (b) $\sqrt[5]{m^{17}} = $ _____

303. Perfect powers can also be factored out of radicals which contain negative powers. That is:

$$\sqrt{x^{-5}} = \sqrt{x^{-4}} \cdot \sqrt{x^{-1}} = x^{-2}\sqrt{x^{-1}}$$

(a) $\sqrt[3]{y^{-5}} = \sqrt[3]{y^{-3}} \cdot \sqrt[3]{y^{-2}} = $ _____

(b) $\sqrt[4]{t^{-11}} = \sqrt[4]{t^{-8}} \cdot \sqrt[4]{t^{-3}} = $ _____

304. A perfect power <u>cannot</u> be factored out of a radical if the absolute value of the exponent in the radicand is smaller than the index (or root). For example:

We cannot factor a perfect fourth out of $\sqrt[4]{x^3}$, $\sqrt[4]{x^2}$, or $\sqrt[4]{x}$.

We cannot factor a perfect cube out of $\sqrt[3]{y^2}$ or $\sqrt[3]{y}$.

Factor out a perfect power if possible:

(a) $\sqrt{x} = $ ____ (b) $\sqrt{y^2} = $ ____ (c) $\sqrt[3]{t^2} = $ ____ (d) $\sqrt[4]{p^2} = $ ____

305. Factor out a perfect power if possible:

(a) $\sqrt{t^{-1}} = $ _____ (b) $\sqrt[4]{b^{-4}} = $ _____ (c) $\sqrt[3]{R^{-2}} = $ _____

Answers:

(1) $\sqrt{a^7} = a^3\sqrt{a}$

(3) $\sqrt[3]{x^{10}} = x^3\sqrt[3]{x}$

a) $t^2\sqrt[4]{t^3}$ b) $m^3\sqrt[5]{m^2}$

a) $y^{-1}\sqrt[3]{y^{-2}}$
b) $t^{-2}\sqrt[4]{t^{-3}}$

a) Not possible.
b) y^1 or y
c) Not possible.
d) Not possible.

306. In the radical $\sqrt{x^4 y}$, "y" is not a perfect square. Therefore, we can only factor out one perfect power. We get:
$$\sqrt{x^4 y} = \sqrt{x^4} \cdot \sqrt{y} = x^2 \sqrt{y}$$

Factor out the perfect powers from each of these radicals:

(a) $\sqrt{ab^{-6}} =$ _____ (b) $\sqrt[3]{p^{12} q^2} =$ _____ (c) $\sqrt[4]{cd^8} =$ _____

| a) Not possible. |
| b) b^{-1} |
| c) Not possible. |

---

307. Though y^5 is not a perfect power in the radical below, we can still factor two perfect powers out of the radical. We get:
$$\sqrt{x^6 y^5} = \sqrt{x^6} \cdot \sqrt{y^4} \cdot \sqrt{y} = x^3 y^2 \sqrt{y}$$

Factor out the perfect powers from each radical below:

(a) $\sqrt{a^{-3} b^4} =$ _____ (b) $\sqrt[3]{s^6 t^8} =$ _____ (c) $\sqrt[4]{16 m^5} =$ _____

| a) $b^{-3} \sqrt{a}$ |
| b) $p^4 \sqrt[3]{q^2}$ |
| c) $d^2 \sqrt[4]{c}$ |

---

308. Neither x^3 nor y^7 is a perfect power in the radical below. However, we can still factor out two perfect powers. Notice that the two remaining radicals are recombined into one in the final form of the simplification.
$$\sqrt{x^3 y^7} = \sqrt{x^2} \cdot \sqrt{x} \cdot \sqrt{y^6} \cdot \sqrt{y} = x\sqrt{x} \cdot y^3 \sqrt{y} = xy^3 \sqrt{xy}$$

Following the procedure above, factor out the perfect powers from each radical below:

(a) $\sqrt{a^9 b^{-5}} =$ _____ (b) $\sqrt[3]{c^5 d^{10}} =$ _____

| a) $a^{-1} b^2 \sqrt{a^{-1}}$ |
| b) $s^2 t^2 \sqrt[3]{t^2}$ |
| c) $2m \sqrt[4]{m}$ |

---

309. Factor out the perfect powers from each of these radicals. <u>Be sure to look at the index first</u>.

(a) $\sqrt[5]{x^5 y^{-10}} =$ _____ (c) $\sqrt[6]{c^6 d^{14}} =$ _____
(b) $\sqrt[4]{pq^{20}} =$ _____ (d) $\sqrt[3]{F^8 T^{-10}} =$ _____

| a) $a^4 b^{-2} \sqrt{ab^{-1}}$ |
| b) $cd^3 \sqrt[3]{c^2 d}$ |

---

310. We can also simplify radicals which contain fractions by factoring out perfect powers. To do so, we apply the law of radicals for fractions and then factor out perfect powers from both the numerator and denominator. For example:
$$\sqrt{\frac{x^4}{y^8}} = \frac{\sqrt{x^4}}{\sqrt{y^8}} = \frac{x^2}{y^4}$$
$$\sqrt[3]{\frac{c^{-6}}{d^9}} = \frac{\sqrt[3]{c^{-6}}}{\sqrt[3]{d^9}} = \frac{c^{-2}}{d^3}$$

Following the procedure above, factor the perfect powers out of:

(a) $\sqrt{\dfrac{16}{R^2}} =$ _____ (c) $\sqrt{\dfrac{4x^{-2}}{y^{10}}} =$ _____
(b) $\sqrt[3]{\dfrac{p^{12}}{27}} =$ _____ (d) $\sqrt[4]{\dfrac{c^4}{d^8 f^{12}}} =$ _____

| a) xy^{-2} |
| b) $q^5 \sqrt[4]{p}$ |
| c) $cd^2 \sqrt[6]{d^2}$ |
| d) $F^2 T^{-3} \sqrt[3]{F^2 T^{-1}}$ |

---

a) $\dfrac{4}{R}$ c) $\dfrac{2x^{-1}}{y^5}$

b) $\dfrac{p^4}{3}$ d) $\dfrac{c}{d^2 f^3}$

364   Exponents and Radicals

311. If we apply the law of radicals for the roots of powers to the radical below, we get:

$$\sqrt{\left(\frac{x}{y}\right)^6} = \left(\frac{x}{y}\right)^{\frac{6}{2}} = \left(\frac{x}{y}\right)^3 = \frac{x^3}{y^3}$$

We can confirm this fact by raising the fraction to the sixth power first and then factoring out perfect squares. We get:

$$\sqrt{\left(\frac{x}{y}\right)^6} = \sqrt{\frac{x^6}{y^6}} = \frac{x^3}{y^3}$$

(a) Apply the law of radicals for the roots of powers to the radical below:

$$\sqrt[3]{\left(\frac{m}{t}\right)^{12}} = \underline{} = \underline{} = \underline{}$$

(b) Now confirm this result by raising the fraction to the twelfth power and then factoring out perfect cubes:

$$\sqrt[3]{\left(\frac{m}{t}\right)^{12}} = \underline{} = \underline{}$$

312. Though the numerators of the fractions below are not perfect powers, we can still factor perfect powers out of each radical. That is:

$$\sqrt{\frac{x}{y^4}} = \frac{\sqrt{x}}{\sqrt{y^4}} = \frac{\sqrt{x}}{y^2} \qquad \sqrt[3]{\frac{p^5}{q^{15}}} = \frac{\sqrt[3]{p^5}}{\sqrt[3]{q^{15}}} = \frac{p\sqrt[3]{p^2}}{q^5}$$

Factor out the perfect powers from each radical:

(a) $\sqrt{\dfrac{25c^{-3}}{d^8}} = $

(b) $\sqrt[4]{\dfrac{t^5}{s^{12}}} = $

313. Though the denominators of the fractions below are not perfect powers, we can still factor out perfect powers from each radical. We get:

$$\sqrt{\frac{x^6}{y}} = \frac{\sqrt{x^6}}{\sqrt{y}} = \frac{x^3}{\sqrt{y}} \qquad \sqrt[3]{\frac{m^9}{p^5}} = \frac{\sqrt[3]{m^9}}{\sqrt[3]{p^5}} = \frac{m^3}{p\sqrt[3]{p^2}}$$

Factor out the perfect powers from each radical:

(a) $\sqrt{\dfrac{t^4}{9a^3}} = $

(b) $\sqrt[4]{\dfrac{16d^8}{c^7}} = $

---

a) $\left(\dfrac{m}{t}\right)^{\frac{12}{3}} = \left(\dfrac{m}{t}\right)^4 = \dfrac{m^4}{t^4}$

b) $\sqrt[3]{\dfrac{m^{12}}{t^{12}}} = \dfrac{m^4}{t^4}$

---

a) $\dfrac{5c^{-1}\sqrt{c^{-1}}}{d^4}$

b) $\dfrac{t\sqrt[4]{t}}{s^3}$

---

a) $\dfrac{t^2}{3a\sqrt{a}}$ b) $\dfrac{2d^2}{c\sqrt[4]{c^3}}$

314. Though neither the numerators nor denominators are perfect powers in the fractions below, we can still factor out perfect powers from each radical. Notice how the two radicals are recombined into one in the final form of the answer.

$$\sqrt{\frac{x^5}{y^7}} = \frac{\sqrt{x^5}}{\sqrt{y^7}} = \frac{x^2\sqrt{x}}{y^3\sqrt{y}} = \frac{x^2}{y^3}\sqrt{\frac{x}{y}}$$

$$\sqrt[3]{\frac{p^8}{q^4}} = \frac{\sqrt[3]{p^8}}{\sqrt[3]{q^4}} = \frac{p^2\sqrt[3]{p^2}}{q\sqrt[3]{q}} = \frac{p^2}{q}\sqrt[3]{\frac{p^2}{q}}$$

Factor out the perfect powers from each radical:

(a) $\sqrt{\dfrac{4a^9}{9b^3}} = $ _____   (b) $\sqrt[4]{\dfrac{c^4 d^5}{p^{12} q^{11}}} = $ _____

315. Factor out the perfect powers from each radical:

(a) $\sqrt{\dfrac{R^{-2}}{T^{-4}}} = $ _____   (c) $\sqrt{\dfrac{49x^2}{16y}} = $ _____

(b) $\sqrt[3]{\dfrac{m^4 p}{q^{10}}} = $ _____   (d) $\sqrt[5]{\dfrac{a^5 b^7}{c^3}} = $ _____

a) $\dfrac{2a^4}{3b}\sqrt{\dfrac{a}{b}}$

b) $\dfrac{cd}{p^3 q^2}\sqrt[4]{\dfrac{d}{q^3}}$

Answer to Frame 315:   a) $\dfrac{R^{-1}}{T^{-2}}$ b) $\dfrac{m}{q^3}\sqrt[3]{\dfrac{mp}{q}}$ c) $\dfrac{7x}{4\sqrt{y}}$ d) $ab\sqrt[5]{\dfrac{b^2}{c^3}}$

SELF-TEST 17 (Frames 291-315)

Simplify the following radicals by factoring out perfect powers:

1. $\sqrt[3]{8a^4} = $ _____   2. $\sqrt{12r^3 t} = $ _____   3. $\sqrt[4]{x^9 y^{-6}} = $ _____   4. $\sqrt[3]{d^{-3} h^2 k^5} = $ _____

Simplify the following radicals by factoring out perfect powers:

5. $\sqrt[5]{\dfrac{a^{-5}}{b^{10}}} = $ _____   6. $\sqrt[3]{\dfrac{64 s^7}{p^4}} = $ _____   7. $\sqrt{\left(\dfrac{w}{t}\right)^3} = $ _____

ANSWERS:
1. $2a\sqrt[3]{a}$ 3. $x^2 y^{-1}\sqrt[4]{xy^{-2}}$ 5. $\dfrac{a^{-1}}{b^2}$ 6. $\dfrac{4s^2}{p}\sqrt[3]{\dfrac{s}{p}}$ 7. $\dfrac{w}{t}\sqrt{\dfrac{w}{t}}$
2. $2r\sqrt{3rt}$ 4. $d^{-1}k\sqrt[3]{h^2 k^2}$

366   Exponents and Radicals

4-18   RAISING RADICALS TO POWERS

It is also possible to raise radicals to powers. In this section, we will discuss the law of radicals which is used to do so. When raising radicals to powers, we will limit the radicals to those in which the radicand is a number, a letter, or a simple power.

---

316. We have already discussed powers in which the base is a number, letter, letter-term, or fraction. It is also possible to have powers in which the base is a radical. For example:

$$\text{In } (\sqrt{3})^2, \text{ the base is } \sqrt{3}.$$

(a) In $(\sqrt[3]{x})^4$, the base is ____. (b) In $(\sqrt[4]{5})^3$, the base is ____.

---

317. Powers whose bases are radicals have the same meaning as other types of powers. That is:

$$(\sqrt{x})^2 \text{ means: raise } \sqrt{x} \text{ to the \underline{second} power.}$$

(a) $(\sqrt[3]{5})^4$ means: raise $\sqrt[3]{5}$ to the _____ power.

(b) $(\sqrt[4]{t})^{-1}$ means: raise $\sqrt[4]{t}$ to the _____ power.

a) $\sqrt[3]{x}$ b) $\sqrt[4]{5}$

---

318. Just like any other base raised to a positive whole-number power greater than 1, a radical raised to such a power stands for a multiplication of identical factors. That is:

$$(\sqrt{x})^2 \text{ means: } \sqrt{x} \cdot \sqrt{x}$$
$$(\sqrt[3]{y})^3 \text{ means: } \sqrt[3]{y} \cdot \sqrt[3]{y} \cdot \sqrt[3]{y}$$
$$(\sqrt[4]{t})^5 \text{ means: } \sqrt[4]{t} \cdot \sqrt[4]{t} \cdot \sqrt[4]{t} \cdot \sqrt[4]{t} \cdot \sqrt[4]{t}$$

Write each of the following as a multiplication of identical factors:

(a) $(\sqrt{7})^3 = $ _____   (b) $(\sqrt[4]{3})^4 = $ _____

a) fourth

b) minus-one

---

319. Write each of the following multiplications of identical radicals as a power:

(a) $\sqrt{x} \cdot \sqrt{x} \cdot \sqrt{x} = $ _____   (c) $\sqrt{2} \cdot \sqrt{2} \cdot \sqrt{2} \cdot \sqrt{2} = $ _____

(b) $\sqrt[3]{5} \cdot \sqrt[3]{5} = $ _____   (d) $\sqrt[4]{p} \cdot \sqrt[4]{p} \cdot \sqrt[4]{p} = $ _____

a) $\sqrt{7} \cdot \sqrt{7} \cdot \sqrt{7}$

b) $\sqrt[4]{3} \cdot \sqrt[4]{3} \cdot \sqrt[4]{3} \cdot \sqrt[4]{3}$

---

320. <u>Any radical raised to a power is equal to the same radical with its radicand raised to that power.</u> For example:

$$(\sqrt{y})^2 = \sqrt{y^2} \qquad (\sqrt[4]{5})^3 = \sqrt[4]{5^3}$$

We can see this fact by converting the original power to a multiplication and then performing the multiplication. That is:

$$(\sqrt{y})^2 = \sqrt{y} \cdot \sqrt{y} = \sqrt{y^2}$$

(a) $(\sqrt{3})^4 = \sqrt{3} \cdot \sqrt{3} \cdot \sqrt{3} \cdot \sqrt{3} = \sqrt{}$

(b) $(\sqrt[3]{t})^2 = \sqrt[3]{t} \cdot \sqrt[3]{t} = \sqrt[3]{}$

(c) $(\sqrt[4]{5})^3 = \sqrt[4]{5} \cdot \sqrt[4]{5} \cdot \sqrt[4]{5} = \sqrt[4]{}$

a) $(\sqrt{x})^3$ (c) $(\sqrt{2})^4$

b) $(\sqrt[3]{5})^2$ (d) $(\sqrt[4]{p})^3$

321. In the last frame, we saw these facts:

$$\left(\sqrt{y}\right)^2 = \sqrt{y^2} \qquad \left(\sqrt[3]{t}\right)^2 = \sqrt[3]{t^2}$$
$$\left(\sqrt{3}\right)^4 = \sqrt{3^4} \qquad \left(\sqrt[4]{5}\right)^3 = \sqrt[4]{5^3}$$

The following general pattern should be obvious:

$$\boxed{\left(\sqrt[n]{a}\right)^m = \sqrt[n]{a^m}}$$

That is: any radical raised to the "m" power is equal to the same radical with its radicand raised to the "m" power.

Using the principle above, complete each of these:

(a) $\left(\sqrt{x}\right)^5 = \sqrt{x^{\square}}$ (c) $\left(\sqrt[4]{t}\right)^9 = \sqrt[4]{t^{\square}}$

(b) $\left(\sqrt[3]{7}\right)^6 = \sqrt[3]{7^{\square}}$ (d) $\left(\sqrt[5]{2}\right)^7 = \sqrt[5]{2^{\square}}$

a) $\sqrt{3^4}$
b) $\sqrt[3]{t^2}$
c) $\sqrt[4]{5^3}$

322. Using the same principle, complete each of these:

(a) $\left(\sqrt{5}\right)^2 = \sqrt{}$ (c) $\left(\sqrt[4]{3}\right)^8 = \sqrt[4]{}$

(b) $\left(\sqrt[3]{p}\right)^9 = \sqrt[3]{}$ (d) $\left(\sqrt{R}\right)^{10} = \sqrt{}$

a) $\sqrt{x^5}$ c) $\sqrt[4]{t^9}$
b) $\sqrt[3]{7^6}$ d) $\sqrt[5]{2^7}$

323. We have already seen the following law for radicals whose radicands are powers:

$$\boxed{\sqrt[n]{a^m} = a^{\frac{m}{n}}}$$

Since any radical raised to the "m" power is equal to a radical whose radicand is raised to the "m" power, we can convert a radical raised to a power to a simple power. To do so, we use the law above. That is:

$$\left(\sqrt{x}\right)^4 = \sqrt{x^4} = x^{\frac{4}{2}} \text{ or } x^2$$

$$\left(\sqrt[3]{7}\right)^9 = \sqrt[3]{7^9} = 7^{\frac{9}{3}} \text{ or } 7^3$$

$$\left(\sqrt[4]{y}\right)^8 = \sqrt[4]{y^8} = \underline{} \text{ or } \underline{}$$

a) $\sqrt{5^2}$ c) $\sqrt[4]{3^8}$
b) $\sqrt[3]{p^9}$ d) $\sqrt{R^{10}}$

324. Here is the general principle we have developed:

$$\boxed{\left(\sqrt[n]{a}\right)^m = \sqrt[n]{a^m} = a^{\frac{m}{n}}}$$

Using this principle, complete each of the following:

(a) $\left(\sqrt{t}\right)^6 = \sqrt{t^6} = \underline{}$ (c) $\left(\sqrt{5}\right)^{10} = \sqrt{5^{10}} = \underline{}$

(b) $\left(\sqrt[3]{p}\right)^{12} = \sqrt[3]{p^{12}} = \underline{}$ (d) $\left(\sqrt[4]{3}\right)^4 = \sqrt[4]{3^4} = \underline{}$

$y^{\frac{8}{4}}$ or y^2

Exponents and Radicals 367

368   Exponents and Radicals

325. Based on the general principle we have developed, we can write the following "law of radicals for raising a radical to a power":

$$\left(\sqrt[n]{a}\right)^m = a^{\frac{m}{n}}$$

Using the law above, complete each of these:

$\left(\sqrt{3}\right)^{12} = 3^{\frac{12}{2}}$ or 3^6 (b) $\left(\sqrt[3]{11}\right)^6 = $ _____ or _____

(a) $\left(\sqrt{m}\right)^8 = $ _____ or _____   (c) $\left(\sqrt[5]{c}\right)^5 = $ _____ or _____

a) t^3 c) 5^5
b) p^4 d) 3^1 or 3

326. There are two laws of radicals which can be used to convert a radical expression to a simple power. They are:

$$\sqrt[n]{a^m} = a^{\frac{m}{n}} \qquad \left(\sqrt[n]{a}\right)^m = a^{\frac{m}{n}}$$

The law on the left is the law of radicals "for the roots of powers".

The law on the right is the law of radicals "for raising a radical to a power".

Using these two laws, convert each of the following to a simple power:
(a) $\sqrt{x^4} = $ ____   (b) $\left(\sqrt{x}\right)^4 = $ ____   (c) $\sqrt[3]{y^9} = $ ____   (d) $\left(\sqrt[3]{y}\right)^9 = $ ____

a) $m^{\frac{8}{2}}$ or m^4
b) $11^{\frac{6}{3}}$ or 11^2
c) $c^{\frac{5}{5}}$ or c^1 or c

327. Convert each of the following radical expressions to a simple power:
(a) $\left(\sqrt{t}\right)^{12} = $ ____   (b) $\sqrt[3]{m^{18}} = $ ____   (c) $\sqrt[4]{R^4} = $ ____   (d) $\left(\sqrt[5]{3}\right)^{10} = $ ____

a) x^2 c) y^3
b) x^2 d) y^3

328. Again, here is the law of radicals for raising a radical to a power:

$$\left(\sqrt[n]{a}\right)^m = a^{\frac{m}{n}}$$

If "m" is a multiple of "n", the simple power on the right has a whole-number exponent. For example:

$\left(\sqrt{t}\right)^4 = t^2$ $\left(\sqrt[3]{p}\right)^{12} = p^4$ $\left(\sqrt[4]{F}\right)^4 = F^1$ or F

If "m" is not a multiple of "n", the simple power on the right has a fractional exponent. For example:

$\left(\sqrt{x}\right)^3 = x^{\frac{3}{2}}$ $\left(\sqrt[3]{y}\right)^5 = y^{\frac{5}{3}}$ $\left(\sqrt[4]{m}\right)^7 = m^{\frac{7}{4}}$

If we convert each of the radical expressions below to a simple power, which of the simple powers would have a whole-number exponent?

(a) $\left(\sqrt{R}\right)^5$ (b) $\left(\sqrt[3]{h}\right)^{15}$ (c) $\left(\sqrt{d}\right)^{10}$ (d) $\left(\sqrt[4]{3}\right)^9$

a) t^6
b) m^6
c) R^1 or R
d) 3^2

Only (b) and (c).

Exponents and Radicals   369

329. The law of radicals for raising a radical to a power also applies when a radical is raised to a <u>negative whole-number power</u>. For example:

$$\left(\sqrt{x}\right)^{-6} = x^{\frac{-6}{2}} \text{ or } x^{-3} \qquad \left(\sqrt[3]{y}\right)^{-6} = y^{\frac{-6}{3}} \text{ or } y^{-2}$$

This fact can be justified by applying the definition of a negative power to each radical expression above, as follows:

Since: $\square^{-n} = \dfrac{1}{\square^n}$: $\left(\sqrt{x}\right)^{-6} = \dfrac{1}{\left(\sqrt{x}\right)^6} = \dfrac{1}{x^3} = x^{-3}$

$$\left(\sqrt[3]{y}\right)^{-6} = \dfrac{1}{\left(\sqrt[3]{y}\right)^6} = \dfrac{1}{y^2} = y^{-2}$$

Let's convert $\left(\sqrt[4]{t}\right)^{-8}$ to a simple power by using both procedures above:

(a) Applying the law of radicals for raising a radical to a power, as shown at the beginning of this frame, we get:

$$\left(\sqrt[4]{t}\right)^{-8} = \underline{\qquad} \text{ or } \underline{\qquad}$$

(b) Applying the definition of a power with a negative exponent, we get:

$$\left(\sqrt[4]{t}\right)^{-8} = \underline{\qquad} = \underline{\qquad} = \underline{\qquad}$$

(c) Do we obtain the same simple power by both procedures? _____

---

330. Use the law of radicals for raising a radical to a power to convert each radical expression below to a simple power:

(a) $\left(\sqrt{5}\right)^{-4} = 5^{\boxed{}}$ (c) $\left(\sqrt{x}\right)^{-2} = x^{\boxed{}}$

(b) $\left(\sqrt[3]{7}\right)^{-9} = 7^{\boxed{}}$ (d) $\left(\sqrt[5]{y}\right)^{-20} = y^{\boxed{}}$

a) $t^{\frac{-8}{4}}$ or t^{-2}

b) $\dfrac{1}{\left(\sqrt[4]{t}\right)^8} = \dfrac{1}{t^2} = t^{-2}$

c) Yes.

---

331. If each of the following radical expressions would be converted to a simple power, which of the simple powers would have a negative <u>fractional</u> exponent? _____

(a) $\left(\sqrt[3]{5}\right)^{-2}$ (b) $\left(\sqrt{v}\right)^{-4}$ (c) $\left(\sqrt[4]{m}\right)^{-8}$ (d) $\left(\sqrt[5]{t}\right)^{-7}$

a) $5^{\boxed{-2}}$ c) $x^{\boxed{-1}}$
b) $7^{\boxed{-3}}$ d) $y^{\boxed{-4}}$

---

332. Any radical raised to the zero power is equal to a simple power raised to the zero power. That is:

$$\left(\sqrt{x}\right)^0 = x^{\frac{0}{2}} = x^0 \qquad \left(\sqrt[3]{y}\right)^0 = y^{\frac{0}{3}} = y^0$$

Therefore, any radical raised to the zero power equals what number? _____

Both (a) and (d).

+1

370 Exponents and Radicals

333. The law of radicals for raising a radical to a power <u>also applies if the radicand is itself a power</u>. For example:

$$\left(\sqrt{x^3}\right)^4 = (x^3)^{\frac{4}{2}} = (x^3)^2 = x^6$$

$$\left(\sqrt[3]{y^2}\right)^6 = (y^2)^{\frac{6}{3}} = (y^2)^2 = y^4$$

This fact can be justified by converting each "radical to a power" to a multiplication of identical radicals. That is:

$$\left(\sqrt{x^3}\right)^4 = \sqrt{x^3} \cdot \sqrt{x^3} \cdot \sqrt{x^3} \cdot \sqrt{x^3} = \sqrt{x^{12}} = x^6$$

$$\left(\sqrt[3]{y^2}\right)^6 = \sqrt[3]{y^2} \cdot \sqrt[3]{y^2} \cdot \sqrt[3]{y^2} \cdot \sqrt[3]{y^2} \cdot \sqrt[3]{y^2} \cdot \sqrt[3]{y^2} = \sqrt[3]{y^{12}} = y^4$$

By using the law of radicals for raising a radical to a power, convert each radical expression below to a simple power:

$$\left(\sqrt{t^5}\right)^6 = (t^5)^{\frac{6}{2}} = (t^5)^3 = t^{15}$$

(a) $\left(\sqrt[3]{R^7}\right)^9 = (R^7)^{\frac{9}{3}} = $ _____ = _____

(b) $\left(\sqrt[4]{p^2}\right)^4 = (p^2)^{\frac{4}{4}} = $ _____ = _____

---

334. Use the same law to convert each of the following to a simple power:

(a) $\left(\sqrt{y^3}\right)^8 = $ _____   (c) $\left(\sqrt[4]{m^3}\right)^{12} = $ _____

(b) $\left(\sqrt[3]{2^5}\right)^3 = $ _____   (d) $\left(\sqrt{d^{10}}\right)^2 = $ _____

a) $(R^7)^3 = R^{21}$
b) $(p^2)^1 = p^2$

---

335. The same law can be used when the radicand is a negative power. For example:

$$\left(\sqrt{x^{-3}}\right)^4 = (x^{-3})^{\frac{4}{2}} = (x^{-3})^2 = x^{-6}$$

This fact can also be justified by converting the original radical expression to a multiplication:

$$\left(\sqrt{x^{-3}}\right)^4 = \sqrt{x^{-3}} \cdot \sqrt{x^{-3}} \cdot \sqrt{x^{-3}} \cdot \sqrt{x^{-3}} = \sqrt{x^{-12}} = x^{-6}$$

Use the law of radicals to convert each of the following to a simple power:

(a) $\left(\sqrt{x^{-2}}\right)^8 = (x^{-2})^{\frac{8}{2}} = $ _____   (c) $\left(\sqrt[3]{p^{-2}}\right)^9 = $ _____

(b) $\left(\sqrt[3]{y^{-1}}\right)^6 = (y^{-1})^{\frac{6}{3}} = $ _____   (d) $\left(\sqrt[4]{t^{-6}}\right)^8 = $ _____

a) y^{12}, from $(y^3)^4$
b) 2^5, from $(2^5)^1$
c) m^9, from $(m^3)^3$
d) d^{10}, from $(d^{10})^1$

---

a) x^{-8}, from $(x^{-2})^4$
b) y^{-2}, from $(y^{-1})^2$
c) p^{-6}, from $(p^{-2})^3$
d) t^{-12}, from $(t^{-6})^2$

336. The same law also applies when the radical is raised to a negative power. That is:

$$\left(\sqrt{t^3}\right)^{-4} = (t^3)^{\frac{-4}{2}} = (t^3)^{-2} = t^{-6}$$

$$\left(\sqrt[3]{R^{-1}}\right)^{-9} = (R^{-1})^{\frac{-9}{3}} = (R^{-1})^{-3} = R^3$$

This fact can be justified by applying the definition of a negative power to each original radical expression. That is:

$$\left(\sqrt{t^3}\right)^{-4} = \frac{1}{\left(\sqrt{t^3}\right)^4} = \frac{1}{t^6} = t^{-6}$$

$$\left(\sqrt[3]{R^{-1}}\right)^{-9} = \frac{1}{\left(\sqrt[3]{R^{-1}}\right)^9} = \frac{1}{R^{-3}} = R^3$$

Use the law of radicals to convert each of the following to a simple power:

(a) $\left(\sqrt{t^5}\right)^{-6} = (t^5)^{\frac{-6}{2}} = $ _____

(b) $\left(\sqrt[3]{7^{-2}}\right)^{-3} = (7^{-2})^{\frac{-3}{3}} = $ _____

(c) $\left(\sqrt{p^{-1}}\right)^{-4} = $ _____

(d) $\left(\sqrt[3]{x^2}\right)^{-6} = $ _____

---

337. Use the law of radicals to convert each of the following to a simple power:

(a) $\left(\sqrt{t^3}\right)^8 = $ _____

(b) $\left(\sqrt[3]{m^{-2}}\right)^3 = $ _____

(c) $\left(\sqrt[6]{p^5}\right)^{-6} = $ _____

(d) $\left(\sqrt{x^{-5}}\right)^{-2} = $ _____

a) t^{-15}, from $(t^5)^{-3}$

b) 7^2, from $(7^{-2})^{-1}$

c) p^2, from $(p^{-1})^{-2}$

d) x^{-4}, from $(x^2)^{-2}$

---

338. Use the law of radicals to convert each of these to a simple power:

(a) $\left(\sqrt{x}\right)^4 = $ _____

(b) $\left(\sqrt[3]{7}\right)^{12} = $ _____

(c) $\left(\sqrt[4]{y}\right)^{-4} = $ _____

(d) $\left(\sqrt[3]{t^2}\right)^6 = $ _____

(e) $\left(\sqrt{m^{-3}}\right)^8 = $ _____

(f) $\left(\sqrt[5]{F^{-1}}\right)^{-10} = $ _____

a) t^{12}, from $(t^3)^4$

b) m^{-2}, from $(m^{-2})^1$

c) p^{-5}, from $(p^5)^{-1}$

d) x^5, from $(x^{-5})^{-1}$

---

a) x^2 d) t^4

b) 7^4 e) m^{-12}

c) y^{-1} f) F^2

372   Exponents and Radicals

SELF-TEST 18 (Frames 316-338)

Complete: 1. $\left(\sqrt[3]{y}\right)^5 = \sqrt[?]{?}$ _____   2. $\sqrt[4]{R^7} = \left(\sqrt[?]{R}\right)^?$ _____

Convert each of the following radical expressions <u>to a simple power</u> which does <u>not</u> contain a radical sign:

3. $\left(\sqrt[4]{x}\right)^8 =$ _____   7. $\left(\sqrt{h}\right)^3 =$ _____   11. $\left(\sqrt{3^4}\right)^{-8} =$ _____

4. $\left(\sqrt{P}\right)^{10} =$ _____   8. $\left(\sqrt[3]{w}\right)^2 =$ _____   12. $\left(\sqrt[3]{t^{-4}}\right)^6 =$ _____

5. $\left(\sqrt[3]{5}\right)^{15} =$ _____   9. $\left(\sqrt{y}\right)^{-2} =$ _____   13. $\left(\sqrt[4]{b^{-2}}\right)^{-12} =$ _____

6. $\left(\sqrt[5]{F}\right)^5 =$ _____   10. $\left(\sqrt[4]{a^3}\right)^8 =$ _____   14. $\left(\sqrt[6]{d^{-3}}\right)^0 =$ _____

ANSWERS:
1. $\sqrt[3]{y^5}$ 3. x^2 7. $h^{\frac{3}{2}}$ 11. 3^{-16}
2. $\left(\sqrt[4]{R}\right)^7$ 4. P^5 8. $w^{\frac{2}{3}}$ 12. t^{-8}
                          5. 5^5 9. y^{-1} 13. b^6
                          6. F^1 or F 10. a^6 14. d^0 or 1

4-19   THE MEANING OF POWERS WITH FRACTIONAL EXPONENTS

In this section, we will define what is meant by a power with a fractional exponent. We will show that any power with a fractional exponent is equivalent to a radical, and vice versa.

339.   The general law of radicals for finding the roots of powers is given below:

$$\boxed{\sqrt[n]{a^m} = a^{\frac{m}{n}}}$$

This law can be used to convert any radical to a power. If "m" is not a multiple of "n", the power has a fractional exponent. For example:

$$\sqrt{3^5} = 3^{\frac{5}{2}} \qquad \sqrt[3]{7^4} = 7^{\frac{4}{3}} \qquad \sqrt[5]{t^3} = t^{\frac{3}{5}}$$

Using the law above, convert each of the following radicals to a power:

(a) $\sqrt{4^3} =$ _____   (b) $\sqrt[3]{p^2} =$ _____   (c) $\sqrt[4]{2^5} =$ _____   (d) $\sqrt[7]{h^4} =$ _____

340. If the exponent of a radicand is not explicitly written, its exponent is "1". For example:

$$\sqrt{3} = \sqrt{3^1} \qquad \sqrt[3]{5} = \sqrt[3]{5^1} \qquad \sqrt[4]{m} = \sqrt[4]{m^1}$$

Using this fact and the law of radicals for the roots of powers, we can convert each of the radicals below to a power. That is:

$$\sqrt{3} = \sqrt{3^1} = 3^{\frac{1}{2}}$$

(a) $\sqrt{t} = \sqrt{t^1} = $ _____

(b) $\sqrt[3]{5} = \sqrt[3]{5^1} = $ _____

(c) $\sqrt[4]{m} = \sqrt[4]{m^1} = $ _____

a) $4^{\frac{3}{2}}$ c) $2^{\frac{5}{4}}$

b) $p^{\frac{2}{3}}$ d) $h^{\frac{4}{7}}$

---

341. Convert each of the following radicals to a power:

(a) $\sqrt{m^7} = $ _____ (b) $\sqrt{d} = $ _____ (c) $\sqrt[3]{p^5} = $ _____ (d) $\sqrt[5]{7} = $ _____

a) $t^{\frac{1}{2}}$ b) $5^{\frac{1}{3}}$

c) $m^{\frac{1}{4}}$

---

342. We have reversed the law of radicals for the roots of powers below:

$$a^{\frac{m}{n}} = \sqrt[n]{a^m}$$

Using this reversed form of the law, we can convert any power with a fractional exponent to a radical. For example:

$$3^{\frac{5}{2}} = \sqrt{3^5} \qquad 7^{\frac{2}{3}} = \sqrt[3]{7^2} \qquad x^{\frac{5}{7}} = \sqrt[7]{x^5}$$

Notice in each case that the <u>denominator</u> of the fractional exponent is the <u>index</u> of the radical.

Convert each of the following powers to a radical:

(a) $y^{\frac{3}{2}} = $ _____ (b) $5^{\frac{4}{3}} = $ _____ (c) $t^{\frac{2}{5}} = $ _____

a) $m^{\frac{7}{2}}$

b) $d^{\frac{1}{2}}$

c) $p^{\frac{5}{3}}$

d) $7^{\frac{1}{5}}$

---

343. Here are some conversions in which the numerator of the fractional exponent is "1":

$$3^{\frac{1}{2}} = \sqrt{3^1} \text{ or } \sqrt{3} \qquad p^{\frac{1}{5}} = \sqrt[5]{p^1} \text{ or } \sqrt[5]{p}$$

Convert each of the following powers to a radical:

(a) $y^{\frac{1}{2}} = $ _____ (b) $7^{\frac{1}{3}} = $ _____ (c) $t^{\frac{1}{4}} = $ _____ (d) $2^{\frac{1}{9}} = $ _____

a) $\sqrt{y^3}$

b) $\sqrt[3]{5^4}$

c) $\sqrt[5]{t^2}$

---

344. Convert each radical to a power and each power to a radical:

(a) $3^{\frac{7}{2}} = $ _____ (b) $\sqrt[3]{x^8} = $ _____ (c) $\sqrt[5]{2^4} = $ _____ (d) $y^{\frac{3}{4}} = $ _____

a) $\sqrt{y^1}$ or $\sqrt{y}$

b) $\sqrt[3]{7^1}$ or $\sqrt[3]{7}$

c) $\sqrt[4]{t^1}$ or $\sqrt[4]{t}$

d) $\sqrt[9]{2^1}$ or $\sqrt[9]{2}$

---

345. Convert each radical to a power and each power to a radical:

(a) $\sqrt{t^9} = $ _____ (b) $5^{\frac{4}{3}} = $ _____ (c) $\sqrt{d^3} = $ _____ (d) $\sqrt[7]{p^2} = $ _____

a) $\sqrt{3^7}$ c) $2^{\frac{4}{5}}$

b) $x^{\frac{8}{3}}$ d) $\sqrt[4]{y^3}$

346. Convert each radical to a power and each power to a radical:

(a) $m^{\frac{1}{2}} =$ _____ (b) $\sqrt[3]{c} =$ _____ (c) $\sqrt[5]{t} =$ _____ (d) $q^{\frac{1}{4}} =$ _____

a) $t^{\frac{9}{2}}$ c) $\sqrt{d^3}$
b) $\sqrt[3]{5^4}$ d) $p^{\frac{2}{7}}$

347. If the base of a fractional power is a number, we can occasionally convert the power to a regular number quite easily. For example:

$$16^{\frac{1}{2}} = \sqrt{16} = 4 \qquad 8^{\frac{2}{3}} = \sqrt[3]{8^2} = \sqrt[3]{64} = 4$$

Convert each of the following to a radical and then to a regular number:

(a) $49^{\frac{1}{2}} =$ _____ = _____   (c) $16^{\frac{1}{4}} =$ _____ = _____

(b) $27^{\frac{1}{3}} =$ _____ = _____   (d) $4^{\frac{3}{2}} =$ _____ = _____

a) $\sqrt{m}$ c) $t^{\frac{1}{5}}$
b) $c^{\frac{1}{3}}$ d) $\sqrt[4]{q}$

348. However, though we can always convert a fractional power with a numerical base to a root of a number, we usually need logarithms to find the root. For example:

$$7^{\frac{1}{5}} = \sqrt[5]{7} \qquad 5^{\frac{2}{3}} = \sqrt[3]{25} \qquad 3^{\frac{3}{4}} = \sqrt[4]{27}$$

Convert each of the following to a radical whose radicand is a number:

(a) $2^{\frac{3}{2}} =$ _____   (b) $3^{\frac{2}{5}} =$ _____   (c) $5^{\frac{1}{7}} =$ _____

a) $\sqrt{49} = 7$
b) $\sqrt[3]{27} = 3$
c) $\sqrt[4]{16} = 2$
d) $\sqrt{4^3} = \sqrt{64} = 8$

349. If we apply the law of radicals to a radical whose radicand is a power with a negative exponent, the radical converts to a negative power. For example:

$$\sqrt{3^{-1}} = 3^{-\frac{1}{2}} \qquad \sqrt[3]{5^{-2}} = 5^{-\frac{2}{3}} \qquad \sqrt[4]{p^{-5}} = p^{-\frac{5}{4}}$$

Use the law of radicals to convert each radical below to a power:

(a) $\sqrt[3]{m^{-7}} =$ _____ (b) $\sqrt{7^{-3}} =$ _____ (c) $\sqrt[4]{5^{-1}} =$ _____ (d) $\sqrt[5]{c^{-2}} =$ _____

a) $\sqrt{8}$
b) $\sqrt[5]{9}$
c) $\sqrt[7]{5}$

350. If we convert a power with a negative fractional exponent to a radical, the exponent of the radicand is always negative. For example:

$$7^{-\frac{1}{2}} = \sqrt{7^{-1}} \qquad p^{-\frac{2}{3}} = \sqrt[3]{p^{-2}} \qquad x^{-\frac{7}{5}} = \sqrt[5]{x^{-7}}$$

Convert each of the following powers to radicals:

(a) $m^{-\frac{1}{3}} =$ _____ (b) $5^{-\frac{3}{2}} =$ _____ (c) $t^{-\frac{5}{4}} =$ _____ (d) $7^{-\frac{8}{3}} =$ _____

a) $m^{-\frac{7}{3}}$ c) $5^{-\frac{1}{4}}$
b) $7^{-\frac{3}{2}}$ d) $c^{-\frac{2}{5}}$

351. Convert each radical to a power and each power to a radical:

(a) $\sqrt{t^{-5}} =$ _____ (b) $y^{-\frac{4}{3}} =$ _____ (c) $5^{-\frac{2}{7}} =$ _____ (d) $\sqrt[4]{10^{-5}} =$ _____

a) $\sqrt[3]{m^{-1}}$ c) $\sqrt[4]{t^{-5}}$
b) $\sqrt{5^{-3}}$ d) $\sqrt[3]{7^{-8}}$

a) $t^{-\frac{5}{2}}$ c) $\sqrt[7]{5^{-2}}$
b) $\sqrt[3]{y^{-4}}$ d) $10^{-\frac{5}{4}}$

352. Convert each radical to a power and each power to a radical:

(a) $x^{\frac{3}{2}} = $ _____ (b) $\sqrt[3]{m^{-2}} = $ _____ (c) $\sqrt[3]{t^7} = $ _____ (d) $h^{-\frac{7}{4}} = $ _____

353. Convert each radical to a power and each power to a radical:

(a) $x^{\frac{1}{2}} = $ _____ (b) $\sqrt[3]{y^{-1}} = $ _____ (c) $t^{-\frac{1}{2}} = $ _____ (d) $p^{\frac{1}{4}} = $ _____

a) $\sqrt{x^3}$ c) $t^{\frac{7}{3}}$

b) $m^{-\frac{2}{3}}$ d) $\sqrt[4]{h^{-7}}$

Answer to Frame 353:   a) $\sqrt{x}$ b) $y^{-\frac{1}{3}}$ c) $\sqrt{t^{-1}}$ d) $\sqrt[4]{p}$

SELF-TEST 19 (Frames 339-353)

Convert each of the following radicals to a power:

1. $\sqrt[4]{r^3} = $ _____   2. $\sqrt{P} = $ _____   3. $\sqrt[3]{h^{-1}} = $ _____   4. $\sqrt{19^{-5}} = $ _____   5. $\sqrt[n]{a^m} = $ _____

Convert each of the following powers to a radical:

6. $d^{\frac{1}{3}} = $ _____   7. $t^{-\frac{1}{2}} = $ _____   8. $7^{\frac{8}{5}} = $ _____   9. $N^{-\frac{7}{4}} = $ _____   10. $b^{\frac{s}{t}} = $ _____

Find the numerical value of each of these powers. Each answer is a whole number:

11. $25^{\frac{1}{2}} = $ _____   12. $64^{\frac{1}{3}} = $ _____   13. $8^{\frac{4}{3}} = $ _____   14. $1^{\frac{3}{4}} = $ _____

ANSWERS:
1. $r^{\frac{3}{4}}$
2. $P^{\frac{1}{2}}$
3. $h^{-\frac{1}{3}}$
4. $19^{-\frac{5}{2}}$
5. $a^{\frac{m}{n}}$
6. $\sqrt[3]{d}$
7. $\sqrt{t^{-1}}$
8. $\sqrt[5]{7^8}$
9. $\sqrt[4]{N^{-7}}$
10. $\sqrt[t]{b^s}$
11. 5
12. 4
13. 16
14. 1

4-20 FRACTIONAL EXPONENTS AND THE LAWS OF EXPONENTS AND RADICALS

In this section, we will begin by using the laws of exponents for powers with fractional exponents. Then we will show that each law of radicals is equivalent to a law of exponents by converting some simple problems from radical form to power form and vice versa.

376　Exponents and Radicals

354. The law of exponents for multiplication is given below:

$$a^m \cdot a^n = a^{m+n}$$

This law also applies to powers with fractional exponents. That is:

$$x^{\frac{1}{2}} \cdot x^{\frac{1}{3}} = x^{\frac{1}{2}+\frac{1}{3}} = x^{\frac{3}{6}+\frac{2}{6}} = x^{\frac{5}{6}}$$

Use the law of exponents for multiplication to find each product below:

(a) $y^{\frac{1}{2}} \cdot y^{\frac{1}{4}} =$ _____

(b) $t^{\frac{2}{3}} \cdot t^{\frac{1}{2}} =$

(c) $m^{\frac{3}{2}} \cdot m^{\frac{5}{2}} =$ _____

(d) $R^{\frac{4}{3}} \cdot R^{\frac{3}{2}} =$

a) $y^{\frac{3}{4}}$　　c) m^4

b) $t^{\frac{7}{6}}$　　d) $R^{\frac{17}{6}}$

355. The law of exponents for division is given below:

$$\frac{a^m}{a^n} = a^{m-n}$$

This law also applies to powers with fractional exponents. That is:

$$\frac{a^{\frac{1}{2}}}{a^{\frac{1}{3}}} = a^{\frac{1}{2}-\frac{1}{3}} = a^{\frac{3}{6}-\frac{2}{6}} = a^{\frac{1}{6}}$$

Use the law of exponents for division to find each quotient below:

(a) $\dfrac{c^{\frac{3}{4}}}{c^{\frac{1}{2}}} =$ _____

(b) $\dfrac{p^{\frac{4}{3}}}{p^{\frac{2}{3}}} =$

(c) $\dfrac{d^{\frac{5}{4}}}{d^{\frac{3}{4}}} =$ _____

(d) $\dfrac{t^{\frac{5}{6}}}{t^{\frac{2}{3}}} =$

a) $c^{\frac{1}{4}}$　　c) $d^{\frac{1}{2}}$

b) $p^{\frac{2}{3}}$　　d) $t^{\frac{1}{6}}$

356. Of course, the laws of exponents for multiplication and division apply only to powers with the same base. They cannot be used to perform multiplications and divisions like those below:

$$a^{\frac{1}{2}} \cdot b^{\frac{1}{3}} \qquad \frac{m^{\frac{3}{4}}}{t^{\frac{1}{2}}} \qquad c^{\frac{2}{3}} \cdot p^{\frac{1}{4}} \qquad \frac{d^{\frac{5}{6}}}{r^{\frac{1}{3}}}$$

Use the law of exponents for either multiplication or division to perform each of the following if a law applies:

(a) $F^{\frac{2}{3}} S^{\frac{3}{4}} =$ _____

(b) $\dfrac{t^{\frac{7}{3}}}{t^{\frac{2}{3}}} =$ _____

(c) $h^{\frac{3}{2}} \cdot h^{\frac{1}{4}} =$ _____

(d) $\dfrac{P^{\frac{5}{4}}}{Q^{\frac{3}{4}}} =$ _____

357. Use the law of exponents for either multiplication or division to complete these:

(a) $\dfrac{c^{\frac{4}{3}}}{c^{\frac{1}{3}}} = $ _____

(b) $x^{\frac{2}{3}} \cdot x^{\frac{1}{3}} = $ _____

(c) $\dfrac{y^{\frac{5}{6}}}{y^{\frac{5}{6}}} = $ _____

(d) $m^{\frac{3}{2}} \cdot m^{\frac{1}{2}} = $ _____

a) Does not apply.
b) $t^{\frac{5}{3}}$
c) $h^{\frac{7}{4}}$
d) Does not apply.

358. The law of exponents for raising a multiplication to a power is given below:

$$(ab)^n = a^n b^n$$

This law also applies when a multiplication is raised to a fractional power. For example:

$(xy)^{\frac{1}{2}} = x^{\frac{1}{2}} y^{\frac{1}{2}} \qquad (cd)^{\frac{2}{3}} = c^{\frac{2}{3}} d^{\frac{2}{3}}$

Use the law above to complete each of these:

(a) $(mt)^{\frac{3}{4}} = m^{\Box} t^{\Box}$

(b) $(RS)^{\frac{1}{3}} = R^{\Box} S^{\Box}$

(c) $(pq)^{\frac{5}{3}} = p^{\Box} q^{\Box}$

a) c^1 or c c) y^0 or $+1$
b) x^1 or x d) m^2

359. The law of exponents for raising a fraction to a power is given below:

$$\left(\dfrac{a}{b}\right)^n = \dfrac{a^n}{b^n}$$

This law also applies when the fraction is raised to a fractional power. For example:

$\left(\dfrac{x}{3}\right)^{\frac{1}{2}} = \dfrac{x^{\frac{1}{2}}}{3^{\frac{1}{2}}} \qquad \left(\dfrac{t}{5}\right)^{\frac{2}{3}} = \dfrac{t^{\frac{2}{3}}}{5^{\frac{2}{3}}}$

Using the law above, complete each of these:

(a) $\left(\dfrac{c}{t}\right)^{\frac{5}{2}} = \dfrac{c^{\Box}}{t^{\Box}}$

(b) $\left(\dfrac{V}{F}\right)^{\frac{1}{4}} = \dfrac{V^{\Box}}{F^{\Box}}$

(c) $\left(\dfrac{x}{y}\right)^{\frac{5}{6}} = \dfrac{x^{\Box}}{y^{\Box}}$

a) $m^{\frac{3}{4}} t^{\frac{3}{4}}$
b) $R^{\frac{1}{3}} S^{\frac{1}{3}}$
c) $p^{\frac{5}{3}} q^{\frac{5}{3}}$

378   Exponents and Radicals

360. If we reverse the law of exponents for raising a multiplication to a power, we get:

$$a^n b^n = (ab)^n$$

By using this reversed form of the law, we can write any multiplication of powers <u>with the same exponent</u> as a single power. For example:

$$x^{\frac{1}{2}} y^{\frac{1}{2}} = (xy)^{\frac{1}{2}} \qquad p^{\frac{4}{3}} q^{\frac{4}{3}} = (pq)^{\frac{4}{3}}$$

Using this reversed form of the law, write each of the following as a single power:

(a) $c^{\frac{1}{3}} t^{\frac{1}{3}} =$ _____ (b) $m^{\frac{6}{5}} p^{\frac{6}{5}} =$ _____ (c) $f^{\frac{2}{5}} h^{\frac{2}{5}} =$ _____

a) $\dfrac{c^{\frac{5}{2}}}{t^{\frac{5}{2}}}$

b) $\dfrac{V^{\frac{1}{4}}}{F^{\frac{1}{4}}}$

c) $\dfrac{x^{\frac{5}{6}}}{y^{\frac{5}{6}}}$

---

361. Don't confuse this reversed form of the law for raising a multiplication to a power with the law for multiplication.

The reversed form of the law for raising a multiplication to a power is used to convert a multiplication of powers <u>with the same exponent</u> to a single power. For example:

$$x^{\frac{2}{3}} y^{\frac{2}{3}} = (xy)^{\frac{2}{3}} \qquad p^{\frac{1}{5}} q^{\frac{1}{5}} = (pq)^{\frac{1}{5}}$$

The law for multiplication is used to convert a multiplication of powers <u>with the same base</u> to a single power. For example:

$$x^{\frac{1}{5}} x^{\frac{3}{5}} = x^{\frac{4}{5}} \qquad p^{\frac{2}{3}} p^{\frac{5}{3}} = p^{\frac{7}{3}}$$

Using the appropriate law, convert each multiplication below to a single power:

(a) $m^{\frac{3}{4}} t^{\frac{3}{4}} =$ _____ (c) $y^{\frac{2}{5}} y^{\frac{4}{5}} =$ _____

(b) $q^{\frac{1}{3}} q^{\frac{4}{3}} =$ _____ (d) $a^{\frac{5}{2}} b^{\frac{5}{2}} =$ _____

a) $(ct)^{\frac{1}{3}}$

b) $(mp)^{\frac{6}{5}}$

c) $(fh)^{\frac{2}{5}}$

---

362. If we reverse the law of exponents for raising a fraction to a power, we get:

$$\dfrac{a^n}{b^n} = \left(\dfrac{a}{b}\right)^n$$

By using this reversed form of the law, we can write any division of powers <u>with the same exponent</u> as a single power. For example:

$$\dfrac{x^{\frac{2}{3}}}{y^{\frac{2}{3}}} = \left(\dfrac{x}{y}\right)^{\frac{2}{3}} \qquad \dfrac{p^{\frac{1}{4}}}{q^{\frac{1}{4}}} = \left(\dfrac{p}{q}\right)^{\frac{1}{4}}$$

Using this reversed form of the law, write each of the following as a single power:

(a) $\dfrac{a^{\frac{1}{2}}}{b^{\frac{1}{2}}} =$ _____ (b) $\dfrac{c^{\frac{5}{3}}}{d^{\frac{5}{3}}} =$ _____ (c) $\dfrac{R^{\frac{3}{4}}}{S^{\frac{3}{4}}} =$ _____

a) $(mt)^{\frac{3}{4}}$ c) $y^{\frac{6}{5}}$

b) $q^{\frac{5}{3}}$ d) $(ab)^{\frac{5}{2}}$

Exponents and Radicals   379

363. Don't confuse this reversed form of the law for raising a fraction to a power with the law for division.

The reversed form of the law for raising a fraction to a power is used to convert a division of powers <u>with the same exponent</u> to a single power. For example:

$$\frac{x^{\frac{2}{5}}}{y^{\frac{2}{5}}} = \left(\frac{x}{y}\right)^{\frac{2}{5}} \qquad \frac{p^{\frac{1}{6}}}{q^{\frac{1}{6}}} = \left(\frac{p}{q}\right)^{\frac{1}{6}}$$

The law for division is used to convert a division of powers <u>with the same base</u> to a single power. For example:

$$\frac{x^{\frac{4}{3}}}{x^{\frac{2}{3}}} = x^{\frac{2}{3}} \qquad \frac{y^{\frac{4}{5}}}{y^{\frac{1}{5}}} = y^{\frac{3}{5}}$$

Using the appropriate law, convert each division below to a single power:

(a) $\dfrac{t^{\frac{5}{6}}}{t^{\frac{1}{6}}} = $ _____

(b) $\dfrac{c^{\frac{5}{6}}}{d^{\frac{5}{6}}} = $ _____

(c) $\dfrac{a^{\frac{1}{3}}}{h^{\frac{1}{3}}} = $ _____

(d) $\dfrac{y^{\frac{7}{5}}}{y^{\frac{4}{5}}} = $ _____

a) $\left(\dfrac{a}{b}\right)^{\frac{1}{2}}$

b) $\left(\dfrac{c}{d}\right)^{\frac{5}{3}}$

c) $\left(\dfrac{R}{S}\right)^{\frac{3}{4}}$

364. The law of exponents for raising a power to a power is given below:

$$\boxed{(a^m)^n = a^{mn}}$$

This law also applies to fractional powers. For example:

$$(b^2)^{\frac{1}{3}} = b^{(2)\left(\frac{1}{3}\right)} = b^{\frac{2}{3}} \qquad \left(t^{\frac{1}{2}}\right)^5 = t^{\left(\frac{1}{2}\right)(5)} = t^{\frac{5}{2}} \qquad \left(V^{\frac{1}{4}}\right)^{\frac{3}{2}} = V^{\left(\frac{1}{4}\right)\left(\frac{3}{2}\right)} = V^{\frac{3}{8}}$$

Using the law above, complete each of these:

(a) $(x^3)^{\frac{1}{4}} = $ _____

(b) $\left(y^{\frac{2}{3}}\right)^2 = $ _____

(c) $\left(p^{\frac{1}{2}}\right)^{\frac{4}{3}} = $ _____

a) $t^{\frac{2}{3}}$ c) $\left(\dfrac{a}{h}\right)^{\frac{1}{3}}$

b) $\left(\dfrac{c}{d}\right)^{\frac{5}{6}}$ d) $y^{\frac{3}{5}}$

365. Using the same law, complete each of these:

(a) $(V^4)^{\frac{1}{4}} = $ _____

(b) $\left(t^{\frac{1}{2}}\right)^2 = $ _____

(c) $\left(m^{\frac{5}{3}}\right)^{\frac{3}{5}} = $ _____

a) $x^{\frac{3}{4}}$

b) $y^{\frac{4}{3}}$

c) $p^{\frac{2}{3}}$

a) V, from V^1

b) t, from t^1

c) m, from m^1

366. The law of radicals for the root of a power was used to define what is meant by a power with a fractional exponent. That law and the law of exponents for raising a power to a power are given below:

$$\sqrt[n]{a^m} = a^{\frac{m}{n}} \qquad (a^m)^n = a^{mn}$$

When "n" in the law of exponents (on the right, above) is a fraction whose numerator is "1", these two laws are equivalent. We can show this fact by converting some problems from power form to radical form. For example:

$$(x^6)^{\frac{1}{3}} = x^2 \qquad (y^5)^{\frac{1}{2}} = y^{\frac{5}{2}}$$
$$\downarrow \qquad \downarrow \qquad \qquad \downarrow \qquad \downarrow$$
$$\sqrt[3]{x^6} = x^2 \qquad \sqrt{y^5} = y^{\frac{5}{2}}$$

Convert each problem below to radical form:

(a) $(t^6)^{\frac{1}{2}} = t^3$ (b) $(y^8)^{\frac{1}{4}} = y^2$ (c) $(m^4)^{\frac{1}{3}} = m^{\frac{4}{3}}$
 $\downarrow \quad \downarrow$ $\qquad \downarrow \quad \downarrow$ $\qquad \downarrow \quad \downarrow$

___ = ___ ___ = ___ ___ = ___

367. Similarly, we can convert problems from radical form to power form. For example:

$$\sqrt{b^{10}} = b^5 \qquad \sqrt[3]{c^7} = c^{\frac{7}{3}}$$
$$\downarrow \quad \downarrow \qquad \qquad \downarrow \quad \downarrow$$
$$(b^{10})^{\frac{1}{2}} = b^5 \qquad (c^7)^{\frac{1}{3}} = c^{\frac{7}{3}}$$

Convert each problem below from radical form to power form:

(a) $\sqrt[3]{t^9} = t^3$ (b) $\sqrt[4]{p^{16}} = p^4$ (c) $\sqrt{d^3} = d^{\frac{3}{2}}$
 $\downarrow \quad \downarrow$ $\qquad \downarrow \quad \downarrow$ $\qquad \downarrow \quad \downarrow$

___ = ___ ___ = ___ ___ = ___

Answers:
a) $\sqrt{t^6} = t^3$
b) $\sqrt[4]{y^8} = y^2$
c) $\sqrt[3]{m^4} = m^{\frac{4}{3}}$

368. The law of exponents for raising a power to a power and the law of radicals for raising a radical to a power are given below:

$$(a^m)^n = a^{mn} \qquad \left(\sqrt[n]{a}\right)^m = a^{\frac{m}{n}}$$

When "m" in the law of exponents (on the left, above) is a fraction whose numerator is "1", these two laws are equivalent. We can see this fact by converting some problems from power form to radical form. That is:

$$\left(x^{\frac{1}{3}}\right)^9 = x^3 \qquad \left(y^{\frac{1}{2}}\right)^5 = y^{\frac{5}{2}}$$
$$\downarrow \quad \downarrow \qquad \qquad \downarrow \quad \downarrow$$
$$\left(\sqrt[3]{x}\right)^9 = x^3 \qquad \left(\sqrt{y}\right)^5 = y^{\frac{5}{2}}$$

Answers:
a) $(t^9)^{\frac{1}{3}} = t^3$
b) $(p^{16})^{\frac{1}{4}} = p^4$
c) $(d^3)^{\frac{1}{2}} = d^{\frac{3}{2}}$

Continued on following page.

368. Continued

Convert each problem below to radical form:

(a) $\left(t^{\frac{1}{2}}\right)^6 = t^3$ ↓ ↓

___ = ___

(b) $\left(c^{\frac{1}{4}}\right)^8 = c^2$ ↓ ↓

___ = ___

(c) $\left(v^{\frac{1}{3}}\right)^2 = v^{\frac{2}{3}}$ ↓ ↓

___ = ___

a) $\left(\sqrt{t}\right)^6 = t^3$

b) $\left(\sqrt[4]{c}\right)^8 = c^2$

c) $\left(\sqrt[3]{v}\right)^2 = v^{\frac{2}{3}}$

---

369. Similarly, we can convert problems from radical form to power form. For example:

$\left(\sqrt{x}\right)^4 = x^2$ ↓ ↓
$\left(x^{\frac{1}{2}}\right)^4 = x^2$

$\left(\sqrt[3]{b}\right)^5 = b^{\frac{5}{3}}$ ↓ ↓
$\left(b^{\frac{1}{3}}\right)^5 = b^{\frac{5}{3}}$

Convert each of these problems from radical form to power form:

(a) $\left(\sqrt[3]{d}\right)^{12} = d^4$ ↓ ↓

___ = ___

(b) $\left(\sqrt[5]{p}\right)^{10} = p^2$ ↓ ↓

___ = ___

(c) $\left(\sqrt{y}\right)^3 = y^{\frac{3}{2}}$ ↓ ↓

___ = ___

a) $\left(d^{\frac{1}{3}}\right)^{12} = d^4$

b) $\left(p^{\frac{1}{5}}\right)^{10} = p^2$

c) $\left(y^{\frac{1}{2}}\right)^3 = y^{\frac{3}{2}}$

---

370. The law of exponents for raising a multiplication to a power and the law of radicals for the root of a multiplication are given below:

$(ab)^n = a^n b^n$

$\sqrt[n]{ab} = \sqrt[n]{a} \cdot \sqrt[n]{b}$

When "n" in the law of exponents (on the left, above) is a fraction whose numerator is "1", these two laws are equivalent. We can show this fact by converting problems from power form to radical form. For example:

$(xy)^{\frac{1}{2}} = x^{\frac{1}{2}} y^{\frac{1}{2}}$ ↓ ↓ ↓
$\sqrt{xy} = \sqrt{x} \cdot \sqrt{y}$

$(cd)^{\frac{1}{3}} = c^{\frac{1}{3}} d^{\frac{1}{3}}$ ↓ ↓ ↓
$\sqrt[3]{cd} = \sqrt[3]{c} \cdot \sqrt[3]{d}$

Convert each of the following from power form to radical form:

(a) $(mt)^{\frac{1}{4}} = m^{\frac{1}{4}} \cdot t^{\frac{1}{4}}$ ↓ ↓ ↓

___ = ___ · ___

(b) $(pq)^{\frac{1}{5}} = p^{\frac{1}{5}} \cdot q^{\frac{1}{5}}$ ↓ ↓ ↓

___ = ___ · ___

---

371. By reversing the procedure in the last frame, we can convert the problems below from radical form to an equivalent power form. Do so:

(a) $\sqrt{3x} = \sqrt{3} \cdot \sqrt{x}$ ↓ ↓ ↓

___ = ___

(b) $\sqrt[3]{5y} = \sqrt[3]{5} \cdot \sqrt[3]{y}$ ↓ ↓ ↓

___ = ___

a) $\sqrt[4]{mt} = \sqrt[4]{m} \cdot \sqrt[4]{t}$

b) $\sqrt[5]{pq} = \sqrt[5]{p} \cdot \sqrt[5]{q}$

---

a) $(3x)^{\frac{1}{2}} = 3^{\frac{1}{2}} x^{\frac{1}{2}}$

b) $(5y)^{\frac{1}{3}} = 5^{\frac{1}{3}} y^{\frac{1}{3}}$

382 Exponents and Radicals

372. If we reverse the law of radicals for the root of a multiplication, we get the law of radicals for multiplying radicals <u>of the same order</u>. The reversed form of that law and the reversed form of the law of exponents for raising a multiplication to a power are given below:

$$\boxed{\sqrt[n]{a} \cdot \sqrt[n]{b} = \sqrt[n]{ab}} \qquad \boxed{a^n b^n = (ab)^n}$$

When "n" in the law of exponents (on the right, above) is a fraction whose numerator is "1", these two laws are equivalent. Here are some examples:

$$x^{\frac{1}{2}} \cdot y^{\frac{1}{2}} = (xy)^{\frac{1}{2}} \qquad p^{\frac{1}{3}} \cdot q^{\frac{1}{3}} = (pq)^{\frac{1}{3}}$$
$$\downarrow \quad \downarrow \quad \downarrow \qquad \qquad \downarrow \quad \downarrow \quad \downarrow$$
$$\sqrt{x} \cdot \sqrt{y} = \sqrt{xy} \qquad \sqrt[3]{p} \cdot \sqrt[3]{q} = \sqrt[3]{pq}$$

Convert each equation below to radical form:

(a) $m^{\frac{1}{2}} \cdot s^{\frac{1}{2}} = (ms)^{\frac{1}{2}}$ (b) $p^{\frac{1}{3}} \cdot q^{\frac{1}{3}} = (pq)^{\frac{1}{3}}$
   $\downarrow \quad \downarrow \quad \downarrow \qquad \qquad \downarrow \quad \downarrow \quad \downarrow$

____ • ____ = ____   ____ • ____ = ____

373. Convert each equation below to power form:

(a) $\sqrt{3} \cdot \sqrt{t} = \sqrt{3t}$ (b) $\sqrt[4]{5} \cdot \sqrt[4]{y} = \sqrt[4]{5y}$
   $\downarrow \quad \downarrow \quad \downarrow \qquad \qquad \downarrow \quad \downarrow \quad \downarrow$

____ • ____ = ____   ____ • ____ = ____

a) $\sqrt{m} \cdot \sqrt{s} = \sqrt{ms}$
b) $\sqrt[3]{p} \cdot \sqrt[3]{q} = \sqrt[3]{pq}$

374. The law of exponents for raising a fraction to a power and the law of radicals for the root of a fraction are given below:

$$\boxed{\left(\frac{a}{b}\right)^n = \frac{a^n}{b^n}} \qquad \boxed{\sqrt[n]{\frac{a}{b}} = \frac{\sqrt[n]{a}}{\sqrt[n]{b}}}$$

When "n" in the law of exponents (on the left, above) is a fraction whose numerator is "1", these two laws are equivalent. For example:

$$\left(\frac{x}{y}\right)^{\frac{1}{2}} = \frac{x^{\frac{1}{2}}}{y^{\frac{1}{2}}} \qquad \left(\frac{t}{5}\right)^{\frac{1}{3}} = \frac{t^{\frac{1}{3}}}{5^{\frac{1}{3}}}$$
$$\downarrow \qquad \downarrow \qquad \qquad \downarrow \qquad \downarrow$$
$$\sqrt{\frac{x}{y}} = \frac{\sqrt{x}}{\sqrt{y}} \qquad \sqrt[3]{\frac{t}{5}} = \frac{\sqrt[3]{t}}{\sqrt[3]{5}}$$

a) $3^{\frac{1}{2}} \cdot t^{\frac{1}{2}} = (3t)^{\frac{1}{2}}$
b) $5^{\frac{1}{4}} \cdot y^{\frac{1}{4}} = (5y)^{\frac{1}{4}}$

Convert each problem below to radical form:

(a) $\left(\frac{c}{d}\right)^{\frac{1}{2}} = \frac{c^{\frac{1}{2}}}{d^{\frac{1}{2}}}$ (b) $\left(\frac{m}{7}\right)^{\frac{1}{4}} = \frac{m^{\frac{1}{4}}}{7^{\frac{1}{4}}}$
   $\downarrow \qquad \downarrow \qquad \qquad \downarrow \qquad \downarrow$

____ = ____   ____ = ____

375. Convert each problem below to power form:

(a) $\sqrt[3]{\dfrac{t}{s}} = \dfrac{\sqrt[3]{t}}{\sqrt[3]{s}}$ ↓ ↓

(b) $\sqrt[5]{\dfrac{y}{3}} = \dfrac{\sqrt[5]{y}}{\sqrt[5]{3}}$ ↓ ↓

= _____   = _____

a) $\sqrt{\dfrac{c}{d}} = \dfrac{\sqrt{c}}{\sqrt{d}}$

b) $\sqrt[4]{\dfrac{m}{7}} = \dfrac{\sqrt[4]{m}}{\sqrt[4]{7}}$

376. If we reverse the law of radicals for the root of a fraction, we get the law of radicals for dividing radicals <u>of the same order</u>. The reversed form of that law and the reversed form of the law of exponents for raising a fraction to a power are given below:

$$\boxed{\dfrac{\sqrt[n]{a}}{\sqrt[n]{b}} = \sqrt[n]{\dfrac{a}{b}}} \qquad \boxed{\dfrac{a^n}{b^n} = \left(\dfrac{a}{b}\right)^n}$$

When "n" in the law of exponents (on the right, above) is a fraction whose numerator is "1", these two laws are equivalent. For example:

$\dfrac{p^{\frac{1}{2}}}{q^{\frac{1}{2}}} = \left(\dfrac{p}{q}\right)^{\frac{1}{2}}$ $\dfrac{t^{\frac{1}{3}}}{5^{\frac{1}{3}}} = \left(\dfrac{t}{5}\right)^{\frac{1}{3}}$
↓ ↓   ↓ ↓
$\dfrac{\sqrt{p}}{\sqrt{q}} = \sqrt{\dfrac{p}{q}}$ $\dfrac{\sqrt[3]{t}}{\sqrt[3]{5}} = \sqrt[3]{\dfrac{t}{5}}$

Convert each problem below to radical form:

(a) $\dfrac{m^{\frac{1}{2}}}{t^{\frac{1}{2}}} = \left(\dfrac{m}{t}\right)^{\frac{1}{2}}$ ↓ ↓

(b) $\dfrac{p^{\frac{1}{4}}}{q^{\frac{1}{4}}} = \left(\dfrac{p}{q}\right)^{\frac{1}{4}}$ ↓ ↓

= _____   = _____

a) $\left(\dfrac{t}{s}\right)^{\frac{1}{3}} = \dfrac{t^{\frac{1}{3}}}{s^{\frac{1}{3}}}$

b) $\left(\dfrac{y}{3}\right)^{\frac{1}{5}} = \dfrac{y^{\frac{1}{5}}}{3^{\frac{1}{5}}}$

377. Convert each problem below to power form:

(a) $\dfrac{\sqrt[4]{s}}{\sqrt[4]{b}} = \sqrt[4]{\dfrac{s}{b}}$ ↓ ↓

(b) $\dfrac{\sqrt[5]{10}}{\sqrt[5]{a}} = \sqrt[5]{\dfrac{10}{a}}$ ↓ ↓

= _____   = _____

a) $\dfrac{\sqrt{m}}{\sqrt{t}} = \sqrt{\dfrac{m}{t}}$

b) $\dfrac{\sqrt[4]{p}}{\sqrt[4]{q}} = \sqrt[4]{\dfrac{p}{q}}$

Answer to Frame 377:   a) $\dfrac{s^{\frac{1}{4}}}{b^{\frac{1}{4}}} = \left(\dfrac{s}{b}\right)^{\frac{1}{4}}$ b) $\dfrac{10^{\frac{1}{5}}}{a^{\frac{1}{5}}} = \left(\dfrac{10}{a}\right)^{\frac{1}{5}}$

384 Exponents and Radicals

378. In the preceding frames, we have shown that each law of radicals is equivalent to a law of exponents. This equivalence is summarized in the table below.

Law of Radicals	Law of Exponents
$\sqrt[n]{a^m} = a^{\frac{m}{n}}$	$(a^m)^n = a^{mn}$
$\left(\sqrt[n]{a}\right)^m = a^{\frac{m}{n}}$	$(a^m)^n = a^{mn}$
$\sqrt[n]{ab} = \sqrt[n]{a} \cdot \sqrt[n]{b}$	$(ab)^n = a^n b^n$
$\sqrt[n]{\dfrac{a}{b}} = \dfrac{\sqrt[n]{a}}{\sqrt[n]{b}}$	$\left(\dfrac{a}{b}\right)^n = \dfrac{a^n}{b^n}$

The equivalence of the laws of radicals and the laws of exponents has been shown by converting some simple problems from radical form to power form and vice versa. By combining the laws above, the equivalence of the two sets of laws could be shown in many more-complicated problems.

---

379. The law of exponents for multiplying powers <u>with the same base</u> is given below:

$$a^m \cdot a^n = a^{m+n}$$

Though this law is not equivalent to any law of radicals, it can be used to multiply radicals in some cases in which the law of radicals for multiplication does not apply. For example:

The law of radicals for multiplication cannot be used below <u>because the radicals have different orders</u>. However, we can perform the multiplication by means of the law of exponents above after converting to power form. We get:

$$\sqrt{x} \cdot \sqrt[3]{x} = \sqrt[6]{x^5}$$
$$\downarrow \quad \downarrow \quad \uparrow$$
$$x^{\frac{1}{2}} \cdot x^{\frac{1}{3}} = x^{\frac{5}{6}}$$

Note: Since each radical has <u>the same radicand</u>, the law of exponents can be used because the powers have <u>the same base</u>.

Perform each multiplication below by converting to power form:

(a) $\sqrt{y} \cdot \sqrt[4]{y} = $ _____
$\quad \downarrow \quad \downarrow \quad \uparrow$

_____ $\cdot$ _____ = _____

(b) $\sqrt[3]{t} \cdot \sqrt[5]{t} = $ _____
$\quad \downarrow \quad \downarrow \quad \uparrow$

_____ $\cdot$ _____ = _____

---

a) $= \sqrt[4]{y^3}$
$\uparrow$
$y^{\frac{1}{2}} \cdot y^{\frac{1}{4}} = y^{\frac{3}{4}}$

b) $= \sqrt[15]{t^8}$
$\uparrow$
$t^{\frac{1}{3}} \cdot t^{\frac{1}{5}} = t^{\frac{8}{15}}$

380. The law of exponents for dividing powers <u>with</u> <u>the</u> <u>same</u> <u>base</u> is given below:

$$\frac{a^m}{a^n} = a^{m-n}$$

Though this law is not equivalent to any law of radicals, it can be used to divide radicals in some cases in which the law of radicals for division does not apply. For example:

The law of radicals for division cannot be used below <u>because</u> <u>the</u> <u>radicals</u> <u>have</u> <u>different</u> <u>orders</u>. However, we can perform the division by means of the law of exponents above after converting to power form. We get:

$$\frac{\sqrt{y}}{\sqrt[3]{y}} = \sqrt[6]{y}$$

$$\frac{y^{\frac{1}{2}}}{y^{\frac{1}{3}}} = y^{\frac{1}{6}}$$

<u>Note</u>: Since each radical has <u>the</u> <u>same</u> <u>radicand</u>, the law of exponents can be used because the powers have <u>the</u> <u>same</u> <u>base</u>.

Perform each division below by converting to power form:

(a) $\dfrac{\sqrt{p}}{\sqrt[4]{p}} =$ _____

(b) $\dfrac{\sqrt[3]{t}}{\sqrt[4]{t}} =$ _____

= _____

= _____

---

a) $\quad = \sqrt[4]{p}$

$\dfrac{p^{\frac{1}{2}}}{p^{\frac{1}{4}}} = p^{\frac{1}{4}}$

b) $\quad = \sqrt[12]{t}$

$\dfrac{t^{\frac{1}{3}}}{t^{\frac{1}{4}}} = t^{\frac{1}{12}}$

386   Exponents and Radicals

SELF-TEST 20 (Frames 354-380)

Complete the following laws of exponents:

1. $a^m \cdot a^n =$ ____
2. $\dfrac{a^m}{a^n} =$ ____
3. $(a^m)^n =$ ____
4. $(ab)^n =$ ____
5. $\left(\dfrac{a}{b}\right)^n =$ ____

Work Problems 6 to 9 using the laws of exponents. Write each answer in power form:

6. $\dfrac{r^{\frac{5}{6}}}{r^{\frac{1}{2}}} =$
7. $\left(x^{\frac{3}{2}}\right)^{\frac{1}{3}} =$ ____
8. $x^{\frac{3}{2}} \cdot x^{\frac{1}{3}} =$ ____
9. $(a^2 b)^{\frac{3}{4}} =$ ____

10. Write this equation in radical form:
$b^{\frac{1}{4}} \cdot d^{\frac{1}{4}} = (bd)^{\frac{1}{4}}$ ____

11. Write this equation in power form:
$\sqrt[3]{\dfrac{t}{w}} = \dfrac{\sqrt[3]{t}}{\sqrt[3]{w}}$ ____

Convert each radical to power form:

12. $\left(\sqrt[3]{G}\right)^5 =$ ____
13. $\left(\sqrt[4]{A^2}\right)^3 =$ ____

Write the product in radical form:

14. $\sqrt{x} \cdot \sqrt[4]{x} =$ ____

Write the quotient in radical form:

15. $\dfrac{\sqrt{y}}{\sqrt[4]{y}} =$ ____

ANSWERS:

1. a^{m+n}
2. a^{m-n}
3. a^{mn}
4. $a^n b^n$
5. $\dfrac{a^n}{b^n}$
6. $r^{\frac{1}{3}}$
7. $x^{\frac{1}{2}}$
8. $x^{\frac{11}{6}}$
9. $a^{\frac{3}{2}} b^{\frac{3}{4}}$
10. $\sqrt[4]{b} \cdot \sqrt[4]{d} = \sqrt[4]{bd}$
11. $\left(\dfrac{t}{w}\right)^{\frac{1}{3}} = \dfrac{t^{\frac{1}{3}}}{w^{\frac{1}{3}}}$
12. $G^{\frac{5}{3}}$
13. $A^{\frac{3}{2}}$
14. $\sqrt[4]{x^3}$
15. $\sqrt[4]{y}$

4-21 POWERS WITH DECIMAL EXPONENTS

In this section, we will briefly discuss the meaning of powers with decimal exponents. We will show that the laws of exponents also apply to powers of this type.

381. We have seen that any radical can be converted to a power with a fractional exponent. Carrying this conversion one more step, we can convert any radical to a power with a "decimal" exponent. For example:

$$\sqrt{4} = 4^{\frac{1}{2}} = 4^{0.5} \qquad\qquad \sqrt[4]{3^5} = 3^{\frac{5}{4}} = 3^{1.25}$$

Following the steps above, convert each radical below to a power with a fractional exponent and then to a power with a decimal exponent:

(a) $\sqrt{27} =$ ____ = ____

(b) $\sqrt[4]{16} =$ ____ = ____

(c) $\sqrt[5]{12^4} =$ ____ = ____

(d) $\sqrt[5]{3^6} =$ ____ = ____

382. When converting from a fractional exponent to a decimal exponent, we usually report the decimal exponent to two places beyond the decimal point. For example:

$$x^{\frac{1}{3}} = x^{0.33} \qquad y^{\frac{2}{3}} = y^{0.67} \qquad p^{\frac{5}{6}} = p^{0.83}$$

Convert each radical below to a power with a fractional exponent and then to a power with a decimal exponent:

(a) $\sqrt[3]{7}$ = _____ = _____

(b) $\sqrt[3]{m^5}$ = _____ = _____

(c) $\sqrt[6]{5^7}$ = _____ = _____

(d) $\sqrt[7]{t^3}$ = _____ = _____

a) $27^{\frac{1}{2}} = 27^{0.5}$

b) $16^{\frac{1}{4}} = 16^{0.25}$

c) $12^{\frac{4}{5}} = 12^{0.8}$

d) $3^{\frac{6}{5}} = 3^{1.2}$

---

383. By reversing the procedure in the last frame, we can convert powers with decimal exponents to radicals. For example:

$$9^{0.5} = 9^{\frac{1}{2}} = \sqrt{9} \qquad\qquad t^{0.67} = t^{\frac{2}{3}} = \sqrt[3]{t^2}$$

Convert each power below to a power with a fractional exponent and then to a radical:

(a) $7^{0.33}$ = _____ = _____

(b) $p^{0.25}$ = _____ = _____

(c) $6^{0.8}$ = _____ = _____

(d) $y^{1.75}$ = _____ = _____

a) $7^{\frac{1}{3}} = 7^{0.33}$

b) $m^{\frac{5}{3}} = m^{1.67}$

c) $5^{\frac{7}{6}} = 5^{1.17}$

d) $t^{\frac{3}{7}} = t^{0.43}$

---

384. Convert each radical to a power with a decimal exponent and vice versa:

(a) $\sqrt{p^3}$ = _____

(b) $t^{2.5}$ = _____

(c) $\sqrt[3]{m^2}$ = _____

(d) $x^{1.33}$ = _____

a) $7^{\frac{1}{3}} = \sqrt[3]{7}$

b) $p^{\frac{1}{4}} = \sqrt[4]{p}$

c) $6^{\frac{4}{5}} = \sqrt[5]{6^4}$

d) $y^{\frac{7}{4}} = \sqrt[4]{y^7}$

---

385. By converting them to radicals first, some powers with numerical bases and decimal exponents can easily be converted to regular numbers. For example:

$$16^{0.5} = 16^{\frac{1}{2}} = \sqrt{16} = 4 \qquad\qquad 8^{0.33} = 8^{\frac{1}{3}} = \sqrt[3]{8} = 2$$

Following the steps above, convert each power below to a regular number:

(a) $36^{0.5}$ = _____ = _____ = _____

(b) $4^{1.5}$ = _____ = _____ = _____

(c) $27^{0.33}$ = _____ = _____ = _____

(d) $8^{0.67}$ = _____ = _____ = _____

a) $p^{1.5}$

b) $\sqrt{t^5}$

c) $m^{0.67}$

d) $\sqrt[3]{x^4}$

---

a) $36^{\frac{1}{2}} = \sqrt{36} = 6$

b) $4^{\frac{3}{2}} = \sqrt{4^3}$ or $\sqrt{64} = 8$

c) $27^{\frac{1}{3}} = \sqrt[3]{27} = 3$

d) $8^{\frac{2}{3}} = \sqrt[3]{8^2}$ or $\sqrt[3]{64} = 4$

388 Exponents and Radicals

386. The laws of exponents also apply to powers with decimal exponents. The laws of exponents for multiplying or dividing powers with the same base are given below:

$$a^m \cdot a^n = a^{m+n} \qquad \frac{a^m}{a^n} = a^{m-n}$$

Using the laws above, complete each of these:

(a) $x^{0.5} \cdot x^{0.4} =$ _____

(b) $\dfrac{y^{0.75}}{y^{0.25}} =$ _____

(c) $t^{1.5} \cdot t^{0.5} =$ _____

(d) $\dfrac{R^{2.5}}{R^{1.2}} =$ _____

---

387. The law of exponents for raising a power to a power is given below:

$$(a^m)^n = a^{mn}$$

Using this law, complete each of these:

(a) $(x^{0.5})^3 =$ _____ (b) $(y^4)^{0.2} =$ _____ (c) $(m^{1.4})^{0.4} =$ _____

a) $x^{0.9}$
b) $y^{0.50}$ or $y^{0.5}$
c) $t^{2.0}$ or t^2
d) $R^{1.3}$

---

388. The laws of exponents for raising a multiplication or fraction to a power are given below:

$$(ab)^n = a^n b^n \qquad \left(\frac{a}{b}\right)^n = \frac{a^n}{b^n}$$

Using the laws above, complete these:

(a) $(3x)^{0.2} = 3^{\boxed{}} x^{\boxed{}}$

(b) $\left(\dfrac{y}{4}\right)^{0.5} = \dfrac{y^{\boxed{}}}{4^{\boxed{}}}$

(c) $(pq)^{1.5} = p^{\boxed{}} q^{\boxed{}}$

(d) $\left(\dfrac{s}{t}\right)^{2.4} = \dfrac{s^{\boxed{}}}{t^{\boxed{}}}$

a) $x^{1.5}$
b) $y^{0.8}$
c) $m^{0.56}$

---

389. The reversed form of each law in the last frame is given below. By using these reversed forms, we can write a multiplication or division of powers with the same exponents but different bases as a single power.

$$a^n b^n = (ab)^n \qquad \frac{a^n}{b^n} = \left(\frac{a}{b}\right)^n$$

Using the laws above, write each multiplication and division below as a single power:

(a) $\dfrac{m^{0.33}}{t^{0.33}} =$ _____

(b) $c^{0.4} \cdot d^{0.4} =$ _____

(c) $\dfrac{x^{2.5}}{y^{2.5}} =$ _____

(d) $p^{1.7} \cdot q^{1.7} =$ _____

a) $3^{\boxed{0.2}} x^{\boxed{0.2}}$

b) $\dfrac{y^{\boxed{0.5}}}{4^{\boxed{0.5}}}$

c) $p^{\boxed{1.5}} q^{\boxed{1.5}}$

d) $\dfrac{s^{\boxed{2.4}}}{t^{\boxed{2.4}}}$

390. By using the appropriate law of exponents, write each multiplication and division below as a single power:

(a) $x^{2.2} \cdot x^{1.7} = $ _____

(b) $\dfrac{V^{0.7}}{R^{0.7}} = $ _____

(c) $c^{1.6} \cdot t^{1.6} = $ _____

(d) $\dfrac{h^{1.9}}{h^{1.2}} = $ _____

a) $\left(\dfrac{m}{t}\right)^{0.33}$ c) $\left(\dfrac{x}{y}\right)^{2.5}$

b) $(cd)^{0.4}$ d) $(pq)^{1.7}$

Answer to Frame 390:   a) $x^{3.9}$ b) $\left(\dfrac{V}{R}\right)^{0.7}$ c) $(ct)^{1.6}$ d) $h^{0.7}$

SELF-TEST 21 (Frames 381–390)

Convert each radical to a power with a decimal exponent:

1. $\sqrt{7} = $ _____
2. $\sqrt[3]{R} = $ _____
3. $\sqrt{8^5} = $ _____
4. $\sqrt[6]{10^5} = $ _____

Convert each power to a radical:

5. $10^{0.75} = $ _____
6. $3^{1.50} = $ _____
7. $V^{0.40} = $ _____
8. $5^{2.67} = $ _____

9. The power $9^{1.50}$ is equal to a whole number. Find it. $9^{1.50} = $ _____

Using the laws of exponents, do these problems. Write each answer as a single power:

10. $r^{0.5} \cdot r^{0.5} = $ _____
11. $\dfrac{x^{2.85}}{x^{1.24}} = $ _____
12. $(t^2)^{1.5} = $ _____
13. $A^{0.58} \cdot B^{0.58} = $ _____

ANSWERS: 1. $7^{0.5}$ 3. $8^{2.5}$ 5. $\sqrt[4]{10^3}$ 7. $\sqrt[5]{V^2}$ 9. 27   10. r^1 or r 12. t^3
2. $R^{0.33}$ 4. $10^{0.83}$ 6. $\sqrt{3^3}$ 8. $\sqrt[3]{5^8}$ 11. $x^{1.61}$ 13. $(AB)^{0.58}$

Chapter 5 LINEAR GRAPHS AND SLOPE

In earlier work, we introduced the coordinate system and graphed both linear and curvilinear equations and formulas on it. In this chapter, we will discuss linear equations and linear graphs in greater detail. The following major topics are included:

(1) A method of identifying linear equations and formulas.
(2) The intercepts of linear graphs.
(3) The slope of linear graphs.
(4) The slope-intercept form of linear equations and formulas.

The chapter is divided into two major parts. In the first part (Sections 5-1 to 5-14), the discussion is limited to linear equations in which the two variables are "x" and "y". In the second part (Sections 5-15 to 5-23), the same concepts are extended to linear equations and formulas in which the two variables are not "x" and "y".

5-1 IDENTIFYING LINEAR EQUATIONS IN "x" AND "y"

The graphs of some equations are <u>linear</u> (or straight lines). The graphs of other equations are <u>curvilinear</u> (or curves). In this section, we will discuss a method of identifying <u>linear</u> equations. The discussion will be limited to two-variable equations in which the variables are "x" and "y".

---

1. The following terms are called "<u>linear</u>" terms. Notice that each contains a <u>single non-squared</u> letter with a numerical coefficient.

    $3x \qquad -5x \qquad 7y \qquad -4y$

    The following terms are called "<u>non-linear</u>" terms. Notice that each contains either a <u>squared letter</u> or <u>two letters</u>.

    $2x^2 \qquad -6y^2 \qquad 9xy \qquad -8xy$

    Which of the following are "<u>linear</u>" terms? _____

    (a) $10x$ (b) $-3y$ (c) $4x^2$ (d) $-2xy$

---

2. Which of the following are "<u>non-linear</u>" terms? _____ | Both (a) and (b).

    (a) $5xy$ (b) $-3y$ (c) $6y^2$ (d) $7x$

| | Both (a) and (c). |

390

Linear Graphs and Slope   391

3. Though not explicitly written, the numerical coefficient of each term below is either +1 or −1.

$$x \quad -y \quad -x^2 \quad y^2 \quad xy$$

Terms of this type can also be classified as either linear or non-linear.

"x" and "−y" are <u>linear</u> terms.

"−x²", "y²", and "xy" are <u>non-linear</u> terms.

Which of the following are <u>linear</u> terms? _____

(a) −x   (b) x²   (c) 5x   (d) xy

4. Which of the following are <u>non-linear</u> terms? _____

(a) y   (b) −xy   (c) −y²   (d) −3y

Both (a) and (c).

5. Draw a box around the <u>linear</u> term or terms in each equation below:

(a) 3x + 4y = 7   (b) xy − x = 9   (c) 3x² = y

Both (b) and (c).

6. Draw a box around the <u>non-linear</u> term or terms in each equation below:

(a) x² − xy = 5   (b) 3y − xy = 0   (c) 4x = 5y²

a) $\boxed{3x} + \boxed{4y} = 7$
b) $xy - \boxed{x} = 9$
c) $3x^2 = \boxed{y}$

7. If each letter-term in an equation in "x" and "y" is <u>linear</u>, the equation is <u>linear</u>. That is, its graph is a <u>straight line</u>. Each equation below, for example, is <u>linear</u>.

3x + 2y = 7   4y − x = 0   y = 5x − 6   7y = x

If one or more letter-terms in an equation in "x" and "y" is <u>non-linear</u>, the equation is <u>non-linear</u>. That is, its graph is a <u>curve</u>. Each equation below, for example, is <u>non-linear</u>.

xy = 10   y = 5x²   3xy − y = 0   y² = 2x − 7

Which of the following equations are <u>linear</u>? _____

(a) 5y − 2x = 3   (b) 2xy = 7   (c) y = 10x² − 1   (d) 3y = 4x

a) $\boxed{x^2} - \boxed{xy} = 5$
b) $3y - \boxed{xy} = 0$
c) $4x = \boxed{5y^2}$

8. Which of the following equations are <u>non-linear</u>? _____

(a) y = 7x − 1   (b) 5xy = 2x   (c) 7x − 2y = 0   (d) y² − 5x² = 0

Both (a) and (d).

9. (a) If an equation in "x" and "y" is <u>linear</u>, its graph is a _____ (straight line/curve).

(b) If an equation in "x" and "y" is <u>non-linear</u>, its graph is a _____ (straight line/curve).

Both (b) and (d).

a) straight line
b) curve

392   Linear Graphs and Slope

10. If the numerical coefficient of a letter-term is a decimal or fraction:

   The term is <u>linear</u> if it contains a single non-squared letter. For example:

   $14.7x \qquad 0.4y \qquad \frac{5}{6}x \qquad \frac{7}{2}y$

   The term is <u>non-linear</u> if it contains a squared letter or two letters. For example:

   $5.8y^2 \qquad \frac{2}{3}x^2 \qquad 0.2xy \qquad \frac{8}{5}xy$

   Draw a box around each <u>linear</u> term or terms in the equations below:

   (a) $0.8x - 0.4y = 10$ (b) $4.2xy = 1.7x$ (c) $15.9y = 10.8x^2$

11. Draw a box around each <u>non-linear</u> term or terms in the equations below:

   (a) $\frac{5}{6}x = \frac{2}{3}xy$ (b) $\frac{3}{4}y^2 = \frac{1}{2}x$ (c) $\frac{7}{8}xy = \frac{1}{4}x^2$

a) $\boxed{0.8x} - \boxed{0.4y} = 10$

b) $4.2xy = \boxed{1.7x}$

c) $\boxed{15.9y} = 10.8x^2$

12. Which of the following equations are <u>linear</u>? _____

   (a) $y = \frac{5}{2}x + 1$ (b) $\frac{5}{4}xy = \frac{1}{2}x$ (c) $\frac{3}{4}y = \frac{2}{3}x$ (d) $\frac{4}{3}y^2 = \frac{1}{2}x - 7$

a) $\frac{5}{6}x = \boxed{\frac{2}{3}xy}$

b) $\boxed{\frac{3}{4}y^2} = \frac{1}{2}x$

c) $\boxed{\frac{7}{8}xy} = \boxed{\frac{1}{4}x^2}$

13. Which of the following equations are <u>non-linear</u>? _____

   (a) $0.8xy = 1.2$ (c) $0.4x - 0.6y = 0$

   (b) $10.2y = 14.7x$ (d) $9.8y = 6.7x^2 + 1.3$

Both (a) and (c).

14. Fractional terms like $\frac{3x}{2}$, $\frac{y}{5}$, $\frac{5xy}{4}$, and $\frac{y^2}{3}$ can easily be written as letter-terms with fractional coefficients. That is:

   $\frac{3x}{2} = \frac{3}{2}x \qquad \frac{y}{5} = \frac{1}{5}y \qquad \frac{5xy}{4} = \frac{5}{4}xy \qquad \frac{y^2}{3} = \frac{1}{3}y^2$

   Therefore, we can determine whether terms like those above are linear or non-linear by simply examining the letter-terms in their numerators.

   $\frac{3x}{2}$ and $\frac{y}{5}$ are <u>linear</u> terms.

   $\frac{5xy}{4}$ and $\frac{y^2}{3}$ are <u>non-linear</u> terms.

   Draw a box around the linear term or terms in each equation below:

   (a) $\frac{3x}{2} - \frac{y}{5} = 1$ (b) $\frac{xy}{7} = \frac{2y}{3}$ (c) $y = \frac{2x^2}{5} - 1$

Both (a) and (d).

Linear Graphs and Slope   393

15. Which of the following are linear equations? _____

   (a) $\dfrac{x}{5} + \dfrac{3y}{2} = 7$ (c) $y = \dfrac{x}{4}$

   (b) $\dfrac{xy}{7} = \dfrac{3y}{4}$ (d) $y = \dfrac{x^2}{3}$

a) $\boxed{\dfrac{3x}{2}} - \boxed{\dfrac{y}{5}} = 1$

b) $\dfrac{xy}{7} = \boxed{\dfrac{2y}{3}}$

c) $\boxed{y} = \dfrac{2x^2}{5} - 1$

Both (a) and (c).

---

16. Since each term on the left below fits the pattern for the addition (or subtraction) of two fractions, it can be written as two simpler terms. For example:

$$\dfrac{3x+4}{2} = \dfrac{3x}{2} + \dfrac{4}{2} = \dfrac{3}{2}x + 2$$

$$\dfrac{5xy-1}{3} = \dfrac{5xy}{3} - \dfrac{1}{3} = \dfrac{5}{3}xy - \dfrac{1}{3}$$

$$\dfrac{y^2-4}{8} = \dfrac{y^2}{8} - \dfrac{4}{8} = \dfrac{1}{8}y^2 - \dfrac{1}{2}$$

Whether the simpler letter term on the far right is linear or non-linear depends on whether the letter term in the original numerator is linear or non-linear. Therefore:

   (a) $\dfrac{3x+4}{2}$ is a _____ (linear/non-linear) term.

   (b) $\dfrac{5xy-1}{3}$ and $\dfrac{y^2-4}{8}$ are _____ (linear/non-linear) terms.

---

17. Which of the following equations are linear? _____

   (a) $y = \dfrac{3x-4}{8}$ (b) $y = \dfrac{xy+1}{3}$ (c) $y = \dfrac{2x^2-1}{5}$

a) linear

b) non-linear

Only (a).

---

18. If a fractional equation contains a letter in its denominator, we must clear the fraction before determining whether the equation is linear or non-linear. For example:

   Since $\dfrac{3y+5}{x} = 7$ is equivalent to $3y + 5 = 7x$, it is linear.

   Since $\dfrac{3}{x} = 10y$ is equivalent to $3 = 10xy$, it is non-linear.

Clear the fraction in each equation below and then determine whether the equation is linear or non-linear:

   (a) $\dfrac{y}{x} = 5$ (b) $11 = \dfrac{2x-7}{y}$ (c) $y = \dfrac{3x-1}{y}$

   _____ = _____   _____ = _____   _____ = _____

   _____   _____   _____

394   Linear Graphs and Slope

19. Clear the fraction in each equation below and then decide whether it is linear or non-linear:

(a) $\dfrac{5}{x} = 2y$ (b) $\dfrac{3y+1}{x} = x$ (c) $\dfrac{y}{x+2} = 3$

___ = ___   ___ = ___   ___ = ___

_____   _____   _____

| a) Linear, since: $y = 5x$
| b) Linear, since: $11y = 2x - 7$
| c) Non-linear, since: $y^2 = 3x - 1$

20. If an equation contains an instance of the distributive principle, you must multiply by the distributive principle before deciding whether or not the equation is linear. For example:

Since $y = 3(x + 5)$ is equivalent to $y = 3x + 15$, it is linear.

Since $y = x(y - 1)$ is equivalent to $y = xy - x$, it is non-linear.

Multiply by the distributive principle in each of these and then decide whether the equation is linear or non-linear:

(a) $y = x(x + 1)$ (b) $x = 3(2y + 7)$ (c) $y(x + 5) = 7$

___ = ___   ___ = ___   ___ = ___

_____   _____   _____

| a) Non-linear, since: $5 = 2xy$
| b) Non-linear, since: $3y + 1 = x^2$
| c) Linear, since: $y = 3x + 6$

21. An equation with many letter-terms is linear if each of the terms is linear. For example:

$3x + 2y - 1 = 5y - x$ is <u>linear</u>.

$5y - 2x = y^2 - xy$ is <u>non-linear</u>.

Which of the equations below are linear? _____

(a) $7y = 5x - y$ (b) $10x = 7y - x^2 + 3x$

| a) Non-linear, since: $y = x^2 + x$
| b) Linear, since: $x = 6y + 21$
| c) Non-linear, since: $xy + 5y = 7$

22. Which of the following equations are linear? _____

(a) $xy = 7$ (c) $y + 3x - 10 = 0$ (e) $7 - x = 5y$
(b) $y = 2x^2 - 9$ (d) $3y^2 = 2x - 1$ (f) $5x - y = 3y - 2$

| Only (a)

23. Which of the following equations are linear? _____

(a) $0.4y^2 = 0.2x$ (c) $5.1x - 3.2y = 0$ (e) $y = \dfrac{6x-5}{3}$
(b) $10.8xy = 19.6$ (d) $\dfrac{3x}{2} = \dfrac{5y}{7} - 1$ (f) $x = \dfrac{xy+2}{4}$

| (c), (e), and (f)

24. Which of the following equations are linear? _____

(a) $10 = \dfrac{xy}{7}$ (c) $1 = \dfrac{y}{x-2}$ (e) $x - 3(y+1) = 0$
(b) $y = \dfrac{x}{5}$ (d) $\dfrac{y^2-1}{5} = 2x$ (f) $\dfrac{x-2}{y} = y$

| (c), (d), and (e)

| (b), (c), and (e)

SELF-TEST 1 (Frames 1-24)

1. Which of the following are "linear" terms? _____

   (a) 8x   (b) -3y   (c) x   (d) y^2 (e) 3xy   (f) 0.37y   (g) $-5x^2$ (h) $\frac{2x}{3}$

---

2. Which of the following are "linear" equations? _____

   (a) $y = x$ (c) $\frac{y}{x} = 5$ (e) $y^2 + 4y = 3x$ (g) $y = \frac{6x + 11}{4}$

   (b) $y = x^2$ (d) $\frac{8}{y} = x$ (f) $\frac{5x}{3} - \frac{2y}{7} = 4$ (h) $2x(x - y) = 9$

ANSWERS: 1. (a), (b), (c), (f), (h)   2. (a), (c), (f), (g)

5-2 THE INTERCEPTS OF LINEAR EQUATIONS

When linear equations are graphed, their straight-line graphs usually cross (intersect) both the horizontal axis and the vertical axis. The points of intersection with the two axes are called "intercepts". In this section, we will briefly review the concept of "intercepts", and then we will show how the coordinates of intercepts can be obtained from the linear equations themselves, without graphing the equations.

---

25. The graphed line on the right intersects both the horizontal axis (the x-axis) and the vertical axis (the y-axis). The coordinates of these points of intersection are given.

    Point A is called the vertical-intercept or y-intercept. Its coordinates are (0,3) where:

         0 is the x-coordinate or abscissa.
         3 is the y-coordinate or ordinate.

    Point B is called the horizontal-intercept or x-intercept. Its coordinates are (2,0) where:

         (a) 2 is the x-coordinate or _____.
         (b) 0 is the y-coordinate or _____.

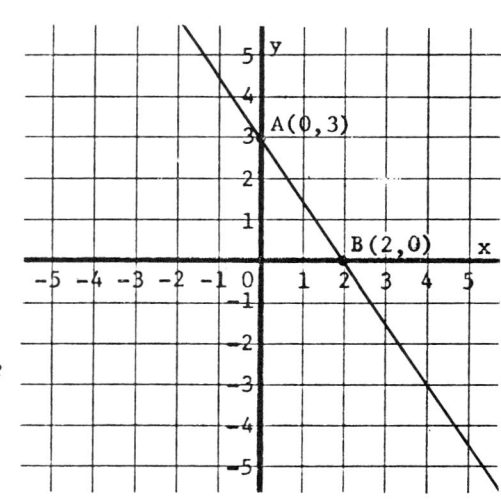

a) abscissa

b) ordinate

396   Linear Graphs and Slope

26. Write another name for each of the following:
    (a) x-axis _____   (c) x-coordinate _____
    (b) y-axis _____   (d) y-coordinate _____

27. (a) The _____ (x-coordinate/y-coordinate) of any point on the vertical (or y-axis) is 0. (b) The _____ (abscissa/ordinate) of any point on the horizontal (or x-axis) is 0.	a) horizontal axis b) vertical axis c) abscissa d) ordinate

28. One of the coordinates of any intercept is always 0.
    (a) If it is an x-intercept, its _____ (abscissa/ordinate) is 0.
    (b) If it is a y-intercept, its _____ (x-coordinate/y-coordinate) is 0.
	a) x-coordinate
	b) ordinate

29. Since the x-coordinate of each point below is "0", each point is a y-intercept. A (0, 5) B (0, −5) C (0, 8) D (0, −8) Which of the y-intercepts lie <u>below</u> the horizontal axis? _____	a) ordinate b) x-coordinate
30. Since the y-coordinate of each point below is "0", each point is an x-intercept. A (7, 0) B (−7, 0) C (3, 0) D (−3, 0) Which of these x-intercepts lie to the <u>left</u> of the vertical axis? _____	Points B and D.
31. Which of the following points are y-intercepts? _____ A (4, 0) B (0, −3) C (−5, 0)	Points B and D.
32. Which of the following points are x-intercepts? _____ A (10, 0) B (−7, 0) C (0, 8)	Only B.
	A and B.

33. The graph of the linear equation below is given on the right.

$$x + 2y = 6$$

The coordinates of each intercept are given on the graph.

We can obtain the coordinates of the intercepts from the equation in the following way:

(1) <u>Vertical intercept (y-intercept)</u>:

Since the x-coordinate (or abscissa) of any vertical intercept is "0", we can plug in "0" for "x" in the equation and compute the corresponding value of "y". We get:

$$x + 2y = 6$$
$$0 + 2y = 6$$
$$2y = 6 \quad \text{or} \quad y = 3$$

Therefore, the coordinates of the vertical intercept are (0, 3).

(2) <u>Horizontal intercept (x-intercept)</u>:

Since the y-coordinate (or ordinate) of any point on the horizontal axis is "0", we can plug in "0" for "y" in the equation and compute the corresponding value of "x". We get:

$$x + 2y = 6$$
$$x + 2(0) = 6$$
$$x = 6$$

Therefore, the coordinates of the horizontal intercept are (6, 0).

---

34. Here is another linear equation: $3x + 2y = 6$

(a) To find the coordinates of its vertical intercept, we can plug in "0" for "x" and compute the corresponding value of "y". Do so.

The coordinates of its vertical intercept are (,   ).

(b) To find the coordinates of its horizontal intercept, we can plug in "0" for "y" and compute the corresponding value of "x". Do so.

The coordinates of its horizontal intercept are (,   ).

---

a) (0, 3)

b) (2, 0)

398   Linear Graphs and Slope

35. Here is another linear equation: $\boxed{2x - 5y = 20}$

   (a) By plugging in "0" for "x" and computing the corresponding value of "y", we can find the coordinates of its y-intercept. Do so.

   The coordinates of its y-intercept are (,).

   (b) By plugging in "0" for "y" and computing the corresponding value of "x", we can find the coordinates of its x-intercept. Do so.

   The coordinates of its x-intercept are (,).

---

36. (a) To find the coordinates of a vertical (or "y") intercept, we must plug in "0" for ____ (x/y).

   (b) To find the coordinates of a horizontal (or "x") intercept, we must plug in "0" for ____ (x/y).

   a) (0, -4)
   b) (10, 0)

---

37. Here is another linear equation: $\boxed{3y = x - 9}$

   Plug in 0's to find the coordinates of its intercepts.

   (a) The coordinates of the x-intercept are (,).
   (b) The coordinates of the y-intercept are (,).

   a) x
   b) y

---

38. Here is another linear equation: $\boxed{7y = 3x - 21}$

   (a) The coordinates of its vertical intercept are (,).

   (b) The coordinates of its horizontal intercept are (,).

   a) (9, 0)
   b) (0, -3)

---

39. Plug in 0's to find the coordinates of the intercepts of: $\boxed{2x + 5y = 3}$

   (a) The coordinates of the vertical intercept are (,).
   (b) The coordinates of the horizontal intercept are (,).

   a) (0, -3)
   b) (7, 0)

---

40. Below each equation, write the coordinates of its <u>horizontal intercept</u>:

   (a) 5x + 3y = 15   (b) 3y = 2x + 18   (c) 2x = 5y - 10

       (,)   (,)   (,)

   a) $\left(0, \dfrac{3}{5}\right)$
   b) $\left(\dfrac{3}{2}, 0\right)$

---

41. Below each equation, write the coordinates of its <u>vertical intercept</u>:

   (a) 4x + 3y = 12   (b) 4y = 5x - 20   (c) 2x = 3y - 6

       (,)   (,)   (,)

   a) (3, 0)
   b) (-9, 0)
   c) (-5, 0)

42. Below each equation, write the coordinates of its x-intercept:

   (a) $2x = 3y + 7$ (b) $7x - 5y = 12$ (c) $y - 5x = 20$

   (,)   (,)   (,)

   a) $(0, 4)$
   b) $(0, -5)$
   c) $(0, 2)$

43. Below each equation, write the coordinates of its y-intercept:

   (a) $7x - 8y = 15$ (b) $x = 5y - 17$ (c) $3y = x - 2$

   (,)   (,)   (,)

   a) $\left(\dfrac{7}{2}, 0\right)$
   b) $\left(\dfrac{12}{7}, 0\right)$
   c) $(-4, 0)$

44. In the equation $\boxed{y = 4x}$, when "x" is 0, "y" is 0, and vice versa. Therefore:

   (a) The coordinates of its vertical intercept must be (,).
   (b) The coordinates of its horizontal intercept must be (,).
   (c) The graph of this equation must pass through the _____.

   a) $\left(0, -\dfrac{15}{8}\right)$
   b) $\left(0, \dfrac{17}{5}\right)$
   c) $\left(0, -\dfrac{2}{3}\right)$

Answer to Frame 44:   a) (0,0)   b) (0,0)   c) origin

SELF-TEST 2 (Frames 25-44)

1. If a point lies on the y-axis, its _____ (abscissa/ordinate) is zero.
2. If the ordinate of a point is zero, the point lies on the _____ (x-axis/y-axis).

3. Which of the following points lie on the x-axis? _____

   (a) (0,1)   (b) (8,0)   (c) (0,-5)   (d) (-2,0)   (e) (0,0)

For each of the following equations, find the coordinates of the x-intercept and the y-intercept:

$\boxed{2x - 5y = 20}$	$\boxed{3y = 4x - 2}$	$\boxed{y - 5x = 0}$
4. x-intercept: _____	6. x-intercept: _____	8. x-intercept: _____
5. y-intercept: _____	7. y-intercept: _____	9. y-intercept: _____

ANSWERS: 1. abscissa   3. (b), (d), (e)   4. (10,0)   6. $\left(\dfrac{1}{2}, 0\right)$ 8. (0,0)
           2. x-axis   5. (0,-4)   7. $\left(0, -\dfrac{2}{3}\right)$ 9. (0,0)

400   Linear Graphs and Slope

5-3   GRAPHING LINEAR EQUATIONS BY PLOTTING INTERCEPTS

Since a straight line is determined by two points, a linear equation can be graphed by plotting only two points. Frequently, the two points which are plotted are its intercepts. In this section, we will graph linear equations by plotting their intercepts.

45. There is a point labeled A on the right. How many straight lines can be drawn through point A?	
46. There are two points labeled A and B on the right. How many <u>straight lines</u> can be drawn that pass through <u>both</u> points?	An infinite number. Here are a few of them:
47. There are two points labeled A and B on the right. How many <u>curved lines</u> can be drawn that pass through both points?	Only <u>one</u>.
48. <u>Two points determine a straight line</u> since only one line can be drawn through <u>both</u> of them. Do <u>two</u> points determine a curved line? _____	An infinite number. Here are some of them:
49. <u>If we know that the graph of an equation is a straight line</u>, we only have to plot two points to determine the line. <u>If we know that the graph of an equation is curvilinear</u>, we must plot many points to determine the exact shape of the curve. (a) $y = x + 6$ is a <u>linear</u> equation. How many points must we plot to graph it? _____ (b) $xy = 50$ is a <u>non-linear</u> equation. How many points must we plot to graph it? _____	No, since many curves can be drawn through the same two points.
50. Which of the following equations are <u>linear</u> and can be graphed by plotting <u>only two points</u>? _____ (a) $3x = y - 7$ (b) $y^2 = x - 2$ (c) $xy = 10$ (d) $5y = x$	a) <u>Two</u> points is enough. b) As many points as are needed to determine the shape of the curve.
51. Which of the following equations are <u>non-linear</u> and can be graphed only by plotting many points? _____ (a) $3y - 4xy = 11$ (b) $y = 2x^2 - x$ (c) $4x - 5y = 20$ (d) $2x - y = 0$	Only (a) and (d), since (b) and (c) are non-linear.
	Both (a) and (b)

52. Since $y = x + 6$ in a linear equation, we can graph it by plotting only two points. The two points below lie on the graph since their coordinates satisfy the equation.

Point A: If $x = 2$, $y = 8$.
Point B: If $x = 4$, $y = 10$.

Plot these two points on the graph at the right and then draw the graph of: $y = x + 6$

(Compare your graph with the graph shown in the next frame.)

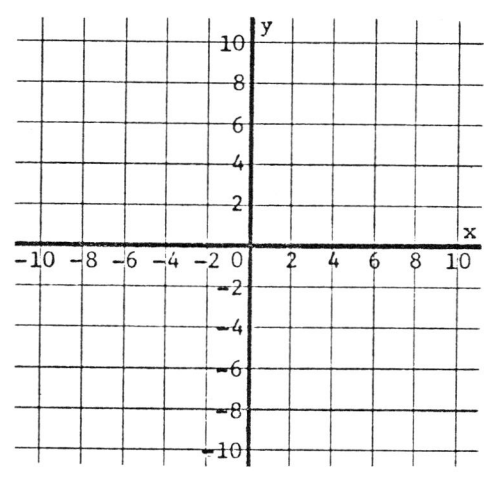

53. The graph of $y = x + 6$ is given on the right.

To check that the graphed line is correct, you should always plot a third point.

If it lies on the line, the graph is correct.

If it does not lie on the line, the graph is incorrect.

The point below should also lie on the graphed line since its coordinates satisfy: $y = x + 6$

Point C: If $x = -2$, $y = 4$.

(a) Does this point lie on the line? _____

(b) Is the graphed line correct? _____

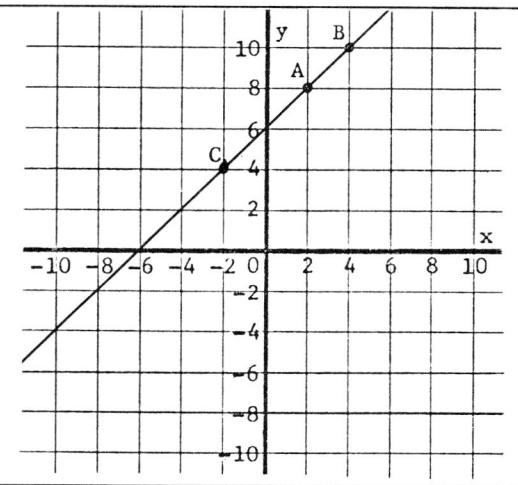

Answer to Frame 53:   a) Yes   b) Yes

54. To graph a linear equation, we can plot any two points (with a third point as a check). However, we frequently graph a linear equation by plotting its two intercepts (with a third point as a check).

Let's graph $6x + 3y = 12$ by plotting its intercepts.

(a) The coordinates of its x-intercept are (,).

(b) The coordinates of its y-intercept are (,).

(c) Graph the equation at the right by plotting these two intercepts and drawing the line.

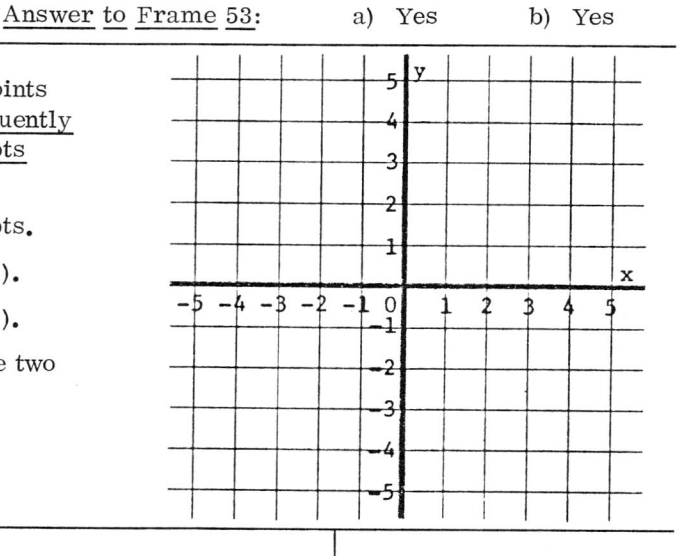

a) (2, 0)
b) (0, 4)
c) See next frame.

55. The graph of $\boxed{6x + 3y = 12}$ is shown at the right. It was graphed by plotting its two intercepts.

To check the correctness of the graph, let's plot a third point. The following pair of values of x and y satisfy the equation:

$$x = 1, \quad y = 2$$

(a) Plot the point (1, 2) on the graph. Does it lie on the graphed line? _____

(b) Is our graph correct? _____

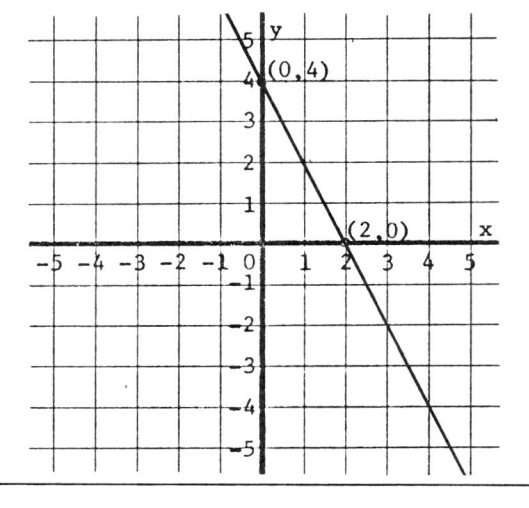

Answer to Frame 55: a) Yes b) Yes

---

56. Both equations below are linear:

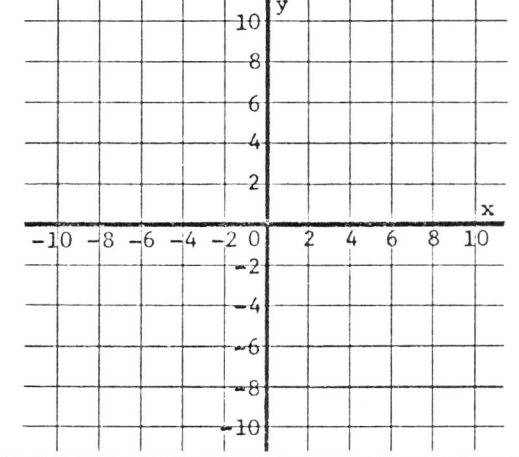

#1: $\boxed{3x = 4y - 24}$ #2: $\boxed{3y = x - 6}$

Determine the intercepts for each equation and use them to graph each equation on the axes at the right. Be sure to plot a third point for each equation as a check.

Answer to Frame 56:

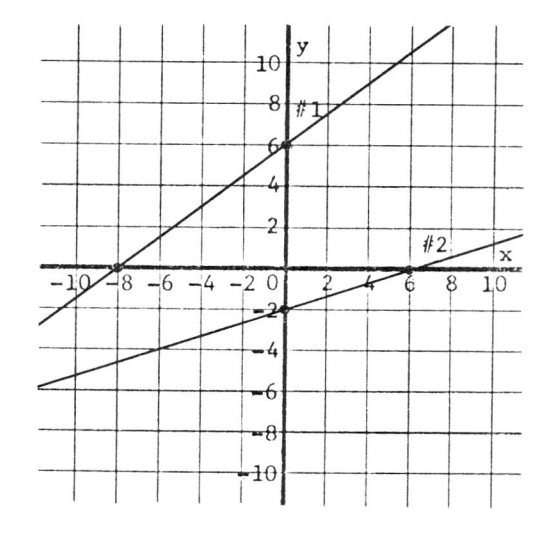

Linear Graphs and Slope 403

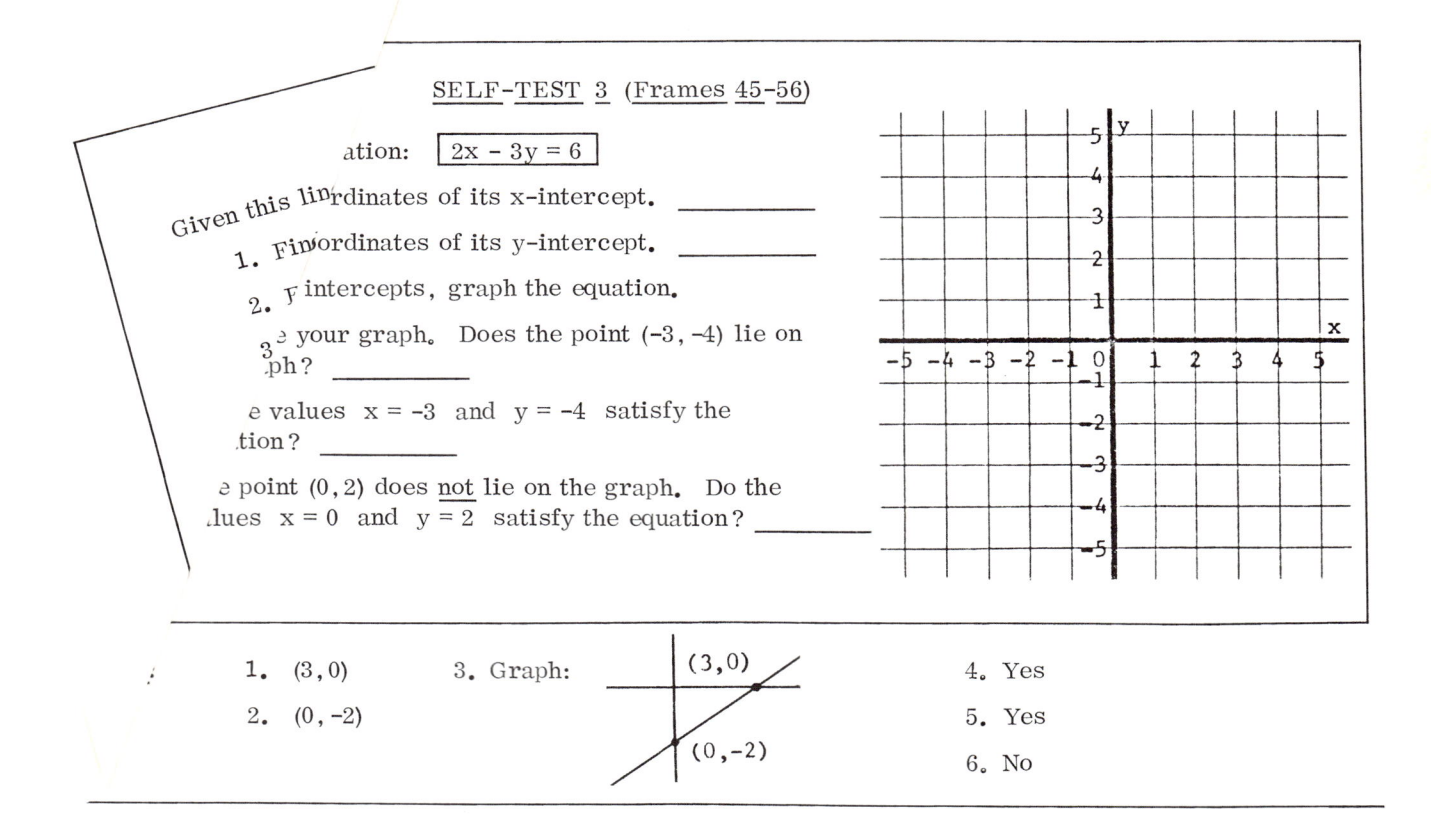

SELF-TEST 3 (Frames 45-56)

Given this linear equation: $2x - 3y = 6$

1. Find the coordinates of its x-intercept. _____
2. Find the coordinates of its y-intercept. _____
3. Using the intercepts, graph the equation.
4. Using your graph. Does the point (-3, -4) lie on the graph? _____
5. Do the values $x = -3$ and $y = -4$ satisfy the equation? _____
6. The point (0, 2) does not lie on the graph. Do the values $x = 0$ and $y = 2$ satisfy the equation? _____

1. (3, 0)   3. Graph: (3, 0), (0, -2)   4. Yes
2. (0, -2)   5. Yes
                                        6. No

5-4 A GRAPHICAL REPRESENTATION OF CHANGES IN "y" CORRESPONDING TO INCREASES IN "x"

In this section, we will examine how y increases or decreases because of a corresponding increase in x in various linear relationships. We will show:

(1) how these changes can be graphically represented by vectors, and

(2) how signed numbers can be used to represent these changes and their corresponding vectors.

This examination of "changes in x" and "changes in y" is needed before we can introduce the concept of the "slope of a line" in the next section.

SELF-TEST 3 (Frames 45-56)

Given this linear equation: $\boxed{2x - 3y = 6}$

1. Find the coordinates of its x-intercept. _____
2. Find the coordinates of its y-intercept. _____
3. Using the intercepts, graph the equation.
4. Examine your graph. Does the point (-3, -4) lie on the graph? _____
5. Do the values $x = -3$ and $y = -4$ satisfy the equation? _____
6. The point (0, 2) does <u>not</u> lie on the graph. Do the values $x = 0$ and $y = 2$ satisfy the equation? _____

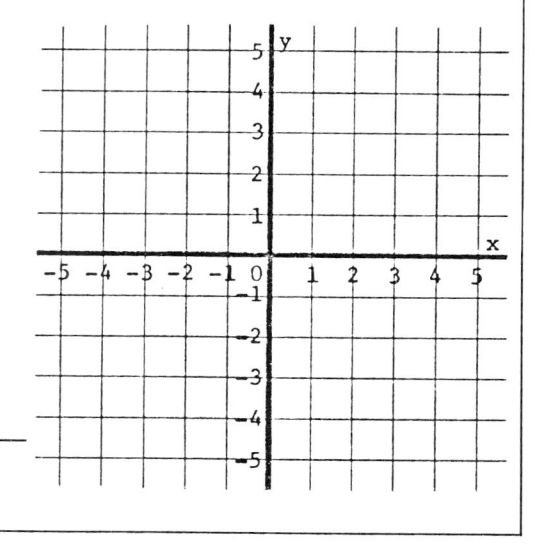

ANSWERS:
1. (3, 0)
2. (0, -2)
3. Graph: (3,0), (0,-2)
4. Yes
5. Yes
6. No

5-4 A GRAPHICAL REPRESENTATION OF CHANGES IN "y" CORRESPONDING TO INCREASES IN "x"

In this section, we will examine how y increases or decreases because of a corresponding increase in x in various linear relationships. We will show:

(1) how these changes can be graphically represented by vectors, and

(2) how signed numbers can be used to represent these changes and their corresponding vectors.

This examination of "changes in x" and "changes in y" is needed before we can introduce the concept of the "slope of a line" in the next section.

404   Linear Graphs and Slope

57. Changes (increases or decreases) in x and y can be graphically represented by horizontal and vertical vectors on the coordinate system. These changes and their corresponding vectors can also be represented by signed numbers. Remember:

   (1) Any vector <u>to</u> <u>the</u> <u>right</u> or <u>upward</u> is <u>positive</u>.

   (2) Any vector <u>to</u> <u>the</u> <u>left</u> or <u>downward</u> is <u>negative</u>.

   We have drawn some vectors on the coordinate system at the right:

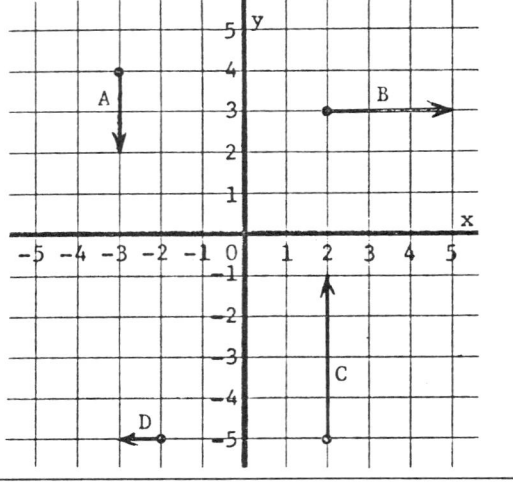

   What signed number represents each vector?

   A: _____   B: _____   C: _____   D: _____

Answer to Frame 57:   A: -2   B: +3   C: +4   D: -1

58. Horizontal and vertical vectors can cross an axis. Do not let this bother you. The same rules apply.

   What signed number represents each vector at the right?

   A: _____   B: _____   C: _____   D: _____

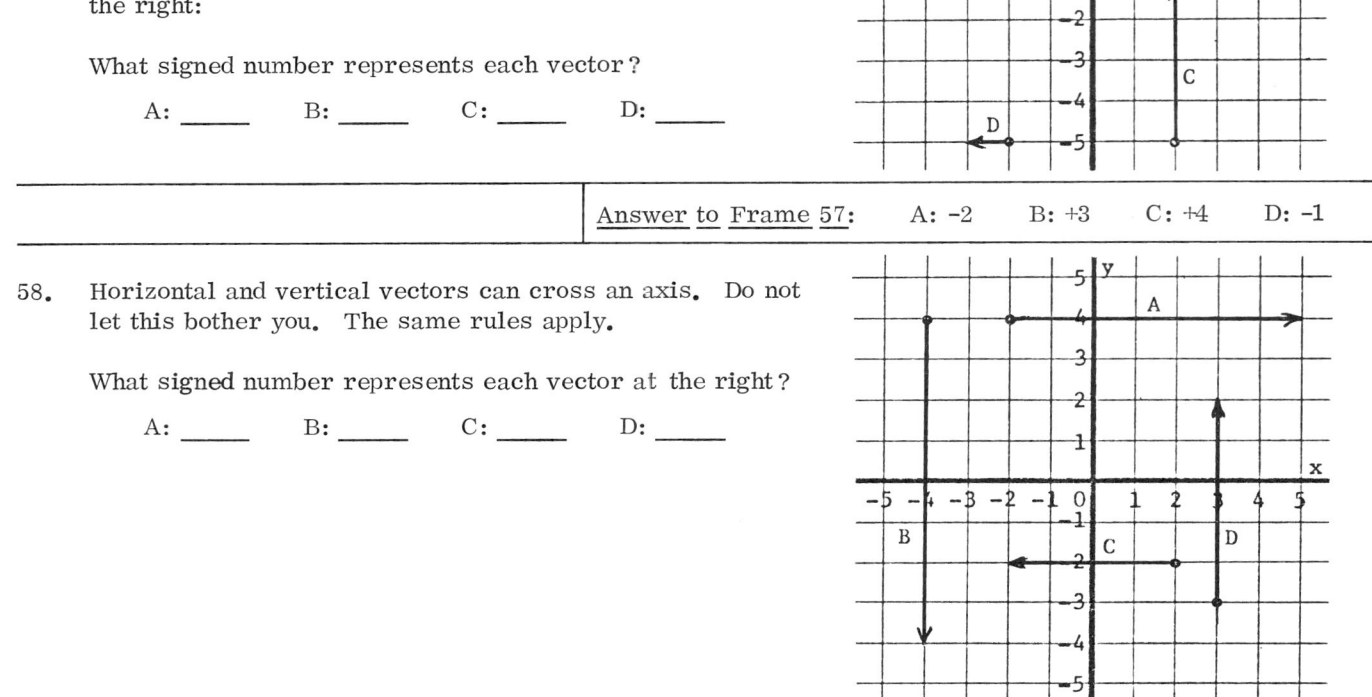

Answer to Frame 58:   A: +7   B: -8   C: -4   D: +5

59. On the coordinate system at the right, we have graphed a line showing a relationship between x and y. Two points, M (2,2) and T (4,5), are plotted on the line. From point M to point T:

   x increases from 2 to 4
   y increases from 2 to 5

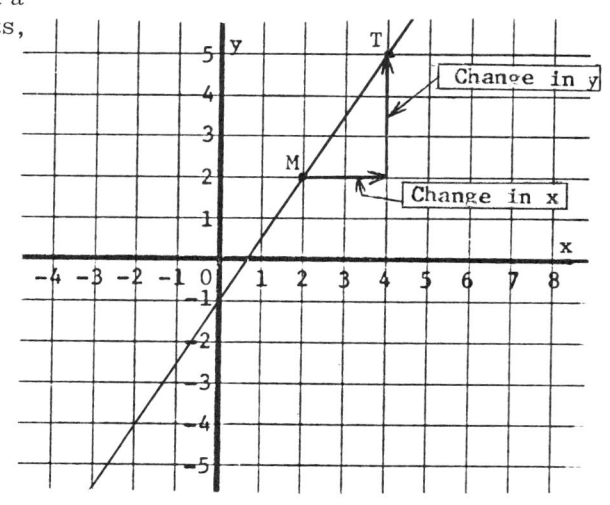

   The increases in x and y are graphically represented by the horizontal and vertical vectors.

   (a) The horizontal vector represents the increase in x. What signed number represents this change and its corresponding vector? _____

   (b) The vertical vector represents the increase in y. What signed number represents this change and its corresponding vector? _____

   (c) Therefore, we can make this statement: With a 2-unit increase in x, we get a ___-unit increase in y.

Linear Graphs and Slope 405

Answer to Frame 59:   a) +2   b) +3   c) 3-unit

60. The graphed line at the right shows another relationship between x and y. The two vectors represent the changes in x and y from point R (-6, 6) to point Q (4, -8).

(a) The horizontal vector represents an <u>increase</u> in x from -6 to +4. It can be represented by what signed number? _____

(b) The vertical vector represents a <u>decrease</u> in y from +6 to -8. It can be represented by what signed number? _____

(c) Therefore, we can make this statement: For a _____-unit increase in x, we get a _____-unit decrease in y.

---

61. Mathematicians use the symbol "Δ" (pronounced "delta") as an abbreviation for the phrase "change in".

Instead of "change in x", they write "Δx".

Instead of "change in y", they write "_____".

a) +10

b) -14, since its direction is <u>downward</u>.

c) 10-unit ... 14-unit

Answer to Frame 61: Δy

62. On the coordinate system at the right, we have graphed another linear relationship between x and y. The vectors which represent the changes in x and y from point L (-5, 3) to point S (-3, -2) are also drawn.

(a) The horizontal vector represents an increase in x from -5 to -3. Therefore, the "change in x" or Δx equals what signed number? Δx = _____

(b) The vertical vector represents a decrease in y from 3 to -2. Therefore, the "change in y" or Δy equals what signed number? Δy = _____

(c) Therefore, with a 2-unit increase in x, we get a corresponding 5-unit _____ (increase/decrease) in y.

a) Δx = +2

b) Δy = -5

c) decrease

406   Linear Graphs and Slope

63. Another linear relationship between x and y is graphed at the right. We have represented the change from point A (−4, −8) to point B (6, −4) by means of vectors.

   (a) Since the increase in x is from −4 to +6,
       $\Delta x =$ _____.

   (b) Since the change in y is from −8 to −4,
       $\Delta y =$ _____.

   (c) With a 10-unit increase in x, we get a corresponding 4-unit _____ (increase/decrease) in y.

---

64. When graphing the change in y for a corresponding increase in x:

   (a) Does the <u>horizontal</u> vector represent the change in x or the change in y? _____

   (b) Does the <u>vertical</u> vector represent the change in x or the change in y? _____

a) $\Delta x = +10$
b) $\Delta y = +4$
c) increase

---

65. When comparing changes in x and y, we always compare the change in y which corresponds to an <u>increase</u> in x. Therefore:

   (a) The direction of the horizontal vector is always to the _____ (right/left).

   (b) Δx is always _____ (positive/negative).

a) Change in x.
b) Change in y.

Answer to Frame 65:   a) right   b) positive

---

66. On the graph at the right, we have shown the changes from A to B on Line #1 and from C to D on Line #2.

   On Line #1, B is <u>lower</u> than A. In this case:

   (a) Δy represents a _____ (increase/decrease) in y.

   (b) Δy is _____ (positive/negative).

   On Line #2, D is <u>higher</u> than C. In this case:

   (c) Δy represents an _____ (increase/decrease) in y.

   (d) Δy is _____ (positive/negative).

---

67. Of Δx and Δy: (a) Which one have we always kept positive? _____

   (b) Which one may be either positive or negative? _____

a) decrease
b) negative
c) increase
d) positive

Answer to Frame 67: a) Δx   b) Δy

68. On any given line, we get identical changes in y with identical increases in x. For example, on the coordinate system at the right, we have graphed the changes from F to G, G to H, and H to J.

In each case, Δx is +3.

What is Δy in each case? Δy = _____

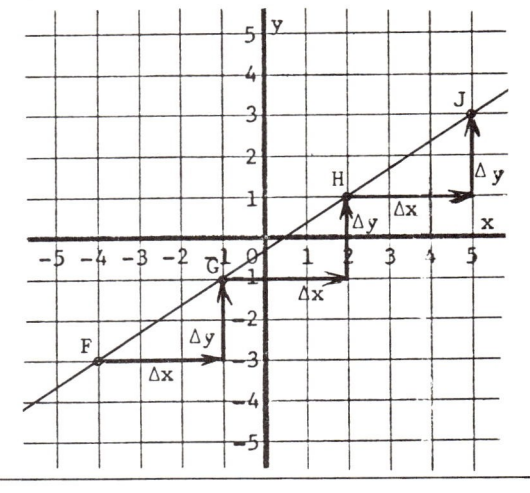

Answer to Frame 68: Δy = +2

69. On the line at the right, we have graphed the changes from A to B, C to D, and E to F.

In each case:

(a) Δx = _____

(b) Δy = _____

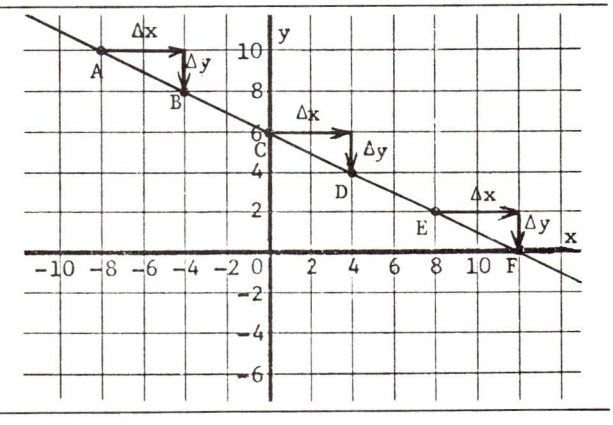

Answer to Frame 69: a) Δx = +4   b) Δy = -2

70. For any line, if two Δx's are different, it should be obvious that their corresponding Δy's will be different. The larger Δx is, the larger the increase or decrease in y will be.

For example, on the coordinate system at the right, we have graphed the changes from M to P, M to Q, and M to R.

(a) In the change from M to P: Δx = +3, Δy = _____

(b) In the change from M to Q: Δx = +6, Δy = _____

(c) In the change from M to R: Δx = +9, Δy = _____

(d) As the increase in x gets larger, the increase in y gets _____ (larger/smaller).

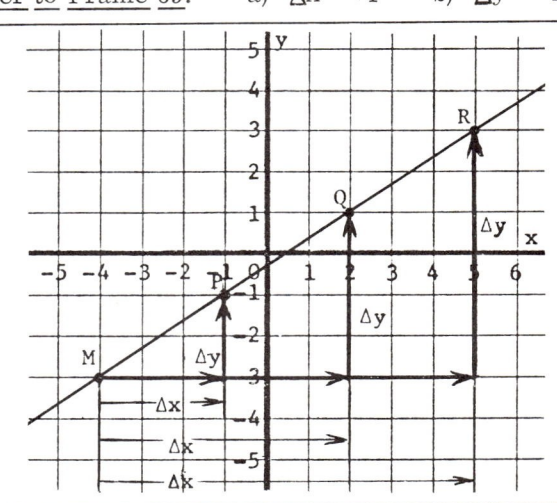

a) Δy = +2
b) Δy = +4
c) Δy = +6
d) larger

408   Linear Graphs and Slope

71. At the right, we have graphed the changes from A to B, A to C, and A to D. In each case, the change in y is a decrease.

(a) In the change from A to B: $\Delta x = $ ___, $\Delta y = $ ___

(b) In the change from A to C: $\Delta x = $ ___, $\Delta y = $ ___

(c) In the change from A to D: $\Delta x = $ ___, $\Delta y = $ ___

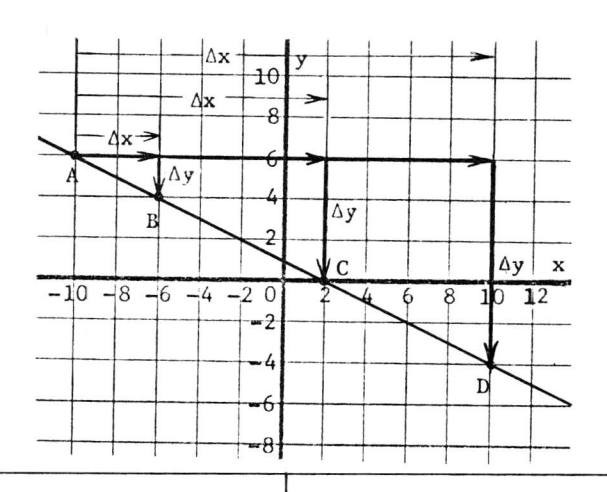

(d) As the increase in x gets larger, the decrease in y gets _____ (larger/smaller).

---

72. In Frame 70, we found the following result: As the increases in x became larger, the increases in y became larger. Here is a table of the corresponding values of Δx and Δy:

CHANGE	Δx	Δy
M to P	+3	+2
M to Q	+6	+4
M to R	+9	+6

Let's compare the size of Δy with the size of Δx in each case. This comparison is a ratio in which Δy is divided by Δx.

CHANGE	RATIO OF Δy TO Δx
M to P	$\frac{\Delta y}{\Delta x} = \frac{+2}{+3}$ or $\frac{2}{3}$
M to Q	$\frac{\Delta y}{\Delta x} = \frac{+4}{+6}$ or $\frac{2}{3}$
M to R	$\frac{\Delta y}{\Delta x} = \frac{+6}{+9}$ or $\frac{2}{3}$

What statement can we make about the ratio of Δy to Δx for these three changes? _____

---

73. In Frame 71, we found the following result: As the increases in x became larger, the decreases in y became larger. Here is a table of the corresponding values of Δx and Δy:

CHANGE	Δx	Δy
A to B	+4	−2
A to C	+12	−6
A to D	+20	−10

Continued on following page.

Answers:

a) $\Delta x = +4$, $\Delta y = -2$

b) $\Delta x = +12$, $\Delta y = -6$

c) $\Delta x = +20$, $\Delta y = -10$

d) larger

All three are equal.

73. Continued

Again, let's examine the ratio of Δy to Δx for these changes:

CHANGE	RATIO OF Δy TO Δx
A to B	$\dfrac{\Delta y}{\Delta x} = \dfrac{-2}{+4}$ or $-\dfrac{1}{2}$
A to C	$\dfrac{\Delta y}{\Delta x} = \dfrac{-6}{+12}$ or $-\dfrac{1}{2}$
A to D	$\dfrac{\Delta y}{\Delta x} = \dfrac{-10}{+20}$ or $-\dfrac{1}{2}$

Even though Δx and Δy differ, is the ratio of Δy to Δx the same in each case? _____

Answer to Frame 73:   Yes

---

74. The points we are making are stated below:

> For any given straight line:
>
> (1) If two Δx's are equal, their corresponding Δy's are equal.
>
> (2) If two Δx's are unequal, their corresponding Δy's are unequal.
>
> (3) Even if two Δx's (and their corresponding Δy's) are unequal, the ratio of Δy to Δx in each case is equal.

SELF-TEST 4 (Frames 57-74)

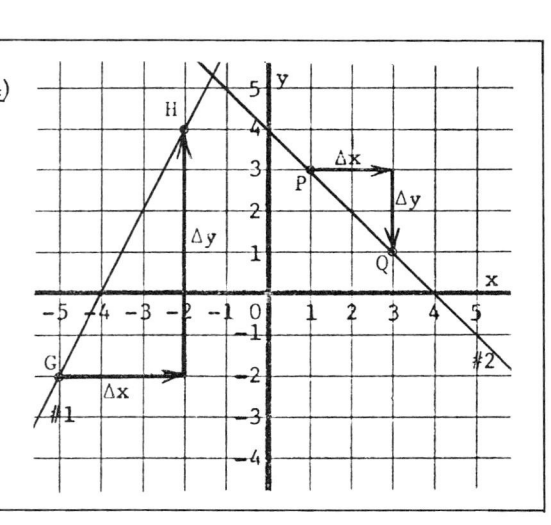

On Line #1, the coordinates of points G and H are (-5, -2) and (-2, 4). In the change from G to H:

1. $\Delta x =$ _____   2. $\Delta y =$ _____   3. $\dfrac{\Delta y}{\Delta x} =$ _____

On Line #2, the coordinates of points P and Q are (1, 3) and (3, 1). In the change from P to Q:

4. $\Delta x =$ _____   5. $\Delta y =$ _____   6. $\dfrac{\Delta y}{\Delta x} =$ _____

ANSWERS:
1. $\Delta x = +3$
2. $\Delta y = +6$
3. $\dfrac{\Delta y}{\Delta x} = \dfrac{+6}{+3} = +2$
4. $\Delta x = +2$
5. $\Delta y = -2$
6. $\dfrac{\Delta y}{\Delta x} = \dfrac{-2}{+2} = -1$

410 Linear Graphs and Slope

5-5 THE SLOPE OF A LINE EXPRESSED AS A SIGNED NUMBER

In the two figures below, we have sketched some graphed lines on the coordinate system:

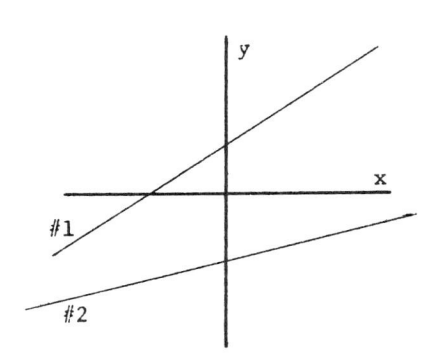

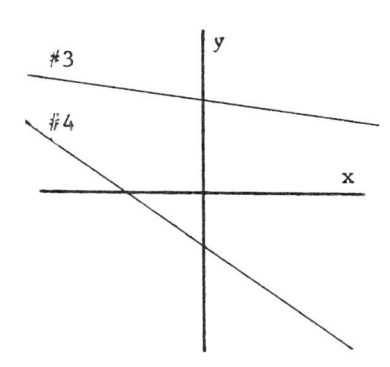

When reading a graph, we always read from left to right.

In the figure on the left, both lines "rise" as we read from left to right.
The "rise" of Line #1 is steeper than the "rise" of Line #2.

In the figure on the right, both lines "fall" as we read from left to right.
The "fall" of Line #4 is steeper than the "fall" of Line #3.

The "steepness of its rise or fall" is called the "slope" of a line. Mathematicians use the ratio "$\frac{\Delta y}{\Delta x}$" as a means of expressing the "slope" of a line as a signed number. When the "$\frac{\Delta y}{\Delta x}$" ratio is used to express the "slope" of a line as a signed number, both the sign and the absolute value of the signed number have a definite meaning. We will discuss this definition of "slope" in this section.

75. Here is the definition of the "slope" of a line:

$$\text{Slope} = \frac{\Delta y}{\Delta x} = \frac{\text{Increase or decrease in y}}{\text{Increase in x}}$$

To compute the slope of a line, we: (1) pick any two points on it, and
(2) then compute the ratio $\frac{\Delta y}{\Delta x}$.

Since the ratio $\frac{\Delta y}{\Delta x}$ is the same for any two points on the same line, we can use any pair of points to compute the slope. On the graphed line at the right, we picked points M and P.

The slope of the line = $\frac{\Delta y}{\Delta x} = \frac{+5}{+2}$ or $\frac{5}{2}$.

(a) A slope of "$\frac{5}{2}$" means that for any 2-unit increase in x, we get a ____-unit increase in y.

(b) The line above "rises" from left to right. Is its slope positive or negative? _____

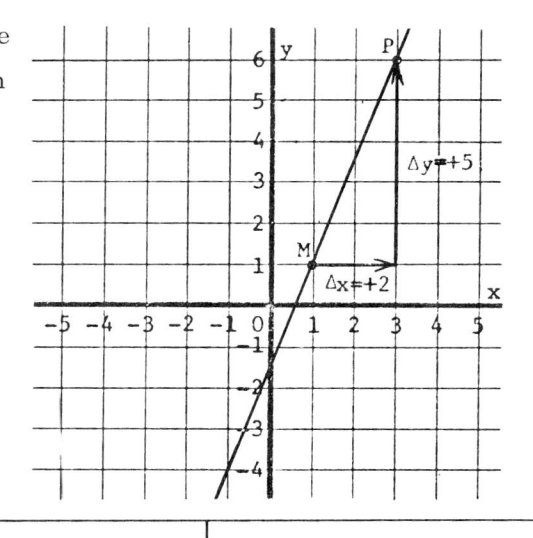

a) 5   b) Positive

76. Here is another graphed line. To compute its slope, we picked the two points R and W.

   For the line at the right: Slope $= \frac{\Delta y}{\Delta x} = \frac{-7}{+5}$ or $-\frac{7}{5}$

   (a) A slope of "$-\frac{7}{5}$" or "$\frac{-7}{+5}$" means that for any 5-unit increase in x, we get a corresponding ____-unit decrease in y.

   (b) The line "falls" from left to right. Is its slope positive or negative? _____

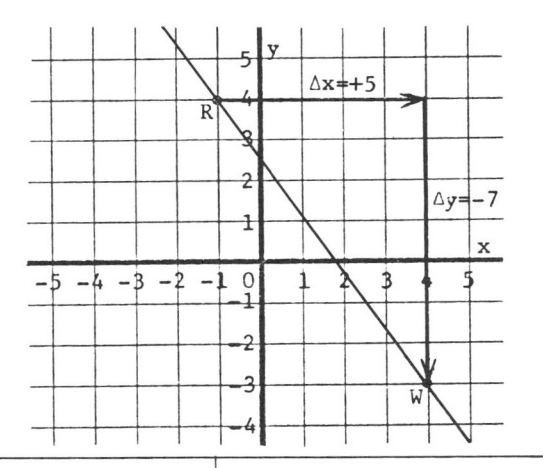

---

77. Slope is defined as the ratio of the "change in y" to a corresponding "increase in x". The "change in y" may be either an increase or decrease.

   Which of the following is the correct formula for "slope"? _____

   (a) Slope $= \frac{\text{change in y}}{\text{increase in x}}$ (b) Slope $= \frac{\text{increase in x}}{\text{change in y}}$

a) 7

b) Negative.

---

78. Which of the following is the correct formula for "slope"? _____

   (a) Slope $= \frac{\Delta x}{\Delta y}$ (b) Slope $= \frac{\Delta y}{\Delta x}$

(a)

---

79. When computing a slope, we always choose Δx so that it is positive. Therefore, the <u>sign</u> of the slope is determined by Δy.

   If Δy is <u>positive</u>, the slope is <u>positive</u>.
   If Δy is <u>negative</u>, the slope is <u>negative</u>.

   (a) If Δy is <u>positive</u>, is the change in y an increase or decrease? _____

   (b) If Δy is <u>negative</u>, is the change in y an increase or decrease? _____

(b)

---

a) Increase.

b) Decrease.

412   Linear Graphs and Slope

80. The sign of a slope tells us whether a line <u>rises</u> or <u>falls</u> from left to right.

   If its sign is <u>positive</u>, the line <u>rises</u> since a <u>positive</u> Δy represents an <u>increase</u> in y.

   If its sign is <u>negative</u>, the line <u>falls</u> since a <u>negative</u> Δy represents a <u>decrease</u> in y.

   On the graph at the right, we have drawn four lines and labeled them #1, #2, #3, and #4.

   (a) Which of these lines have a positive slope? _____
   (b) Which of these lines have a negative slope? _____

---

81. (a) If the slope of a line is <u>positive</u>, does it "rise" or "fall" from left to right? _____

    (b) If the slope of a line is <u>negative</u>, does it "rise" or "fall" from left to right? _____

   a) Lines #3 and #4.
   b) Lines #1 and #2.

---

82. If the slope of a line is $\frac{7}{3}$:

    (a) $\Delta y =$ _____
    (b) The line _____ (rises/falls) from left to right.

   a) Rise.
   b) Fall.

---

83. If the slope of a line is $-\frac{5}{3}$:

    (a) $\Delta y =$ _____
    (b) The line _____ (rises/falls) from left to right.

   a) $\Delta y = +7$
   b) rises

---

84. A slope of "$\frac{5}{4}$" means: For any 4-unit increase in x, there is a corresponding _____-unit _____ (increase/decrease) in y.

   a) $\Delta y = -5$
   b) falls

---

85. A slope of "$-\frac{4}{7}$" means: For any 7-unit increase in x, there is a corresponding _____-unit _____ (increase/decrease) in y.

   5-unit increase

---

86. Slope is a ratio or fraction. When the slope is computed, <u>the fraction</u> <u>should always be reduced to lowest terms</u>.

    (a) If $\Delta x = 10$ and $\Delta y = 8$, the slope of the line is _____.
    (b) If $\Delta y = -6$ and $\Delta x = 9$, the slope of the line is _____.

   4-unit decrease

---

a) $\frac{4}{5}$ (from $\frac{8}{10}$)

b) $-\frac{2}{3}$ (from $\frac{-6}{9}$)

87. When computing a slope, sometimes the ratio (or fraction) reduces to a whole number. Here is an example:

    To compute the slope of the line at the right, we plotted the two points L and R.

    Since $\Delta x = +4$ and $\Delta y = +8$,

    the slope $= \dfrac{+8}{+4} = \dfrac{2}{1}$ or 2.

Show that we also get "2" as the value of the slope if we use points M and R to compute it.

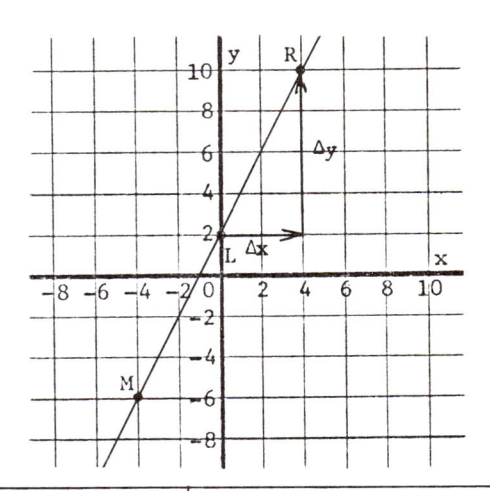

---

88. The slope of the line in the last frame was "2". When the slope reduces to a whole number, it is easy to forget that the whole number stands for a ratio (or fraction).

    If we think of "2" as $\dfrac{"2"}{1}$, we will not forget the fact that the slope stands for a ratio.

A slope of 2 (or $\dfrac{2}{1}$) means: For every 1-unit increase in x, there is a corresponding 2-unit _____ (increase/decrease) in y.

Slope $= \dfrac{+16}{+8} = \dfrac{2}{1}$ or 2

---

89. If $\Delta y = -20$ and $\Delta x = 5$, the slope $= \dfrac{-20}{5} = \dfrac{-4}{1} = -4$.

A slope of -4 ($-\dfrac{4}{1}$) means: For every 1-unit increase in x, there is a _____ (4-unit increase/4-unit decrease) in y.

increase

---

90. If the slope of a line is "10":

(a) Does the line rise or fall from left to right? _____

(b) For every 1-unit increase in x, what happens to y? _____

4-unit decrease

---

91. If the slope of a line is "-7":

(a) Does the line rise or fall from left to right? _____

(b) What does this slope mean in terms of changes in x and y? _____

a) It rises.

b) A 10-unit <u>increase</u>.

---

92. If $\Delta x = 5$ and $\Delta y = 5$, the slope $= \dfrac{5}{5} = \dfrac{1}{1} = 1$.

What does a slope of "+1" mean in terms of changes in x and y? _____

a) It falls.

b) For every 1-unit <u>increase</u> in x, there is a 7-unit <u>decrease</u> in y.

414   Linear Graphs and Slope

93. If $\Delta y = -9$ and $\Delta x = 9$, the slope $= \dfrac{-9}{9} = \dfrac{-1}{1} = -1$.

   What does a slope of "-1" mean? _____

   | For every 1-unit <u>increase</u> in x, there is a 1-unit <u>increase</u> in y.

   <u>Answer</u> <u>to</u> <u>Frame</u> <u>93</u>: For every 1-unit <u>increase</u> in x, there is a 1-unit <u>decrease</u> in y.

94. The <u>sign</u> of a slope tells us whether the line <u>rises</u> or <u>falls</u> from left to right. The point of the next few frames is to show that the <u>absolute value</u> of the slope also has a meaning. That is, its absolute value tells us <u>how steep the rise or fall is</u>.

   We have graphed three lines at the right. The rise of Line #1 is steeper than the rise of Line #2, and the rise of Line #2 is steeper than the rise of Line #3.

   (a) Using points A and D, the slope of Line #1 is _____.

   (b) Using points A and C, the slope of Line #2 is _____.

   (c) Using points A and B, the slope of Line #3 is _____.

   (d) The steeper the rise of a line, the _____ (larger/smaller) is the absolute value of its slope.

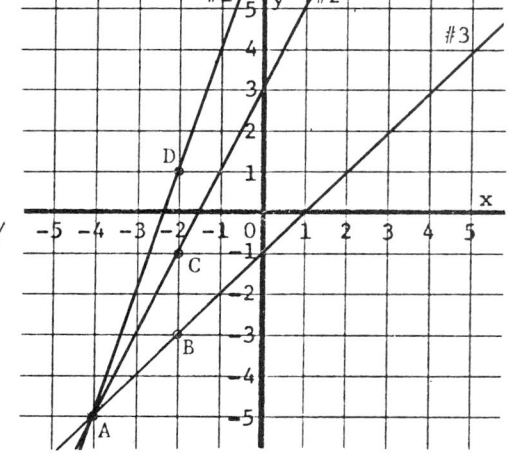

   <u>Answer</u> <u>to</u> <u>Frame</u> <u>94</u>:   a) $\dfrac{6}{2} = +3$ b) $\dfrac{4}{2} = +2$ c) $\dfrac{2}{2} = +1$ d) larger

95. On the right, we have graphed three lines with <u>different</u> <u>negative slopes</u>. Of the three lines, the fall of Line #1 is the steepest.

   (a) Using points E and F, the slope of Line #3 is _____.

   (b) Using points E and G, the slope of Line #2 is _____.

   (c) Using points E and H, the slope of Line #1 is _____.

   (d) The steeper the fall of a line, the _____ (larger/smaller) is the absolute value of its slope.

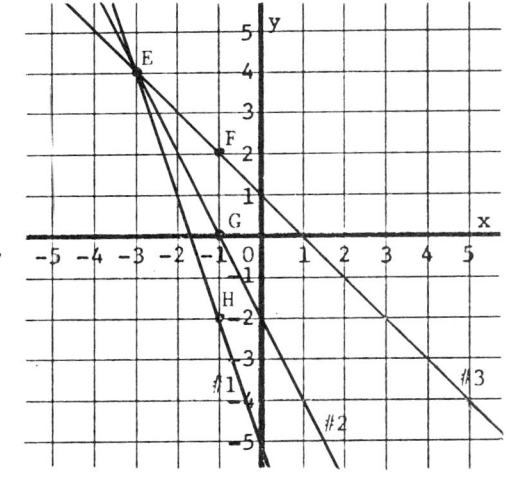

   a) $\dfrac{-2}{2} = -1$

   b) $\dfrac{-4}{2} = -2$

   c) $\dfrac{-6}{2} = -3$

   d) larger

Linear Graphs and Slope 415

96. If the slope of a line (call it #1) is "10":

For every 1-unit increase in x, there is a 10-unit <u>increase</u> in y.

If the slope of another line (call it #2) is "5":

For every 1-unit increase in x, there is a 5-unit <u>increase</u> in y.

The <u>rise</u> of which line (#1 or #2) is steeper? _____

---

97. If the slope of a line (call it #1) is "-6": | #1

For every 1-unit increase in x, there is a 6-unit <u>decrease</u> in y.

If the slope of another line (call it #2) is "-3":

For every 1-unit increase in x, there is a 3-unit <u>decrease</u> in y.

The <u>fall</u> of which line (#1 or #2) is steeper? _____

---

98. If one line (#1) has a slope of $\frac{7}{8}$ and another line (#2) has a slope of $\frac{11}{8}$, which line has the steeper <u>rise</u>, #1 or #2? _____ | #1

---

99. If one line (#1) has a slope of $-\frac{5}{3}$ and another line (#2) has a slope of $-\frac{2}{3}$, which line has the steeper <u>fall</u>, #1 or #2? _____ | #2, since:

$\frac{11}{8}$ is larger than $\frac{7}{8}$.

---

100. (a) If Line #1 has a slope of $\frac{7}{4}$ and Line #2 has a slope of $\frac{1}{3}$, which line has the steeper rise? _____ | #1, since:

$\frac{5}{3}$ is larger than $\frac{2}{3}$.

(b) If Line #3 has a slope of $-\frac{6}{5}$ and Line #4 has a slope of -2, which line has the steeper fall? _____

---

101. Both the <u>sign</u> and the <u>absolute value</u> of its slope give us information about a line.

The <u>sign</u> tells us <u>whether</u> the line rises or falls (from left to right).

The <u>absolute value</u> tells us <u>how much</u> the line rises or falls (from left to right).

| a) #1, since:

$\frac{7}{4}$ is larger than $\frac{1}{3}$.

b) #4, since:

2 is larger than $\frac{6}{5}$.

416 Linear Graphs and Slope

SELF-TEST 5 (Frames 75-101)

1. Using Δx and Δy, write the formula for slope.

Find the slope of each of these lines:

2. If $\Delta x = 4$, $\Delta y = -12$. Slope = _____

3. If $\Delta x = 6$, $\Delta y = 2$. Slope = _____

Refer to the three lines graphed at the right:

4. Find the slope of Line #1. Slope = _____
5. Find the slope of Line #2. Slope = _____
6. Find the slope of Line #3. Slope = _____

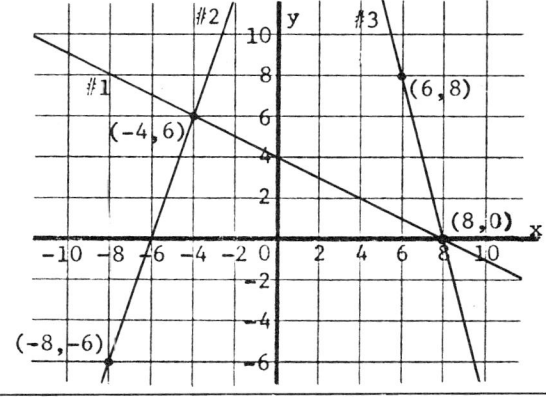

Four lines are sketched on the graph at the right:

7. Which lines have a positive slope? _____
8. Which lines have a negative slope? _____
9. Which line has the steepest rise? _____
10. Which line has the steepest fall? _____

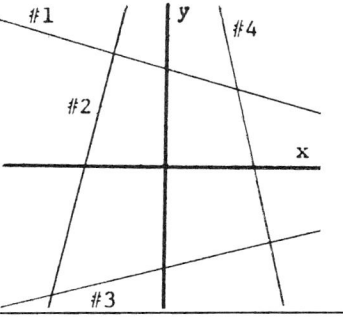

If the slope of a line is -2:

11. From left to right, the line _____ (rises/falls).
12. For every 1-unit increase in x, there is a ____-unit _____ (increase/decrease) in y.

ANSWERS:

1. Slope = $\dfrac{\Delta y}{\Delta x}$
2. Slope = -3
3. Slope = $\dfrac{1}{3}$
4. Slope = $-\dfrac{1}{2}$
5. Slope = 3
6. Slope = -4
7. Lines #2 and #3.
8. Lines #1 and #4.
9. Line #2.
10. Line #4.
11. falls
12. 2-unit decrease

5-6 THE SLOPE OF A LINE AS A RATIO OF VARIOUS PAIRS OF CHANGES IN "x" AND "y"

In the last section, we defined the "slope" concept and discussed the use of signed numbers to represent the slope of lines. In this section, we will show that the signed number used to represent the slope of a line is actually a ratio which represents various pairs of changes in "x" and "y".

102. The corresponding changes in "x" and "y" between any two points on a line can be used to compute its slope. Therefore, the pairs of values of Δx and Δy which are used can differ. Since we always reduce the ratio $\frac{\Delta y}{\Delta x}$ to lowest terms, we can lose sight of the fact that the slope represents various pairs of changes. For example:

If the slope of a line is reported as $"\frac{1}{2}"$, this slope could have been obtained from any of the following equivalent $\frac{\Delta y}{\Delta x}$ ratios, since all of them reduce to $"\frac{1}{2}"$:

$$\frac{1}{2}, \frac{2}{4}, \frac{3}{6}, \frac{5}{10}, \frac{10}{20}, \frac{50}{100}$$

Therefore, a slope of $"\frac{1}{2}"$ really represents all of the following pairs of changes.

If x increases 2 units, y increases 1 unit.

If x increases 4 units, y increases 2 units.

If x increases 20 units, y increases _____ units.

103. If the slope of a line is reported as $"-\frac{2}{3}"$, this slope could have been obtained from any of the following equivalent $\frac{\Delta y}{\Delta x}$ ratios, since all of them reduce to $"-\frac{2}{3}"$:

$$\frac{-2}{3}, \frac{-4}{6}, \frac{-6}{9}, \frac{-20}{30}, \frac{-40}{60}, \frac{-80}{120}$$

Therefore, a slope of $"-\frac{2}{3}"$ really represents all of the following pairs of changes:

If x increases 3 units, y decreases 2 units.

(a) If x increases 9 units, y decreases _____ units.

(b) If x increases 120 units, y decreases _____ units.

---

10 units

a) 6 units

b) 80 units

418 Linear Graphs and Slope

104. If the slope of a line is reported as "5", this slope could have been obtained from any of the following equivalent $\frac{\Delta y}{\Delta x}$ ratios:

$$\frac{5}{1}, \frac{10}{2}, \frac{20}{4}, \frac{40}{8}, \frac{100}{20}, \frac{200}{40}$$

Therefore, a slope of "5" represents all of the following pairs of changes:

  If x increases 1 unit, y increases 5 units.

  (a) If x increases 2 units, y increases _____ units.
  (b) If x increases 20 units, y increases _____ units.

---

105. Similarly, a slope reported as "-3" could have been obtained from any of the following equivalent $\frac{\Delta y}{\Delta x}$ ratios:

$$\frac{-3}{1}, \frac{-6}{2}, \frac{-9}{3}, \frac{-15}{5}, \frac{-45}{15}, \frac{-90}{30}$$

Therefore, a slope of "-3" represents all of the following pairs of changes:

  If x increases 1 unit, y decreases 3 units.

  (a) If x increases 3 units, y decreases _____ units.
  (b) If x increases 15 units, y decreases _____ units.

a) 10 units
b) 100 units

---

106. A slope reported as "1.7" could have been obtained from any of the following equivalent $\frac{\Delta y}{\Delta x}$ ratios:

$$\frac{1.7}{1}, \frac{3.4}{2}, \frac{5.1}{3}, \frac{8.5}{5}, \frac{17}{10}, \frac{170}{100}$$

Therefore, a slope of "1.7" represents all of the following pairs of changes:

  If x increases 1 unit, y increases 1.7 units.

  (a) If x increases 2 units, y increases _____ units.
  (b) If x increases 10 units, y increases _____ units.

a) 9 units
b) 45 units

---

a) 3.4 units
b) 17 units

107. A slope of $"\frac{3}{4}"$ stands for any $\frac{\Delta y}{\Delta x}$ ratio which reduces to $\frac{3}{4}$. Therefore, to find the Δy corresponding to a specific Δx of a line whose slope is $"\frac{3}{4}"$, we can substitute the specific Δx in the following proportion:

$$\boxed{\frac{3}{4} = \frac{\Delta y}{\Delta x}}$$

That is: If $\Delta x = 8$, $\frac{3}{4} = \frac{\Delta y}{8}$. Therefore: $\Delta y = 8\left(\frac{3}{4}\right) = 6$

(a) If $\Delta x = 20$, $\frac{3}{4} = \frac{\Delta y}{20}$. Therefore: $\Delta y = 20\left(\frac{3}{4}\right) = $ _____

(b) If $\Delta x = 40$, $\frac{3}{4} = \frac{\Delta y}{40}$. Therefore: $\Delta y = 40\left(\frac{3}{4}\right) = $ _____

108. To find the Δy corresponding to a specific Δx for a line whose slope is $"-\frac{1}{5}"$, we can substitute the specific Δx in the following proportion.

a) 15

b) 30

$$\boxed{-\frac{1}{5} = \frac{\Delta y}{\Delta x}}$$

That is: If $\Delta x = 10$, $-\frac{1}{5} = \frac{\Delta y}{10}$. Therefore: $\Delta y = 10\left(-\frac{1}{5}\right) = -2$

(a) If $\Delta x = 15$, $-\frac{1}{5} = \frac{\Delta y}{15}$. Therefore: $\Delta y = 15\left(-\frac{1}{5}\right) = $ _____

(b) If $\Delta x = 50$, $-\frac{1}{5} = \frac{\Delta y}{50}$. Therefore: $\Delta y = 50\left(-\frac{1}{5}\right) = $ _____

109. By setting up a proportion, find the corresponding change in "y" for each of the following:

a) -3

b) -10

(a) If the slope is $\frac{2}{5}$ and $\Delta x = 10$, $\Delta y = $ _____.

(b) If the slope is $\frac{7}{2}$ and $\Delta x = 10$, $\Delta y = $ _____.

(c) If the slope is $-\frac{1}{3}$ and $\Delta x = 9$, $\Delta y = $ _____.

(d) If the slope is $-\frac{5}{4}$ and $\Delta x = 28$, $\Delta y = $ _____.

a) $\Delta y = 4$, from: $\frac{2}{5} = \frac{\Delta y}{10}$

b) $\Delta y = 35$, from: $\frac{7}{2} = \frac{\Delta y}{10}$

c) $\Delta y = -3$, from: $-\frac{1}{3} = \frac{\Delta y}{9}$

d) $\Delta y = -35$, from: $-\frac{5}{4} = \frac{\Delta y}{28}$

420   Linear Graphs and Slope

110. (a) If the slope is $\frac{1}{4}$ and x increases 12 units,

　　　　y _____ (increases/decreases) _____ units.

(b) If the slope is $\frac{6}{5}$ and x increases 25 units,

　　　　y _____ (increases/decreases) _____ units.

(c) If the slope is $-\frac{5}{7}$ and x increases 14 units,

　　　　y _____ (increases/decreases) _____ units.

(d) If the slope is $-\frac{8}{3}$ and x increases 12 units,

　　　　y _____ (increases/decreases) _____ units.

a) y increases 3 units.
b) y increases 30 units.
c) y decreases 10 units.
d) y decreases 32 units.

111. When the slope of a line is a whole number or decimal, the slope is still a ratio. For example:

A slope of "3" means $\frac{3}{1}$.　　　　A slope of "2.7" means $\frac{2.7}{1}$.

A slope of "-5" means $\frac{-5}{1}$.　　　A slope of "-14.3" means $\frac{-14.3}{1}$.

Therefore, to find Δy corresponding to a specific Δx for a line whose slope is "3", we can substitute the specific Δx in the following proportion.

$$\boxed{\frac{3}{1} = \frac{\Delta y}{\Delta x}}$$

However, it is easier to write the proportion with the given slope in whole-number form as we have done below:

$$\boxed{3 = \frac{\Delta y}{\Delta x}}$$

That is:　　If $\Delta x = 2$, $3 = \frac{\Delta y}{2}$. Therefore: $\Delta y = 2(3) = 6$

(a) If $\Delta x = 4$, $3 = \frac{\Delta y}{4}$. Therefore: $\Delta y =$ _____ = _____

(b) If $\Delta x = 10$, $3 = \frac{\Delta y}{10}$. Therefore: $\Delta y =$ _____ = _____

a)　$4(3) = 12$
b)　$10(3) = 30$

112. To find the Δy corresponding to a specific Δx for a line whose slope is "-5", we can substitute into the following proportion:

$$\boxed{\frac{-5}{1} = \frac{\Delta y}{\Delta x}}$$

However, it is easier to write the proportion with the given slope in whole-number form as we have done below:

$$\boxed{-5 = \frac{\Delta y}{\Delta x}}$$

That is: If $\Delta x = 3$, $-5 = \frac{\Delta y}{3}$. Therefore: $\Delta y = 3(-5) = -15$

(a) If $\Delta x = 4$, $-5 = \frac{\Delta y}{4}$. Therefore: $\Delta y =$ _____ = _____

(b) If $\Delta x = 8$, $-5 = \frac{\Delta y}{8}$. Therefore: $\Delta y =$ _____ = _____

113. To find the Δy corresponding to a specific Δx for a line whose slope is "2.7", we can substitute into the following proportion:

$$\boxed{\frac{2.7}{1} = \frac{\Delta y}{\Delta x}}$$

a) $4(-5) = -20$

b) $8(-5) = -40$

However, it is easier to write the proportion with the given slope in decimal form as we have done below:

$$\boxed{2.7 = \frac{\Delta y}{\Delta x}}$$

That is: If $\Delta x = 2$, $2.7 = \frac{\Delta y}{2}$. Therefore: $\Delta y = 2(2.7) = 5.4$

(a) If $\Delta x = 3$, $2.7 = \frac{\Delta y}{3}$. Therefore: $\Delta y =$ _____ = _____

(b) If $\Delta x = 10$, $2.7 = \frac{\Delta y}{10}$. Therefore: $\Delta y =$ _____ = _____

114. By setting up a proportion, find the corresponding change in "y" for each of the following:

(a) If the slope is 2 and $\Delta x = 7$, $\Delta y =$ _____.

(b) If the slope is 10 and $\Delta x = 12$, $\Delta y =$ _____.

(c) If the slope is -4 and $\Delta x = 5$, $\Delta y =$ _____.

(d) If the slope is 1.6 and $\Delta x = 3$, $\Delta y =$ _____.

a) $3(2.7) = 8.1$

b) $10(2.7) = 27$

422   Linear Graphs and Slope

115. (a) If the slope of a line is 7 and x increases 3 units,
    y _____ (increases/decreases) _____ units.

    (b) If the slope of a line is -6 and x increases 5 units,
    y _____ (increases/decreases) _____ units.

    (c) If the slope of a line is 0.125 and x increases 10 units,
    y _____ (increases/decreases) _____ units.

    (d) If the slope of a line is -1.25 and x increases 5 units,
    y _____ (increases/decreases) _____ units.

a) $\Delta y = 14$, from:
   $2 = \dfrac{\Delta y}{7}$

b) $\Delta y = 120$, from:
   $10 = \dfrac{\Delta y}{12}$

c) $\Delta y = -20$, from:
   $-4 = \dfrac{\Delta y}{5}$

d) $\Delta y = 4.8$, from:
   $1.6 = \dfrac{\Delta y}{3}$

---

116. (a) The slope of a line is $\dfrac{5}{7}$. What happens to y if x increases 21 units? _____

    (b) The slope of a line is $-\dfrac{9}{2}$. What happens to y if x increases 10 units? _____

    (c) The slope of a line is -10. What happens to y if x increases 3 units? _____

    (d) The slope of a line is 0.02. What happens to y if x increases 6 units? _____

a) y increases 21 units.
b) y decreases 30 units.
c) y increases 1.25 units.
d) y decreases 6.25 units.

---

117. We are frequently interested in the change in y when x increases 1 unit (that is, when $\Delta x = 1$). If the slope is a whole number, remember that the whole number stands for a ratio. That is:

   A slope of "2" means $\dfrac{"2"}{1}$.   A slope of "-5" means $\dfrac{"-5"}{1}$.

    (a) The ratio $\dfrac{"2"}{1}$ means that for every 1-unit increase in x, y increases _____ units.

    (b) The ratio $\dfrac{"-5"}{1}$ means that for every 1-unit increase in x, y decreases _____ units.

a) y increases 15 units.
b) y decreases 45 units.
c) y decreases 30 units.
d) y increases 0.12 unit.

---

118. If the slope is a decimal, remember that the decimal stands for a ratio. That is:

   A slope of "3.5" means $\dfrac{"3.5"}{1}$.   A slope of "-0.8" means $\dfrac{"-0.8"}{1}$.

    (a) The ratio $\dfrac{3.5}{1}$ means that for every 1-unit increase in x, y increases _____ units.

    (b) The ratio $\dfrac{-0.8}{1}$ means that for every 1-unit increase in x, y decreases _____ unit.

a) 2 units
b) 5 units

119. If the slope of a line is a whole number or decimal, <u>the slope itself is</u> the Δy corresponding to a Δx of +1. Using this fact, complete the following:

    (a) If the slope is 10, what happens to y if x increases 1 unit?
    _____

    (b) If the slope is -7, what happens to y if x increases 1 unit?
    _____

    (c) If the slope is 0.25, what happens to y if Δx is +1?
    _____

    (d) If the slope is -15.8, what happens to y if Δx is +1?
    _____

a) 3.5 units
b) 0.8 unit

---

120. Up to this point, we have given the following meaning for a slope of "$\frac{7}{4}$":

    For every 4-unit increase in x, y increases 7 units.

    However, since $\frac{7}{4} = \frac{\frac{7}{4}}{1}$, we can also give the following new meaning for a slope of $\frac{7}{4}$:

    For every 1-unit increase in x, y increases _____ units.

a) y increases 10 units.
b) y decreases 7 units.
c) y increases 0.25 unit.
d) y decreases 15.8 units.

---

121. Up to this point, we have given the following meaning for a slope of "$-\frac{3}{5}$":

    For every 5-unit increase in x, y decreases 3 units.

    However, since $-\frac{3}{5} = \frac{-\frac{3}{5}}{1}$, we can also give the following new meaning for a slope of $-\frac{3}{5}$:

    For every 1-unit increase in x, y decreases _____ unit.

$\frac{7}{4}$ units

---

122. Just like slopes which are whole numbers or decimals, a fractional slope <u>is itself the Δy</u> when Δx is +1. That is:

    If the slope is $\frac{6}{5}$, for every 1-unit increase in x, we get a $\frac{6}{5}$-unit increase in y.

    If the slope is $-\frac{3}{4}$, for every 1-unit increase in x, we get a $\frac{3}{4}$-unit decrease in y.

    (a) If the slope is $\frac{9}{4}$, what happens to y for every 1-unit increase in x?
    _____

    (b) If the slope is $-\frac{7}{5}$, what happens to y for every 1-unit increase in x?
    _____

$\frac{3}{5}$ unit

123. Any fractional slope can be converted to a decimal. That is:

A slope of $\frac{8}{5}$ equals a slope of 1.6.

Therefore, if the slope is $\frac{8}{5}$ (or 1.6), for every 1-unit increase in x, y increases $\frac{8}{5}$ units or 1.6 units.

(a) If the slope is $\frac{5}{4}$ (or 1.25), for every 1-unit increase in x,

y increases _____ units or _____ units.

(b) If the slope is $-\frac{2}{5}$ (or -0.4), for every 1-unit increase in x,

y decreases _____ unit or _____ unit.

a) y increases $\frac{9}{4}$ units.

b) y decreases $\frac{7}{5}$ units.

Answer to Frame 123:   a) $\frac{5}{4}$ units or 1.25 units   b) $\frac{2}{5}$ unit or 0.4 unit

---

SELF-TEST 6 (Frames 102-123)

In Problems 1 to 6, find Δy. In each problem, the slope and Δx are given.

1. Slope = $\frac{7}{3}$ $\Delta x = 6$ $\Delta y =$ ___
2. Slope = $-\frac{5}{2}$ $\Delta x = 10$ $\Delta y =$ ___
3. Slope = 4 $\Delta x = 1$ $\Delta y =$ ___
4. Slope = -1 $\Delta x = 3$ $\Delta y =$ ___
5. Slope = -1.9 $\Delta x = 2$ $\Delta y =$ ___
6. Slope = 0.57 $\Delta x = 1$ $\Delta y =$ ___

7. If the slope of a line is $\frac{3}{2}$ and x increases 2 units, y _____ (increases/decreases) ____ units.

8. If the slope of a line is -0.25 and x increases 5 units, y _____ (increases/decreases) ____ units.

9. If the slope of a line is $-\frac{9}{4}$ and x increases 1 unit, y _____ (increases/decreases) ____ units.

10. The slope of a line is -8. For every 3-unit increase in x, what happens to y?

11. The slope of a line is 1. What happens to y if x increases 1 unit?

ANSWERS:

1. $\Delta y = 14$
2. $\Delta y = -25$
3. $\Delta y = 4$
4. $\Delta y = -3$
5. $\Delta y = -3.8$
6. $\Delta y = 0.57$
7. y increases 3 units.
8. y decreases 1.25 units.
9. y decreases $\frac{9}{4}$ units.
10. y decreases 24 units.
11. y increases 1 unit.

5-7 THE SLOPE-INTERCEPT FORM OF LINEAR EQUATIONS

Any linear equation in "x" and "y" can be written in the following form:

$$\boxed{y = mx + b}$$ where "x" and "y" are the <u>variables</u> and "m" and "b" are <u>constants</u>.

This form of a linear equation is called its "<u>slope-intercept</u>" form because the two constants "m" and "b" have a definite meaning. That is:

"m" is the <u>slope of the line</u>, and
"b" is the <u>vertical intercept</u> of the line.

In this section, we will confirm the fact that "m" and "b" in the slope-intercept form of a linear equation represent the slope and the vertical intercept of the graphed line.

---

124. Any linear equation in "x" and "y" can be rearranged to the following form:

$$\boxed{y = mx + b}$$

Here are some examples of linear equations written in the form above:

$$y = 5x + 8 \qquad y = 7x + 10$$

Notice these points:
(1) In each case, "y" is the "solved for" variable.
(2) In each case, "m" and "b" are numerical constants.

In $y = 5x + 8$: $m = 5$ and $b = 8$.

In $y = 7x + 10$: $m = \underline{}$ and $b = \underline{}$.

---

125. When a linear equation is written in $\boxed{y = mx + b}$ form, the numerical constants "m" and "b" can be decimals or fractions. For example:

(a) In $y = 7.5x + 0.8$, $m = \underline{}$ and $b = \underline{}$.

(b) In $y = \frac{3}{4}x + \frac{1}{9}$, $m = \underline{}$ and $b = \underline{}$.

---

126. When a linear equation is written in $\boxed{y = mx + b}$ form, "m" and "b" can be either positive or <u>negative</u>. For example:

(a) In $y = -3x + 7$, $m = \underline{}$ and $b = \underline{}$.

(b) In $y = \frac{5}{2}x + (-1)$, $m = \underline{}$ and $b = \underline{}$.

Answers:

$m = 7$ and $b = 10$

a) $m = 7.5$ and $b = 0.8$

b) $m = \frac{3}{4}$ and $b = \frac{1}{9}$

a) $m = -3$ and $b = 7$

b) $m = \frac{5}{2}$ and $b = -1$

127. When a linear equation is written in the form $\boxed{y = mx + b}$, we say that it is written in "slope-intercept" form since:

"m" is the slope of the line, and
"b" is the vertical intercept of the line.

In the next few frames, we will confirm the fact that "m" and "b" are the slope and the vertical intercept by examining the graphs of some specific linear equations.

The equation $\boxed{y = 2x + 1}$ is graphed on the right. The coordinates of two points (G and H) on the line are given. Note the following:

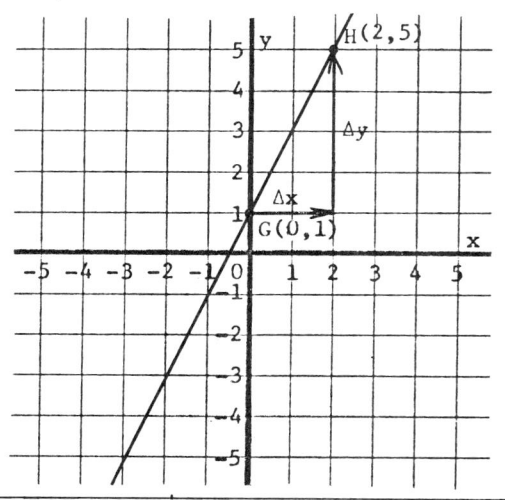

(1) In $y = 2x + 1$, $m = 2$. We can show that "2" is the slope of the line by using the Δx and Δy from G to H to compute the slope:

Since $\Delta x = 2$, and $\Delta y = 4$, the slope is $\frac{\Delta y}{\Delta x} = \frac{4}{2} = 2$.

(2) In $y = 2x + 1$, $b = 1$. We can see that "1" is the ordinate of the vertical intercept by examining the graph:

Since point G is the vertical intercept, the coordinates of the vertical intercept are (0,1).

128. The equation $\boxed{y = -\frac{2}{3}x + 3}$ is graphed on the right. The coordinates of two points (R and T) on the line are given.

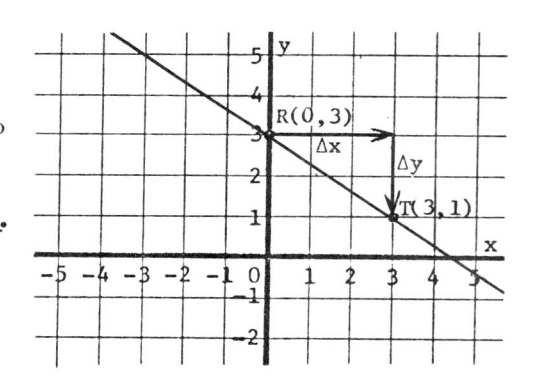

(a) In $y = -\frac{2}{3}x + 3$, $m = -\frac{2}{3}$. We can see that "$-\frac{2}{3}$" is the slope of the line by using the Δx and Δy from R to T to compute the slope:

Since $\Delta x = 3$ and $\Delta y = -2$, the slope $= \frac{\Delta y}{\Delta x} = $ _____.

(b) In $y = -\frac{2}{3}x + 3$, $b = 3$. We can see that "3" is the ordinate of the vertical intercept by examining the graph:

Since point R is the vertical intercept, the coordinates of the vertical intercept are (,).

---

a) $\frac{-2}{3}$ or $-\frac{2}{3}$

b) (0,3)

129. The equation $\boxed{y = \frac{5}{2}x + (-4)}$ is graphed on the right. The coordinates of two points (C and E) on the line are given.

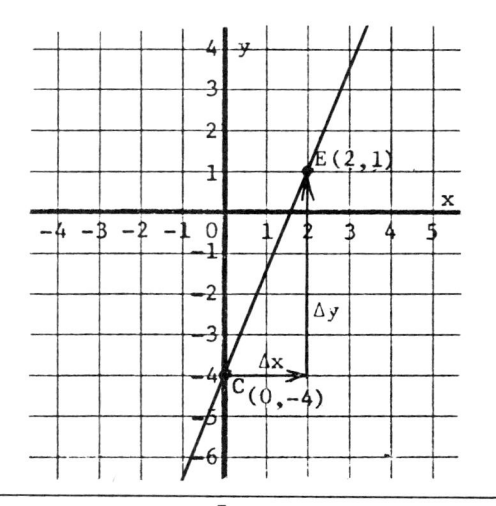

(a) In $y = \frac{5}{2}x + (-4)$, $m = \frac{5}{2}$. We can see that "$\frac{5}{2}$" is the <u>slope</u> of the line by using the Δx and Δy from C to E to compute the slope:

Since $\Delta x = 2$ and $\Delta y = 5$, the slope $= \frac{\Delta y}{\Delta x} =$ _____.

(b) In $y = \frac{5}{2}x + (-4)$, $b = -4$. We can see that "-4" is the ordinate of the <u>vertical intercept</u> by examining the graph:

Since C is the vertical intercept, the coordinates of the vertical intercept are (,).

Answer to Frame 129: a) $\frac{5}{2}$ b) $(0, -4)$

130. The equation $\boxed{y = -3x + (-1)}$ is graphed on the right. Two points on the line are labeled D and F.

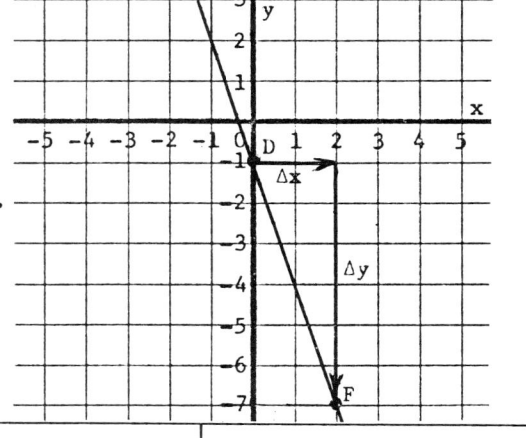

(a) In $y = -3x + (-1)$, $m = -3$. We can see that "-3" is the <u>slope</u> of the line by using the Δx and Δy from D to F to compute the slope:

Since $\Delta x = 2$ and $\Delta y = -6$, the slope $= \frac{\Delta y}{\Delta x} =$ _____.

(b) In $y = -3x + (-1)$, $b = -1$. We can see that "-1" is the ordinate of the <u>vertical intercept</u> by examining the graph:

Since point D is the vertical intercept, the coordinates of the vertical intercept are (,).

131. When a linear equation is written in the form $\boxed{y = mx + b}$:

(1) "m" is the slope of the line, and
(2) "b" is the ordinate of the y-intercept.

Since the constants in a linear equation in this form tell you its slope and y-intercept, this form is called the <u>slope-intercept form</u> of a linear equation.

Each equation below is linear. Which of them are in slope-intercept form? _____

(a) $y = 5x + 1$ (b) $2x + 3y = 7$ (c) $5 = x - y$ (d) $y = -\frac{5}{4}x + 6$

a) -3 From: $\frac{-6}{2}$

b) $(0, -1)$

Only (a) and (d).

428   Linear Graphs and Slope

132. In $y = mx + b$, we have seen that "m" is the slope of the line. Mathematicians use the symbol "m" instead of the word "slope". The formula for the "slope" or "m" is:

$$m = \frac{\Delta y}{\Delta x}$$

If an equation is written in slope-intercept form, we do not need a graph in order to find its slope. The slope is immediately given in the equation.

Find "m", the slope of each of the following linear equations:

(a) $y = 7x + 10$ m = _____   (c) $y = \frac{3}{4}x + (-2)$ m = _____

(b) $y = -5x + 3$ m = _____   (d) $y = -\frac{8}{3}x + 1$ m = _____

---

133. The constant "m" is not explicitly written in the equations below:

$y = x + 5$ $y = -x + 3$

However, since "x = 1x" and "-x = -1x", we can write the same equations in the following form:

$y = 1x + 5$ $y = -1x + 3$

Therefore: (a) The slope of $y = x + 5$ is: m = _____

(b) The slope of $y = -x + 3$ is: m = _____

a) m = 7   c) m = $\frac{3}{4}$

b) m = -5   d) m = $-\frac{8}{3}$

---

134. In $y = mx + b$, we have seen that "b" is the <u>ordinate of the vertical intercept</u> (or <u>y-intercept</u>). Therefore, when an equation is written in slope-intercept form, we can write the coordinates of the vertical intercept immediately. That is:

For $y = 3x + 7$, the coordinates of the vertical intercept are (0, 7).

Using "b" from the equation, write the coordinates of the y-intercept for each of these equations:

(a) $y = -x + 9$ (b) $y = \frac{3}{5}x + (-6)$ (c) $y = -3x + (-11)$

   (,)   (,)   (,)

a) +1   b) -1

---

135. Examine "b" in each equation below to answer each of the following:

(a) Does $y = -3x + 4$ cut the vertical axis above or below the origin? _____

(b) Does $y = x + (-3)$ cut the vertical axis above or below the origin? _____

a) (0, 9)
b) (0, -6)
c) (0, -11)

---

a) Above.
b) Below.

136. (a) $\boxed{y = mx + b}$ is called the _____ form of a linear equation.

(b) The abscissa (x-coordinate) of any y-intercept is always what number? _____

a) slope-intercept

b) 0

---

137. Write the slope-intercept form of the linear equations with the following slopes and y-intercepts:

	Slope	y-intercept	Equation
(a)	5	$(0, -7)$	_____
(b)	$-\frac{7}{2}$	$(0, 2)$	_____
(c)	1	$(0, -\frac{3}{2})$	_____
(d)	-1	$(0, \frac{5}{7})$	_____

---

138. When an equation is written in slope-intercept form, the vertical intercept is given by "b". However, the horizontal intercept is not immediately given in the equation. To find the coordinates of the horizontal intercept, we must plug in "0" for "y" and compute the corresponding value of "x". For the equation below, we get:

$$y = 5x + 9$$
$$0 = 5x + 9$$
$$x = -\frac{9}{5}$$

Therefore, the coordinates of the horizontal intercept of $y = 5x + 9$ are (,).

a) $y = 5x + (-7)$

b) $y = -\frac{7}{2}x + 2$

c) $y = 1x + \left(-\frac{3}{2}\right)$

or

$y = x + \left(-\frac{3}{2}\right)$

d) $y = -1x + \frac{5}{7}$

or

$y = -x + \frac{5}{7}$

---

139. For $y = 2x + (-6)$: (a) The coordinates of the vertical intercept are (,).

(b) The coordinates of the horizontal intercept are (,).

$\left(-\frac{9}{5}, 0\right)$

---

140. $\boxed{y = -2x + 4}$

(a) The y-intercept of the above equation is (,).

(b) The x-intercept of the above equation is (,).

(c) Graph the equation on the axes at the right by plotting the two intercepts and then drawing the line.

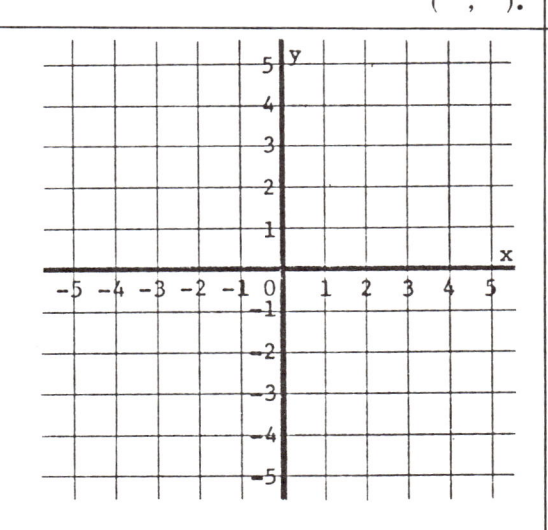

a) $(0, -6)$

b) $(3, 0)$

430   Linear Graphs and Slope

141. In $y = 5x + 7$, the slope is "5". A slope of "5" means:

    (a) For every 1-unit increase in x, there is a _____-unit increase in y.

    (b) For every 3-unit increase in x, there is a _____-unit increase in y.

    (c) For every 7-unit increase in x, there is a _____-unit increase in y.

a) (0,4)
b) (2,0)
c) Graph: (line through (0,4) and (2,0))

---

142. In $y = \frac{5}{2}x + 1$, the slope is $\frac{5}{2}$. A slope of $\frac{5}{2}$ means:

    (a) When x increases 1 unit, y increases _____ units.

    (b) When x increases 6 units, y increases _____ units.

    (c) When x increases 50 units, y increases _____ units.

a) 5
b) 15
c) 35

---

143. In $y = 0.4x + 1.2$, the slope is 0.4. A slope of 0.4 means:

    (a) When x increases 1 unit, y increases _____ units.

    (b) When x increases 50 units, y increases _____ units.

    (c) When x increases 100 units, y increases _____ units.

a) $\frac{5}{2}$ or 2.5 units
b) 15 units
c) 125 units

---

144. In $y = -\frac{3}{5}x + (-1)$, the slope is $-\frac{3}{5}$. With a slope of $-\frac{3}{5}$, what change in y occurs for each of the following:

    (a) A 1-unit increase in x? _____

    (b) A 5-unit increase in x? _____

    (c) A 10-unit increase in x? _____

    (d) A 25-unit increase in x? _____

a) 0.4 units
b) 20 units
c) 40 units

---

a) y decreases $\frac{3}{5}$ or 0.6 unit.
b) y decreases 3 units.
c) y decreases 6 units.
d) y decreases 15 units.

Linear Graphs and Slope   431

SELF-TEST 7 (Frames 124-144)

Find the slope "m" and the y-intercept "b" of each of these linear equations:

1. $y = 5x + (-8)$ m = _____
   b = _____

2. $y = -\frac{3}{4}x + \frac{1}{2}$ m = _____
   b = _____

3. $y = x + (-1)$ m = _____
   b = _____

4. Write the linear equation whose slope is 9 and whose y-intercept is at (0, -5). _____

Given: $y = 4x + 6$

5. Find the coordinates of its y-intercept. _____
6. Find the coordinates of its x-intercept. _____
7. Find its slope. _____
8. If x increases 2 units, y _____ (increases/decreases) _____ units.

ANSWERS:
1. m = 5, b = -8
2. m = $-\frac{3}{4}$, b = $\frac{1}{2}$
3. m = 1, b = -1
4. y = 9x + (-5)
5. (0, 6)
6. $\left(-\frac{3}{2}, 0\right)$
7. Slope = 4
8. y increases 8 units

5-8 WRITING LINEAR EQUATIONS IN SLOPE-INTERCEPT FORM

Linear equations are not always written in slope-intercept form. However, when they are not in slope-intercept form, the constants in the equations do not represent the slopes and y-intercepts of their graphed lines. Therefore, linear equations in some other form are frequently rearranged to slope-intercept form so that their constants do represent their slopes and y-intercepts. We will discuss the rearrangement process in this section.

---

145. Each equation below is in slope-intercept form. Notice that in each case, the x-term and number on the right are <u>added</u>:

$y = 3x + 7$ $y = -2x + 5$ $y = 4x + (-1)$ $y = -5x + (-4)$

Neither equation below is in slope-intercept form because in each case the x-term and number are <u>subtracted</u>. However, each can be written in slope-intercept form by simply converting the <u>subtraction</u> to <u>addition</u>. Do so:

(a) $y = 3x - 7$
    $y = 3x +$ _____

(b) $y = -2x - 5$
    $y = -2x +$ _____

a) $y = 3x + (-7)$
b) $y = -2x + (-5)$

432 Linear Graphs and Slope

146. (a) Since "x = 1x", the slope of y = x + 9 is _____.

(b) Since "-x = -1x", the slope of y = -x + 5 is _____.

147. Convert each equation below to slope-intercept form by converting the subtraction to addition. Then identify its slope and the coordinates of its y-intercept.

(a) y = x - 3   y = _____   m = _____   (,)

(b) y = -x - 7   y = _____   m = _____   (,)

a) +1

b) -1

148. To convert $\boxed{y = 7 - 3x}$ to slope-intercept form:

(1) We convert the subtraction to addition: y = 7 + (-3x)
(2) and then commute the two terms on the right: y = -3x + 7

Therefore: (a) m = _____   (b) Its vertical intercept is (,).

a) y = x + (-3)
    m = 1   (0, -3)

b) y = -x + (-7)
    m = -1   (0, -7)

149. Following the steps in the last frame, convert each equation below to slope-intercept form. Then identify its slope and the coordinates of its vertical intercept.

(a) y = 1 - 5x   y = _____   m = _____   (,)

(b) y = 5 - x   y = _____   m = _____   (,)

a) m = -3

b) (0, 7)

150. In $\boxed{y = mx + b}$, "y" is the "solved for" variable. Therefore, the equation below is not in slope-intercept form because "-y" is "solved for":

-y = 2x + 7

We can, however, solve for "y" and put the equation in slope-intercept form by replacing each side with its opposite.

Since the opposite of "-y" is "y"
and the opposite of 2x + 7 is (-2x) + (-7),

we get: y = -2x + (-7)

Therefore: (a) m = _____   (b) Its y-intercept is (,).

a) y = -5x + 1
    m = -5   (0, 1)

b) y = -x + 5
    m = -1   (0, 5)

151. Convert each equation below to slope-intercept form by replacing each side with its opposite:

(a) -y = 7x + 1   (b) -y = x + 5

___ = _____   ___ = _____

a) m = -2

b) (0, -7)

152. The equation on the right can be converted to slope-intercept form by:

(1) converting the subtraction on the right to an addition.

(2) and then replacing each side with its opposite.

Therefore: (a) m = _____   (b) Its vertical intercept is (,).

$\boxed{-y = 3x - 5}$

-y = 3x + (-5)

y = -3x + 5

a) y = (-7x) + (-1)

b) y = (-x) + (-5)

153. Following the steps in the last frame, convert each equation below to slope-intercept form:

    (a) $-y = 4x - 9$ (b) $-y = x - 10$ (c) $-y = 8 - 5x$

    ___ = _____   ___ = _____   ___ = _____

a) -3
b) $(0, 5)$

---

154. Convert each equation below to slope-intercept form:

    (a) $y = 10 - x$ $y = $ _____   (c) $-y = x - 3$ $y = $ _____

    (b) $-y = 7x + 8$ $y = $ _____   (d) $-y = 7 - 2x$ $y = $ _____

a) $y = -4x + 9$
b) $y = -x + 10$
c) $y = 5x + (-8)$

---

155. Convert each equation below to slope-intercept form. Then identify its slope and the coordinates of its vertical intercept.

    (a) $y = 7 - 5x$ $y = $ _____ $m = $ ____   (,)

    (b) $-y = x + 1$ $y = $ _____ $m = $ ____   (,)

    (c) $-y = 9 - 4x$ $y = $ _____ $m = $ ____   (,)

    (d) $-y = 10 - x$ $y = $ _____ $m = $ ____   (,)

a) $y = -x + 10$
b) $y = -7x + (-8)$
c) $y = -x + 3$
d) $y = 2x + (-7)$

---

156. To put the equation on the right in slope-intercept form, we must solve for "y". The steps are shown.

$$x = y + 8$$
$$x + (-8) = y$$
or
$$y = x + (-8)$$

Therefore: (a) $m = $ ____   (b) The y-intercept is (,).

a) $y = -5x + 7$
   $m = -5$ $(0, 7)$
b) $y = -x + (-1)$
   $m = -1$ $(0, -1)$
c) $y = 4x + (-9)$
   $m = 4$ $(0, -9)$
d) $y = x + (-10)$
   $m = 1$ $(0, -10)$

---

157. Put each equation below in slope-intercept form by solving for "y":

    (a) $x = y - 5$ (b) $2x = 1 + y$ (c) $5x = 10 - y$

    $y = $ _____ $y = $ _____ $y = $ _____

a) $m = 1$
b) $(0, -8)$

---

158. Put each equation below in slope-intercept form by solving for "y":

    (a) $x + y = 5$ (b) $y - 3x = 7$ (c) $5x - y = 9$

    $y = $ _____ $y = $ _____ $y = $ _____

a) $y = x + 5$
b) $y = 2x + (-1)$
c) $y = -5x + 10$

---

a) $y = -x + 5$
b) $y = 3x + 7$
c) $y = 5x + (-9)$

434 Linear Graphs and Slope

159. The equation on the right is not in slope-intercept form since "y" has a numerical coefficient.

$$2y = 3x + 4$$

To put it in slope-intercept form, we solve for "y" by multiplying both sides by $\frac{1}{2}$, the reciprocal of the coefficient of "y". We have done so on the right.

$$\frac{1}{2}(2y) = \frac{1}{2}(3x + 4)$$

$$(1)y = \frac{1}{2}(3x) + \frac{1}{2}(4)$$

$$y = \frac{3}{2}x + 2$$

Therefore: (a) m = _____   (b) The vertical intercept is (,).

160. Put each equation below in slope-intercept form by solving for "y":

(a) $5y = 15x + 4$ (b) $6y = 3x - 12$

y = _____   y = _____

a) $m = \frac{3}{2}$

b) $(0, 2)$

161. To put the equation on the right in slope-intercept form, we:

$$2x + 3y = 6$$

(1) Isolate the y-term by adding "-2x" to both sides.

$$3y = 6 + (-2x)$$

(2) Commute the terms on the right side.

$$3y = -2x + 6$$

(3) Then multiply both sides by $\frac{1}{3}$.

$$\frac{1}{3}(3y) = \frac{1}{3}(-2x + 6)$$

$$(1)y = \frac{1}{3}(-2x) + \frac{1}{3}(6)$$

$$y = -\frac{2}{3}x + 2$$

Therefore: (a) m = _____   (b) The y-intercept is (,).

a) $y = 3x + \frac{4}{5}$

b) $y = \frac{1}{2}x + (-2)$

162. Put each equation below in slope-intercept form:

(a) $5y - 4x = 15$ (b) $x + 7y = 3$

y = _____ (b) y = _____

a) $m = -\frac{2}{3}$

b) $(0, 2)$

a) $y = \frac{4}{5}x + 3$

b) $y = -\frac{1}{7}x + \frac{3}{7}$

163. To put the equation on the right in slope-intercept form, we:

$\boxed{3x - 5y = 15}$

(1) Add "-3x" to both sides, and commute the right side.

$-5y = 15 + (-3x)$
$-5y = -3x + 15$

(2) Replace each side with its opposite.

$5y = 3x + (-15)$

(3) Then multiply both sides by "$\frac{1}{5}$".

$\frac{1}{5}(5y) = \frac{1}{5}\left[3x + (-15)\right]$

$(1)y = \frac{1}{5}(3x) + \frac{1}{5}(-15)$

$y = \frac{3}{5}x + (-3)$

Therefore: (a) m = _____ (b) The y-intercept is (,).

164. Write each equation below in slope-intercept form:

(a) $7x - 4y = 8$ (b) $x - 6y = 7$

y = _____   y = _____

a) $m = \frac{3}{5}$

b) $(0, -3)$

165. Write each equation below in slope-intercept form:

(a) $5y - 2x + 10 = 0$ (b) $3x - 2y - 12 = 0$

y = _____   y = _____

a) $y = \frac{7}{4}x + (-2)$

b) $y = \frac{1}{6}x + \left(-\frac{7}{6}\right)$

166. To put each equation below in slope-intercept form, we must begin by clearing the fraction. Do so:

(a) $x = \frac{7y + 6}{5}$ (b) $3y = \frac{2x - 1}{4}$ (c) $11 = \frac{2x - 7}{y}$

y = _____   y = _____   y = _____

a) $y = \frac{2}{5}x + (-2)$

b) $y = \frac{3}{2}x + (-6)$

167. Though "y" is solved for in the equation on the right, the equation is not in slope-intercept form since the right side fits the pattern for the sum of two fractions. By breaking the fraction apart, we have put the equation in standard form.

$$y = \frac{3x+4}{2}$$

$$y = \frac{3x}{2} + \frac{4}{2}$$

$$y = \frac{3}{2}x + 2$$

a) $y = \frac{5}{7}x + \left(-\frac{6}{7}\right)$

b) $y = \frac{1}{6}x + \left(-\frac{1}{12}\right)$

c) $y = \frac{2}{11}x + \left(-\frac{7}{11}\right)$

Write each equation below in standard form:

(a) $y = \frac{7x+16}{4}$ (b) $y = \frac{5x-11}{10}$ (c) $x = \frac{5y-1}{7}$

y = _____   y = _____   y = _____

168. After rearranging an equation to slope-intercept form, it is easy to identify its slope and the coordinates of its vertical intercept. For example:

Since $\boxed{7x - 4y = 8}$ is equivalent to $\boxed{y = \frac{7}{4}x + (-2)}$:

Its slope is $\frac{7}{4}$.

Its vertical intercept is (0, -2).

However, to find its horizontal intercept, we must plug in "0" for "y" in one of the two equations above.

If we plug in "0" for "y" in
7x - 4y = 8, we get:

$7x - 4(0) = 8$

$7x = 8$

$x = \frac{8}{7}$

If we plug in "0" for "y" in
$y = \frac{7}{4}x + (-8)$, we get:

$0 = \frac{7}{4}x + (-2)$

$2 = \frac{7}{4}x$

$8 = 7x$

$x = \frac{8}{7}$

a) $y = \frac{7}{4}x + 4$

b) $y = \frac{1}{2}x + \left(-\frac{11}{10}\right)$

c) $y = \frac{7}{5}x + \frac{1}{5}$

Using either equation, the coordinates of the horizontal intercept are (,).

169. (a) Write the equation on the right in slope-intercept form.

$\boxed{30 + 10y = 5x}$

$\left(\frac{8}{7}, 0\right)$

(b) Its slope is _____.

(c) Its vertical intercept is (,).

(d) Its horizontal intercept is (,).

y = _____

170. An equation can only be written in slope-intercept form if it is <u>linear</u>. If an equation contains an x^2-term, a y^2-term, or an xy-term, it is <u>non-linear</u>. Therefore, it <u>cannot</u> be written in slope-intercept form.

Which of the following equations cannot be written in slope-intercept form because they are non-linear? _____

(a) $xy = 50$ (b) $x + y^2 = 1$ (c) $3x - 4y = 24$ (d) $7y = 5x^2 - 1$

a) $y = \frac{1}{2}x + (-3)$

b) $m = \frac{1}{2}$

c) $(0, -3)$

d) $(6, 0)$

Answer to Frame 170:   (a), (b), and (d)

SELF-TEST 8 (Frames 145-170)

By algebraic rearrangement, put each of the following equations in slope-intercept form:

1. $2x - 5y = 10$

2. $3y + 9x + 8 = 0$

3. $\frac{8y - 1}{2} = 6x$

For each of the following equations, find the slope, y-intercept, and x-intercept:

4. Slope = _____ $3x + 4y = 24$

5. y-intercept:
   (,)

6. x-intercept:
   (,)

7. Slope = _____ $8x - 5y + 12 = 0$

8. y-intercept:
   (,)

9. x-intercept:
   (,)

ANSWERS: 1. $y = \frac{2}{5}x + (-2)$ 4. $-\frac{3}{4}$ 7. $\frac{8}{5}$

2. $y = -3x + \left(-\frac{8}{3}\right)$ 5. $(0, 6)$ 8. $\left(0, \frac{12}{5}\right)$

6. $(8, 0)$

3. $y = \frac{3}{2}x + \frac{1}{8}$ 9. $\left(-\frac{3}{2}, 0\right)$

438 Linear Graphs and Slope

5-9 DETERMINING THE SLOPE OF A LINE THROUGH TWO KNOWN POINTS

If we know the coordinates of two points on a line, we can determine the slope of the line by means of a formula. We will show the method in this section.

171. A line passes through the following two points: M (−3,−4) and T (2,5)

To find the slope of the line, we can use the following steps:

(1) Roughly plot the points on a graph.
(2) Draw the line through them.
(3) Draw vectors representing Δx and Δy.
(4) Then use the numerical values of Δx and Δy to compute the slope.

We have done the first three steps for the two points above on the graph at the right.

(a) Since Δx represents a change from "−3" to "+2", Δx = _____ .

(b) Since Δy represents a change from "−4" to "+5", Δy = _____ .

(c) The slope of the line is _____ .

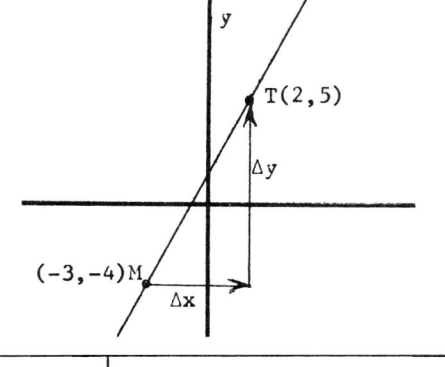

---

172. Use the graph at the right to find the slope of the line through these two points: S(2,−3) , W(−4,4). Follow the steps in the last frame.

(a) Δx = _____
(b) Δy = _____
(b) m = _____

a) 5
b) 9
c) $\dfrac{9}{5}$

---

173. Use the graph at the right to find the slope of the line through these two points: Q(15,50) , P(25,70).

(a) Δx = _____
(b) Δy = _____
(c) m = _____

a) 6 (from −4 to 2)
b) −7 (from 4 to −3)
c) $-\dfrac{7}{6}$

---

a) 10 (from 15 to 25)
b) 20 (from 50 to 70)
c) $\dfrac{20}{10} = 2$

174. If we know the coordinates of two points, their coordinates can be used to determine Δx, Δy, and m. On the right, we have graphed a line through points L(-3,-4) and R(2,-1). The vectors for Δx and Δy are also drawn.

In the example on the right, $\Delta x = +5$, $\Delta y = +3$, and $m = \frac{3}{5}$.

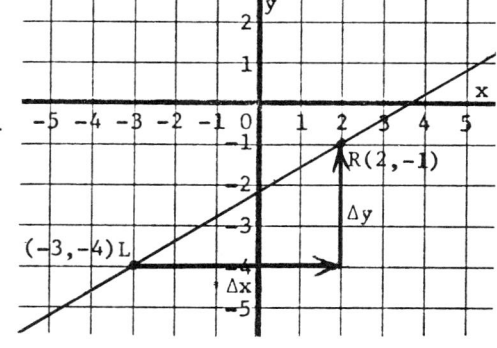

We can get Δx and Δy <u>by subtracting</u> the <u>abscissa</u> <u>and</u> <u>ordinate</u> <u>of</u> <u>the</u> <u>left</u> <u>point</u> <u>(L)</u> <u>from</u> <u>the</u> <u>abscissa</u> <u>and</u> <u>ordinate</u> <u>of</u> <u>the</u> <u>right</u> <u>point</u> <u>(R)</u>. That is:

$$\Delta x = x_R - x_L = 2 - (-3) = 2 + 3 = +5$$

$$\Delta y = y_R - y_L = (-1) - (-4) = (-1) + 4 = +3$$

Therefore: $m = \frac{\Delta y}{\Delta x} = \frac{y_R - y_L}{x_R - x_L} = \frac{3}{5}$

---

175. The point of the last frame was to introduce the following formulas:

$$\Delta y = y_R - y_L$$
$$\Delta x = x_R - x_L$$

$$m = \frac{\Delta y}{\Delta x} = \frac{y_R - y_L}{x_R - x_L}$$

Using these formulas, we can find the slope of a line without graphing it if we know the coordinates of two points on it.

Let's use the formulas to find the slope of the line through the following two points:

$$(4,1) \text{ and } (2,5)$$

Before plugging into the formulas above, we must determine which is the <u>right point</u> and which is the <u>left point</u>. To do so, we draw a rough sketch like the one on the right.

Since (4,1) is the <u>right</u> point, $x_R = 4$ and $y_R = 1$.
Since (2,5) is the <u>left</u> point, $x_L = 2$ and $y_L = 5$.
Therefore: $\Delta x = x_R - x_L = 4 - 2 = 2$

$$\Delta y = y_R - y_L = 1 - 5 = -4$$

$$m = \frac{\Delta y}{\Delta x} = \frac{y_R - y_L}{x_R - x_L} = \underline{}$$

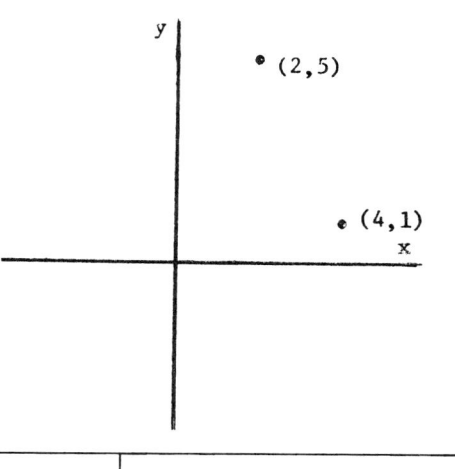

$$m = \frac{-4}{2} = -2$$

440   Linear Graphs and Slope

176. To use the formulas, we must be able to determine which point is <u>on the right</u> and which one is <u>on the left</u>. The easiest way to do so is <u>to draw a rough sketch</u> of the two points on the coordinate system.

Draw a rough sketch of each of the following pairs of points to determine which of the two is <u>on the right</u>:

(a) C(5, 2) and D(1, -4)   (b) G(-2, 4) and H(1, 3)   (c) T(-5, 10) and S(-15, 20)   (d) M(-10, -15) and R(-8, -20)

Right point:_____   Right point:_____   Right point:_____   Right point:_____

Answer to Frame 176:   a) C(5, 2)   b) H(1, 3)   c) T(-5, 10)   d) R(-8, -20)

---

177. Once we have decided which point is on the right, we can plug the coordinates into the slope formula to determine the slope. (Note: A rough sketch of the points and the line will save you from "sign" errors. From your sketch, you can easily see whether the slope is positive or negative.)

Use the slope formula to determine the slope of the lines through each of the following pairs of points. (Draw a rough sketch to avoid errors.)

(a) B(7, 5) and D(2, 2)

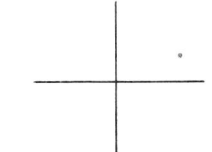

(b) T(10, -20) and S(15, 5)

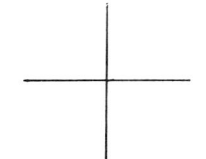

$m = \dfrac{y_R - y_L}{x_R - x_L} =$ _____

$m = \dfrac{y_R - y_L}{x_R - x_L} =$ _____

---

178. If the coordinates of the points are decimals, Δx and Δy can be decimal numbers. Here is an example:

D(5.7, 1.4)   and   F(10.3, 6.6)

The right point is F. Therefore:

$m = \dfrac{\Delta y}{\Delta x} = \dfrac{y_R - y_L}{x_R - x_L} = \dfrac{6.6 - 1.4}{10.3 - 5.7} = \dfrac{5.2}{4.6} =$ _____

When Δx and Δy are decimals, we ordinarily convert the slope ratio to a decimal by performing the division. Do so above.

a) $m = \dfrac{5 - 2}{7 - 2} = \dfrac{3}{5}$

b) $m = \dfrac{5 - (-20)}{15 - 10}$

  $= \dfrac{5 + 20}{5} = \dfrac{25}{5} = 5$

m = 1.13 or 1.1

179. Find the slope of the line through each pair of points below. (Make a rough sketch to avoid gross errors.)

   (a) J(9.8, 5.1) and K(3.1, 1.7) (b) P(15.9, 9.7) and Q(26.2, 5.6)

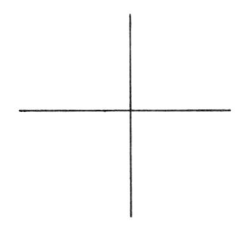

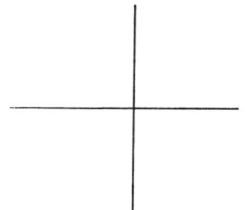

$$m = \frac{y_R - y_L}{x_R - x_L} = \underline{\hspace{2in}}$$ $$m = \frac{y_R - y_L}{x_R - x_L} = \underline{\hspace{2in}}$$

Answer to Frame 179: a) $\frac{5.1 - 1.7}{9.8 - 3.1} = \frac{3.4}{6.7} = 0.507$ or 0.51 b) $\frac{5.6 - 9.7}{26.2 - 15.9} = \frac{-4.1}{10.3} = -0.398$ or -0.40

180. Both of the points listed below lie on an axis. We have sketched them at the right and drawn the line through them.

$$(0, -3) \text{ and } (4, 0)$$

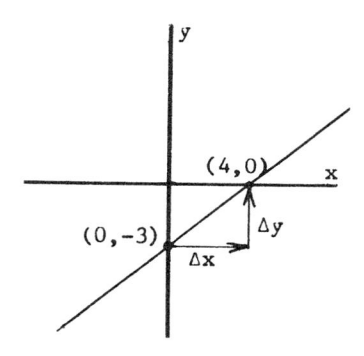

As you can see from the graph:

$$\Delta x = 4, \quad \Delta y = 3, \quad m = \frac{3}{4}$$

When using the formula to compute the slope of a line through points on an axis, you must be especially careful of the 0's and the signs. Of the two points above, (4, 0) is the one to the right. Therefore:

$$m = \frac{\Delta y}{\Delta x} = \frac{y_R - y_L}{x_R - x_L} = \frac{0 - (-3)}{4 - 0} = \frac{0 + 3}{4} \text{ or } \frac{3}{4}$$

Is this the same slope we obtained by examining the graph above? _____

Answer to Frame 180: Yes.

181. Find the slope of the line through each pair of points below. Roughly sketch the points on the axes to avoid gross errors.

   (a) (0, 5) and (10, 0) (b) (0, -5) and (-5, -10)

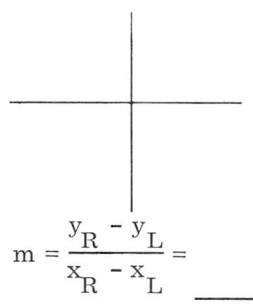

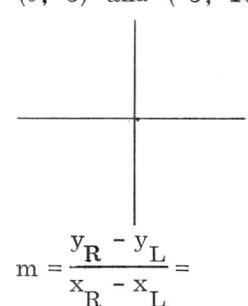

$$m = \frac{y_R - y_L}{x_R - x_L} = \underline{\hspace{2in}}$$ $$m = \frac{y_R - y_L}{x_R - x_L} = \underline{\hspace{2in}}$$

Answer to Frame 181: a) $m = \frac{0 - 5}{10 - 0} = \frac{-5}{10} = -\frac{1}{2}$ b) $m = \frac{(-5) - (-10)}{0 - (-5)} = \frac{(-5) + 10}{0 + 5} = \frac{5}{5} = 1$

442   Linear Graphs and Slope

SELF-TEST 9 (Frames 171-181)

Using the formula, find the slope of the line through each pair of points below:

1. (-6, 5) and (3, 2)

   m = _____

2. (36, 67) and (-24, -83)

   m = _____

3. (-8, 0) and (0, -18)

   m = _____

ANSWERS: 1. $m = -\frac{1}{3}$ 2. $m = \frac{5}{2}$ 3. $m = -\frac{9}{4}$

5-10 DETERMINING THE EQUATION OF A LINE THROUGH TWO KNOWN POINTS

In the last section, we showed how to determine the slope of a line through two known points. In this section, we will show how to determine the y-intercept of a line through two points. Once we know "m" and "b", we can then write the equation of the line.

---

182. If you know both the slope "m" and the y-intercept "b" of a line, you can write the equation of the line in slope-intercept form. For example:

    If m = 5 and b = 3, the equation of the line is: y = 5x + 3

    If $m = -\frac{2}{3}$ and b = 1, the equation of the line is: _____

$y = -\frac{2}{3}x + 1$

183. At the right, we have drawn a line through the two points T(1, 2) and Z(4, 3).

The slope of the line is: $m = \dfrac{y_R - y_L}{x_R - x_L} = \dfrac{3-2}{4-1} = \dfrac{1}{3}$

By examining the point at which the line crosses the y-axis, we can estimate the coordinates of the y-intercept. They are approximately (0, 1.7).

Since "m" is $\dfrac{1}{3}$ and "b" is approximately 1.7, the equation of the line is approximately: y = _____

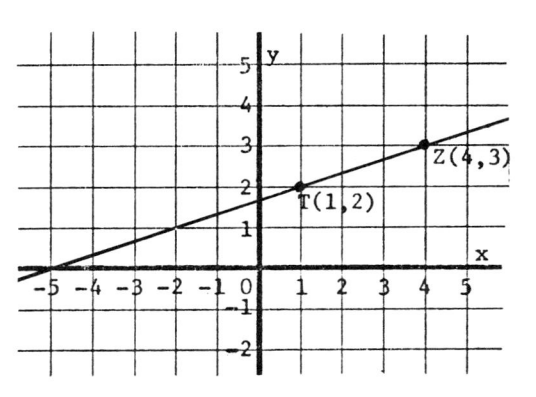

Answer to Frame 183: $y = \dfrac{1}{3}x + 1.7$

---

184. Since we only estimated "b" in the last frame, we could not write the exact equation of the line. In this frame, we will show an exact method of obtaining the value of "b" for the same line. Then we can write the equation of the line exactly.

We found that the slope of the line through T(1, 2) and Z(4, 3) was $\dfrac{1}{3}$. Therefore, we can substitute $\dfrac{1}{3}$ for "m" in y = mx + b. We get:

$$y = \dfrac{1}{3}x + b$$

The coordinates of both points (T and Z) must satisfy the equation of the line since they lie on it. If we plug either pair of coordinates into the equation above, we can solve for b as follows:

(1) Using T(x = 1, y = 2):

$y = \dfrac{1}{3}x + b$

$2 = \dfrac{1}{3}(1) + b$

$2 = \dfrac{1}{3} + b$

$b = 2 - \dfrac{1}{3} = \dfrac{5}{3}$ (or 1.67)

(2) Using Z(x = 4, y = 3):

$y = \dfrac{1}{3}x + b$

$3 = \dfrac{1}{3}(4) + b$

$3 = \dfrac{4}{3} + b$

$b = 3 - \dfrac{4}{3} = \dfrac{5}{3}$ (or 1.67)

(a) Did we get the same value for "b" with the coordinates of each point? _____

(b) If $m = \dfrac{1}{3}$ and $b = \dfrac{5}{3}$, the equation of the line is: y = _____

---

a) Yes. $\dfrac{5}{3}$ (or 1.67)

b) $y = \dfrac{1}{3}x + \dfrac{5}{3}$

185. The slope of a line is $\frac{1}{4}$ and the line passes through the point (7,5). We want to find the equation of the line. Since we know the slope, we can substitute it for "m" and get:

(a) Since we know that the coordinates (7,5) must satisfy the equation, we can plug them in and solve for "b". Do so:

$$y = \frac{1}{4}x + b$$

b = _____

(b) The exact equation of the line is: y = _____

---

186. The slope of a line is $-\frac{1}{2}$ and the line passes through (-10, -6).

(a) Find "b" by plugging the coordinates above into $y = -\frac{1}{2}x + b$.

(b) Write the equation of the line: y =

a) $b = \frac{13}{4}$, since:

$5 = \frac{1}{4}(7) + b$

$b = 5 - \frac{7}{4} = \frac{13}{4}$

b) $y = \frac{1}{4}x + \frac{13}{4}$

b = _____

Answer to Frame 186:   a) b = -11   b) $y = -\frac{1}{2}x + (-11)$ or $y = -\frac{1}{2}x - 11$

---

187. We want to find the equation of the line through the following two points which are plotted on the right:

A(-6, 2) and B(4, -6)

(a) $m = \dfrac{y_R - y_L}{x_R - x_L} = $ _____

(b) Reading from the graph, it appears that "b" lies between "-2" and "-3". We can, however, find the exact value of "b" by substituting the numerical value of "m" into y = mx + b, and then plugging in the coordinates of either point A or point B. Do so:

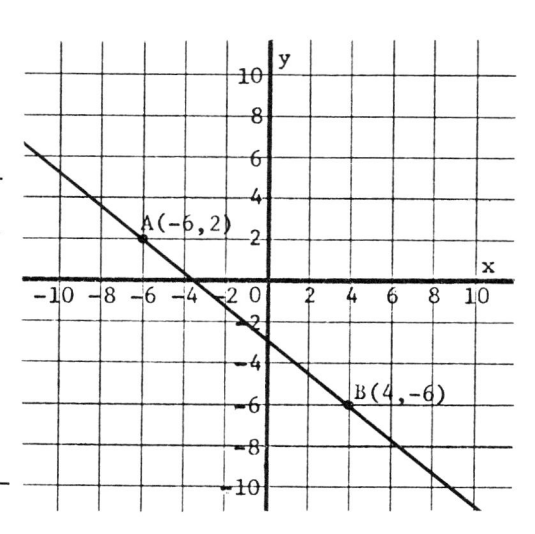

b = _____

(c) The equation of the line is: y = _____

Answer to Frame 187:   a) $m = \dfrac{(-6) - 2}{4 - (-6)} = \dfrac{-8}{10} = -\dfrac{4}{5}$ b) $b = -\dfrac{14}{5}$ (or -2.8)   c) $y = -\dfrac{4}{5}x + \left(-\dfrac{14}{5}\right)$

or $y = -\dfrac{4}{5}x - \dfrac{14}{5}$

188. To find the equation of the line through the two points D(-1, -4) and T(2, 5), draw a rough sketch of the line at the right.

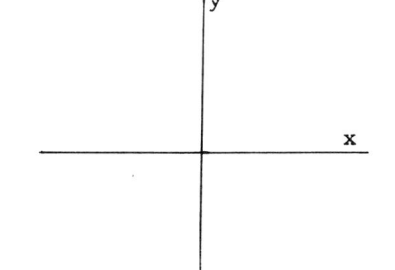

   (a) Find "m": m = _____

   (b) Find "b": b = _____

   (c) The equation of the line is: y = _____

189. To find the equation of the line through the two points: (5, 10) and (10, 0), draw a rough sketch of the line at the right.

   (a) Find "m": m = _____

   (b) Find "b": b = _____

   (c) The equation of the line is: y = _____

a) $m = \dfrac{5 - (-4)}{2 - (-1)} = \dfrac{9}{3} = 3$

b) $b = -1$

c) $y = 3x + (-1)$
   or
   $y = 3x - 1$

190. After finding the equation, its correctness should be checked by substituting the coordinates of each of the two given points into it.

   For example, in the last frame it was found that the equation of the line through (5, 10) and (10, 0) is: $y = -2x + 20$

   (a) Show that the equation is satisfied by (5, 10).

   (b) Show that the equation is satisfied by (10, 0).

a) $m = \dfrac{0 - 10}{10 - 5} = \dfrac{-10}{5} = -2$

b) $b = 20$

c) $y = -2x + 20$

Linear Graphs and Slope

When finding "m", "b", and the equation of a line through two given points, students are tempted to use formulas and equations without drawing a rough sketch. <u>Always draw a rough sketch</u> because you can use it to make sure that:

(1) The "sign" and "absolute value" of the slope make sense.

(2) The computed value of "b" makes sense.

A rough sketch can be used to avoid gross errors.

a) $y = -2x + 20$
$10 = -2(5) + 20$
$10 = -10 + 20$
$10 = 10$

b) $y = -2x + 20$
$0 = -2(10) + 20$
$0 = -20 + 20$
$0 = 0$

SELF-TEST 10 (Frames 182-190)

1. Find the equation of the line which passes through these two points:

   (3, 0) and (0, -3)

   y = _____

2. Find the equation of the line which passes through these two points:

   (-3, 7) and (5, 3)

   y = _____

ANSWERS: 1. $y = x - 3$ 2. $y = -\dfrac{1}{2}x + \dfrac{11}{2}$ or $y = -0.5x + 5.5$

5-11 THE EQUATIONS OF LINES THROUGH THE ORIGIN

In discussing the slope-intercept form of linear equations up to this point, we have avoided equations whose graphed lines pass through the origin. We will discuss the slope-intercept form of such equations in this section.

---

191. The graphed line on the right passes through the origin. Therefore, its vertical intercept is (0,0).

    Using the changes from point A to point B, we can compute its slope:

    $\Delta x = 1$ and $\Delta y = 3$ $m = \dfrac{3}{1} = 3$

    Since $m = 3$ and $b = 0$, the slope-intercept form of the equation of the line is:

    $y = 3x + 0$
    or $y = 3x$

    There is only one constant in $y = 3x$. Does this constant represent the slope or the vertical intercept of the line? _____

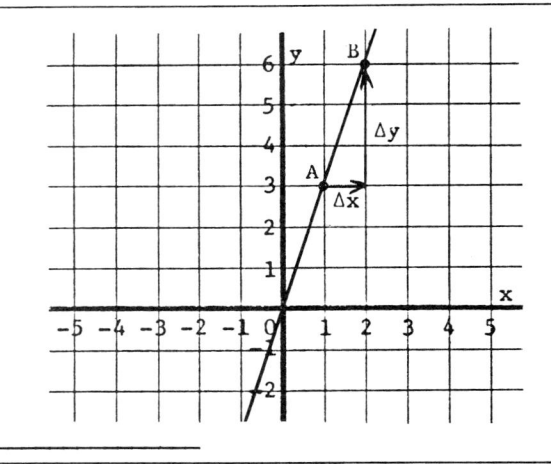

Answer to Frame 191: Slope.

---

192. The general slope-intercept form of linear equations is:

    $$\boxed{y = mx + b}$$

    However, the vertical intercept of all lines through the origin is (0,0). Since "b" is "0" in the slope-intercept form of the equations of all lines through the origin, their slope-intercept form is:

    $y = mx + 0$

    or

    $$\boxed{y = mx}$$

    As you can see, the only constant in this last form of the equation is "m" (the slope).

    Which equations below graph as lines through the origin? _____

    (a) $y = 9x$ (b) $y = 9x + 1$ (c) $y = -2x$ (d) $y = -2x + 7$

---

193. $y = 5x$ is the equation of a line through the origin.

    (a) The vertical intercept of the line is (,).
    (b) The horizontal intercept of the line is (,).

(a) and (c)

---

a) (0,0)
b) (0,0)

448   Linear Graphs and Slope

194. A linear equation can be graphed by plotting only two points. If we know that its line passes through the origin, we can use the origin as one of the points.

   For example, to graph: $y = 2x$

   (1) We can use the origin as one point.

   (2) To find a second point, we merely plug in some other value for "x". For example:

   If $x = 4$, $y = 8$

   By plotting $(0,0)$ and $(4,8)$, we have graphed $y = 2x$ at the right.

   Using the same method, graph and label each equation below on the same set of axes.

   $y = -3x$ $y = \frac{1}{2}x$

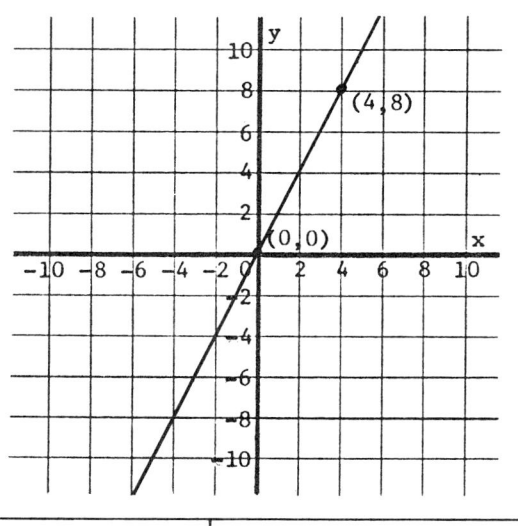

195. The graphs of $y = -3x$, $y = \frac{1}{2}x$, and $y = 2x$ are shown at the right.

   (a) The graph of $y = -3x$ lies in Quadrants ____ and ____.

   (b) The graph of $y = \frac{1}{2}x$ lies in Quadrants ____ and ____.

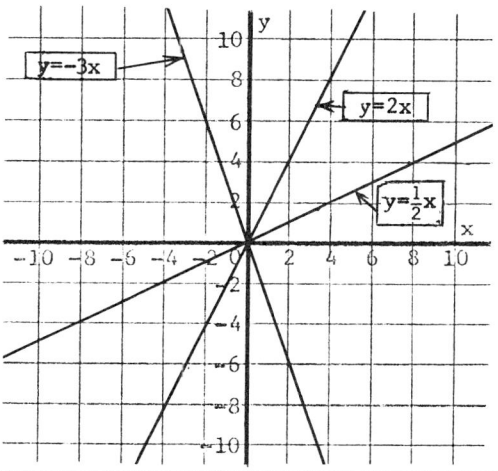

   See next frame.

196. If we know the slope of a line through the origin, we can write its equation immediately. That is:

   If $m = \frac{1}{2}$, $y = \frac{1}{2}x$

   (a) If $m = -\frac{5}{4}$, $y = $ _____

   (b) If $m = 1.75$, $y = $ _____

   a) II and IV
   b) I and III

   ---

   a) $y = -\frac{5}{4}x$
   b) $y = 1.75x$

197. We can easily find the slope and equation of a line through the origin if we know the coordinates of any other point on the line. Here is an example:

The line at the right passes through the origin (0,0) and the point (4,8). By examining the graph, complete these:

(a) $\Delta x =$ _____

(b) $\Delta y =$ _____

(c) $m =$ _____

(d) The equation of the line is: $y =$ _____

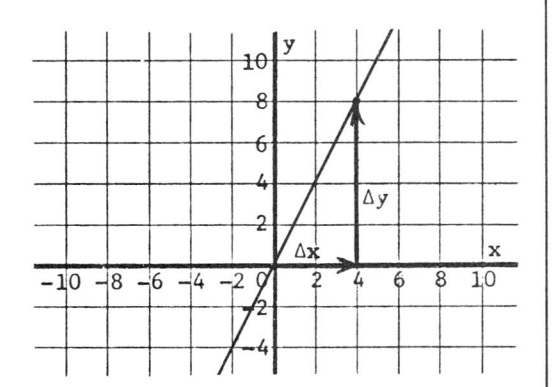

198. The line at the right passes through the origin and the point (-3,-5). By examining the graph, complete these:

(a) $\Delta x =$ _____

(b) $\Delta y =$ _____

(c) $m =$ _____

(d) The equation of the line is: $y =$ _____

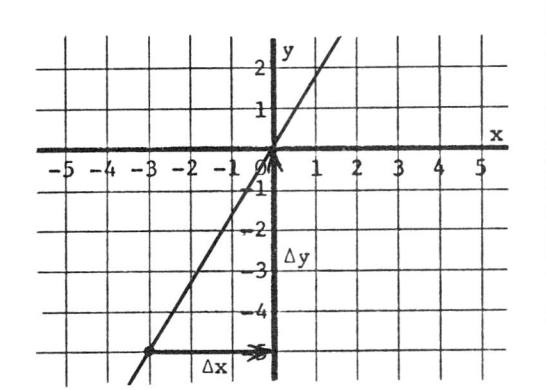

a) 4

b) 8

c) $\frac{8}{4} = 2$

d) $y = 2x$

Answer to Frame 198:   a) 3   b) 5   c) $\frac{5}{3}$ d) $y = \frac{5}{3}x$

---

199. To determine the equation of a line through the origin and one other known point, we must find its slope. To find its slope, we roughly sketch the line and determine the values of Δx and Δy.

Find the equation of the line passing through the origin and each of the following points. Sketch each line first:

(a)   (2,6)   (b)   (9,4)   (c)   (-8,6)   (d)   (-3,-5)

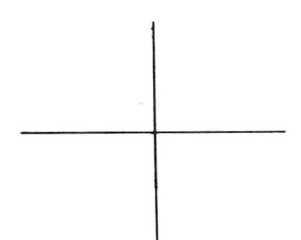

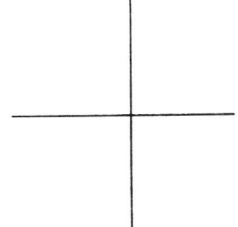

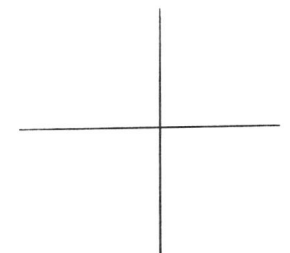

$y =$ _____ $y =$ _____ $y =$ _____ $y =$ _____

Linear Graphs and Slope

200. $\boxed{y = mx}$ is the general slope-intercept form for an equation whose graphed line <u>passes through the origin</u>. Not every equation whose graphed line passes through the origin is written in slope-intercept form.

Here is one that is not: $y - 2x = 0$

(a) Write the equation in slope-intercept form: $y =$ _____

(b) The slope of the line is _____.

a) $y = 3x$

b) $y = \frac{4}{9}x$

c) $y = -\frac{3}{4}x$

d) $y = \frac{5}{3}x$

---

201. Determine the slope of each of the following by writing the equation in slope-intercept form:

(a) $3y = 2x$ $m =$ _____   (c) $5y - 2x = 0$ $m =$ _____

(b) $y + 4x = 0$ $m =$ _____   (d) $7x + 9y = 0$ $m =$ _____

a) $y = 2x$

b) $m = 2$

---

202. The equation $y = x$ can be written $y = 1x$.
    Therefore, its slope is "+1".

What is the slope of the following equation?

$y = -x$ $m =$ _____

a) $m = \frac{2}{3}$ c) $m = \frac{2}{5}$

b) $m = -4$ d) $m = -\frac{7}{9}$

---

203. Which of the following lines pass through the origin? _____

(a) $y = 2x - 1$ (b) $x = 3y$ (c) $x + y = 0$ (d) $\frac{x}{2} = \frac{y}{5}$

$m = -1$, from $y = -1x$

(b), (c), and (d)

SELF-TEST 11 (Frames 191-203)

1. Find the slope of the line whose equation is $y = 6x$.   m = _____

2. Find the coordinates of the y-intercept of $y = -4x$.   _____

A line passes through the origin and the point (-3, -3). Find the following:

3. Its slope.   m = _____

4. Its equation.   _____

A line passes through the origin and the point (40, -25). Find the following:

5. Its slope.   m = _____

6. Its equation.   _____

Find the slope of each of the following lines:

7. $3x = 2y$ m = _____

8. $4x + 5y = 0$ m = _____

9. $3y - x = 0$ m = _____

ANSWERS:
1. m = 6
2. (0, 0)
3. m = 1
4. $y = 1x$ or $y = x$
5. $m = -\dfrac{5}{8}$
6. $y = -\dfrac{5}{8}x$
7. $m = \dfrac{3}{2}$
8. $m = -\dfrac{4}{5}$
9. $m = \dfrac{1}{3}$

452 Linear Graphs and Slope

5-12　SLOPE AND PARALLEL LINES

In this section, we will discuss the slope of parallel lines. We will show that parallel lines are lines <u>with the same slope</u> but <u>different y-intercepts</u>.

204. On the graph at the right, we have graphed the following three equations. Notice that all three graph as straight lines <u>with the same slope</u> but <u>different y-intercepts</u>.

　　#1　　y = 2x + 3
　　#2　　y = 2x + 1
　　#3　　y = 2x + (-3)

If we graph equations with the same slopes but different y-intercepts, the graphed lines are _____ .

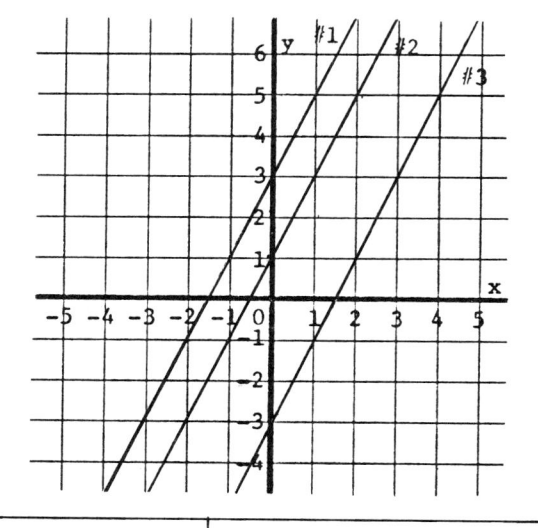

205. When linear equations are written in slope-intercept form, you can tell immediately whether their graphed lines will be parallel or not. If their slopes are the same, they are <u>parallel</u>; if their slopes are different, they are <u>not parallel</u>.

Which of the following equations graph as parallel lines? _____

　(a)　$y = \frac{3}{2}x$　　(b)　$y = \frac{3}{2}x + 7$　　(c)　y = 3x + 1　　(d)　$y = \frac{3}{2}x + (-4)$

parallel

206. To decide whether the graphed lines of the equations below are parallel or not, we must put them in slope-intercept form first so that we can compare their slopes. Do so:

　　#1　　3y - 2x = 12　　　　#2　　2x - 3y = 18

(a), (b), and (d)

　(a)　y = _____　　(b)　y = _____
(c) Are the graphed lines parallel? _____

207. Which of the following pairs of equations will graph as a pair of parallel lines? To decide, write each equation in slope-intercept form.

Pair A $\begin{bmatrix} x = 2y - 7 \\ 2y - 4x = 9 \end{bmatrix}$ Pair B $\begin{bmatrix} 5x - 2y + 10 = 0 \\ y = \frac{5}{2}x + 11 \end{bmatrix}$

a) $y = \frac{2}{3}x + 4$

b) $y = \frac{2}{3}x + (-6)$

c) Yes. They are parallel because they have the same slope, $m = \frac{2}{3}$.

---

208. Which of the following lines are parallel? _____

(a) $y = x$ (c) $y = 4x + 5$ (e) $4y = 4x - 5$

(b) $y = 5x$ (d) $2y = x$ (f) $y - x = 3$

Only Pair B. The slope of each equation in Pair B is $\frac{5}{2}$.

The slopes of the two equations in Pair A are $\frac{1}{2}$ and 2, respectively.

Answer to Frame 208:   (a), (e), and (f), since the slope of each is "+1".

SELF-TEST 12 (Frames 204-208)

By using slope-intercept form, list the slope of each of these linear equations:

1. $y = 2x + 5$	2. $y = -2x - 1$	3. $2y - 4x = 9$	4. $4y - 2x = 0$
m = ____	m = ____	m = ____	m = ____

5. In Problems 1 to 4 above, which equations have graphs which are parallel? _____

ANSWERS: 1. m = 2   3. m = 2   5. (1) $y = 2x + 5$ and (3) $2y - 4x = 9$ are parallel,
          2. m = -2   4. $m = \frac{1}{2}$ since each has a slope of +2.

454 Linear Graphs and Slope

5-13 THE EQUATIONS OF HORIZONTAL AND VERTICAL LINES

In this section, we will discuss the equations of horizontal and vertical lines on the coordinate system. Then in the next section, we will discuss the slope of horizontal and vertical lines.

209. A horizontal line is graphed on the right.

(a) Write the coordinates of the following points:

A(,) B(,) C(,) D(,)

(b) Though the abscissas (or x-coordinates) vary, the ordinate (or y-coordinate) of any point on this line is _____.

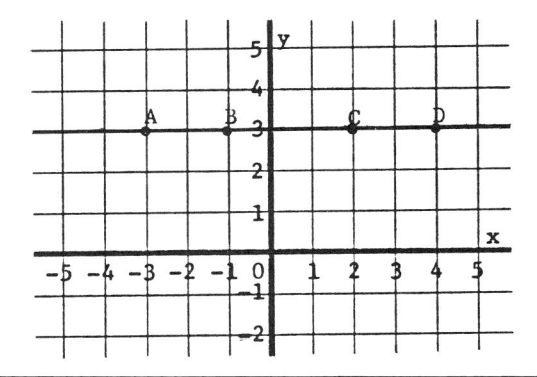

Answer to Frame 209:   a) A(-3,3)   B(-1,3)   C(2,3)   D(4,3)   b) 3

210. On Line #1 at the right, the y-coordinate (or ordinate) of each point is "3", no matter what the x-coordinate (or abscissa) is. Therefore, the equation of the line is:

$y = 3$ "y = 3" means: For every x-value, y = 3.

On Line #2 at the right, y = -4 for every x-value. Therefore, the equation of the line is: _____

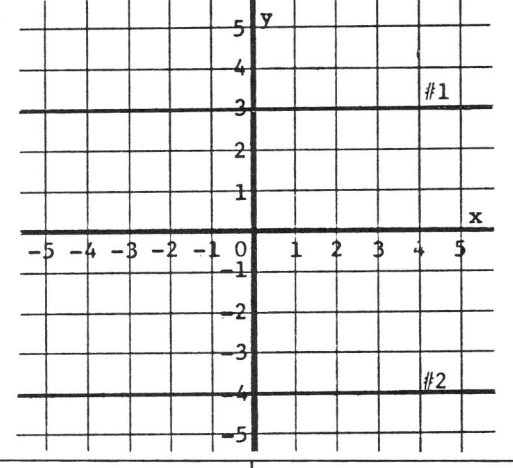

211. The equation $y = 6$ is graphed at the right:

(a) If x = -4, y = _____.

(b) If x = 8, y = _____.

(c) On the line, for any value of "x", y = _____.

y = -4

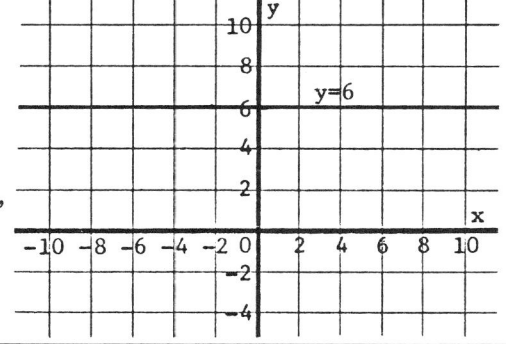

a) 6
b) 6
c) 6

212. On the graph at the right, we have drawn and labeled four horizontal lines.

Write the equation of each line:

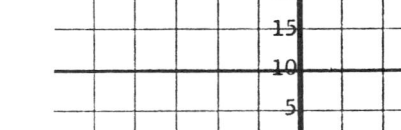

(a) #1 _____

(b) #2 _____

(c) #3 _____

(d) #4 _____

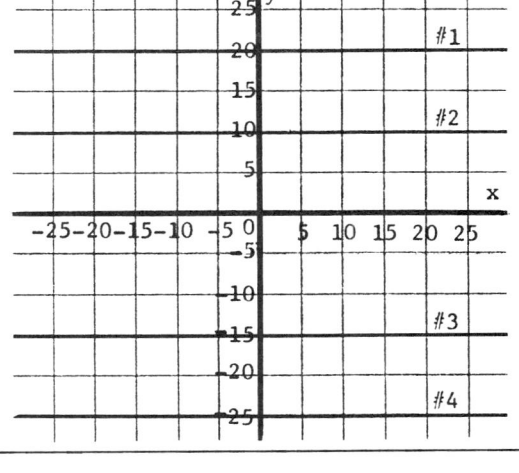

a) $y = 20$ c) $y = -15$
b) $y = 10$ d) $y = -25$

---

213. The x-axis is also a horizontal line on the coordinate system.

(a) On the x-axis, for every value of "x", y = _____.

(b) The equation of the x-axis must be _____.

a) 0

b) $y = 0$

---

214. Describe the graph of $y = 11$. _____

The graph is a horizontal line which cuts the vertical axis at "11". The <u>ordinate</u> of any point on it is "11".

---

215. At what point does the graph of $y = -7$:

(a) Intercept the y-axis? (,)

(b) Intercept the x-axis? (,)

a) $(0, -7)$

b) There is no x-intercept. The line is horizontal and therefore does not cross the x-axis.

---

216. (a) $y = 0$ is the equation of what special line? _____

(b) The equation $y = -4$ means this: For any value of "x", "y" is _____.

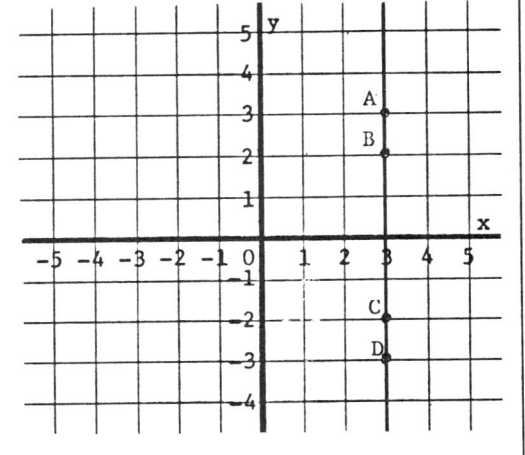

a) The x-axis (or horizontal axis).

b) -4

---

217. A vertical line is graphed at the right.

(a) Write the coordinates of the following points:

A(,)   C(,)

B(,)   D(,)

(b) On this graphed line, no matter what the value of "y" is, "x" is always _____.

456   Linear Graphs and Slope

218. At the right, we have graphed another vertical line.

(a) What are the coordinates of the following points?

A(,)   B(,)

(b) On this graphed line, the <u>abscissa</u> or (x-coordinate) is always ____ no matter what the value of "y" is.

(c) What are the coordinates of the following points on the y-axis?

C(,)   D(,)

(d) On the y-axis, no matter what the value of "y" is, "x" is always ____.

a) A(3,3)   C(3,-2)
   B(3,2)   D(3,-3)

b) 3

---

219. <u>For a horizontal line</u>: If "y" is 10 no matter what the value of "x" is, the equation of the line is: $y = 10$

<u>For a vertical line</u>: If "x" is 3 no matter what the value of "y" is, the equation of the line is: $x = $ ____

a) A(-4,5)   B(-4,-2)

b) -4

c) C(0,3)   D(0,-4)

d) 0

---

220. At the right, we have graphed and labeled two lines:

(a) On Line #1, "x" is 8 no matter what the value of "y" is. Therefore, the equation of line #1 is _____.

(b) On Line #2, "x" is -6 no matter what the value of "y" is. Therefore, the equation of line #2 is _____.

(c) On the y-axis, "x" is 0 no matter what the value of "y" is. Therefore, the equation of the y-axis is _____.

$x = 3$

---

221. On the graph at the right, we have drawn and labeled four lines.

Write the equation of each line:

#1 _____

#2 _____

#3 _____

#4 _____

a) $x = 8$

b) $x = -6$

c) $x = 0$

222. (a) The equation of the x-axis is _____.

(b) The equation of the y-axis is _____.

| #1 y = 4 | #3 x = -2 |
| #2 y = -3 | #4 x = 4 |

Answer to Frame 222:   a) y = 0   b) x = 0

SELF-TEST 13 (Frames 209-222)

Write the equation of each of the following lines:

1. The horizontal line which is 8 units below the x-axis. _____

2. The horizontal line which passes through (-5, 9). _____

3. The x-axis. _____

4. The vertical line which is 4 units to the right of the y-axis. _____

5. The vertical line which passes through (-15, -2). _____

6. The y-axis. _____

ANSWERS: 1. y = -8   2. y = 9   3. y = 0   4. x = 4   5. x = -15   6. x = 0

5-14   THE SLOPE OF HORIZONTAL AND VERTICAL LINES

In the last section, we discussed the equations of horizontal and vertical lines. In this section, we will discuss the slope of such lines.

223. Slope is a measure of the rise or fall of a line from left to right.

(a) Does a horizontal line rise or fall? _____

(b) What do you think the numerical value of the slope of a horizontal line is? _____

a) Neither.

b) 0 (meaning "no rise or fall")

458   Linear Graphs and Slope

224. Let's confirm the fact that the slope of any horizontal line is "0" by examining the line at the right.

    (a) From point A to point B, the horizontal change (Δx) is _____.

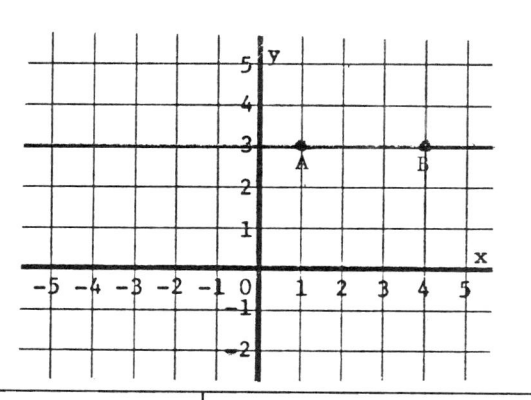

    (b) From point A to point B, the vertical change (Δy) is _____.

    (c) Therefore: $m = \dfrac{\Delta y}{\Delta x} =$ _____ = _____

---

225. Let's consider the slope of the vertical line at the right.

From point A to point B:

    $\Delta y = 2$ units
    $\Delta x = 0$ units (it does not change)

Therefore: $m = \dfrac{\Delta y}{\Delta x} = \dfrac{2}{0}$

But division by "0" is undefined since:

    $\dfrac{2}{0} = 2$(the reciprocal of 0), and "0" has no reciprocal.

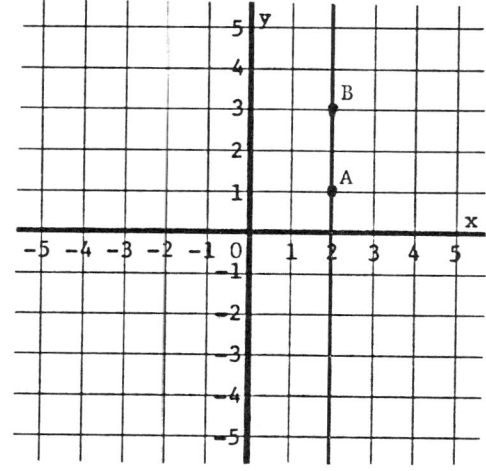

Since division by "0" is undefined and Δx is always "0" for a vertical line, the slope of a vertical line is undefined.

a) 3

b) 0

c) $\dfrac{0}{3} = 0$

---

226. (a) The slope of a vertical line is _____.

    (b) The slope of a horizontal line is _____.

Go to next frame.

---

227. (a) Since $y = 10$ is a horizontal line, its slope is _____.

    (b) Since $x = -5$ is a vertical line, its slope is _____.

a) undefined

b) 0

---

228. What are the slopes of the following lines?

    (a) $y = 7$ $m =$ _____   (c) $y = -4$ $m =$ _____

    (b) $x = -3$ $m =$ _____   (d) $x = 19$ $m =$ _____

a) 0

b) undefined

---

229. (a) The slope of the x-axis is _____.

    (b) The slope of the y-axis is _____.

a) 0   c) 0

b) undefined d) undefined

---

230. (a) The equation of the x-axis is: _____

    (b) The equation of the y-axis is: _____

a) 0

b) undefined

---

a) $y = 0$

b) $x = 0$

Linear Graphs and Slope   459

SELF-TEST 14 (Frames 223-230)

What is the numerical value of the slope of:

| 1. Any horizontal line? _____ | 2. Any vertical line? _____ |

What is the numerical value of the slope of each of the following lines?

| 3. $x = 0$ | 4. $y = -2$ | 5. $y = 10$ | 6. $x = 3$ |
| _____ | _____ | _____ | _____ |

7. Write the equation of the line for which $m = 0$ and $b = 7$. _____

ANSWERS:
1. 0
2. undefined
3. undefined
4. 0
5. 0
6. undefined
7. $y = 7$
From: $y = 0x + 7$

5-15 IDENTIFYING LINEAR EQUATIONS WITH VARIABLES OTHER THAN "x" AND "y"

The graphs of many equations with variables other than "x" and "y" are linear. Equations of this type can also be identified by the fact that each letter-term in the equation is linear. We will discuss the method of identifying these linear equations in this section.

231. Letter-terms containing letters other than "x" and "y" can also be identified as either linear or non-linear.

Each term below is a <u>linear</u> term. Notice that each term contains a <u>single non-squared</u> letter with a numerical coefficient.

$-5p$ $-6t$ $\frac{7}{3}m$ $-0.4a$

Each term below is a <u>non-linear</u> term. Notice that each contains either a <u>squared letter</u> or <u>two letters</u>:

$7b^2$ $-3cd$ $-\frac{5}{2}pq$ $1.6R^2$

Which of the following are linear terms? _____

(a) $10d$ (b) $-\frac{1}{2}t^2$ (c) $4.7h$ (d) $-\frac{6}{5}F$ (e) $5st$

232. Though not explicitly written, the coefficient of each term below is either "+1" or "-1":

t $-F$ gh r^2 $-I^2$

Which of the five terms are linear? _____

(a), (c), and (d)

Only "t" and "-F"

460   Linear Graphs and Slope

233. Any two-variable equation is <u>linear</u> if each letter-term in it is <u>linear</u>. However, if one or more letter-terms in it are <u>non-linear</u>, the equation is <u>non-linear</u>.

Which of the following equations are <u>linear</u>? _____

(a) $4s - 3t = 12$ (b) $7pq = 10$ (c) $m = 5R^2 + 1$ (d) $10c = 11d$

---

234. Which of the following equations are <u>non-linear</u>? _____

(a) $EI = 10$ (b) $F = 0.14s$ (c) $A = 3.14r^2$ (d) $s = \frac{5}{2}t$

Both (a) and (d).

---

235. (a) If an equation is <u>linear</u>, its graph is a _____.

(b) If an equation is <u>non-linear</u>, it graph is a _____.

Both (a) and (c).

---

236. Which of the following equations graph as straight lines? _____

(a) $5p - 6q = 60$ (b) $E + I = 30$ (c) $PV = 100$ (d) $t = 2d^2 - 9$

a) straight line

b) curve

---

237. Fractional terms like $\frac{5p}{3}$, $\frac{t}{10}$, $\frac{7cd}{5}$, and $\frac{m^2}{50}$ can be written as letter-terms with a fractional coefficient. That is:

$$\frac{5p}{3} = \frac{5}{3}p \qquad \frac{t}{10} = \frac{1}{10}t \qquad \frac{7cd}{5} = \frac{7}{5}cd \qquad \frac{m^2}{50} = \frac{1}{50}m^2$$

Therefore, terms of this type are linear if their numerator is linear and non-linear if their numerator is non-linear.

Which of the following terms are linear? _____

(a) $\frac{pq}{7}$ (b) $\frac{10F^2}{3}$ (c) $\frac{R}{6}$ (d) $\frac{7m}{4}$

Only (a) and (b).

---

238. Which of the following equations are linear? _____

(a) $\frac{3t}{2} + \frac{s}{5} = 1$ (b) $\frac{a}{3} = \frac{bc}{7}$ (c) $F = \frac{R^2}{3}$ (d) $d = \frac{h}{7}$

Only (c) and (d).

---

239. Each term below can be written as two simpler terms. That is:

$$\frac{7p + 15}{5} = \frac{7p}{5} + \frac{15}{5} = \frac{7}{5}p + 3$$

$$\frac{10F^2 - 14}{7} = \frac{10F^2}{7} - \frac{14}{7} = \frac{10}{7}F^2 - 2$$

Therefore: (a) $\frac{7p + 15}{5}$ is a _____ (linear/non-linear) term.

(b) $\frac{10F^2 - 14}{7}$ is a _____ (linear/non-linear) term.

Only (a) and (d).

---

a) linear

b) non-linear

240. Which of the following equations are linear? _____

   (a) $h = \dfrac{cd - 1}{7}$ (b) $t = \dfrac{5s + 7}{3}$ (c) $R = \dfrac{0.4F^2 + 0.8}{1.3}$

   Only (b).

241. If a fractional equation contains a letter in its denominator, we must clear the fraction before deciding whether or not the equation is linear.

   Clear the fraction in each of these and then determine whether the equation is linear or non-linear:

   (a) $\dfrac{E}{R} = 10$ (b) $V = \dfrac{100}{P}$ (c) $m = \dfrac{3t + 1}{m}$ (d) $10 = \dfrac{5Q + 7}{R}$

   _____ = _____   _____ = _____   _____ = _____   _____ = _____

   _____   _____   _____   _____

   a) Linear, since:
      $E = 10R$
   b) Non-linear, since:
      $VP = 100$
   c) Non-linear, since:
      $m^2 = 3t + 1$
   d) Linear, since:
      $10R = 5Q + 7$

242. If an equation contains an instance of the distributive principle, we must multiply by the distributive principle before deciding whether or not the equation is linear.

   (a) Since $p = 3(q + 1)$ is equivalent to $p = 3q + 3$, the equation is _____.

   (b) Since $t = s(s - 1)$ is equivalent to $t = s^2 - s$, the equation is _____.

243. Which of the following equations are linear? _____

   (a) $\dfrac{m^2}{V} = 5$ (b) $1.6a - 0.7b = 0$ (c) $H = \dfrac{5E - 7}{3}$ (d) $\dfrac{x^2}{9} - \dfrac{y^2}{16} = 1$

   a) linear
   b) non-linear

244. Which of the following equations are linear? _____

   (a) $p = \dfrac{q - 7}{p}$ (b) $R = R(F + 3)$ (c) $\dfrac{h + 5}{v} = 1$ (d) $7(t - 1) = s$

   Only (b) and (c).

   Only (c) and (d).

462   Linear Graphs and Slope

SELF-TEST 15 (Frames 231-244)

State whether each equation is "linear" or "non-linear":

1. $I = \dfrac{6}{R}$ _____
3. $C = 3.14d$ _____
5. $s = \dfrac{2(r+3)}{5}$ _____

2. $v = 32t$ _____
4. $F = w(w - 2)$ _____
6. $h = \dfrac{1-p}{p}$ _____

7. Which of the following equations graph as straight lines? _____

   (a) $\dfrac{r}{s} = 8$ (b) $W = \dfrac{2P}{W}$ (c) $\dfrac{1}{R} = G$ (d) $v = \dfrac{t}{4}$

ANSWERS: 1. Non-linear.   3. Linear.   5. Linear.   7. (a) and (d)
             2. Linear.   4. Non-linear.   6. Non-linear.

5-16   THE INTERCEPTS OF "NON-xy" LINEAR EQUATIONS

When an "xy" equation is graphed, "y" is <u>always</u> scaled on the <u>vertical</u> axis; "x" is <u>always</u> scaled on the <u>horizontal</u> axis. However, when equations containing two other variables are graphed, <u>either variable can be scaled on either axis</u>. The coordinates of the intercepts depend upon the choice of axes for the two variables. We will discuss the intercepts of "non-xy" linear equations in this section.

245. The variables in the equation below are not "x" and "y". Therefore, either "p" or "q" can be scaled on either axis.

     $\boxed{5p + 3q = 7}$

  (a) If you are told to plot "p" as "x", "p" is scaled on the
    _____ (horizontal/vertical) axis.

  (b) If you are told to plot "p" as "y", "p" is scaled on the
    _____ (horizontal/vertical) axis.

246. In $E + 10I = 20$, the variables are not "x" and "y".

  (a) If "E" is scaled on the vertical axis, it is plotted as the
    _____ (abscissa/ordinate).

  (b) If "E" is scaled on the horizontal axis, it is plotted as the
    _____ (abscissa/ordinate).

        a) horizontal
        b) vertical

247. For $d = 5t$, you are told to plot "t" as the abscissa. Therefore, which variable should be scaled:

  (a) On the horizontal axis? _____   (b) On the vertical axis? _____

        a) ordinate
        b) abscissa

        a) t   b) d

248. The point of this frame is to show that the coordinates of the intercepts of "non-xy" equations depend upon which variable is scaled on which axis.

We have graphed the following formula two different ways below. (Note: Only the first quadrant is used because only positive values of E and I are graphed.)

$$E + 10I = 20$$

In Figure 1, "E" is scaled on the horizontal axis.
In Figure 2, "E" is scaled on the vertical axis.

Figure 1

Figure 2

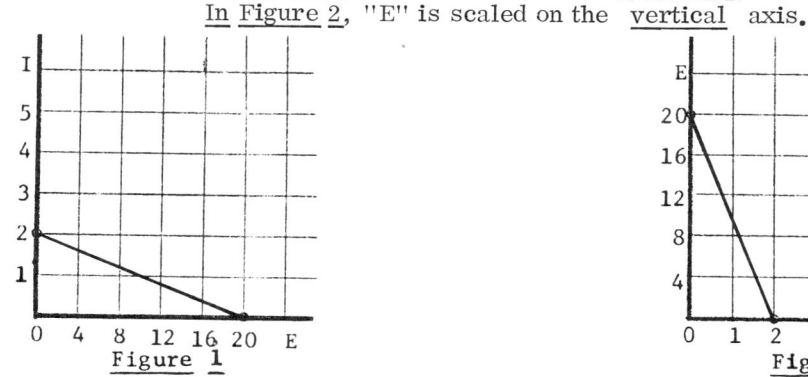

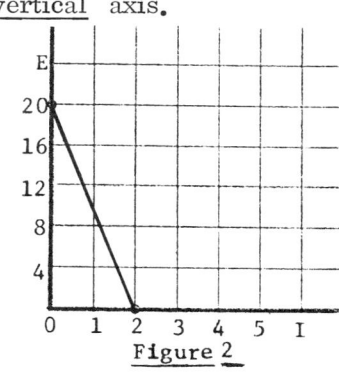

In Figure 1: (a) The coordinates of the horizontal (or E) intercept are (,   ).

(b) The coordinates of the vertical (or I) intercept are (,   ).

In Figure 2: (c) The coordinates of the horizontal (or I) intercept are (,   ).

(d) The coordinates of the vertical (or E) intercept are (,   ).

Answer to Frame 248:   a) (20,0)   b) (0,2)   c) (2,0)   d) (0,20)

249. Though the same numbers ("0", "2", "20") were involved in the coordinates of the intercepts in the last frame, whether they were abscissas or ordinates depended on which variable was scaled on which axis.

Here is another equation which does not contain "x" and "y": $5M + 2P = 20$

A sketch of the axes with "M" plotted as the abscissa (on the horizontal axis) is given on the right.

(a) On the horizontal axis, "P" is 0. Therefore, to find the coordinates of the horizontal (or M) intercept, we must plug in "0" for "P" in the equation and solve for the corresponding value of M. Do so.

The coordinates of the horizontal (or M) intercept are (,   ).

(b) On the vertical axis, "M" is 0. Therefore, to find the coordinates of the vertical (or P) intercept, we must plug in "0" for "M" and solve for the corresponding value of P. Do so:

The coordinates of the vertical (or P) intercept are (,   ).

---

a) (4,0)
b) (0,10)

464   Linear Graphs and Slope

250. Let's find the intercepts for $\boxed{4S + 3R = 24}$ when "R" is plotted as the abscissa. A sketch of the axes with "R" as the abscissa is given on the right.

Refer to the sketch for these:

(a) Which variable is "0" on the horizontal axis? _____

(b) The coordinates of the horizontal (or R) intercept are (,).

(c) Which variable is "0" on the vertical axis? _____

(d) The coordinates of the vertical (or S) intercept are (,).

---

251. Draw your own sketch of the axes in the following frames.

If "v" is plotted as the abscissa for $\boxed{7v + 2t = 28}$:

(a) The coordinates of the v-intercept are (,).

(b) The coordinates of the t-intercept are (,).

a) S
b) (8, 0)
c) R
d) (0, 6)

---

252. If "Q" is plotted as the abscissa for $\boxed{3W + 5Q = 45}$:

(a) The coordinates of the horizontal intercept are (,).

(b) The coordinates of the vertical intercept are (,).

a) (4, 0)
b) (0, 14)

---

253. If "S" is plotted as the abscissa for $\boxed{4S = 5R - 20}$:

(a) The coordinates of the S-intercept are (,).

(b) The coordinates of the R-intercept are (,).

a) (9, 0)
b) (0, 15)

---

254. If "P" is plotted as the abscissa for $\boxed{5M = 2P + 20}$:

(a) The coordinates of the horizontal intercept are (,).

(b) The coordinates of the vertical intercept are (,).

a) (−5, 0)
b) (0, 4)

---

255. If we know that the graph of an equation or formula is a straight line, we can graph it by plotting <u>two</u> points. Frequently, the two intercepts are used for these points. Let's graph the following linear formula by the intercept method.

$\boxed{E + 5I = 10}$

If E is plotted as the abscissa:

(a) The coordinates of the horizontal intercept are (,).

(b) The coordinates of the vertical intercept are (,).

(c) Plot these two intercepts on the graph at the right and draw the line through them.

a) (−10, 0)
b) (0, 4)

256. The graph of $\boxed{E + 5I = 10}$ is given on the right.

When graphing a line by the intercept method, we plot a third point as a check. From the formula: When E is 5, I is 1.

(a) Plot (5,1) on the graph. Does it lie on the graphed line? _____

(b) Is the line drawn correctly? _____

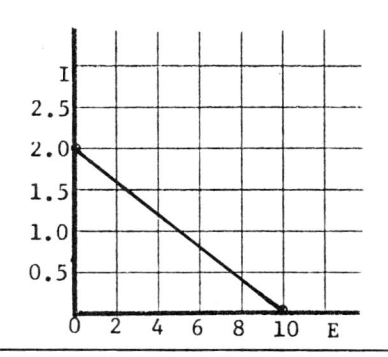

a) Yes.
b) Yes.

---

257. When graphing equations and formulas which contain variables other than "x" and "y", we frequently use only the first quadrant. Whether all four quadrants or only the first quadrant should be used depends on <u>whether negative values of the variables occur</u>.

If negative values <u>occur</u>, we use all four quadrants.

If negative values <u>do not occur</u>, we use only the first quadrant.

Therefore, we have to know something about the variables before deciding how many quadrants to use.

SELF-TEST 16 (Frames 245-257)

Given: $\boxed{2t + 5h = 10}$ If "t" is plotted as the <u>ordinate</u>, which variable is scaled:

1. On the horizontal axis? _____   2. On the vertical axis? _____

Given: $\boxed{P + 4W = 20}$ "P" is plotted as the abscissa.

3. Find the coordinates of the horizontal intercept. _____
4. Find the coordinates of the vertical intercept. _____
5. Graph the equation on the axes at the right.

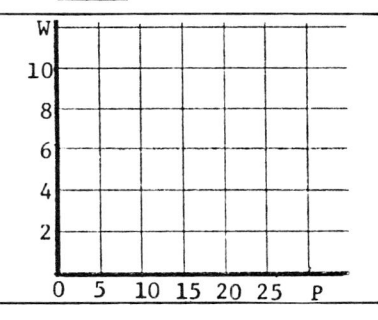

ANSWERS: 1. h   2. t   3. (20,0)   4. (0,5)   5.

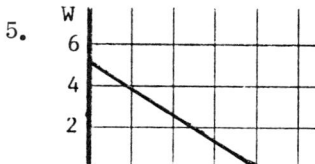

5-17 THE SLOPE OF "NON-xy" LINEAR EQUATIONS

Slope is a measure of the steepness of the rise or fall of a line. When the variables are "x" and "y", the slope of the line is defined by the following ratio:

$$m = \frac{\Delta y}{\Delta x}$$

In this section, we will examine the slope of lines when the variables are not "x" and "y". Defining slope with Δy and Δx does not make sense for other variables. However, since Δy is a "vertical change" and Δx is a "horizontal increase", we will show that we can use the following definition which is more general:

$$m = \frac{\text{Vertical Change}}{\text{Horizontal Increase}}$$

Furthermore, we will see that the slope of a "non-xy" line depends on the choice of axes for the variables.

---

258. We have graphed a line on the right. The two variables are "F" and "T", with "F" plotted as the abscissa. We want to find the slope of the line.

To find the slope, we have given the coordinates of two points, (-6, -4) and (2, 6). In this case:

(a) ΔF is the horizontal increase. $\Delta F =$ _____

(b) ΔT is the vertical change. $\Delta T =$ _____

(c) $m = \dfrac{\text{Vertical Change}}{\text{Horizontal Increase}} = \dfrac{\Delta T}{\Delta F} =$ _____

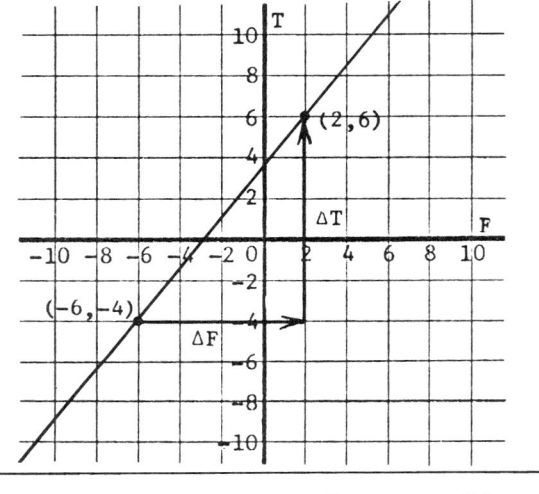

Answer to Frame 258: a) +8   b) +10   c) $\dfrac{5}{4}$ From: $\dfrac{+10}{+8}$

---

259. The two variables on the right are "K" and "S", with "S" plotted as the abscissa. The graphed line appears only in the first quadrant. The coordinates of the intercepts are (5, 0) and (0, 4), as shown.

In this case, the formula for the slope is:

$$m = \frac{\text{Vertical Change}}{\text{Horizontal Increase}} = \frac{\Delta K}{\Delta S}$$

(a) Since the line "falls" from left to right, we know that the slope must be _____ (positive/negative).

(b) $\Delta S =$ _____   (c) $\Delta K =$ _____   (d) $m =$ _____

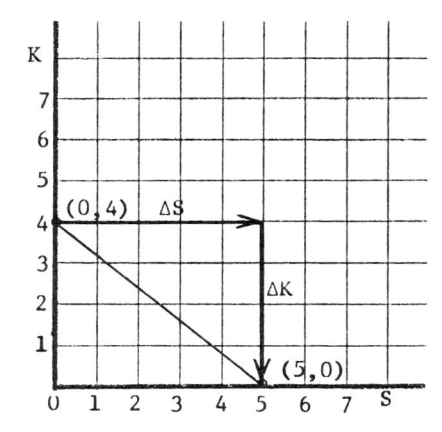

Answer to Frame 259:   a) negative   b) $\Delta S = +5$ c) $\Delta K = -4$ d) $m = \dfrac{-4}{+5} = -\dfrac{4}{5}$

260. In Figures 1 to 4 below, we have sketched the graphs of four lines:

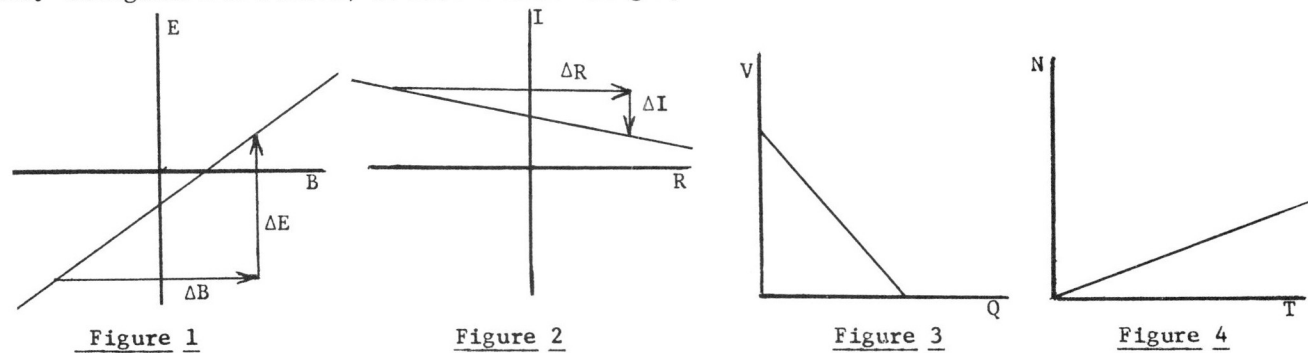

Figure 1   Figure 2   Figure 3   Figure 4

In Figure 1, the formula for the slope of the line is $m = \dfrac{\Delta E}{\Delta B}$, since ΔE is the vertical change and ΔB is the horizontal increase.

Write the formula for the slope of each of the other three lines:

(a) In Figure 2, m = _____   (b) In Figure 3, m = _____   (c) In Figure 4, m = _____

Answer to Frame 260:   a) $m = \dfrac{\Delta I}{\Delta R}$ b) $m = \dfrac{\Delta V}{\Delta Q}$ c) $m = \dfrac{\Delta N}{\Delta T}$

---

261. When an equation does not contain "x" and "y" as its variables, the slope depends on which variable is scaled on which axis. Below we have graphed the following equation two different ways by the intercept method.

$$5t - 3d = 15$$

In Figure 1, "t" is plotted on the horizontal axis, and "d" is plotted on the vertical axis.
In Figure 2, "d" is plotted on the horizontal axis, and "t" is plotted on the vertical axis.
You can see that the slope of the line in Figure 1 is steeper.

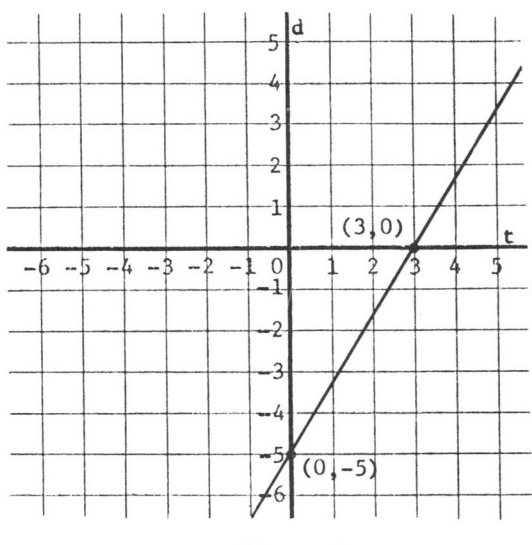

Figure 1 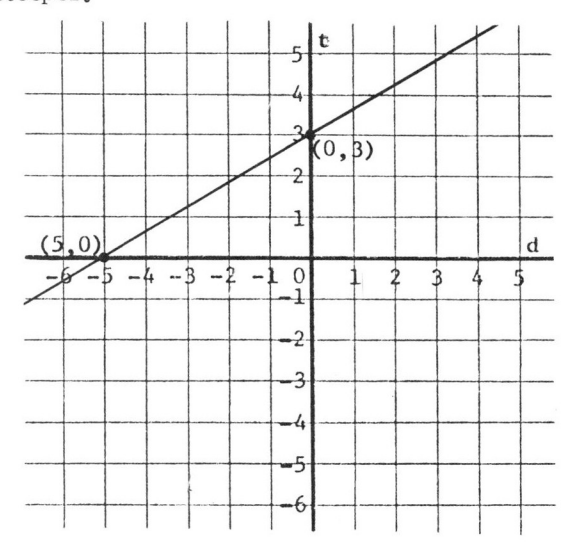   Figure 2

Continued on following page.

468 Linear Graphs and Slope

261. Continued.

In Figure 1:

The formula for the slope is: $m = \frac{\Delta d}{\Delta t}$.

In Figure 2:

The formula for the slope is: $m = \frac{\Delta t}{\Delta d}$.

Using the coordinates of the intercepts in each case:

(a) The slope in Figure 1 is: m = _____

(b) The slope in Figure 2 is: m = _____

Answer to Frame 261:   a) $m = \frac{+5}{+3} = \frac{5}{3}$ b) $m = \frac{+3}{+5} = \frac{3}{5}$

---

262. On the graphs below, we have graphed the following equation two different ways by the intercept method. (Note: Because only positive values of "K" and "R" are needed, the graphs appear only in the first quadrant.)

$$2R + 3K = 12$$

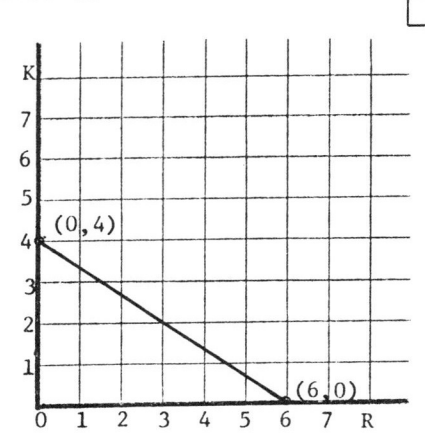

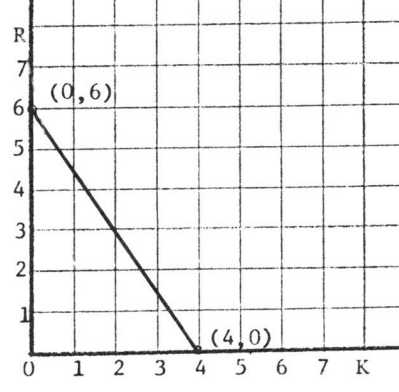

Figure 1

Figure 2

(a) By examining the graphs, you can see that the line in _____ (Figure 1/Figure 2) has the steeper slope.

In Figure 1: (b) The slope formula is: m = _____

(c) The slope is: m = _____

In Figure 2: (d) The slope formula is: m = _____

(e) The slope is: m = _____

Answer to Frame 262:   a) Figure 2   b) $\frac{\Delta K}{\Delta R}$ c) $\frac{-4}{+6} = -\frac{2}{3}$ d) $\frac{\Delta R}{\Delta K}$ e) $\frac{-6}{+4} = -\frac{3}{2}$

263. Earlier we showed that we can find the slope of a straight line if we know the coordinates of any two points on it. Here are two pairs of values of the variables I and P.

$$\text{If } I = 5, \ P = 10. \qquad \text{If } I = 20, \ P = 15.$$

In Figure 1 and 2 below, plot the points representing these values and draw the straight lines through them. (Be sure to examine which variable goes on which axis.)

Figure 1

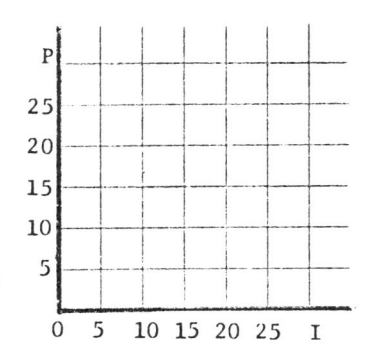

Figure 2

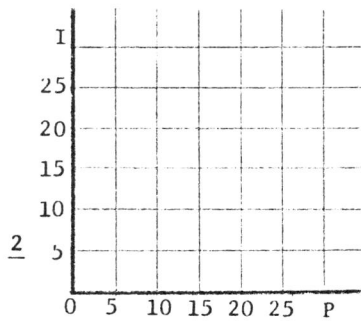

In Figure 1: (a) The slope formula is: m = _____

(b) The slope is: m = _____

In Figure 2: (c) The slope formula is: m = _____

(d) The slope is: m = _____

Answer to Frame 263: a) $\dfrac{\Delta P}{\Delta I}$ b) $\dfrac{1}{3}$ From: $\dfrac{+5}{+15}$ c) $\dfrac{\Delta I}{\Delta P}$ d) 3 From: $\dfrac{+15}{+5}$

264. Here are two pairs of relationships between two variables: If $N = 10$, $P = 40$.
If $N = 40$, $P = 0$.

Plot the two points on each graph at the right, and draw the two lines.

Find the numerical value of the slope of the line in:

(a) Figure 1: m = _____

(b) Figure 2: m = _____

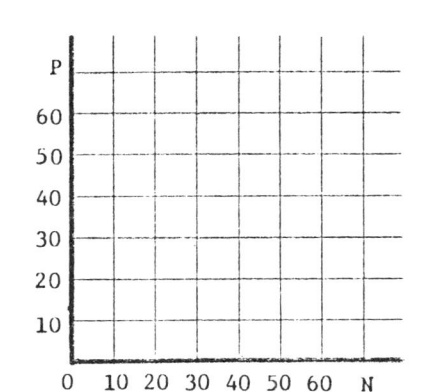

Figure 1

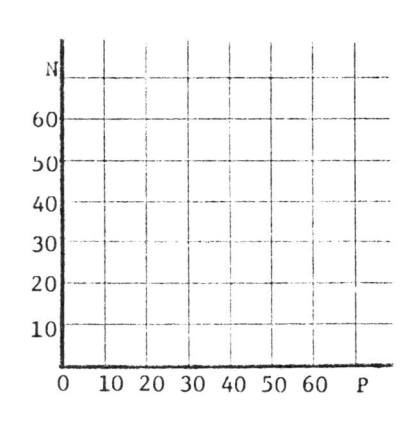

Figure 2

Answer to Frame 264: a) $m = -\dfrac{4}{3}$ From: $m = \dfrac{\Delta P}{\Delta N} = \dfrac{-40}{+30} = -\dfrac{4}{3}$ b) $m = -\dfrac{3}{4}$ From: $m = \dfrac{\Delta N}{\Delta P} = \dfrac{-30}{+40} = -\dfrac{3}{4}$

SELF-TEST 17 (Frames 258-264)

1. Using "Δs" and "Δt", write the formula for finding the slope of a line on the graph at the right. m = _____

2. Find the slope of Line #1. m = _____

3. Find the slope of Line #2. m = _____

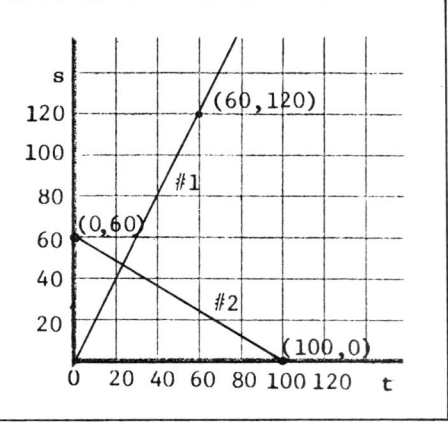

ANSWERS: 1. $m = \dfrac{\Delta s}{\Delta t}$ 2. $m = 2$ 3. $m = -\dfrac{3}{5}$

5-18 THE SLOPE-INTERCEPT FORM OF "NON-xy" EQUATIONS

Any two-variable linear equation can be written in slope-intercept form.

If the two variables are "x" and "y", the slope-intercept form is:

$$y = mx + b$$

If the two variables are different than "x" and "y", the more general slope-intercept form is:

$$\bigcirc = m\,\square + b$$

Note: Any two variables can be plugged into the circle and square.

In this section, we will discuss the slope-intercept form of "non-xy" equations.

265. The general slope-intercept form of "non-xy" linear equations is:

$$\bigcirc = m\,\square + b$$

Here are some examples of "non-xy" equations in this form:

(#1) $F = \dfrac{9}{5}C + 32$ (#2) $p = 3q + 5$ (#3) $t = \dfrac{5}{3}d + \left(-\dfrac{4}{3}\right)$

(a) In #1, "m" is _____.

(b) In #2, "b" is _____.

(c) In #3, the two variables are _____ and _____.

a) $\dfrac{9}{5}$

b) 5

c) t and d

266. Here is an equation whose graph is a straight line: $\boxed{2k + 3t = 6}$

Though not in slope-intercept form, it can be rearranged to slope-intercept form <u>by solving for either of the variables</u> as follows:

<u>Solving for k, we get</u>:

$2k + 3t = 6$

$2k + 3t + (-3t) = 6 + (-3t)$

$2k = (-3t) + 6$

$\boxed{\frac{1}{2}}(2k) = \frac{1}{2}[(-3t) + 6]$
↓
$(1)(k) = \frac{1}{2}(-3t) + \frac{1}{2}(6)$

$k = -\frac{3}{2}t + 3$

<u>Solving for t, we get</u>:

$2k + 3t = 6$

$(-2k) + 2k + 3t = 6 + (-2k)$

$3t = (-2k) + 6$

$\boxed{\frac{1}{3}}(3t) = \frac{1}{3}[(-2k) + 6]$
↓
$(1)(t) = \frac{1}{3}(-2k) + \frac{1}{3}(6)$

$t = -\frac{2}{3}k + 2$

(a) In the slope-intercept form in which "k" is "solved for", m = _____.

(b) In the slope-intercept form in which "t" is "solved for", b = _____.

<u>Answer to Frame 266</u>: a) $m = -\frac{3}{2}$ b) $b = 2$

267. The graph of the following equation is a straight line: $\boxed{5m - 3q = 7}$

To put the equation in slope-intercept form, we must solve for either "m" or "q".

(a) Solve for "m": $5m - 3q = 7$ (b) Solve for "q": $5m - 3q = 7$

m = _____   q = _____

<u>Answer to Frame 267</u>: a) $m = \frac{3}{5}q + \frac{7}{5}$ b) $q = \frac{5}{3}m + \left(-\frac{7}{3}\right)$

472 Linear Graphs and Slope

268. Here is a linear equation in slope-intercept form: $\boxed{F = 2D + 4}$

We have graphed the equation below in two different ways:

In Figure 1, "F" is scaled on the horizontal axis.
In Figure 2, "F" is scaled on the vertical axis.

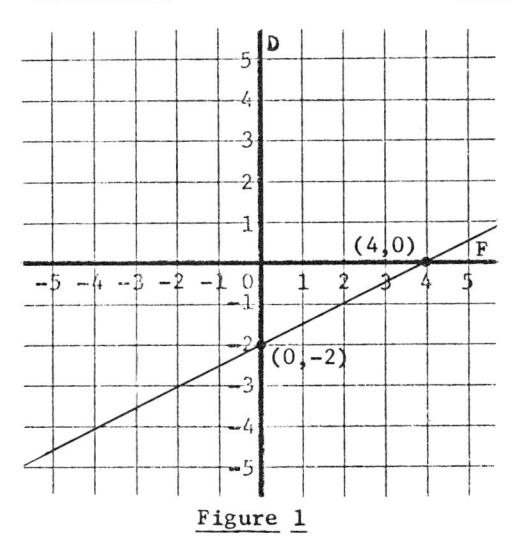

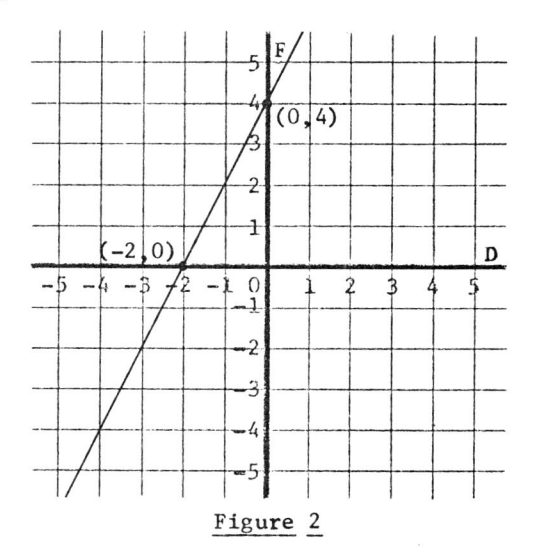

Figure 1

Figure 2

In the equation $\boxed{F = 2D + 4}$, m = 2 and b = 4.

"m = 2" says that the slope of the line is 2.
"b = 4" says that the coordinates of the vertical intercept are (0,4).

Let's determine the slope and vertical intercept of each line graphed above.

In Figure 1:

(a) The slope is ____.

(b) The vertical intercept is (,).

In Figure 2:

(c) The slope is ____.

(d) The vertical intercept is (,).

(e) "m" and "b" in the equation tell us the slope and vertical intercept only if ____ (F or D) is scaled on the vertical axis.

---

269. When an "xy" equation is written in slope-intercept form:

(1) "y" is always scaled on the vertical axis.

(2) "m" and "b" always tell us the slope and vertical intercept of the line.

When a "non-xy" equation is written in slope-intercept form:

(1) Either variable can be scaled on the vertical axis.

(2) "m" and "b" tell us the slope and vertical intercept only if the "solved for" variable is scaled on the vertical axis.

The following equation is in slope-intercept form: $\boxed{F = \frac{9}{5}C + 32}$

The slope will be $\frac{9}{5}$ and the vertical intercept will be (0,32) only if ____
(F or C) is scaled on the vertical axis.

a) $\frac{1}{2}$ c) 2

b) (0,-2)   d) (0,4)

e) F

270.	The constants "m" and "b" in the slope-intercept form of a "non-xy" equation tell us the slope and vertical intercept only if the _____ variable is scaled on the vertical axis.	F, the "solved for" variable.

271.	When $\boxed{F = \frac{9}{5}C + 32}$ is graphed, either "F" or "C" can be scaled on the vertical axis.
	
	If F is scaled on the vertical axis,
	    the slope is $\frac{9}{5}$ and the vertical intercept is $(0, 32)$.
	
	If C is scaled on the vertical axis,
	    do the constants in the equation above give you the slope and vertical intercept of the line? _____
		"solved for"

272.	(a) If $\boxed{V = 17D + 9}$, are $(0, 9)$ the coordinates of the vertical intercept if "V" is scaled on the vertical axis? _____ (b) If $\boxed{P = 9Q + (-11)}$, is "9" the slope if "Q" is scaled on the vertical axis?	No.
273.	(a) If $\boxed{H = 3G + (-7)}$, can you immediately tell from this formula the slope if "H" is scaled on the vertical axis? _____ (b) If $\boxed{M = 5P + 9}$, can you immediately tell from this formula the coordinates of the vertical intercept if "P" is scaled on the vertical axis? _____	a) Yes. b) No. "9" is the slope only if "P" is scaled on the vertical axis.

274.	If the "solved for" variable is scaled on the vertical axis, the vertical intercept is given by "b" in the equation. However, the <u>horizontal</u> intercept is not given by a constant in the equation. To find the horizontal intercept, we must plug in "0" for the "solved for" variable.
	
	Find the coordinates of the horizontal intercept for each equation below if the "solved for" variable is plotted on the vertical axis.
	
	    (a) $V = \frac{3}{5}t + 9$ (b) $P = \frac{2}{7}Q + (-6)$
	
	            (,) (,)
		a) Yes. It is "3".
		
		b) No.

		a) $(-15, 0)$ b) $(21, 0)$

474   Linear Graphs and Slope

275. The equation at the right is linear: $\boxed{5p - 6q = 60}$

To determine its slope and vertical intercept if "p" is scaled on the vertical axis, proceed as follows:

(a) Solve for "p" and write the solution in slope-intercept form.

p = _____

(b) Its slope is ____.   (c) Its vertical intercept is (,).

---

276. The equation at the right is linear: $\boxed{7W + 4R = 28}$

a) $p = \dfrac{6}{5}q + 12$

To determine its slope and vertical intercept if "R" is scaled on the vertical axis, proceed as follows:

b) $m = \dfrac{6}{5}$

(a) Solve for "R" and write the solution in slope-intercept form.

c) $(0, 12)$

R = _____

(b) Its slope is ____.   (c) Its vertical intercept is (,).

---

277. If E is scaled on the vertical axis when graphing: $\boxed{E = -5I + 30}$

a) $R = -\dfrac{7}{4}W + 7$

The slope is -5.
The vertical intercept is (0, 30).

b) $m = -\dfrac{7}{4}$

To find its slope and vertical intercept when "I" is scaled on the vertical axis:

c) $(0, 7)$

(a) Solve for "I" and write the solution in slope-intercept form.

I = _____

(b) Its slope is ____.   (c) Its vertical intercept is (,).

---

278. Of course, an equation can be written in slope-intercept form <u>only if it is linear</u>. Which of the equations below <u>cannot</u> be written in slope-intercept form because they are non-linear? _____

a) $I = -\dfrac{1}{5}E + 6$

(a) IR = 10   (b) $t = 2d^2 - 9$ (c) R + 3S - 10 = 0   (d) $3h^2 = 6p + 9$

b) $m = -\dfrac{1}{5}$

c) $(0, 6)$

(a), (b), and (d)

SELF-TEST 18 (Frames 265-278)

Given: $\boxed{s = 2t + 8}$ If "s" is scaled on the vertical axis, find the following:

1. Slope: m = _____
2. Vertical intercept: (,)
3. Horizontal intercept: (,)

Given: $\boxed{E + 8I = 12}$ If "I" is scaled on the vertical axis, find the following:

4. Slope-intercept form of the equation: _____

5. Slope: m = _____
6. Vertical intercept: (,)
7. Horizontal intercept: (,)

Given: $\boxed{2P - 3F = 12}$ If "F" is scaled on the vertical axis, find the following:

8. Slope-intercept form of the equation: _____

9. Slope: m = _____
10. Vertical intercept: (,)
11. Horizontal intercept: (,)

ANSWERS:
1. m = 2
2. (0, 8)
3. (−4, 0)
4. $I = -\frac{1}{8}E + \frac{3}{2}$
5. $m = -\frac{1}{8}$
6. $(0, \frac{3}{2})$
7. (12, 0)
8. $F = \frac{2}{3}P + (-4)$
9. $m = \frac{2}{3}$
10. (0, −4)
11. (6, 0)

5-19 "NON-xy" EQUATIONS OF LINES THROUGH THE ORIGIN

When a line passes through the origin, its vertical intercept is (0, 0). Therefore, "b" in the slope-intercept form of its equation is "0".

If the variables are "x" and "y", the slope-intercept form reduces to:

$$\boxed{y = mx}$$

If the variables are not "x" and "y", the more general slope-intercept form is:

$$\boxed{\bigcirc = m \square}$$

Note: Any two variables can be plugged into the circle and square.

In this section, we will discuss the slope-intercept form of "non-xy" equations whose graphed lines pass through the origin.

---

279. The general slope-intercept form of equations whose lines pass through the origin is:

$$\boxed{\bigcirc = m \square}$$

Here are some examples: d = 50t   E = 10R   C = 3.14d

Both the vertical and horizontal intercept of each equation is (,).

(0, 0)

476 Linear Graphs and Slope

280. The graph of $\boxed{d = 50t}$ is a straight line through the origin. If "d" is plotted as the vertical coordinate:

   The slope of the line is 50.
   The vertical intercept is the origin or (0,0).

   To find the slope when "t" is plotted as the vertical coordinate, we must solve for "t". We get: $\boxed{t = \frac{1}{50}d}$. Therefore, with "t" plotted as the vertical coordinate:

   (a) The slope is _____.   (b) The vertical intercept is still (,).

---

281. Here are the two forms of the equation in the last frame:

   $\boxed{d = 50t}$ $\boxed{t = \frac{1}{50}d}$

   Both lines pass through the origin, but <u>their slopes are a pair of reciprocals</u>.

   The graph of $E = 10R$ is also a line through the origin:

   If E is plotted as "y", the slope is 10.

   If R is plotted as "y", the slope is the "reciprocal of 10" or _____.

a) $\frac{1}{50}$ b) (0,0)

---

282. The graph of the relationship between two variables "p" and "q" is a straight line through the origin.

   (a) If "p" is plotted as "y", the slope is 5. Write the equation of the line in slope-intercept form: _____

   (b) If "q" is plotted as "y", the equation in slope-intercept form must be _____.

$\frac{1}{10}$

---

283. Here is the formula relating the length of the circumference (C) of a circle with the length of its diameter:

   $\boxed{C = \pi d \text{ or } C = 3.14d}$ (since $\pi = 3.14$)

   Its graph is linear. If "C" is scaled on the vertical axis:

   (a) The slope is _____.

   (b) The coordinates of both the vertical and horizontal intercepts are (,).

a) $p = 5q$

b) $q = \frac{1}{5}p$

---

284. If the temperature of a 50 inch aluminum rod is increased from 70° Fahrenheit, its increase in length is given by this formula:

   $\boxed{L = 0.00065t}$

   Where "t" is the <u>increase</u> in temperature.
   "L" is the <u>increase</u> in inches.

   (a) If this formula were graphed, would the graphed relationship be a straight line? _____

   (b) The slope of the line is 0.00065 if _____ ("L" or "t") is scaled on the vertical axis.

a) 3.14

b) (0,0), since the line passes through the origin.

Linear Graphs and Slope   477

285. The graph of the relationship between two variables "E" and "I" is a straight line through the origin.

When E is 15, I = 20.

We want to find the slope and equation of the line if "I" is scaled as "y". To do so, we use a sketch.

(a) The slope of the line is _____.

(b) The equation of the line is _____.

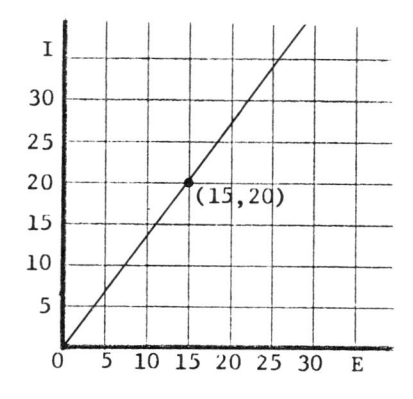

a) Yes, it would be a straight line <u>passing through the origin</u>.

b) "L"

---

286. The graph of the relationship between "R" and "S" is a straight line through the origin.

When R = 30, S = 25.

If "R" is plotted as the ordinate, the equation of the line must be _____. (Note: Make your own sketch.)

a) $\frac{4}{3}$ (from $\frac{20}{15}$)

b) $I = \frac{4}{3} E$

---

287. Here is the formula relating distance "d" traveled at a constant velocity "v" in a given amount of time "t":

$$d = vt$$

There are <u>three variables</u> in this formula. <u>Only two variables</u> can be graphed on the coordinate system since there are only two axes. Therefore, we cannot graph all three variables at one time. However, we can graph the relationship between two of the three variables <u>by replacing the third variable with some fixed value (or constant)</u>.

For example, we can graph the relationship between "d" and "t" if we replace "v" by the constant 25.

Then "d = vt" changes to "d = 25t".

(a) What would the graph of "d = 25t" look like?

(b) For d = 25t, the slope is _____.

$R = \frac{6}{5} S$

---

a) A straight line through the origin. (First quadrant only – negative values are ordinarily not used.)

b) 25

478 Linear Graphs and Slope

288. When substituting a constant for "v" in $\boxed{d = vt}$ so that we can graph the relationship between "d" and "t", we can substitute any constant for "v". For example, we could graph:

$$d = 25t \qquad d = 50t \qquad d = 75t \qquad d = 100t$$

Therefore, we can graph a whole family of relationships. In each case, however:

(a) The graphed line begins at the _____.

(b) The specific constant substituted for "v" is the _____ of the line.

---

289. Beginning with the same formula $\boxed{d = vt}$, we can graph the relationship between "d" and "v" if we replace "t" by a constant or various constants. If we replace "t" by 5, we get:

$$d = v(5) \quad \text{or} \quad d = 5v$$

(a) If we graphed $d = 5v$, would we get a straight line through the origin? _____

(b) The slope of the line will be "5" if we plot _____ ("d" or "v") on the vertical axis.

(c) What would be the slope of the line if we plot "v" on the vertical axis? _____

a) origin
b) slope

---

290. Based on the relationship $\boxed{d = vt}$:

(a) If we graph $d = 35t$, we should recognize that the slope "35" is a specific value of what variable? _____

(b) If we graph $d = 10v$, we should recognize that the slope "10" is a specific value of what variable? _____

a) Yes.
b) "d"
c) $\frac{1}{5}$ or 0.2

---

291. To graph the relationship between "v" and "t" in $\boxed{d = vt}$, we must replace "d" by some specific constant. If we replace "d" by 120, we get:

$$120 = vt \quad \text{or} \quad vt = 120$$

Is the graph of $vt = 120$ a straight line? _____

a) v   b) t

---

292. The following formula states a relationship between the voltage (E), the current (I), and the resistance (R) in an electrical circuit:

$$\boxed{E = IR}$$

(a) To graph the relationship between "E" and "R", we must replace "I" with some constant. Then, this fixed value of "I" becomes the _____ of the line.

(b) When we graph a relationship between "E" and "I" (such as $E = 8I$), we should immediately recognize that the slope "8" is a fixed value of which variable (E, I, or R)? _____

No. Its graph is a curve, since $vt = 120$ is not a linear equation.

---

a) slope
b) R

293. When substituting a constant in order to graph the relationship between two of the variables in $\boxed{E = IR}$, the graph is a <u>curve</u>:

   (a) If we substitute a constant for _____,

   (b) And then graph the relationship between _____ and _____.

Answer to Frame 293:   a) E   b) I and R

SELF-TEST 19 (Frames 279-293)

1. The graphs of which of the following equations pass through (0,0)? _____

   (a) $h = 0.15d$ (b) $t - 2p = 3$ (c) $P - 2S = 0$ (d) $y = x$ (e) $3v = 1 - d$

The graph of an equation is a straight line through the origin and the point (8,4). If the variables are "r" and "s", and if "r" is scaled on the vertical axis, find the following:

2. The slope of the line. m = _____

3. The equation of the line. _____

Refer to the graph at the right. Write the equation of:

4. Line #1: _____

5. Line #2: _____

6. Line #3: _____

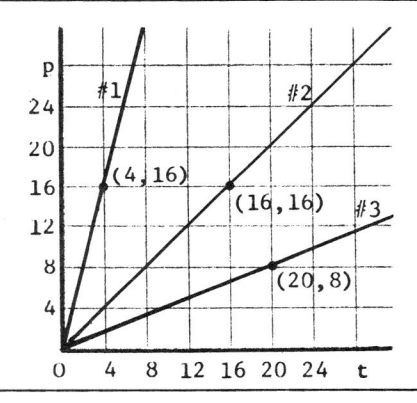

In the formula $\boxed{P = \dfrac{F}{A}}$, "A" is replaced by the constant "30".

7. Write the resulting two-variable equation. _____

8. If "F" is scaled on the vertical axis, find the slope of the graph. m = _____

9. In the formula $\boxed{v = rt}$, if "v" is replaced by a constant, is the resulting equation linear or non-linear? _____

ANSWERS: 1. (a), (c), (d)   2. $m = \dfrac{1}{2}$ 4. $p = 4t$ 7. $P = \dfrac{F}{30}$ 9. Non-linear.

3. $r = \dfrac{1}{2}s$ 5. $p = t$ 8. $m = 30$

6. $p = \dfrac{2}{5}t$

480 Linear Graphs and Slope

5-20 THE SLOPE AS A RATIO OF VARIOUS PAIRS OF CHANGES IN "NON-xy" VARIABLES

In an earlier section, we saw that the slope of an "xy" equation is a ratio which represents various pairs of changes in "x" and "y". In this section, we will show that the slope of a "non-xy" equation represents various pairs of changes in the "non-xy" variables.

294. When a linear equation containing variables other then "x" and "y" is graphed, the slope is a ratio of the vertical change to the horizontal increase between any two points on the line. That is:

$$\text{Slope} = \frac{\text{Vertical Change}}{\text{Horizontal Increase}}$$

If the equation is written in slope-intercept form and the "solved for" variable is scaled on the vertical axis, the slope is a ratio of the change in the "solved for" variable to any increase in the other variable. For example:

For $p = 3q + 7$, the slope $= \frac{\Delta p}{\Delta q}$.

(a) For $t = 5m + 1$, the slope = _____.

(b) For $d = \frac{7}{3}h$, the slope = _____.

295. The following formula shows the relationship between distance traveled "d" and time "t" at a constant velocity of 50 miles per hour:

$$d = 50t$$

If "d" is plotted as the ordinate, the slope of the line is "50". Remember that this slope represents the ratio of the change in "d" (Δd) to the change in "t" (Δt) between any two points on the graphed line. That is, all of the following ratios of Δd to Δt would lead to a slope of "50" when reduced to lowest terms.

$$\frac{\Delta d}{\Delta t} = \frac{50}{1}, \frac{100}{2}, \frac{200}{4}, \frac{500}{10}, \text{ etc.}$$

Therefore in the equation above, a slope of "50" really means all of the following:

(a) For any 1-unit increase in time, the distance traveled increases _____ units.

(b) For any 2-unit increase in time, the distance traveled increases _____ units.

(c) For any 10-unit increase in time, the distance traveled increases _____ units.

a) $\frac{\Delta t}{\Delta m}$

b) $\frac{\Delta d}{\Delta h}$

a) 50
b) 100
c) 500

296. If $d = 50t$, we can find the increase in "d" corresponding to any specific increase in "t" by plugging the specific Δt into the following proportion:

$$\frac{50}{1} = \frac{\Delta d}{\Delta t} \quad \text{or simply} \quad 50 = \frac{\Delta d}{\Delta t}$$

That is: For any 3-unit increase in "t", "d" increases 150 units, since:

$$50 = \frac{\Delta d}{3} \quad \text{and} \quad \Delta d = 3(50) \text{ or } 150$$

Using the same procedure, complete each of the following:

(a) For any 5-unit increase in "t", "d" increases _____ units.

(b) For any 20-unit increase in "t", "d" increases _____ units.

---

297. The following formula shows the relationship between the circumference (C) and diameter (d) of a circle:

$$\boxed{C = \pi d} \quad \text{or} \quad \boxed{C = 3.14d}$$

If C is scaled on the vertical axis, the slope of the line is 3.14. This slope stands for any of the following ratios of ΔC to Δd:

$$\frac{\Delta C}{\Delta d} = \frac{3.14}{1}, \frac{6.28}{2}, \frac{9.42}{3}, \frac{31.4}{10}, \text{etc.}$$

Therefore in the equation above, a slope of 3.14 really means all of the following:

(a) For any 1-unit increase in diameter, the circumference increases _____ units.

(b) For any 3-unit increase in diameter, the circumference increases _____ units.

(c) For any 10-unit increase in diameter, the circumference increases _____ units.

Answers:
a) 250 units
b) 1,000 units

---

298. If $C = 3.14d$, we can find the increase in "C" corresponding to any specific increase in "d" by plugging the specific Δd into the following proportion:

$$\frac{3.14}{1} = \frac{\Delta C}{\Delta d} \quad \text{or simply} \quad 3.14 = \frac{\Delta C}{\Delta d}$$

That is: For any 4-unit increase in "d", "C" increases 12.56 units, since:

$$3.14 = \frac{\Delta C}{4} \quad \text{and} \quad \Delta C = 4(3.14) \text{ or } 12.56$$

Using the same procedure, complete each of the following:

(a) For any 6-unit increase in "d", "C" increases _____ units.

(b) For any 100-unit increase in "d", "C" increases _____ units.

Answers:
a) 3.14
b) 9.42
c) 31.4

---

a) 18.84
b) 314

482 Linear Graphs and Slope

299. Here is the formula showing the relationship between degrees-Fahrenheit (F) and degrees-Centigrade (C):

$$\boxed{F = \frac{9}{5}C + 32}$$

If F is scaled on the vertical axis, the slope of the line is $\frac{9}{5}$. This slope stands for any of the following ratios of ΔF to ΔC:

$$\frac{\Delta F}{\Delta C} = \frac{9}{5}, \frac{18}{10}, \frac{27}{15}, \frac{45}{25}, \frac{90}{50}, \text{ etc.}$$

Therefore in the equation above, a slope of $\frac{9}{5}$ really means all of the following:

(a) For any 5-degree increase on the Centigrade scale, we get a _____-degree increase on the Fahrenheit scale.

(b) For any 10-degree increase on the Centigrade scale, we get a _____-degree increase on the Fahrenheit scale.

(c) For any 50-degree increase on the Centigrade scale, we get a _____-degree increase on the Fahrenheit scale.

---

300. If $F = \frac{9}{5}C + 32$, we can find the increase in "F" corresponding to any specific increase in "C" by plugging the specific ΔC into the following proportion:

$$\frac{9}{5} = \frac{\Delta F}{\Delta C}$$

That is: For any 20-degree increase in "C", "F" increases 36 degrees, since:

$$\frac{9}{5} = \frac{\Delta F}{20} \quad \text{and} \quad \Delta F = 20\left(\frac{9}{5}\right) \text{ or } 4(9) \text{ or } 36$$

Using the same procedure, complete each of the following:

(a) For any 30-degree increase in "C", "F" increases _____ degrees.

(b) For any 40-degree increase in "C", "F" increases _____ degrees.

a) 9-degree
b) 18-degree
c) 90-degree

---

301. In $F = \frac{9}{5}C + 32$, $\frac{9}{5}$ can be written as the ratio $\frac{\frac{9}{5}}{1}$.

Therefore, for any 1-degree increase in C, there is a _____-degree increase in F.

a) 54 degrees
b) 72 degrees

---

302. If "p" is plotted as "y" when graphing $\boxed{p = 3q + 1}$, what change occurs in "p":

(a) For any 1-unit increase in "q"? _____

(b) For any 10-unit increase in "q"? _____

(c) For any 5.5-unit increase in "q"? _____

$\frac{9}{5}$ or 1.8-degree

303. If "t" is scaled on the vertical axis when graphing $\boxed{t = 1.8v}$, what happens to "t":

    (a) For any 1-unit increase in "v"? _____

    (b) For any 5-unit increase in "v"? _____

    (c) For any 1.5-unit increase in "v"? _____

a) A 3-unit increase.
b) A 30-unit increase.
c) A 16.5-unit increase.

---

304. If "R" is plotted as "y" when graphing $\boxed{R = -\frac{3}{2}F}$, what happens to "R":

    (a) For any 1-unit increase in "F"? _____

    (b) For any 2-unit increase in "F"? _____

    (c) For any 10-unit increase in "F"? _____

a) A 1.8-unit increase.
b) A 9-unit increase.
c) A 2.7-unit increase.

Answer to Frame 304:   a) A $\frac{3}{2}$ or 1.5-unit decrease.   b) A 3-unit decrease.   c) A 15-unit decrease.

SELF-TEST 20 (Frames 294-304)

In graphing $\boxed{v = 32t}$, "v" is plotted on the vertical axis.

1. Using Δt and Δv, write the formula for the slope of the line. m = _____

2. The numerical value of the slope is _____.

3. If $\Delta t = 2$ units, $\Delta v =$ _____ units.

4. For any 5-unit increase in "t", "v" increases _____ units.

In graphing $\boxed{W = -\frac{3}{2}H + 4}$, "W" is plotted as the ordinate.

5. Using ΔH and ΔW, write the formula for the slope of the line. m = _____

6. The numerical value of the slope is _____.

7. If $\Delta H = 8$ units, $\Delta W =$ _____ units.

8. For any 2-unit increase in "H", "W" decreases _____ units.

ANSWERS: 1. $m = \frac{\Delta v}{\Delta t}$ 3. 64 units   5. $m = \frac{\Delta W}{\Delta H}$ 7. -12 units

                 2. 32   4. 160 units   6. $-\frac{3}{2}$ 8. 3 units

484 Linear Graphs and Slope

5-21 ESTIMATING THE EQUATIONS OF GRAPHED LINES

If we know the exact coordinates of two points on a line, we can determine the exact equation of the line. However, we occasionally want to determine the equation of a graphed line when the exact coordinates of two points are not given. In such cases, we can only estimate the equation of the line by estimating its slope and y-intercept. We will discuss the method in this section.

305. If the exact coordinates of two points on a line are given, we can find the exact value of "m", the exact value of "b", and the exact equation of the line. Here is an example:

On the graph at the right, the exact coordinates of two points are given. Therefore, we can determine the exact "xy" equation of the line by:

(1) Determining its exact slope "m" by means of the slope formula:

$$m = \frac{1 - (-3)}{3 - 1} = \frac{1 + 3}{2} = \frac{4}{2} = 2$$

(2) Determining its exact y-intercept "b" by:

(a) Plugging in "2" for "m" in the slope-intercept equation, $y = mx + b$:

$$y = 2x + b$$

(b) Solving for "b" by plugging in the coordinates of one of the known points. We will use (3,1):

$$1 = 2(3) + b$$
$$1 = 6 + b$$
$$b = -5$$

Therefore, the exact equation of the line is: _____

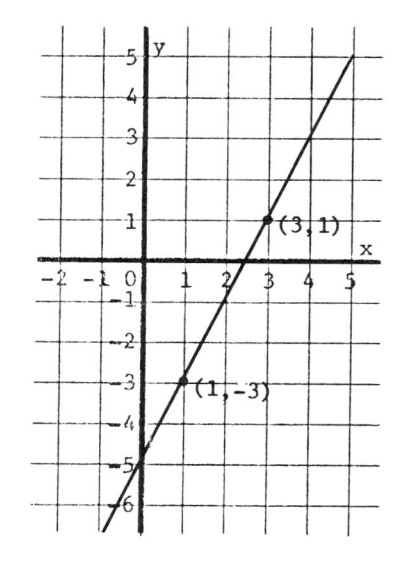

Answer to Frame 305: $y = 2x + (-5)$ or $y = 2x - 5$

---

306. Since the exact coordinates of two points on the line at the right are given, we can determine the exact "non-xy" equation of the line by the same method. Note that the variables are "p" and "h". The steps are:

(1) Determine its exact slope "m" by means of the slope formula:

$$m = \frac{0 - 4}{4 - (-6)} = \frac{-4}{4 + 6} = \frac{-4}{10} = -\frac{2}{5}$$

(2) Determine its exact vertical intercept "b" by:

(a) Substituting $-\frac{2}{5}$ for "m" in the slope-intercept equation, $h = mp + b$:

$$h = -\frac{2}{5}p + b$$

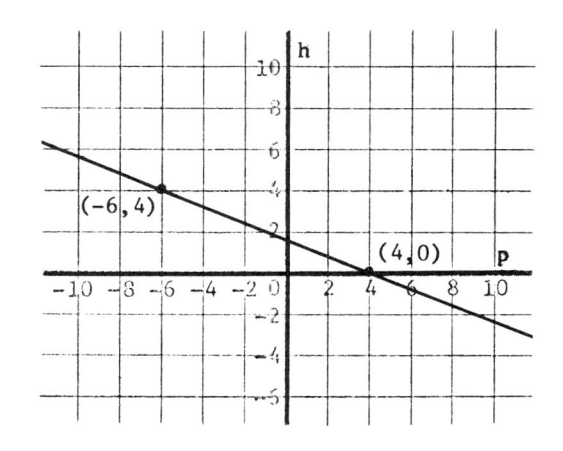

Continued on the following page.

306. Continued.

(b) Solving for "b" by plugging in the coordinates of one of the known points. We will use (4,0).

$$0 = -\frac{2}{5}(4) + b$$

$$0 = -\frac{8}{5} + b$$

$$b = \frac{8}{5}$$

Therefore, the exact equation of the line is: _____

| | Answer to Frame 306: $h = -\frac{2}{5}p + \frac{8}{5}$ or $h = -0.4p + 1.6$ |

307. The line in each graph below passes through the origin. Besides the exact coordinates of the origin (0,0), the exact coordinates of one other point on each line is given.

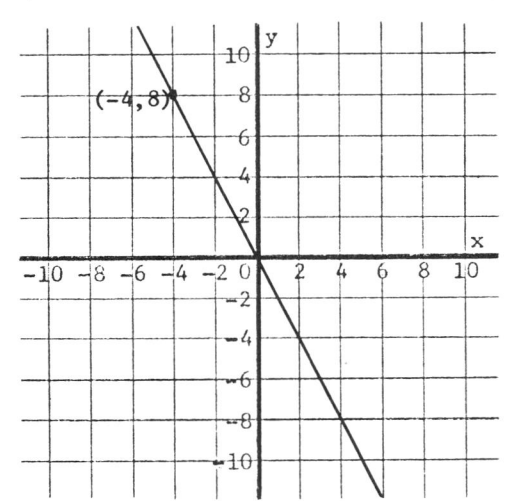

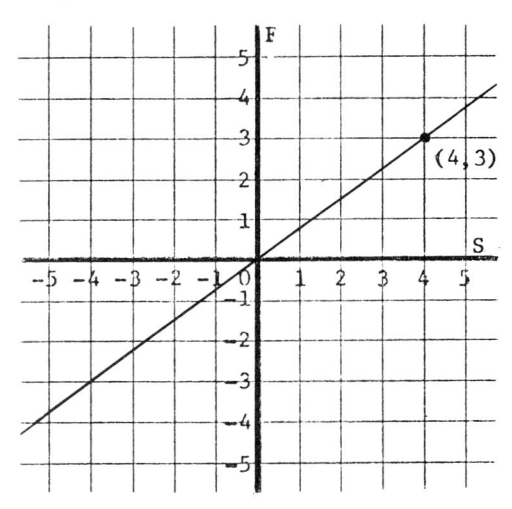

We can determine the exact equation of each line by simply determining its slope. In doing so, <u>be sure to use the variables labeled on the axes.</u>

(a) The equation of the line on the left is: _____

(b) The equation of the line on the right is: _____

	a) $y = -2x$
	b) $F = \frac{3}{4}S$
	or
	$F = 0.75S$

308. We want to determine the equation of the line at the right. Since we are not given the exact coordinates of any points, we can only estimate the vertical intercept and slope of the line. Therefore, we can only estimate the equation of the line.

   (1) To estimate the vertical intercept, we estimate the coordinates of point B where the graphed line crosses the vertical axis. The coordinates of point B are approximately (0,1), and "b" is approximately "1".

   (2) To estimate the slope, we try to locate two points where the graphed line seems to cross an intersection of the grid lines, like points A and C. The coordinates of such points are approximately whole numbers.

   Since the coordinates of point A are approximately (2,6) and the coordinates of point C are approximately (-2,-4),

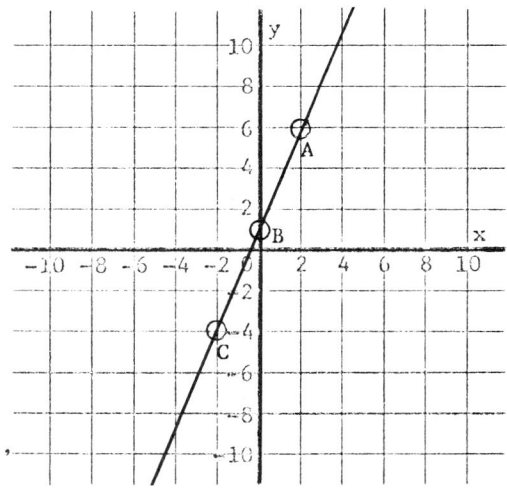

   the slope "m" is approximately: $\frac{6 - (-4)}{2 - (-2)} = \frac{6 + 4}{2 + 2} = \frac{10}{4} = \frac{5}{2}$

Therefore, an estimate of the equation of the line is:

Answer to Frame 308: $y = \frac{5}{2}x + 1$

---

309. Since we are not given the coordinates of any points on the line at the right, we can only estimate its equation.

   (1) Since the vertical intercept is approximately $(0, 3.5)$ or $(0, \frac{7}{2})$, "b" is approximately 3.5 or $\frac{7}{2}$.

   (2) The coordinates of points F and T are approximately whole numbers since the graphed line seems to cross an intersection of the grid lines at those points.

   The coordinates of point F are approximately (-3, 4).
   The coordinates of point T are approximately (2, 3).

   Therefore, "m" is approximately: $\frac{3 - 4}{2 - (-3)} = \frac{-1}{2 + 3} = \frac{-1}{5}$ or $-\frac{1}{5}$ or -0.2

Therefore, the equation of the line is approximately:

Answer to Frame 309: $y = -\frac{1}{5}x + \frac{7}{2}$ or $y = -0.2x + 3.5$

310. Two lines are drawn on the graph at the right. On each line, we have encircled two points whose coordinates are approximately whole numbers. Let's estimate the equation of each line.

For Line #1:

(a) The vertical intercept "b" is approximately _____.

(b) The slope "m" is approximately _____.

(c) The equation is approximately: _____

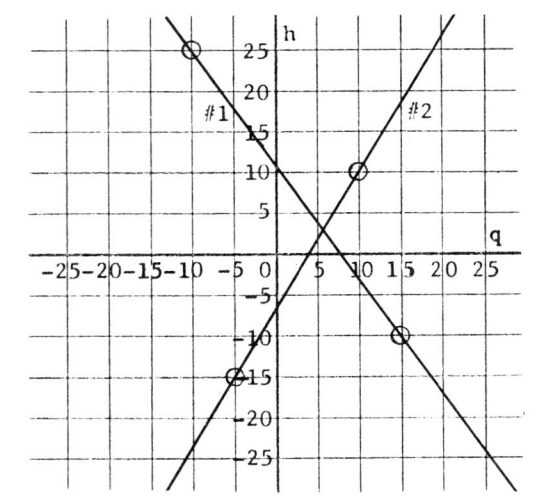

For Line #2:

(d) The vertical intercept "b" is approximately _____.

(e) The slope "m" is approximately _____.

(f) The equation is approximately: _____

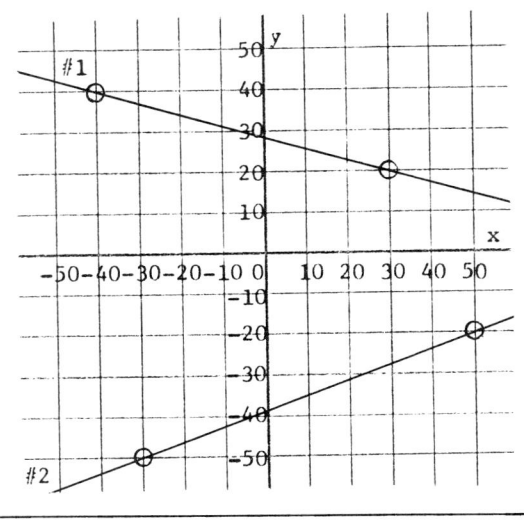

Answer to Frame 310:

a) 28

b) $-\frac{2}{7}$, from $\frac{-20}{70}$

c) $y = -\frac{2}{7}x + 28$

d) -39

e) $\frac{3}{8}$, from $\frac{30}{80}$

f) $y = \frac{3}{8}x + (-39)$ or $y = \frac{3}{8}x - 39$

---

311. On each line at the right, we have encircled two points whose coordinates are approximately whole numbers. Let's estimate the equation of each line.

Note: The variables on the axes are "h" and "q".

For Line #1:

(a) "b" is approximately _____.

(b) "m" is approximately _____.

(c) The equation is approximately: _____

For Line #2:

(d) "b" is approximately _____.

(e) "m" is approximately _____.

(f) The equation is approximately: _____

Answer to Frame 311:

a) 11

b) $-\frac{7}{5}$, from $\frac{-35}{25}$

c) $h = -\frac{7}{5}q + 11$

d) -7

e) $\frac{5}{3}$, from $\frac{25}{15}$

f) $h = \frac{5}{3}q + (-7)$ or $h = \frac{5}{3}q - 7$

488   Linear Graphs and Slope

312. Both lines on the right apparently pass through the origin. We have encircled one point on each line whose coordinates are approximately whole numbers. We can estimate the equation of each line by estimating its slope.

(a) The equation of Line #1 is approximately:

_____

(b) The equation of Line #2 is approximately:

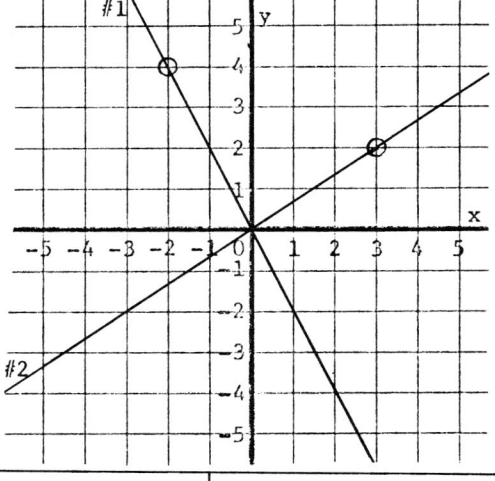

_____

313. Both lines at the right apparently start at the origin. One point, whose coordinates are approximately whole numbers, is encircled on each line. Let's estimate the equation of each line.

Note: The variables on the axes are "t" and "d".

(a) The equation of Line #1 is approximately: _____

(b) The equation of Line #2 is approximately: _____

a) $y = -2x$

b) $y = \frac{2}{3}x$

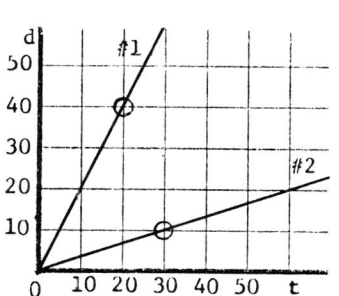

Answer to Frame 313:   a) $d = 2t$ b) $d = \frac{1}{3}t$

314. Let's estimate the equation of each line at the right by:

(1) Estimating its vertical intercept.
(2) Estimating its slope by locating two points whose coordinates are approximately whole numbers.

Notice that the variables on the axes are "v" and "t".

(a) The equation of Line #1 is approximately:

_____

(b) The equation of Line #2 is approximately:

_____

(c) The equation of Line #3 is approximately:

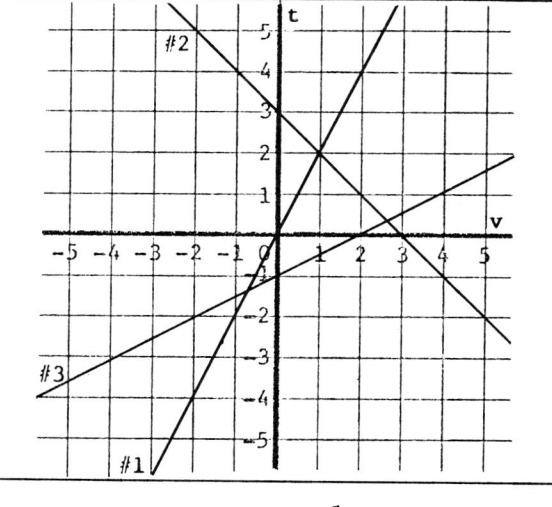

_____

Answer to Frame 314:   a) $t = 2v$ b) $t = -1v + 3$ c) $t = \frac{1}{2}v + (-1)$
or
$t = -v + 3$
or
$t = \frac{1}{2}v - 1$

315. On the line at the right, there are no points which cross an intersection of grid lines. Therefore, there are no points with both coordinates approximately whole numbers. We want to estimate the equation of the line.

We can estimate "b" in the usual way. It is approximately $\frac{1}{2}$ or 0.5.

To estimate "m", however, we use the following steps:

(1) Pick two points whose <u>abscissas</u> are approximately whole numbers. (These points should be fairly far apart so that any errors in estimating the ordinates will only affect the estimated value of the slope slightly.)

We picked: Point A (whose abscissa is approximately -3)
Point B (whose abscissa is approximately 3)

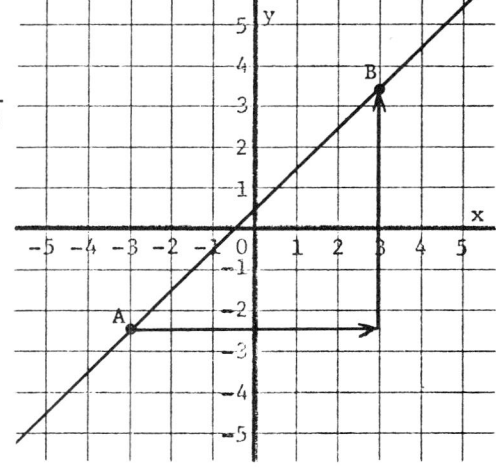

Therefore, Δx is approximately 6, <u>a whole number</u>.

(2) Estimate the <u>ordinates</u> of the two points.

The ordinate of A is approximately -2.4.
The ordinate of B is approximately 3.4.

Therefore, Δy is approximately 5.8.

The slope "m" is approximately $\frac{5.8}{6}$ or 0.97.

Therefore, the equation of the line is approximately: _____

Answer to Frame 315:   y = 0.97x + 0.5

316. The line at the right does not pass through any points which lie at intersections of grid lines. Let's estimate the equation of the line.

Note: The variables are "H" and "F".

The vertical intercept is approximately "3". To estimate the slope:

(1) We picked two points (far apart) whose <u>abscissas</u> are approximately whole numbers so that ΔH will be approximately a whole number.

The abscissa of point A is approximately -4.
The abscissa of point B is approximately 6.

Therefore, ΔH is approximately 10.

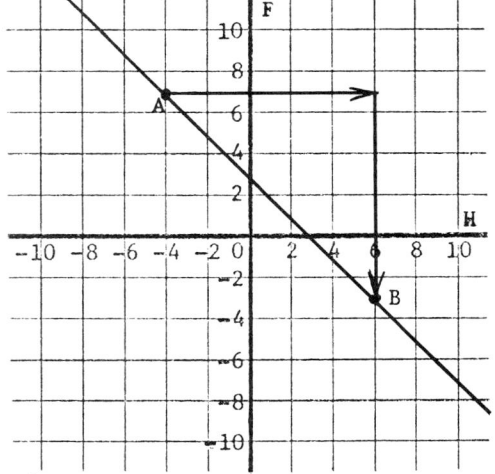

(2) By estimating the ordinates of A and B, we can estimate ΔF.

The ordinate of A $\doteq$ 7.
The ordinate of B $\doteq$ -3.

Therefore, ΔF is approximately -10.

(a) The slope of the line is approximately _____.

(b) The equation of the line is approximately: _____

490 Linear Graphs and Slope

317. Though the line at the right apparently begins at the origin, it seems to have no other point with two whole-number coordinates. Let's estimate its equation. (Note that the variables are "p" and "t".)

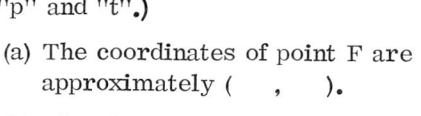

   (a) The coordinates of point F are approximately (,).

   (b) The slope is approximately _____.

   (c) The equation is approximately: _____

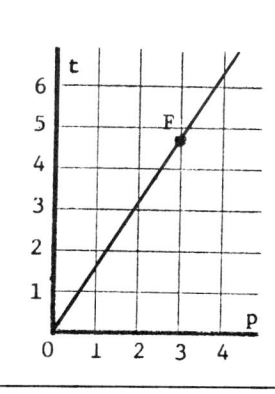

a) -1, from $\dfrac{-10}{10}$

b) $F = (-1)H + 3$
   or
   $F = -H + 3$

Answer to Frame 317:   a) (3, 4.8)   b) 1.6   c) $t = 1.6p$

SELF-TEST 21 (Frames 305-317)

On the graph at the right, the variables are "F" and "P". Write the equation of:

1. Line #1: _____

2. Line #2: _____

3. Line #3: _____

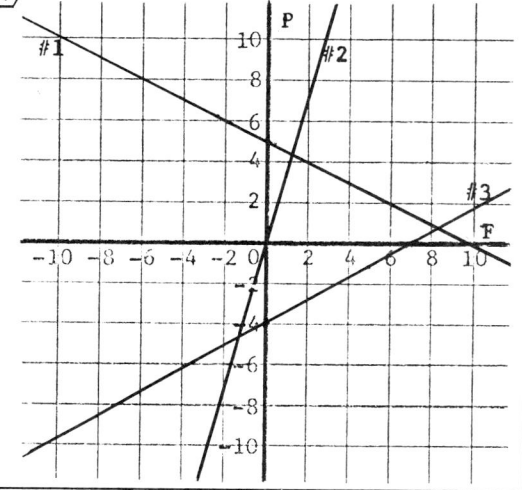

The variables are "E" and "I" on the graph at the right. Write the equation of:

4. Line #1: _____

5. Line #2: _____

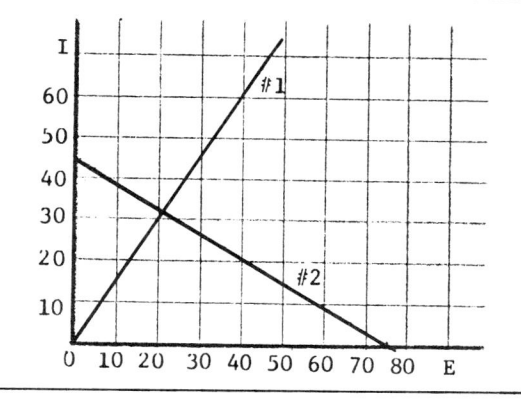

ANSWERS: 1. $P = -0.5F + 5$ 4. $I = 1.5E$
                2. $P = 3.5F$ 5. $I = -0.6E + 45$
                3. $P = 0.57F - 4$

5-22 THE SLOPE OF LINES WHEN THE SCALES ON THE AXES ARE DIFFERENT

In discussing the slope of graphed lines up to this point, we have avoided axes with different scales. In this section, we will briefly discuss the slope of graphed lines on axes with different scales. We will show that the actual slope of such lines can be quite different than their apparent slope.

318. When graphing on the "xy" system, the scales on both axes are usually the same. We have graphed both equations below on the axes at the right. The scales on the axes are the same.

   Graph #1: $y = x$ (or $y = 1x$)
   (with a slope of +1)

   Graph #2: $y = -x$ (or $y = -1x$)
   (with a slope of -1)

   Notice that the angle between the graphed line and the x-axis is 45° for each equation.

   When the scales on both axes are the same, if a graphed line forms a 45° angle with the x-axis, the numerical value of its slope must be either _____ or _____.

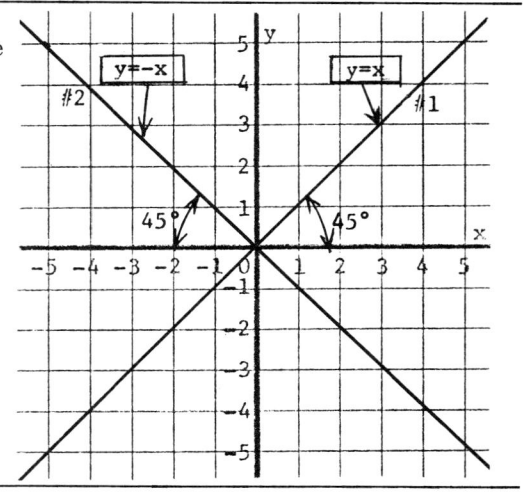

Answer to Frame 318: +1 or -1

319. When graphing on the "xy" system, <u>we do not always use the same scales on both axes</u>. Here are some equations graphed on a system with different scales on the axes:

   Notice that both lines form a 45° angle with the x-axis.

   (a) What is the slope of line #1? _____

   (b) What is the slope of line #2? _____

   (c) On an xy-system <u>with different scales on the axes</u>, is the slope of a line which forms a 45° angle with the x-axis either +1 or -1? _____

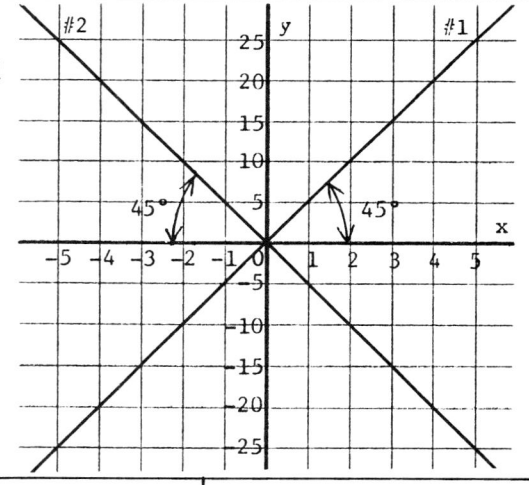

a) +5

b) -5

c) No.

492   Linear Graphs and Slope

320. "Eyeballing" a graphed line to estimate its slope can be very deceiving unless the scales on the axes are taken into account. At the right, we have graphed two formulas on axes with different scales. Both graphed lines form a 45° angle with the horizontal axis.

From the formulas:

(a) The slope of Line #1 is _____.

(b) The slope of Line #2 is _____.

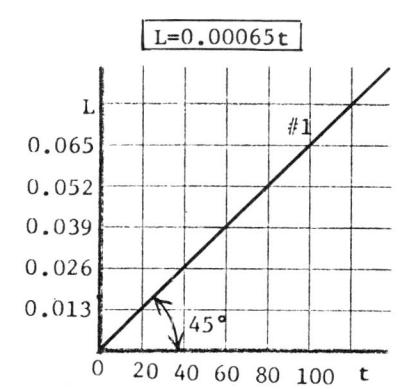

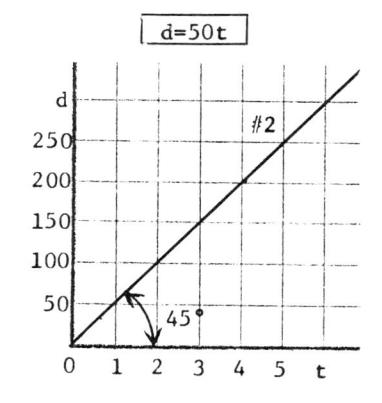

---

321. If a graphed line forms a 45° angle with the horizontal axis, its slope is +1 or −1 only if what condition is true?

a) 0.00065

b) 50

Answer to Frame 321:   Only if the scales on each axis are the same.

322. The scales on the axes at the right are different.

(a) Using the circled points, estimate the slope of Line #1.

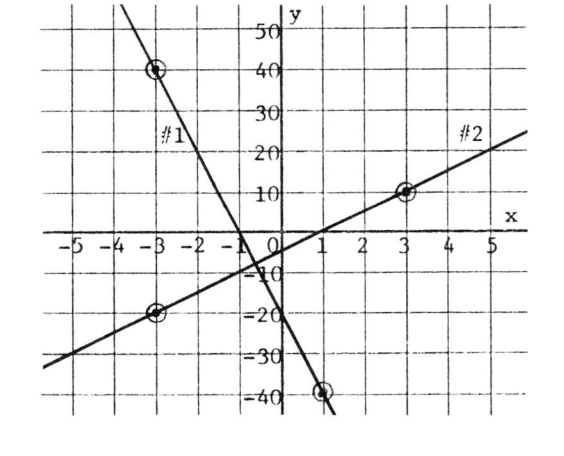

$$m = \underline{\qquad}$$

(b) The equation of Line #1 is approximately: _____

(c) Using the circled points, estimate the slope of Line #2.

$$m = \underline{\qquad}$$

(d) The equation of Line #2 is approximately: _____

---

323. The scales on each set of axes below are different. Let's estimate the slope of each line.

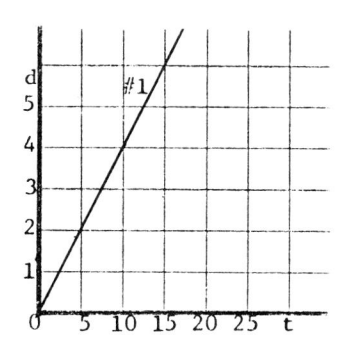

Continued on following page.

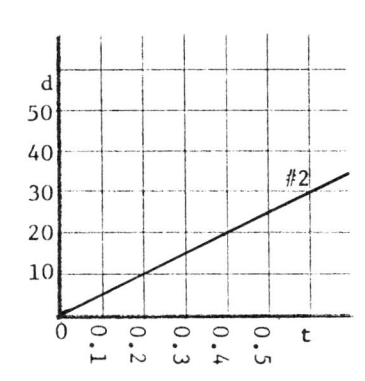

a) −20, from $\dfrac{-80}{4}$

b) $y = -20x + (-20)$
or
$y = -20x - 20$

c) +5, from $\dfrac{30}{6}$

d) $y = 5x + (-5)$
or
$y = 5x - 5$

323. Continued.

(a) For Line #1,   m ≐ _____

(b) For Line #2,   m ≐ _____

(c) Which line appears to have the greatest slope? _____

(d) Which line actually has the greatest slope? _____

Answer to Frame 323:   a) $m \doteq \frac{2}{5}$ or 0.4   b) $m \doteq 50$ c) Line #1.   d) Line #2.

SELF-TEST 22 (Frames 318-323)

Refer to the graph at the right. Find the following:

1. Slope of Line #1: _____

2. Equation of Line #1: _____

3. Slope of Line #2: _____

4. Equation of Line #2: _____

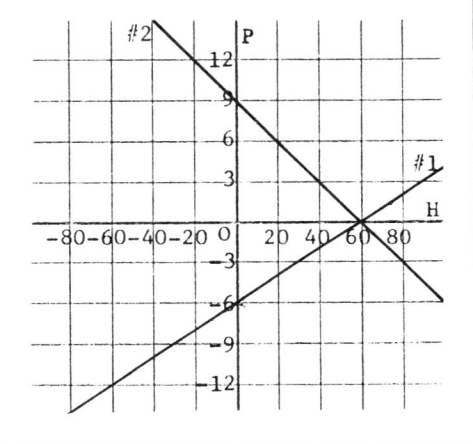

Referring to the graph below, write the equation of:

5. Line #1: _____
6. Line #2: _____

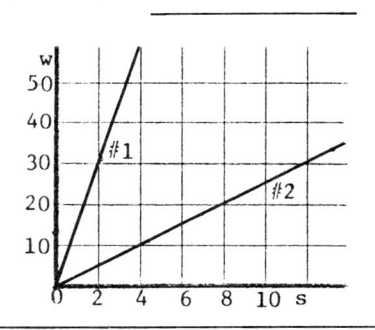

7. The graph of a relationship between the variables "F" and "s" is a straight line through the origin. When F = 50, s = 0.2. If "F" is plotted on the vertical axis, find the equation of the line.

_____

ANSWERS: 1. $m = \frac{1}{10}$ or 0.1   3. $m = -\frac{3}{20}$ or -0.15   5. w = 15s   7. F = 250s

2. P = 0.1H - 6   4. P = -0.15H + 9   6. $w = \frac{5}{2}s$ or w = 2.5s

---

5-23   THE SLOPE OF CURVES

In this section, we will give a brief introduction to the slope of curves. The main point we want to make is this: <u>Curves also have slopes</u>. The concept of slope is not limited to straight-line figures.

494 Linear Graphs and Slope

324. "Slope" is a concept which is defined in terms of a straight line. To apply this concept to curved figures, we must somehow introduce a straight line. The straight line which is introduced is called the tangent to the curve at a particular point.

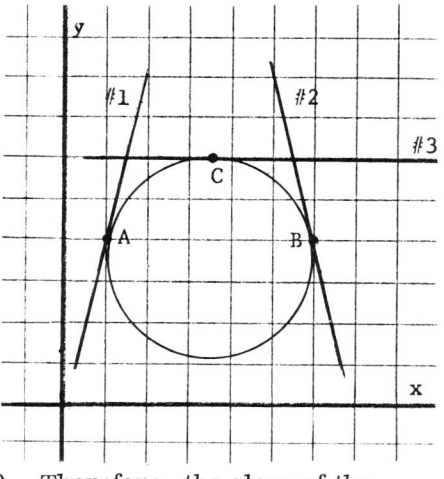

At the right, we have graphed a circle in the first quadrant. Note the following:

(1) We have labeled three points on the circle with the letters A, B, and C.

(2) We have drawn tangents at these three points and labeled them Lines #1, #2, and #3.

Line #1 is the tangent to the curve at point A. The slope of the tangent IS the slope of the curve at that point. Since the slope of Line #1 is positive, the slope of the curve at point A is positive.

(a) The slope of Line #2 is _____ (positive/negative). Therefore, the slope of the curve at point B is _____ (positive/negative).

(b) What is the slope of Line #3? _____ Therefore, the slope of the curve at point C is _____.

Answer to Frame 324:   a) negative ... negative   b) 0 ... 0 (Since it is a horizontal line.)

---

325. We have graphed another curve at the right:

(a) The slope of Line #1 is _____ (positive/negative). The slope of the curve at point A is _____ (positive/negative).

(b) The slope of Line #2 is _____ (positive/negative). The slope of the curve at point B is _____ (positive/negative).

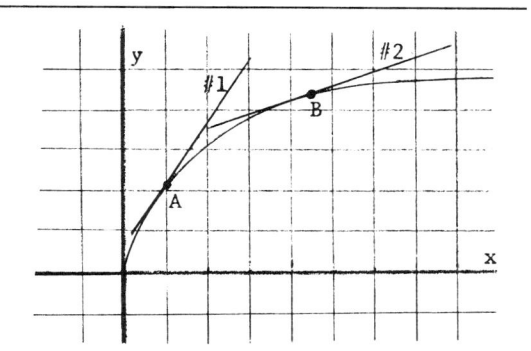

Answer to Frame 325:   a) positive ... positive   b) positive ... positive

---

326. Here is another graph. The curve is called a parabola.

State whether the slope is positive, negative, or 0 at:

(a) Point A: _____

(b) Point B: _____

(c) Point C: _____

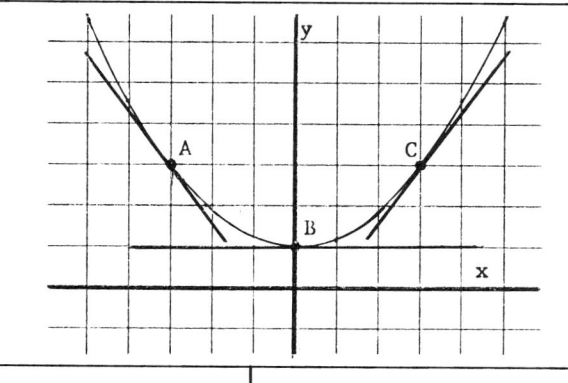

a) Negative.

b) 0

c) Positive.

327. The curve at the right is called a hyperbola.

State whether the slope is positive or negative at:

(a) Point A: _____

(b) Point B: _____

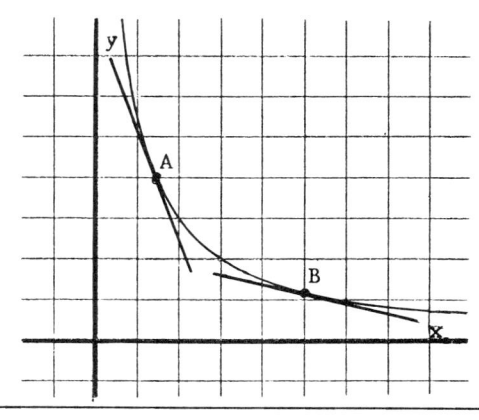

Answer to Frame 327:   a) Negative.   b) Negative.

328. Another parabola is shown at the right.

Tangents are shown at points A, B, and C.

(a) The slope is 0 at which point? _____

(b) The slope is negative at which point? _____

(c) The slope is positive at which point? _____

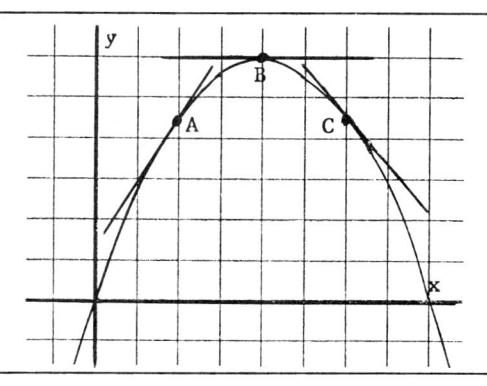

Answer to Frame 328:   a) B   b) C   c) A

329. The graph at the right is another parabola.

(a) At point A, the slope is positive. Reading from left to right, is the curve rising or falling at this point? _____

(b) At point C, the slope is negative. Is the curve rising or falling at this point? _____

(c) At point B, the slope is 0. What is happening to the curve at this point in terms of rising or falling? _____

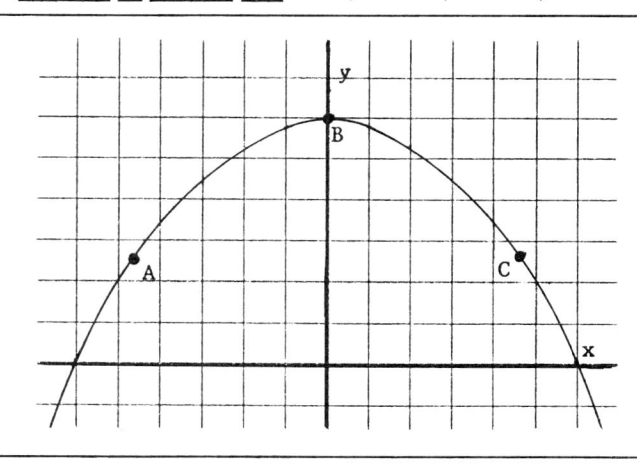

---

a) Rising.

b) Falling.

c) It is neither rising nor falling.

496   Linear Graphs and Slope

330. A hyperbola in the first quadrant is shown at the right.

    (a) The slope at both points A and B is negative. A negative slope means that the curve is _____ (rising/falling) at these points.

    (b) When the slope of a curve is negative, the curve is falling. This means that when "x" is increasing, "y" is _____ (increasing/decreasing).

Answer to Frame 330:   a) falling   b) decreasing

---

331. Another parabola is shown at the right.

At points A and B the slope is negative.

    (a) When the slope is negative, the curve is _____ (rising/falling).

    (b) When the slope is negative, as "x" increases, "y" _____ (increases/decreases).

At points C and D the slope is positive.

    (c) When the slope is positive, the curve is _____ (rising/falling).

    (d) When the slope is positive, as "x" increases, "y" _____ (increases/decreases).

---

332. When the slope of a curve is positive, "y" is increasing as "x" increases. This means that the curve is _____ (rising/falling).

a) falling   c) rising
b) decreases d) increases

333. In terms of changes in "x" and "y", what does a falling curve mean?
_____

rising

As "x" increases, "y" decreases.

334. On the graph at the right, the slope is positive at both points A and B.

    (a) At which of the two points is the slope steeper? _____

    (b) At which of the two points is "y" increasing more with each unit increase in "x"? _____

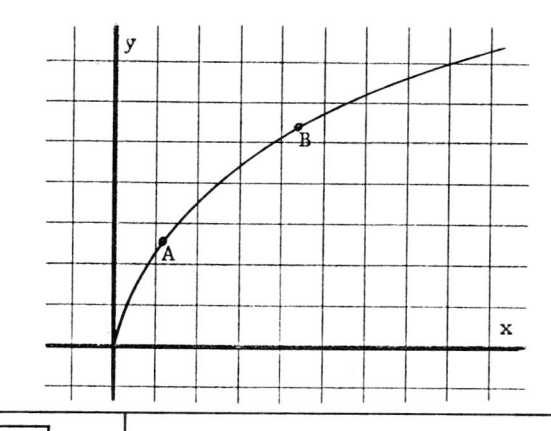

Note: The numerical value of the slope of a curve at various points can be computed. We will not present a method at this time.

a) At point A.
   (Draw the tangents.)

b) At point A.
   (This is what a steeper slope means.)

Chapter 6 COMMON AND NATURAL LOGARITHMS

In this chapter, we will review common (base "10") logarithms and introduce natural (base "e") logarithms. The meaning of logarithms and logarithmic notation will be emphasized. Evaluations will be performed with "log", "e", and "ln" formulas. Semi-log, log-log, and exponential graphs will also be discussed.

6-1   THE POWER-OF-TEN FORM OF NUMBERS GREATER THAN 1

In earlier work, we showed that any positive number can be written in power-of-ten form. In this section, we will briefly review the power-of-ten form of numbers greater than 1.

1. When numbers like 10, 100, 1,000, and so on, are written in power-of-ten form, the exponents are whole numbers. For example: $$10 = 10^1 \qquad 100 = 10^2 \qquad 1,000 = 10^3$$ Write the power-of-ten form for each of the following: (a) 10,000 _____ (b) 1,000,000 = _____ (c) 1 = _____	
2. When most numbers are written in power-of-ten form, the exponents are decimal numbers. If the regular number lies between 1 and 10, the exponent of its power-of-ten form is given in the "exponent table" on pages 499 and 500. (The table is titled "COMMON LOGARITHMS OF NUMBERS".) The exponents in the following two examples were obtained from the table: $$2.73 = 10^{0.4362} \qquad 8.60 = 10^{0.9345}$$ Using the exponent table, write the proper exponent in each box: (a) $4.93 = 10^{\boxed{}}$ (b) $6.88 = 10^{\boxed{}}$ (c) $9.04 = 10^{\boxed{}}$	a) 10^4 b) 10^6 c) 10^0
3. Using the exponent table, write the power-of-ten form for each regular number below: (a) 7.01 = _____ (b) 3.20 = _____ (c) 1.06 = _____	a) $10^{0.6928}$ b) $10^{0.8376}$ c) $10^{0.9562}$
	a) $10^{0.8457}$ b) $10^{0.5051}$ c) $10^{0.0253}$

Common and Natural Logarithms 501

4. The exponent table only contains the power-of-ten exponents for numbers between 1 and 10. Therefore, in order to use the table to convert a number greater than 10 to power-of-ten form, we must write the number in standard notation first. (Note: The first factor in standard-notation form is always a number between 1 and 10).

Write each number below in standard notation:

(a) 49 = _____   (c) 9,380 = _____

(b) 649,000 = _____   (d) 563 = _____

a) 4.9×10^1
b) 6.49×10^5
c) 9.38×10^3
d) 5.63×10^2

5. The following steps are used to convert any number greater than 10 (like 48,100) to power-of-ten form:

(1) Write it in standard notation: $48,100 = 4.81 \times 10^4$

(2) Use the exponent table to convert the first factor to power-of-ten form: $= 10^{0.6821} \times 10^4$

(3) Then use the law of exponents for multiplication to combine the two power-of-ten factors: $= 10^{4.6821}$

Therefore, the power-of-ten form of 48,100 is: _____

$10^{4.6821}$

6. Convert each number below to power-of-ten form:

(a) $583,000 = 10^{\boxed{}}$ (b) $6,940 = 10^{\boxed{}}$ (c) $10,500,000 = 10^{\boxed{}}$

a) $10^{5.7657}$
b) $10^{3.8414}$
c) $10^{7.0212}$

7. If a number contains more than three non-zero digits, we round to three non-zero digits in order to use the exponent table. An example is given at the right.

$7,128 = 7.13 \times 10^3$
$= 10^{0.8531} \times 10^3$
$= 10^{3.8531}$

Write the correct exponent in each box.

(a) $25.48 = 10^{\boxed{}}$ (b) $19,721 = 10^{\boxed{}}$ (c) $7.3669 = 10^{\boxed{}}$

a) $10^{1.4065}$
b) $10^{4.2945}$
c) $10^{0.8675}$

502   Common and Natural Logarithms

8. If the exponent of a power of ten is a whole number, we can convert the power of ten to a regular number without using the exponent table. Convert each power of ten below to a regular number:

 (a) $10^3 = $ _____ (c) $10^5 = $ _____

 (b) $10^1 = $ _____ (d) $10^0 = $ _____

---

9. If the exponent of a power of ten is a decimal number between 0 and 1, we can convert the power of ten to a regular number directly by using the exponent table. Convert each power of ten below to a regular number:

 (a) $10^{0.5623} = $ _____ (b) $10^{0.6998} = $ _____ (c) $10^{0.8751} = $ _____

a) 1,000
b) 10
c) 100,000
d) 1

---

10. The process of converting "737" to power-of-ten form is given below:

     Mantissa Characteristic
     ↓   ↓

$$737 = 7.37 \times 10^2 = 10^{\boxed{0.8675}} \times 10^{\boxed{2}} = 10^{0.8675 + 2} = 10^{2.8675}$$

The special names "mantissa" and "characteristic" are given to two exponents which appear in this process:

 The "mantissa" is the exponent of the power-of-ten form of 7.37.

 The "characteristic" is the whole-number exponent obtained when 737 is converted to standard notation.

We have converted 14.7 to power-of-ten form below:

 $14.7 = 1.47 \times 10^1 = 10^{0.1673} \times 10^1 = 10^{1.1673}$

In this conversion: (a) The mantissa is _____.

      (b) The characteristic is _____.

a) 3.65
b) 5.01
c) 7.50

---

11. Any power of ten with a decimal exponent greater than 1 can be factored into mantissa-characteristic form. For example:

$$10^{4.6899} = 10^{0.6899} \times 10^4$$
$$10^{2.5066} = 10^{0.5066} \times 10^2$$
$$10^{1.0857} = 10^{0.0857} \times 10^1$$

Notice in each case that:

 (1) The mantissa is the decimal part of the original exponent. It is a decimal number between 0 and 1.

 (2) The characteristic is the whole-number part of the original exponent.

Factor each power of ten into mantissa-characteristic form:

 (a) $10^{3.1067} = $ _____ (b) $10^{5.9011} = $ _____

a) 0.1673
b) 1

---

a) $10^{0.1067} \times 10^3$
b) $10^{0.9011} \times 10^5$

12. In order to convert a power of ten with a decimal exponent greater than 1 to a regular number, we must write it in mantissa-characteristic form first. Then the exponent table can be used to convert the mantissa-factor to a regular number. For example:

$$10^{3.1584} = 10^{0.1584} \times 10^3$$
$$= 1.44 \times 10^3$$
$$= 1,440$$

Convert each power of ten below to a regular number:

(a) $10^{3.9460} =$ _____

(b) $10^{5.6294} =$ _____

(c) $10^{1.8293} =$ _____

(d) $10^{6.0453} =$ _____

---

13. When converting a power of ten back to a regular number, sometimes the exact exponent does not appear in the exponent table. Here is an example:
$$10^{0.3549}$$
The exponent "0.3549" is not in the table. In such a case, we use the closest exponent, which is "0.3541".

Therefore: $10^{0.3549} =$ _____

a) 8,830
b) 426,000
c) 67.5
d) 1,110,000

---

14. Suppose we want to convert $10^{0.8528}$ to a regular number. In the exponent table, "8528" lies exactly halfway between "8525" and "8531":

$$10^{0.8525} = 7.12$$
$$10^{0.8531} = 7.13$$

To be consistent, we always pick the regular number which ends in an even digit.

Therefore: $10^{0.8528} =$ _____

2.26

---

15. Convert the following to regular numbers:

(a) $10^{0.2940} =$ _____   (c) $10^{0.6582} =$ _____

(b) $10^{0.6498} =$ _____   (d) $10^{0.7122} =$ _____

↓
7.12 (Since "2" is even.)

---

16. Convert the following to regular numbers:

(a) $10^{3.5077} =$ _____

(b) $10^{5.6400} =$ _____

a) 1.97
b) 4.46 (not 4.47)
c) 4.55
d) 5.16 (not 5.15)

---

a) 3,220
b) 436,000
   (not 437,000)

17. In $2 = 10^{0.3010}$: "2" is a <u>regular</u> <u>number</u>.
    "10" is a <u>base</u>.
    "0.3010" is an <u>exponent</u>.

   In $564 = 10^{2.7513}$: (a) The base is _____.

   (b) The exponent is _____.

18. When a regular number is written in power-of-ten form, the <u>base</u> is always _____.

a) 10
b) 2.7513

<u>Answer</u> <u>to</u> <u>Frame</u> <u>18</u>: 10

SELF-TEST 1 (Frames 1-18)

Convert each number to power-of-ten form:

1. $3.49 =$ _____
2. $8{,}170 =$ _____
3. $164 =$ _____
4. $52{,}637 =$ _____

Convert each power of ten to a regular number:

5. $10^{3.1931} =$ _____
6. $10^{1.5827} =$ _____
7. $10^{6.8666} =$ _____

Given: $7{,}410 = 10^{3.8698}$ Write the: 8. Characteristic: _____   9. Mantissa: _____

ANSWERS:
1. $10^{0.5428}$
2. $10^{3.9122}$
3. $10^{2.2148}$
4. $10^{4.7210}$
5. 1,560
6. 38.3
7. 7,360,000
8. 3
9. 0.8698

6-2   COMMON LOGARITHMS OF NUMBERS GREATER THAN 1

When a regular number is written in power-of-ten form, the <u>exponent</u> of the power of ten is called the "logarithm" of the number. In this section, we will review the logarithms of numbers greater than 1. In doing so, we will also review the meaning of logarithmic or "log" notation.

19. The word "logarithm" is another name for the word "exponent". Therefore, in the statement: $20 = 10^{1.3010}$, "1.3010" can be called either the _____ or _____ .	
20. THE LOGARITHM OF A NUMBER IS THE EXPONENT OF ITS POWER-OF-TEN FORM. It is the exponent only; it is not the exponent with the base. For example: Since $37.2 = 10^{1.5705}$: The logarithm of 37.2 is 1.5705 (only the exponent). The logarithm of 37.2 is not $10^{1.5705}$ (the exponent with the base). Since $48.9 = 10^{1.6893}$, the logarithm of 48.9 is _____ .	exponent or logarithm
21. Remember: A logarithm is an exponent without the base. (a) Since $104,000 = 10^{5.0170}$, the logarithm of 104,000 is _____ . (b) Since $429 = 10^{2.6325}$, the logarithm of 429 is _____ .	1.6893 (not $10^{1.6893}$)
22. (a) Since $1,000 = 10^3$, the logarithm of 1,000 is _____ . (b) Since $1 = 10^0$, the logarithm of 1 is _____ .	a) 5.0170 (not $10^{5.0170}$) b) 2.6325 (not $10^{2.6325}$)
23. Instead of the words "the logarithm of", the abbreviation "log" is generally used. Therefore: (a) Since $389 = 10^{2.5899}$, log 389 = _____ . (b) Since $10 = 10^1$, log 10 = _____ .	a) 3 (or 3.0000) b) 0 (or 0.0000)
24. Any statement in "power-of-ten" notation can be written in "log" notation. For example: $775 = 10^{2.8893}$ IS EQUIVALENT TO: log 775 = 2.8893 Similarly: (a) $8,970 = 10^{3.9528}$ is equivalent to: log 8,970 = _____ (b) $14,300 = 10^{4.1553}$ is equivalent to: log 14,300 = _____	a) 2.5899 b) 1 (or 1.0000)
25. Write the following statements in "log" notation: (a) $14.6 = 10^{1.1644}$ (b) $386 = 10^{2.5866}$	a) 3.9528 b) 4.1553
	a) log 14.6 = 1.1644 b) log 386 = 2.5866

506 Common and Natural Logarithms

26. Any statement in "log" notation can be written in "power-of-ten" notation. For example:

> log 506 = 2.7042,
> IS EQUIVALENT TO:
> 506 = 10$^{2.7042}$

(a) If log 642 = 2.8075, then 642 = 10$^{\boxed{}}$

(b) If log 7.06 = 0.8488, then 7.06 = 10$^{\boxed{}}$

a) 10$^{\boxed{2.8075}}$
b) 10$^{\boxed{0.8488}}$

27. Write these statements in "power-of-ten" notation:

(a) log 279 = 2.4456   (b) log 15.7 = 1.1959

28. Convert each equation in power-of-ten notation to log notation and vice versa:

(a) 100 = 10^2 (c) 1 = 10^0

(b) log 1,000 = 3   (d) log 10 = 1

a) 279 = 10$^{2.4456}$
b) 15.7 = 10$^{1.1959}$

29. In each case below, the "log" of the number is a whole number. Write the "log" of each:

(a) log 100 = _____   (b) log 100,000 = _____   (c) log 10 = _____

a) log 100 = 2
b) 1,000 = 10^3
c) log 1 = 0
d) 10 = 10^1

30. The exponent table we have been using is really a "log" table. Therefore, the "log" of any number between 1 and 10 is given directly in the table. Using the log table, complete these:

(a) log 6.40 = _____   (b) log 1.82 = _____   (c) log 9.04 = _____

a) 2   b) 5   c) 1

31. The exponent of the power-of-ten form of any number greater than 10 is the sum of a mantissa and a characteristic. For example:

524 = 5.24 × 10^2 = 10$^{0.7193}$ × 10^2 = 10$^{2.7193}$

Therefore, the logarithm of any number greater than 10 is the sum of a mantissa and a characteristic. For example:

In log 524 = 2.7193, the logarithm is the sum of:

(a) the mantissa _____ and   (b) the characteristic _____.

a) 0.8062
b) 0.2601
c) 0.9562

a) 0.7193   b) 2

32. Here are the steps needed to find the logarithm of any number greater than 10. We will find log 47.8 as an example:

   (1) We write 47.8 in standard notation: 4.78×10^1

   (2) We find the <u>mantissa</u> of its logarithm in the log table. Its mantissa is 0.6794.

   (3) Its <u>characteristic</u> is "1", the whole-number exponent of the second factor.

   Adding the mantissa and characteristic, log 47.8 = _____.

33. Let's find log 96,400 by the same method:

   (a) The mantissa is _____.

   (b) The characteristic is _____.   (c) log 96,400 = _____

	1.6794

34. Find the following logarithms:

   (a) log 32.4 = _____   (c) log 2.140 = _____

   (b) log 984,000 = _____   (d) log 794,000,000 = _____

   a) 0.9841
   b) 4   c) 4.9841

35. In the equation log 100 = w :

   "100" is a regular number.
   "w" is its logarithm (or the exponent in its power-of-ten form).

   Therefore, the numerical value of "w" is _____.

   a) 1.5105   c) 0.3304
   b) 5.9930   d) 8.8998

36. (a) If log 1,000 = t,   (b) If log 7.62 = V,

       t = _____.   V = _____.

   2

37. (a) If log 3,762 = P,   (b) If log 27,100 = d,

       P = _____.   d = _____.

   a) t = 3
   b) V = 0.8820

38. In the equation log t = 4 :

   "t" stands for some regular number.
   "4" is its logarithm (or the exponent in its power-of-ten form).

   Therefore: $t = 10^4$ or _____

   a) P = 3.5752
   b) d = 4.4330

39. In the equation log y = 0.4594 :

   "y" stands for some regular number.
   "0.4594" is the logarithm of that number.

   Therefore: $y = 10^{0.4594}$ = _____

   10,000

40. (a) If log d = 2,   (b) If log B = 0.7340,

       d = _____.   B = _____.

   2.88

508   Common and Natural Logarithms

41. In the equation $\boxed{\log x = 3.9462}$:

"x" stands for some regular number.
"3.9462" is the logarithm of that number.

Therefore: $x = 10^{3.9462} = 10^{0.9462} \times 10^3 = 8.83 \times 10^3 = 8,830$

(Notice that we began by converting $10^{3.9462}$ to mantissa-characteristic form.)

Following the steps above, let's find "N" in: $\boxed{\log N = 2.3034}$

$N = 10^{2.3034} = 10^{\boxed{}} \times 10^{\boxed{}}$

$= \underline{} \times 10^{\boxed{}} = \underline{}$

a) d = 100 (from 10^2)
b) B = 5.42 (from $10^{0.7340}$)

---

42. (a) If log K = 5.2993,   (b) If log H = 2.5730,

K = _____ .   H = _____ .

$N = 10^{0.3034} \times 10^2$
$= 2.01 \times 10^2 = 201$

---

43. (a) In log 729 = t, does "t" stand for a logarithm or a regular number? _____

(b) In log w = 3.4622, does "w" stand for a logarithm or a regular number? _____

a) K = 199,000
b) H = 374

---

44. Solve each of the following equations:

(a) log 6.433 = D   D = _____   (c) log 1,739 = M   M = _____

(b) log T = 1.9584   T = _____   (d) log S = 4.9784   S = _____

a) A logarithm.
b) A regular number.

---

45. Solve each of these equations:

(a) If log 1,000 = x, x = _____   (c) If log t = 0, t = _____

(b) If log h = 2, h = _____   (d) If log 10 = y, y = _____

a) D = 0.8082
b) T = 90.9
c) M = 3.2405
d) S = 95,200 (not 95,100)

---

a) x = 3   c) t = 1
b) h = 100   d) y = 1

Common and Natural Logarithms 509

SELF-TEST 2 (Frames 19-45)

1. Write the following statement in logarithmic notation:

    $85.4 = 10^{1.9315}$ _____

2. Write the following statement in power-of-ten notation:

    $\log 2,870 = 3.4579$ _____

In $\log 709 = 2.8506$, list the: 3. Logarithm: _____ 4. Mantissa: _____

In $67.3 = 10^{1.8280}$, list the: 5. Logarithm: _____ 6. Characteristic: _____

7. Find: $\log 49,200 =$ _____

8. If $\log 14.7 = P$, $P =$ _____

9. If $G = \log 6,630$, $G =$ _____

10. If $\log Q = 2.7370$, $Q =$ _____

11. If $\log x = 4.6454$, $x =$ _____

12. If $\log R = 6.0000$, $R =$ _____

ANSWERS:
1. $\log 85.4 = 1.9315$
2. $2,870 = 10^{3.4579}$
3. 2.8506
4. 0.8506
5. 1.8280
6. 1
7. 4.6920
8. $P = 1.1673$
9. $G = 3.8215$
10. $Q = 546$
11. $x = 44,200$
12. $R = 1,000,000$ (From: $R = 10^6$)

6-3 THE POWER-OF-TEN FORM OF NUMBERS BETWEEN 0 AND 1

In this section, we will review the power-of-ten form of numbers between 0 and 1. Since this topic frequently causes difficulties for students, pay special attention in this section.

46. When numbers like 0.1, 0.01, 0.001, and so on, are written in power-of-ten form, the exponents are <u>negative</u> whole numbers. For example:

$0.1 = 10^{-1}$ $0.01 = 10^{-2}$ $0.001 = 10^{-3}$

Write the power-of-ten form for each of the following:

(a) $0.0001 =$ _____   (b) $0.000001 =$ _____   (c) $1 =$ _____

a) 10^{-4}
b) 10^{-6}
c) 10^{0}

510   Common and Natural Logarithms

47. Some powers of ten with whole-number exponents are given in the table on the right. Since $1 = 10^0$, the number "1" is the dividing point between numbers whose exponents in power-of-ten form are positive or negative. That is:

$$100 = 10^2$$
$$10 = 10^1$$
$$1 = 10^0$$
$$0.1 = 10^{-1}$$
$$0.01 = 10^{-2}$$

(a) If a positive number is greater than "1", its exponent in power-of-ten form is _____ (positive/negative).

(b) If a positive number is less than "1", its exponent in power-of-ten form is _____ (positive/negative).

48. Any decimal number between 0 and 1 has a negative exponent in its power-of-ten form. As an example, we will convert 0.56 to power-of-ten form.

Step 1: We write 0.56 in standard notation: $0.56 = 5.6 \times 10^{-1}$

Step 2: We use the exponent table to convert "5.6" to power-of-ten form: $= 10^{0.7482} \times 10^{-1}$

Step 3: We combine the two powers of ten by adding their exponents: $= 10^{0.7482 + (-1)}$

$$= 10^{-0.2518}$$

Therefore, the power-of-ten form of 0.56 is _____.

a) positive

b) negative

49. When converting a decimal number between 0 and 1 to power-of-ten form, we begin by writing it in standard notation.

Write the following numbers in standard notation:

(a) $0.88 =$ _____   (c) $0.017 =$ _____

(b) $0.193 =$ _____   (d) $0.00732 =$ _____

$10^{-0.2518}$

50. Write these numbers in standard notation:

(a) $0.000567 =$ _____   (c) $0.0000005 =$ _____

(b) $0.0000067 =$ _____   (d) $0.6 =$ _____

a) 8.8×10^{-1}
b) 1.93×10^{-1}
c) 1.7×10^{-2}
d) 7.32×10^{-3}

51. When a number between 0 and 1 is converted to power-of-ten form, the mantissa is always a positive decimal number and the characteristic is always a negative whole number. For example:

Mantissa   Characteristic
  ↓   ↓
$0.0652 = 6.52 \times 10^{-2} = 10^{\boxed{0.8142}} \times 10^{\boxed{-2}} = 10^{-1.1858}$

a) 5.67×10^{-4}
b) 6.7×10^{-6}
c) 5×10^{-7}
d) 6×10^{-1}

Continued on following page.

51. Continued.

Here are the steps needed to add two exponents of this type:

(1) To find the <u>absolute value</u> of the sum, we subtract the smaller absolute value from the larger. That is, we subtract the mantissa from the characteristic. We get:

    2.0000   (<u>Note</u>: "2" is written as "2.0000"
    0.8142   to make the subtraction
    1.1858   easier to perform.)

(2) The <u>sign</u> of the sum is the same as the sign of the number with the larger absolute value. Since the sign of the characteristic "-2" is <u>negative</u>, the sum is <u>negative</u>. Therefore:

$$0.8142 + (-2) = -1.1858$$

Using the steps above, perform each addition below:

(a) $0.6021 + (-3) = $ _____   (c) $0.9146 + (-2) = $ _____

(b) $0.3907 + (-4) = $ _____   (d) $0.8888 + (-1) = $ _____

---

52. Convert each of the following to power-of-ten form:

(a) $0.259 = $ _____   (b) $0.0061 = $ _____

a) -2.3979 c) -1.0854
b) -3.6093 d) -0.1112

---

53. Convert each of these to power-of-ten form:

(a) $0.000123 = $ _____   (b) $0.00009 = $ _____

a) $10^{-0.5867}$
b) $10^{-2.2147}$

---

a) $10^{-3.9101}$
b) $10^{-4.0458}$

512   Common and Natural Logarithms

54. Convert each of these to power-of-ten form:

   (a) $0.1 = $ _____ (c) $0.0001 = $ _____

   (b) $0.86 = $ _____ (d) $0.05 = $ _____

---

55. Convert each of the following to power-of-ten form:
(Note: To use the exponent table, you must round the first standard-notation factor to three digits.)

   (a) $0.01019 = $ _____ (b) $0.00007144 = $ _____

a) 10^{-1}
b) $10^{-0.0655}$
c) 10^{-4}
d) $10^{-1.3010}$

---

56. If a power of ten has a negative whole-number exponent, it can easily be converted to a regular number. For example:

$$10^{-1} = 0.1 \qquad 10^{-3} = 0.001 \qquad 10^{-5} = 0.00001$$

Convert each of the following to regular numbers:

   (a) $10^{-4} = $ _____ (b) $10^{-2} = $ _____ (c) $10^{-6} = $ _____

a) $10^{-1.9914}$
b) $10^{-4.1463}$

---

57. Four numbers are converted to power-of-ten form below. We have drawn a box around the <u>mantissa-characteristic forms</u> in the various conversions:

$$758 = 7.58 \times 10^2 = \boxed{10^{0.8797} \times 10^2} = 10^{2.8797}$$
$$63.2 = 6.32 \times 10^1 = \boxed{10^{0.8007} \times 10^1} = 10^{1.8007}$$
$$0.249 = 2.49 \times 10^{-1} = \boxed{10^{0.3962} \times 10^{-1}} = 10^{-0.6038}$$
$$0.0571 = 5.71 \times 10^{-2} = \boxed{10^{0.7566} \times 10^{-2}} = 10^{-1.2434}$$

By examining the mantissa-characteristic forms above, answer each of the following:

   (a) Is a mantissa ever <u>greater than</u> 1? _____

   (b) Is a mantissa ever <u>negative</u>? _____

   (c) Can a characteristic be a <u>non-whole number</u>? _____

   (d) Is a characteristic ever <u>negative</u>? _____

a) 0.0001
b) 0.01
c) 0.000001

Common and Natural Logarithms 513

58. Below, 27.3 is converted to power-of-ten form. Notice that the exponent of the "10" is positive.

$$27.3 = 2.73 \times 10^1 = 10^{0.4362} \times 10^1 = 10^{1.4362}$$

As you can see from the example, the mantissa and characteristic are readily apparent in the final exponent. That is:

The <u>mantissa</u> is the decimal part of the final exponent.

The <u>characteristic</u> is the whole-number part of the final exponent.

Using these facts, write each power of ten below in mantissa-characteristic form:

(a) $10^{3.7011} = 10^{\boxed{}} \times 10^{\boxed{}}$ (b) $10^{2.6599} = 10^{\boxed{}} \times 10^{\boxed{}}$

a) No. A mantissa is always a number between 0 and 1.

b) No.

c) No.

d) Yes.

59. 0.0852 is converted to power-of-ten form below. Notice that the exponent of the "10" is negative.

$$0.0852 = 8.52 \times 10^{-2} = 10^{0.9304} \times 10^{-2} = 10^{-1.0696}$$

As you can see from the example, the mantissa and characteristic are <u>not</u> readily apparent in the final exponent. They are obscured by the addition of a positive mantissa and a negative characteristic.

To convert $10^{-1.0696}$ back to mantissa-characteristic form, we must get a mantissa. That is, we must get <u>a positive exponent between 0.0000 and 0.9996</u>. To do so, we add "+2" to "-1.0696" and then add "-2" to that sum. We get:

$$10^{-1.0696} = 10^{[2 + (-1.0696)] + (-2)}$$
$$= 10^{0.9304 + (-2)}$$
$$= 10^{0.9304} \times 10^{-2}$$

Note: Adding both "+2" and "-2" is equivalent to adding "0", since (+2) + (-2) = 0. Therefore: -1.0696 = 0.9304 + (-2).

Is $10^{0.9304} \times 10^{-2}$ the same mantissa-characteristic form we obtained above when converting 0.0852 to power-of-ten form? _____

a) $10^{0.7011} \times 10^3$

b) $10^{0.6599} \times 10^2$

Yes.

514   Common and Natural Logarithms

60. To convert a power of ten with a negative exponent (such as $10^{-3.1568}$) back to mantissa-characteristic form, we add some instance of +n and -n to it at the same time. This addition of a number and its opposite is always equivalent to adding "0" to the exponent.

To determine the specific number to add, we examine the whole-number part of the exponent:

$$10^{-3.1568}$$

In this case, the whole number part is -3. The absolute value of this whole-number part is 3. To determine the positive number to use, we simply add 1 to this absolute value (3 + 1 = 4). Therefore, we add +4 to -3.1568 and then add -4 to that result. We get:

$$10^{-3.1568} = 10^{[4 + (-3.1568)] + (-4)}$$
$$= 10^{0.8432 + (-4)}$$
$$= 10^{0.8432} \times 10^{-4}$$

In the mantissa-characteristic form: (a) The mantissa is _____.

(b) The characteristic is _____.

---

61. Let's convert this power of ten back to mantissa-characteristic form:

$$10^{-0.0991}$$

Since the whole-number part of the exponent is "-0", we must add +1 to -0.0991 and then add -1 to this sum. (Note: The absolute value of +1 is one more than that of 0.)

We get: $10^{-0.0991} = 10^{[1 + (-0.0991)] + (-1)}$
$= 10^{\boxed{}} \times 10^{\boxed{}}$

a) 0.8432

b) -4

---

62. What pair of numbers would you add in each case to convert to mantissa-characteristic form?

(a) $10^{-1.5655}$ ____ and ____   (c) $10^{-0.1099}$ ____ and ____

(b) $10^{-4.7622}$ ____ and ____   (d) $10^{-7.4766}$ ____ and ____

$10^{\boxed{0.9009}} \times 10^{\boxed{-1}}$

---

63. When converting a power of ten with a negative exponent to mantissa-characteristic form, there is a shortcut you can use. Here is an example:

$$\phantom{10^{-2.5659}}3.0000$$
$$10^{-2.5659 + (-3)}$$
$$10^{0.4341} \times 10^{-3}$$

Note: We wrote the +3 as 3.0000 immediately above the -2.5659 so that we could easily perform the addition.

Use the shortcut to convert this one
to mantissa-characteristic form: $10^{-1.2644}$

a) +2 and -2

b) +5 and -5

c) +1 and -1

d) +8 and -8

64. Convert each of the following to mantissa-characteristic form:

(a) $10^{-5.2199} = 10^{\boxed{}} \times 10^{\boxed{}}$ (c) $10^{-0.2318} = 10^{\boxed{}} \times 10^{\boxed{}}$

(b) $10^{-1.7651} = 10^{\boxed{}} \times 10^{\boxed{}}$ (d) $10^{-3.1055} = 10^{\boxed{}} \times 10^{\boxed{}}$

$$= 10^{-1.2644 + (-2)}\quad\overset{2.0000}{}$$
$$= 10^{0.7356} \times 10^{-2}$$

a) $10^{\boxed{0.7801}} \times 10^{\boxed{-6}}$
b) $10^{\boxed{0.2349}} \times 10^{\boxed{-2}}$
c) $10^{\boxed{0.7682}} \times 10^{\boxed{-1}}$
d) $10^{\boxed{0.8945}} \times 10^{\boxed{-4}}$

65. After writing the power of ten below in mantissa-characteristic form, it is easy to convert it to standard notation and then to a regular number. We use the exponent table to do so.

$$10^{-1.2069} = 10^{0.7931} \times 10^{-2}$$

(a) = _____ $\times 10^{-2}$

(b) = _____

66. Complete this conversion back to a regular number:

(a) $10^{-2.1752} = 10^{\boxed{}} \times 10^{\boxed{}}$

(b) = _____ $\times 10^{\boxed{}}$

(c) = _____

a) 6.21×10^{-2}
b) 0.0621

67. Convert the following to regular numbers:

(a) $10^{-3.3625} =$ _____   (c) $10^{-1.1203} =$ _____

(b) $10^{-0.9281} =$ _____   (d) $10^{-2.0610} =$ _____

a) $10^{\boxed{0.8248}} \times 10^{\boxed{-3}}$
b) $6.68 \times 10^{\boxed{-3}}$
c) 0.00668

68. Write each of the following as a regular number. Be careful of the sign of the exponent:

(a) $10^{4.5972} =$ _____   (c) $10^{-0.7777} =$ _____

(b) $10^{-4.5972} =$ _____   (d) $10^{0.7777} =$ _____

a) 0.000434
 (from 4.34×10^{-4})
b) 0.118
 (from 1.18×10^{-1})
c) 0.0758
 (from 7.58×10^{-2})
d) 0.00869
 (from 8.69×10^{-3})

516   Common and Natural Logarithms

69. Write each of the following as a regular number:

(a) 10^1 = _____   (b) 10^{-1} = _____   (c) 10^3 = _____   (d) 10^{-3} = _____

| a) 39,600 | c) 0.167 |
| b) 0.0000253 | d) 5.99 |

Answer to Frame 69:   a) 10   b) 0.1   c) 1,000   d) 0.001

SELF-TEST 3 (Frames 46-69)

Convert each of the following numbers to power-of-ten form:

1. 0.00261 = _____   2. 0.848 = _____   3. 0.0574 = _____   4. 5,740 = _____

Convert each of the following powers of ten to a regular number:

5. $10^{-1.3841}$ = _____   6. $10^{-0.7258}$ = _____   7. $10^{-4.8906}$ = _____   8. $10^{4.8906}$ = _____

ANSWERS: 1. $10^{-2.5834}$ 3. $10^{-1.2411}$ 5. 0.0413   7. 0.0000129
           2. $10^{-0.0716}$ 4. $10^{3.7589}$ 6. 0.188   8. 77,700

6-4 COMMON LOGARITHMS OF NUMBERS BETWEEN 0 AND 1

In this section, we will review what is meant by the logarithms of numbers between 0 and 1. In doing so, we will again review the meaning of logarithmic or "log" notation.

70. When a number is written in power-of-ten form, the exponent is called the "logarithm" of the number. This same term is used even when the exponent is negative. For example:

Since $0.789 = 10^{-0.1029}$, the logarithm of 0.789 is -0.1029.

Since $0.0512 = 10^{-1.2907}$, the logarithm of 0.0512 is _____.

71. (a) Since $0.1 = 10^{-1}$, log 0.1 = _____

(b) Since $0.01 = 10^{-2}$, log 0.01 = _____

-1.2907

a) -1 (or -1.0000)
b) -2 (or -2.0000)

72. Each of the following power-of-ten equations is equivalent to a log equation. That is:

$0.784 = 10^{-0.1057}$ is equivalent to $\log 0.784 = -0.1057$

(a) $0.0192 = 10^{-1.7167}$ is equivalent to _____

(b) $0.001 = 10^{-3}$ is equivalent to _____

---

73. Each of the following log equations is equivalent to a power-of-ten equation. That is:

$\log 0.00398 = -2.4001$ is equivalent to $0.00398 = 10^{-2.4001}$

(a) $\log 0.000483 = -3.3161$ is equivalent to _____

(b) $\log 0.00001 = -5$ is equivalent to _____

a) $\log 0.0192 = -1.7167$

b) $\log 0.001 = -3$
  (or -3.0000)

---

74. Convert each power-of-ten equation to an equivalent log equation and vice versa:

(a) $0.1 = 10^{-1}$ (c) $\log 0.001 = -3$

_____

(b) $\log 0.678 = -0.1688$ (d) $0.00289 = 10^{-2.5391}$

_____

a) $0.000483 = 10^{-3.3161}$

b) $0.00001 = 10^{-5}$

---

75. In each case below, the "log" of the number is a <u>negative</u> whole number. Write the "log" of each:

(a) $\log 0.01 = $ _____ (b) $\log 0.1 = $ _____ (c) $\log 0.0001 = $ _____

a) $\log 0.1 = -1$

b) $0.678 = 10^{-0.1688}$

c) $0.001 = 10^{-3}$

d) $\log 0.00289 = -2.5391$

---

76. The logarithm of 0.0265 is the sum of a positive mantissa and a negative characteristic. The steps needed to find $\log 0.0265$ are given below:

(1) We write 0.0265 in standard notation: 2.65×10^{-2}

(2) We find its <u>mantissa</u> in the log (or exponent) table. Its mantissa is 0.4232.

(3) Its <u>characteristic</u> is "-2", the negative whole-number exponent of the second factor.

Therefore: $\log 0.0265 = 0.4232 + (-2) = $ _____

a) -2

b) -1

c) -4

---

77. Let's find $\log 0.159$ by the same method. Since $0.159 = 1.59 \times 10^{-1}$:

(a) The mantissa is _____.

(b) The characteristic is ____. (c) $\log 0.159 = $ _____

-1.5768

---

a) 0.2014

b) -1

c) -0.7986

518   Common and Natural Logarithms

78. Using the same method, complete each of the following:

    (a) log 0.761 = _____   (c) log 0.000963 = _____

    (b) log 0.0102 = _____   (d) log 0.0000032 = _____

---

79. Complete each of these:

    (a) log 0.001 = _____   (c) log 0.000001 = _____

    (b) log 0.07 = _____   (d) log 0.000000005 = _____

a) -0.1186
b) -1.9914
c) -3.0164
d) -5.4949

---

80. In the equation $\boxed{\log 0.123 = t}$:

    "0.123" is a regular number.
    "t" is the logarithm of "0.123".

    Therefore, the numerical value of "t" is _____.

a) -3 (or -3.0000)
b) -1.1549
c) -6 (or -6.0000)
d) -8.3010

---

81. (a) If log 0.01 = x, x = _____.

    (b) If log 0.00000437 = V, V = _____.

-0.9101

---

82. In the equation $\boxed{\log x = -4}$:

    "x" stands for a regular number.
    "-4" is the logarithm of the number.

    Therefore: $x = 10^{-4}$ or _____

a) $x = -2$
b) $V = -5.3595$

0.0001

83. In the equation $\boxed{\log y = -1.1694}$:

"y" stands for a regular number.
"-1.1694" is the logarithm of that number.

Therefore: $y = 10^{-1.1694} = 10^{0.8306} \times 10^{-2} = 6.77 \times 10^{-2} = 0.0677$

Notice that we began by converting $10^{-1.1694}$ to mantissa-characteristic form. To do so, we added both +2 and -2 to -1.1694.

Following the steps above, let's solve: $\boxed{\log x = -0.6635}$

$x = 10^{-0.6635} = 10^{\boxed{}} \times 10^{\boxed{}}$

$= \underline{} \times 10^{\boxed{}} = \underline{}$

---

84. Let's solve: $\boxed{\log t = -2.3478}$

$t = 10^{-2.3478} = 10^{\boxed{}} \times 10^{\boxed{}}$

$= \underline{} \times 10^{\boxed{}} = \underline{}$

$= 10^{0.3365} \times 10^{-1}$
$= 2.17 \times 10^{-1} = 0.217$

---

85. Let's solve: $\boxed{\log y = -3.4962}$

$y = 10^{-3.4962} = 10^{\boxed{}} \times 10^{\boxed{}}$

$= \underline{} \times 10^{\boxed{}} = \underline{}$

$= 10^{0.6522} \times 10^{-3}$
$= 4.49 \times 10^{-3} = 0.00449$

---

86. Solve each of the following equations:

(a) If $\log d = -0.4311$, d = _____   (b) If $\log a = -2.7709$, a = _____

$= 10^{0.5038} \times 10^{-4}$
$= 3.19 \times 10^{-4} = 0.000319$

---

87. In which equations below does "x" stand for a <u>logarithm</u>? _____

(a) $\log x = -0.8651$ (b) $\log 0.067 = x$ (c) $\log x = -4.1659$

a) $d = 3.71 \times 10^{-1} = 0.371$
b) $a = 1.69 \times 10^{-3} = 0.00169$

---

88. In which equations below does "y" stand for a regular number? _____

(a) $\log 0.569 = y$ (b) $\log 0.007 = y$ (c) $\log y = -1.6233$

Only in (b).

---

89. Answer "regular number" or "logarithm" for each of these:

(a) In $\log P = -3$, "P" stands for a _____.

(b) In $\log 0.1 = t$, "t" stands for a _____.

(c) In $\log 0.001 = m$, "m" stands for a _____.

(d) In $\log H = -1$, "H" stands for a _____.

Only in (c).

520   Common and Natural Logarithms

90. In the equations below, the letters can stand for either logarithms or regular numbers. Solve each equation:

(a) log 0.01 = x   x = _____
(b) log h = -1   h = _____
(c) log P = -3   P = _____
(d) log 0.0001 = y y = _____

a) regular number
b) logarithm
c) logarithm
d) regular number

91. Solve each equation below:

(a) log 0.0132 = B   B = _____
(b) log f = -1.9747   f = _____
(c) log R = -0.4841   R = _____
(d) log 0.000989 = V   V = _____

a) $x = -2$
b) $h = 10^{-1} = 0.1$
c) $P = 10^{-3} = 0.001$
d) $y = -4$

92. In P = log 378, "P" stands for a <u>logarithm</u>.
In 3.6122 = log x, "x" stands for a <u>regular number</u>.

(a) In -0.7168 = log y, "y" stands for a _____.
(b) In m = log 0.66, "m" stands for a _____.

a) $B = -1.8794$
b) $f = 0.0106$
c) $R = 0.328$
d) $V = -3.0048$

93. Convert each equation to power-of-ten form. Do not solve.

(a) log 77 = t
(b) x = log 0.01
(c) log R = -1.6257
(d) 3.0768 = log V

a) regular number
b) logarithm

a) $77 = 10^t$
b) $10^x = 0.01$
c) $R = 10^{-1.6257}$
d) $10^{3.0768} = V$

94. Solve each equation below:

    (a) h = log 100   h = _____   (c) D = log 0.1   D = _____

    (b) 4 = log V   V = _____   (d) -3 = log q   q = _____

---

95. Solve each equation below:

    (a) P = log 27.1   P = _____   (c) t = log 0.06   t = _____

    (b) 2.5514 = log y   y = _____   (d) -0.2636 = log x   x = _____

---

a) h = 2 (or 2.0000)
b) V = 10,000 (from 10^4)
c) D = -1 (or -1.0000)
d) q = 0.001 (from 10^{-3})

---

a) P = 1.4330
b) y = 356
c) t = -1.2218
d) x = 0.545

SELF-TEST 4 (Frames 70-95)

1. Convert this power-of-ten equation to a logarithmic equation:

   $10^{-2.3690} = 0.00428$ _____

2. Convert this logarithmic equation to a power-of-ten equation:

   $\log 0.751 = -0.1244$ _____

Find the following logarithms:

3. $\log 0.0186 =$ _____

4. $\log 0.000953 =$ _____

5. $\log 0.000001 =$ _____

In $0.0287 = 10^{-1.5421}$, list the: 6. Logarithm: _____   7. Characteristic: _____   8. Mantissa: _____

9. If $\log N = -3.3212$, $N =$ _____

10. If $\log P = -0.8705$, $P =$ _____

Solve these equations:

11. $h = \log 0.870$

    $h =$ _____

12. $\log T = -1.2998$

    $T =$ _____

13. $\log R = 1.2998$

    $R =$ _____

ANSWERS:

1. $\log 0.00428 = -2.3690$
2. $0.751 = 10^{-0.1244}$
3. -1.7305
4. -3.0209
5. -6
6. -1.5421
7. -2
8. 0.4579
9. $N = 0.000477$
10. $P = 0.135$
11. $h = -0.0605$
12. $T = 0.0501$
13. $R = 19.9$

Common and Natural Logarithms   523

6-5 EVALUATING "LOG" FORMULAS FOR "SOLVED-FOR" VARIABLES

There are various formulas which contain "log" expressions. Some examples are given below:

$$D = 10 \log R \qquad \log P = A - \frac{B}{T}$$

In these formulas, the "solved-for" variable is either a letter (like "D") or the log of a letter (like "log P"). In this section, we will do some evaluations with formulas of this type. In each case, we will find the value of the "solved-for" variable when the values of all other variables are given.

---

96. The formula below contains a "log" expression:

    $$\boxed{D = 10 \log R}$$

    There are two factors ("10" and "log R") on the right side of the formula. To show this fact, we have inserted parentheses on the right side below.

    $$\boxed{D = (10)(\log R)}$$

    If we are given a specific value for "R", we can find the corresponding value of "D" by a simple evaluation. For example:

    If R = 847, then: D = (10)(log R)

    $\qquad\qquad\qquad\quad = (10)(\log 847)$

    $\qquad\qquad\qquad\quad = 10(2.9279) = $ _____

---

97. Here is the same formula: $\boxed{D = 10 \log R}$ Let's find "D" when R = 0.0527.

    D = 10(log 0.0527)

    (a) = 10(   )

    (b) = _____

D = 29.3

    In this section, we will round all answers to three digits. Therefore, 29.279 was rounded to 29.3 .

---

98. Here is a different formula: $\boxed{H = w \log T}$

    If we are given specific values for both "w" and "T", we can find the corresponding value of "H". For example:

    If w = 100 and T = 27.6 , H = 100 log 27.6

    (a) = 100(   )

    (b) = _____

    a) 10(-1.2782)

    b) -12.8
        (rounded to three digits)

---

a) = 100 (1.4409)

b) = 144

524   Common and Natural Logarithms

99. Here is a different formula: $\boxed{D = 10 \log\left(\dfrac{P_2}{P_1}\right)}$

In this formula, $\log\left(\dfrac{P_2}{P_1}\right)$ is the log of a division or fraction. If we know the values of "P_2" and "P_1", we can find the corresponding value of "D". For example:

If $P_2 = 15.5$ and $P_1 = 12.7$: $D = 10 \log\left(\dfrac{15.5}{12.7}\right)$

Performing the division, we get: $\dfrac{15.5}{12.7} = 1.22$

Therefore: $D = 10 \log\left(\dfrac{15.5}{12.7}\right)$

$\qquad = 10 \log 1.22$

(a) $\qquad = 10(\qquad)$

(b) $\qquad = \underline{\qquad}$

---

100. Let's do another one with the same formula: $\boxed{D = 10 \log\left(\dfrac{P_2}{P_1}\right)}$

a) $10(0.0864)$
b) 0.864

If $P_2 = 840$ and $P_1 = 5270$: $D = 10 \log\left(\dfrac{840}{5270}\right)$

(a) $\qquad = 10 \log(\qquad)$

(b) $\qquad = 10(\qquad)$

(c) $\qquad = \underline{\qquad}$

---

101. In $\boxed{M = 2.5 \log\left(\dfrac{I_1}{I}\right)}$, find M when $I_1 = 86.2$ and $I = 15.8$.

a) $= 10 \log 0.159$
b) $= 10(-0.7986)$
c) $= -7.99$
  (three digits)

$M = 2.5 \log\left(\dfrac{86.2}{15.8}\right)$

(a) $\qquad = 2.5 \log(\qquad)$

(b) $\qquad = 2.5(\qquad)$

(c) $\qquad = \underline{\qquad}$

---

102. Here is a different type of formula: $\boxed{P = A - K \log Q}$

a) $2.5 \log 5.46$
b) $2.5(0.7372)$
c) 1.84

If $A = 56$, $K = 17$, and $Q = 0.172$, we can find the corresponding value of "P" by plugging in the known values.

$P = 56 - 17 \log 0.172$

$\quad = 56 - \underline{17(-0.7645)}$
$\qquad\qquad\qquad\downarrow$
$\quad = 56 - (-13.0) \quad = 56 + (+13) = \underline{\qquad}$

103. Do this one with the same formula: $\boxed{P = A - K \log Q}$

If $A = 97$, $K = 20$, and $Q = 2.6$, find "P":

$P =$ _____

---

104. Here is another type: $\boxed{K \log(W + H) = B}$

In the formula above, "log(W + H)" is one of the two factors on the left side. It is the log of an addition. For any specific values of "K", "W", and "H", we can find the corresponding value of "B". For example:

If $K = 3.11$, $W = 19.6$, and $H = 37.5$, we get:

$(3.11)\log(19.6 + 37.5) = B$

$3.11 \log(57.1) \qquad = B$

(a) $3.11(\qquad) \qquad = B$

(b) $\qquad\qquad = B$

$P = 97 - 20 \log 2.6$
$\quad = 97 - 20(0.4150)$
$\quad = 97 - 8.3$
$\quad = 88.7$

---

105. Here is another type: $\boxed{\log\left(\dfrac{H - B}{G - B}\right) = K}$

On the left side of the formula above, there is a log of a division or fraction. Both the numerator and denominator of the fraction are subtractions. Though the formula looks complicated, it is easy to find the value of "K" for specific values of "H", "B", and "G". For example:

If $H = 1190$, $B = 2.3$, and $G = 16.9$, we get: $\log\left(\dfrac{1190 - 2.3}{16.9 - 2.3}\right) = K$

Performing the subtractions, we get: $\log\left(\dfrac{1187.7}{14.6}\right) = K$

Performing the division, we get: $\log 81.4 = K$

Therefore: $K =$ _____

a) $3.11(1.7566)$

b) $B = 5.46$

---

106. The formula on the right also looks complicated but it is easy to find "C" for specific values of "D" and "R".

$C = \dfrac{0.02}{\log\left(\dfrac{D - R}{R}\right)}$

If $D = 570$ and $R = 0.122$:

$C = \dfrac{0.02}{\log\left(\dfrac{570 - 0.122}{0.122}\right)}$

$= \dfrac{0.02}{\log\left(\dfrac{569.878}{0.122}\right)}$

$= \dfrac{0.02}{\log 4{,}670}$

$= \dfrac{0.02}{3.6693} =$ _____

$K = 1.91$ (from 1.9106)

526   Common and Natural Logarithms

107. In the formula below, the "-" in front of $\log A_H$ can be confusing:

$$P_H = -\log A_H$$

Just as $-x = -1x$, $-\log A_H = -1 \log A_H$. Therefore, it helps to write the formula in the following form:

$$P_H = (-1)(\log A_H)$$

Therefore: If $A_H = 0.000067$, $P_H = (-1)(\log 0.000067)$
$= (-1)(-4.1739)$
$= +4.1739$ or 4.17

(a) If $A_H = 0.00057$, $P_H = (-1)(\log 0.00057)$
$= (-1)(\quad\quad)$
$= \underline{\quad\quad\quad}$

(b) If $A_H = 0.000027$, $P_H = (-1)(\log 0.000027)$
$= (-1)(\quad\quad)$
$= \underline{\quad\quad\quad}$

C = 0.00545

108. In this formula, a "log" expression is "solved for": $\log R = \dfrac{D}{10}$

If we are given a specific value for "D", we can find the corresponding values of "log R" and "R". For example:

If $D = 18.7$, $\log R = \dfrac{18.7}{10}$

$\log R = 1.87$ or 1.8700
(Note: By adding two 0's after 1.87, it is easier to use the log table.)

Therefore: R = _____

a) $= (-1)(-3.2441)$
$= +3.2441$ or 3.24

b) $= (-1)(-4.5686)$
$= +4.5686$ or 4.57

109. Here is the same formula: $\log R = \dfrac{D}{10}$

Let's find the value of "R" when $D = 31.7$:

$\log R = \dfrac{31.7}{10}$

(a) $\log R = $ _____

(b) $R = $ _____

R = 74.1

a) 3.17 or 3.1700

b) 1,480

110. Here is another formula in which a "log" expression is "solved for":

$$\log P = A - \frac{B}{T}$$

If we are given specific values for "A", "B", and "T", we can find "log P" and "P". For example:

If $A = 4.79$, $B = 50$ and $T = 37$: $\log P = 4.79 - \frac{50}{37}$

$= 4.79 - 1.35$

$= 3.44$

Therefore: P = _____

111. A log expression is also "solved for" in this formula:

$$\log G = \frac{H - T}{H}$$

P = 2,750

If we are given specific values for "H" and "T", we can find the corresponding values of "log G" and "G". For example:

If $H = 500$ and $T = 300$: $\log G = \frac{500 - 300}{500}$

$\log G = \frac{200}{500}$

(a) log G = _____

(b) G = _____

Answer to Frame 111: a) 0.4 or 0.4000   b) 2.51

SELF-TEST 5 (Frames 96-111)

1. $D = 10 \log Q$

   If $Q = 0.0514$, D = _____

2. $P_H = -\log A_H$

   If $A_H = 0.0001$, P_H = _____

3. $N = 20 \log\left(\frac{E_2}{E_1}\right)$

   If $E_1 = 1.88$ and $E_2 = 73.5$,

   N = _____

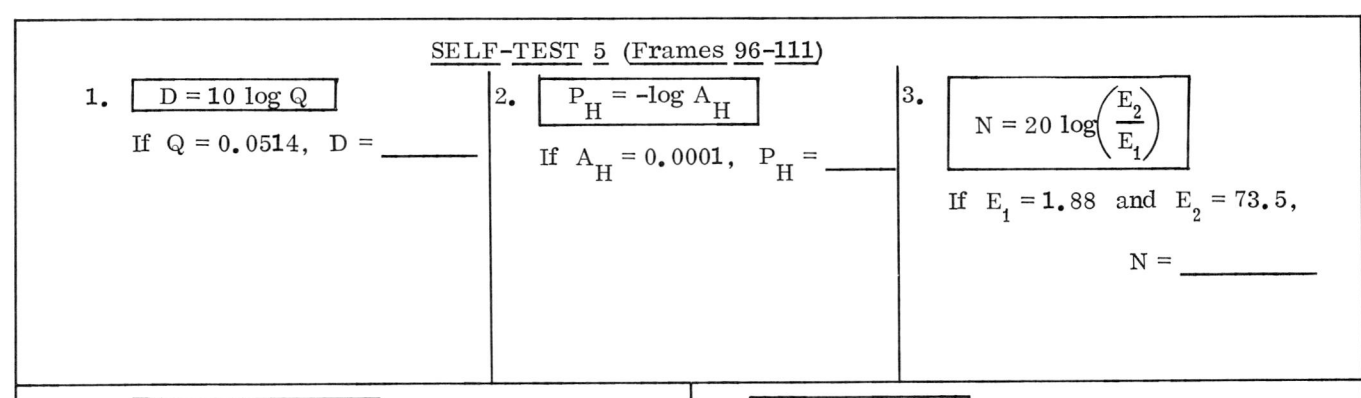

4. $R = A - H \log P$

   If $P = 631$, $H = 0.25$, and $A = 3.7$,

   R = _____

5. $\log N = \frac{E - G}{E}$

   If $E = 16.2$ and $G = 93.7$, N = _____

ANSWERS:

1. D = -12.9   2. $P_H = 4$ 3. N = 31.8   4. R = 3.0   5. $N = 1.66 \times 10^{-5}$ or 0.0000166
(from $\log N = -4.78$)

528   Common and Natural Logarithms

6-6 EVALUATING "LOG" FORMULAS FOR "NON-SOLVED FOR" VARIABLES

In the last section, we evaluated "log" formulas for "solved for" variables. In this section, we will evaluate the same type of formulas for "non-solved for" variables by simply plugging the known values into the formulas as they stand. In later work, we will perform the same type of evaluations by rearranging the log formulas first.

112. In the formula at the right, the variable "log R" is not "solved for". To find "R" when $D = 17.9$, we can use the steps below:

    $\boxed{D = 10 \log R}$

    (1) Plug in 17.9 for "D": $17.9 = 10 \log R$

    (2) Solve for "log R":
    $$\frac{1}{10}(17.9) = \frac{1}{10}(10)(\log R)$$
    $$\frac{17.9}{10} = 1 \cdot \log R$$
    $$1.79 = \log R$$

    (3) Use the log table to find the value of R:

    Since $\log R = 1.79$ (or 1.7900), R = _____

113. Following the steps in the last frame, do this evaluation:

    If $\boxed{D = 10 \log R}$, find R when $D = -17.7$

    (a) $\log R =$ _____
    (b) $R =$ _____

    R = 61.7

114. "w" is not "solved for" in $\boxed{H = w \log T}$. To find the value of "w" when $H = 150$ and $T = 23.8$, we can use the steps below:

    (1) Plug in the known values: $150 = w \log 23.8$

    (2) Replace log 23.8 with its numerical value:
    $150 = w(1.3766)$
    or
    $150 = 1.3766w$

    (3) Now solve for "w" in the usual way:
    $$w = \frac{150}{1.3766}$$
    $w =$ _____

    a) $\log R = -1.7700$
    b) $R = 0.0170$

115. If $\boxed{H = w \log T}$, find the value of "T" when $H = 178$ and $w = 100$.

    (a) $\log T =$ _____
    (b) $T =$ _____

    w = 109

    a) $\log T = 1.7800$
       From: $178 = 100 \log T$
       $\log T = \frac{178}{100}$
    b) $T = 60.3$

116. "Q" is not "solved for" in $\boxed{P = A - K \log Q}$. To find the value of "Q" when $P = 690$, $A = 560$, and $K = 170$, we can use the steps below:

    (1) Plug in the known values: $690 = 560 - 170 \log Q$

    (2) Add "-560" to both sides: $130 = -170 \log Q$

    (3) Replace each side with its opposite: $-130 = 170 \log Q$

    (4) Solve for "log Q" and then for "Q" in the usual way: $\log Q = \dfrac{-130}{170}$

        (a) $\log Q = $ _____

        (b) $Q = $ _____

---

117. In $\boxed{P = A - K \log Q}$, we want to find the value of "A" when $P = 37.8$, $K = 9.20$, and $Q = 14.7$. The steps are given below:

    (1) Plug in the known values: $37.8 = A - 9.20 \log 14.7$

    (2) Replace log 14.7 with its numerical value: $37.8 = A - 9.20(1.1673)$

    (3) Perform the multiplication in the last term on the right: $37.8 = A - 10.7$

    (4) Complete the solution in the usual way: $A = $ _____

a) $\log Q = -0.7650$
b) $Q = 0.172$

---

118. In $\boxed{K \log(W + H) = B}$, we want to find the value of "K" when $W = 14.6$, $H = 17.3$, and $B = 5.19$. To do so, we use the steps below:

    (1) Plug in the known values: $K \log(14.6 + 17.3) = 5.19$

    (2) Perform the addition in the grouping: $K \log 31.9 = 5.19$

    (3) Replace log 31.9 with its numerical value: $K(1.5038) = 5.19$
         or
         $1.5038K = 5.19$

    (4) Complete the solution in the usual way: $K = $ _____

$A = 48.5$

$K = 3.45$ (from $\dfrac{5.19}{1.5038}$)

530 Common and Natural Logarithms

119. In $\boxed{\log P = A - \dfrac{B}{T}}$, we want to find the value of "A" when $P = 23,600$, $B = 67.1$, and $T = 5.88$. The steps below can be used:

(1) Plug in the known values: $\quad \log 23,600 = A - \dfrac{67.1}{5.88}$

(2) Replace log 23,600 with its numerical value: $\quad 4.3729 = A - \dfrac{67.1}{5.88}$

(3) Perform the division in the term on the right: $\quad 4.3729 = A - 11.4$

(4) Complete the solution in the usual way: $\quad A = \rule{2cm}{0.4pt}$

---

120. In $\boxed{\log P = A - \dfrac{B}{T}}$, we want to find the value of "T" when $P = 25,700$, $A = 14.0$, and $B = 62.8$. The steps below can be used:

(1) Plug in the known values: $\quad \log 25,700 = 14.0 - \dfrac{62.8}{T}$

(2) Replace log 25,700 with its numerical value: $\quad 4.4099 = 14.0 - \dfrac{62.8}{T}$

(3) Add "-14.0" to both sides: $\quad -9.5901 = -\dfrac{62.8}{T}$

(4) Replace each side with its opposite: $\quad 9.5901 = \dfrac{62.8}{T}$

(5) Complete the solution: $\quad T = \rule{2cm}{0.4pt}$

$A = 15.8$

---

121. In $\boxed{\log G = \dfrac{H - T}{H}}$, we want to find the value of "T" when $H = 600$ and $G = 2.60$. The steps below can be used:

(1) Plug in the known values: $\quad \log 2.60 = \dfrac{600 - T}{600}$

(2) Replace log 2.60 with its numerical value: $\quad 0.4150 = \dfrac{600 - T}{600}$

(3) Multiply both sides by 600 to clear the fraction: $\quad 249 = 600 - T$

(4) Complete the solution: $\quad T = \rule{2cm}{0.4pt}$

$T = 6.55$
From: $\dfrac{62.8}{9.5901}$

$T = 351$

122. In $\boxed{\log G = \dfrac{H - T}{H}}$, we want to find the value of "H" when $G = 2.47$ and $T = 400$. The steps below can be used:

    (1) Plug in the known values: $\log 2.47 = \dfrac{H - 400}{H}$

    (2) Replace $\log 2.47$ with its numerical value: $0.3927 = \dfrac{H - 400}{H}$

    (3) Multiply both sides by "H" to clear the fraction: $0.3927H = H - 400$

    (4) Add "$-1H$" to both sides: $-0.6073H = -400$

    (5) Replace each side with its opposite: $0.6073H = 400$

    (6) Complete the solution: $H = $ _____

123. "$\log A_H$" is not "solved for" in $\boxed{P_H = -\log A_H}$ since its coefficient is <u>negative</u>. Let's find "A_H" when $P_H = 9.16$:

    (1) Plugging in the known value: $9.16 = -\log A_H$

    (2) Taking the opposite of both sides: $-9.16 = \log A_H$
                                 or
                                 $\log A_H = -9.16$

    (3) Complete the solution:

                                $A_H = $ _____

124. Here's the same formula: $\boxed{P_H = -\log A_H}$

    Find the value of "A_H" when $P_H = 4.90$.

                                $A_H = $ _____

Answers:

$H = 659$

$A_H = 6.92 \times 10^{-10}$

$A_H = 1.26 \times 10^{-5}$

(from $\log A_H = -4.9000$)

SELF-TEST 6 (Frames 112-124)

1. $\boxed{P = 20 \log H}$

   If $P = 42.8$, $H = \underline{}$

2. $\boxed{T = -\log Q}$

   If $T = 1.65$, $Q = \underline{}$

3. $\boxed{W = k \log R}$

   If $W = -40$ and $R = 0.01$, $k = \underline{}$

4. $\boxed{F = B - P \log A}$

   If $B = 500$, $F = 200$, and $P = 150$,
   $A = \underline{}$

5. $\boxed{\log E = \dfrac{t+1}{t}}$ If $E = 15.7$, $t = \underline{}$

ANSWERS: 1. $H = 138$ 2. $Q = 0.0224$ 3. $k = 20$ 4. $A = 100$ 5. $t = 5.10$

6-7 EVALUATING "LOG" FORMULAS WHEN A CONVERSION TO POWER-OF-TEN FORM IS REQUIRED

When performing evaluations to find the values of variables in log formulas, sometimes the unknown variable is part of a complicated "log" expression. In such cases, we encounter equations like those below in the evaluation process:

$$\log\left(\frac{P}{1700}\right) = 0.59 \qquad \log(W + 11.1) = 1.66 \qquad \log\left(\frac{650 - R}{R}\right) = 3.67$$

To solve for the letter in each equation above, we must convert the log equation to power-of-ten form. We will discuss the method in this section.

125. The variable "P" is part of a complicated log expression in the equation below:

$$\log\left(\frac{P}{1700}\right) = 0.59$$

To solve for "P", we must convert the log equation to power-of-ten form. That is:

Just as $\log x = 0.73$ means: $x = 10^{0.73}$

$\log\left(\frac{P}{1700}\right) = 0.59$ means: $\frac{P}{1700} = 10^{0.59}$

Having converted the log equation to power-of-ten form, we can solve the new equation by replacing $10^{0.59}$ by a regular number. Since $10^{0.59}$ or $10^{0.5900} = 3.89$, we get:

$$\frac{P}{1700} = 10^{0.59}$$

$$\frac{P}{1700} = 3.89$$

$$P = \underline{\qquad}$$

---

126. Two equations containing complicated log expressions have been converted to power-of-ten form below:

If $\log(M + 3) = 2.56$, then $M + 3 = 10^{2.56}$

If $\log\left(\frac{D - 15.8}{17.9}\right) = -0.67$, then $\frac{D - 15.8}{17.9} = 10^{-0.67}$

Notice in each case that the <u>whole quantity in parentheses</u> is equal to the power of ten.

Convert each log equation below to power-of-ten form:

(a) $\log\left(\frac{300}{t}\right) = 1.33$ (b) $\log(x - 7.88) = -1.66$

---

127. Convert each log equation below to power-of-ten form:

(a) $\log\left(\frac{650 - R}{R}\right) = 3.67$ (b) $\log\left(\frac{988 - B}{16.1 - B}\right) = -1.91$

P = 3.89(1700)
= 6,610

a) $\frac{300}{t} = 10^{1.33}$

b) $x - 7.88 = 10^{-1.66}$

a) $\frac{650 - R}{R} = 10^{3.67}$

b) $\frac{988 - B}{16.1 - B} = 10^{-1.91}$

534 Common and Natural Logarithms

128. To solve for "W" in the equation below, we use the following steps:

$$\log(W + 6.72) = 1.89$$

(1) Convert the equation to power-of-ten form:

$W + 6.72 = 10^{1.89}$ or $10^{1.8900}$

(2) Replace $10^{1.8900}$ with a regular number:

$W + 6.72 = 77.6$

(3) Solve in the usual way:

W = _____

---

129. Let's solve this equation:

$$\log\left(\frac{D - 15.8}{17.9}\right) = -0.67$$

Converting to power-of-ten form, we get:

$\frac{D - 15.8}{17.9} = 10^{-0.67}$ or $10^{-0.6700}$

Replacing $10^{-0.6700}$ with a regular number, we get:

$\frac{D - 15.8}{17.9} = 0.214$

Multiplying both sides by 17.9 to clear the fraction, we get:

$D - 15.8 = 0.214(17.9)$

$D - 15.8 = 3.83$

Complete the solution:

D = _____

W = 70.88 or 70.9

---

130. Let's solve this equation:

$$\log\left(\frac{988 - B}{16.1 - B}\right) = 1.91$$

Converting to power-of-ten form, we get:

$\frac{988 - B}{16.1 - B} = 10^{1.9100}$

Replacing the power of ten with a regular number, we get:

$\frac{988 - B}{16.1 - B} = 81.3$

Multiplying both sides by (16.1 - B) to clear the fraction, we get:

$988 - B = 81.3(16.1 - B)$

$988 - B = 1,310 - 81.3B$

Complete the solution:

B = _____

D = 19.63 or 19.6

B = 4.01

From: 80.3B = 322

131. Here is an evaluation in which the unknown variable is part of a complicated log expression.

In $\boxed{D = 10 \log\left(\dfrac{P_2}{P_1}\right)}$, find the value of "$P_2$" when $D = 25.4$ and $P_1 = 710$.

(1) Plug in the known values: $\qquad 25.4 = 10 \log\left(\dfrac{P_2}{710}\right)$

(2) Isolate the log expression: $\qquad \dfrac{1}{10}(25.4) = \left(\dfrac{1}{10}\right)10 \log\left(\dfrac{P_2}{710}\right)$

$$2.54 = \log\left(\dfrac{P_2}{710}\right)$$

or $\log\left(\dfrac{P_2}{710}\right) = 2.54$

(3) Convert to power-of-ten form and solve for "P_2":

$P_2 = $ _____

$P_2 = 246{,}000$ or 2.46×10^5

Here are the steps:

$\dfrac{P_2}{710} = 10^{2.5400}$

$\dfrac{P_2}{710} = 347$

$P_2 = (710)(347)$

$P_2 = 246{,}000$ (or 2.46×10^5)

---

132. Here is another evaluation in which the unknown variable is part of a complicated log expression: $\boxed{K \log(W + H) = B}$

Find "H" when $K = 14.8$, $W = 11.1$, and $B = 17.6$.

Step 1: Plugging in the known values: $\quad 14.8 \log(11.1 + H) = 17.6$

Step 2: Isolating $\log(11.1 + H)$: $\quad \log(11.1 + H) = \dfrac{17.6}{14.8}$

$\log(11.1 + H) = 1.19$

Now convert to power-of-ten form and solve for "H":

H = _____

$H = 4.4$

Here are the steps:

$11.1 + H = 10^{1.1900}$

$11.1 + H = 15.5$

$H = 4.4$

536   Common and Natural Logarithms

133. Find "R" when C = 0.00545 and D = 570:

$$C = \dfrac{0.02}{\log\left(\dfrac{D-R}{R}\right)}$$

Step 1: Plugging in the known values, we get:

$$0.00545 = \dfrac{0.02}{\log\left(\dfrac{570-R}{R}\right)}$$

Step 2: Solving for the log expression, we get:

$$\log\left(\dfrac{570-R}{R}\right) = \dfrac{0.02}{0.00545}$$

$$\log\left(\dfrac{570-R}{R}\right) = 3.67$$

Now convert to power-of-ten form and solve for "R":

R = _____

---

134. Find "B" when H = 1190, G = 16.9, and K = 2.15:

$$\log\left(\dfrac{H-B}{G-B}\right) = K$$

Plugging in the known values, we get:

$$\log\left(\dfrac{1190-B}{16.9-B}\right) = 2.15$$

Convert to power-of-ten form and complete the solution:

B = _____

R = 0.122

Here are the steps:

$$\dfrac{570-R}{R} = 10^{3.6700}$$

$$\dfrac{570-R}{R} = 4{,}680$$

$$570 - R = 4{,}680R$$

$$570 = 4{,}681R$$

$$R = \dfrac{570}{4{,}681}$$

$$R = 0.122$$

Answer to Frame 134:   B = 8.50   Here are the steps:

(1) $\dfrac{1190-B}{16.9-B} = 10^{2.1500}$ (3) $1190 - B = 141(16.9 - B)$

(2) $\dfrac{1190-B}{16.9-B} = 141$ (4) $1190 - B = 2380 - 141B$

(5) $140B = 1190$

(6) $B = \dfrac{1190}{140} = 8.50$

SELF-TEST 7 (Frames 125-134)

1. Find "P_1" when:
   $A = 16.0$
   $P_2 = 765$

   $$A = 10 \log\left(\frac{P_2}{P_1}\right)$$

   $P_1 = $ _____

2. Find "H" when:
   $A = 250$
   $G = 67.5$
   $R = 21.3$

   $$R \log(A + H) = G$$

   $H = $ _____

ANSWERS: 1. $P_1 = 19.2$ 2. $H = 1,230$

6-8 LOGARITHMIC SCALES

In this section, we will show what is meant by a logarithmic scale. We will discuss the method of generating a logarithmic scale of any length, and then show that the C and D scales on the slide rule are really logarithmic scales. Then in the next section, we will show how logarithmic scales are used in graphing.

135. Here are two different types of scales:

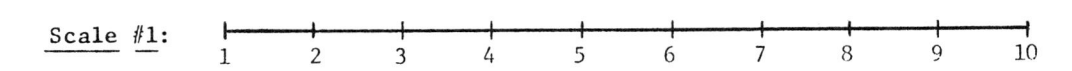

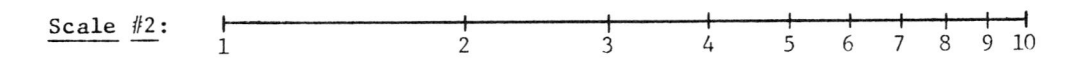

   Scale #1: |—|—|—|—|—|—|—|—|—|
              1 2 3 4 5 6 7 8 9 10

   Scale #2: |——————|————|———|——|—|—|—|—|
              1   2   3   4 5 6 7 8 9 10

Scale #1 is a uniform scale. It is the most common type of scale. It is called a "uniform" scale because each division has the same length. For example, the distances between 1 and 2, 2 and 3, and so on, are equal.

Scale #2 is a logarithmic scale. The C and D scales on the slide rule are logarithmic scales. Logarithmic scales are not uniform scales because each division does not have the same length. For example, the distance between 1 and 2 is longer than the distance between 2 and 3.

On a logarithmic scale, the numbers are crowded together on the _____ (right/left) side.

right

538   Common and Natural Logarithms

136. Any logarithmic scale is generated by plotting the logarithms of the numbers rather than the numbers themselves. To show what we mean, we will construct a logarithmic scale below.

Here is a table of logarithms for the whole numbers between 1 and 10. We have rounded each logarithm to hundredths:

log 1 = 0.00	log 6 = 0.78
log 2 = 0.30	log 7 = 0.85
log 3 = 0.48	log 8 = 0.90
log 4 = 0.60	log 9 = 0.95
log 5 = 0.70	log 10 = 1.00

Since the logs range from 0.00 to 1.00, the scale below ranges from 0 to 1. We have plotted and labeled each log in the table on the scale below:

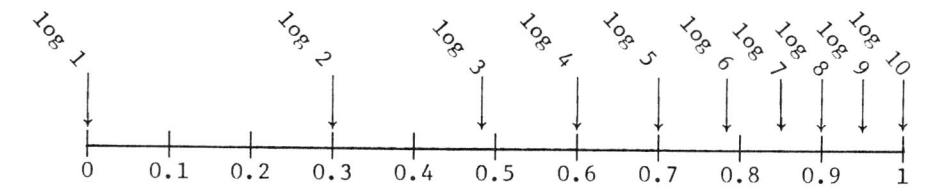

Here is the same scale redrawn. In redrawing it, we have not written the logs (0 to 1) along the scale. Instead, we have written each whole number (1 to 10) at the point where its logarithm lies:

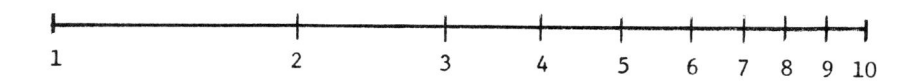

This redrawn scale is a logarithmic scale. Though it is generated by plotting the logarithms of the whole numbers, the whole numbers rather than their logarithms are written along the scale.

(Go to next frame.)

137. There is a difference between plotting intermediate values on a uniform scale and a logarithmic scale.

Here is a uniform scale. To plot 1.5 and 2.5, we simply put them halfway between the whole numbers, as we did below:

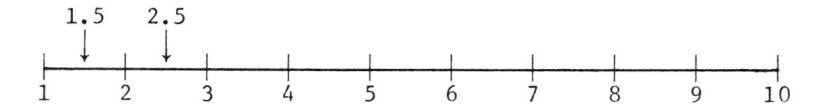

When plotting 1.5 and 2.5 on a logarithmic scale, there is a temptation to simply put them halfway between the whole numbers. However, doing so overlooks the fact that numbers are plotted on a logarithmic scale according to the size of their logarithms and not according to their own size.

Continued on following page.

137. Continued.

To determine where to plot 1.5 and 2.5 on a logarithmic scale, we must return to our original scale on which the logarithms are given. On the scale below, we have plotted:

$$\log 1.5 = 0.18$$
$$\log 2.5 = 0.40$$

Note: (1) $\log 1.5 = 0.18$. Its log is not halfway between $\log 1$ and $\log 2$ which are 0.00 and 0.30 respectively.

(2) $\log 2.5 = 0.40$. Its log is not halfway between $\log 2$ and $\log 3$ which are 0.30 and 0.48 respectively.

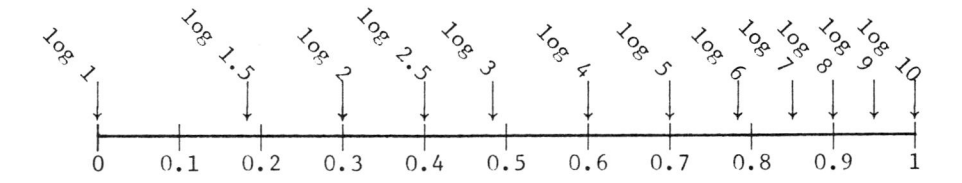

On the logarithmic scale below, we have used the scale above to plot the whole numbers according to the size of their logarithms. Using the scale above, include 1.5 and 2.5 on the scale below.

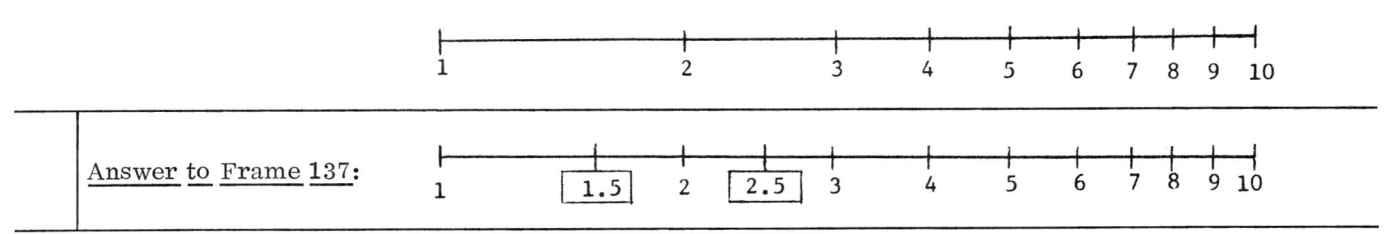

Answer to Frame 137:

138. As an aid in understanding logarithmic scales, we will show a method for constructing one of any length. As an example, we will construct and calibrate a log scale which is <u>5 inches</u> long.

Step 1: We look up the logarithms of the whole numbers between 1 and 10 in a table.

log 1 = 0.0000	log 6 = 0.7782
log 2 = 0.3010	log 7 = 0.8451
log 3 = 0.4771	log 8 = 0.9031
log 4 = 0.6021	log 9 = 0.9542
log 5 = 0.6990	log 10 = 1.0000

Step 2: We multiply each logarithm in the table by <u>5 inches</u>. These products are then rounded to hundredths so that they are easier to plot.

5(log 1) = 0.0000 = 0.00	5(log 6) = 3.8910 ≐ 3.89
5(log 2) = 1.5050 ≐ 1.50	5(log 7) = 4.2255 ≐ 4.23
5(log 3) = 2.3855 ≐ 2.39	5(log 8) = 4.5155 ≐ 4.52
5(log 4) = 3.0105 ≐ 3.01	5(log 9) = 4.7710 ≐ 4.77
5(log 5) = 3.4950 ≐ 3.50	5(log 10) = 5.0000 = 5.00

Continued on following page.

540 Common and Natural Logarithms

138. Continued.

Step 3: We plot whole numbers on a 5-inch scale in the following way:

From the table, 5(log 2) ≐ 1.50". Therefore, the calibration mark for the point "2" on the log scale is located 1.50" from the left end of the log scale. (See sketch below.)

From the table, 5(log 3) ≐ 2.39". Therefore, the calibration mark for the point "3" is located 2.39" from the left end of the log scale. (See sketch below.)

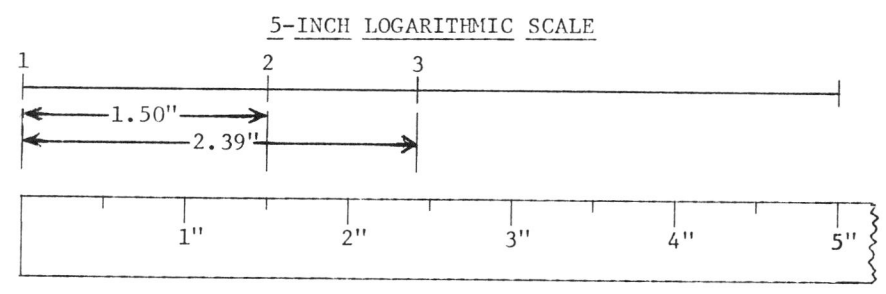

5-INCH LOGARITHMIC SCALE

(a) The calibration mark for point "4" should be located how far from the left end of the scale? _____

(b) The calibration mark for point "____" should lie 4.23" from the left end of the scale.

Answer to Frame 138:   a) 3.01"   b) "7"

---

139. Using the procedure and data of the last frame, calibrate and label this 5-inch log scale. (For convenience, a ruler calibrated in inches and tenths of an inch is shown.)

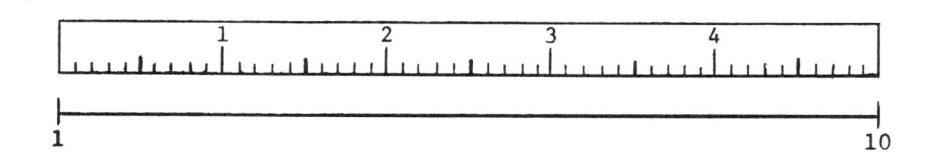

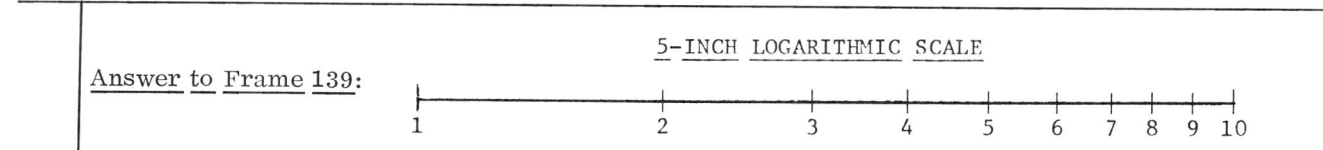

Answer to Frame 139:

5-INCH LOGARITHMIC SCALE

---

140. The method used in calibrating a 5-inch logarithmic scale consisted of these steps:

Step 1: Multiply log 1, log 2, log 3, ... , log 9, log 10 by "5", the scale length.

Step 2: Using the distances calculated in Step 1, locate the calibration marks for 1, 2, 3, ... , 9, 10. These distances were layed off from the left end of the 5" scale.

To calibrate a 10-inch logarithmic scale, a similar procedure would be used. For example, on a 10-inch logarithmic scale, the calibration mark for the point "2" would be located 3.01" from the left end of the scale, since:

$$10"(\log 2) = 10"(0.3010) = 3.01"$$

For a 10-inch logarithmic scale, determine how far each point below should be located from the left end of the scale:

(a) Point "3" _____   (b) Point "5" _____   (c) Point "1" _____   (d) Point "10" _____

Answer to Frame 140:   a) 10"(log 3) = 4.77"   c) 10"(log 1) = 0.00"
                       b) 10"(log 5) = 6.99"   d) 10"(log 10) = 10.00"

141. The C and D scales on the slide rule are logarithmic scales. When you perform a multiplication or division by means of these scales, you are really performing the operation by means of logarithms. However, the logarithms are "built-into" the scales, and so you do not have to look them up in a log table.

The slide-rule setting below illustrates how log scales are used to perform the following multiplication:

$$2 \times 3 = 6$$

$$\log(2 \times 3) = \log 6$$

$$\log 2 + \log 3 = \log 6$$

Note from the diagram that: log 2 (on D) + log 3 (on C) = log 6 (on D)

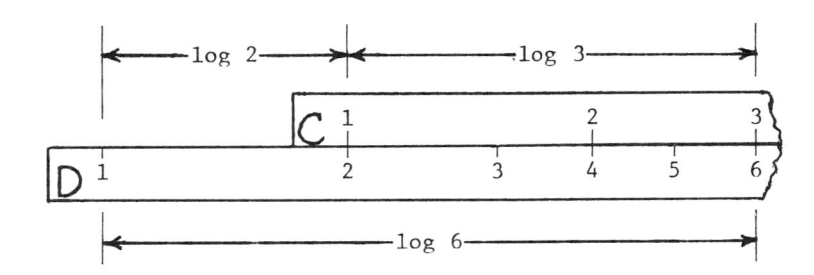

This diagram also illustrates the following division problem:

$$\frac{6}{3} = 2$$

$$\log\left(\frac{6}{3}\right) = \log 2$$

$$\log 6 - \log 3 = \log 2$$

Examine the diagram carefully. From the diagram, is the following relation true? _____

log 6 (on D) − log 3 (on C) = log 2 (on D)

Answer to Frame 141:   Yes.

SELF-TEST 8 (Frames 135-141)

We want to calibrate an 8-inch logarithmic scale. The left end of the scale is labeled "1", and the right end is labeled "10". Determine how far each of the following calibration marks should be located from the left end of the scale. Report each answer to the nearest hundredth of an inch:

1. "2" _____   2. "3" _____   3. "5" _____   4. "9" _____

ANSWERS: 1. 2.41"   2. 3.82"   3. 5.59"   4. 7.63"

542 Common and Natural Logarithms

6-9 SEMI-LOG AND LOG-LOG GRAPHS

In all of the graphs discussed up to this point, the scales on both axes have been uniform scales. Sometimes, however, the scale on one or both of the axes is a logarithmic scale. Graphs of this type are called "semi-log" or "log-log" graphs. Semi-log and log-log graphs are used because they show the relationship between two variables in a more useful and meaningful way than could be shown on a graph with two uniform scales. We will discuss semi-log and log-log graphs in this section.

142. On each graph below, <u>one of the axes has a uniform</u> scale <u>and the other has a logarithmic scale</u>. A graph of either type is called a <u>SEMI-LOG GRAPH</u>. It is called a "semi-log" graph or "half-log" graph because only one of the two scales is a logarithmic scale.

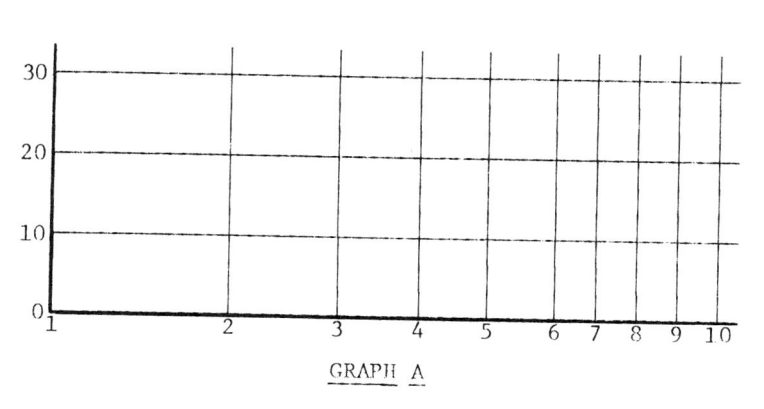

GRAPH A

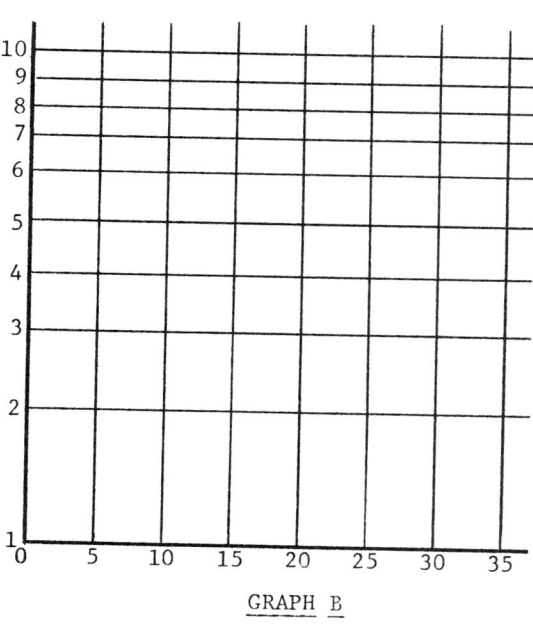

GRAPH B

(a) On graph A above, the logarithmic scale is on the _____ (horizontal/vertical) axis.

(b) On graph B above, the logarithmic scale is on the _____ (horizontal/vertical) axis.

Answer to Frame 142:   a) horizontal   b) vertical

---

143. On the graph at the right, both axes have logarithmic scales. A graph of this type is called a <u>LOG-LOG GRAPH</u>.

(a) If a graph has a logarithmic scale on only one axis, it is called a _____ graph.

(b) If a graph has logarithmic scales on both axes, it is called a _____ graph.

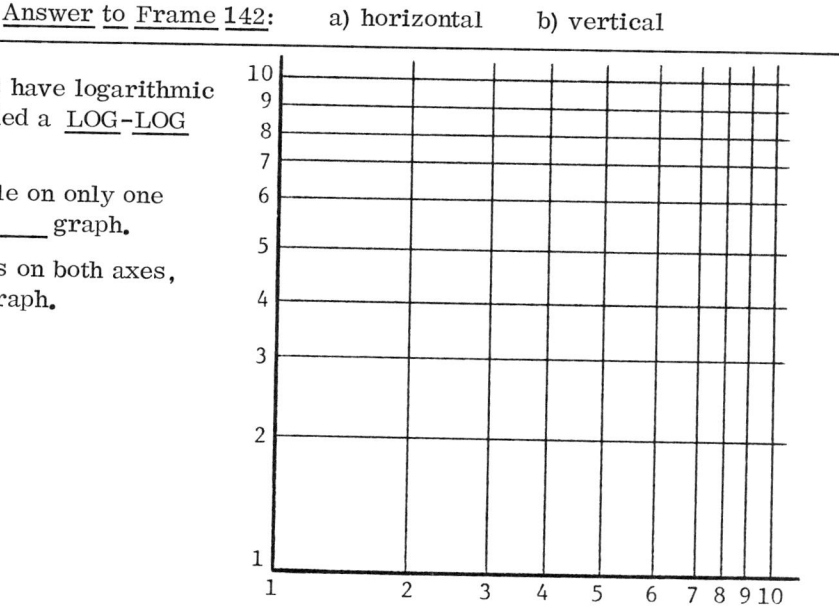

Answer to Frame 143:   a) semi-log   b) log-log

144. Just like any other graph, each point on a semi-log or log-log graph represents one pair of coordinates. Here is a semi-log graph:

The coordinates of point A are (5, 200). Point A represents this fact: When t is 5, w is 200.

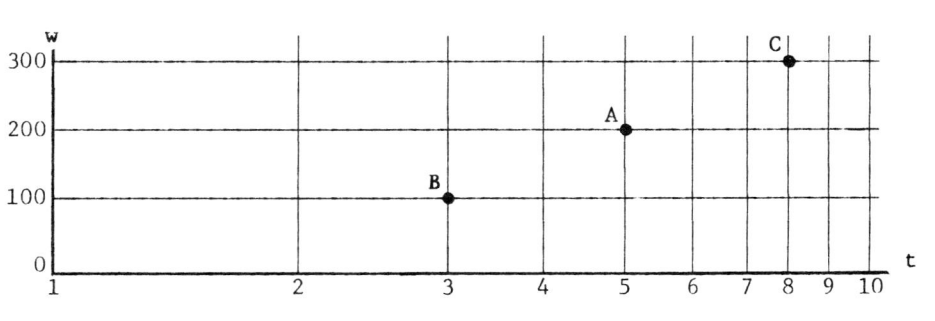

(a) The coordinates of point B are (,   ).

(b) Point C represents this fact: When t is _____, w is _____.

Answer to Frame 144:   a) (3, 100)   b) When t is 8, w is 300.

145. A log-log graph is shown at the right.

The coordinates of point H are (6, 4). Point H represents this fact: When R is 6, D is 4.

(a) Point F represents this fact: When R is _____, D is _____.

(b) The coordinates of point P are (,   ).

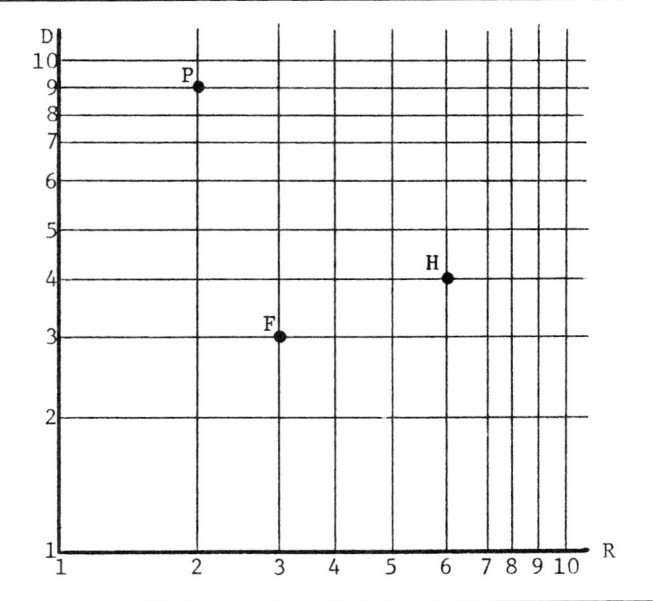

Answer to Frame 145:   a) When R is 3, D is 3.   b) (2, 9)

146. When a log scale is used in graphing, we usually want more values than 1 to 10 on it. Therefore, a single logarithmic scale sometimes consists of two or more compressed log scales. The scale below is a two-cycle log scale. There is one cycle from 1 to 10 and another cycle from 10 to 100. Each cycle is the same length.

On this scale, what number is associated with:

(a) Point A? ____   (c) Point C? ____
(b) Point B? ____   (d) Point D? ____

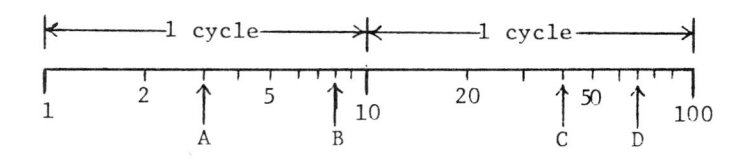

a) 3   c) 40
b) 8   d) 70

544 Common and Natural Logarithms

147. Since a log scale is based on the logarithms of numbers rather than on the numbers themselves, the dividing points for cycles are numbers whose logarithms are whole numbers.

There are three log cycles on the scale below. The dividing points are 1, 10, 100, and 1,000 since:

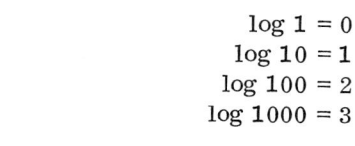

$$\log 1 = 0$$
$$\log 10 = 1$$
$$\log 100 = 2$$
$$\log 1000 = 3$$

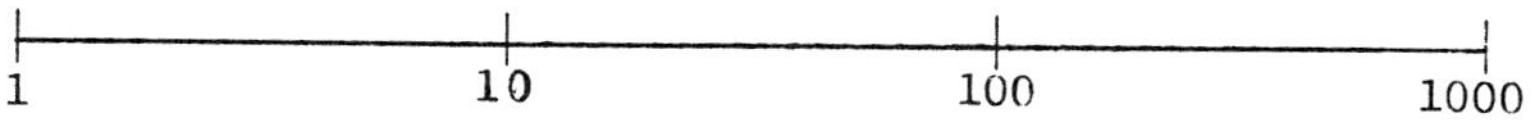

|—————|—————|—————|
1   10   100   1000

A scale of this type is a _____-cycle log scale.

---

148. A log scale does not have to begin at "1". It can begin at any number whose logarithm is a positive or negative whole number. Here are two examples:

Scale #1: |—————|—————|—————|
         10   100   1,000 10,000

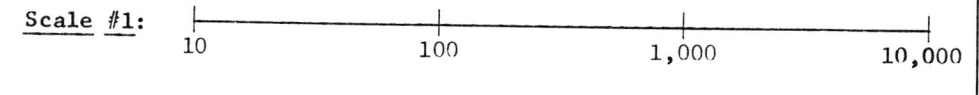

Scale #2: |—————|—————|—————|—————|
         0.01 0.1   1   10   100

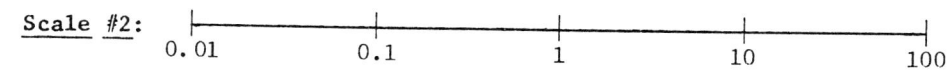

(a) Scale #1 begins at 10, whose log is 1. It is a _____-cycle scale.

(b) Scale #2 begins at 0.01, whose log is -2. It is a _____-cycle scale.

three

---

149. In order to include more cycles on a log scale, the length of each cycle has to be made smaller. When the length of each cycle is made smaller, it is difficult to calibrate and label all of the subdivisions within a cycle.

Here is an example of one common type of calibration and labeling. The scale is a two-cycle log scale. All eight subdivisions (like 2, 3, 4, 5, 6, 7, 8, 9) within each cycle are calibrated. However, only the even subdivisions are labeled since there is not enough room for all of the labels.

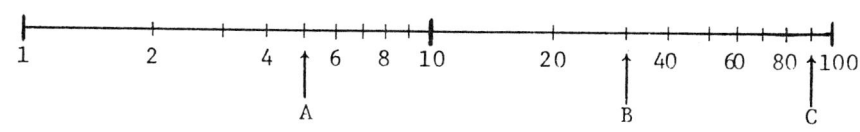

|——|——|—|—|—|—|—|—|——|——|—|—|—|—|—|—|
1   2   4 ↑ 6 8 10   20 ↑ 40 60 80 ↑100
             A   B   C

If they were included, what number would be written at:

(a) Point A? _____   (b) Point B? _____   (c) Point C? _____

a) three
b) four

---

a) 5
b) 30
c) 90

150. Here is an example of another common type of calibration and labeling. The scale is a four-cycle log scale. Only the four-even subdivisions (like 2, 4, 6, 8) are calibrated within each cycle. Though we have labeled the subdivisions within the first cycle, frequently even this labeling is not done.

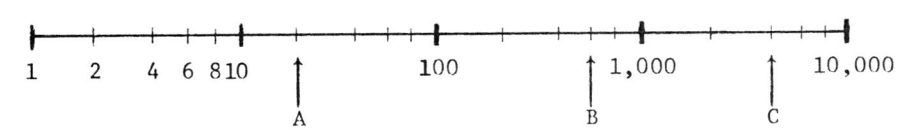

   If they were included, what number would be written at:

   (a) Point A? _____ (b) Point B? _____ (c) Point C? _____

151. Here is a third common type of calibration. The scale is a five-cycle log scale. Only one subdivision (like 2, 20, etc.) is given within each cycle.

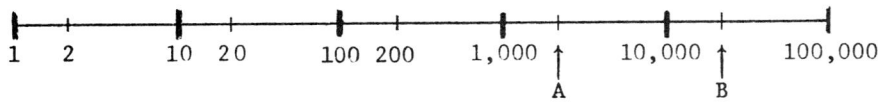

   a) 20
   b) 600
   c) 4,000

   If they were included, what numbers would be written at:

   (a) Point A? _____ (b) Point B? _____

152. This four-cycle log scale begins at 10. Only the even subdivisions within each cycle are calibrated.

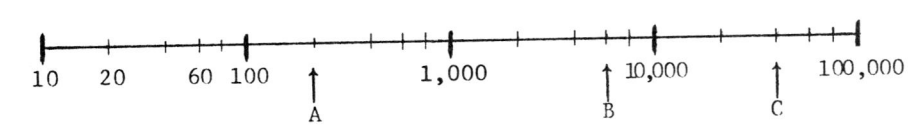

   a) 2,000   b) 20,000

   If they were included, what numbers would be written at:

   (a) Point A? _____ (b) Point B? _____ (c) Point C? _____

153. This five-cycle log scale begins at 0.01. Only one subdivision is given within each cycle.

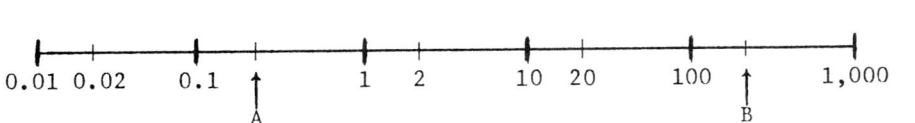

   a) 200
   b) 6,000
   c) 40,000

   If they were included, what numbers would be written at:

   (a) Point A? _____ (b) Point B? _____

   a) 0.2   b) 200

546   Common and Natural Logarithms

154. At the right is a <u>semi-log</u> graph. Note the following:

The horizontal axis is a <u>logarithmic</u> scale with four cycles.

The vertical axis is a uniform scale.

Write the coordinates of the following points. (<u>Note</u>: On this graph and the following ones, you may have to estimate the value on an axis.)

A: _____
B: _____
C: _____

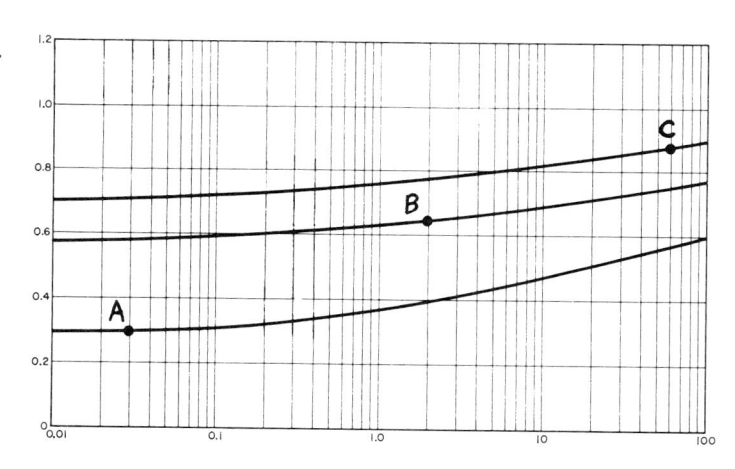

Answer to Frame 154:   A: (0.03, 0.3)   B: (2, 0.65)   C: (60, 0.88)

---

155. In the semi-log graph at the right, the <u>logarithmic scale</u> is again on the <u>horizontal axis</u>.

Note that each cycle on the horizontal axis is divided into <u>five</u> subdivisions.

Write the coordinates of the following points:

A: _____
B: _____
C: _____

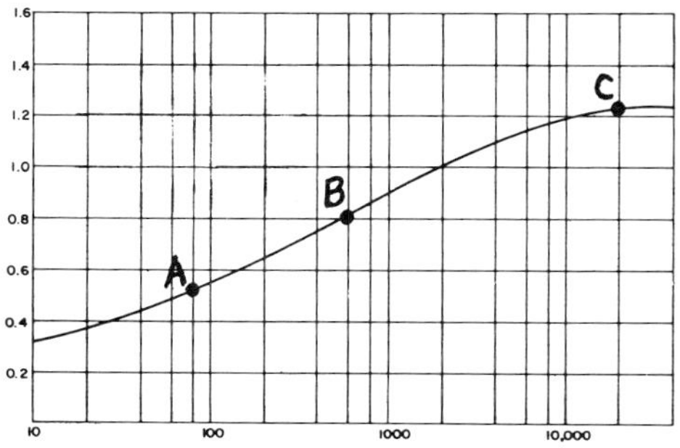

Answer to Frame 155:   A: (80, 0.52)   B: (600, 0.81)   C: (20,000, 1.22)

---

156. In this semi-log graph, the <u>logarithmic scale</u> is on the <u>vertical axis</u>.

Write the coordinates of the following points:

A: _____   B: _____

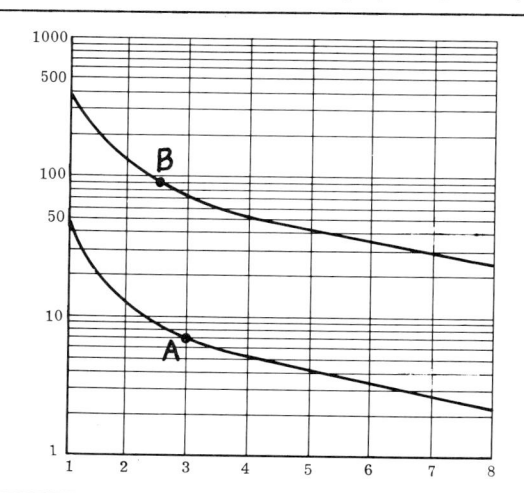

Answer to Frame 156:   A: (3, 7)   B: (2.5, 90)

Common and Natural Logarithms   547

157. This graph is a <u>log-log graph</u>. That is, both axes are <u>logarithmic scales</u>. There are two <u>cycles</u> on each axis.

Write the coordinates of the following points:

A: _____   B: _____

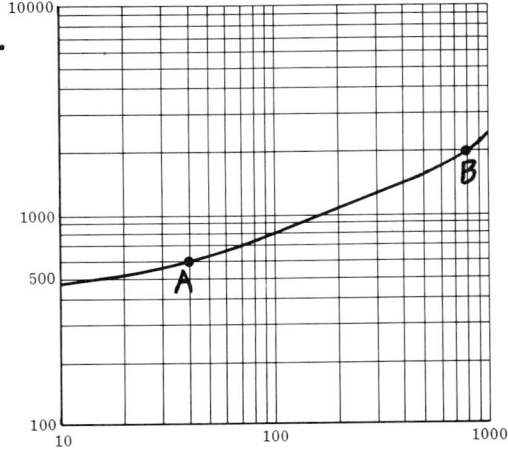

Answer to Frame 157:   A: (40, 600)   B: (800, 2,000)

158. Another <u>log-log</u> graph is shown at the right.

Write the coordinates of the following points:

A: _____

B: _____

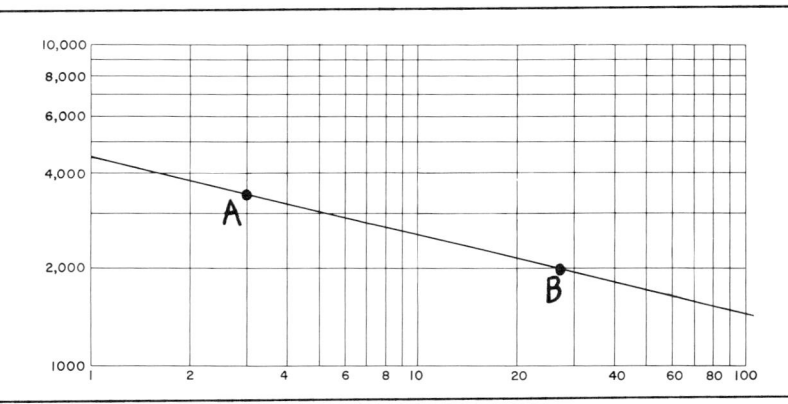

Answer to Frame 158:   A: (3, 3,500)   B: (27, 2,000)

159. In the log-log graph at the right, note that the horizontal axis is labeled with powers of ten. In reading the horizontal axis, you should mentally convert each power of ten to decimal form (such as $10^{-3} = 0.001$). For example, the coordinates of point P are (0.003, 50).

Write the coordinates of the following points:

Q: _____   R: _____

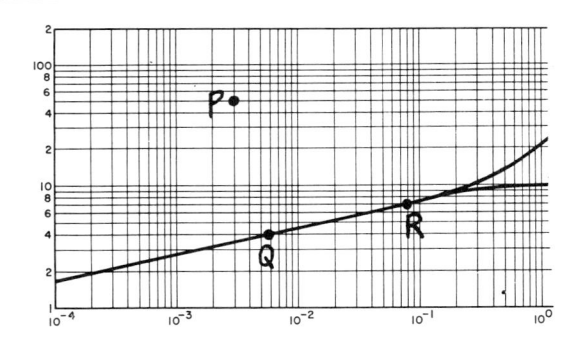

Answer to Frame 159:   Q: (0.006, 4)   R: (0.08, 7)

548 Common and Natural Logarithms

SELF-TEST 9 (Frames 142-159)

1. If both the horizontal and vertical axes of a graph are logarithmic scales, the graph is called a _____ graph.

Refer to the graph shown in Frame 158. On this graph:

2. The horizontal axis is a _____-cycle log scale.

3. The vertical axis is a _____-cycle log scale.

Refer to the graph shown in Frame 155.	Refer to the graph shown in Frame 157.
4. Write the coordinates of the point whose abscissa is 2,000. _____	6. For an abscissa of 500, is the ordinate greater or less than 2,000? _____
5. Write the coordinates of the point whose ordinate is 1.19. _____	7. Write the coordinates of the point whose ordinate is 700. _____

ANSWERS: 1. log-log   2. two-cycle   4. (2,000 , 1.0)   6. Less.
           3. one-cycle   5. (10,000 , 1.19)   7. (70, 700)

6-10 FINDING COMMON LOGARITHMS ON THE SLIDE RULE

If the logarithm of a number is needed, we usually find it by referring to a logarithm table. However, the logarithm of a number can also be found on the slide rule. We will briefly show the method in this section.

160. To find logarithms on the slide rule, we need a logarithmic scale and a uniform scale. The two scales used are the D-scale which is <u>logarithmic</u> and the L-scale which is <u>uniform</u>. The letter "L" on the L-scale refers to the fact that the logarithms of numbers are read on this scale.

The general procedure for finding the logarithm of a number is:

> (1) Set the <u>number</u> on the D-scale.
> (2) Read its <u>logarithm</u> on the L-scale.

Depending upon the type of slide rule you have, the L-scale may be either on the front or back of it. There is a different procedure for each type. Read the procedure below which fits your slide rule.

<u>If the L-scale is on the front of your slide rule</u>:

To find the logarithm of "4": (1) Line up the C and D scales.
                                (2) Set "4" on the D-scale.
                                (3) Read the logarithm of 4 on the L-scale.

Here is a diagram of the settings:

[log 4 = 0.602]

```
L ├──┬──┬──┬──┬──┬──┬──┬──┬──┤
   1  2  3  4  5  6  7  8  9
D ├────┬────┬──┬──┬─┬─┬─┬─┬─┤
   1    2    3  4  5 6 7 8 9 1
           [4]
```

Continued on following page.

Common and Natural Logarithms 549

160. Continued.

If the L-scale is on the back of your slide rule:

There are two hairlines on the back of the slide rule, one on each end. These hairlines are used to read the logarithms of numbers on the L-scale.

To find the logarithm of "4":

(1) Move the left end of C to "4" on D.

(2) Turn the slide rule over. Read the logarithm of "4" on the L-scale under the right back hairline. Again, the reading is "602". Therefore:

$$\boxed{\log 4 = 0.602}$$

Note: The L-scale is "backwards". That is, values increase from right to left.

---

161. Let's review the procedure by finding log 2.4: | Go to next frame.

If the L-scale is on the front of your slide rule:

(1) Set "24" on D.

(2) Read the logarithm on L. log 2.4 = _____

If the L-scale is on the back of your slide rule:

(1) Move the left end of C to "24" on D.

(2) Turn your slide rule over and read the logarithm on L under the right back hairline.   log 2.4 = _____

---

162. | Read this frame only if the L-scale is on the back of your slide rule. Otherwise, go on to the next frame. | log 2.4 = 0.380

Note: Be alert to the fact that your L-scale is backwards. That is, values increase as you move to the left.

Sometimes the logarithm is read under the left back hairline. For example, to find the logarithm of 8.2:

(1) Move the right end of C to "82" on D.

(2) Turn your slide rule over and read the logarithm under the left back hairline.   log 8.2 = _____

---

163. Find these logarithms on your slide rule: | log 8.2 = 0.914

(a) log 2 = _____   (b) log 1.3 = _____

---

164. Find these logarithms on your slide rule: | a) 0.301   b) 0.114

(a) log 5.7 = _____   (b) log 4.1 = _____   (c) log 1.27 = _____

---

165. Do these problems: | a) 0.756
| b) 0.613
(a) log 6.73 = _____   (b) log π = log 3.14 = _____ | c) 0.104

550   Common and Natural Logarithms

166. Just like your log-table, the L-scale contains only the <u>mantissa</u> logarithms. You must supply the <u>characteristics</u>. Do so in the following problems:

(a) log 77,800 = _____   (c) log 0.0525 = _____

(b) log 236 = _____   (d) log 0.000922 = _____

| a) 0.828   b) 0.497 |

167. If you know the logarithm of a number, you can find the number by reversing the procedure. For example:

To find N, if log N = 0.773:

(1) Set "773" on the L-scale.
(2) Read 5.93 on the D-scale.

Using the same procedure, complete these:

(a) Find N, if log N = 0.575.   N = _____
(b) Find P, if log P = 0.787.   P = _____

| a) 4.891
| b) 2.373
| c) −1.280
| From: 0.720 + (−2)
| d) −3.035
| From: 0.965 + (−4)

168. Do these problems on your slide rule:

(a) If log R = 3.358, R = _____   (c) If log Q = 1.874, Q = _____

(b) If log H = 2.722, H = _____   (d) If log P = −1.580, P = _____

| a) N = 3.76
| b) P = 6.12

<u>Answer</u> to <u>Frame</u> <u>168</u>:   a) R = 2,280   b) H = 527   c) Q = 74.8   d) P = 0.0263

When finding the logarithms of numbers, the slide rule is not as accurate as log tables. The slide rule gives logarithms only to <u>three</u> digits; our log table gives them to <u>four</u> digits and other log tables are even more accurate. Therefore, <u>it is foolish to use the slide rule when a log table is available</u>.

Common and Natural Logarithms 551

SELF-TEST 10 (Frames 160-168)

Do the following problems on your slide rule:

1. log 1,840 = _____   4. If log P = 2.143, P = _____

2. log 32.5 = _____   5. If log A = 5.882, A = _____

3. log 0.00624 = _____   6. If log N = -1.325, N = _____

ANSWERS:
1. 3.265
2. 1.512
3. -2.205   From: 0.795 + (-3)
4. P = 139
5. A = 762,000
6. N = 0.0473   From: log N = 0.675 + (-2)

6-11   THE NUMBER "e" AS A BASE

Many formulas contain a number which is called "e". The numerical value of "e" is:

$$e = 2.718281828459\ldots$$

This number "e" appears in formulas related to various physical phenomena, such as radioactive decay, change of voltage or current in a circuit, cooling of objects, rocket flight, chemical reactions, gas expansion, and many others.

Mathematicians have found that many formulas can be written in a simpler form if "e" is used. When "e" appears in a formula, it is used as a <u>base</u> with some exponent. Here are some examples:

$$R = e^{-ft} \qquad A = ke^{-ct}$$

In this section, we will discuss the number "e" and a table of values of e^x and e^{-x}. Then in a later section, we will do some evaluations with formulas in which "e" appears.

---

169. The number "e" is an unending decimal number. The three periods below mean that the decimal part of the number does not end.

$$e = 2.7182818\ldots$$

However, when working with the number "e", we usually round it to 3 digits. Rounded to 3 digits, e = _____.

2.72

552 Common and Natural Logarithms

170. In this and the following frames, round "e" to three digits when using it. That is, use the value: $\boxed{e = 2.72}$

(a) Just as $10^1 = 10$, $e^1 =$ _____. (b) Just as $10^0 = 1$, $e^0 =$ _____.

---

171. The number "e" can be converted to power-of-ten form. That is:

If $e = 2.72$ and $2.72 = 10^{0.4346}$, then $e = 10^{\boxed{}}$

a) $e = 2.72$

b) $e^0 = 1$
(Any base raised to the "0" power is +1.)

---

172. In the last frame, we saw that: $e = 2.72 = 10^{0.4346}$

However, 2.72 is a rounded value for "e". Therefore, there is a slight error in the power-of-ten form above. It is more accurate to use the following power-of-ten form for "e":

$$\boxed{e = 10^{0.4343}}$$

Note: We will always use this more accurate form in our work.

$e = 10^{0.4346}$

(See next frame.)

---

173. By replacing "e" with $10^{0.4343}$, we can convert any power of "e" to power-of-ten form. That is:

$e^5 = (10^{0.4343})^5$ $e^{0.7} = (10^{0.4343})^{0.7}$
$ = 10^{(0.4343)(5)}$ $\phantom{e^{0.7}} = 10^{(0.4343)(0.7)}$
$ = 10^{2.1715}$ $\phantom{e^{0.7}} = 10^{\boxed{}}$

Go to next frame.

---

174. Using the steps in the last frame, convert each of the following to power-of-ten form:

(a) $e^{2.7} = 10^{\boxed{}}$ (b) $e^{-1.4} = 10^{\boxed{}}$

$10^{0.30401}$ or $10^{0.3040}$

---

175. By converting powers of "e" to power-of-ten form, we can find the numerical value of any power of "e". To do so, we simply use the log table.

$$e^{0.7} = 10^{0.3040} = 2.01$$

(a) $e^{2.7} = 10^{1.1726} =$ _____   (b) $e^{-1.4} = 10^{-0.6080} =$ _____

a) $10^{1.1726}$

b) $10^{-0.6080}$

---

a) 14.9   b) 0.247

Common and Natural Logarithms 553

176. Though any power of "e" can be converted to a regular number by converting it to power-of-ten form first, this method is tedious. Since powers of "e" appear in many formulas, mathematicians have prepared a table of powers of "e". A condensed form of this table is given on Page 554. It is entitled TABLE OF e^x AND e^{-x}.

There are three columns of numbers in the table. The "x-column" on the left is a column of exponents.

Here is one entry:

x	e^x	e^{-x}
1.5	4.48	0.223

It means: $e^{1.5} = 4.48$ and $e^{-1.5} = 0.223$

Here is another entry:

x	e^x	e^{-x}
4.5	90.0	0.0111

It means: (a) $e^{4.5} = $ _____ (b) $e^{-4.5} = $ _____

177. Refer to the table to find the following powers of "e":

(a) $e^{3.3} = $ _____   (c) $e^{7.1} = $ _____

(b) $e^{-3.3} = $ _____ (d) $e^{-7.1} = $ _____

a) 90.0   b) 0.0111

178. Examine the table. As the exponent "x" increases:

(a) Does e^x increase or decrease? _____

(b) Does e^{-x} increase or decrease? _____

a) 27.1   c) 1210
b) 0.0369 d) 0.000825

179. As the exponent "x" increases from 0.0 to 10.0:

(a) e^x increases from _____ to _____.

(b) e^{-x} decreases from _____ to _____.

a) It increases.
b) It decreases.

180. Any adjacent pair of values for e^x and e^{-x} in the table are a pair of reciprocals. (That is, their product is +1.)

We can show this fact in two ways:

(1) By means of the law of exponents for multiplication:

$e^{2.1} \times e^{-2.1} = e^{2.1 + (-2.1)} = e^0 = 1$

(2) By multiplying the two values shown in the table:

$(e^{2.1})(e^{-2.1}) = (8.17)(0.122) = $ _____

a) 1.00 to 22,000
b) 1.00 to 0.0000454

181. From the table: $e^{8.5} = 4,910$ and $e^{-8.5} = 0.000204$

(a) $(4,910)(0.000204) = $ _____

(b) 4,910 and 0.000204 are a pair of _____.

0.997, which is approximately "1".

a) 1.00 or 1
b) reciprocals

TABLE OF e^x AND e^{-x}

x	e^x	e^{-x}	x	e^x	e^{-x}	x	e^x	e^{-x}
0.0	1.00	1.00	3.3	27.1	0.0369	6.6	735.	0.00136
0.1	1.11	0.905	3.4	30.0	0.0334	6.7	812.	0.00123
0.2	1.22	0.819				6.8	898.	0.00111
0.3	1.35	0.741	3.5	33.1	0.0302	6.9	992.	0.00101
0.4	1.49	0.670	3.6	36.6	0.0273			
			3.7	40.4	0.0247	7.0	1100	0.000912
0.5	1.65	0.607	3.8	44.7	0.0224	7.1	1210	0.000825
0.6	1.82	0.549	3.9	49.4	0.0202	7.2	1340	0.000747
0.7	2.01	0.497				7.3	1480	0.000676
0.8	2.23	0.449	4.0	54.6	0.0183	7.4	1640	0.000611
0.9	2.46	0.407	4.1	60.3	0.0166			
			4.2	66.7	0.0150	7.5	1810	0.000553
1.0	2.72	0.368	4.3	73.7	0.0136	7.6	2000	0.000500
1.1	3.00	0.333	4.4	81.5	0.0123	7.7	2210	0.000453
1.2	3.32	0.301				7.8	2440	0.000410
1.3	3.67	0.273	4.5	90.0	0.0111	7.9	2700	0.000371
1.4	4.06	0.247	4.6	99.5	0.0101			
			4.7	110.	0.00910	8.0	2980	0.000336
1.5	4.48	0.223	4.8	122.	0.00823	8.1	3290	0.000304
1.6	4.95	0.202	4.9	134.	0.00745	8.2	3640	0.000275
1.7	5.47	0.183				8.3	4020	0.000248
1.8	6.05	0.165	5.0	148.	0.00674	8.4	4450	0.000225
1.9	6.69	0.150	5.1	164.	0.00610			
			5.2	181.	0.00552	8.5	4910	0.000204
2.0	7.39	0.135	5.3	200.	0.00499	8.6	5430	0.000184
2.1	8.17	0.122	5.4	221.	0.00452	8.7	6000	0.000167
2.2	9.02	0.111				8.8	6630	0.000151
2.3	9.97	0.100	5.5	245.	0.00409	8.9	7330	0.000136
2.4	11.0	0.0907	5.6	270.	0.00370			
			5.7	299.	0.00335	9.0	8100	0.000123
2.5	12.2	0.0821	5.8	330.	0.00303	9.1	8960	0.000112
2.6	13.5	0.0743	5.9	365.	0.00274	9.2	9900	0.000101
2.7	14.9	0.0672				9.3	10900	0.0000914
2.8	16.4	0.0608	6.0	403.	0.00248	9.4	12100	0.0000827
2.9	18.2	0.0550	6.1	446.	0.00224			
			6.2	493.	0.00203	9.5	13400	0.0000749
3.0	20.1	0.0498	6.3	545.	0.00184	9.6	14800	0.0000677
3.1	22.2	0.0450	6.4	602.	0.00166	9.7	16300	0.0000613
3.2	24.5	0.0408				9.8	18000	0.0000555
			6.5	665.	0.00150	9.9	19900	0.0000502
						10.0	22000	0.0000454

182. (a) If $x = 5.2$, $e^x = e^{5.2} =$ _____

(b) If $x = 4.1$, $e^{-x} = e^{-4.1} =$ _____

(c) If $x = 1.1$, $e^x = e^{\boxed{}} =$ _____

(d) If $x = 2.7$, $e^{-x} = e^{\boxed{}} =$ _____

a) = 181
b) = 0.0166
c) $e^{1.1} = 3.00$
d) $e^{-2.7} = 0.0672$

---

183. The e^x-table presented in this book is a condensed form of a more complete e^x-table. In a more complete e^x-table, the exponents in the x-column are generally rounded to hundredths. In our condensed table, the exponents in the x-column are rounded to tenths.

Suppose you are asked to find e^x when $x = 1.43$. In a more complete e^x-table, the value "1.43" will appear in the x-column, and the value of $e^{1.43}$ will appear in the e^x-column. However, to use the condensed table in this book, we must round "1.43" to tenths.

If $x = 1.43$, we round to 1.4. Since $e^{1.4} = 4.06$, we say that $e^{1.43} = 4.06$.

Round the following values of "x" to tenths:

(a) 4.66 rounds to _____. (b) 2.437 rounds to _____.

---

184. Round to tenths and evaluate by using the condensed table:

(a) If $x = 9.27$, $e^x = e^{\boxed{}} =$ _____

(b) If $x = 3.655$, $e^{-x} = e^{\boxed{}} =$ _____

(c) If $x = 8.073$, $e^x = e^{\boxed{}} =$ _____

(d) If $x = 1.014$, $e^{-x} = e^{\boxed{}} =$ _____

a) 4.7
b) 2.4

a) $e^{9.3} = 10,900$
b) $e^{-3.7} = 0.0247$
c) $e^{8.1} = 3,290$
d) $e^{-1.0} = 0.368$

---

185. We can also use the e^x-table to convert regular numbers to power-of-"e" form. For example:

(a) To convert 9.97 to power-of-"e" form, we locate 9.97 in the e^x-column and then use the corresponding value of "x".

Therefore: $9.97 = e^{\boxed{}}$

(b) To convert 0.0608 to power-of-"e" form, we locate 0.0608 in the e^{-x}-column and then use the corresponding value of "-x".

Therefore: $0.0608 = e^{\boxed{}}$

a) $e^{2.3}$
b) $e^{-2.8}$

556 Common and Natural Logarithms

186. When using the e^x-table to convert a regular number to power-of-"e" form, be careful to use the correct sign with the exponent.

Since $\boxed{1.00 = e^0}$:
(a) If a number is greater than 1.00, it appears in the e^x-column. Therefore, its exponent in power-of-"e" form is _____ (positive/negative).

(b) If a number is between 0 and 1.00, it appears in the e^{-x}-column. Therefore, its exponent in power-of-"e" form is _____ (positive/negative).

---

187. Use the table to convert each number below to power-of-"e" form. (<u>Be careful of the signs of the exponents.</u>)

(a) $3.32 = e^{\boxed{}}$ (b) $0.135 = e^{\boxed{}}$ (c) $164 = e^{\boxed{}}$

a) positive
b) negative

---

188. Convert each number below to power-of-"e" form:

(a) $0.0166 =$ _____ (b) $4{,}450 =$ _____ (c) $0.000553 =$ _____

a) $e^{1.2}$
b) $e^{-2.0}$ or e^{-2}
c) $e^{5.1}$

---

189. Some regular numbers do not appear exactly as an entry in either the e^x-column or the e^{-x}-column. In such cases, we use the entry closest to the regular number in order to convert it to power-of-"e" form. For example:

(a) 5.01 does not appear as an entry in the table. Since it lies between 4.95 and 5.47 <u>and is closer to 4.95</u>, we say:

$$5.01 = e^{\boxed{}}$$

(b) 0.118 does not appear as an entry in the table. Since it lies between 0.111 and 0.122 <u>and is closer to 0.122</u>, we say:

$$0.118 = e^{\boxed{}}$$

a) $e^{-4.1}$
b) $e^{8.4}$
c) $e^{-7.5}$

---

190. Use the e^x-table to convert each number below to power-of-"e" form:

(a) $13.1 =$ _____ (b) $0.0486 =$ _____ (c) $376 =$ _____

a) $e^{1.6}$
b) $e^{-2.1}$

---

191. Convert each number below to power-of-"e" form:

(a) $0.0108 =$ _____ (b) $10{,}100 =$ _____ (c) $0.000701 =$ _____

a) $e^{2.6}$
b) $e^{-3.0}$ or e^{-3}
c) $e^{5.9}$

---

a) $e^{-4.5}$
b) $e^{9.2}$
c) $e^{-7.3}$

SELF-TEST 11 (Frames 169-191)

1. Rounded to three digits, the value of the number "e" is _____.

2. Complete: $e^{0.6} = (10^{0.4343})^{0.6} = 10^{\boxed{}} = $ _____ (Use log table.)

3. If $e^{5.7}$ is multiplied by $e^{-5.7}$, the product is _____.

Work Problems 4 to 7 by referring to the e^x-table:

| 4. $e^{4.8}=$ _____ | 5. $e^{-4.8}=$ _____ | 6. $e^{0.9}=$ _____ | 7. $e^{-0.9}=$ _____ |

Work Problems 8 to 11 by referring to the e^x-table:

8. If $x = 2.2$, $e^x =$ _____

9. If $x = 6.7$, $e^{-x}=$ _____

10. If $x = 5.48$, $e^x =$ _____

11. If $x = 0.813$, $e^{-x} =$ _____

In Problems 12 to 15, use the e^x-table to convert each number to power-of-"e" form:

| 12. $24.5 =$ _____ | 13. $0.0037 =$ _____ | 14. $1,290 =$ _____ | 15. $0.159 =$ _____ |

ANSWERS:
1. 2.72
2. $10^{0.2606} = 1.82$
3. 1 (from e^0)
4. 122
5. 0.00823
6. 2.46
7. 0.407
8. 9.02
9. 0.00123
10. 245
11. 0.449
12. $e^{3.2}$
13. $e^{-5.6}$
14. $e^{7.2}$
15. $e^{-1.8}$

6-12 EVALUATIONS WITH FORMULAS WHICH CONTAIN "e"

In this section, we will perform some evaluations with formulas which contain "e" as a base. The evaluations in this section are limited to those which can be performed by using the "e^x-table". That is, we will avoid evaluations in which the unknown variable is the exponent, or part of the exponent, of "e". The latter type of evaluation will be examined in a later section.

192. In $\boxed{T = e^x}$, we can find the value of "T" corresponding to any given value of "x". For example:

If $x = 3.5$, $T = e^{3.5} = 33.1$

If $x = 7.6$, $T = e^{7.6} = 2,000$

In $\boxed{R = e^p}$, we can find the value of "R" corresponding to any given value of "p". That is:

(a) If $p = 1.8$, $R = e^{\boxed{}} =$ _____

(b) If $p = 9.2$, $R = e^{\boxed{}} =$ _____

a) $e^{1.8} = 6.05$

b) $e^{9.2} = 9,900$

558   Common and Natural Logarithms

193. In the formula $\boxed{S = e^{-x}}$, the exponent of "e" is "-x".

$\boxed{\text{"-x" means: the \underline{opposite} of x}}$

That is: If $x = 2$, $-x =$ the opposite of $2 = -2$

Since we can obtain the opposite of any quantity by multiplying it by "-1", we can replace "-x" in the formula with "(-1)(x)". That is:

$\boxed{S = e^{-x}}$ is equivalent to $\boxed{S = e^{(-1)(x)}}$

Therefore: If $x = 3.1$, $S = e^{(-1)(3.1)} = e^{-3.1} = 0.0450$

(a) If $x = 1.5$, $S = e^{(-1)(1.5)} = e^{\boxed{}} = \underline{}$

(b) If $x = 0.8$, $S = e^{(-1)(0.8)} = e^{\boxed{}} = \underline{}$

a) $e^{-1.5} = 0.223$

b) $e^{-0.8} = 0.449$

194. In $\boxed{V = e^{-p}}$, we can find the value of "V" corresponding to any given value of "p".

Since $\boxed{V = e^{-p}}$ is equivalent to $\boxed{V = e^{(-1)(p)}}$

(a) If $p = 4$, $V = e^{\boxed{}} = \underline{}$ (b) If $p = 1.2$, $V = e^{\boxed{}} = \underline{}$

a) $e^{-4} = 0.0183$

b) $e^{-1.2} = 0.301$

195. Here is a formula which contains "e" twice: $\boxed{H = \dfrac{e^x + e^{-x}}{2}}$

If $x = 1.4$, we can find the corresponding value of "H" by using the e^x-table. We get:

$$H = \frac{e^{1.4} + e^{-1.4}}{2} = \frac{4.06 + 0.247}{2} = \frac{4.307}{2} = 2.153 \text{ or } 2.15$$

Find the value of "H" when: $x = 0.9$

$H = \underline{}$

196. Here is another type of formula: $\boxed{P = 14.7e^{-0.2h}}$

If $h = 2$, we can find the corresponding value of "P". The steps are:

$P = 14.7e^{(-0.2)(2)}$

$= 14.7e^{-0.4} = 14.7(0.670) = \underline{}$

$H = 1.43$

(Note: We will report only three digits in our answers.)

$P = 9.85$

197. In $\boxed{R = e^{-ft}}$, we can substitute "-1" for the "-" in the exponent. That is:

$\boxed{R = e^{-ft}}$ is equivalent to $\boxed{R = e^{(-1)(ft)}}$

Substitute "-1" for the "-" in each exponent below:

(a) $y = e^{-x}$ (b) $A = ke^{-ct}$ (c) $i = I_o e^{-\frac{Rt}{L}}$

_____

198. In this formula, note that there are two letters in the exponent of "e": $\boxed{R = e^{-ft}}$

Writing it with a "-1" instead of a "-" alone, we get: $R = e^{(-1)(ft)}$

If $f = 0.5$ and $t = 11.6$, we get: $R = e^{(-1)(0.5)(11.6)}$

$= e^{-5.8} =$ _____

a) $y = e^{(-1)(x)}$

b) $A = ke^{(-1)(ct)}$

c) $i = I_o e^{(-1)\left(\frac{Rt}{L}\right)}$

199. Here is the same formula: $\boxed{R = e^{-ft}}$

Find "R" when: $f = 0.1$ and $t = 22$

R = _____

R = .00303

200. In $\boxed{A = ke^{-ct}}$, let's find "A" when: $k = 4$, $c = 0.02$, and $t = 140$

Writing the formula with a "-1" in the exponent, we get: $A = ke^{(-1)(ct)}$

Substituting the known values, we get: $A = 4e^{(-1)(0.02)(140)}$

$= 4e^{-2.8}$

(a) $= 4($ $)$

(b) $=$ _____

R = 0.111

From: $R = e^{(-1)(0.1)(22)}$

$= e^{-2.2} = 0.111$

201. Here is the same formula: $\boxed{A = ke^{-ct}}$

Find "A" when: $k = 6$, $c = 0.04$, and $t = 50$

A = _____

a) $4(0.0608)$

b) 0.243

A = 0.810

From:
$A = 6e^{(-1)(0.04)(50)}$
$= 6e^{-2.0}$
$= 6(0.135)$
$= 0.810$

560 Common and Natural Logarithms

202. Though some formulas which contain "e" look complicated, don't let that bother you. Evaluations with them are merely a matter of computation. Here is an example:

In $\boxed{i = I_o e^{-\frac{Rt}{L}}}$, find "i" when: $I_o = 60$
$R = 15.7$
$t = 3.5$
$L = 22.6$

Writing the exponent with a "-1" instead of the "-", we get: $i = I_o e^{(-1)\left(\frac{Rt}{L}\right)}$

Plugging in the known values, we get: $i = 60 e^{(-1)\left[\frac{(-15.7)(3.5)}{22.6}\right]}$

$= 60 e^{(-1)(2.43)}$

$= 60 e^{-2.43}$

(a) $= 60(\quad\quad)$

(b) $= \underline{\hphantom{xxxx}}$

---

203. In $\boxed{V = E e^{-\frac{t}{RC}}}$, let's find "V" when: $E = 200$
$t = 0.85$
$R = 12$
$C = 0.05$

a) $60(0.0907)$
b) 5.44

Writing an explicit "-1" in the exponent and substituting the known values, we get: $V = 200 e^{(-1)\left[\frac{0.85}{12(0.05)}\right]}$

$= 200 e^{(-1)(1.42)}$

$= 200 e^{-1.42}$

(a) $= 200(\quad\quad)$

(b) $= \underline{\hphantom{xxxx}}$

---

204. In $\boxed{i = \frac{E}{R}\left(1 - e^{-\frac{Rt}{L}}\right)}$, let's find "i" when: $E = 12$
$R = 16$
$t = 0.02$
$L = 0.80$

a) $200(0.247)$
b) 49.4

We get: $i = \frac{12}{16}\left(1 - e^{(-1)\left(\frac{(16)(0.02)}{0.80}\right)}\right)$

$= \frac{3}{4}(1 - e^{-0.4})$

$= (0.75)(1 - 0.670) = (0.75)(0.330) = \underline{\hphantom{xxxx}}$

$i = 0.248$

205. In $\boxed{A = ke^{-ct}}$, "k" is not "solved for". However, it is easy to find the corresponding value of "k" if values for A, c, and t are given. Here is an example:

Find "k" when: $A = 0.810$, $c = 0.04$, and $t = 50$

Plugging in the known values, and simplifying the exponent, we get:
$$0.810 = ke^{(-1)(0.04)(50)}$$
$$0.810 = ke^{-2.0}$$

Using the e^x-table, we can substitute 0.135 for $e^{-2.0}$:
$$0.810 = k(0.135)$$

Complete the solution:

$k = $ _____

---

206. Here is the same formula: $\boxed{A = ke^{-ct}}$

Let's find "k" when: $A = 5.36$, $c = 0.08$, and $t = 5$

Plugging in the known values, we get: $5.36 = ke^{(-1)(0.08)(5)}$
$$5.36 = ke^{-0.4}$$

Substitute a regular number for $e^{-0.4}$ and complete the solution:

$k = $ _____

$k = 6.00$

From: $k = \dfrac{0.810}{0.135}$

---

207. In $\boxed{i = I_o e^{-\frac{Rt}{L}}}$, let's find "$I_o$" when:
$i = 25$
$R = 3.1$
$t = 0.40$
$L = 0.65$

Plugging in the known values, we get: $25 = I_o e^{(-1)\left(\frac{(3.1)(0.40)}{0.65}\right)}$
$$25 = I_o e^{-1.91}$$

Complete the solution:

$I_o = $ _____

$5.36 = k(0.670)$
$k = 8.00$

$I_o = \dfrac{25}{0.150}$

$I_o = 167$

208.

In $\boxed{i = \dfrac{E}{R}\left(1 - e^{-\frac{Rt}{L}}\right)}$, let's find "E" when:
$i = 4.06$
$R = 28.8$
$t = 3.75$
$L = 66.7$

Plugging in the known values, we get:

$$4.06 = \dfrac{E}{28.8}\left(1 - e^{(-1)\left(\frac{(28.8)(3.75)}{66.7}\right)}\right)$$

$$4.06 = \dfrac{E}{28.8}(1 - e^{-1.62})$$

$$4.06 = \dfrac{E}{28.8}(1 - 0.202)$$

$$4.06 = \dfrac{E}{28.8}(0.798)$$

Complete the solution:

$E = $ _____

$E = 147$

From: $E = \dfrac{(4.06)(28.8)}{0.798}$

SELF-TEST 12 (Frames 192-208)

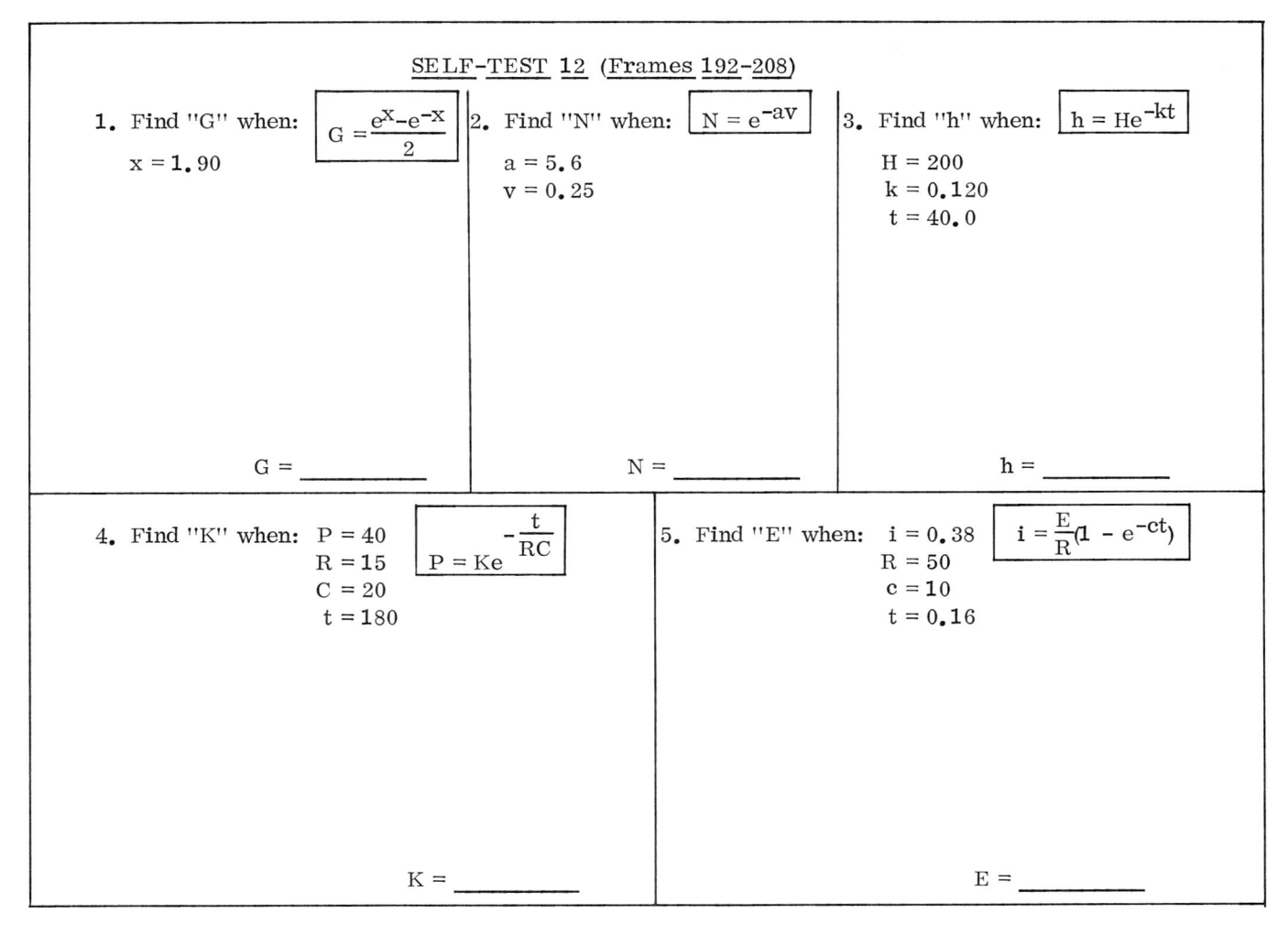

1. Find "G" when: $G = \dfrac{e^x - e^{-x}}{2}$

   $x = 1.90$

   G = _____

2. Find "N" when: $N = e^{-av}$

   $a = 5.6$
   $v = 0.25$

   N = _____

3. Find "h" when: $h = He^{-kt}$

   $H = 200$
   $k = 0.120$
   $t = 40.0$

   h = _____

4. Find "K" when: $P = Ke^{-\frac{t}{RC}}$

   $P = 40$
   $R = 15$
   $C = 20$
   $t = 180$

   K = _____

5. Find "E" when: $i = \dfrac{E}{R}(1 - e^{-ct})$

   $i = 0.38$
   $R = 50$
   $c = 10$
   $t = 0.16$

   E = _____

ANSWERS:
1. G = 3.27
2. N = 0.247
3. h = 1.65
4. K = 73
5. E = 23.8 or 24

6-13 NATURAL (BASE "e") LOGARITHMS

The logarithms we have discussed up to this point are called <u>common</u> logarithms. <u>Common</u> logarithms are based on powers of ten. The <u>common</u> <u>logarithm</u> of a number is simply <u>the</u> <u>exponent</u> when it is written in <u>power-of-ten</u> form. For example:

$$\text{Since } 221 = 10^{2.3444}, \quad \log 221 = 2.3444$$

Any positive number can also be written in power-of-"e" form. Therefore, there is a second type of logarithm called "<u>natural</u>" logarithms. <u>Natural</u> logarithms are based on powers of "e". The <u>natural</u> <u>logarithm</u> of a number is simply <u>the</u> <u>exponent</u> when it is written in <u>power-of-"e"</u> form. We will discuss <u>natural</u> logarithms in this section.

564 Common and Natural Logarithms

209. We have seen that we can write any positive number in power-of-ten form. For example:

$$11 = 10^{1.0414} \qquad 90 = 10^{1.9542} \qquad 221 = 10^{2.3444}$$

Similarly, we can write any positive number in power-of-"e" form. We use the e^x-table to do so. That is:

(a) $11 = e^{\boxed{}}$ (b) $90 = e^{\boxed{}}$ (c) $221 = e^{\boxed{}}$

---

210. Here is the same number written in power-of-ten form and power-of-"e" form:

$$221 = 10^{2.3444} \qquad \text{and} \qquad 221 = e^{5.4}$$

(a) The <u>common</u> logarithm of a number is the <u>exponent</u> when it is written in <u>power-of-ten</u> form.

The <u>common</u> logarithm of 221 is _____.

(b) The <u>natural</u> logarithm of a number is the <u>exponent</u> when it is written in <u>power-of-"e"</u> form.

The <u>natural</u> logarithm of 221 is _____.

a) $11 = e^{2.4}$
b) $90 = e^{4.5}$
c) $221 = e^{5.4}$

---

211. Since $33.1 = e^{3.5}$ and $33.1 = 10^{1.5198}$

(a) The <u>common</u> logarithm of 33.1 is _____.

(b) The <u>natural</u> logarithm of 33.1 is _____.

a) 2.3444
b) 5.4

---

212. (a) <u>Common</u> logarithms are based on powers of _____.

(b) <u>Natural</u> logarithms are based on powers of _____.

a) 1.5198
b) 3.5

---

213. Any power-of-ten or power-of-"e" statement can be written in logarithmic notation.

Here is a base-ten statement: $602 = 10^{2.7796}$

Here is the same statement in logarithmic notation:

The <u>common</u> logarithm of 602 is 2.7796, or simply:

$$\log 602 = 2.7796$$

Here is a base-"e" statement: $602 = e^{6.4}$

Here is the same statement in logarithmic notation:

The <u>natural</u> logarithm of 602 is 6.4, or simply:

$$\ln 602 = 6.4$$

Notice the different meaning of the symbols "log" and "ln":

"<u>log</u>" means "the <u>common</u> logarithm of".

"<u>ln</u>" means "the _____ logarithm of".

a) ten or 10
b) "e"

---

natural

214. The symbol "log" is used when referring to common or base-ten logarithms.

The symbol "ln" is used when referring to natural or base-e logarithms.

Write each of the following in logarithmic notation:
  (a) $11 = 10^{1.0414}$ _____
  (b) $11 = e^{2.4}$ _____

215. Write each of the following in logarithmic notation:
  (a) $2.01 = e^{0.70}$ _____
  (b) $e^{-1.80} = 0.165$ _____
  (c) $10^{-2.2790} = 0.00526$ _____

a) $\log 11 = 1.0414$
b) $\ln 11 = 2.4$

216. Any logarithmic statement can be written in exponential form. For example:

If $\log 3.81 = 0.5809$, then: $3.81 = 10^{0.5809}$.

If $\ln 73.7 = 4.30$, then: $73.7 = e^{4.30}$.

Write each of the following in exponential form:
  (a) $\ln 21.0 = 3.0445$ _____
  (b) $\ln 4.41 = 1.4839$ _____
  (c) $\ln 0.0910 = -2.3969$ _____
  (d) $\log 4.41 = 0.6444$ _____

a) $\ln 2.01 = 0.70$
b) $\ln 0.165 = -1.80$
c) $\log 0.00526 = -2.2790$

217. Convert each exponential statement to a logarithmic statement and vice versa:
  (a) $e^{4.6052} = 100$ _____
  (b) $\log 0.782 = -0.1068$ _____
  (c) $\ln 0.900 = -0.1054$ _____
  (d) $0.102 = e^{-2.2828}$ _____

a) $21.0 = e^{3.0445}$
b) $4.41 = e^{1.4839}$
c) $0.0910 = e^{-2.3969}$
d) $4.41 = 10^{0.6444}$

566   Common and Natural Logarithms

218. Base-ten logarithms are called "common" logarithms.
Base-e logarithms are called "natural" logarithms.

(a) Is log 418 = 2.6212 a common or natural logarithm statement? _____

(b) Is ln 43.0 = 3.7612 a common or natural logarithm statement? _____

a) ln 100 = 4.6052
b) 0.782 = 10$^{-0.1068}$
c) 0.900 = e$^{-0.1054}$
d) ln 0.102 = -2.2828

---

219. (a) Since 1 = e^0, ln 1 = _____.

(b) Since 2.72 = e$^{1.0}$, ln 2.72 = _____.

(c) Since 20.1 = e$^{3.0}$, ln 20.1 = _____.

a) Common.
b) Natural.

---

220. Mathematicians have prepared a lengthy table of natural logarithms. Such a table can be found in mathematical handbooks.

A condensed version of such a table is included on Page 567 and is entitled TABLE OF NATURAL LOGARITHMS.

Examine the table and notice these points:

(1) There are two columns headed "N" and "ln N".

"N" stands for "number".

"ln N" stands for the "natural logarithm" of the number.

(2) The numbers in the "ln N" column are the entire natural logarithms. This table is not like the common log table in which mantissas are given and you have to determine the characteristic.

From the table, you can see that the natural logarithm of 1 is 0.0000. This entry reflects the fact that 1 = e^0 or e$^{0.0000}$.

(a) If a regular number is less than 1, its natural logarithm is _____ (positive/negative).

(b) If a regular number is greater than 1, its natural logarithm is _____ (positive/negative).

a) 0
b) 1.0 or 1
c) 3.0 or 3

---

a) negative
b) positive

TABLE OF NATURAL LOGARITHMS

N	ln N	N	ln N	N	ln N	N	ln N
0.00	-------	5.00	1.6094	10.0	2.3026	60.0	4.0943
0.10	-2.3026	5.10	1.6292	11.0	2.3979	61.0	4.1109
0.20	-1.6094	5.20	1.6487	12.0	2.4849	62.0	4.1271
0.30	-1.2040	5.30	1.6677	13.0	2.5650	63.0	4.1431
0.40	-0.9163	5.40	1.6864	14.0	2.6391	64.0	4.1589
0.50	-0.6932	5.50	1.7048	15.0	2.7080	65.0	4.1744
0.60	-0.5108	5.60	1.7228	16.0	2.7726	66.0	4.1896
0.70	-0.3567	5.70	1.7405	17.0	2.8332	67.0	4.2047
0.80	-0.2231	5.80	1.7579	18.0	2.8904	68.0	4.2195
0.90	-0.1054	5.90	1.7750	19.0	2.9444	69.0	4.2341
1.00	0.0000	6.00	1.7918	20.0	2.9957	70.0	4.2485
1.10	0.0953	6.10	1.8083	21.0	3.0445	71.0	4.2627
1.20	0.1823	6.20	1.8246	22.0	3.0910	72.0	4.2767
1.30	0.2624	6.30	1.8406	23.0	3.1355	73.0	4.2905
1.40	0.3365	6.40	1.8563	24.0	3.1780	74.0	4.3041
1.50	0.4055	6.50	1.8718	25.0	3.2189	75.0	4.3175
1.60	0.4700	6.60	1.8871	26.0	3.2581	76.0	4.3307
1.70	0.5306	6.70	1.9021	27.0	3.2958	77.0	4.3438
1.80	0.5878	6.80	1.9169	28.0	3.3322	78.0	4.3567
1.90	0.6418	6.90	1.9315	29.0	3.3673	79.0	4.3694
2.00	0.6932	7.00	1.9459	30.0	3.4012	80.0	4.3820
2.10	0.7419	7.10	1.9601	31.0	3.4340	81.0	4.3944
2.20	0.7885	7.20	1.9741	32.0	3.4657	82.0	4.4067
2.30	0.8329	7.30	1.9879	33.0	3.4965	83.0	4.4188
2.40	0.8755	7.40	2.0015	34.0	3.5264	84.0	4.4308
2.50	0.9163	7.50	2.0149	35.0	3.5554	85.0	4.4426
2.60	0.9555	7.60	2.0282	36.0	3.5835	86.0	4.4544
2.70	0.9932	7.70	2.0412	37.0	3.6109	87.0	4.4659
2.80	1.0296	7.80	2.0541	38.0	3.6376	88.0	4.4773
2.90	1.0647	7.90	2.0669	39.0	3.6636	89.0	4.4886
3.00	1.0986	8.00	2.0794	40.0	3.6889	90.0	4.4998
3.10	1.1314	8.10	2.0919	41.0	3.7136	91.0	4.5109
3.20	1.1632	8.20	2.1041	42.0	3.7377	92.0	4.5218
3.30	1.1939	8.30	2.1163	43.0	3.7612	93.0	4.5326
3.40	1.2238	8.40	2.1282	44.0	3.7842	94.0	4.5433
3.50	1.2528	8.50	2.1401	45.0	3.8067	95.0	4.5539
3.60	1.2809	8.60	2.1518	46.0	3.8286	96.0	4.5644
3.70	1.3083	8.70	2.1633	47.0	3.8502	97.0	4.5747
3.80	1.3350	8.80	2.1748	48.0	3.8712	98.0	4.5850
3.90	1.3610	8.90	2.1860	49.0	3.8918	99.0	4.5951
4.00	1.3863	9.00	2.1972	50.0	3.9120	100.	4.6052
4.10	1.4110	9.10	2.2083	51.0	3.9318	150.	5.0106
4.20	1.4351	9.20	2.2192	52.0	3.9512	200.	5.2983
4.30	1.4586	9.30	2.2300	53.0	3.9703	250.	5.5215
4.40	1.4816	9.40	2.2407	54.0	3.9890	300.	5.7038
4.50	1.5041	9.50	2.2513	55.0	4.0073	350.	5.8579
4.60	1.5261	9.60	2.2618	56.0	4.0254	400.	5.9915
4.70	1.5476	9.70	2.2721	57.0	4.0430	450.	6.1092
4.80	1.5686	9.80	2.2824	58.0	4.0604	500.	6.2146
4.90	1.5892	9.90	2.2925	59.0	4.0775		

568   Common and Natural Logarithms

221. As we said, the natural log table in this book is a condensed form of a table found in mathematics handbooks. In our condensed version:

(1) "N" increases by <u>tenths</u> from 0.00 to 10.00.

(2) "N" increases by <u>units</u> (<u>ones</u>) from 10.0 to 100.

(3) "N" increases by <u>50's</u> from 100 to 500.

One entry in the table is: | If N = 18, ln N = 2.8904 |

This entry means: ln 18 = 2.8904 or $18 = e^{2.8904}$

Another entry in the table is: | If N = 0.7, ln N = -0.3567 |

This entry means: ln ____ = _____ or ____ = e^____

---

ln 0.7 = -0.3567

or

$0.7 = e^{-0.3567}$

---

222. Using the table, find these natural logarithms:

(a) ln 0.1 = _____   (b) ln 8.4 = _____   (c) ln 350 = _____

---

223. To use our condensed version of the table, we round to the nearest entry. For example:

ln 74.8 ≐ ln 75

≐ 4.3175

The values we obtain when rounding to the nearest entry are only approximate. However, they are accurate enough for the purposes of this book. From now on, we will not use the "is approximately equal to" or "≐" symbol.

Complete these: (a) ln 14.2 = _____   (b) ln 77.6 = _____

---

a) -2.3026

b) 2.1282

c) 5.8579

---

224. If you know the natural log of a number, you can use the table to find the number.

If ln N = 3.3673, N = 29

(a) If ln A = 1.9879, A = _____   (b) If ln P = -0.5108, P = _____

---

a) 2.6391
(Same as ln 14.)

b) 4.3567
(Same as ln 78.)

---

225. When looking up entries in the "ln N" column, choose the one which is closest to the given value:

(a) If ln N = 3.2190, N = _____   (c) If ln K = -1.6021, K = _____

(b) If ln T = 5.9433, T = _____   (d) If ln Q = 1.6021, Q = _____

---

a) 7.30   b) 0.60

---

226. From the table, we know this fact: | If N = 0.80, ln N = -0.2231. |

Convert the statement above to exponential (base "e") form:

---

a) 25.0   c) 0.20

b) 400   d) 5.00

$0.80 = e^{-0.2231}$

227. The same information is contained in both the Table of Natural Logarithms and the Table of e^x and e^{-x}. Since this fact may not be clear to you, we will give a few examples:

In the e^x-table, we find this entry:

$$\boxed{\text{If } x = 4.5, \ e^x = 90.0}$$

This entry means either: (1) $\quad 90 = e^{4.5}$
or (2) $\ \ln 90 = 4.5$
or (3) If $N = 90$, $\ln N = 4.5$

In the natural-log table, we find this entry:

$$\boxed{\text{If } N = 90, \ \ln N = 4.4998}$$

The two tables give these values (4.5 and 4.4998) for the natural logarithm of 90. Are these two values approximately equal? _____

---

228. Here is another example comparing the two tables:

In the natural-log table, we find this entry:

$$\boxed{\text{If } N = 200, \ \ln N = 5.2983}$$

This entry means either: (1) $\ln 200 = 5.2983$
or (2) $\quad 200 = e^{5.2983}$

In the e^x-table, we find this entry:

$$\boxed{\text{If } e^x = 200, \ x = 5.3}$$

This entry means either: (1) $\quad 200 = e^{5.3}$
or (2) $\ln 200 = 5.3$

The two tables give these values (5.2983 and 5.3) for the natural log of 200. Are the two values approximately equal? _____

Yes.

Yes.

570   Common and Natural Logarithms

SELF-TEST 13 (Frames 209-228)

Convert each statement to <u>logarithmic</u> form:

1. $83.7 = 10^{1.9227}$ _____

2. $299 = e^{5.70}$ _____

Convert each statement to <u>exponential</u> form:

3. $\ln 43.0 = 3.7612$ _____

4. $\log 3{,}960 = 3.5977$ _____

Work Problems 5 to 12 by referring to the "Table of Natural Logarithms":

5. $\ln 7.90 =$ _____ | 6. $\ln 0.40 =$ _____ | 7. $\ln 152 =$ _____ | 8. $\ln 17.9 =$ _____

9. If $\ln Q = -0.2231$, $Q =$ _____

10. If $\ln P = 3.800$, $P =$ _____

11. If $\ln A = 1.905$, $A =$ _____

12. If $\ln G = -1.5000$, $G =$ _____

13. Write this "ln table" entry in <u>exponential</u> form: [If $N = 30.0$, $\ln N = 3.4012$] _____

14. Write this "e^x-table" entry in <u>logarithmic</u> form: [If $x = 3.4$, $e^x = 30$] _____

ANSWERS:

1. $\log 83.7 = 1.9227$
2. $\ln 299 = 5.70$
3. $43.0 = e^{3.7612}$
4. $3{,}960 = 10^{3.5977}$
5. 2.0669
6. -0.9163
7. 5.0106
8. 2.8904
9. $Q = 0.80$
10. $P = 45.0$
11. $A = 6.70$
12. $G = 0.20$
13. $e^{3.4012} = 30.0$
14. $\ln 30 = 3.4$

6-14   EVALUATING "Ln" FORMULAS

Though the "e^x-table" and the "natural-log table" really contain the same information in a different form, there is a reason for having both tables. The reason is this: Some formulas contain "e" as a base and others contain the symbol "ln". Just as the "e^x-table" is more useful when evaluating formulas in which "e" appears, the "natural-log table" is more useful when evaluating formulas in which "ln" appears.

We will do some evaluations with "ln" formulas in this section. However, we will avoid evaluations in which the variable is part of a complicated "ln" expression. Evaluations of the latter type will be discussed in the next section.

229. Here is a formula in which the symbol "ln" appears: $\boxed{D = k \ln P}$

Let's find the value of "D" when k = 10 and P = 40. The steps are:

(1) Plug in the known values: D = 10 ln 40

(2) Use the natural-log table to substitute a regular number for "ln 40":   D = 10(3.6889)

(3) Complete the solution: D = _____

---

230. In $\boxed{D = k \ln P}$, find "D" when: k = 5.60, P = 7.80

D = _____

Answer: D = 36.889 or 36.9

(We will report three digits in our answers.)

---

231. Here is a somewhat more complicated formula: $\boxed{v = c \ln\left(\dfrac{M}{m}\right)}$

Find "v" when: c = 1.92, M = 850, and m = 118

Plugging in the known values, we get: $v = 1.92 \ln\left(\dfrac{850}{118}\right)$

$= 1.92 \ln 7.20$

$= 1.92(1.9741) =$ _____

Answer:
D = 5.60 ln 7.80
= 5.60(2.0541)
= 11.5

---

232. In $\boxed{v = c \ln\left(\dfrac{M}{m}\right)}$, find "v" when: c = 1.87, M = 874, m = 92

v = _____

Answer: v = 3.79

---

233. Here is a different type:

Let's find "t" when: $\boxed{t = \dfrac{\ln\left(\dfrac{A_o}{A}\right)}{k}}$

$A_o = 40$, A = 20, k = 0.33

Plugging in the known values, we get: $t = \dfrac{\ln\left(\dfrac{40}{20}\right)}{0.33}$

$= \dfrac{\ln 2}{0.33}$

$= \dfrac{0.6932}{0.33} =$ _____

Answer:
$v = 1.87 \ln\left(\dfrac{874}{92}\right)$
= 1.87 ln(9.50)
= 1.87(2.2513)
= 4.21

---

t = 2.1

234. In $\boxed{k = \dfrac{\ln\left(\dfrac{A_o}{A}\right)}{t}}$, find "k" when: $A_o = 32.7$
$A = 8.35$
$t = 0.545$

k = _____

235. In $\boxed{D = k \ln P}$, let's find "k" when: $D = 40$ and $P = 50$

Plugging in the known values, we get: $40 = k \ln 50$

Substituting a regular number for "ln 50", we get: $40 = k(3.9120)$
or
$40 = 3.912k$

Complete the solution:

k = _____

$k = \dfrac{\ln\left(\dfrac{32.7}{8.35}\right)}{0.545}$

$= \dfrac{\ln 3.92}{0.545}$

$= \dfrac{1.3610}{0.545}$

$= 2.50$

236. In $\boxed{w = \dfrac{d}{\ln P}}$, let's find the value of "d" when: $w = 0.91$ and $P = 80$

Plugging in the known values, we get: $0.91 = \dfrac{d}{\ln 80}$

Substituting a regular number for "ln 80", we get: $0.91 = \dfrac{d}{4.3820}$

Complete the solution:

d = _____

$k = \dfrac{40}{3.912}$

$= 10.2$

237. In $\boxed{\ln\left(\dfrac{A_o}{A}\right) = kt}$,

let's find "k" when:

$A_o = 50$
$A = 10$
$t = 1.85$

The steps are given on the right.

$\ln\left(\dfrac{50}{10}\right) = k(1.85)$

$\ln 5 = 1.85k$

$1.6094 = 1.85k$

k = _____

$d = (0.91)(4.3820)$
$= 3.99$

$k = 0.870$ or 0.87

238. In $\boxed{\ln R = t}$, we can find "ln R" and then use the natural-log table to find "R" for any specific value of "t". That is:

(a) If t = 3.7377, ln R = 3.7377 and R = _____.

(b) If t = 1.3083, ln R = 1.3083 and R = _____.

---

239. In $\boxed{\ln h = b}$, we can find "ln h" and "h" for any specific value of "b". That is:

(a) If b = 3.91 (or 3.9100), ln h = 3.9100 and h = _____.

(b) If b = 1.4 (or 1.4000), ln h = 1.4000 and h = _____.

a) R = 42.0 or 42
b) R = 3.70 or 3.7

---

240. In $\boxed{\ln A = \dfrac{h}{t}}$, we can find "ln A" and "A" for any specific values of "h" and "t". That is:

(a) If h = 60 and t = 20,
$\ln A = \dfrac{60}{20} = 3$ (or 3.0000) and A = _____

(b) If h = 14.9 and t = 4.86,
$\ln A = \dfrac{14.9}{4.86} = 3.07$ (or 3.0700) and A = _____

a) h = 50
b) h = 4.1

---

241. In $\boxed{D = k \ln P}$, let's find "P" when D = 40 and k = 20. The steps are:

Plug in the known values: 40 = 20 ln P

Solve for "ln P" and then "P":

ln P = _____

P = _____

a) A = 20
b) A = 22

---

242. In $\boxed{w = \dfrac{d}{\ln P}}$, let's find the value of "P" when w = 0.88 and d = 4.27. The steps are:

Plug in the known values: $0.88 = \dfrac{4.27}{\ln P}$

Now solve for "ln P" and then "P":

ln P = _____

P = _____

ln P = 2 or (2.0000)
P = 7.4

$\ln P = \dfrac{4.27}{0.88}$

$= 4.85$ (or 4.8500)

P = 150

574 Common and Natural Logarithms

243. In the formula $\boxed{Q = -\ln B}$, "$-\ln B$" means "the opposite of $\ln B$". To find the value of "B" when $Q = 1.1$, we use the following steps:

(1) Plug in the known value: $1.1 = -\ln B$

(2) Take the opposite of both sides: $-1.1 = \ln B$
or
$\ln B = -1.1000$

(3) Use the "ln" table to find "B": $B = 0.30$ or 0.3

Following the steps above, find B when $Q = -0.8$:

B = _____

---

244. Here is the same formula: $\boxed{Q = -\ln B}$ To find "Q" for any specific value of "B", it is easier to rewrite the equation in the following equivalent way:

$$Q = (-1)(\ln B)$$

Then: (a) If $B = 7.5$, $Q = (-1)(\ln 7.5) = (-1)($ _____ $) = $ _____

(b) If $B = 80$, $Q = (-1)(\ln 80) = (-1)($ _____ $) = $ _____

$B = 2.2$, from:

$-0.8 = -\ln B$
$0.8 = \ln B$
or
$\ln B = 0.8000$

a) $(-1)(2.0149) = -2.0149$
    or -2.01

b) $(-1)(4.3820) = -4.3820$
    or -4.38

SELF-TEST 14 (Frames 229-244)

1. Find "H" when: $c = 20.0$, $B = 5.90$

   $\boxed{H = c \ln B}$

   H = _____

2. Find "w" when: $d = 3.88$, $P = 83.0$

   $\boxed{w \ln P = d}$

   w = _____

3. Find "Q" when: $G = 0.710$

   $\boxed{G = -\ln Q}$

   Q = _____

4. Find "a" when: $M = 400$, $R = 48.6$, $G = 12.8$

   $\boxed{a = M \ln\left(\dfrac{R}{G}\right)}$

   a = _____

5. Find "A" when: $h = 15.3$, $t = 5.65$

   $\boxed{t = \dfrac{h}{\ln A}}$

   A = _____

ANSWERS: 1. $H = 35.5$ 2. $w = 0.878$ 3. $Q = 0.50$ 4. $a = 534$ 5. $A = 15.0$

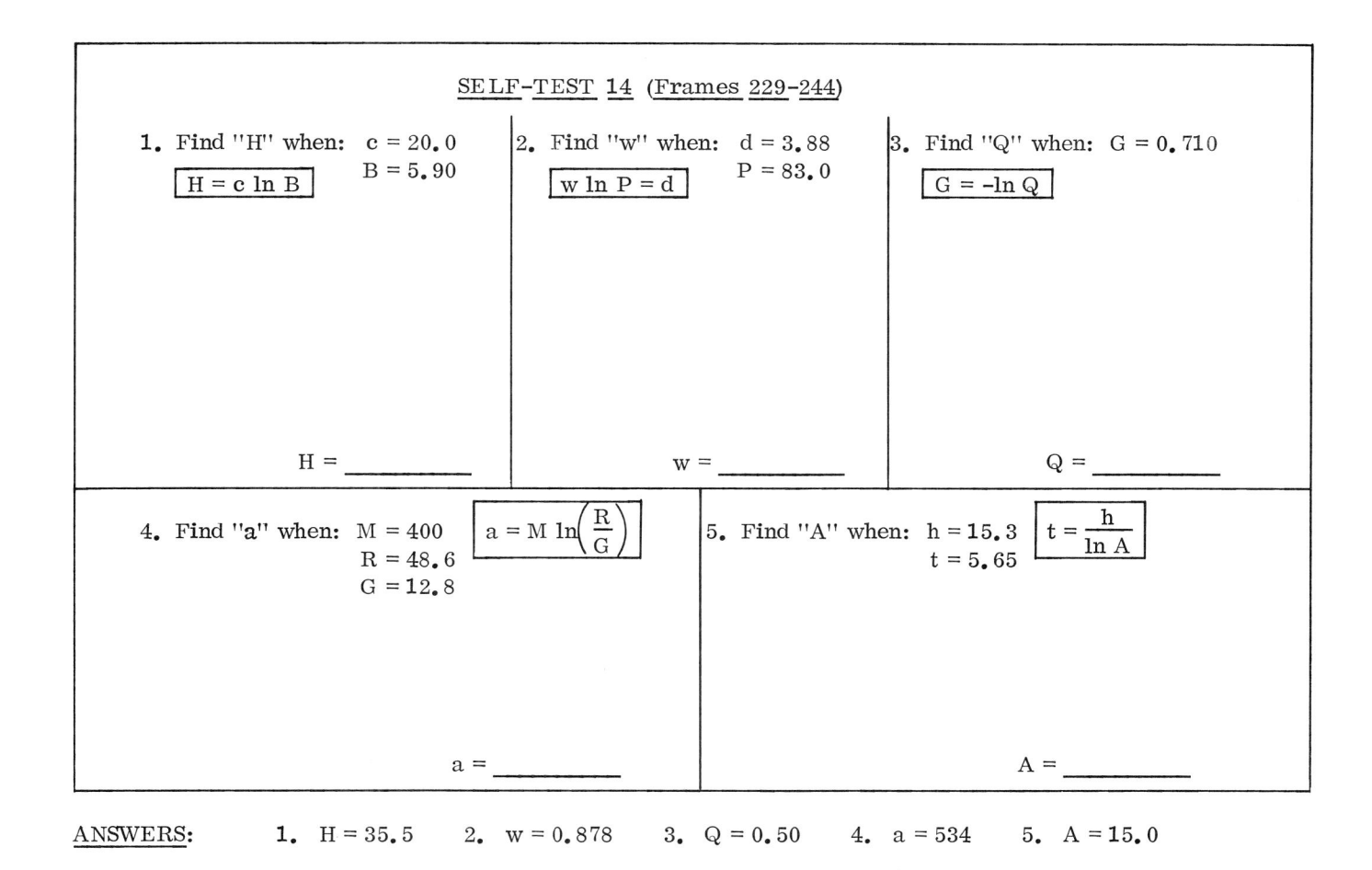

6-15 EVALUATING "Ln" FORMULAS WHEN A CONVERSION TO POWER-OF-"e" FORM IS REQUIRED

When performing evaluations with "ln" formulas, sometimes the unknown variable is part of a complicated "ln" expression. In such cases, we encounter equations like those below in the evaluation process:

$$\ln\left(\frac{A_o}{4.01}\right) = 2.77 \qquad \ln\left(\frac{825}{m}\right) = 3.17$$

To solve for the letter in each equation above, we will convert the "ln" equation to power-of-"e" form. We will discuss the method in this section.

576 Common and Natural Logarithms

245. The variable "T" is part of a complicated "ln" expression in the equation below:

$$\ln\left(\frac{T}{50}\right) = 2.5$$

To solve for "T", we must convert the log equation to power-of-"e" form. That is:

Just as $\ln P = 2.5$ means: $P = e^{2.5}$

$\ln\left(\frac{T}{50}\right) = 2.5$ means: $\frac{T}{50} = e^{2.5}$

Having converted the "ln" equation to power-of-"e" form, we can solve the new equation by replacing $e^{2.5}$ with a regular number. From the e^x-table, $e^{2.5} = 12.2$. Therefore, we get:

$$\frac{T}{50} = e^{2.5}$$

$$\frac{T}{50} = 12.2$$

$$T = (12.2)(50) = \underline{}$$

---

246. Two equations containing complicated "ln" expressions have been converted to power-of-"e" form below:

If $\ln\left(\frac{40}{d}\right) = 1.5$, then $\frac{40}{d} = e^{1.5}$

If $\ln(20 + b) = -2.7$, then $20 + b = e^{-2.7}$

Notice in each case that the whole quantity in parentheses is equal to the power-of-"e".

Convert each "ln" equation below to power-of-"e" form:

(a) $\ln\left(\frac{P}{20}\right) = 1.3$ (b) $\ln\left(\frac{100}{R}\right) = -0.8$

_____ _____

T = 610

---

247. Convert each "ln" equation below to power-of-"e" form:

(a) $\ln\left(\frac{a}{50}\right) = -1$ (b) $\ln(C + 17.8) = 2.9$

_____ _____

a) $\frac{P}{20} = e^{1.3}$

b) $\frac{100}{R} = e^{-0.8}$

---

a) $\frac{a}{50} = e^{-1}$

b) $C + 17.8 = e^{2.9}$

248. In $\boxed{\ln\left(\dfrac{A_o}{A}\right) = kt}$, "$A_o$" is part of a complicated "ln" expression.

To find the value of "A_o" when $A = 3.32$, $k = 0.540$ and $t = 3.91$, we use the following steps:

(1) Plug in the known values and simplify:
$$\ln\left(\dfrac{A_o}{3.32}\right) = (0.540)(3.91)$$
$$\ln\left(\dfrac{A_o}{3.32}\right) = 2.11$$

(2) Convert the "ln" equation to power-of-"e" form:
$$\dfrac{A_o}{3.32} = e^{2.11}$$

(3) Replace $e^{2.11}$ with a regular number by using the e^x-table:
$$\dfrac{A_o}{3.32} = 8.17$$

(4) Complete the solution:

$$A_o = \underline{\qquad}$$

---

249. In $\boxed{\ln\left(\dfrac{A}{A_o}\right) = -kt}$, let's find "$A_o$" when: $A = 16.2$, $k = 0.0488$, $t = 14.2$

Plugging in the known values and simplifying, we get:
$$\ln\left(\dfrac{16.2}{A_o}\right) = (-1)(0.0488)(14.2)$$
$$\ln\left(\dfrac{16.2}{A_o}\right) = -0.693$$

Converting to power-of-"e" form, we get:
$$\dfrac{16.2}{A_o} = e^{-0.693}$$

Substituting a regular number for $e^{-0.693}$ (or $e^{-0.7}$), we get:
$$\dfrac{16.2}{A_o} = 0.497$$

Now complete the solution:

$$A_o = \underline{\qquad}$$

$A_o = (3.32)(8.17)$
$ = 27.1$

$A_o = \dfrac{16.2}{0.497} = 32.6$

578   Common and Natural Logarithms

250. In $\boxed{v = c \ln\left(\dfrac{M}{m}\right)}$, find "M" when: $v = 4.11$
  $c = 1.87$
  $m = 92$

Plugging in the known values, we get: $4.11 = 1.87 \ln\left(\dfrac{M}{92}\right)$

Isolating the "ln" expression, we get: $\dfrac{4.11}{1.87} = \ln\left(\dfrac{M}{92}\right)$
    or
    $\ln\left(\dfrac{M}{92}\right) = 2.20$

Convert to exponential form and complete the solution:

M = _____

---

251. In $\boxed{\ln(c + t) = -k}$, find "t" when $c = 0.800$ and $k = 0.100$.

Plugging in the known values, we get: $\ln(0.800 + t) = -0.100$

Now convert to power-of-"e" form and complete the solution:

M = 830

Here are the steps:

$\dfrac{M}{92} = e^{2.20}$

$\dfrac{M}{92} = 9.02$

$M = 92(9.02) = 830$

t = _____

Answer to Frame 251: $t = 0.105$ From: $0.8 + t = e^{-0.100}$
    $0.8 + t = 0.905$

SELF-TEST 15 (Frames 245-251)

1. Solve for "R":
   $\boxed{\ln(R + 20) = 4.50}$

   R = _____

2. Find "H" when: $\boxed{\ln\left(\dfrac{F}{H}\right) = bt}$
   $F = 120$
   $b = 0.125$
   $t = 15.2$

   H = _____

3. Find "T" when: $\boxed{w = k \ln\left(\dfrac{T}{t}\right)}$
   $t = 2.35$
   $k = 0.332$
   $w = 1.86$

   T = _____

ANSWERS: 1. $R = 70$ 2. $H = 17.9$ 3. $T = 635$

Common and Natural Logarithms

6-16 EVALUATING "e" FORMULAS WHEN A CONVERSION TO "Ln" NOTATION IS REQUIRED

When performing evaluations with "e" formulas, sometimes the unknown variable is the exponent (or part of the exponent) of the base "e". In such cases, we encounter equations like those below in the evaluation process:

$$950 = e^h \qquad 5.90 = e^{-0.3t} \qquad 0.197 = e^{-25p}$$

To solve for the letter in each exponent above, we can convert the "e" equation to "ln" notation. We will discuss the method in this section.

---

252. The unknown variable "q" is the exponent of "e" in the equation below:

$$\boxed{3.50 = e^q}$$

To solve for "q", we can convert the power-of-"e" equation to "ln" notation. We get:

$$\boxed{\ln 3.50 = q}$$

Now we can solve for "q" by using the natural-log table to substitute a regular number for ln 3.5.

Since $\ln 3.50 = 1.2528$, q = _____.

---

253. The variable "p" is part of the exponent of "e" in the equation below:

$$\boxed{40 = e^{-p}}$$

To solve for "p", we can convert the power-of-"e" equation to "ln" notation. We get:

$$\boxed{\ln 40 = -p}$$

Now we can solve for "p" in the following steps.

(1) Substitute a regular number for ln 40: $3.6889 = -p$

(2) Solve for "p" by taking the opposite of both sides: p = _____

---

q = 1.2528 or 1.25

p = -3.6889 or -3.69

254. The variable "t" is part of the exponent of "e" in the equation below:

$$5.10 = e^{-0.3t}$$

We can solve for "t" by means of the following steps:

    (1) Convert the equation to "ln" notation: $\ln 5.10 = -0.3t$

    (2) Substitute a regular number for ln 5.10: $1.6292 = -0.3t$

    (3) Take the opposite of both sides: $-1.6292 = 0.3t$
                                    or
                                      $0.3t = -1.6292$

    (4) Complete the solution:

                         t = _____

Answer: $t = \dfrac{-1.6292}{0.3}$
    = -5.43

255. Let's solve for "x" in: $350 = 10e^{-50x}$

    Isolating the power of "e", we get: $35 = e^{-50x}$

    Converting to "ln" notation, we get: $\ln 35 = -50x$

    Substituting a regular number for ln 35, we get: $3.5554 = -50x$

    Taking the opposite of both sides, we get: $-3.5554 = 50x$
                                    or
                                  $50x = -3.5554$

    Complete the solution:

                         x = _____

Answer: $x = \dfrac{-3.5554}{50}$
    = -0.0711

256. When the variable in an equation is the exponent (or part of the exponent) of "e", we can solve for the variable by converting the equation to "ln" notation.

Convert each equation below to "ln" notation:

    (a) $47 = e^{-m}$ (b) $1.5 = e^{-0.1p}$

Answer:
a) $\ln 47 = -m$
b) $\ln 1.5 = -0.1p$

257. If the power of "e" has a numerical coefficient, we must isolate the power of "e" before converting to "ln" notation.

In each case below, isolate the power of "e" and then convert to "ln" notation:

(a) $250 = 100e^{-b}$ (b) $75 = 10e^{-20x}$

_____ _____

_____ _____

---

258. In $\boxed{P = e^h}$, let's find "h" when $P = 21.9$.

Plugging in the known value, we get: $21.9 = e^h$

Converting to "ln" notation, we get: $\ln 21.9 = h$

By substituting a regular number for $\ln 21.9$, complete the solution:

$h = $ _____

a) $2.5 = e^{-b}$
$\ln 2.5 = -b$

b) $7.5 = e^{-20x}$
$\ln 7.5 = -20x$

---

259. In $\boxed{L = e^{-t}}$, let's solve for "t" when $L = 0.70$.

Plugging in the known value, we get: $0.70 = e^{-t}$

Converting to "ln" notation, we get: $\ln 0.70 = -t$

Substituting a regular number for $\ln 0.70$, we get: $-0.3567 = -t$

Complete the solution by taking the opposite of both sides:

$t = $ _____

$h = 3.0910$ or 3.09

---

260. If $\boxed{P = 14.7e^{-0.2h}}$, let's find "h" when $P = 4.42$. The steps are:

Plug in the known value: $4.42 = 14.7e^{-0.2h}$

Isolate the power of "e": $\dfrac{4.42}{14.7} = e^{-0.2h}$

$0.300 = e^{-0.2h}$

Convert to "ln" notation: $\ln 0.300 = -0.2h$

Substitute a regular number for $\ln 0.300$ and complete the solution:

$h = $ _____

$t = 0.3567$ or 0.36

582 Common and Natural Logarithms

261. In $\boxed{A = ke^{-ct}}$, find "c" when:
$A = 0.267$
$k = 2.59$
$t = 100$

Plugging in the known values, we get:

$$0.267 = 2.59e^{(-1)(c)(100)}$$
or
$$0.267 = 2.59e^{-100c}$$

Isolating the power of "e" and simplifying, we get:

$$\frac{0.267}{2.59} = e^{-100c}$$

$$0.103 = e^{-100c}$$

Converting to "ln" notation, we get:

$$\ln 0.103 = -100c$$

Substitute a regular number for ln 0.103 and complete the solution:

$c = $ _____

---

262. In $\boxed{A = A_o e^{-kt}}$, find "t" if:
$A = 2.92$
$A_o = 14.6$
$k = 50$

Plugging in the known values, we get:

$$2.92 = 14.6e^{(-1)(50)(t)}$$
or
$$2.92 = 14.6e^{-50t}$$

Isolating the power of "e" and simplifying, we get:

$$\frac{2.92}{14.6} = e^{-50t}$$

$$0.200 = e^{-50t}$$

Complete the solution after converting to "ln" notation:

$t = $ _____

$h = 6.02$

From:
$-1.2040 = -0.2h$
$1.2040 = 0.2h$
$$h = \frac{1.2040}{0.2}$$
$h = 6.02$

$c = 0.023$

From:
$-2.3026 = -100c$
$2.3026 = 100c$
$$c = \frac{2.3026}{100}$$
$c = 0.023$

$t = 0.0322$

From:
$\ln 0.200 = -50t$
$-1.6094 = -50t$
$1.6094 = 50t$
$$t = \frac{1.6094}{50} = 0.0322$$

SELF-TEST 16 (Frames 252-262)

1. Convert this power-of-"e" equation to "ln" notation: $87 = e^{2w}$ _____

2. Solve for "t": $\boxed{47 = e^t}$

   t = _____

3. Solve for "h": $\boxed{300 = 200e^{5h}}$

   h = _____

4. Solve for "p": $\boxed{100 = e^{-0.5p}}$

   p = _____

5. Find "x" when: $\boxed{P = P_o e^{cx}}$

   P = 60
   P_o = 25
   c = 5

   x = _____

6. Find "t" when: $\boxed{A = A_o e^{-kt}}$

   A = 210
   A_o = 350
   k = 0.25

   t = _____

ANSWERS:
1. ln 87 = 2w
2. t = 3.85
3. h = 0.0811
4. p = -9.21
5. x = 0.175
6. t = 2.04

6-17 A REVIEW OF THE STRATEGIES USED IN EVALUATIONS WITH "Log", "e", AND "Ln" FORMULAS

In performing evaluations with "log", "e", and "ln" formulas, three different tables are used. Furthermore, in order to complete some of the evaluations, a logarithmic equation must be converted to exponential form or vice versa. In this section, we will briefly review the strategies used in these evaluations. Emphasis will be given to the choice of the correct table and to the type of evaluations in which a conversion to another form is needed.

584 Common and Natural Logarithms

263. In <u>all</u> <u>evaluations</u> with formulas which contain "log" expressions, the <u>common-log</u> table is used. Therefore, the common-log table would be used to perform all evaluations with which formula below? _____

(a) $A = ke^{-ct}$ (b) $D = 10 \log R$ (c) $D = k \ln P$

---

264. When evaluating to find the value of a variable which is part of a complicated "log" expression, a conversion to power-of-ten form is needed. For example, we must eventually convert to power-of-ten form to find the value of either "P_2" or "P_1" in the formula below.

$$D = 10 \log\left(\frac{P_2}{P_1}\right)$$

Must we eventually convert to power-of-ten form:

(a) to find the value of "K" in $K \log(W + H) = B$? _____

(b) to find the value of "W" in $K \log(W + H) = B$? _____

(c) to find the value of "H" in $\log\left(\dfrac{H - B}{G - B}\right) = K$? _____

(b) $D = 10 \log R$

---

265. When performing evaluations with "e" formulas, the e^x-<u>table</u> is used to find the value of:

(1) a "solved for" variable, like "R" in $R = e^{-ft}$.

(2) a "non-solved for" variable which is not part of the exponent, like "k" in $A = ke^{-ct}$.

Identify the table needed to find the value of each variable below:

(a) "A" in $A = ke^{-ct}$ _____

(b) "I_0" in $i = I_0 e^{-\frac{Rt}{L}}$ _____

(c) "K" in $P = A - K \log Q$ _____

a) No.
b) Yes.
c) Yes.

---

a) e^x-table
b) e^x-table
c) Common-log table

266. However, to find the value of a variable <u>which is part of the exponent of "e"</u>, we must eventually convert the equation from power-of-"e" to "ln" notation. For example, we must eventually convert to "ln" notation to find the value of either "c" or "t" in the formula below:

$$A = ke^{-ct}$$

Must we eventually convert to "ln" notation:

(a) to find the value of "f" in $\boxed{R = e^{-ft}}$? _____

(b) to find the value of "R" in $\boxed{i = I_o e^{-\frac{Rt}{L}}}$? _____

(c) to find the value of "E" in $\boxed{V = Ee^{-\frac{t}{RC}}}$? _____

---

267. Of course, when an "e" formula is converted to "ln" notation, the <u>natural-log</u> table is used to complete the evaluation. For example, we eventually use the <u>natural-log</u> table to find the value of either "c" or "t" in the formula below:

$$A = ke^{-ct}$$

Identify the table needed to find the value of each variable below:

(a) "i" in $\boxed{i = \frac{E}{R}\left(1 - e^{-\frac{Rt}{L}}\right)}$ _____

(b) "R" in $\boxed{i = I_o e^{-\frac{Rt}{L}}}$ _____

(c) "A_H" in $\boxed{P_H = -\log A_H}$ _____

a) Yes.

b) Yes.

c) No.

---

268. When performing evaluations with "ln" formulas, the <u>natural-log</u> table is used to find the value of:

(1) a "solved for" variable, like "D" in $\boxed{D = k \ln P}$.

(2) a "non-solved for" variable <u>which is not part of a complicated</u> "ln" expression, like "k" in $\boxed{\ln\left(\frac{A_o}{A}\right) = kt}$.

Identify the table needed to find the value of each variable below:

(a) "d" in $\boxed{w = \dfrac{d}{\ln P}}$ _____

(b) "A" in $\boxed{\ln A = \dfrac{h}{t}}$ _____

(c) "H" in $\boxed{P = 20 \log H}$ _____

a) e^x-table

b) Natural-log table

c) Common-log table

269. However, to find the value of a variable which is part of a complicated "ln" expression, we must eventually convert the "ln" equation to power-of-"e" form. For example, we must eventually convert to power-of-"e" form to find the value of either "M" or "m" in the formula below:

$$v = c \ln\left(\frac{M}{m}\right)$$

Must we eventually convert to power-of-"e" form:

(a) to find the value of "P" in $\boxed{w = \dfrac{d}{\ln P}}$? _____

(b) to find the value of "A" in $\boxed{\ln\left(\dfrac{A_o}{A}\right) = kt}$? _____

(c) to find the value of "k" in $\boxed{D = k \ln\left(\dfrac{t}{s}\right)}$? _____

a) Natural-log table

b) Natural-log table

c) Common-log table

270. Of course, when an "ln" equation is converted to power-of-"e" form, the e^x-table is used to complete the evaluation. For example, we eventually use the e^x-table to find the value of either "G" or "P" in the formula below:

$$r \ln\left(\frac{G}{P}\right) = v$$

Identify the table needed to find the value of each variable below:

(a) "h" in $\boxed{\ln A = \dfrac{h}{t}}$ _____

(b) "m" in $\boxed{v = c \ln\left(\dfrac{M}{m}\right)}$ _____

(c) "p" in $\boxed{T = e^{-kp}}$ _____

a) No.

b) Yes.

c) No.

271. Identify the table needed to find the value of each variable below:

(a) "P_2" in $\boxed{D = 10 \log\left(\dfrac{P_2}{P_1}\right)}$ _____

(b) "t" in $\boxed{V = Ee^{-\frac{t}{RC}}}$ _____

(c) "V" in $\boxed{R = c \ln\left(\dfrac{V}{T}\right)}$ _____

a) Natural-log table

b) e^x-table

c) Natural-log table

a) Common-log table

b) Natural-log table

c) e^x-table

SELF-TEST 17 (Frames 263-271)

In solving for the listed variable in each formula below, state whether the <u>common-log table</u>, the <u>natural-log table</u>, or the e^x-<u>table</u> would be used. (<u>Note</u>: In each formula, the numerical values of all variables are known except the variable which is to be solved for.)

1. Find "H" in: $\boxed{H \log A = R}$ _____

2. Find "P" in: $\boxed{\dfrac{d}{\ln P} = r}$ _____

3. Find "F" in: $\boxed{w = b \ln(F + G)}$ _____

4. Find "I" in: $\boxed{i = Ie^{-\frac{t}{RC}}}$ _____

5. Find "E_2" in: $\boxed{D = 20 \log\left(\dfrac{E_2}{E_1}\right)}$ _____

6. Find "t" in: $\boxed{v = V(1 - e^{-kt})}$ _____

7. In Problems 1 to 6 above, there is <u>one</u> problem whose solution will involve converting the formula to <u>power-of-ten form</u>. It is Problem _____.

8. In Problems 1 to 6 above, there is <u>one</u> problem whose solution will involve converting the formula to <u>power-of-"e" form</u>. It is Problem _____.

ANSWERS:
1. Common-log table.
2. Natural-log table.
3. e^x-table.
4. e^x-table.
5. Common-log table.
6. Natural-log table.
7. Problem 5.
8. Problem 3.

6-18 THE GRAPHS OF EXPONENTIAL EQUATIONS IN "x" AND "y"

In each equation below, there is an exponent which contains a variable (or letter):

$$y = 10^x \qquad y = e^{-x} \qquad y = 10e^{0.4x} \qquad y = k(1 - e^x)$$

Equations of this type are called <u>exponential equations</u>. Their graphs are called <u>exponential graphs</u>. In this section, we will discuss the graphs of exponential equations in which the variables are "x" and "y". Major emphasis will be given to exponential equations which contain the letter "e".

272. An <u>exponential</u> <u>equation</u> <u>is</u> <u>one</u> <u>in</u> <u>which</u> <u>an</u> <u>exponent</u> <u>contains</u> <u>a</u> <u>variable</u> (<u>or</u> <u>letter</u>).

$y = 3^x$ <u>is</u> an exponential equation since the exponent "x" <u>is</u> a variable.

$y = x^3$ <u>is not</u> an exponential equation since the exponent "3" <u>is not</u> a variable.

Which of the following equations are <u>exponential equations</u>? _____

(a) $y = 10^x$ (b) $y = x^{10}$ (c) $A = 5e^{-0.4}$ (d) $A = 5e^{-0.4x}$

588 Common and Natural Logarithms

273. Here is an exponential equation: $\boxed{y = 10^x}$

To graph this equation, we must find pairs of values for "x" and "y" which satisfy the equation.

If "x" is a whole number, it is easy to find the corresponding value of "y". For example:

If $x = 1$, $y = 10^1$ or 10.
If $x = 0$, $y = 10^0$ or 1.
If $x = -1$, $y = 10^{-1}$ or 0.1 .

If "x" is a decimal number, we must use the common log table to find the corresponding value of "y". For example:

If $x = 0.4$, $y = 10^{0.4000} = 2.51$

If $x = -0.4$, $y = 10^{-0.4000} = 0.398$

(a) If $x = 0.6$, $y = 10^{0.6000} = $ _____

(b) If $x = -0.2$, $y = 10^{-0.2000} = $ _____

Only (a) and (d).

In (b) and (c), there is no variable in the exponent.

---

274. In the exponential equation below, there is a "-" in the exponent:

$$\boxed{y = 10^{-x}}$$

"-x" means "the opposite of x". Since we get the opposite of any quantity by multiplying it by "-1", it is helpful to write the equation this way:

$$\boxed{y = 10^{(-1)(x)}}$$

Then we can find pairs of values for "x" and "y" which satisfy the equation. Complete these:

(a) If $x = 1$, $y = 10^{(-1)(1)}$ $= 10^{-1}$ = _____

(b) If $x = -1$, $y = 10^{(-1)(-1)}$ $= 10^{+1}$ = _____

(c) If $x = 0.5$, $y = 10^{(-1)(0.5)}$ $= 10^{-0.5000}$ = _____

(d) If $x = -0.5$, $y = 10^{(-1)(-0.5)}$ $= 10^{+0.5000}$ = _____

a) 3.98

b) 0.631
(From $10^{0.8000} \times 10^{-1}$)

---

a) 0.1

b) 10

c) 0.316
(From $10^{0.5000} \times 10^{-1}$)

d) 3.16

Common and Natural Logarithms 589

275. At the right are two tables of pairs of values for each of the exponential equations in the last two frames. (As an aid in plotting the points, we have rounded all "y" values to tenths.)

$y = 10^x$	
x	y
-1.0	0.1
-0.5	0.3
-0.3	0.5
0.0	1.0
0.2	1.6
0.5	3.2
0.7	5.0
1.0	10.0

$y = 10^{-x}$	
x	y
-1.0	10.0
-0.7	5.0
-0.5	3.2
-0.2	1.6
0.0	1.0
0.3	0.5
0.5	0.3
1.0	0.1

Using these pairs of values, we have graphed both equations at the right:

Notice these points about the graphs:

(1) Both curves are called <u>exponential curves</u>.

   The graph of $y = 10^x$ is called an <u>ascending</u> exponential curve.

   The graph of $y = 10^{-x}$ is called a <u>descending</u> exponential curve.

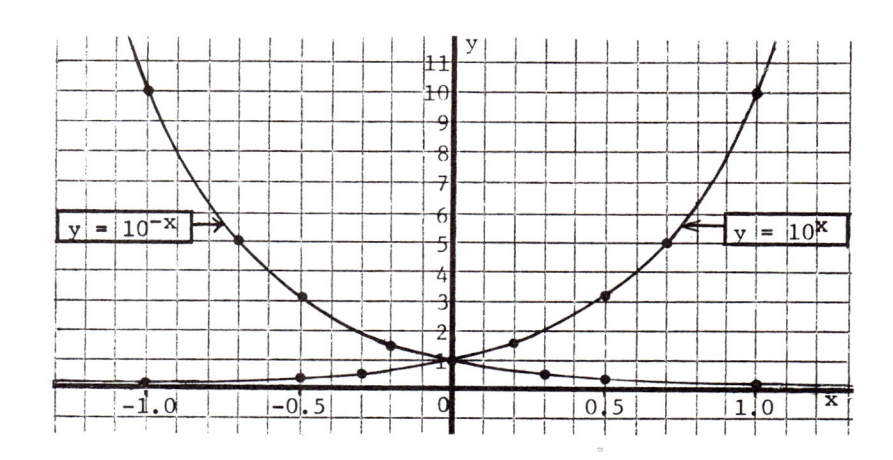

(2) Since powers of ten <u>always</u> stand for <u>positive</u> numbers, "y" cannot be negative. Therefore, <u>both curves</u> <u>appear</u> <u>only</u> <u>in Quadrants I and II</u>.

(3) Both curves cross the y-axis at (0,1). The y-intercept is (0,1) because $y = 10^0$ or 1 when "x" is 0.

(4) The graph of $y = 10^x$ rises slowly in Quadrant II but its ascent is rapid in Quadrant I. The graph of $y = 10^{-x}$ descends very rapidly in Quadrant II and then its descent tapers off in Quadrant I.

276. Here is another exponential equation: $y = e^x$

Go to next frame.

To find pairs of values for "x" and "y", we use the "e^x-table". For example:

   If $x = 5$, $y = e^5 = 148$
   If $x = 0.8$, $y = e^{0.8} = 2.23$
   If $x = -1.2$, $y = e^{-1.2} = 0.301$

Use the e^x-table to complete these:

   (a) If $x = 1.9$, $y = e^{1.9} =$ _____

   (b) If $x = -0.4$, $y = e^{-0.4} =$ _____

   (c) If $x = -3$, $y = e^{-3} =$ _____

a) 6.69

b) 0.67

c) 0.0498

277. The following exponential equation contains "-x" as the exponent:

$$y = e^{-x}$$

As a help in finding pairs of values, we can rewrite the equation this way:

$$y = e^{(-1)(x)}$$

Therefore: (a) If $x = 3$, $\quad y = e^{(-1)(3)} \quad = e^{-3} = $ _____

(b) If $x = 0.6$, $\quad y = e^{(-1)(0.6)} = e^{-0.6} = $ _____

(c) If $x = -2.9$, $\quad y = e^{(-1)(-2.9)} = e^{+2.9} = $ _____

(d) If $x = -7$, $\quad y = e^{(-1)(-7)} = e^{+7} = $ _____

Answer to Frame 277:   a) 0.0498   b) 0.549   c) 18.2   d) 1,100

278. Here is a table of pairs of values for each of the exponential equations in the last two frames. (Again, we have rounded "y" to tenths.)

$y = e^x$

x	y
-2.0	0.1
-1.0	0.4
0.0	1.0
0.5	1.6
1.0	2.7
1.5	4.5
2.0	7.4
2.5	12.2
3.0	20.1

$y = e^{-x}$

x	y
-3.0	20.1
-2.5	12.2
-2.0	7.4
-1.5	4.5
-1.0	2.7
-0.5	1.6
0.0	1.0
1.0	0.4
2.0	0.1

Using the pairs of values in the tables, we have graphed both equations at the right:

(a) The graph of $y = e^x$ is called an <u>ascending</u> exponential curve. The graph of $y = e^{-x}$ is called a _____ exponential curve.

(b) Since $y = e^0$ or 1 when "x" is 0, both curves cross the y-axis at (,).

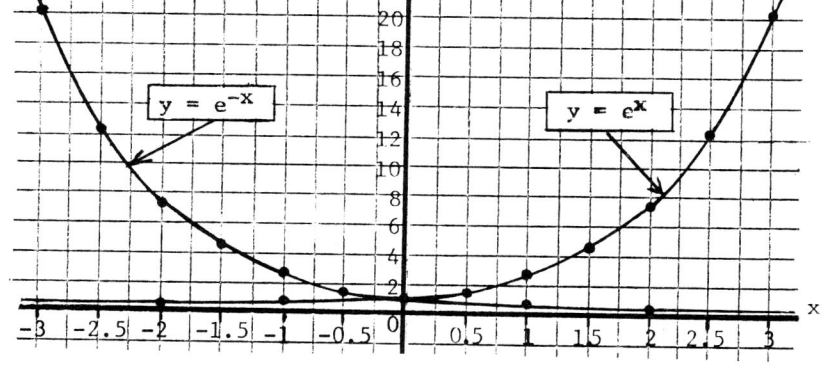

(c) The graph of $y = e^x$ rises rapidly in Quadrant ____. The graph of $y = e^{-x}$ descends rapidly in Quadrant ____.

a) descending

b) (0,1)

c) Quadrant I
   Quadrant II

279. Here is a more complicated exponential equation: $y = ke^{cx}$

There are four variables in this equation: "x", "y", "k", and "c". If we want to graph the relationship between "x" and "y", we must replace "k" and "c" by some numbers since only two variables can be graphed at one time on the coordinate system.

If we replace "k" with 10 and "c" with "0.2", we get: $y = 10e^{0.2x}$

(a) If $x = 1$, $y = 10e^{0.2(1)} = 10e^{0.2} = 10(1.22) = $ _____

(b) If $x = 0$, $y = 10e^{0.2(0)} = 10e^{0} = 10(1) = $ _____

(c) If $x = -0.5$, $y = 10e^{0.2(-0.5)} = 10e^{-0.1} = 10(0.905) = $ _____

---

280. Here is the same basic equation: $y = ke^{cx}$

If we replace "k" with 5 and "c" with -0.4, we get: $y = 5e^{-0.4x}$

Then: If $x = 1.5$, $y = 5e^{-0.4(1.5)} = 5e^{-0.6} = 5(0.549) = 2.745$

If $x = -2$, $y = 5e^{-0.4(-2)} = 5e^{+0.8} = 5(2.23) = 11.15$

Find the value of "y" when:

(a) $x = 0.5$ (b) $x = -5$

y = _____ y = _____

a) 12.2
b) 10
c) 9.05

Answer to Frame 280:
a) $y = 5e^{-0.4(0.5)} = 5e^{-0.2} = 5(0.819) = 4.095$ b) $y = 5e^{-0.4(-5)} = 5e^{+2.0} = 5(7.39) = 36.95$

---

281. On the graph below, we have graphed the following four specific instances of $y = ke^{cx}$:

$y = 10e^{0.4x}$ $y = 10e^{-0.4x}$ $y = 5e^{0.3x}$ $y = 5e^{-0.3x}$

Notice these points about the graphs:

(1) The number plugged in for "k" determines the point at which the curve crosses the y-axis:

Both $y = 10e^{0.4x}$ and $y = 10e^{-0.4x}$ cross the y-axis at "10".

Both $y = 5e^{0.3x}$ and $y = 5e^{-0.3x}$ cross the y-axis at "5".

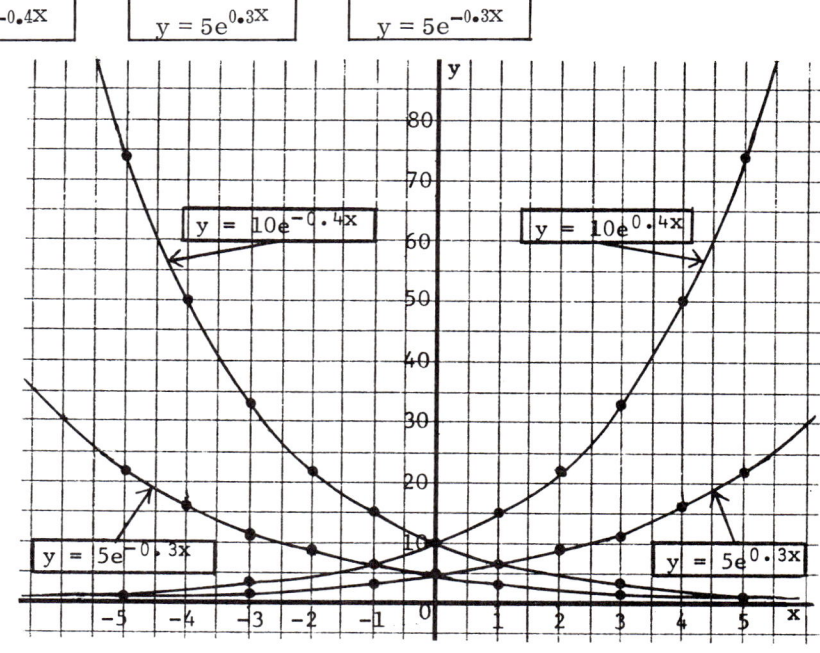

Continued on following page.

592 Common and Natural Logarithms

281. Continued.

(2) The number plugged in for "c" determines how fast the curve will ascend or descend. The larger the absolute value of the number, the faster the ascent or descent.

$y = 10e^{0.4X}$ ascends faster than $y = 5e^{0.3X}$.

$y = 10e^{-0.4X}$ descends faster than $y = 5e^{-0.3X}$.

If we graphed the following two exponential equations:

#1 $\boxed{y = 20e^{0.1X}}$ #2 $\boxed{y = 15e^{0.5X}}$

(a) Curve #1 would cross the y-axis at the point (,).

(b) Curve #2 would cross the y-axis at the point (,).

(c) Both curves would ascend. Which would ascend faster? _____

---

282. Here is another more complicated exponential equation:

$\boxed{y = k(1 - e^X)}$

To graph the relationship between "x" and "y", we must replace "k" with some number. If $k = 10$, we get: $\boxed{y = 10(1 - e^X)}$

(a) If $x = 2$, $y = 10(1 - e^2)$
$= 10(1 - 7.39) = 10(-6.39) = $ _____

(b) If $x = -1.5$, $y = 10(1 - e^{-1.5})$
$= 10(1 - 0.223) = 10(0.777) = $ _____

(c) If $x = 0$, $y = 10(1 - e^0) = 10(1 - 1) = $ _____

a) (0, 20)

b) (0, 15)

c) Curve #2, since $0.5 > 0.1$

---

283. In this equation, the exponent is "-x":

$\boxed{y = k(1 - e^{-X})}$

To graph the relationship between "x" and "y", let's replace "k" with 20 and "-x" with "(-1)(x)". We get:

$\boxed{y = 20\left(1 - e^{(-1)(x)}\right)}$

(a) If $x = 1$, $y = 20(1 - e^{(-1)(1)})$
$= 20(1 - e^{-1})$
$= 20(1 - 0.368) = $ _____ = _____

(b) If $x = 0$, $y = 20\left(1 - e^{(-1)(0)}\right)$
$= 20(1 - e^0)$
$= 20(1 - 1) = $ _____ = _____

(c) If $x = -2.5$, $y = 20(1 - e^{(-1)(-2.5)})$
$= 20(1 - e^{+2.5})$
$= 20(1 - 12.2) = $ _____ = _____

a) −63.9

b) 7.77

c) 0, since $10(0) = 0$

---

a) $20(0.632) = 12.64$

b) $20(0) = 0$

c) $20(-11.2) = -224$

284. We have graphed two instances of the general equation at the right. Both of the curves are <u>ascending</u> <u>exponential</u> <u>curves</u>.

$$y = k(1 - e^{-x})$$

Notice these points:

(1) Both of the curves cross the y-axis at the <u>origin</u> (0, 0).

(2) "k" in this case does not give the y-intercept. Instead, it gives the value of "y" which the curve approaches but never quite reaches.

The graph of $y = 10(1 - e^{-x})$ approaches the horizontal line $y = 10$ as "x" increases.

The graph of $y = 20(1 - e^{-x})$ approaches the horizontal line $y = 20$ as "x" increases.

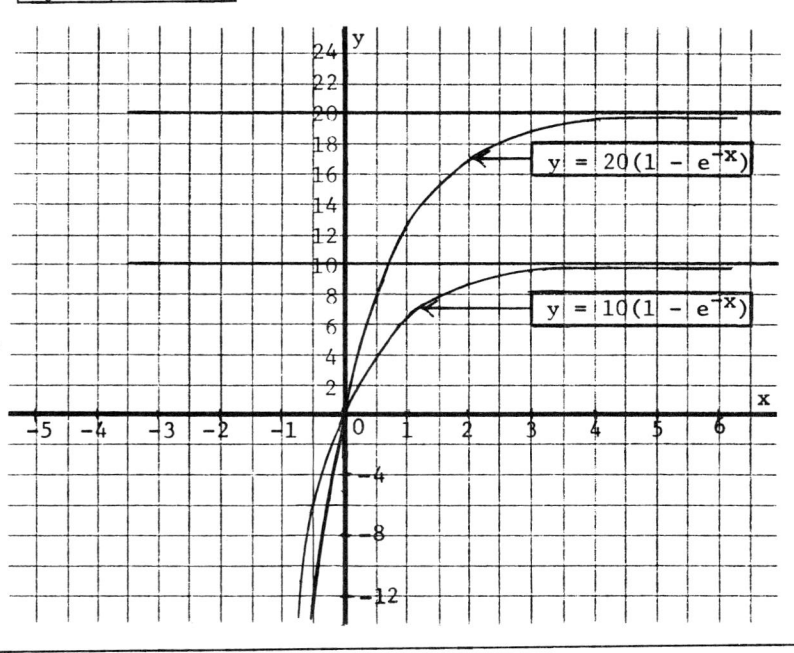

---

285. Here are two general exponential equations which are very useful in science. We have given a sketch of each below:

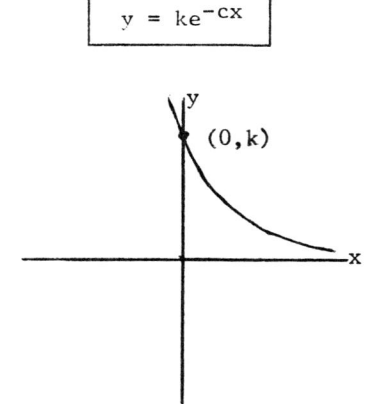

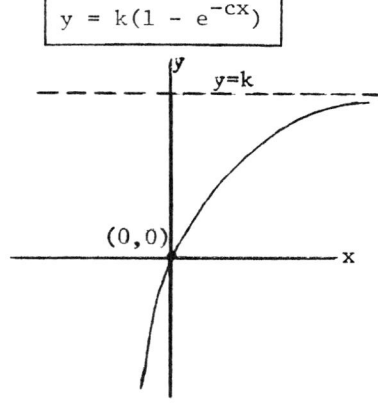

$$y = ke^{-cx}$$

$$y = k(1 - e^{-cx})$$

In each case, "c" is related to how fast the curve <u>ascends</u> or <u>descends</u>.

The main points of interest, however, are the y-intercepts and the value of "y" which the curve approaches as "x" increases.

For $y = ke^{-cx}$: (1) The y-intercept is given by "k".
(2) The curve approaches the line $y = 0$ (the horizontal axis) as x increases.

For $y = k(1 - e^{-cx})$: (1) The y-intercept is the <u>origin</u>.
(2) The curve approaches the line $y = k$ as x increases.

(a) If $y = 25e^{-0.7x}$ were graphed, the curve would cross the y-axis at the point (,   ).

(b) If $y = 25(1 - e^{-0.7x})$ were graphed, the curve would cross the y-axis at the point (,   ).

a) (0, 25)
b) (0, 0)

286. (a) If $y = 1.9e^{-0.2x}$ were graphed, it would approach, but not reach, what value of "y" as "x" increases? _____

(b) If $y = 1.9(1 - e^{-0.2x})$ were graphed, it would approach, but not reach, what value of "y" as "x" increases? _____

287. If $\boxed{y = 10e^{-x}}$ were graphed:

(a) The y-intercept would be (,).

(b) The curve would _____ (ascend/descend).

(c) The curve would approach what value of "y" as "x" increases? _____

(d) Roughly sketch the curve on the right:

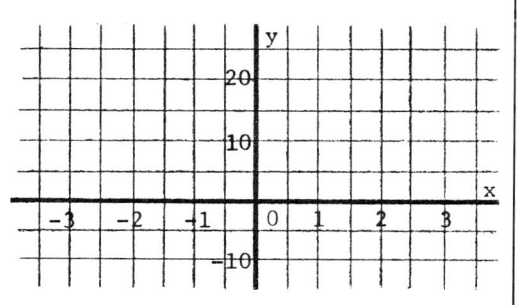

a) 0 (the line $y = 0$)

b) 1.9 (the line $y = 1.9$)

288. If $\boxed{y = 15(1 - e^{-x})}$ were graphed:

(a) The y-intercept would be (,).

(b) The curve would _____ (ascend/descend).

(c) The curve would approach what value of "y" as "x" increases? _____

(d) Roughly sketch the curve on the right:

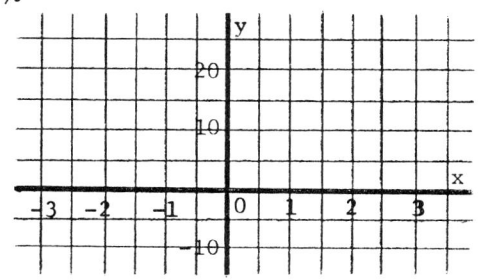

a) (0, 10)

b) descend

c) 0 (the line $y = 0$)

d) Sketch:

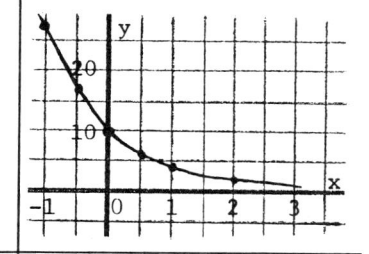

a) (0, 0)

b) ascend

c) 15 (the line $y = 15$)

d) Sketch:

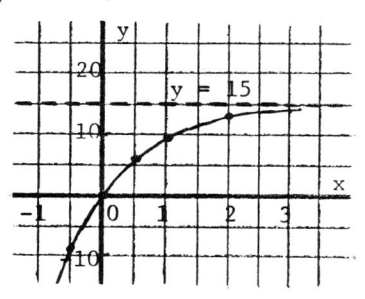

SELF-TEST 18 (Frames 272-288)

If $y = 10^x$ is graphed:

1. As "x" increases, does the graph rise or fall? _____
2. Find the coordinates of the y-intercept.

If $y = 40e^{-x}$ is graphed:

3. As "x" increases, does the graph rise or fall? _____
4. Find the coordinates of the y-intercept. _____

If $y = 75(1 - e^{-x})$ is graphed:

5. As "x" increases, does the graph rise or fall? _____
6. Find the coordinates of the y-intercept. _____
7. As "x" increases and reaches large positive values, the graph approaches what value of "y"? _____

ANSWERS:
1. It rises.
2. (0,1)
3. It falls.
4. (0,40)
5. It rises.
6. (0,0)
7. $y = 75$

6-19 THE GRAPHS OF EXPONENTIAL FORMULAS

When exponential equations in "x" and "y" are graphed, negative values for "x" and "y" can occur. Therefore, the graphed relationships lie in various quadrants. As we saw in the last section, for example:

The graph of $y = 10e^{-x}$ lies in Quadrants I and II.

The graph of $y = 15(1 - e^{-x})$ lies in Quadrants I and III.

When graphing formulas which contain "e", however, negative values for the variables do not generally need to be considered. Therefore, their graphs generally appear only in the first quadrant. We will discuss the graphs of a few exponential formulas in this section.

596   Common and Natural Logarithms

289.

Here is an exponential formula: $i = I_m e^{-\frac{t}{RC}}$

(a) List the variables in the formula: _____

(b) In a particular problem, values for three of the five variables are known. They are:

$I_m = 50$, $R = 1.00$, $C = 0.50$

Plug these values into the formula, and write the resulting formula: $i =$ _____

(c) How many variables does the new formula contain? _____

---

290. The two-variable formula we obtained in the last frame is:

$i = 50e^{-2t}$

The formula in this form can be graphed since it contains only two variables, "i" and "t". To construct the graph, it is necessary to find pairs of values of "t" and "i" as follows:

If $t = 0.25$, $i = 50e^{-0.50} = 50(0.607) = 30.35$

If $t = 1.50$, $i = 50e^{-3} = 50(0.0498) = 2.49$

Complete the following: (a) If $t = 0$, $i =$ _____

(b) If $t = 0.50$, $i =$ _____

(c) If $t = 1$, $i =$ _____

a) i, I_m, t, R, and C

(Note: "e" is not a variable. It stands for the number 2.718....)

b) $i = 50e^{-\frac{t}{0.5}}$

or

$i = 50e^{-2t}$

c) Two (i and t)

Answer to Frame 290:   a) $i = 50$ b) $i = 18.4$ c) $i = 6.75$

---

291. At the right are some pairs of values which satisfy the formula:

$i = 50e^{-2t}$

t	i
0	50
0.25	30.35
0.50	18.40
0.75	11.15
1	6.75

t	i
1.25	4.10
1.50	2.49
1.75	1.51
2	0.92

Using these pairs of values, we have graphed the formula at the right. Since the table contains only positive values of "t" and "i", the graph lies entirely in the first quadrant.

(a) Write the coordinates of the vertical intercept of the graph.

_____

(b) Does the graph descend (fall) more rapidly between $t = 0.25$ and $t = 0.50$, or between $t = 1.25$ and $t = 1.50$?

_____

(c) Does the graph touch the t-axis?

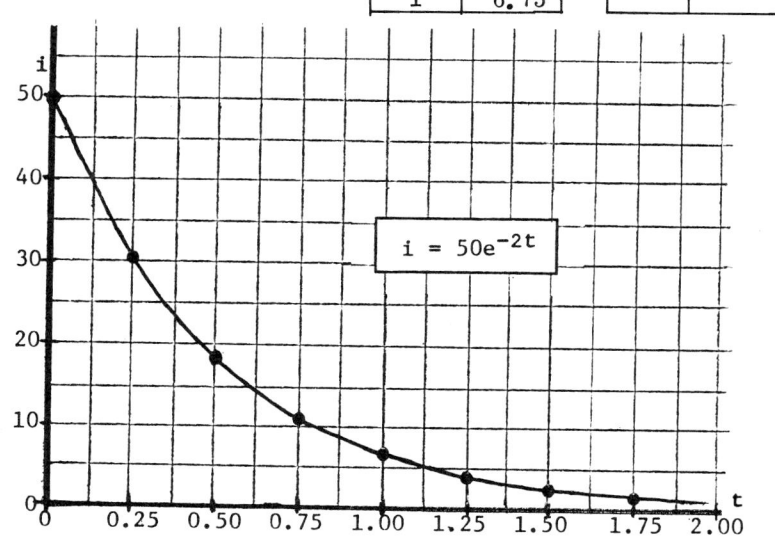

_____

292.

Here is another exponential formula: $v = V\left(1 - e^{-\frac{t}{RC}}\right)$

(a) List the variables in this formula. _____

(b) In a particular problem, the values of three of the five variables are known. They are:

$V = 100$, $R = 5.00$, $C = 0.40$

Plug these values into the formula, and write the resulting formula: $v = $ _____

(c) How many variables does the new formula contain? _____

a) (0, 50)

b) It descends more rapidly between $t = 0.25$ and $t = 0.50$.

c) No. The graph will approach, but never touch, the t-axis.

---

293. The two-variable formula we obtained in the last frame is:

$v = 100(1 - e^{-0.5t})$

The formula in this form can be graphed since it contains only two variables, "v" and "t". To construct the graph, we find pairs of values of "v" and "t" as follows:

If $t = 3$, $v = 100(1 - e^{-1.5})$
$ = 100(1 - 0.223)$
$ = 100(0.777)$
$ = 77.7$

If $t = 6$, $v = 100(1 - e^{-3})$
$ = 100(1 - 0.0498)$
$ = 100(0.9502)$
$ = 95.02$

Complete the following: (a) If $t = 0$, $v = $ _____

(b) If $t = 1$, $v = $ _____

(c) If $t = 2$, $v = $ _____

a) v, V, t, R, and C

b) $v = 100(1 - e^{-0.5t})$

c) Two. (v and t)

Answer to Frame 293: a) $v = 0$ b) $v = 39.3$ c) $v = 63.2$

---

294. Here are some pairs of values which satisfy the formula:

$v = 100(1 - e^{-0.5t})$

t	v
0	0
0.50	22.1
1	39.3
2	63.2
3	77.7

t	v
4	86.5
5	91.8
6	95.0
7	97.0

Using these pairs of values, we have graphed the formula at the right. Since the table contains only positive values of "t" and "v", the graph lies entirely in the first quadrant.

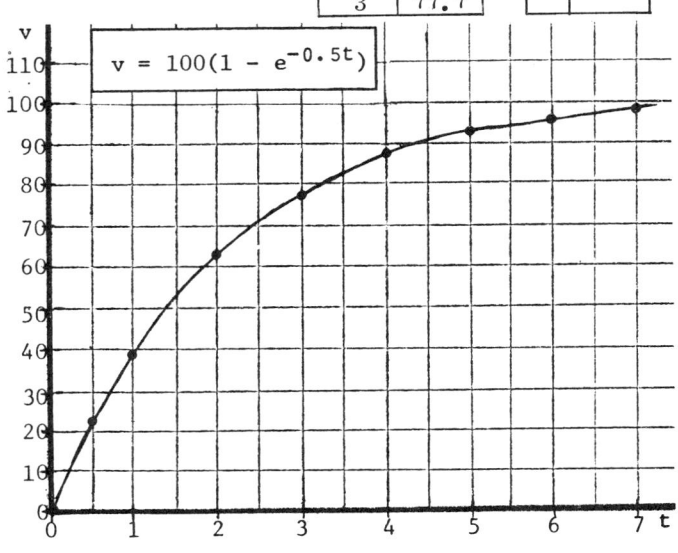

(a) Write the coordinates of the vertical intercept. _____

(b) Does the graph ascend (rise) more rapidly between $t = 0$ and $t = 1$, or $t = 4$ and $t = 5$? _____

(c) As "t" increases, what value of "v" does the graph approach? _____

(d) Complete the following by reading the graph: When $t = 3.5$, $v = $ _____

598 Common and Natural Logarithms

Answer to Frame 294:

a) (0,0)   b) It ascends more rapidly between $t = 0$ and $t = 1$.   c) 100 (It approaches the line $v = 100$.)   d) $v = 82$ or 83 (approximately)

SELF-TEST 19 (Frames 289-294)

In $\boxed{n = Ne^{-kt}}$, $k = 0.15$ and $N = 30$.

1. Plug these values of "k" and "N" into the formula. The new formula is: _____
2. List the variables in the new formula. _____

In $\boxed{i = \dfrac{E}{R}\left(1 - e^{-\tfrac{Rt}{L}}\right)}$, $E = 12$, $L = 25$, and $R = 100$.

3. Plug these values of "E", "L", and "R" into the formula. The new formula is: _____
4. List the variables in the new formula. _____   _____

Here is an exponential formula:

$\boxed{v = 20(1 - e^{-5t})}$

5. If $t = 0$, $v =$ _____
6. If $t = 1$, $v =$ _____
7. If "t" increases from $t = 1$ to $t = 5$, does the value of "v" increase or decrease? _____

Here is another exponential formula:

$\boxed{h = 400e^{-0.4w}}$

8. If $w = 0$, $h =$ _____
9. If $w = 10$, $h =$ _____
10. If "w" increases from $w = 1$ to $w = 5$, does the value of "h" increase or decrease? _____

ANSWERS: 1. $n = 30e^{-0.15t}$ 3. $i = 0.12(1 - e^{-4t})$ 5. $v = 0$ 8. $h = 400$
           2. "n" and "t"   4. "i" and "t"   6. $v = 19.9$ 9. $h = 7.32$
                                                                   7. Increases.   10. Decreases.

6-20 A NOTE ON "exp" NOTATION FOR BASE-"e" EXPONENTIAL FORMULAS

In exponential formulas which contain "e" as a base, the exponent of "e" can be quite complicated. Therefore, an alternate notation involving the abbreviation "exp" is frequently used to write these formulas. We will briefly discuss the "exp" notation in this section.

295. Instead of writing the base "e" in an exponential formula, the abbreviation "exp" is frequently used. For example:

$R = e^{-ft}$ is written: $R = \exp(-ft)$

$P = 14.7e^{-0.2h}$ is written: $P = 14.7 \exp(-0.2h)$

Notice these points about the "exp" notation:
(1) The base "e" is not explicitly shown.
(2) There is no period after "exp".
(3) The exponent of "e" is written in parentheses after the "exp" and on the same line as the "exp".

Write each of the following formulas in "exp" notation:

(a) $M = e^{-pt}$ _____ (b) $A = ke^{-ct}$ _____

296. The "exp" notation is especially useful when the exponent of "e" is quite complicated. For example:

$i = I_0 e^{-\frac{Rt}{L}}$ is written $i = I_0 \exp\left(-\frac{Rt}{L}\right)$

Write each formula below in "exp" notation:

(a) $V = Ee^{-\frac{t}{RC}}$ (b) $i = \frac{E}{R}\left(1 - e^{-\frac{Rt}{L}}\right)$

a) $M = \exp(-pt)$
b) $A = k \exp(-ct)$

297. In "exp" notation, the "exp" is an abbreviation for the word "exponential". Of course, the exponential is a base-e exponential. Therefore, the following words are used when reading a formula in this form.

$y = \exp(-kt)$ is read: "y equals exponential (-kt)"

$A = k \exp(-ct)$ is read: "A equals k times _____ (-ct)"

a) $V = E \exp\left(-\frac{t}{RC}\right)$
b) $i = \frac{E}{R}\left[1 - \exp\left(-\frac{Rt}{L}\right)\right]$

298. Any exponential formula in "exp" notation can be written in regular base-"e" form. That is:

$T = \exp(-0.1m)$ is the same as $T = e^{-0.1m}$

$P = 10 \exp(-50x)$ is the same as $P = 10e^{-50x}$

Convert each formula below to regular base-"e" form:

(a) $d = \exp(-at)$ (b) $V = 1.8 \exp(-0.7p)$ (c) $i = I_0 \exp\left(-\frac{Rt}{L}\right)$

exponential

a) $d = e^{-at}$
b) $V = 1.8e^{-0.7p}$
c) $i = I_0 e^{-\frac{Rt}{L}}$

Chapter 7 LOGARITHMS AND EXPONENTIALS: LAWS AND FORMULAS

In this chapter, we will discuss the laws of both common and natural logarithms. We will also discuss the rearrangement of "log", "ln", and "base-e" formulas. Some of these rearrangements require the use of the laws of logarithms. The relationship between logarithmic and exponential formulas is emphasized. The chapter is concluded with an introduction to the "log principle for equations" and its use in evaluating exponential formulas which contain bases other than "10" and "e".

7-1 THE LAWS OF COMMON LOGARITHMS

In an earlier chapter, we introduced the laws of common logarithms for multiplication, division, and raising a base to a power. In this section, we will briefly review these three laws.

1. Here is the law of logarithms for multiplication: $\boxed{\log(ab) = \log a + \log b}$

   Put into words, it means: The logarithm of a multiplication is equal to the sum of the logarithms of each individual factor.

   We can justify the law by means of numerical examples:

   Example #1: $\log(3 \times 4) = \log 3 + \log 4$

   Since: $\log(3 \times 4) = \log 12 = \underline{1.0792}$

   and: $\log 3 + \log 4 = 0.4771 + 0.6021 = \underline{1.0792}$

   Example #2: $\log(25 \times 64) = \log 25 + \log 64$

   Since: $\log(25 \times 64) = \log 1,600 = \underline{3.2041}$

   and: $\log 25 + \log 64 = 1.3979 + 1.8062 = \underline{3.2041}$

2. Using the law of logarithms for multiplication, complete these:

   (a) $\log(7 \times 8) = \log 7 + \underline{\qquad}$

   (b) $\log(0.1 \times 0.9) = \underline{\qquad} + \log 0.9$

   (c) $\log(MN) = \underline{\qquad} + \underline{\qquad}$

   Go to next frame.

   a) $\log 7 + \underline{\log 8}$

   b) $\underline{\log 0.1} + \log 0.9$

   c) $\underline{\log M} + \underline{\log N}$

3. This law also applies when the multiplication contains more than two factors. Here is an example:

$$\log(5 \times 4 \times 3) = \log 5 + \log 4 + \log 3$$

Since: $\log(5 \times 4 \times 3) = \log 60 = \underline{1.7782}$

and: $\log 5 + \log 4 + \log 3 = 0.4771 + 0.6021 + 0.6990 = \underline{1.7782}$

Complete these using the law of logarithms for multiplication:

(a) $\log(7 \times 8 \times 9) = \log 7 + \underline{\qquad} + \underline{\qquad}$

(b) $\log(pqt) = \underline{\qquad} + \underline{\qquad} + \underline{\qquad}$

Answer to Frame 3: a) $\log 7 + \underline{\log 8} + \underline{\log 9}$ b) $\underline{\log p} + \underline{\log q} + \underline{\log t}$

---

4. Here is the law of logarithms for <u>a fraction</u> (or <u>division</u>): $\boxed{\log\left(\dfrac{a}{b}\right) = \log a - \log b}$

Put into words, it means: <u>The logarithm of a fraction (or division) is obtained by subtracting the logarithm of the denominator from the logarithm of the numerator.</u>

We can justify this law by means of numerical examples:

<u>Example #1</u>: $\log\left(\dfrac{12}{3}\right) = \log 12 - \log 3$

Since: $\log\left(\dfrac{12}{3}\right) = \log 4 = \underline{0.6021}$

and: $\log 12 - \log 3 = 1.0792 - 0.4771 = \underline{0.6021}$

<u>Example #2</u>: $\log\left(\dfrac{1600}{25}\right) = \log 64 - \log 25$

Since: $\log\left(\dfrac{1600}{25}\right) = \log 64 = \underline{1.8062}$

and: $\log 1600 - \log 25 = 3.2041 - 1.3979 = \underline{1.8062}$

---

5. Using the law of logarithms for division, complete these:

(a) $\log\left(\dfrac{27.8}{1.59}\right) = \log 27.8 - \underline{\qquad}$

(b) $\log\left(\dfrac{0.65}{3.21}\right) = \underline{\qquad} - \log 3.21$

(c) $\log\left(\dfrac{T}{V}\right) = \underline{\qquad} - \underline{\qquad}$

Go to next frame.

---

6. Complete each of these using the laws of logarithms:

(a) $\log\left(\dfrac{P_1}{P_2}\right) = \underline{\qquad}$

(b) $\log(RS) = \underline{\qquad}$

a) $\log 27.8 - \underline{\log 1.59}$

b) $\underline{\log 0.65} - \log 3.21$

c) $\underline{\log T} - \underline{\log V}$

---

a) $\log P_1 - \log P_2$

b) $\log R + \log S$

7. Complete each of these using the laws of logarithms:

   (a) $\log(0.5t) = $ _____

   (b) $\log \frac{V}{15} = $ _____

   (c) $\log(15pq) = $ _____

Answer to Frame 7: a) $\log 0.5 + \log t$ b) $\log V - \log 15$ c) $\log 15 + \log p + \log q$

---

8. Here is the law of logarithms for raising a base to a power: $\boxed{\log(b^a) = (a)(\log b)}$

   Put into words, it means: The <u>logarithm</u> of a <u>base-exponent quantity</u> is <u>equal to the exponent times the logarithm of the base.</u>

   We can justify this law by means of numerical examples:

   Example #1: $\log(3^2) = 2\log 3$

   Since: $\log(3^2) = \log 9 = \underline{0.9542}$

   and: $2 \log 3 = 2(0.4771) = \underline{0.9542}$

   Example #2: $\log(49^{\frac{1}{2}}) = \frac{1}{2}\log 49$

   Since: $\log(49^{\frac{1}{2}}) = \log 7 = \underline{0.8451}$

   and: $\frac{1}{2}\log 49 = \frac{1}{2}(1.6902) = \underline{0.8451}$

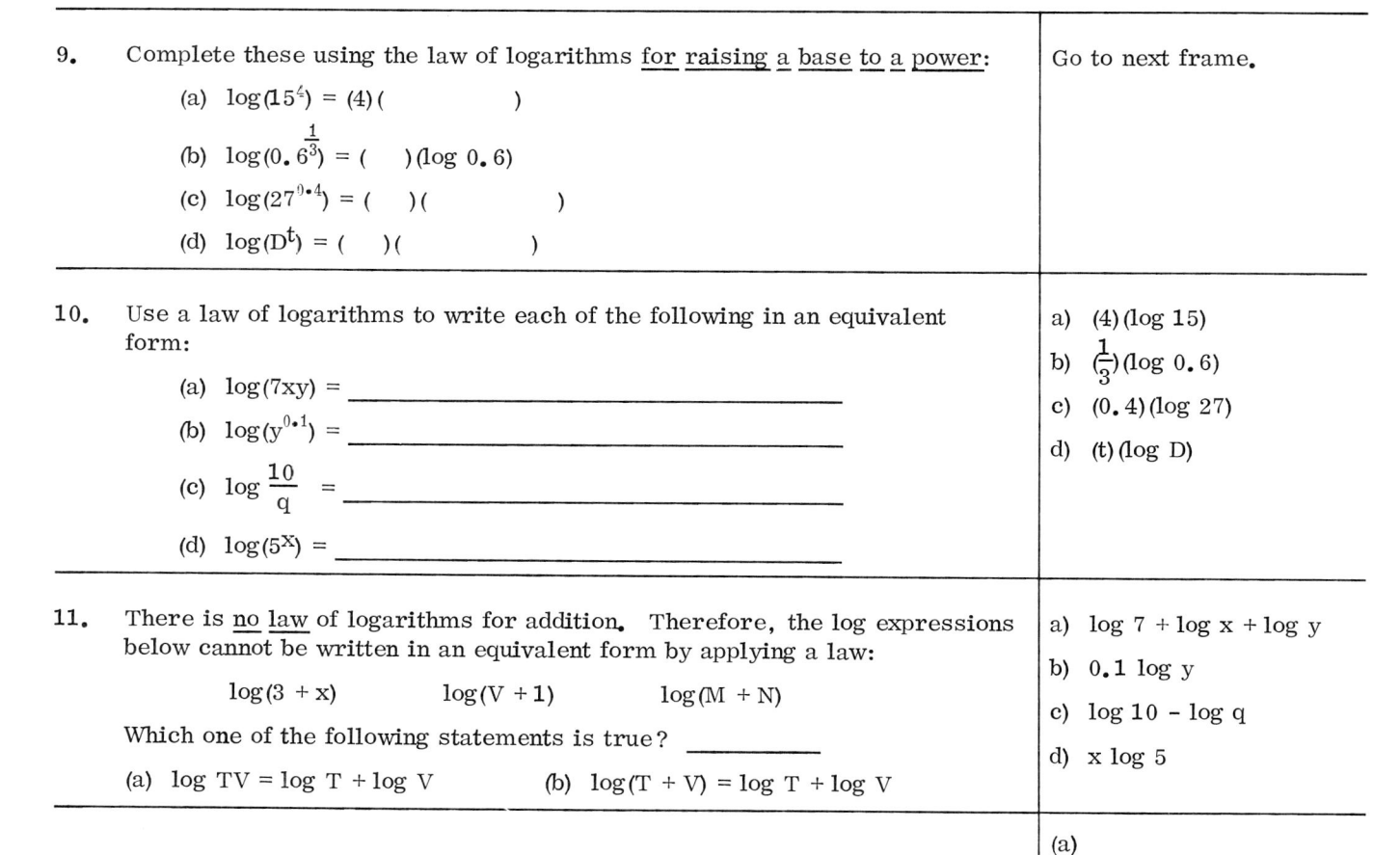

---

9. Complete these using the law of logarithms <u>for raising a base to a power</u>:

   (a) $\log(15^4) = (4)($ $)$

   (b) $\log(0.6^{\frac{1}{3}}) = ($ $)(\log 0.6)$

   (c) $\log(27^{0.4}) = ($ $)($ $)$

   (d) $\log(D^t) = ($ $)($ $)$

   Go to next frame.

---

10. Use a law of logarithms to write each of the following in an equivalent form:

    (a) $\log(7xy) = $ _____

    (b) $\log(y^{0.1}) = $ _____

    (c) $\log \frac{10}{q} = $ _____

    (d) $\log(5^x) = $ _____

    a) $(4)(\log 15)$
    b) $(\frac{1}{3})(\log 0.6)$
    c) $(0.4)(\log 27)$
    d) $(t)(\log D)$

---

11. There is <u>no law</u> of logarithms for addition. Therefore, the log expressions below cannot be written in an equivalent form by applying a law:

    $\log(3 + x)$ $\log(V + 1)$ $\log(M + N)$

    Which one of the following statements is true? _____

    (a) $\log TV = \log T + \log V$ (b) $\log(T + V) = \log T + \log V$

    a) $\log 7 + \log x + \log y$
    b) $0.1 \log y$
    c) $\log 10 - \log q$
    d) $x \log 5$

(a)

12. There is no law of logarithms for subtraction. Therefore, the log expressions below cannot be written in an equivalent form by applying a law:

$$\log(7 - y) \qquad \log(C - 5) \qquad \log(h - t)$$

Which one of the following statements is true? _____

(a) $\log(D - F) = \log D - \log F$ (b) $\log \dfrac{D}{F} = \log D - \log F$

---

13. If possible, use a law of logarithms to write each expression below in an equivalent form:

    (a) $\log\left(\dfrac{p}{q}\right) = $ _____

    (b) $\log(T - P) = $ _____

    (c) $\log(a + b) = $ _____

    (d) $\log(RS) = $ _____

(b)

Answer to Frame 13: a) $\log p - \log q$ c) Not possible.
                    b) Not possible.   d) $\log R + \log S$

---

14. Here is a summary of the three basic laws of logarithms:

    (1) **Multiplication.** The logarithm of a multiplication equals the sum of the logarithms of the factors.

    $$\boxed{\log(ab) = \log a + \log b}$$

    (2) **Fraction (or Division).** The logarithm of a fraction (or division) equals the logarithm of the numerator minus the logarithm of the denominator.

    $$\boxed{\log\left(\dfrac{a}{b}\right) = \log a - \log b}$$

    (3) **Raising to a Power.** The logarithm of a base raised to a power equals the power (or exponent) times the logarithm of the base.

    $$\boxed{\log(b^a) = a \log b}$$

SELF-TEST 1 (Frames 1-14)

Using the laws of logarithms, write each of these expressions in an equivalent form:

1. $\log(B^x) = $ _____
2. $\log\left(\dfrac{R}{5}\right) = $ _____
3. $\log(3y) = $ _____
4. $\log(18^{2.35}) = $ _____

Given: $\log 2 = \underline{0.3010}$ and $\log 3 = \underline{0.4771}$ Using the values of log 2 and log 3 and the laws of logarithms, find the numerical value of each of the following logarithms:

5. $\log 6 = $ _____
    Note: $\log 6 = \log(2 \times 3)$

6. $\log 1.5 = $ _____
    Note: $\log 1.5 = \log\left(\dfrac{3}{2}\right)$

7. $\log 9 = $ _____
    Note: $\log 9 = \log(3^2)$

ANSWERS:
1. $x \log B$
2. $\log R - \log 5$
3. $\log 3 + \log y$
4. $2.35 \log 18$
5. $\log 6 = 0.7781$ (From: $\log 2 + \log 3$)
6. $\log 1.5 = 0.1761$ (From: $\log 3 - \log 2$)
7. $\log 9 = 0.9542$ (From: $2 \log 3$)

7-2 REARRANGING "LOG" FORMULAS - WITHOUT USING THE LAWS OF LOGARITHMS

Like other formulas, "log" formulas can be rearranged to solve for a variable or the logarithm of a variable. Some of these rearrangements require the use of the laws of common logarithms; others do not. In this section, we will discuss the rearrangements which do not require the use of the laws of logarithms.

15. In $\boxed{D = 10 \log R}$, the coefficient of log R is 10. We can solve for "log R" by multiplying both sides by the reciprocal of 10. We get:

$$\frac{1}{10}(D) = \frac{1}{10}(10) \log R$$

$$\frac{D}{10} = 1 \cdot \log R$$

$$\log R = \frac{D}{10}$$

Using the steps above, solve for $\log\left(\frac{P_2}{P_1}\right)$ in the formula $\boxed{D = 20 \log\left(\frac{P_2}{P_1}\right)}$ on the right:

$$\log\left(\frac{P_2}{P_1}\right) = \underline{\qquad}$$

16. When rearranging a "log" formula, any "log" expression is treated as a single quantity. For example, "log(W + H)" should be treated as a single quantity when rearranging the formula below:

$$\boxed{k \log(W + H) = B}$$

(a) Solve for "log(W + H)":   (b) Solve for "k"

log(W + H) = _____   k = _____

17. We can solve for "log A" in $\boxed{P = -\log A}$ by replacing each side with its opposite. We get:

$$-P = \log A \quad \text{or} \quad \log A = -P$$

Here is another formula of the same type: $\boxed{T = -\log m}$

Solve for "log m":   log m = _____

$\log\left(\frac{P_2}{P_1}\right) = \frac{D}{20}$

a) $\log(W + H) = \frac{B}{k}$

b) $k = \frac{B}{\log(W + H)}$

log m = -T

606   Logarithms and Exponentials: Laws and Formulas

18. To solve for either "k" or "log M" in $\boxed{T = B + k \log M}$, we must isolate the "log" term first. We get:

$$k \log M = T - B$$

(a) Solve for "k": (b) Solve for "log M":

k = _____   log M = _____

---

19. To solve for either "k" or "log Q" in $\boxed{P = A - k \log Q}$, we must also isolate the "log" term first. To do so, we proceed as follows:

(1) Add "-A" to both sides: $P - A = -k \log Q$

(2) Replace each side with its opposite: $A - P = k \log Q$

Note: The opposite of $P - A$ is $A - P$.

Now we can easily solve for "k" or "log Q".

(a) Solve for "k".   (b) Solve for "log Q":

k = _____   log Q = _____

a) $k = \dfrac{T - B}{\log M}$

b) $\log M = \dfrac{T - B}{k}$

---

20. To solve for "A" in $\boxed{P = A - k \log Q}$, the steps are:

(1) Convert the subtraction to addition: $P = A + (-k \log Q)$

(2) Add the opposite of "-k log Q" to both sides:

$$P + (k \log Q) = A + \underbrace{(-k \log Q) + (k \log Q)}$$

$$P + k \log Q = A + \quad 0$$

$$A = P + k \log Q$$

Solve for "B" in: $\boxed{T = B + k \log M}$

B = _____

a) $k = \dfrac{A - P}{\log Q}$

b) $\log Q = \dfrac{A - P}{k}$

$B = T - k \log M$

From:

$B = T + (-k \log M)$

Logarithms and Exponentials: Laws and Formulas 607

21. To solve for "A" in $\boxed{\log P = A - \dfrac{B}{T}}$, the easiest steps are:

    (1) Convert the subtraction to an addition:

$$\log P = A + \left(-\dfrac{B}{T}\right)$$

    (2) Add the opposite of $"-\dfrac{B}{T}"$ to both sides:

$$\log P + \dfrac{B}{T} = A + \left(-\dfrac{B}{T}\right) + \dfrac{B}{T}$$

$$\log P + \dfrac{B}{T} = A + 0$$

$$A = \log P + \dfrac{B}{T}$$

$$\text{or } A = \dfrac{B}{T} + \log P$$

Solve for "log V" in: $\boxed{\dfrac{a}{h} + \log V = d}$

log V = _____

---

22. To solve for either "B" or "T" in $\boxed{\log P = A - \dfrac{B}{T}}$, the easiest method is to solve for $\dfrac{"B"}{T}$ first. The steps are:

    (1) Add "-A" to both sides:

$$\log P - A = -\dfrac{B}{T}$$

    (2) Replace each side with its opposite:

$$A - \log P = \dfrac{B}{T}$$

$$\text{or } \dfrac{B}{T} = A - \log P$$

(a) Solve for "B":   (b) Solve for "T":

B = _____   T = _____

$\log V = d - \dfrac{a}{h}$

From:

$\log V = d + \left(-\dfrac{a}{h}\right)$

a) $B = T(A - \log P)$

b) $T = \dfrac{B}{A - \log P}$

From:

$B = T(A - \log P)$

23. To solve for either "a" or "h" in $\boxed{\dfrac{a}{h} + \log V = d}$, the easiest method is to solve for "$\dfrac{a}{h}$" first. We get:

$$\dfrac{a}{h} = d - \log V$$

(a) Solve for "a":   (b) Solve for "h":

a = _____   h = _____

a) $a = h(d - \log V)$

b) $h = \dfrac{a}{d - \log V}$

24. To solve for "$\log\left(\dfrac{D - R}{R}\right)$" in: $\boxed{C = \dfrac{0.020}{\log\left(\dfrac{D - R}{R}\right)}}$

(a) Clear the fraction and get:

(b) Multiply both sides by the _____ reciprocal of the coefficient of $\log\left(\dfrac{D - R}{R}\right)$ and get: _____

a) $C \log\left(\dfrac{D - R}{R}\right) = 0.020$

b) $\log\left(\dfrac{D - R}{R}\right) = \dfrac{0.020}{C}$

25. When performing an evaluation to find the value of a "non-solved for" variable in a "log" formula, we can use either of two methods. Here is an example:

In $\boxed{R = P + k \log T}$, let's find "k" when: $R = 150$, $P = 50$, $T = 200$

Method 1: Plugging the known values into the formula as it stands and then solving for "k". We get:

$$150 = 50 + k \log 200$$
$$100 = k(2.3010)$$
$$k = \dfrac{100}{2.3010} = 43.5$$

Method 2: Rearranging the formula to solve for "k" first and then plugging in the known values. We get:

$$k = \dfrac{R - P}{\log T} = \dfrac{150 - 50}{2.3010} = \dfrac{100}{2.3010} = 43.5$$

Did we obtain the same value for "k" in both methods? _____

Yes.

SELF-TEST 2 (Frames 15-25)

1. Solve for "A":

   $$B = W - A \log R$$

   A = _____

2. Solve for "log N":

   $$T = H + F \log N$$

   log N = _____

3. Solve for "log(P + 1)":

   $$s = b - d \log(P + 1)$$

   log(P + 1) = _____

4. Solve for "t": $r = \dfrac{a}{t} - \log K$

   t = _____

5. In $r = \dfrac{a}{t} - \log K$, find "t" when: r = 22, a = 12, K = 100

   t = _____

ANSWERS: 1. $A = \dfrac{W - B}{\log R}$ 2. $\log N = \dfrac{T - H}{F}$ 3. $\log(P + 1) = \dfrac{b - s}{d}$ 4. $t = \dfrac{a}{r + \log K}$ 5. t = 0.5

7-3 REARRANGING "LOG" FORMULAS - USING THE LAWS OF LOGARITHMS

When a variable is part of a complicated log expression, we can frequently solve for the variable (or its "log") by applying a law of logarithms. We will discuss rearrangements of this type in this section.

610   Logarithms and Exponentials: Laws and Formulas

26. To solve for "log P_2" in $\boxed{D = 10 \log\left(\dfrac{P_2}{P_1}\right)}$, we must use the law of logarithms for division. The steps are:

(1) Isolate $\log\left(\dfrac{P_2}{P_1}\right)$ and get: $\log\left(\dfrac{P_2}{P_1}\right) = \dfrac{D}{10}$

(2) Apply the law of logs for division to $\log\left(\dfrac{P_2}{P_1}\right)$ and get: $\log P_2 - \log P_1 = \dfrac{D}{10}$

(3) Complete the solution: $\log P_2 = \dfrac{D}{10} + \log P_1$

Following the steps above, solve for "log I_1" in the formula on the right: $\boxed{M = 2.5 \log\left(\dfrac{I_1}{I}\right)}$

$\log I_1 = $ _____

$\log I_1 = \dfrac{M}{2.5} + \log I$

27. To solve for "log P_1" in $\boxed{D = 10 \log\left(\dfrac{P_2}{P_1}\right)}$, we must also use the law of logs for division. The steps are:

(1) Isolate $\log\left(\dfrac{P_2}{P_1}\right)$: $\log\left(\dfrac{P_2}{P_1}\right) = \dfrac{D}{10}$

(2) Apply the law of logs for division: $\log P_2 - \log P_1 = \dfrac{D}{10}$

(3) Add "$-\log P_2$" to both sides: $-\log P_1 = \dfrac{D}{10} - \log P_2$

(4) Take the opposite of both sides: $\log P_1 = \log P_2 - \dfrac{D}{10}$

Note: The opposite of "$\dfrac{D}{10} - \log P_2$" is "$\log P_2 - \dfrac{D}{10}$".

Following the steps above, solve for "log I" in the formula on the right: $\boxed{M = 2.5 \log\left(\dfrac{I_1}{I}\right)}$

$\log I = $ _____

$\log I = \log I_1 - \dfrac{M}{2.5}$

Logarithms and Exponentials: Laws and Formulas   611

28. To solve for "log M" in $\boxed{\log(MN) = T}$, we must apply the law of logs for multiplication to log(MN). The steps are:

  (1) Apply the law of logs to log(MN): $\log M + \log N = T$

  (2) Complete the solution:

  log M = _____

---

29. To solve for "q" in $\boxed{R = \log(p^q)}$, we must apply the law of logs for raising a base to a power to log(p^q). The steps are:

  (1) Apply the law of logs to log(p^q): $R = q \log p$

  (2) Complete the solution:

  q = _____

$\log M = T - \log N$

---

30. Solve for "log b" in the formula on the right: $\boxed{\log(b^r) = ak}$

  log b = _____

$q = \dfrac{R}{\log p}$

---

31. When a law must be used to solve for a variable in a "log" formula, <u>the log expression should be isolated first</u>. Here is an example:

  Let's solve for "log d" if: $\boxed{h \log\left(\dfrac{d}{t}\right) = 2r}$

  (1) Isolating the log expression, we get: $\log\left(\dfrac{d}{t}\right) = \dfrac{2r}{h}$

  (2) Applying the law of logs for division, we get: $\log d - \log t = \dfrac{2r}{h}$

  (3) Completing the solution, we get: $\log d = \dfrac{2r}{h} + \log t$

  Following the steps above, solve for "log V" in the formula on the right: $\boxed{M = k \log\left(\dfrac{S}{V}\right)}$

  log V = _____

$\log b = \dfrac{ak}{r}$

32. In order to solve for "t" or "log G" in the formula on the right, we isolate $\log(G^t)$ before applying the law of logs. The steps are shown:

$$H = m \log(G^t)$$

$$\frac{H}{m} = \log G^t$$

$$\frac{H}{m} = t \log G$$

$$\frac{1}{t}\left(\frac{H}{m}\right) = \log G$$

$$\log G = \frac{H}{mt}$$

$\log V = \log S - \frac{M}{k}$

From:

$-\log V = \frac{M}{k} - \log S$

Using the steps above, solve for "r" in the formula on the right:

$$k \log(A^r) = 1.4b$$

r = _____

33. Solving for the "log" expression in $\boxed{k \log(W + H) = B}$, we get:

$$\log(W + H) = \frac{B}{k}$$

Can we solve for "log W" by using a law of logs? _____

$r = \dfrac{1.4b}{k \log A}$

34. When performing an evaluation to find the value of a variable which is part of a complicated "log" expression, either of two methods can be used. Here is an example:

In $\boxed{D = 10 \log\left(\dfrac{P_2}{P_1}\right)}$, find "$P_2$" when: $D = 1.17$, $P_1 = 13.7$

No. Since there is no law of logs for addition, $\log(W + H)$ cannot be broken up into a simpler form.

    Method 1: Plugging the known values into the formula as it stands. This method requires a conversion to power-of-ten form as one step. The steps are:

$$1.17 = 10 \log\left(\frac{P_2}{13.7}\right)$$

$$0.117 = \log\left(\frac{P_2}{13.7}\right)$$

$$10^{0.117} = \frac{P_2}{13.7}$$

$$P_2 = 13.7(10^{0.117}) = 13.7(1.31) = 17.9$$

Continued on following page.

34. Continued.

  Method 2: Rearranging the formula to solve for "log P_2" first and then plugging in the known values. The steps are:

  $$\log P_2 = \frac{D}{10} + \log P_1$$

  $$\log P_2 = \frac{1.17}{10} + \log 13.7$$

  $$\log P_2 = 0.117 + 1.1367 = 1.2537$$

  $$P_2 = 10^{1.2537} = 17.9$$

  Did we obtain the same value for "P_2" in both methods? _____

35. The law of logs for raising a base to a power can be used to write the formula on the right in a simpler form. The steps are shown:

$N = 10 \log(E^2)$

$N = 10(2 \log E)$

$N = 20 \log E$

Following the steps above, write the formula on the right in a simpler form:

$D = 10 \log\left(\dfrac{P_2}{P_1}\right)^2$

$D = $ _____

Yes.

Answer to Frame 35: $D = 20 \log\left(\dfrac{P_2}{P_1}\right)$

SELF-TEST 3 (Frames 26–35)

1. Solve for "log W":

   $F = A \log(BW)$

   log W = _____

2. Solve for "log V_1":

   $r = d - \log\left(\dfrac{V_1}{V_2}\right)$

   log V_1 = _____

3. Solve for "t":

   $h = c \log(A^t)$

   t = _____

4. Solve for "log E_1": $D = 10 \log\left(\dfrac{E_2}{E_1}\right)^2$

   log E_1 = _____

5. In $D = 10 \log\left(\dfrac{E_2}{E_1}\right)^2$, find "$E_1$" when: $D = 30$, $E_2 = 140$

   E_1 = _____

ANSWERS: 1. $\log W = \dfrac{F}{A} - \log B$ 3. $t = \dfrac{h}{c \log A}$ 4. $\log E_1 = \log E_2 - \dfrac{D}{20}$

  2. $\log V_1 = d - r + \log V_2$ 5. $E_1 = 4.43$

614   Logarithms and Exponentials: Laws and Formulas

7-4 THE RELATIONSHIP BETWEEN "LOG" FORMULAS AND BASE-10 EXPONENTIAL FORMULAS

Because of the relationship between "log" and "power-of-ten" expressions, there is a corresponding relationship between "log" and "power-of-ten" formulas. We will briefly discuss that relationship in this section in order to reemphasize the basic meaning of "log" formulas, even though the base-10 exponential form of such formulas is not generally used.

36. The conversion of exponential formulas with "10" as the base to "log" formulas is based on the following definition of a common logarithm:

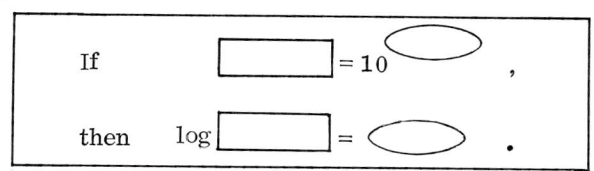

$$\text{If } \boxed{} = 10^{\bigcirc},$$
$$\text{then } \log \boxed{} = \bigcirc.$$

Here are two examples: (1) If $P = 10^a$,   (2) If $\left(\dfrac{M}{N}\right) = 10^{4.5t}$
                       then $\log P = a$ then $\log\left(\dfrac{M}{N}\right) = 4.5t$

Convert each of the following exponential formulas to "log" form:

(a) $Q = 10^{0.1m}$ (b) $10^{a-b} = R$ (c) $\dfrac{L}{t} = 10^{\frac{H-S}{H}}$

---

37. Similarly, "log" formulas can be converted to an equivalent exponential form which contains "10" as the base. Here are two examples:

(1) If $\log t = 0.9R$,   (2) If $\log\left(\dfrac{p}{q}\right) = R - D$,
    then $t = 10^{0.9R}$ then $\dfrac{p}{q} = 10^{R-D}$

Convert each of the following "log" formulas to exponential form:

(a) $\log\left(\dfrac{E_2}{E_1}\right) = \dfrac{N}{20}$ (b) $\log A = \dfrac{P - E}{K}$

a) $\log Q = 0.1m$	
b) $a - b = \log R$	
c) $\log\left(\dfrac{L}{t}\right) = \dfrac{H - S}{H}$	

---

38. Convert each of the following to exponential form:

(a) $\log P = A - \dfrac{B}{T}$ (b) $\log\left(\dfrac{H - B}{G - B}\right) = K$

a) $\dfrac{E_2}{E_1} = 10^{\frac{N}{20}}$	
b) $A = 10^{\frac{P-E}{K}}$	

39. To convert a "log" formula to exponential form, we must isolate the "log" expression first. Here is an example:

$$D = 20 \log Q$$

Step 1: Isolating the "log" expression, we get:
$$\log Q = \frac{D}{20}$$

Step 2: Converting to exponential form, we get:
$$Q = 10^{\frac{D}{20}}$$

Let's convert $\boxed{K \log(W + H) = B}$ to exponential form:

(a) Isolate the "log" expression: _____

(b) Convert to exponential form: _____

a) $P = 10^{A - \frac{B}{T}}$

b) $\frac{H - B}{G - B} = 10^K$

40. After isolating the "log" expression, convert each of the following to exponential form:

(a) $P = -\log A$ (b) $M = 2.5 \log\left(\frac{I_1}{I}\right)$

a) $\log(W + H) = \frac{B}{K}$

b) $W + H = 10^{\frac{B}{K}}$

41. In an earlier section, we showed that "log T" in the formula at the right can be solved for by the ordinary algebraic method. That is:

$$P = k \log T$$
$$\log T = \frac{P}{k}$$

By converting the bottom formula to exponential form, we can solve for "T" instead of "log T". Do so:

T = _____

a) $A = 10^{-P}$, from: $\log A = -P$

b) $\frac{I_1}{I} = 10^{\frac{M}{2.5}}$, from: $\log\left(\frac{I_1}{I}\right) = \frac{M}{2.5}$

42. In an earlier section, we solved for "log b" in the formula at the right by applying the law of logs for raising a base to a power. That is:

$$\log(b^r) = ak$$
$$r \log b = ak$$
$$\log b = \frac{ak}{r}$$

By converting the bottom formula to exponential form, we can solve for "b" instead of "log b". Do so:

b = _____

$T = 10^{\frac{P}{k}}$

$b = 10^{\frac{ak}{r}}$

Logarithms and Exponentials: Laws and Formulas 615

616   Logarithms and Exponentials: Laws and Formulas

43. In an earlier section, we solved for "log P_2" in the formula at the right by isolating $\log\left(\dfrac{P_2}{P_1}\right)$ and applying the law of logs for division. That is:

$$D = 10 \log\left(\dfrac{P_2}{P_1}\right)$$

$$\log\left(\dfrac{P_2}{P_1}\right) = \dfrac{D}{10}$$

$$\log P_2 - \log P_1 = \dfrac{D}{10}$$

$$\log P_2 = \dfrac{D}{10} + \log P_1$$

After isolating $\log\left(\dfrac{P_2}{P_1}\right)$, we can solve for "$P_2$" instead of "log P_2" by converting the formula to exponential form. We get:

$$\log\left(\dfrac{P_2}{P_1}\right) = \dfrac{D}{10}$$

$$\dfrac{P_2}{P_1} = 10^{\frac{D}{10}} \quad \text{or} \quad P_2 = \underline{\qquad\qquad}$$

44. Because of the definition of common logarithms, each formula below represents the same basic relationship:

$$D = 20 \log R \qquad R = 10^{\frac{D}{20}}$$

However, since the base-10 exponential form is not generally used to express relationships in mathematics and science, which of the two ways of stating the relationship above would generally be used? _____

$P_2 = P_1\left(10^{\frac{D}{10}}\right)$

45. In $\boxed{R = \log(p^q)}$, we can solve for either "log p" or "p". Both solutions are shown below. Notice that the steps are identical except for the fact that solving for "p" requires a final conversion to exponential form:

Solving for "log p":

$R = q \log p$

$\log p = \dfrac{R}{q}$

Solving for "p":

$R = q \log p$

$\log p = \dfrac{R}{q}$

$p = 10^{\frac{R}{q}}$

However, since base-10 exponential formulas are not generally used, would we ordinarily solve for "log p" or "p"? _____

$D = 20 \log R$

We would solve for "log p".

46. In $\boxed{M = k \log\left(\dfrac{S}{V}\right)}$, we can solve for either "log S" or "S". Both solutions are shown below:

Solving for "log S":
$$\log\left(\dfrac{S}{V}\right) = \dfrac{M}{k}$$
$$\log S - \log V = \dfrac{M}{k}$$
$$\log S = \dfrac{M}{k} + \log V$$

Solving for "S":
$$\log\left(\dfrac{S}{V}\right) = \dfrac{M}{k}$$
$$\dfrac{S}{V} = 10^{\frac{M}{k}}$$
$$S = V\left(10^{\frac{M}{k}}\right)$$

Notice this major difference in the two solutions:

When solving for "log S", we apply the law of logs for division in the second step.

When solving for "S", we convert to exponential form in the second step.

(a) Are both solutions mathematically correct? _____

(b) Would we ordinarily solve for "log S" or "S"? _____

Answer to Frame 46:   a) Yes.   b) For "log S", since the solution for "S" leads to a base-10 exponential formula.

SELF-TEST 4 (Frames 36-46)

Convert each formula to exponential form:

1. $\boxed{-\log Q = t}$
2. $\boxed{s = \log\left(\dfrac{a}{b}\right)}$
3. $\boxed{G \log P = A}$
4. $\boxed{w = h - \log R}$

Convert each formula to logarithmic form:

5. $\boxed{10^{-h} = N}$
6. $\boxed{AF = 10^{bt}}$
7. $\boxed{10^{a+b} = K + R}$
8. $\boxed{10^{\frac{c}{s}} = \dfrac{B}{E}}$

ANSWERS:
1. $Q = 10^{-t}$
2. $\dfrac{a}{b} = 10^s$
3. $P = 10^{\frac{A}{G}}$
4. $R = 10^{h-w}$
5. $\log N = -h$
6. $\log AF = bt$
7. $\log(K + R) = a + b$
8. $\log\left(\dfrac{B}{E}\right) = \dfrac{c}{s}$

7-5 THE LAWS OF NATURAL LOGARITHMS

There are three basic laws of common (base-10) logarithms. These same three laws are also true for natural (base-"e") logarithms. In this section, we will justify the laws for natural logarithms with some numerical examples.

47. Here is the law of common logarithms for multiplication: $\boxed{\log(ab) = \log a + \log b}$

This same law also holds for natural logarithms. That is: $\boxed{\ln(ab) = \ln a + \ln b}$

We can justify this fact with a numerical example: (Note: The natural log table is used to do so.)

$$\ln(5 \times 10) = \ln 5 + \ln 10$$

Since: $\ln(5 \times 10) = \ln 50 = \underline{3.9120}$

and: $\ln 5 + \ln 10 = 1.6094 + 2.3026 = \underline{3.9120}$

Use the natural-log table to justify the following example of the same law:

$$\ln(3 \times 9) = \ln 3 + \ln 9$$

(a) $\ln(3 \times 9) = \ln 27 = $ _____

(b) $\ln 3 + \ln 9 = $ _____ + _____ = _____

Answer to Frame 47:   a) 3.2958   b) 1.0986 + 2.1972 = 3.2958

48. Here is the law of common logarithms for a fraction (or division): $\boxed{\log\left(\dfrac{M}{N}\right) = \log M - \log N}$

This same law also holds for natural logarithms. That is: $\boxed{\ln\left(\dfrac{M}{N}\right) = \ln M - \ln N}$

We can justify this fact with a numerical example:

$$\ln\left(\frac{30}{5}\right) = \ln 30 - \ln 5$$

Since: $\ln\left(\dfrac{30}{5}\right) = \ln 6 = 1.7918$

and: $\ln 30 - \ln 5 = 3.4012 - 1.6094 = 1.7918$

Use the natural log table to justify the following example of the same law:

$$\ln\left(\frac{48}{12}\right) = \ln 48 - \ln 12$$

(a) $\ln\left(\dfrac{48}{12}\right) = \ln 4 = $ _____

(b) $\ln 48 - \ln 12 = $ _____ - _____ = _____

---

a) 1.3863

b) 3.8712 - 2.4849 = 1.3863

49. Here is the law of common logarithms <u>for raising a base to a power</u>: $\boxed{\log P^a = a \log P}$

    This law also holds for natural logarithms. That is: $\boxed{\ln P^a = a \ln P}$

    We can justify this fact with a numerical example:
    $$\ln 3^2 = 2 \ln 3$$
    Since: $\ln 3^2 = \ln 9 = \underline{2.1972}$
    and: $2 \ln 3 = 2(1.0986) = \underline{2.1972}$

    Use the natural log table to justify the following example of the same law:
    $$\ln 4^3 = 3 \ln 4$$
    (a) $\ln 4^3 = \ln 64 = $ _____ (b) $3 \ln 4 = 3($ _____ $) = $ _____

50. A law of logs for natural logarithms can be used to write each of the following log expressions in an equivalent form. Do so:

    (a) $\ln(0.01t) = $ _____
    (b) $\ln\left(\dfrac{D}{0.6}\right) = $ _____
    (c) $\ln R^{0.9} = $ _____
    (d) $\ln(MT) = $ _____

    a) 4.1589
    b) 3(1.3863) = 4.1589

51. Use a law of logs to write each "ln" expression below in an equivalent form:

    (a) $\ln\left(\dfrac{P}{Q}\right) = $ _____
    (b) $\ln t^a = $ _____
    (c) $\ln(abc) = $ _____
    (d) $\ln\left(\dfrac{8}{d}\right) = $ _____

    a) $\ln 0.01 + \ln t$
    b) $\ln D - \ln 0.6$
    c) $0.9 \ln R$
    d) $\ln M + \ln T$

52. Use a law of logs to write the right side of each formula below in an equivalent form:

    (a) $kt = \ln\left(\dfrac{A_o}{A}\right)$

    $kt = $ _____

    (b) $D = \ln P^k$

    $D = $ _____

    a) $\ln P - \ln Q$
    b) $a \ln t$
    c) $\ln a + \ln b + \ln c$
    d) $\ln 8 - \ln d$

    a) $= \ln A_o - \ln A$
    b) $= k \ln P$

620   Logarithms and Exponentials: Laws and Formulas

SELF-TEST 5 (Frames 47-52)

Using the laws of logarithms, write each expression below in an equivalent form:

1. $\ln(2w) =$ _____
2. $\ln\left(\dfrac{P_2}{P_1}\right) =$ _____
3. $\ln 5^r =$ _____

Using the laws of logarithms, write the right side of each formula below in an equivalent form:

4. $R = \ln AFG$ $R =$ _____
5. $H = 3 \ln M^r$ $H =$ _____

ANSWERS:

1. $\ln 2 + \ln w$ 2. $\ln P_2 - \ln P_1$ 3. $r \ln 5$ 4. $R = \ln A + \ln F + \ln G$ 5. $H = 3r \ln M$

7-6   REARRANGING "ln" FORMULAS

Like "log" formulas, "ln" formulas can be rearranged to solve for a variable or the logarithm of a variable. Some of the rearrangements require the use of the laws of logarithms. We will discuss rearrangements of "ln" formulas in this section.

53. Here is a simple "ln" formula: $D = k \ln P$ When rearranging this formula, "ln P" is treated as a single quantity.

(a) Solve for "ln P":   (b) Solve for "k".

$\ln P =$ _____ $k =$ _____

54. To solve for "ln T" in $w = \dfrac{V}{\ln T}$, the steps are:

(1) Clear the fraction and get: $w \ln T = V$

(2) Complete the solution: $\ln T =$ _____

a) $\ln P = \dfrac{D}{k}$

b) $k = \dfrac{D}{\ln P}$

$\ln T = \dfrac{V}{w}$

55. (a) Solve for "t" in: $\boxed{\ln A = \dfrac{h}{t}}$ (b) Solve for "S" in: $\boxed{V = \dfrac{S}{\ln B}}$

   t = _____   S = _____

---

56. (a) Solve for "$\ln\left(\dfrac{M}{m}\right)$" in: (b) Solve for "k" in:

   $\boxed{v = c \ln\left(\dfrac{M}{m}\right)}$ $\boxed{D = k \ln\left(\dfrac{a}{b}\right)}$

   Answers:
   a) $t = \dfrac{h}{\ln A}$
   b) $S = V \ln B$

   $\ln\left(\dfrac{M}{m}\right) = $ _____   k = _____

---

57. To solve for "K" in $\boxed{G - K \ln PV = W}$, the steps are:

   (1) Add "-G" to both sides to isolate the "ln" term: $-K \ln PV = W - G$

   (2) Replace each side with its opposite: $K \ln PV = G - W$

   (3) Complete the solution:

   K = _____

   Answers:
   a) $\ln\left(\dfrac{M}{m}\right) = \dfrac{v}{c}$
   b) $k = \dfrac{D}{\ln\left(\dfrac{a}{b}\right)}$

---

58. (a) Solve for "ln T" in: (b) Solve for "M" in:

   $\boxed{A = B + D \ln T}$ $\boxed{M - t \ln C = R}$

   $K = \dfrac{G - W}{\ln PV}$

   ln T = _____   M = _____

---

59. To solve for "$\ln A_o$" in $\boxed{\ln\left(\dfrac{A_o}{A}\right) = kt}$, we must apply the law of logs for division. The steps are:

   (1) Apply the law of logs for division to $\ln\left(\dfrac{A_o}{A}\right)$: $\ln A_o - \ln A = kt$

   (2) Complete the solution:

   Answers:
   a) $\ln T = \dfrac{A - B}{D}$
   b) $M = R + t \ln C$

   $\ln A_o = $ _____

622 Logarithms and Exponentials: Laws and Formulas

60. To solve for "ln t" in $\boxed{\ln\left(\frac{c}{t}\right) = b}$, the steps are:

(1) Apply the law of logs for division to $\ln\left(\frac{c}{t}\right)$: $\ln c - \ln t = b$

(2) Add the opposite of "ln c" to both sides: $-\ln t = b - \ln c$

(3) Replace each side with its opposite: $\ln t = $ _____

$\ln A_0 = kt + \ln A$

---

61. To solve for either "ln a" or "ln b" in $\boxed{D = k \ln\left(\frac{a}{b}\right)}$, we must isolate the "ln" expression first. We get:

$$\frac{D}{k} = \ln\left(\frac{a}{b}\right) \quad \text{or} \quad \ln\left(\frac{a}{b}\right) = \frac{D}{k}$$

Now by applying the law of logs for division:

(a) Solve for "ln a":   (b) Solve for "ln b":

$\ln a = $ _____ $\ln b = $ _____

$\ln t = \ln c - b$

---

62. To solve for "ln t" in $\boxed{RV = \ln(st)}$, the steps are:

(a) Apply the law of logs for multiplication to $\ln(st)$:   _____

(b) Complete the solution: $\ln t = $ _____

a) $\ln a = \frac{D}{k} + \ln b$

b) $\ln b = \ln a - \frac{D}{k}$

---

63. To solve for "ln A" in $\boxed{0.1 \ln(AB) = V}$, the steps are:

(1) Isolate $\ln(AB)$ and get: $\ln(AB) = \frac{V}{0.1}$

(2) Apply the law of logs for multiplication to $\ln(AB)$: $\ln A + \ln B = \frac{V}{0.1}$

(3) Complete the solution: $\ln A = $ _____

a) $RV = \ln s + \ln t$

b) $\ln t = RV - \ln s$

$\ln A = \frac{V}{0.1} - \ln B$

64. To solve for "t" in $\boxed{\ln R^t = A - B}$, the steps are:

    (a) Apply the law of logs for raising a base to a power to $\ln R^t$: _____

    (b) Complete the solution: t = _____

---

65. To solve for "ln P" in $\boxed{s - \ln P^t = w}$, the steps are:

    (1) Isolate "$\ln P^t$": $-\ln P^t = w - s$

                        $\ln P^t = s - w$

    (2) Apply the law for raising a base to a power to $\ln P^t$: $t \ln P = s - w$

    (3) Complete the solution: ln P = _____

a) $t \ln R = A - B$

b) $t = \dfrac{A - B}{\ln R}$

---

66. When performing an evaluation to find the value of a "non-solved for" variable in an "ln" formula, we can use either of two methods. Here is an example:

    In $\boxed{G - K \ln P = W}$, let's find "K" when: $G = 100$
                                                   $P = 35$
                                                   $W = 20$

    Method 1: Plugging the known values into the formula as it stands and then solving for "K". We get:

$$100 - K \ln 35 = 20$$
$$100 - K(3.5554) = 20$$
$$-3.5554K = -80$$
$$3.5554K = 80$$
$$K = 22.5$$

    Method 2: Rearranging the formula to solve for "K" first and then plugging in the known values. We get:

$$K = \frac{G - W}{\ln P}$$

$$K = \frac{100 - 20}{\ln 35}$$

$$K = \frac{80}{3.5554} = 22.5$$

Did we obtain the same value for "K" in both methods? _____

$\ln P = \dfrac{s - w}{t}$

Yes.

624   Logarithms and Exponentials: Laws and Formulas

SELF-TEST 6 (Frames 53-66)

1. Solve for "ln B":

$$F = R - \ln(AB)$$

ln B = _____

2. Solve for "ln m":

$$c \ln\left(\frac{M}{m}\right) = v$$

ln m = _____

3. Solve for "ln R":

$$W = H \ln R^{0.4}$$

ln R = _____

4. In $\boxed{k \ln A^t = r}$, find "t" when: k = 60
A = 20
r = 45

t = _____

5. In $\boxed{N = D \ln(BG)}$, find "D" when: N = 176
B = 29
G = 73

D = _____

ANSWERS:

1. $\ln B = R - F - \ln A$ 2. $\ln m = \ln M - \dfrac{v}{c}$ 3. $\ln R = \dfrac{W}{0.4H}$ or $\ln R = \dfrac{2.5W}{H}$ 4. t = 0.25   5. D = 23

7-7 THE RELATIONSHIP BETWEEN "Ln" FORMULAS AND BASE-"e" EXPONENTIAL FORMULAS

Because of the relationship between "ln" and "base-e" expressions, there is a corresponding relationship between "ln" and "base-e" formulas. We will discuss that relationship in this section. Unlike base-10 exponential formulas, base-"e" exponential formulas do frequently occur in mathematics and science. Therefore, conversions to exponential form are often used to solve for variables in formulas containing natural logarithms.

67. The conversion of exponential formulas containing "e" to "ln" formulas is based on the following definition of a natural logarithm.

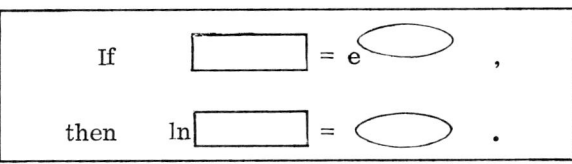

$$\text{If } \boxed{} = e^{\bigcirc}, \text{ then } \ln\boxed{} = \bigcirc.$$

Here are two examples: (1) If $P = e^{-x}$, (2) If $\dfrac{A_o}{A} = e^{kt}$,

then $\ln P = -x$ then $\ln\left(\dfrac{A_o}{A}\right) = kt$

Convert each of these formulas to "ln" formulas:

(a) $R = e^{-ft}$ (b) $\dfrac{P}{k} = e^{-t}$ (c) $\dfrac{V}{E} = e^{-\frac{t}{RC}}$

---

68. The conversion of "ln" formulas to exponential formulas containing "e" is based on the same principle. Here are two examples:

(1) If $\ln Q = -t$,   (2) If $\ln\left(\dfrac{a}{b}\right) = cy$,

then $Q = e^{-t}$ then $\dfrac{a}{b} = e^{cy}$

Convert each of these "ln" formulas to exponential form:

(a) $\ln R = -ft$ (b) $\ln\left(\dfrac{V}{D}\right) = \dfrac{R}{k}$

Answers:

a) $\ln R = -ft$

b) $\ln\left(\dfrac{P}{k}\right) = -t$

c) $\ln\left(\dfrac{V}{E}\right) = -\dfrac{t}{RC}$

---

69. Before converting to exponential form, we must isolate the "ln" expression first. Here is an example:

$$\boxed{v = c \ln\left(\dfrac{M}{m}\right)}$$

Step 1: Isolating the "ln" expression, we get: $\dfrac{v}{c} = \ln\left(\dfrac{M}{m}\right)$

Step 2: Converting to exponential form, we get: $e^{\frac{v}{c}} = \dfrac{M}{m}$

Convert each of these formulas to exponential form:

(a) $D = k \ln P$ (b) $t = \dfrac{\ln\left(\dfrac{A_o}{A}\right)}{k}$

Answers:

a) $R = e^{-ft}$

b) $\dfrac{V}{D} = e^{\frac{R}{k}}$

626   Logarithms and Exponentials: Laws and Formulas

70. If an "ln" formula contains the natural log of a power, we apply the law of logs for raising a base to a power before converting to exponential form. Here are two examples:

$$\boxed{k = \ln V^t}$$

$$k = t \ln V$$

$$\frac{k}{t} = \ln V$$

$$e^{\frac{k}{t}} = V$$

$$\boxed{R = c \ln M^a}$$

$$\frac{R}{c} = \ln M^a$$

$$\frac{R}{c} = a \ln M$$

$$\frac{R}{ac} = \ln M$$

$$e^{\frac{R}{ac}} = M$$

Using the steps above, convert each formula below to exponential form:

(a) $\ln A^p = -1.5$ (b) $0.1 \ln T^b = -L$

a) $P = e^{\frac{D}{k}}$, from:

$$\ln P = \frac{D}{k}$$

b) $\frac{A_o}{A} = e^{kt}$, from:

$$\ln\left(\frac{A_o}{A}\right) = kt$$

---

71. In an earlier section, we solved for "ln P" in the formula at the right. We obtained:

$$\boxed{D = k \ln P}$$

$$\ln P = \frac{D}{k}$$

By converting the bottom formula to exponential form, we can solve for "P" instead of "ln P". Do so:

P = _____

a) $A = e^{-\frac{1.5}{p}}$

b) $T = e^{-\frac{L}{0.1b}}$

or $T = e^{-\frac{10L}{b}}$

72. After solving for "ln T" in the formula on the right, we can solve for "T" by converting to exponential form. Do so:

$$\boxed{w = \frac{V}{\ln T}}$$

T = _____

$P = e^{\frac{D}{k}}$

$T = e^{\frac{V}{w}}$, from:

$$\ln T = \frac{V}{w}$$

Logarithms and Exponentials: Laws and Formulas   627

73. The last step in each solution below is a conversion to exponential form:

    (a) Solve for "T" in: (b) Solve for "C" in:

    $\boxed{A = B + D \ln T}$ $\boxed{M - t \ln C = R}$

    T = _____   C = _____

---

74. We can solve for "Q" in the formula on the right by solving for "ln Q" and then converting to exponential form. Do so:

    $\boxed{\ln Q^p = -t}$

    Q = _____

a) $T = e^{\frac{A-B}{D}}$, from: $\ln T = \frac{A-B}{D}$

b) $C = e^{\frac{M-R}{t}}$, from: $\ln C = \frac{M-R}{t}$

---

75. We can solve for "S" in the formula on the right by solving for "ln S" and then converting to exponential form. Do so:

    $\boxed{V = k \ln S^{-L}}$

    S = _____

$Q = e^{-\frac{t}{p}}$, from: $\ln Q = -\frac{t}{p}$

---

76. To solve for "A_0" in $\boxed{\ln\left(\frac{A_0}{A}\right) = kt}$, the steps are:

    (1) Convert the formula to exponential form: $\frac{A_0}{A} = e^{kt}$

    (2) Complete the solution:

    A_0 = _____

$S = e^{-\frac{V}{kL}}$, from: $\ln S = -\frac{V}{kL}$

---

77. To solve for "t" in the formula on the right, we begin by converting the formula to exponential form. Do so and solve for "t":

    $\boxed{\ln\left(\frac{c}{t}\right) = b}$

    t = _____

$A_0 = A e^{kt}$

78. To solve for either "a" or "b" in the formula at the right, we isolate the "ln" expression and then convert to exponential form as shown:

$$D = k \ln\left(\frac{a}{b}\right)$$

$$\ln\left(\frac{a}{b}\right) = \frac{D}{k}$$

$$\frac{a}{b} = e^{\frac{D}{k}}$$

Therefore: (a) a = _____ (b) b = _____

$t = \dfrac{c}{e^b}$, from:

$\dfrac{c}{t} = e^b$

---

79. To solve for "t" in the formula on the right, we begin by converting to exponential form. Complete the solution:

$$RV = \ln(st)$$
$$e^{RV} = st$$

t = _____

a) $a = be^{\frac{D}{k}}$

b) $b = \dfrac{a}{e^{\frac{D}{k}}}$

---

80. To solve for "A" in the formula on the right, we isolate the "ln" expression and then convert to exponential form. Complete the solution:

$$0.1 \ln(AB) = V$$

$$\ln(AB) = \frac{V}{0.1}$$

$$AB = e^{\frac{V}{0.1}}$$

A = _____

$t = \dfrac{e^{RV}}{s}$

---

81. (a) Solve for "Q" in:

$$-0.7 = \ln(PQ)$$

Q = _____

(b) Solve for "P_2" in:

$$\ln\left(\frac{P_1}{P_2}\right) = m$$

$P_2 =$ _____

$A = \dfrac{e^{\frac{V}{0.1}}}{B}$

or

$A = \dfrac{e^{10V}}{B}$

---

a) $Q = \dfrac{e^{-0.7}}{P}$

b) $P_2 = \dfrac{P_1}{e^m}$

82. (a) Solve for "t" in:

$$k \ln(at) = -0.3$$

(b) Solve for "V_1" in:

$$h \ln\left(\frac{V_1}{V_2}\right) = -d$$

t = _____ V$_1$ = _____

a) $t = \dfrac{e^{-\frac{0.3}{k}}}{a}$

b) $V_1 = V_2 e^{-\frac{d}{h}}$

---

83. In the preceding frames, we obtained each of the solutions below:

$$t = \frac{e^{RV}}{s} \qquad A = \frac{e^{\frac{V}{0.1}}}{B} \qquad Q = \frac{e^{-\frac{0.3}{k}}}{a}$$

None of these solutions is in the preferred form. That is:

Instead of $t = \dfrac{e^{RV}}{s}$, we would write $t = \dfrac{1}{s}(e^{RV})$

Instead of $A = \dfrac{e^{\frac{V}{0.1}}}{B}$, we would write $A = \dfrac{1}{B}\left(e^{\frac{V}{0.1}}\right)$

Instead of $Q = \dfrac{e^{-\frac{0.3}{k}}}{a}$, we would write Q = _____

$Q = \dfrac{1}{a}\left(e^{-\frac{0.3}{k}}\right)$

---

84. In the preceding frames, we also obtained each of the solutions below:

$$t = \frac{c}{e^b} \qquad P_2 = \frac{P_1}{e^m} \qquad b = \frac{a}{e^{\frac{D}{k}}}$$

These solutions are also not in the preferred form. That is:

Instead of $t = \dfrac{c}{e^b}$, we would write: $t = ce^{-b}$

Note: $\dfrac{c}{e^b} = c$ (the reciprocal of e^b) $= ce^{-b}$

The reciprocal of e^b is e^{-b}, since: $e^b \cdot e^{-b} = e^0 = 1$

Write each solution below in the preferred form:

(a) $P_2 = \dfrac{P_1}{e^m}$ _____

(b) $b = \dfrac{a}{e^{\frac{D}{k}}}$ _____

a) $P_2 = P_1 e^{-m}$

b) $b = ae^{-\frac{D}{k}}$

630   Logarithms and Exponentials: Laws and Formulas

85. Write each solution below in the preferred form:

(a) $V = \dfrac{R}{e^{-t}}$ (b) $S = \dfrac{e^{-0.7}}{T}$ (c) $d = \dfrac{0.3Q}{e^{ct}}$

Answer to Frame 85:   a) $V = Re^{t}$ b) $S = \dfrac{1}{T}(e^{-0.7})$ c) $d = 0.3Qe^{-ct}$

---

86. When performing an evaluation to find the value of a variable which is in an "ln" expression in a formula, we can use any of three methods. Here is an example:

In $\boxed{D = k \ln P}$, let's find "P" when: $D = 34$ and $k = 10$

Method 1: Plugging in the known values, solving for "ln P", and using the natural log table:

$34 = 10 \ln P$

$\ln P = \dfrac{34}{10} = 3.4$ (or 3.4000)

$P = \underline{30.0}$

Method 2: Solving for "ln P" first, then plugging in the known values, and using the natural log table:

$\ln P = \dfrac{D}{k}$

$\ln P = \dfrac{34}{10} = 3.4$ (or 3.4000)

$P = \underline{30.0}$

Method 3: Solving for "P", plugging in the known values, and using the e^x-table:

$P = e^{\frac{D}{k}}$

$P = e^{\frac{34}{10}} = e^{3.4}$

$P = \underline{30.0}$

Note that we obtained the same value for "P" in all three methods.

87. Here is another example of the three methods which can be used to find the value of a variable in an "ln" expression in a formula:

$$\boxed{\ln\left(\frac{A_o}{A}\right) = kt}, \quad \text{find "}A_o\text{" when:} \quad \begin{aligned} A &= 1.60 \\ k &= 0.630 \\ t &= 4.30 \end{aligned}$$

Method 1: Plugging in the known values, converting to exponential form, and using the e^x-table:

$$\ln\left(\frac{A_o}{1.60}\right) = (0.630)(4.30)$$

$$\ln\left(\frac{A_o}{1.60}\right) = 2.71$$

$$\frac{A_o}{1.60} = e^{2.71}$$

$$\frac{A_o}{1.60} = 14.9$$

$$A_o = (1.60)(14.9) = \underline{23.8}$$

Method 2: Solving for "ln A_o", plugging in the known values, and using the natural log table:

$$\ln A_o = kt + \ln A$$

$$\ln A_o = (0.630)(4.30) + \ln 1.60$$

$$\ln A_o = 2.71 + 0.4700$$

$$\ln A_o = 3.1800$$

$$A_o = \underline{24.0}$$

Method 3: Solving for "A_o", plugging in the known values, and using the e^x-table:

$$A_o = Ae^{kt}$$

$$A_o = 1.60e^{(0.630)(4.30)}$$

$$A_o = 1.60e^{2.71}$$

$$A_o = 1.60(14.9) = \underline{23.8}$$

Note that we obtained essentially the same value of "A_o" in all three methods. The slight difference in the answers was caused by using our condensed "ln" table and e^x-table. If we had used the more extensive tables available in mathematics handbooks, the slight difference in the answers would not have occurred.

632 Logarithms and Exponentials: Laws and Formulas

SELF-TEST 7 (Frames 67-87)

Convert to logarithmic form:

1. $R = e^{-cw}$ _____

2. $e^{bh} = \dfrac{G}{P}$ _____

Convert to exponential form:

3. $s = c \ln F$ _____

4. $\ln\left(\dfrac{B}{H}\right) = -pt$ _____

To solve Problems 5 to 8, first isolate the "ln" expression and then convert to exponential form:

5. Solve for "A":
$t = \dfrac{\ln A}{c}$

A = _____

6. Solve for "P":
$r = \ln P^h$

P = _____

7. Solve for "V":
$-d = \ln\left(\dfrac{V}{P}\right)$

V = _____

8. Solve for "R":
$p \ln(HR) = -kt$

R = _____

In each solution below, put the right side of the equation in preferred form:

9. $h = \dfrac{m}{e^{-ct}}$ h = _____

10. $r = \dfrac{e^{ks}}{a}$ r = _____

11. In $\ln\left(\dfrac{N_1}{N_2}\right) = -kt$, find "$N_2$" when: $k = 0.25$, $t = 9.2$, $N_1 = 15$

$N_2 =$ _____

ANSWERS:

1. $\ln R = -cw$
2. $\ln\left(\dfrac{G}{P}\right) = bh$
3. $F = e^{\frac{s}{c}}$
4. $\dfrac{B}{H} = e^{-pt}$
5. $A = e^{ct}$
6. $P = e^{\frac{r}{h}}$
7. $V = Pe^{-d}$
8. $R = \dfrac{1}{H}\left(e^{-\frac{kt}{p}}\right)$
9. $h = me^{ct}$
10. $r = \dfrac{1}{a}(e^{ks})$
11. $N_2 = 150$

7-8 COMBINED USE OF THE LAWS OF LOGARITHMS

Sometimes more than one of the laws of common or natural logarithms can be applied to the same logarithmic expression in order to write it in an equivalent form. We will discuss the procedure in this section.

88. The "log" expression below can be written in an equivalent form by applying the laws of common logarithms for both division and multiplication. The two-step process is shown:

    Step 1: $\log\left(\dfrac{pq}{t}\right) = \underline{\log pq} - \log t$

    Step 2: $= \overline{\log p + \log q} - \log t$

    Note: In Step 1, we applied the law for division.

    In going from Step 1 to Step 2, we applied the law for multiplication to "log pq".

    Following the steps above, write the expression below in an equivalent form:

    $\log\left(\dfrac{7x}{y}\right) = $ _____

89. The "ln" expression below can be written in an equivalent form by applying the laws of natural logarithms for multiplication and raising to a power. The two-step process is shown:

    Step 1: $\ln(bc^2) = \ln b + \underline{\ln c^2}$

    Step 2: $= \ln b + \overline{2 \ln c}$

    Note: In Step 1, we applied the law for a multiplication.

    In going from Step 1 to Step 2, we applied the law for raising to a power to "log c^2".

    Following the steps above, write the expression below in an equivalent form:

    $\ln KT^{1.6} = $ _____

    Answer: $\log 7 + \log x - \log y$

90. Two laws of common logarithms are used in the process below.

    Step 1: $\log\left(\dfrac{t^2}{R}\right) = \underline{\log t^2} - \log R$

    Step 2: $= \overline{2 \log t} - \log R$

    Note: In Step 1, we applied the law for division.

    In going from Step 1 to Step 2, we applied the law for raising to a power to "log t^2".

    Using the steps above, write each expression below in an equivalent form:

    (a) $\log\left(\dfrac{p}{q^5}\right) = $ _____

    (b) $\log\left(\dfrac{D^2}{S^3}\right) = $ _____

    Answer: $\ln K + 1.6 \ln T$

634   Logarithms and Exponentials: Laws and Formulas

91. Here is another use of two laws of <u>natural</u> logarithms:

Step 1: $\ln\left(\dfrac{G}{H}\right)^4 = 4\ln\left(\dfrac{G}{H}\right)$

Step 2: $= 4(\ln G - \ln H)$

Note: In Step 1, we applied the law for raising to a power.

In going from Step 1 to Step 2, we applied the law for division to "$\ln\left(\dfrac{G}{H}\right)$".

Using the steps above, write each expression below in an equivalent form:

(a) $\ln\left(\dfrac{a}{b}\right)^{1.24} = $ _____

(b) $\ln\left(\dfrac{17}{D}\right)^m = $ _____

a) $\log p - 5\log q$

b) $2\log D - 3\log S$

---

92. Here is another use of two laws of <u>common</u> logarithms:

Step 1: $\log p(1+q)^n = \log p + \log(1+q)^n$

Step 2: $= \log p + n\log(1+q)$

Note: In Step 1, we applied the law for multiplication.

In going from Step 1 to Step 2, we applied the law for raising to a power to "$\log(1+q)^n$".

Using two laws of <u>natural</u> logarithms, write the expression below in an equivalent form:

$\ln 5(a+b)^{\frac{1}{2}} = $

a) $1.24(\ln a - \ln b)$

b) $m(\ln 17 - \ln D)$

---

93. Two laws of <u>natural</u> logarithms have been used to write the expression below in an equivalent form:

$\ln(3t)^5 = 5\ln(3t)$

$= 5(\ln 3 + \ln t)$ or $5\ln 3 + 5\ln t$

Note: $5(\ln 3 + \ln t)$ is an instance of the distributive principle.

Using two laws of <u>common</u> logarithms, write the expression below in an equivalent form:

$\log(ab)^7 = $

$\ln 5 + \dfrac{1}{2}\ln(a+b)$

$7(\log a + \log b)$

or

$7\log a + 7\log b$

94. Two laws of <u>common</u> logarithms have been used to write the expression below in an equivalent form:

$$\log\left(\frac{m}{pq}\right) = \log m - \log pq$$
$$= \log m - (\log p + \log q)$$
$$= \log m - \log p - \log q$$

Using two laws of <u>natural</u> logarithms, write the expression below in an equivalent form:

$$\ln\left(\frac{t}{5v}\right) = \underline{\hspace{3cm}}$$

---

95. When applying the laws of logarithms to write expressions in equivalent forms:

    (1) The laws of <u>common</u> logarithms are used with "<u>log</u>" expressions.

    (2) The laws of <u>natural</u> logarithms are used with "<u>ln</u>" expressions.

Using as many laws as possible, write each expression below in an equivalent form:

(a) $\log t^2 V = $ _____

(b) $\ln\left(\dfrac{RS}{T}\right) = $ _____

(c) $\log\left(\dfrac{7}{D}\right)^{1.5} = $ _____

$\ln t - \ln 5 - \ln v$

---

96. Using as many laws as possible, write each expression below in an equivalent form:

(a) $\ln(7d)^3 = $ _____

(b) $\log S(1 - q)^{\frac{1}{5}} = $ _____

(c) $\ln\left(\dfrac{T}{SV}\right) = $ _____

a) $2 \log t + \log V$

b) $\ln R + \ln S - \ln T$

c) $1.5(\log 7 - \log D)$
    or
$1.5 \log 7 - 1.5 \log D$

---

a) $3(\ln 7 + \ln d)$
    or
$3 \ln 7 + 3 \ln d$

b) $\log S + \dfrac{1}{5} \log(1 - q)$

c) $\ln T - \ln S - \ln V$

97. Here is a case in which three laws of logarithms can be applied successively to write the expression in an equivalent form:

$$\log\left(\frac{R^2S}{M^3}\right) = \underline{\log R^2S} - \log M^3$$
$$= \overline{\log R^2 + \log S} - \log M^3$$
$$= 2\log R + \log S - 3\log M$$

Using as many laws as possible, write each expression below in an equivalent form:

(a) $\log\left(\dfrac{ab^2}{c}\right) = $ _____

(b) $\ln\left(\dfrac{m^4}{t^5}\right)^{3.7} = $ _____

Answer to Frame 97: a) $\log a + 2\log b - \log c$ b) $3.7(4\ln m - 5\ln t)$ or $14.8\ln m - 18.5\ln t$

SELF-TEST 8 (Frames 88-97)

Using the laws of logarithms, write each expression in an equivalent form:

1. $\log\left(\dfrac{R}{TW}\right) = $ _____

2. $\ln(\pi r^2) = $ _____

3. $\ln\left(\dfrac{at^2}{2}\right) = $ _____

4. $\log A(1+r)^n = $ _____

5. $\ln(Mv^2)^3 = $ _____

6. $\log\left(\dfrac{x^2}{y}\right)^{1.5} = $ _____

7. $\log\left(\dfrac{1}{G}\right) = $ _____

8. $\ln\left(\dfrac{a}{b}\right)^{-1} = $ _____

ANSWERS:

1. $\log R - \log T - \log W$
2. $\ln \pi + 2\ln r$
3. $\ln a + 2\ln t - \ln 2$
4. $\log A + n\log(1+r)$
5. $3\ln M + 6\ln v$
6. $3\log x - 1.5\log y$
7. $\log 1 - \log G$ or $0 - \log G$ or $-\log G$
8. $(-1)(\ln a - \ln b)$ or $\ln b - \ln a$

7-9 REVERSING THE LAWS OF LOGARITHMS

By reversing the laws of common and natural logarithms, we can write various "log" and "ln" expressions in an equivalent form. We will discuss the method in this section.

98. The reversed direction of each of the three basic laws of logarithms is given in a box below. Though the laws of common logarithms are used, the same statements apply to the laws of natural logarithms.

(1) FOR MULTIPLICATION:

Just as: $\log(ab) = \log a + \log b$,

$\log a + \log b = \log(ab)$

(2) FOR A FRACTION (OR DIVISION):

Just as: $\log\left(\dfrac{M}{N}\right) = \log M - \log N$

$\log M - \log N = \log\left(\dfrac{M}{N}\right)$

(3) FOR RAISING TO A POWER:

Just as: $\log T^b = b \log T$

$b \log T = \log T^b$

By using a law of logarithms in the reversed direction, we can write the left side of each formula below in an equivalent form. Do so:

(a) $\log C - \log R = T$ (b) $T \ln P = S$ (c) $\log D + \log V = Q$

Answer to Frame 98:   a) $\log\left(\dfrac{C}{R}\right) = T$ b) $\ln P^T = S$ c) $\log(DV) = Q$

99. Use a law of logarithms to write each formula below in an equivalent form:

(a) $\ln 0.5 + \ln H = A - B$ (b) $L = \ln 0.02 - \ln c$ (c) $5 \log D = P$

---

100. Sometimes we can use two laws to write a complicated expression in an equivalent form. For example, the law for raising to a power and the law for division are used below:

$\log R - b \log V = \log R - \log V^b = \log\left(\dfrac{R}{V^b}\right)$

Note: The law for raising to a power was used first.

Using the same two laws, write each expression below in an equivalent form:

(a) $5 \log t - \log m =$

(b) $3 \ln S - 4 \ln V =$

a) $\ln(0.5H) = A - B$

b) $L = \ln\left(\dfrac{0.02}{c}\right)$

c) $\log D^5 = P$

638 Logarithms and Exponentials: Laws and Formulas

101. Both the law for raising to a power and the law for multiplication are used to write the expression below in an equivalent form:

$$c \ln a + \ln b = \ln a^c + \ln b = \ln(a^c b)$$

Note: The law for raising to a power was used first.

Using the same two laws, write each expression below in an equivalent form:

(a) $\ln D + 3 \ln F =$ _____

(b) $m \log t + 5 \log y =$ _____

a) $\log\left(\dfrac{t^5}{m}\right)$

b) $\ln\left(\dfrac{S^3}{V^4}\right)$

102. Both the law for multiplication and the law for division are used to write the expression below in an equivalent form:

$$\log a + \log b - \log c = \log(ab) - \log c = \log\left(\dfrac{ab}{c}\right)$$

(Notice that we applied the laws from left to right.)

Using the same two laws, write each expression below in an equivalent form:

(a) $\log 0.1 + \log t - \log V =$ _____

(b) $\ln P + \ln Q - \ln R =$ _____

a) $\ln(DF^3)$

b) $\log(t^m y^5)$

103. We have applied the laws for division and multiplication (from left to right) to the expression below:

$$\ln a - \ln b + \ln c = \ln\left(\dfrac{a}{b}\right) + \ln c = \ln\left[\left(\dfrac{a}{b}\right)(c)\right] = \ln\left(\dfrac{ac}{b}\right)$$

Following the steps above, write each expression below in an equivalent form:

(a) $\ln 7 - \ln D + \ln F =$ _____

(b) $\log T - \log S + \log 1.5 =$ _____

a) $\log\left(\dfrac{0.1t}{V}\right)$

b) $\ln\left(\dfrac{PQ}{R}\right)$

a) $\ln\left(\dfrac{7F}{D}\right)$

b) $\log\left(\dfrac{1.5T}{S}\right)$

104. We have applied the law for division twice (from left to right) to the expression below:

$$\log M - \log P - \log Q = \log\left(\frac{M}{P}\right) - \log(Q) = \log\left[\frac{\frac{M}{P}}{Q}\right] = \log\left(\frac{M}{PQ}\right)$$

Note: $\log\left[\dfrac{\frac{M}{P}}{Q}\right] = \log\left[\left(\dfrac{M}{P}\right)\left(\dfrac{1}{Q}\right)\right] = \log\left(\dfrac{M}{PQ}\right)$

Using the steps above, write each expression below in an equivalent form:

(a) $\log T - \log 0.8 - \log V =$ _____

(b) $\ln V - \ln A - \ln 15 =$ _____

105. Here is a case in which all three laws are used:

$$\log t + b \log v - \log p = \log t + \log v^b - \log p$$
$$= \log(tv^b) - \log p$$
$$= \log\left(\frac{tv^b}{p}\right)$$

Note: (1) The law for raising to a power was applied first.
(2) Then the other two laws were applied from left to right.

Using all three laws, write the expression below in an equivalent form:

$3 \ln H - \ln F + \ln T =$ _____

106. Use as many laws as possible to write each formula below in an equivalent form:

(a) $\boxed{D = 0.02 \log q - \log(p - a)}$ (b) $\boxed{\ln M - \ln F - \ln K = R}$

107. Use as many laws as possible to write each formula below in an equivalent form:

(a) $\boxed{\ln(H - B) - \ln(C - B) = H}$ (b) $\boxed{P = \log b + 2 \log d - \log t}$

---

a) $\log\left(\dfrac{T}{0.8V}\right)$

b) $\ln\left(\dfrac{V}{15A}\right)$

$\ln\left(\dfrac{H^3 T}{F}\right)$

---

a) $D = \log\left(\dfrac{q^{0.02}}{p - a}\right)$

b) $\ln\left(\dfrac{M}{FK}\right) = R$

---

a) $\ln\left(\dfrac{H - B}{C - B}\right) = H$

b) $P = \log\left(\dfrac{bd^2}{t}\right)$

640 Logarithms and Exponentials: Laws and Formulas

108. When the law of logarithms for division is applied in the reverse direction, be sure to reduce the resulting fraction to lowest terms if necessary. For example:

$$\log H - \log AH = \log\left(\frac{H}{AH}\right) = \log\left(\frac{1}{A}\right)$$

Use the law for division to write each expression below in an equivalent form:

(a) $\log ab - \log b = $ _____

(b) $\ln PQ - \ln P^2 = $ _____

---

109. Use as many laws as possible to write each expression below in an equivalent form:

(a) $\log T - \log TS + \log R = $ _____

(b) $2 \ln F - \ln FH = $ _____

a) $\log a$, from:
$$\log\left(\frac{ab}{b}\right)$$

b) $\ln \frac{Q}{P}$, from:
$$\ln\left(\frac{PQ}{P^2}\right)$$

---

110. Though the formula $\boxed{P = -\log A}$ is in a simple form, we can write it in an equivalent form which does not contain a "−" symbol. The steps are:

(1) Substitute a "−1" for the "−": $P = (-1)(\log A)$

(2) Apply the reverse of the law for raising to a power: $P = \log A^{-1}$

(3) Apply the definition of a power with a negative exponent: $P = \log\left(\frac{1}{A}\right)$

Following the steps above, write each formula below in an equivalent form which does not contain a "−" symbol:

(a) $T = -3 \log m$

(b) $R = -\ln p$

a) $\log\left(\frac{R}{S}\right)$

b) $\ln\left(\frac{F}{H}\right)$

There are many more-complicated manipulations which can be done by means of the laws of logarithms. We have considered only a few of the basic types. If you remember the basic laws, you should be able to follow more complicated manipulations when they occur.

a) $T = \log\left(\frac{1}{m^3}\right)$, from:
$$T = \log m^{-3}$$

b) $R = \ln\left(\frac{1}{p}\right)$, from:
$$R = \ln p^{-1}$$

SELF-TEST 9 (Frames 98-110)

By using the reversed laws of logarithms, write each of the following formulas in an equivalent form:

1. $N = \log P + 2 \log R$ $N = $ _____

2. $V = c \ln K - \ln T$ $V = $ _____

3. $G = 2 \ln N - \ln\left(\dfrac{1}{V}\right)$ $G = $ _____

4. $W = \log kt - \log k + \log t$ $W = $ _____

5. Write $F = -\log H$ in an equivalent form without a "-" sign. _____

ANSWERS: 1. $N = \log(PR^2)$ 2. $V = \ln\left(\dfrac{K^c}{T}\right)$ 3. $G = \ln(N^2 V)$ 4. $W = \log t^2$ 5. $F = \log\left(\dfrac{1}{H}\right)$

7-10 REARRANGING BASE-"e" EXPONENTIAL FORMULAS

In this section, we will discuss the rearrangement of base-"e" exponential formulas. Since these rearrangements are usually done to solve for a variable which is in the exponent of "e", we will emphasize that type. We will see that this common type of rearrangement requires a conversion to "ln" form.

111. In the formula at the right, "k" is the coefficient of e^{-x}. We have solved for "k". The solution has been put in the <u>preferred form</u>, as shown in the bottom equation.

$P = ke^{-x}$

$k = \dfrac{P}{e^{-x}}$

$k = Pe^{x}$

In each case below, solve for the indicated variable and write the solution in the preferred form:

(a) Solve for "k": $A = ke^{-ct}$ (b) Solve for "I_0": $i = I_0 e^{-\frac{Rt}{L}}$

$k = $ _____ $I_0 = $ _____

a) $k = Ae^{ct}$, from:
$$k = \dfrac{A}{e^{-ct}}$$

b) $I_0 = ie^{\frac{Rt}{L}}$, from:
$$I_0 = \dfrac{i}{e^{-\frac{Rt}{L}}}$$

112. When base-"e" exponential formulas are rearranged, the rearrangement is usually done to solve for a variable in the exponent of "e". Rearrangements of this type require a conversion to "ln" form. Here is an example:

We can solve for "x" in the formula below by simply converting the formula to "ln" form. We get:

$$P = e^x$$

$\ln P = x$ (or $x = \ln P$)

Solve for the indicated variable in each case below by simply converting to "ln" form:

(a) Solve for "t": $\boxed{Q = e^t}$ (b) Solve for "k": $\boxed{T = e^k}$

t = _____   k = _____

113. To solve for "t" in $\boxed{D = e^{ct}}$, the steps are:

(1) Convert the formula to "ln" form: $\ln D = ct$

(2) Complete the solution: $t = \dfrac{\ln D}{c}$

Following the steps above, solve for the indicated variable in each of these:

(a) Solve for "p": $\boxed{R = e^{kp}}$ (b) Solve for "m": $\boxed{S = e^{mt}}$

p = _____   m = _____

114. To solve for "x" in $\boxed{R = e^{-x}}$, the steps are:

(1) Convert to "ln" form: $\ln R = -x$

(2) Take the opposite of each side: $-\ln R = x$
      or
      $x = -\ln R$

Using the steps above, solve for the indicated letter in each formula below:

(a) Solve for "t": $\boxed{S = e^{-t}}$ (b) Solve for "v": $\boxed{M = e^{-v}}$

t = _____   v = _____

---

a) $t = \ln Q$
b) $k = \ln T$

---

a) $p = \dfrac{\ln R}{k}$
b) $m = \dfrac{\ln S}{t}$

---

a) $t = -\ln S$
b) $v = -\ln M$

115. In the last frame, we obtained these solutions:

$$x = -\ln R \qquad t = -\ln S \qquad v = -\ln m$$

Ordinarily, we prefer to write these solutions in a form which does not contain a "-" symbol. To do so, we apply the "reversed" form of the law of logarithms for raising to a power, and then use the definition of a power with a negative exponent. For example:

$$\boxed{x = -\ln R} \quad \text{is written as} \quad \boxed{x = \ln\left(\frac{1}{R}\right)},$$

since: $-\ln R = \ln R^{-1} = \ln\left(\frac{1}{R}\right)$

Put each of the other solutions in the preferred form:

(a) $t = -\ln S$ is written as _____

(b) $v = -\ln M$ is written as _____

---

116. To solve for "t" in $\boxed{R = e^{-ft}}$, the steps are:

(1) Convert to "ln" form: $\qquad \ln R = -ft$

(2) Replace each side with its opposite: $\qquad -\ln R = ft$

(3) Complete the solution: $\qquad -\dfrac{\ln R}{f} = t$

$$\text{or} \quad t = -\dfrac{\ln R}{f}$$

Using the steps above, solve for the indicated letter in each formula below:

(a) Solve for "v": $\boxed{D = e^{-kv}}$ \qquad (b) Solve for "h": $\boxed{V = e^{-ht}}$

v = _____ \qquad h = _____

a) $t = \ln\left(\dfrac{1}{S}\right)$

b) $v = \ln\left(\dfrac{1}{M}\right)$

---

117. In the last frame, we obtained these solutions:

$$t = -\dfrac{\ln R}{f} \qquad v = -\dfrac{\ln D}{k} \qquad h = -\dfrac{\ln V}{t}$$

These solutions are also in a non-preferred form. The preferred form of the formula on the left is:

$$\boxed{t = \dfrac{1}{f} \ln\left(\dfrac{1}{R}\right)}$$

The preferred form is obtained in these steps:

$$t = -\dfrac{\ln R}{f} = \dfrac{-\ln R}{f} = \dfrac{\ln R^{-1}}{f} = \dfrac{\ln\left(\dfrac{1}{R}\right)}{f} = \dfrac{1}{f}\ln\left(\dfrac{1}{R}\right)$$

Write each solution below in the preferred form:

(a) $v = -\dfrac{\ln D}{k}$ _____ \qquad (b) $h = -\dfrac{\ln V}{t}$ _____

a) $v = -\dfrac{\ln D}{k}$

b) $h = -\dfrac{\ln V}{t}$

644 Logarithms and Exponentials: Laws and Formulas

118. Solve for the indicated letter and write each solution in the preferred form:

(a) Solve for "y": $\boxed{M = e^{-y}}$ (b) Solve for "x": $\boxed{y = e^{-ax}}$

y = _____   x = _____

a) $v = \frac{1}{k}\ln\left(\frac{1}{D}\right)$

b) $h = \frac{1}{t}\ln\left(\frac{1}{V}\right)$

---

119. Before converting a base-"e" exponential formula to "ln" form, we must isolate the power-of-"e" first. For example:

To convert $\boxed{P = ke^{-x}}$ to "ln" form:

(1) We isolate the power-of-"e": $\frac{P}{k} = e^{-x}$

(2) Then convert to "ln" form: $\ln\left(\frac{P}{k}\right) = -x$

Let's convert $\boxed{V = Ee^{-\frac{t}{RC}}}$ to "ln" form:

(a) Isolate the power-of-"e": _____

(b) Convert to "ln" form: _____

a) $y = \ln\left(\frac{1}{M}\right)$, from:
$y = -\ln M$

b) $x = \frac{1}{a}\ln\left(\frac{1}{y}\right)$, from:
$x = -\frac{\ln y}{a}$

---

120. Convert each formula below to "ln" form:

(a) $R = ke^{-mt}$ (b) $i = I_0 e^{-\frac{Rt}{L}}$ (c) $0.5e^{cx} = V$

a) $\frac{V}{E} = e^{-\frac{t}{RC}}$

b) $\ln\left(\frac{V}{E}\right) = -\frac{t}{RC}$

---

121. Let's solve for "t" in $\boxed{M = ke^{t}}$:

(a) Isolate the power-of-"e": _____

(b) Convert to "ln" form: _____

a) $\ln\left(\frac{R}{k}\right) = -mt$

b) $\ln\left(\frac{i}{I_0}\right) = -\frac{Rt}{L}$

c) $cx = \ln\left(\frac{V}{0.5}\right)$ or $\ln(2v)$

---

a) $\frac{M}{k} = e^t$

b) $\ln\left(\frac{M}{k}\right) = t$

or $t = \ln\left(\frac{M}{k}\right)$

Logarithms and Exponentials: Laws and Formulas   645

122. To solve for "x" in $\boxed{P = ke^{-x}}$, the steps are:

 (1) Isolate the power-of-"e": $\dfrac{P}{k} = e^{-x}$

 (2) Convert to "ln" form: $\ln\left(\dfrac{P}{k}\right) = -x$

 (3) Replace each side with its opposite: _____

Answer to Frame 122: $x = -\ln\left(\dfrac{P}{k}\right)$ (See the next frame for a preferred form of this solution.)

---

123. In the last frame, we obtained the following solution: $\boxed{x = -\ln\left(\dfrac{P}{k}\right)}$

This solution is not in the preferred form because of the "-" in front of "ln". The preferred form is: $\boxed{x = \ln\left(\dfrac{k}{P}\right)}$

(Note: The "-" is gone and the fraction was replaced by its reciprocal.)

We can show that $\boxed{-\ln\left(\dfrac{P}{k}\right) = \ln\left(\dfrac{k}{P}\right)}$ by means of the following steps:

 (1) Reversing the law of logarithms for raising a base to a power, we get: $-\ln\left(\dfrac{P}{k}\right) = (-1)\ln\left(\dfrac{P}{k}\right)$

                 $= \ln\left(\dfrac{P}{k}\right)^{-1}$

 (2) Using the definition of a negative power, we get: $= \ln\left(\dfrac{1}{\frac{P}{k}}\right)$

 (3) Performing the division of "1" by a fraction, we get: $= \ln\left(\dfrac{k}{P}\right)$

---

124. List the steps needed to show that: $\boxed{-\ln\left(\dfrac{a}{b}\right) = \ln\left(\dfrac{b}{a}\right)}$

 Go to next frame.

$-\ln\left(\dfrac{a}{b}\right) = \ln\left(\dfrac{a}{b}\right)^{-1}$

$= \ln\left(\dfrac{1}{\frac{a}{b}}\right)$

$= \ln\left(\dfrac{b}{a}\right)$

125. Write each of these solutions in the preferred form:

 (a) $R = -\ln\left(\dfrac{T}{V}\right)$ (b) $M = -\ln\left(\dfrac{R}{Q}\right)$

126. To solve for "t" in $\boxed{A = ke^{-ct}}$, the steps are:

(1) Isolate the power-of-"e": $\dfrac{A}{k} = e^{-ct}$

(2) Convert to "ln" form: $\ln\left(\dfrac{A}{k}\right) = -ct$

(3) Replace each side with its opposite: $-\ln\left(\dfrac{A}{k}\right) = ct$

(4) Complete the solution: $t = $ _____

a) $R = \ln\left(\dfrac{V}{T}\right)$

b) $M = \ln\left(\dfrac{Q}{R}\right)$

---

127. Here is the solution from the last frame: $t = \dfrac{-\ln\left(\dfrac{A}{k}\right)}{c}$

We can eliminate the "−" in the numerator by substituting $\ln\left(\dfrac{k}{A}\right)$ for $-\ln\left(\dfrac{A}{k}\right)$. We get:

$t = \dfrac{\ln\left(\dfrac{k}{A}\right)}{c}$

or

$t = \dfrac{1}{c}\ln\left(\dfrac{k}{A}\right)$

Write each solution below in the preferred form:

(a) $x = -\dfrac{\ln\left(\dfrac{P}{T}\right)}{k}$

(b) $m = -\dfrac{\ln\left(\dfrac{V}{0.1}\right)}{P}$

x = _____   m = _____

$t = \dfrac{-\ln\left(\dfrac{A}{k}\right)}{c}$

(See the next frame for a preferred form of this solution.)

---

128. To solve for "t" in $\boxed{i = I_0 e^{-\frac{Rt}{L}}}$, the steps are:

(1) Isolate the power-of-"e": $\dfrac{i}{I_0} = e^{-\frac{Rt}{L}}$

(2) Convert to "ln" form: $\ln\left(\dfrac{i}{I_0}\right) = -\dfrac{Rt}{L}$

(3) Replace each side with its opposite: $-\ln\left(\dfrac{i}{I_0}\right) = \dfrac{Rt}{L}$

(4) Clear the fraction by multiplying both sides by "L": $-L\ln\left(\dfrac{i}{I_0}\right) = Rt$

(5) Complete the solution: $t = $ _____

a) $x = \dfrac{\ln\left(\dfrac{T}{P}\right)}{k}$ or $\dfrac{1}{k}\ln\left(\dfrac{T}{P}\right)$

b) $m = \dfrac{\ln\left(\dfrac{0.1}{V}\right)}{P}$ or $\dfrac{1}{P}\ln\left(\dfrac{0.1}{V}\right)$

$t = \dfrac{-L\ln\left(\dfrac{i}{I_0}\right)}{R}$

(See the next frame for the preferred form of this solution.)

129.

Here is the solution from the last frame:

$$t = \frac{-L \ln\left(\frac{i}{I_o}\right)}{R}$$

We can eliminate the "−" in front of the "ln" by means of the steps on the right:

$$t = \frac{L\left[-\ln\left(\frac{i}{I_o}\right)\right]}{R}$$

$$t = \frac{L \ln\left(\frac{I_o}{i}\right)}{R}$$

$$t = \frac{L}{R} \ln\left(\frac{I_o}{i}\right)$$

Write each solution below in the preferred form:

(a) $t = -GH \ln\left(\frac{P}{Q}\right)$ $t = $ _____

(b) $V = -0.3 \ln\left(\frac{a}{b}\right)$ $V = $ _____

---

130. Solve for "t" in the formula on the right and write the solution in the preferred form:

$$V = Ee^{-\frac{t}{RC}}$$

$t = $ _____

a) $t = GH \ln\left(\frac{Q}{P}\right)$

b) $V = 0.3 \ln\left(\frac{b}{a}\right)$

---

131. To solve for "L" in $\boxed{i = I_o e^{-\frac{Rt}{L}}}$, the steps are:

(1) Isolate the power-of-"e":

$$\frac{i}{I_o} = e^{-\frac{Rt}{L}}$$

(2) Convert to "ln" form:

$$\ln\left(\frac{i}{I_o}\right) = -\frac{Rt}{L}$$

(3) Multiply both sides by "L":

$$L \ln\left(\frac{i}{I_o}\right) = -Rt$$

(4) Complete the solution: $L = $ _____

$t = RC \ln\left(\frac{E}{V}\right)$, from:

$$t = -RC \ln\left(\frac{V}{E}\right)$$

$$L = \frac{-Rt}{\ln\left(\frac{i}{I_o}\right)} \text{ or } -\frac{Rt}{\ln\left(\frac{i}{I_o}\right)}$$

(See the next frame for the preferred form of this solution.)

648 Logarithms and Exponentials: Laws and Formulas

132. The solution in the last frame was:

$$L = -\frac{Rt}{\ln\left(\frac{i}{I_o}\right)}$$

We can eliminate the "-" in the solution by the steps on the right:

$$L = \frac{Rt}{-\ln\left(\frac{i}{I_o}\right)}$$

$$L = \frac{Rt}{\ln\left(\frac{I_o}{i}\right)}$$

Write each solution below in the preferred form:

(a) $R = \dfrac{-bc}{\ln\left(\frac{V}{T}\right)}$ R = _____

(b) $M = -\dfrac{0.6V}{\ln\left(\frac{P}{15}\right)}$ M = _____

---

133. To solve for "t" in $\boxed{v = V(1 - e^{-\frac{t}{TC}})}$, the steps are:

(1) Multiply both sides by $\dfrac{"1"}{V}$: $\dfrac{v}{V} = 1 - e^{-\frac{t}{RC}}$

(2) Add "-1" to both sides: $\dfrac{v}{V} - 1 = -e^{-\frac{t}{RC}}$

(3) Replace each side with its opposite: $1 - \dfrac{v}{V} = e^{-\frac{t}{RC}}$

or

$\dfrac{V - v}{V} = e^{-\frac{t}{RC}}$

(4) Convert to "ln" form: $\ln\left(\dfrac{V - v}{V}\right) = -\dfrac{t}{RC}$

(5) Multiply both sides by "RC": $RC \ln\left(\dfrac{V - v}{V}\right) = -t$

(6) Replace each side with its opposite: $-RC \ln\left(\dfrac{V - v}{V}\right) = t$

or

$t = -RC \ln\left(\dfrac{V - v}{V}\right)$

(7) Write the solution in the preferred form: t = _____

---

a) $R = \dfrac{bc}{\ln\left(\frac{T}{V}\right)}$

b) $M = \dfrac{0.6V}{\ln\left(\frac{15}{P}\right)}$

$t = RC \ln\left(\dfrac{V}{V - v}\right)$

SELF-TEST 10 (Frames 111-133)

In Problems 1 to 4, put the right side of each equation in the preferred form:

1. $w = -\dfrac{\ln K}{H}$

    $w = $ _____

2. $p = -\ln\left(\dfrac{A}{A_o}\right)$

    $p = $ _____

3. $t = \dfrac{-\ln\left(\dfrac{P}{F}\right)}{k}$

    $t = $ _____

4. $s = -\dfrac{A \ln\left(\dfrac{H}{V}\right)}{B}$

    $s = $ _____

In Problems 5 to 9, solve each formula for the indicated variable, and put each solution in the preferred form:

5. Solve for "s": $h = ce^{-s}$

    $s = $ _____

6. Solve for "D": $d = De^{-cr}$

    $D = $ _____

7. Solve for "x": $w = ae^{kx}$

    $x = $ _____

8. Solve for "t" $b = Be^{-\dfrac{Gt}{H}}$

    $t = $ _____

9. Solve for "v": $p = A(1 - e^{-kv})$

    $v = $ _____

ANSWERS:

1. $w = \dfrac{1}{H} \ln\left(\dfrac{1}{K}\right)$

2. $p = \ln\left(\dfrac{A_o}{A}\right)$

3. $t = \dfrac{1}{k} \ln\left(\dfrac{F}{P}\right)$

4. $s = \dfrac{A}{B} \ln\left(\dfrac{V}{H}\right)$

5. $s = \ln\left(\dfrac{c}{h}\right)$

6. $D = de^{cr}$

7. $x = \dfrac{1}{k} \ln\left(\dfrac{w}{a}\right)$

8. $t = \dfrac{H}{G} \ln\left(\dfrac{B}{b}\right)$

9. $v = \dfrac{1}{k} \ln\left(\dfrac{A}{A - p}\right)$

650 Logarithms and Exponentials: Laws and Formulas

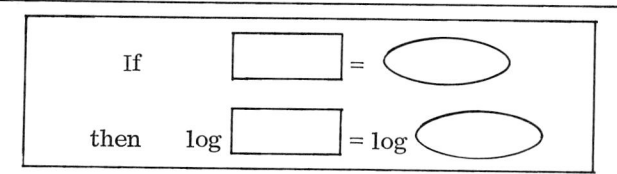

7-11 THE LOG PRINCIPLE FOR EQUATIONS

When solving equations, we use various axioms and principles in which <u>the same operation is performed on both sides of the equation</u>. Up to this point, we have used the addition and multiplication axioms for equations and the squaring and square root principles for equations. There is another basic principle which is used to solve exponential equations which contain bases other than "10" or "e". This principle, which is called the "<u>log principle for equations</u>", involves taking the logarithm of both sides of an equation. In this section, we will introduce the "log principle" for equations and justify it with various examples.

---

134. The "log principle for equations" is shown symbolically at the right:

$$\text{If } \boxed{} = \bigcirc$$
$$\text{then } \log \boxed{} = \log \bigcirc$$

The "log principle for equations" says this:

<u>If we take the logarithm of both sides of an equation, the new equation is equivalent to the original equation. By "equivalent", we mean that the two equations have the same root.</u>

Here are some examples of the "log principle for equations":

If $t = 3^{1.4}$,   If $65 = 8^m$,   If $T = a^p$,

then $\log t = \log 3^{1.4}$ then $\log 65 = \log 8^m$ then $\log T = \log a^p$

In the next five frames, we will justify the "log principle for equations" with various examples. <u>The main point of this justification is to show that the new equation is equivalent to the original equation when the log principle is applied.</u>

---

135. The root of the equation on the right is "9".

$$\boxed{x = 3^2}$$

To apply the log principle for equations, we take the logarithm of <u>both sides</u>. We get:

$\log x = \log 3^2$

Now we can apply the law of logarithms for raising to a power to the right side. We get:

$\log x = 2 \log 3$

To show that "9" is also the root of this new equation, we can plug in "9" for "x" and then use the common logarithm table to evaluate each side. We get:

$\log 9 = 2 \log 3$

$0.9542 = 2(0.4771)$

$0.9542 = 0.9542$

Since the two equations on the right have the same root, we say that they are _____ equations.

$$\boxed{x = 3^2}$$
$$\boxed{\log x = 2 \log 3}$$

Go to next frame.

---

equivalent

136. The root of the equation on the right is "2", since $49 = 7^2$.

$$49 = 7^y$$

Applying the log principle for equations, we get:

$$\log 49 = \log 7^y$$

Now applying the law of logarithms for raising to a power, we get:

$$\log 49 = y \log 7$$

To show that "2" is also the root of this new equation, we can plug in "2" for "y" and evaluate each side. We get:

$$\log 49 = 2 \log 7$$
$$1.6902 = 2(0.8451)$$
$$1.6902 = 1.6902$$

The two equations on the right are equivalent. Why are they equivalent?

$$49 = 7^y$$

$$\log 49 = y \log 7$$

---

137. The root of the equation on the right is "4", since $64 = 4^3$.

$$64 = t^3$$

Because they have the same root.

Applying the log principle for equations, we get:

$$\log 64 = \log t^3$$

Applying the law of logarithms for raising to a power to the right side, we get:

$$\log 64 = 3 \log t$$

Show that "4" is also the root of this new equation by plugging in "4" for "t" and evaluating each side:

---

138. The root of the equation on the right is "15".

$$\frac{D}{5} = 3$$

$\log 64 = 3 \log t$
$\log 64 = 3 \log 4$
$1.8062 = 3(0.6021)$
$1.8062 = 1.8063$

Applying the log principle, we get:

$$\log\left(\frac{D}{5}\right) = \log 3$$

(Note: There is a slight difference because the entries in the log table have been rounded.)

Applying the law of logarithms for division to the left side, we get:

$$\log D - \log 5 = \log 3$$

By plugging in "15" for "D" and evaluating each side, show that "15" is also the root of this new equation:

$\log D - \log 5 = \log 3$
$\log 15 - \log 5 = \log 3$
$1.1761 - 0.6990 = 0.4771$
$0.4771 = 0.4771$

Logarithms and Exponentials: Laws and Formulas 651

139. The root of the equation on the right is "6", since $\left(\dfrac{12}{6}\right)^3 = 2^3 = 8$.

$$\left(\dfrac{12}{x}\right)^3 = 8$$

Applying the log principle, we get:

$$\log\left(\dfrac{12}{x}\right)^3 = \log 8$$

Applying two laws of logarithms (for raising to a power and for division) to the left side, we get:

$$3 \log\left(\dfrac{12}{x}\right) = \log 8$$

$$3(\log 12 - \log x) = \log 8$$

Show that "6" is also the root of this new equation:

---

140. In the last five frames, we justified the log principle for equations. After the log principle is applied to an equation, we <u>always</u> use the laws of logarithms to simplify each side <u>when a simplification is possible</u>. Here is an example:

Applying the log principle to the equation on the right, we get:

$$14 = m^{1.3}$$

$$\log 14 = \log m^{1.3}$$

Now we can apply the law of logarithms for raising to a power to the right side. We get:

$$\log 14 = 1.3 \log m$$

Let's use the same procedure with:

$$7^{1.4} = \dfrac{37}{t}$$

(a) Applying the log principle, we get: _____

(b) Applying the appropriate law of logarithms to each side, we get: _____

$3(\log 12 - \log x) = \log 8$
$3(\log 12 - \log 6) = \log 8$
$3(1.0792 - 0.7782) = 0.9031$
$\qquad 3(0.3010) = 0.9031$
$\qquad 0.9030 = 0.9031$

---

141. The same procedure can be used with formulas. In each case below, apply the log principle and then apply the appropriate law of logarithms to each side if possible:

(a) $\boxed{M = t^a}$ (b) $\boxed{D^b = \dfrac{x}{y}}$

a) $\log 7^{1.4} = \log\left(\dfrac{37}{t}\right)$

b) $1.4 \log 7 = \log 37 - \log t$

---

a) $\log M = \log t^a$
$\log M = a \log t$

b) $\log D^b = \log\left(\dfrac{x}{y}\right)$
$b \log D = \log x - \log y$

142. Sometimes two laws of logarithms can be applied to one side of an equation after the log principle is applied. Here is an example:

Applying the log principle to the equation on the right, we get:

$$D = \left(\frac{a}{b}\right)^{1.9}$$

$$\log D = \log\left(\frac{a}{b}\right)^{1.9}$$

Two laws of logarithms can be applied to the right side.

(1) Applying the law for raising to a power, we get:

$$\log D = 1.9 \log\left(\frac{a}{b}\right)$$

(2) Applying the law for division, we get:

$$\log D = 1.9(\log a - \log b)$$

We have applied the log principle to each equation below. Now apply as many laws of logarithms as possible to each side:

(a) $\dfrac{R_1}{R_2} = \left(\dfrac{T_1}{T_2}\right)^{1.24}$

$\log\left(\dfrac{R_1}{R_2}\right) = \log\left(\dfrac{T_1}{T_2}\right)^{1.24}$

(b) $\left(\dfrac{P_2}{P_1}\right)^{0.8} = \left(\dfrac{T_2}{T_1}\right)^{1.3}$

$\log\left(\dfrac{P_2}{P_1}\right)^{0.8} = \log\left(\dfrac{T_2}{T_1}\right)^{1.3}$

---

143. We have applied the log principle to the equation on the right and then applied as many laws of logarithms as possible to each side.

$$TV^{0.4} = b$$

$\log TV^{0.4} = \log b$

$\log T + \log V^{0.4} = \log b$

$\log T + 0.4 \log V = \log b$

Following the steps above, apply the log principle to each formula below and then apply as many laws of logarithms to each side as possible:

(a) $H = KT^n$

(b) $KP^{1.5} = I$

---

a) $\log R_1 - \log R_2 =$
  $1.24(\log T_1 - \log T_2)$

b) $0.8(\log P_2 - \log P_1) =$
  $1.3(\log T_2 - \log T_1)$

---

a) $\log H = \log K + n \log T$

b) $\log K + 1.5 \log P = \log I$

654   Logarithms and Exponentials: Laws and Formulas

144. We have applied the log principle and then as many laws of logarithms as possible to the equation on the right.

$$A = P(1 + i)^n$$
$$\log A = \log P(1 + i)^n$$
$$\log A = \log P + \log(1 + i)^n$$
$$\log A = \log P + n \log(1 + i)$$

Following the steps above, apply the log principle to each formula below and then apply as many laws of logarithms as possible to each side:

(a) $\boxed{M = T(a + b)^{0.1}}$ (b) $\boxed{PQ^a = R}$

---

145. When applying the log principle to an equation, we must take the log of both sides. Students frequently make the common error of taking the log of only that side of the equation which can be further rearranged by a law (or laws) of logarithms. Here is an example of this common error.

$\boxed{T = P^a}$
ERROR — $\boxed{T = \log P^a}$
           $\boxed{T = a \log P}$

Apply the log principle correctly to: $\boxed{T = P^a}$

a) $\log M = \log T + 0.1 \log(a + b)$

b) $\log P + a \log Q = \log R$

---

146. Don't confuse the log principle for equations and the use of the laws of logarithms.

The <u>log principle for equations</u> involves taking the log of both sides of an equation.

The <u>laws of logarithms</u> are used to write a logarithmic expression in an equivalent form.

In which case below have we applied the log principle to an equation?

(a) $F = \left(\dfrac{S}{T}\right)^{0.1}$ (b) $t = \log\left(\dfrac{m}{p}\right)^a$

$\log F = \log\left(\dfrac{S}{T}\right)^{0.1}$ $t = a \log\left(\dfrac{m}{p}\right)$

$\log F = 0.1 \log\left(\dfrac{S}{T}\right)$ $t = a(\log m - \log p)$

$\log F = 0.1(\log S - \log T)$

$T = P^a$
$\log T = \log P^a$
$\log T = a \log P$

Answer to Frame 146:   Only in (a). In (b), we have merely used two laws of logarithms to write the logarithmic expression on the right side in an equivalent form.

SELF-TEST 11 (Frames 134-146)

Apply the "log principle for equations" to each formula below; then apply as many laws of logarithms to each side as possible:

1. $A = GP^3$

2. $\dfrac{V_1}{V_2} = \left(\dfrac{P_2}{P_1}\right)^{0.95}$

3. $D = K(t+1)^n$

ANSWERS:

1. $\log A = \log G + 3 \log P$ 2. $\log V_1 - \log V_2 = 0.95(\log P_2 - \log P_1)$ 3. $\log D = \log K + n \log(t+1)$

7-12 EXPONENTIAL FORMULAS WITH OTHER BASES

Each of the formulas below contains a power or exponential quantity. However, the bases of the exponentials are not "e" or "10".

$$M = T^a \qquad I = KP^{1.5} \qquad A = P(1+i)^n$$

Formulas of this type are also called "exponential" formulas. We will discuss evaluations with exponential formulas of this type in this section. We will see that such evaluations lead to exponential equations which are solved by applying the log principle for equations.

147. Here is an exponential formula: $M = T^a$

To find the value of "M" when $T = 19$ and $a = 1.4$, we use the following steps:

(1) Plug in the known values: $\qquad M = 19^{1.4}$

(2) Apply the log principle for equations and then use a law of logarithms to write the right side in an equivalent form:

$\log M = \log 19^{1.4}$

$\log M = 1.4 \log 19$

(3) Use the common log table to replace log 19 with a regular number and then simplify:

$\log M = 1.4(1.2788)$

$\log M = 1.79$

(4) Find the value of "M" by using the log table:

If $\log M = 1.79$ (or 1.7900)

$M = \underline{\qquad}$

$M = 61.7$

656 Logarithms and Exponentials: Laws and Formulas

148. In $\boxed{P = V^k}$, let's find the value of "k" when $P = 89.4$ and $V = 17.6$.
The steps are:

(1) Plug in the known values: $89.4 = 17.6^k$

(2) Apply the log principle and then the law of logarithms for raising to a power:
$\log 89.4 = \log 17.6^k$
$\log 89.4 = k \log 17.6$

(3) Replace $\log 89.4$ and $\log 17.6$ with regular numbers: $1.9513 = k(1.2455)$

(4) Complete the solution: $k = \dfrac{1.9513}{1.2455} = $ _____

---

149. In $\boxed{t = R^m}$, let's find the value of "R" when $t = 57.1$ and $m = 2.7$.
The steps are:

(a) Plug in the known values: _____

(b) Apply the log principle and then the law of logarithms for raising to a power: _____

(c) Replace $\log 57.1$ with its numerical value: _____

(d) Complete the solution: $\log R = $ _____
$R = $ _____

$k = 1.56$

(Note: All answers in this section will be reported only <u>to three digits</u>.

---

150. In $\boxed{I = KP^{1.5}}$, let's find "I" when $K = 0.001$ and $P = 16$:

(1) Plug in the known values: $I = (0.001)(16)^{1.5}$

(2) Apply the log principle and then the appropriate laws of logarithms:
$\log I = \log (0.001)(16)^{1.5}$
$\log I = \log 0.001 + \log 16^{1.5}$
$\log I = \log 0.001 + 1.5 \log 16$

(3) Replace the logarithms of numbers with their numerical values and simplify:
$\log I = (-3.0000) + 1.5(1.2041)$
$= (-3.0000) + (1.8062)$
$= -1.1938$

(4) Complete the solution: If $\log I = -1.1938$,
$I = $ _____

a) $57.1 = R^{2.7}$

b) $\log 57.1 = \log R^{2.7}$
$\log 57.1 = 2.7 \log R$

c) $1.7566 = 2.7 \log R$

d) $\log R = \dfrac{1.7566}{2.7} = 0.6500$
$R = 4.47$

$I = 0.0640$

151. In $\boxed{TV^{0.4} = b}$, let's find "T" when $V = 100$ and $b = 3{,}160$. The steps are:

(1) Plug in the known values:	$T(100)^{0.4} = 3{,}160$

(2) Apply the log principle and the appropriate laws of logarithms:

$\log T(100)^{0.4} = \log 3{,}160$

$\log T + \log 100^{0.4} = \log 3{,}160$

$\log T + 0.4 \log 100 = \log 3{,}160$

(3) Replace the logarithms of numbers with their numerical values and simplify:

$\log T + 0.4(2.0000) = 3.4997$

$\log T + 0.8000 = 3.4997$

$\log T = 2.6997$

(4) Complete the solution:	If $\log T = 2.6997$,

$T = \underline{\qquad}$

---

152. In $\boxed{TV^{0.4} = b}$, let's find "V" when $T = 800$ and $b = 3{,}160$.	$T = 501$

(a) Plug in the known values:	$\underline{\qquad\qquad\qquad}$

(b) Apply the log principle and the appropriate laws of logarithms:

$\underline{\qquad\qquad\qquad}$

(c) Replace the logs of numbers with their numerical values, simplify, and solve for log V:

$\log V = \underline{\qquad\qquad}$

(d) Complete the solution:	$V = \underline{\qquad}$

---

a) $800V^{0.4} = 3{,}160$

b) $\log 800V^{0.4} = \log 3{,}160$

$\log 800 + \log V^{0.4} = \log 3{,}160$

$\log 800 + 0.4 \log V = \log 3{,}160$

c) $2.9031 + 0.4 \log V = 3.4997$

$0.4 \log V = 0.5966$

$\log V = \dfrac{0.5966}{0.4} = 1.4915$

d) $V = 31.0$ or 31

658 Logarithms and Exponentials: Laws and Formulas

153. After plugging the known values into an exponential formula, always look for opportunities to simplify before applying the log principle. Here is an example:

In $\boxed{\dfrac{R_1}{R_2} = \left(\dfrac{T_1}{T_2}\right)^{1.24}}$, let's find "$R_1$" when: $R_2 = 36.2$, $T_1 = 2{,}650$, $T_2 = 1{,}920$

Plugging in the known values:

$$\dfrac{R_1}{36.2} = \left(\dfrac{2{,}650}{1{,}920}\right)^{1.24}$$

Now we can simplify by performing the division on the right before applying the log principle:

$$\dfrac{R_1}{36.2} = (1.38)^{1.24}$$

Applying the log principle and the appropriate laws of logarithms, we get:

$$\log\left(\dfrac{R_1}{36.2}\right) = \log 1.38^{1.24}$$

$$\log R_1 - \log 36.2 = 1.24 \log 1.38$$

Replace the log of each number with its numerical value and complete the solution:

$\log R_1 = $ _____

$R_1 = $ _____

---

154. Here is another case where we can simplify before applying the log principle.

In $\boxed{\left(\dfrac{P_2}{P_1}\right)^{0.67} = \left(\dfrac{T_2}{T_1}\right)^{1.67}}$, find "$P_1$" when: $P_2 = 37.6$, $T_2 = 342$, $T_1 = 295$

Plugging in the known values and simplifying, we get:

$$\left(\dfrac{37.6}{P_1}\right)^{0.67} = \left(\dfrac{342}{295}\right)^{1.67}$$

$$\left(\dfrac{37.6}{P_1}\right)^{0.67} = (1.16)^{1.67}$$

Applying the log principle and the appropriate laws of logarithms, we get:

$$\log\left(\dfrac{37.6}{P_1}\right)^{0.67} = \log 1.16^{1.67}$$

$$0.67 \log\left(\dfrac{37.6}{P_1}\right) = 1.67 \log 1.16$$

$$0.67(\log 37.6 - \log P_1) = 1.67 \log 1.16$$

Replace each log of a number with its numerical value and complete the solution:

$$\log 37.6 - \log P_1 = \dfrac{1.67 \log 1.16}{0.67}$$

$\log P_1 = $ _____

$P_1 = $ _____

$\log R_1 - 1.5587 = 1.24(0.1399)$

$\log R_1 - 1.5587 = 0.1735$

$\log R_1 = 1.7322$

$R_1 = 54.0$

155. In $\boxed{A = P(1 + i)^n}$, let's find "A" when:
$P = 1,000$
$i = 0.06$
$n = 20$

Plugging in the known values and simplifying, we get:

$A = 1,000(1 + 0.06)^{20}$
$A = 1,000(1.06)^{20}$

Applying the log principle and the appropriate laws of logarithms, we get:

$\log A = \log(1,000)(1.06)^{20}$
$\log A = \log 1,000 + 20 \log 1.06$

Replace the logs of numbers with their numerical values and complete the solution:

$\log A = \underline{\qquad}$
$A = \underline{\qquad}$

$1.5751 - \log P_1 = \dfrac{1.67(0.0645)}{0.67}$

$1.5752 - \log P_1 = 0.1608$

$-\log P_1 = -1.4144$

$\log P_1 = 1.4144$

$P_1 = 26.0$

---

156. In $\boxed{A = P(1 + i)^n}$, let's find "i" when:
$A = 22.8$
$P = 10$
$n = 21$

Plugging in the known values and simplifying, we get:

$22.8 = 10(1 + i)^{21}$

$\dfrac{22.8}{10} = (1 + i)^{21}$

$2.28 = (1 + i)^{21}$

Applying the log principle and the appropriate law of logarithms, we get:

$\log 2.28 = \log(1 + i)^{21}$

$\log 2.28 = 21 \log(1 + i)$

Replace log 2.28 with its numerical value and complete the solution:

$\log(1 + i) = \underline{\qquad}$
$1 + i = \underline{\qquad}$
$i = \underline{\qquad}$

$\log A = 3 + 20(0.0253)$
$= 3 + 0.506$
$= 3.5060$

$A = 3,210$

$\log(1 + i) = \dfrac{0.3579}{21}$
$= 0.0170$
$1 + i = 1.04$
$i = 0.04$

157. In $\boxed{A = P(1 + i)^n}$, let's find "n" when:
$A = 27,500$
$P = 10,000$
$i = 0.07$

Plugging in the known values and simplifying, we get:

$27,500 = 10,000(1 + 0.07)^n$

$\dfrac{27,500}{10,000} = (1.07)^n$

$2.75 = (1.07)^n$

Applying the log principle and the appropriate law of logarithms, we get:

$\log 2.75 = \log 1.07^n$

$\log 2.75 = n \log 1.07$

Replace each log of a number with its numerical value and complete the solution:

n = _____

Answer to Frame 157: $n = 14.9$ From: $n = \dfrac{0.4393}{0.0294}$

SELF-TEST 12 (Frames 147-157)

1. $\boxed{F = R^w}$ Find "w" when: $F = 708$
   $R = 12.6$

   w = _____

2. $\boxed{H = KE^{1.60}}$ Find "E" when: $H = 56.5$
   $K = 4.38$

   E = _____

3. $\boxed{CP^t = A}$ Find "t" when: $A = 96.0$
   $C = 2.28$
   $P = 2.46$

   t = _____

4. $\boxed{A = P(1 + i)^n}$ Find "P" when: $A = 100,000$
   $i = 0.08$
   $n = 50$

   P = _____

ANSWERS: 1. $w = 2.59$ 2. $E = 4.94$ 3. $t = 4.15$ or 4.16 4. $P = 2,140$